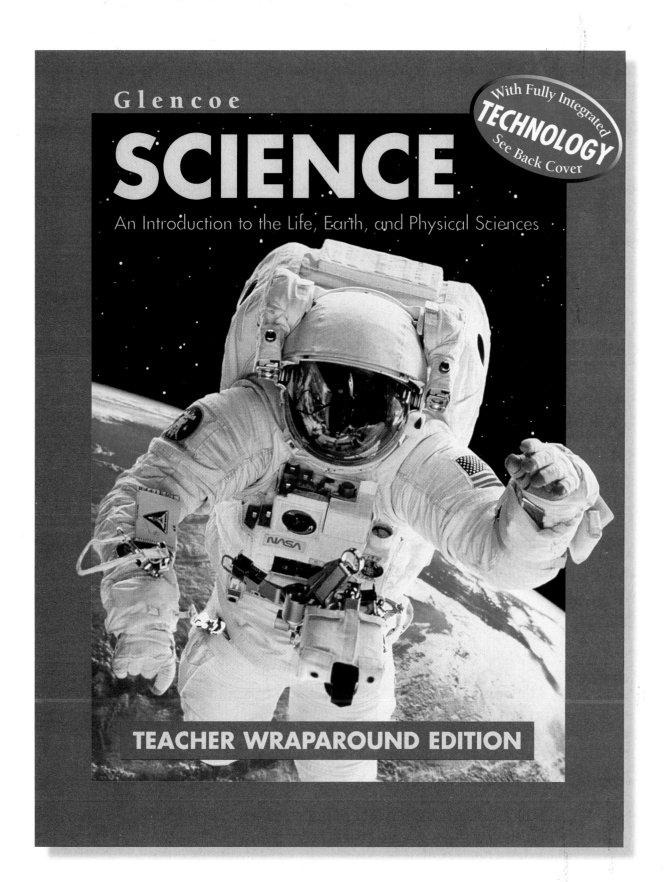

Glencoe

SCIENCE

An Introduction to the Life, Earth, and Physical Sciences

With Fully Integrated
TECHNOLOGY
See Back Cover

TEACHER WRAPAROUND EDITION

 Glencoe
McGraw-Hill

New York, New York   Columbus, Ohio   Woodland Hills, California   Peoria, Illinois

GLENCOE Science

An Introduction to the Life, Earth, and Physical Sciences

Student Edition

Teacher Wraparound Edition

Laboratory Manual: SE

Laboratory Manual: TE

Study Guide for Content Mastery: SE

Study Guide for Content Mastery: TE

Teaching Transparency Package

Section Focus Transparency Package

Science Integration Transparency Package

Chapter Review

Assessment

Performance Assessment

Computer Test Bank Windows and Macintosh Versions

MindJogger Videoquizzes

English/Spanish Audiocassettes

Electronic Teacher Classroom Resources

Glencoe/McGraw-Hill

A Division of The **McGraw·Hill** *Companies*

Send all inquiries to:

Glencoe/McGraw-Hill
8787 Orion Place
Columbus, OH 43240
ISBN 0-02-828316-3

Printed in the United States of America.
3 4 5 6 7 8 9 071/046 04 03 02 01 00

Teacher Guide

Table of Contents

Student Edition

Table of Contents

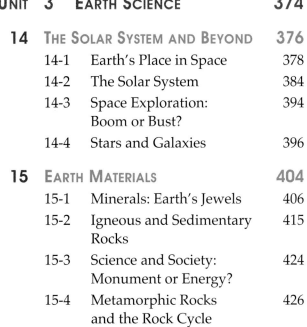

About Our Authors

INTRODUCING THE AUTHOR TEAM

DANIEL BLAUSTEIN teaches physical science and physical anthropology at Waukegan High School in Waukegan, Illinois. He has a B.A. in zoology and anthropology from the University of Wisconsin at Madison and an M.A. in physical anthropology from the State University of New York at Stony Brook. Mr. Blaustein is an active member of several professional organizations, including the American Association of Physical Anthropologists and the National Science Teachers Association.

LOUISE BUTLER was a sixth-grade teacher for 26 years before becoming a school administrator. She is a principal in the Normandy School District in St. Louis, Missouri, where she is an integral part of the district's NSF grant to improve minority achievement in science and math. Ms. Butler is a graduate of the University of Northern Colorado, and has Masters degrees in Administration from the University of Missouri-St. Louis, and in Economic Education from the University of Delaware. She is a member of the National Science Teachers Association and the Science Teachers of Missouri. She is a past Chairman of the NSTA Advisory Board for Talented and Gifted Science Education and presently sits on the NSTA Materials Review Board. Ms. Butler has published articles in *Science and Children, Science Scope, Principal Magazine,* and *Missouri Science Notes.*

BRYCE HIXSON is an author, illustrator, teacher, speaker, and entrepreneur—all in the name of making science meaningful for students and their teachers. Mr. Hixson graduated from Oregon State University at Corvalis with a degree in Science Education and has taught both elementary and junior high school science. Mr. Hixson has written and illustrated many hands-on science books. He is the founder of two companies that promote engaging science and math curriculum materials and also is president of the Utah Museum of Science and Industry.

WANDA MATTHIAS is a retired middle school science department chair with ten years experience as a Grade 6 physical science teacher in Garland, Texas. Ms. Matthias holds a B.S. in Physics from Lamar University, Beamont, Texas. She is a former in-service presenter and technical specialist for WICAT (World-wide Instructional Computer-Assisted Training). Ms. Matthias was given Life Membership in the Parent Teacher Student Association for involving local businesses in science fairs at her school. She has been a spokesperson for programs designed by the *Dallas Morning News* to encourage teachers to develop science fairs.

CONTRIBUTING AUTHORS

CAROLYN FAIR RANDOLPH is Coordinator of Science, Social Studies, and Health Programs for Lexington County, South Carolina, School District Two. Dr. Randolph is a graduate of Fisk University and has a Master's degree in Botany from Howard University and a Doctor of Philosophy in Science Education Curriculum from the University of South Carolina, Columbia. She has taught science in elementary school, junior high school, high school, and at the college level. She is past president of the South Carolina Science Council and the South Carolina Science Supervisors Association, and has held positions in the National Science Teachers Association. Most recently, she has been a member of National Science Foundation Review Panels and Site Review Teams.

LEONARD G. RODRIGUEZ is Assistant Principal at First Avenue Middle School, Arcadia Unified School District, Arcadia, California. Mr. Rodriguez received his bachelor of science degree in biology from Xavier University in Cincinnati, Ohio, and holds a Master's degree in Education Curriculum and Supervision from California State University, Los Angeles. Mr. Rodriguez taught for 20 years at the junior high, middle school, and high school levels. He has been a middle school administrator for 16 years. In 1997, he was received as an Associate into the California School Leadership Academy.

Responding to Changes in
Science Education

The need for new directions in science education

By today's projections, seven out of every ten American jobs will be related to science, mathematics, or electronics in the 21st century. And according to the experts, if students haven't grasped the fundamentals, they probably won't go further in science and may not have a future in a global job market. Studies also reveal that students are avoiding taking "advanced" science classes.

The time for action is now!

In the 1990s, educators, public policy makers, corporate America, and parents have recognized the need for reform in science education. These groups united in a call to action to solve this national problem. As a result, three important projects published reports to point the way for America:

- **Project on National Science Standards** . . . by the National Research Council.
- **Benchmarks for Science Literacy** . . . by the American Association for the Advancement of Science.
- **Scope, Sequence, and Coordination of Secondary School Science** . . . by the National Science Teachers Association (NSTA).

Together, these reports spell out unified guiding principles for new directions in U.S. science education. *Glencoe Science* is based on these guiding principles:

- developing scientific, technological, and mathematical literacy in all students.
- educating students to use scientific principles and processes appropriately in making personal decisions.
- helping students to experience the richness and excitement of knowing about and understanding the natural world.
- teaching students how to engage intelligently in public discourse and debate about matters of scientific and technological concern.

GLENCOE SCIENCE answers the challenge!

At Glencoe Publishing, we believe that *Glencoe Science* will help you bring science reform to the front lines—the classrooms of America. But more importantly, we believe it will help students succeed in science so that they will choose to continue learning about science in high school and into adulthood.

In response to the goals of science curriculum reform, *Glencoe Science* promotes:

- solid, accurate content,
- frequent practice of science process skills,
- hands-on activities that facilitate learning through student-planned activities,
- integration of science concepts across the curriculum,
- numerous opportunities for performance assessment, and
- opportunities for decision making and problem solving.

GLENCOE SCIENCE is loaded with activities.

You'll choose from dozens of activities in the *Student Edition.* These activities are easy to set up and manage and will allow you to teach using a hands-on, inquiry-based approach to learning.

MiniLABs and **Activities** in every chapter involve students in learning and applying scientific methods to practice thinking skills and construct science concepts. They provide an engaging, diverse, active program. **MiniLABs** require a minimum of equipment, and students may take responsibility for organization and execution.

Activities develop and reinforce concepts and develop process skills. Activity formats are structured to guide students to make their own discoveries.

In each chapter, there is one open-ended Activity called **Design Your Own Experiment** that gives students a broad topic to explore and guides them with leading questions to the point where they can determine the direction of the investigation themselves. Students are asked to brainstorm hypotheses, make a decision to investigate one that can be tested, plan procedures, and in the end, think about why their hypotheses were or were not supported.

GLENCOE SCIENCE integrates science and math.

Mathematics is a tool that all students, regardless of their career goals, will use throughout their lives. *Glencoe Science* provides opportunities to hone mathematics skills while learning about the natural world. **Using Math, Practice Problems, MiniLABs,** and **Activities** offer numerous instances for practicing math and math-related skills, including making and using tables and graphs, measuring in SI, and calculating. **Across the Curriculum** strategies in the *Teacher Wraparound Edition* provide additional connections between science and mathematics.

GLENCOE SCIENCE integrates the sciences for understanding.

No subject exists in isolation. At appropriate points throughout *Glencoe Science*, attention is called to other sciences—in the *Student Edition*, in the *Teacher Wraparound Edition*, and in the *Teacher Classroom Resources*.

Science Integration Activities are full-page activities in the *Teacher Classroom Resources* that relate physics, chemistry, Earth science, and life science to chapter topics. *Science Integration Transparencies* also relate the sciences to one another using full-color illustrations that can be used with an overhead projector. These opportunities for integrating the sciences will help your students discover the science behind things they see every day.

It's time for new directions in science education.

The need for new directions in science education has been established. America's students must prepare themselves for the high-tech jobs of the future. We at Glencoe believe that *Glencoe Science* is one answer in preparing for this challenge. Middle school is not too early to begin preparing students for real-world situations. We believe *Glencoe Science* will assist you better in preparing your students for a lifetime of science learning.

Glencoe Science
The National Science Education Standards

The *National Science Education Standards*, published by the National Research Council and representing the contribution of thousands of educators and scientists, offer a comprehensive vision of a scientifically literate society. The standards not only describe what students should know, but also offer guidelines for science teaching and assessment. If you are using, or plan to use, the standards to guide changes in your science curriculum, you can be assured that *Glencoe Science* aligns with the *National Science Education Standards*.

Glencoe Science reflects how Glencoe's commitment to science education is changing the materials used in science classrooms today. *Glencoe Science* is a program that provides numerous opportunities for students, teachers, and school districts to meet the *National Science Education Standards*.

Content Standards

The table on pages 10-11T shows the close alignment between *Glencoe Science* and the grade-appropriate content standards. *Glencoe Science* allows students to discover concepts within each of the content standards, giving them opportunities to make connections between the science disciplines. Our hands-on activities and inquiry-based lessons reinforce the science processes emphasized in the standards.

Teaching Standards

Alignment with the *National Science Education Standards* requires much more than alignment with the outcomes in the content standards. The way in which concepts are presented is critical to effective learning. The teaching standards within the *National Science Education Standards* recommend an inquiry-based program facilitated and guided by teachers.

Glencoe Science provides such opportunities through activities and discussions that allow students to discover by inquiry critical concepts and apply the knowledge they've constructed to their own lives. Throughout the program, students are building critical skills that will be available to them for lifelong learning. *The Teacher Wraparound Edition* helps you make the most of every instructional moment. It offers an abundance of effective strategies and suggestions for guiding students as they explore science.

Assessment Standards

The assessment standards are supported by many of the components that make up the *Glencoe Science* program. *The Teacher Wraparound Edition* and *Teacher Classroom Resources* provide multiple chances to assess students' understanding of important concepts, as well as their ability to perform a wide range of skills. Ideas for portfolios, performance activities, written reports, and other assessment activities accompany every lesson. Rubrics and Performance Task Assessment Lists can be found in Glencoe's Professional Series booklet *Performance Assessment in the Science Classroom.*

Program Coordination

The scope of the content standards requires students to meet the outcomes over the course of their education. The correlation on the following pages demonstrates the close alignment of this course of *Glencoe Science* with the content standards.

Correlation of Glencoe Science *to the National Science Standards, Content Standards, Grades 5–8*

Content Standard	Page Numbers
(UCP) UNIFYING CONCEPTS AND PROCESSES	
1. Systems, order, and organization	36, 45-47, 49-53, 62, 74-77, 80-81, 118, 147-148, 151-155, 157-159, 175, 200-201, 275-279, 351-359, 362, 366, 377, 384-393, 396-400, 411, 430-431, 456-461, 468-469, 476-477, 481-488, 504-506, 516
2. Evidence, models, and explanation	10-12, 17, 24, 26, 69, 78-79, 126-137, 188-189, 196, 214-218, 220-221, 257, 263, 273, 275-283, 291, 298, 311, 326, 330-331, 343-345, 352, 363, 381, 397, 429, 437, 442, 444, 448, 450, 456-458, 477, 481-486, 498, 504, 512-513, 516-517
3. Change, constancy, and measurement	9, 36, 124-125, 129-133, 139-142, 158, 183, 239-240, 246-247, 252-257, 263, 266-268, 271, 274, 280-281, 319, 326, 330-331, 338-339, 345, 356-357, 377, 382-386, 415-431, 438-447, 450-451, 456-461, 468-469, 475-480, 482, 495, 511-517
4. Evolution and equilibrium	124-125, 128-133, 166-167
5. Form and function	49-52, 54, 62-73, 75-77, 89, 121-122
(A) SCIENCE AS INQUIRY	
1. Abilities necessary to do scientific inquiry	6, 8-9, 14-23, 40-41, 46-48, 61, 71, 75, 78-79, 92, 94-95, 104, 126-127, 150, 160-161, 183, 200-201, 215, 220-221, 226, 228, 239, 246-247, 255, 257, 267, 271, 280-281, 326, 330-331, 333-345, 348, 352, 356-357, 361, 412, 418-419, 429, 458, 472-473, 477-478, 495, 501-502, 508-509
2. Understandings about scientific inquiry	4, 6-9, 18-21, 23, 40-41, 47, 83, 101, 183, 200-201, 224, 246-247, 267-268, 271, 280-281, 294, 326, 330-331, 333, 343, 352, 356-357, 361, 368, 394-395, 398, 412, 418-419, 452, 458-459, 477, 495, 501-502, 508-509, 516-517
(B) PHYSICAL SCIENCE	
1. Properties and changes of properties in matter	154, 210-214, 216-219, 223-229, 232, 237-248, 250-258, 360, 405-423, 495, 501-502
2. Motions and forces	220, 254, 263-287, 292-301, 326, 345-350, 482-487, 507-511
3. Transfer of energy	43-44, 155, 168-169, 320-343, 366, 438, 448-449
(C) LIFE SCIENCE	
1. Structure and function in living systems	36-37, 49-52, 61-73, 75-77, 80-81, 103, 122
2. Reproduction and heredity	34, 37-38, 89-93, 96-99, 102-103, 106-111, 134-135
3. Regulation and behavior	35, 39, 42-43, 120, 321-322
4. Populations and ecosystems	134-135, 140-142, 147-155, 157-164, 166-170, 175-202, 250-251, 366-367, 413, 424-425, 467, 471, 482, 488-489, 496-497, 514-517
5. Diversity and adaptations of organisms	98, 110, 117-123, 124-125
(D) EARTH AND SPACE SCIENCE	
1. Structure of the earth system	254-255, 338-339, 405-431, 437-461, 471-474, 476-477, 481-487, 496-500
2. Earth's history	136-142, 452-454, 456-461, 481-482, 518
3. Earth in the solar system	265-266, 378-393, 396-400, 486-487, 503-504

(E) SCIENCE AND TECHNOLOGY	
1. Abilities of technological design	180, 188-189, 266, 305, 308, 352, 358, 361, 390-391
2. Understandings about science and technology	3, 6, 9, 82-83, 101, 105, 137, 180, 188-189, 200-201, 225, 230, 244, 250-251, 265-266, 271-275, 296-311, 312-313, 325, 335, 338-339, 346, 349, 351, 359, 364, 366-367, 390-391, 394-395, 398, 406, 413, 441, 445-447, 485, 512-513

(F) SCIENCE IN PERSONAL AND SOCIAL PERSPECTIVES	
1. Personal health	55, 244, 284-285
2. Populations, resources, and environments	117, 119, 134-135, 166-170, 175-202, 250-251, 255, 366-367, 387-392, 413, 424-425, 467, 471, 482, 488-489, 496-497, 502-507, 514-517
3. Natural hazards	242, 245, 248, 350, 440, 443, 475-477, 492, 498, 507-511
4. Risks and benefits	176-177, 200-201, 249, 250-251, 366-367
5. Science and technology in society	3-4, 9, 26, 82-83, 98, 100-101, 105, 131, 134-135, 137, 178, 181, 200-201, 225, 230, 232, 244, 250-251, 269, 271-275, 282-285, 296-313, 335, 338-339, 346, 349, 351, 359, 364, 366-367, 394-395, 398, 406, 413, 441, 445-447, 470, 485, 488-489, 512-513

(G) HISTORY AND NATURE OF SCIENCE	
1. Science as a human endeavor	4, 13, 84, 100-101, 112, 123, 134-135, 165, 186, 200-201, 212-216, 231, 249, 275-279, 284-286, 305, 340, 408, 432, 446-447, 452-455, 490, 500, 512-513, 516
2. Nature of science	4, 6-8, 238-239, 368, 517
3. History of science	212-214, 216-217, 223, 264-265, 452-455

Proficiency Tests
Helping You Prepare Your Students

FOCUSING ON MEASURABLE GOALS

Proficiency? Competency? In the wake of increased concern on the part of educators and parents about students' ability to read and write, these terms have taken on new meaning and new weight in classrooms across the country. Throughout the United States and the rest of the world, students at various grade levels, from elementary through high school, are being tested to demonstrate their competency in reading, writing skills, computer skills, mathematics, and science.

STANDARDIZED TESTS

Standardized achievement tests have been a part of American curriculum for many years, but their importance is increasing. A scientifically literate population is of critical importance to our society's future, and a certain amount of testing can measure our students' progress toward that goal. Standardized tests that are used frequently are:

- Stanford Achievement Tests
- Iowa Tests of Basic Skills (ITBS)
- Metropolitan Achievement Tests (MAT)
- Comprehensive Test of Basic Skills (CTBS)
- TerraNova (TN)—an expanded version of CTBS
- California Achievement Tests (CAT)

HOW DOES *GLENCOE SCIENCE* HELP?

On the following page, you will see that *Glencoe Science,* a program formulated for the beginning middle school student, satisfies important science criteria in standardized tests. This correlation chart shows that *Glencoe Science* has the content and skills needed to perform well on standardized tests. *Glencoe Science* contains numerous opportunities for preparing your students to successfully master important content and process skill objectives. That success will be demonstrated on these nationally recognized achievement tests and will prepare your students for your state and/or local testing program.

The Student Edition of *Glencoe Science* has also been correlated to the National Science Content Standards, Grades 5-8 (See pages 9T through 11T of this *Teacher Wraparound Edition.*). This means that chapter objectives throughout the text are in line with the broad learning objectives required for Grade 6.

In addition, there are many places in the *Student Edition,* the *Teacher Wraparound Edition,* and the *Teacher Classroom Resources* of *Glencoe Science* where students can practice for state and standardized tests. For instance, the Section Wrap-up questions and questions in the Chapter Review for each chapter are correlated to chapter objectives. **Chapter Tests** and **Study Guide for Content Mastery** in the *Teacher Classroom Resources* present more opportunities for review of basics. A comprehensive **Computer Test Bank,** available in both Macintosh and Windows formats, provides up to 50 additional questions per chapter for practice. *MindJogger Videoquizzes* are a fun way for students to review content. Use them also to assess prior knowledge. You can feel confident that **Glencoe Science** supports your efforts to prepare your students thoroughly.

Correlation to Achievement Tests

Chapter	Achievement Tests					
	Stanford	ITBS	MAT	CTBS	TN	CAT
1	✔	✔	✔	✔	✔	✔
2	✔	✔	✔	✔	✔	✔
3	✔	✔	✔	✔	✔	✔
4	✔	✔	✔	✔	✔	✔
5	✔	✔	✔	✔	✔	✔
6	✔	✔	✔	✔	✔	✔
7	✔	✔	✔	✔	✔	✔
8	✔	✔	✔	✔	✔	✔
9	✔	✔	✔	✔	✔	✔
10	✔	✔	✔	✔	✔	✔
11	✔	✔	✔	✔	✔	✔
12	✔	✔	✔	✔	✔	✔
13	✔	✔	✔	✔	✔	✔
14	✔	✔	✔	✔	✔	✔
15	✔	✔	✔	✔	✔	✔
16	✔	✔	✔	✔	✔	✔
17	✔	✔	✔	✔	✔	✔
18	✔	✔	✔	✔	✔	✔

Themes & Scope & Sequence

Our society is becoming more aware of the inter-relationship of the disciplines of science. It is also necessary to recognize the precarious nature of the stability of some systems and the ease with which this stability can be disturbed so that the system changes. For most people, then, the ideas that unify the sciences and make connections between them are the most important.

Themes provide a construct for unifying the sciences. Woven throughout *Glencoe Science*, themes integrate facts and concepts. Like story lines, themes provide "big ideas" that can link the science disciplines. Themes are an important part of any teaching strategy because they help students see the importance of truly understanding concepts rather than simply memorizing isolated facts. While there are many possible themes around which to unify science, we have chosen four: Energy, Systems and Interactions, Scale and Structure, and Stability and Change.

Energy

Energy is a central concept of the physical sciences that pervades the biological and Earth sciences. In physical terms, energy is the ability of an object to change itself or its surroundings—the capacity to do work. In chemical terms, it forms the basis of reactions between compounds. In biological terms, it gives living systems the ability to maintain themselves, to grow, and to reproduce. Energy sources are crucial in the interactions among science, technology, and society.

Systems and Interactions

A system can be incredibly tiny, such as an atom's nucleus and electrons; extremely complex, such as an ecosystem; or unbelievably large, as the stars in a galaxy. By defining the boundaries of the system, one can study the interactions among its parts. The interactions may be a force of attraction between the positively charged nucleus and negatively charged electron. In an ecosystem, on the other hand, the interactions may be between the predator and its prey, or among the plants and animals.

Scale and Structure

Used as a theme, Structure emphasizes the relationship among different structures. Scale defines the focus of the relationship. As the focus is shifted from a system to its components, the properties of the structure may remain constant. In other systems—an ecosystem for example, which includes a change in scale from interactions between prey and predator to the interactions among systems inside an animal—the structure changes drastically. In *Glencoe Science*, the authors have tried to stress how we know what we know and why we believe it to be so. Thus, explanations remain on the macroscopic level until students have the background needed to understand how the microscopic structure operates.

Stability and Change

A system that is stable is constant. Often, the stability is the result of a system being in equilibrium. If a system is not stable, it undergoes change. Changes in an unstable system may be characterized as trends (position of falling objects), cycles (the motion of planets around the sun), or irregular changes (radioactive decay).

Theme Development

These four major themes are developed within the student material and discussed throughout the *Teacher Wraparound Edition.* Each chapter of *Glencoe Science* incorporates at least one of these themes. These themes are interwoven throughout and are developed as appropriate to the topic presented.

The *Teacher Wraparound Edition* includes a **Theme Connection** section for each unit opener. This section provides a framework for integrating concepts discussed in the unit opener. Each chapter opener includes a **Theme Connection** section to explain the chapter's themes and to point out the major chapter concepts supporting those themes. Throughout the chapters, **Theme Connections** show specifically how a topic in the *Student Edition* relates to the themes.

Theme connections stressed in each chapter of *Glencoe Science* are indicated in the chart on page 15T.

GLENCOE SCIENCE Chapters

Themes

Chapters	Scale and Structure	Energy	Stability and Change	Systems and Interactions
1 The Nature of Science				✔
2 What is life?	✔			
3 Cells	✔			✔
4 Heredity and Reproduction		✔	✔	
5 Diversity and Adaptations			✔	
6 Ecology				✔
7 Resources				✔
8 What is matter?	✔			
9 Properties and Changes			✔	
10 Forces and Motion			✔	
11 Work and Machines				✔
12 Energy		✔	✔	✔
13 Electricity and Magnetism		✔	✔	✔
14 The Solar System and Beyond	✔			
15 Earth Materials	✔		✔	
16 Earth's Structures		✔		
17 Water				✔
18 Earth's Atmosphere	✔		✔	

Using the Internet

Getting Started

If you're already familiar with the Internet, skip to the sites listed on the next page. If you need some tips on how to get started, keep reading.

The Internet is an enormous reference library and a communication tool. You can use it to retrieve information quickly from computers around the globe. Like any good reference, it has an index so you can locate the right piece of information. An Internet index entry is called a Universal Resource Locator, or URL. Here's an example:

http://www.glencoe.com /sec/science

The first part of the URL tells the computer how to display the information. The second part, after the double slash, names the organization and the computer where the information is stored. The part after the first single slash tells the computer which directory to search and which file to retrieve. File locations change frequently. If you can't find what you're looking for, use the first part of the address only, and follow links to what you need.

The World Wide Web

The World Wide Web (WWW), a subset of the Internet, began in 1992. Unlike regular text files, Web files can have links to other text files, images, and sound files. By clicking on a link, you can see or hear the linked information.

How do I get access?

To use the Internet, you need a computer, a modem, a telephone line, and a connection to the Internet. If your school doesn't have a connection, contact your local public library or a university; they often give free access to students and educators.

Glencoe Online Science

The Glencoe Science Web site at www.glencoe.com/sec/science/glencoescience provides students and teachers a wide range of materials. On the Glencoe Science Web site are links to other web sites on the Internet that provide more information on the topics students are studying. Internet projects give students an opportunity to research and share data with other students from around the country and even around the world. Professional development materials give teachers additional resources for teaching science. A complete list of what is on the Glencoe Science Web site is provided below.

CAUTION: Contents may shift!

The sites referenced on any portion of Glencoe's Web site are not under Glencoe's control. Therefore, Glencoe can make no representation concerning the content of these sites. Extreme care has been taken to list only reputable links by using educational and government sites whenever possible. Internet searches have been used that return only sites that contain no content intended for mature audiences.

Glencoe Science: An Introduction to the Life, Earth, and Physical Sciences Internet Site Features	
Student Site	**Teacher Site**
Web Links - Links to other sites on the Internet that provide information on science topics	**Web Links -** Links to other sites on the Internet that provide information on science topics
Interactive Tutor - Online worksheets for study, practice, and review along with answers	**Interactive Tutor -** Online worksheets for study, practice, and review along with answers
Skill Handbook - A handbook describing how to use basic science skills	**Skill Handbook -** A handbook describing how to use basic science skills
Unit Internet Projects - Internet based projects	**Unit Internet Projects -** Internet based projects with additional information to help students carry out the projects
	Teacher Corner • Professional Development • Teacher Forum - a place to get and give information on teaching science

Using Technology in the Classroom

Technology helps you adapt your teaching methods to the needs of your students. Glencoe classroom technology products provide many pathways to help you match students' different learning styles. To make your lesson planning easier, all of the technology products listed below are correlated to the student text.

MindJogger

▶ A videoquiz for each chapter can be used to assess prior knowledge or review content before an exam. Student teams work cooperatively to answer three rounds of questions posed by the video host. Each round requires higher-level thinking skills than the previous round.

NATIONAL GEOGRAPHIC SOCIETY

▶ Glencoe is proud to offer Videodiscs and CD-ROMs from the *National Geographic Society.* These interactive learning tools remove the walls from your classroom and transport your students to the far corners of the world. Award-winning titles from National Geographic give students exciting, easy-to-use resources. With the click of a mouse, students can explore the world—and all that's in it—through video, photographs, sound, and text.

Science and Technology Videodisc Series

Infinite Voyage Videodisc and Videotape Series

Videodiscs

▶ Videodiscs are designed to be used interactively in the classroom. Bar codes in this book allow you to step through the programs and pause to discuss and answer on-screen questions. Bar codes for the following videodiscs also appear throughout this Teacher Edition:

▶ Science & Technology Videodiscs
▶ Infinite Voyage Videodiscs
▶ Glencoe Life, Earth, and Physical Science Interactive Videodiscs
▶ National Geographic Videodiscs

Glencoe Software Support Hotline   1-800-437-3715

Should you encounter any difficulty when setting up or running Glencoe software, contact the Software Support Center at Glencoe Publishing between 8:30 A.M. and 4:30 P.M. Eastern Time.

Computer Test Bank

▶Glencoe's Test Generator for Macintosh and for Windows makes creating, editing, and printing tests quick and easy. You also can edit questions or add your own favorite questions and graphics.

English/Spanish Audiocassettes

▶Audio chapter summaries in English and in Spanish are a way for auditory learners, students with reading difficulties, and ELL students to review key chapter concepts. Students can listen individually during class or check out tapes and use them at home. You may find them useful for reviewing the chapter as you plan lessons.

Electronic Teacher Classroom Resources

Glencoe Science Teacher Classroom Resources come in print form or easy-to-use CD-ROM software.

▶Carries easily
▶Comprehensive
▶Convenient
▶Customizes easily

Glencoe Interactive CD-ROMs

▶Glencoe has an interacive CD-ROM for each of the textbooks in the *Glencoe Life, Earth, and Physical Science* series. These can be used with *Glencoe Science* as a library resource, as a whole-class presentation, or as a review for individual students. Each CD-ROM can be used with its corresponding content unit in *Glencoe Science*.

To order, or for more information about Glencoe Technology, call Customer Service at
1-800-334-7344.

TEACHER Classroom Resources

Color Transparencies
▶ Teaching Transparencies
▶ Science Integration Transparencies
▶ Section Focus Transparencies

A Total of 118 Transparencies

Review and Reinforcement
▶ Chapter Review
▶ Reinforcement
▶ Study Guide for Content Mastery

Enrichment and Application
▶ Enrichment
▶ Cross-Curricular Integration
▶ Science and Society/Technology Integration
▶ Multicultural Connections

Assessment Variety

▶ Assessment (Chapter and Unit Tests)
▶ Performance Assessment
▶ Computer Test Bank Manual

Innovative Hands-On Learning

▶ Activity Worksheets
▶ Science Integration Activities
▶ Laboratory Manual, SE and TE

Teaching Resources

▶ Alternate Assessment in the Science Classroom
▶ Performance Assessment in the Science Classroom
▶ Lab and Safety Skills
▶ Spanish Resources
▶ Cooperative Learning
▶ Exploring Environmental Issues

Facilitated Learning

Strategies suggested in *Glencoe Science* support a facilitated learning approach to science education. The role of the teacher as facilitator is to provide an atmosphere in which students design and direct activities. To develop the idea that science investigation is not made up of closed-end questions, the teacher should ask guiding questions and be prepared to help students draw meaningful conclusions when results do not match predictions. Through the numerous activities, cooperative learning opportunities, and a variety of critical thinking exercises in *Glencoe Science,* you can feel comfortable taking a facilitated learning approach to science in your classroom.

Activities

A facilitated learning approach to science is rooted in an activities-based plan. Students must be provided with sensorimotor experiences as a base for developing abstract ideas. *Glencoe Science* utilizes a variety of learning-by-doing opportunities. **MiniLABs** allow students to consider questions about the concepts to come, make observations, and share prior knowledge. **MiniLABs** require a minimum of equipment, and students may take responsibility for organization and execution.

Design Your Own Activities develop and reinforce or restructure concepts, as well as develop the ability to use process skills. **Design Your Own Activity** formats guide students to make their own discoveries. Students collect real evidence and are encouraged through open-ended questions to reflect and reformulate their ideas based on this evidence.

Three **Unit Internet Projects** enable students to practice their computer and research skills. Links to sites that are sources of data related to each project are provided on the *Glencoe Science* Web site, **www.glencoe.com/sec/science.** Each project also comes with directions for doing the project by using the Internet.

Cooperative Learning

Cooperative learning adds the element of social interaction to science learning. Group learning allows students to verbalize ideas and encourages the reflection that leads to active construction of concepts. It allows students to recognize the inconsistencies in their own perspectives and the strengths of others'. By presenting the idea that there is no one, "ready-made" answer, all students may gain the courage to try to find a viable solution. **Cooperative Learning** strategies appear in the *Teacher Wraparound Edition* margins whenever appropriate.

And More . . .

Flex Your Brain—a self-directed, critical-thinking matrix—is introduced in Chapter 1, The Nature of Science. This activity, referenced wherever appropriate in the *Teacher Wraparound Edition* margins, assists students in identifying what they already know about a subject, then in developing independent strategies to investigate further. A reproducible version of **Flex Your Brain** is provided on page 5 of the *Activity Worksheets* booklet.

Students are encouraged to discover the pleasure of solving a problem through a variety of features. The **Science and Society** features in each chapter invite students to confront real-life problems while practicing useful skills, such as distinguishing fact from opinion, outlining, making generalizations, and analyzing news media. In addition, the **Skill Handbook** and the **Technology Handbook** give specific examples to guide students through the steps of developing thinking and process skills and technology skills such as using E-mail and spreadsheets. **Developing Skills** and **Thinking Critically** sections of the **Chapter Review** allow the teacher to assess and encourage progress in a variety of thinking skills.

Developing and Applying
Thinking Skills

Science is not just a collection of facts for students to memorize. Rather, it is a process of applying those observations and intuitions to situations and problems, formulating hypotheses, and drawing conclusions. This interaction of the thinking process with the content of science is the core of science and should be the focus of science study. Students, much like scientists, will plan and conduct research or experiments based on observations, evaluate their findings, and draw conclusions to explain their results. This process then begins again, using the new information for further investigation.

Basic Process Skills

Observing

The most basic process is observing. Through observation—seeing, hearing, touching, smelling, tasting—the student begins to acquire information about an object or event. Observation allows a student to gather information regarding size, shape, texture, or quantity of an object or event.

Classifying

One of the simplest ways to organize information gathered through observation is by classifying. Classifying involves the sorting and grouping of objects according to similarities or differences.

Using Numbers

Numbers are used to quantify information, including variables and measurements. Quantified information is useful for making comparisons and classifying data or objects.

Communicating

Communicating information is an important part of science. Once all the information is gathered, it is necessary to organize the observations so that the findings can be considered and shared by others. Information can be presented in tables, charts, a variety of graphs, or models.

Developing Thinking Skills

Measuring

Measuring is a way of quantifying information in order to classify or order it by magnitude, such as area, length, volume, or mass. Measuring can involve the use of instruments and the necessary skills to manipulate them.

Inferring

Inferences are logical conclusions based on observations and are made after careful evaluation of all the available facts or data. Inferences are a means of explaining or interpreting observations and are based on making judgments. Therefore, inferences can be invalid.

Predicting

Predicting involves suggesting future events based on observations and inferences about current events. Reliable predictions are based on making accurate observations and measurements and interpreting them accurately.

Using Space/Time Relationships

This process skill involves describing the spatial relationships of objects and how those relationships change with time. For example, a student may be required to describe the motion, direction, symmetry, or shape of an object or objects.

Sequencing

Sequencing involves arranging objects or events in a particular order and may imply a hierarchy or a chronology. Developing a sequence involves the skill of identifying relationships among objects or events. A sequence may be in the form of a numbered list or a series of objects or ideas with directional arrows.

Comparing and Contrasting

Comparing is a way of identifying similarities among objects or events, while contrasting identifies their differences. Comparing and contrasting provides a way of describing and evaluating what is known about something.

Recognizing Cause and Effect

Recognizing cause and effect involves observing actions or events and making logical inferences about why they occur. Recognizing cause and effect can lead to further investigation to isolate a specific cause of a particular event.

Interpreting Scientific Illustrations

Illustrations provide examples that clarify difficult concepts or give additional information about the topic being studied. Interpreting illustrations involves processing visual information into a conceptual construct.

Integrated Process Skills

Interpreting Data

Interpreting data involves synthesizing information (collected data) to make generalizations about the problem under study and apply those generalizations to new problems. Interpreting data may therefore involve many of the other process skills, such as predicting, inferring, classifying, and using numbers.

Defining Operationally

Forming operational definitions involves stating the meaning of something in terms of its function. Operational definitions are formed through observation.

Experimenting

Experimenting involves gathering data to test a hypothesis. For data to be reliable, the experiment must be conducted under controlled conditions with a limited number of variables.

Controlling Variables

Controlling variables requires understanding and regulating all the possible variables that might affect the outcome of an experiment. Failure to control variables can lead to unreliable results.

Formulating Hypotheses

Formulating a hypothesis involves making observations followed by some kind of inference as to what those observations mean. The hypothesis is then tested via experimentation to determine its validity.

Formulating Models

A model is a way to concretely represent abstract ideas or relationships. Models can also be used to predict the outcome of future events or relationships. Models may be expressed physically, verbally, or mentally.

Concept Mapping

Concept Maps are visual representations or graphic organizers of relationships among concepts. Concept mapping may involve sequencing temporal events in which a final outcome is observed (events chain), or in which the last event relates back to the beginning of the sequence (cycle map). Hierarchies can be represented by network tree concept maps. Central and supporting ideas that are nonhierarchical can be represented by spider concept maps.

Making and Using Tables

Making tables involves understanding cause and effect and relationships among ideas, and representing those relationships in a tabular format.

Making and Using Graphs

Making graphs involves interpreting data and representing that data in a visual format.

Interaction of Content and Process

Glencoe Science encourages the interaction between science content and thinking processes. We've known for a long time that hands-on activities are a way of providing a bridge between science content and student comprehension. *Glencoe Science* encourages the interaction between content and thinking processes by offering literally hundreds of hands-on activities that are easy to set up and do. In the student text, the **MiniLABs** require students to make observations and collect and record a variety of data. Full-page **Activities** connect hands-on experiences to the content information.

At the end of each chapter, students use thinking processes as they complete **Developing Skills, Thinking Critically,** and **Performance Assessment** questions.

Skill and Technology Handbooks

The **Skill Handbook** provides students with opportunities to practice the science processes relevant to the material they are studying. The **Technology Handbook** provides guidance in using technology skills such as using E-mail, setting up a multimedia presentation, and using a computerized catalog in a library.

Thinking Skills in the *Teacher Wraparound Edition*

These processes are also featured throughout the *Teacher Wraparound Edition.* In the margins are suggestions for students to write in a journal. Keeping a journal encourages students to communicate their ideas, a key process in science.

Flex Your Brain is a self-directed activity designed to assist students in developing thinking processes as they investigate content areas. Suggestions for using this decision-making matrix are in the margins of the *Teacher Wraparound Edition* whenever appropriate for the topic. Further discussion of **Flex Your Brain** can be found in the *Activity Worksheets* booklet of the *Teacher Classroom Resources.*

Thinking Skills Map

On page 26T, you will find a Thinking Skills Map that indicates how frequently thinking skills are encouraged and developed in *Glencoe Science.*

Developing Thinking Skills

Thinking Skills

Thinking Skills	1	2	3	4	5	6	7	8	9	10	11	12	13	14	15	16	17	18
ORGANIZING INFORMATION																		
Communicating	✓	✓	✓	✓	✓	✓	✓	✓	✓	✓	✓	✓	✓	✓	✓	✓	✓	✓
Classifying		✓	✓			✓	✓	✓	✓					✓	✓			
Sequencing	✓	✓	✓	✓	✓	✓		✓			✓	✓		✓	✓	✓		✓
Concept Mapping				✓			✓		✓	✓	✓	✓		✓				
Making and Using Tables	✓	✓	✓	✓		✓	✓	✓				✓		✓	✓	✓	✓	✓
Making and Using Graphs			✓	✓		✓		✓		✓		✓		✓				✓
THINKING CRITICALLY																		
Observing and Inferring	✓	✓	✓	✓	✓	✓	✓	✓	✓	✓	✓	✓	✓	✓	✓	✓	✓	
Comparing and Contrasting	✓	✓	✓	✓	✓		✓		✓	✓	✓	✓	✓	✓		✓		
Recognizing Cause and Effect	✓	✓		✓		✓	✓	✓	✓	✓	✓				✓	✓		✓
PRACTICING SCIENTIFIC PROCESSES																		
Forming a Hypothesis	✓	✓	✓				✓	✓	✓	✓	✓	✓		✓	✓	✓	✓	✓
Designing an Experiment to Test a Hypothesis	✓	✓	✓	✓	✓	✓	✓	✓	✓	✓	✓	✓	✓	✓	✓	✓	✓	✓
Separating and Controlling Variables	✓	✓	✓	✓						✓			✓			✓	✓	✓
Interpreting Data	✓			✓	✓	✓	✓	✓	✓	✓			✓		✓		✓	✓
REPRESENTING AND APPLYING DATA																		
Interpreting Scientific Illustrations	✓		✓	✓	✓		✓		✓	✓	✓		✓	✓			✓	✓
Making Models	✓			✓		✓	✓	✓		✓	✓	✓		✓		✓		✓
Measuring	✓	✓			✓	✓	✓		✓	✓	✓	✓					✓	✓
Predicting and Estimating				✓	✓			✓	✓	✓	✓	✓		✓		✓	✓	
Using Numbers and Calculating				✓		✓	✓		✓	✓	✓			✓		✓		✓

Flex Your Brain

A key element in the coverage of problem-solving and critical-thinking skills in *Glencoe Science* is a critical-thinking matrix called **Flex Your Brain.**

Flex Your Brain provides students with an opportunity to explore a topic in an organized, self-checking way, and then identify how they arrived at their responses during each step of their investigation. The activity incorporates many of the skills of critical thinking. It helps students to consider their own thinking and learn about thinking from their peers.

Where is Flex Your Brain found?

Chapter 1, The Nature of Science, is an introduction to critical thinking and problem solving. **Flex Your Brain** accompanies the text section in the introductory chapter. A worksheet for **Flex Your Brain** appears on page 5 of the *Activity Worksheets* book and as a transparency (Teaching Transparency 1) in the *Teacher Classroom Resources.* The *Activity Worksheets* version provides spaces for students to write in their responses.

In the *Teacher Wraparound Edition,* suggested topics are given in each chapter for the use of **Flex Your Brain.** You can photocopy the worksheet master from the *Teacher Classroom Resources.*

Using Flex Your Brain

Flex Your Brain can be used as a whole-class activity or in cooperative groups, but is primarily designed to be used by individual students within the class. There are three basic steps.

1. Teachers assign a class topic to be investigated using **Flex Your Brain.**
2. Students use **Flex Your Brain** to guide them in their individual explorations of a topic.
3. After students have completed their explorations, teachers guide them in a discussion of their experiences with **Flex Your Brain,** bridging content and thinking processes.

Flex Your Brain can be used at many different points in the lesson plan.

Introduction: Ideal for introducing a topic, **Flex Your Brain** elicits students' prior knowledge and identifies misconceptions, enabling the teacher to formulate plans specific to student needs.

Development: Flex Your Brain leads students to find out more about a topic on their own, and develops their research skills while increasing their knowledge. Students actually pose their own questions to explore, making their investigations relevant to their personal interests and concerns.

Review and Extension: Flex Your Brain allows teachers to check student understanding while allowing students to explore aspects of the topic that go beyond the material presented in class.

Concept Maps

In science, concept maps make abstract information concrete and useful, improve retention of information, and show students that thought has shape.

Concept maps are visual representations or graphic organizers of relationships among particular concepts. Concept maps can be generated by individual students, small groups, or an entire class. *Glencoe Science* develops and reinforces four types of concept maps—the **network tree, events chain, cycle concept map,** and **spider concept map**—that are most applicable to studying science. Examples of the four types and their applications are shown on this page and page 29T. Students can learn how to construct each of these types of concept maps by referring to the **Skill Handbook.** To further develop concept mapping skills, each chapter of the *Study Guide for Content Mastery* in the *Teacher Classroom Resources* has a concept map.

Building Concept Mapping Skills

The **Developing Skills** section of the **Chapter Review** provides opportunities for practicing concept mapping. A variety of concept mapping approaches is also used throughout the text. Students may be directed to make a specific type of concept map and may be provided the terms to use. At other times, students may be given only general guidelines. For example, concept terms to be used may be provided and students may be required to select the appropriate model to apply, or vice versa. Finally, students may be asked to provide both the terms and type of concept map to explain relationships among concepts. When students are given this flexibility, it is important for you to recognize that while sample answers are provided, student responses may vary. Look for the conceptual strength of student responses, not absolute accuracy. You'll notice that most network tree maps provide connecting words that explain the relationships between concepts. We recommend that you not require all students to supply these words, but many students may be challenged by this aspect.

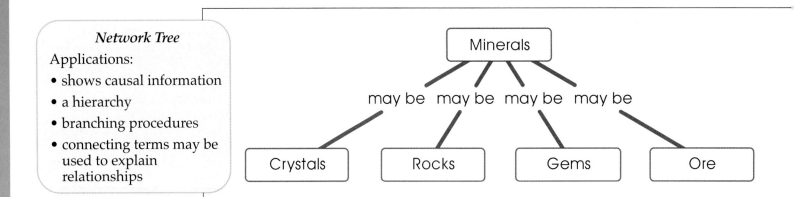

Network Tree

Applications:
- shows causal information
- a hierarchy
- branching procedures
- connecting terms may be used to explain relationships

Study Guide for Content Mastery

The *Study Guide for Content Mastery* book of the *Teacher Classroom Resources,* too, provides a developmental approach for students to practice concept mapping.

As a teaching strategy, generating concept maps can be used to preview a chapter's content by visually relating the concepts to be learned and allowing the students to read with purpose. Using concept maps for previewing is especially useful when there are many new key science terms for students to learn. As an assessment and review strategy, constructing concept maps reinforces main ideas and clarifies their relationships. Construction of concept maps using cooperative learning strategies as described in this Teacher Guide will allow students to practice both interpersonal and process skills.

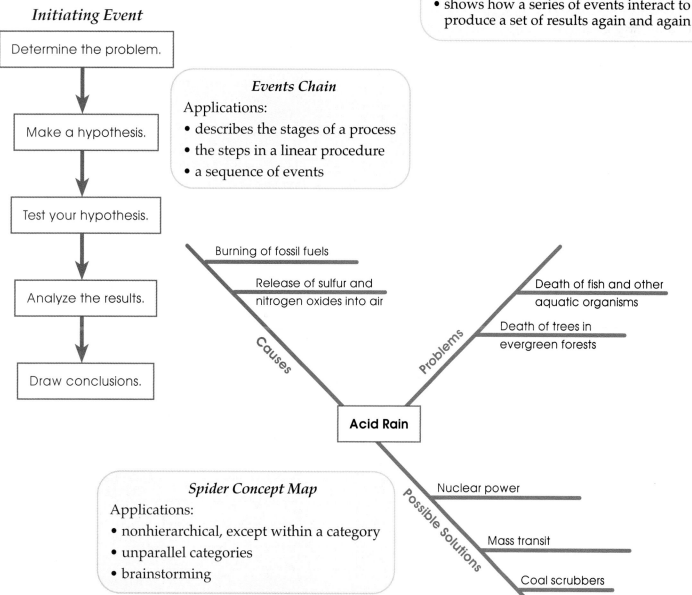

CO_2 + H_2O — Sunlight/Chlorophyll

Respiration — Energy released — Energy stored — Photosynthesis

O_2 + $C_6H_{12}O_6$

Cycle Concept Map
Application:
- shows how a series of events interact to produce a set of results again and again

Initiating Event

Determine the problem.

Make a hypothesis.

Test your hypothesis.

Analyze the results.

Draw conclusions.

Events Chain
Applications:
- describes the stages of a process
- the steps in a linear procedure
- a sequence of events

Burning of fossil fuels

Release of sulfur and nitrogen oxides into air

Death of fish and other aquatic organisms

Death of trees in evergreen forests

Causes

Problems

Acid Rain

Possible Solutions

Nuclear power

Mass transit

Coal scrubbers

Spider Concept Map
Applications:
- nonhierarchical, except within a category
- unparallel categories
- brainstorming

Using Trade Books

ENHANCING LESSONS

A full-curriculum program such as *Glencoe Science* provides you with many transparencies and supplemental booklets to enhance your presentation. But there are numerous outside sources for both teacher and student that can expand science into what may be more familiar settings. Trade books—books for broad readership—can serve that purpose. Listed below are trade books for both teachers (Category A) and students (Category B) that can compliment your *Glencoe Science* presentation. These books have been correlated to *Glencoe Science* in a chart at the bottom of the page. They are not provided as part of the *Glencoe Science* program.

A1. Sarquis, Jerry, Lynn Hogue, Mickey Sarquis, and Linda Woodward. *Investigating Solids, Liquids, and Gases with TOYS.* NewYork: The McGraw-Hill Companies, Inc., 1997.

A2. Sumners, Carolyn. *TOYS in Space.* New York: The McGraw-Hill Companies, Inc., 1997.

A3. Iritz, Maxine Haren. *Super Science Fair Sourcebook.* New York: The McGraw-Hill Companies, Inc., 1996.

A4. Taylor, Beverly A. P., James Poth, and Dwight J. Portman. *Teaching Physics with TOYS.* New York: The McGraw-Hill Companies, Inc., 1995.

A5. Sarquis, Jerry L., Mickey Sarquis, and John P. Williams. *Teaching Chemistry with TOYS.* New York: The McGraw-Hill Companies, Inc., 1995.

A6. Daley, Michael. *Amazing Sun Fun Activities.* New York: The McGraw-Hill Companies, Inc., 1998.

A7. Gertz, Susan E. *Teaching Physical Science Through Children's Literature.* New York: The McGraw-Hill Companies, Inc., 1996.

A8. Taylor, Beverly A. P. *Exploring Energy with TOYS.* New York: The McGraw-Hill Companies, Inc., 1998.

A9. Carrow, Robert. *Put a Fan in Your Hat!* New York: The McGraw-Hill Companies, Inc., 1997.

A10. Sobey, Ed. *Car Smarts.* New York: The McGraw-Hill Companies, Inc., 1997.

A11. Sarquis, Mickey. *Exploring Matter with TOYS.* New York: The McGraw-Hill Companies, Inc., 1997.

B1. Benoit, Margaret. *Who Killed Olive Souffle?* New York: The McGraw-Hill Companies, Inc., 1997.

B2. Lloyd, Emily. *Forest Slump—The Case of the Pilfered Pine Needles.* New York: The McGraw-Hill Companies, Inc., 1997.

B3. Silver, Donald M. *One Small Square: Coral Reef.* New York: The McGraw-Hill Companies, Inc., 1998.

B4. Silver, Donald M. *One Small Square: Swamp.* New York: The McGraw-Hill Companies, Inc., 1997.

B5. Aronson, Billy. *Betting on Forever.* New York: The McGraw-Hill Companies, Inc., 1997.

B6. Ball, Jackie, and Kit Carlson. *The Leopard Son.* The McGraw-Hill Companies, Inc., 1997.

CHAPTER	TRADE BOOK	CHAPTER	TRADE BOOK
1	A3, B1	10	A4, A7, A9, A10
2		11	A4, A8
3	B3	12	A4, A6, A7, A8
4	B3, B4	13	A4, A6, A7
5	B3, B4, B5, B6	14	A2, A6
6	B2, B3, B4, B5, B6	15	A10
7	A3, A6, B2	16	A10
8	A1, A5, A11	17	
9	A1, A5, A7, A11, B1	18	A7, A10

Planning Your Course

Traditional Scheduling

Glencoe Science provides flexibility in the selection of topics and content, which allows teachers to adapt the text to the needs of individual students and classes. In this regard, the teacher is in the best position to decide what topics to present, the pace at which to cover the content, and what material to give the most emphasis. To assist the teacher in planning the course, a planning guide has been provided.

Glencoe Science may be used in a full-year, two-semester course that is comprised of 180 periods of approximately 45 minutes each. This type of schedule is represented in the table under the heading of Single-Class Scheduling. As an alternative, the table also outlines a plan for block scheduling, which is described in the next section.

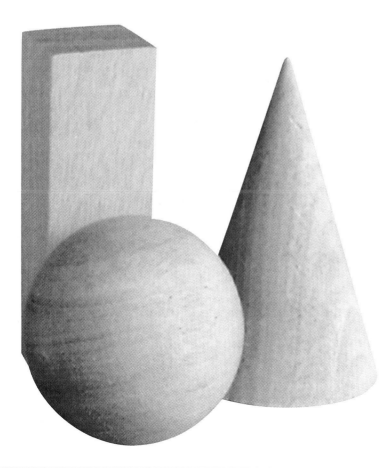

Block Scheduling

To build flexibility into the curriculum, many schools are introducing a block scheduling approach. Block scheduling often involves covering the same information in fewer days with longer class periods. Thus, a course that lasts an entire year may take only one semester under a block scheduling approach. This type of approach allows curriculum supervisors and teachers to tailor the curriculum to meet students' needs while achieving local and/or state curriculum goals. Long, concentrated periods of study can facilitate the learning of complex material. Furthermore, students may be able to take a wider variety of course-work under a block scheduling plan than under a traditional full-year plan, thus enriching their middle-school experience and giving them a broader foundation for high school-level work.

If you follow a block schedule, you may want to consider either combining lessons or eliminating certain topics and spending more time on the topics you do cover. *Glencoe Science* provides the flexibility that allows you to tailor the program to your needs. *Glencoe Science* also provides a wide variety of support materials that will help you and your students, whether you follow a block schedule or full-year schedule. A block schedule may enable you to complete some of the longer **Activities** in one class period. Have students fill out the accompanying *Activity Worksheets* from the *Teacher Classroom Resources* as they complete the activities. The *Teacher Classroom Resources* provide a wealth of materials that provide classroom management support.

In the table shown on the following pages, it is assumed that for block scheduling, the course will be taught for one semester and include 90 periods of approximately 90 minutes each.

Please remember that the planning guide is provided as an aid in planning the best course for your students. You should use the planning guide in relation to your curriculum and the ability levels of the classes you teach, the materials available for activities, and the time allotted for teaching.

Planning Guide

Unit	Chapter/Section	Single-Class (180 days)	Block (90 days)
1	THE NATURE OF SCIENCE	6	3
1-1	What is science?	2	1
1-2	Doing Science	3	1.5
	Chapter Review	1	0.5
UNIT 1	LIFE SCIENCE	62	31
2	WHAT IS LIFE?	10	5
2-1	Alive or Not	5	2.5
2-2	Classifying Life	3	1.5
2-3	Science and Society: Viruses	1	0.5
	Chapter Review	1	0.5
3	CELLS	10	5
3-1	Cells—Building Blocks of Life	5	2.5
3-2	The Different Jobs of Cells	3	1.5
3-3	Science and Society: Test-Tube Tissues	1	0.5
	Chapter Review	1	0.5
4	HEREDITY AND REPRODUCTION	10	5
4-1	Continuing Life	4	1.5
4-2	Science and Society: Cloning	1.5	1
4-3	Genetics: The Study of Inheritance	3.5	2
	Chapter Review	1	1.5
5	DIVERSITY AND ADAPTATIONS	10	5
5-1	Diversity of Life	3	1
5-2	Unity of Life	3	2
5-3	Science and Society: Captive-Breeding Programs	1	0.5
5-4	History of Life	2	1
	Chapter Review	1	0.5

Unit	Chapter/Section	Single-Class (180 days)	Block (90 days)
6	ECOLOGY	10	5
6-1	What is an ecosystem?	3	1
6-2	Relationships Among Living Things	3	2
6-3	Science and Society: Gators at the Gate!	1	0.5
6-4	Energy Through the Ecosystem	2	1
	Chapter Review	1	0.5
7	RESOURCES	12	6
7-1	Natural Resource Use	4	2
7-2	People and the Environment	4	2
7-3	Protecting the Environment	2	1
7-4	Science and Society: A Tool for the Environment	1	0.5
	Chapter Review	1	0.5
UNIT 2	PHYSICAL SCIENCE	60	30
8	WHAT IS MATTER?	10	5
8-1	Matter and Atoms	5	3
8-2	Types of Matter	3	1
8-3	Science and Society: Synthetic Elements	1	0.5
	Chapter Review	1	0.5
9	PROPERTIES AND CHANGES	10	5
9-1	Physical and Chemical Properties	4.5	2
9-2	Science and Society: Road Salt	1	0.5
9-3	Physical and Chemical Changes	3.5	2
	Chapter Review	1	0.5

Planning Guide for GLENCOE SCIENCE

Unit	Chapter/Section	Scheduling Single-Class (180 days)	Block (90 days)
10	FORCES AND MOTION	10	5
	10-1 What is gravity?	2	1
	10-2 How fast is "fast"?	2.5	1.5
	10-3 How do things move?	3.5	1.5
	10-4 Science and Society: Air-Bag Safety	1	0.5
	Chapter Review	1	0.5
11	WORK AND MACHINES	10	5
	11-1 What is work?	3	1.5
	11-2 Simple Machines	3	1.5
	11-3 What is a compound machine?	2	1
	11-4 Science and Society: Access for All	1	0.5
	Chapter Review	1	0.5
12	ENERGY	10	5
	12-1 How does energy change?	4	2
	12-2 How do you use thermal energy?	4	2
	12-3 Science and Society: Sun Power	1	0.5
	Chapter Review	1	0.5
13	ELECTRICITY AND MAGNETISM	10	5
	13-1 What is an electric charge?	2	1
	13-2 How do circuits work?	3	1.5
	13-3 What is a magnet?	2	1
	13-4 Science and Society: Electricity Sources	2	1
	Chapter Review	1	0.5
UNIT 3	EARTH SCIENCE	52	26
14	THE SOLAR SYSTEM AND BEYOND	10	5
	14-1 Earth's Place in Space	3	1.5
	14-2 The Solar System	4	2
	14-3 Science and Society: Space Exploration: Boom or Bust?	1	0.5

Planning Guide for GLENCOE SCIENCE

Unit	Chapter/Section	Scheduling Single-Class (180 days)	Block (90 days)
	14-4 Stars and Galaxies	1	0.5
	Chapter Review	1	0.5
15	EARTH MATERIALS	11	5.5
	15-1 Minerals: Earth's Jewels	4	2
	15-2 Igneous and Sedimentary Rocks	2.5	1.5
	15-3 Science and Society: Monument or Energy?	1	0.5
	15-4 Metamorphic Rocks and the Rock Cycle	2.5	1
	Chapter Review	1	0.5
16	EARTH'S STRUCTURE	12	6
	16-1 What's shaking?	3.5	1.5
	16-2 Science and Society: Predicting Earthquakes	1	0.5
	16-3 A Journey to Earth's Center	2.5	1.5
	16-4 Crashing Continents	4	2
	Chapter Review	1	0.5
17	WATER	9	4.5
	17-1 Recycling Water	2.5	1
	17-2 Earth Shaped by Water	2	1
	17-3 Oceans	2.5	1.5
	17-4 Science and Society: Ocean Pollution	1	0.5
	Chapter Review	1	0.5
18	EARTH'S ATMOSPHERE	10	5
	18-1 What's in the air?	3	1.5
	18-2 Weather	3.5	2
	18-3 Science and Society: Forecasting the Weather	1	0.5
	18-4 Climate	1.5	1
	Chapter Review	1	0.5

Content Mastery

The *Study Guide for Content Mastery* was developed for students in need of additional help with reading, for English Language Learners, and for any student who may require extra help. The exercises in the booklet help beginning middle school students master the basic concepts presented in *Glencoe Science,* and reinforce reading and study skills. A Study Skills section precedes the section-by-section Study Guide pages. This section gives students ideas and tips on how to improve their reading and vocabulary skills and to help them learn from visuals.

Every numbered section in *Glencoe Science* is represented in the *Study Guide for Content Mastery* with a full-page exercise, many of which contain concept maps. Concept maps have been found to be helpful for visual learners. Concept maps also help to focus students' attention on the big picture—to give them an overview of the main points of a topic. Other visuals will give students practice in identifying science concepts in context, thereby extending their knowledge to a real situation.

Opportunities for practicing reading and writing and for exercising skills in classifying and sequencing are throughout. In all, *Glencoe Science Study Guide for Content Mastery* will make a difference in your students' ability to appreciate science concepts.

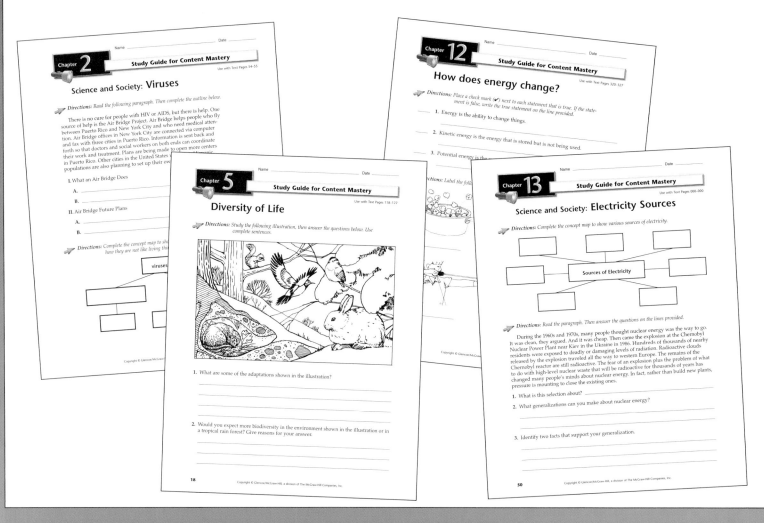

Meeting Individual Needs

Each student brings his or her own unique set of abilities, perceptions, and needs into the classroom. It is important that the teacher try to make the classroom environment as receptive to these differences as possible and to ensure a good learning environment for all students.

It is important to recognize that individual learning styles are different and that learning style does not reflect a student's ability level. While some students learn primarily through visual or auditory senses, others are kinesthetic learners and do best if they have hands-on exploratory interaction with materials. Some students work best alone and others learn best in a group environment. While some students seek to understand the "big picture" in order to deal with specifics, others need to understand the details first in order to put the whole concept together.

In an effort to provide all students with a positive science experience, this text offers a variety of ways for students to interact with materials so that they can utilize their preferred method of learning the concepts. The variety of approaches allows students to become familiar with other learning approaches, as well.

Ability Levels

The activities are broken down into three levels to accommodate students of all ability levels. *Glencoe Science Teacher Wraparound Edition* designates the activities and strategies as follows:

L1 activities are basic activities designed to be within the ability range of low-level students. These activities reinforce the concepts presented.

L2 activities are application activities designed for all students. These activities give students an opportunity for practical application of the concepts presented.

L3 activities are challenging activities designed for the students who are able to go beyond the basic concepts presented. These activities allow students to expand their perspectives on the basic concepts.

The chart on pages 36T-37T gives tips you may find useful in structuring the learning environment in your classroom to meet students' special needs. In addition, **Inclusion Strategies** that address special needs are provided in the bottom margins of the *Teacher Wraparound Edition.*

English Language Learners

In providing for the student with limited English proficiency, the focus needs to be on overcoming a language barrier. Once again, it is important not to confuse ability in speaking/reading English with academic ability or "intelligence." In general, the best method for dealing with ELL variations in learning styles and ability levels is to provide all students with a variety of ways to learn, apply, and be assessed on the concepts. Look for this symbol **ELL** in the teacher margin for specific strategies for students with limited English proficiency.

Learning Styles

We at Glencoe believe it is our responsibility to provide you with a program that allows you to apply diverse instructional strategies to a population of students with diverse learning styles. Several learning styles are emphasized in the *Teacher Wraparound Edition:* Kinesthetic, Visual-Spatial, Logical-Mathematical, Interpersonal, Intrapersonal, Linguistic, Naturalist, and Auditory-Musical. A student with a kinesthetic style learns from touch, movement, and manipulating objects. Visual-spatial learners respond to images and illustrations. Using numbers and reasoning are characteristics of the logical-mathematical learner. A student with an interpersonal style has confidence in social settings, while a student with an intrapersonal style may prefer to learn on his or her own. Linguistic learning involves the use and understanding of words. Finally, auditory-musical learning involves listening to the spoken word and to tones and rhythms.

Any student may display any or all of these styles. The *Student Edition* and *Teacher Wraparound Edition* provide a number of strategies for encouraging students with diverse learning styles. These include **Using Math, Activities, MiniLABs,** and **Science Journal** features. The *Teacher Wraparound Edition* contains **Visual Learning** and **Science Journal,** as well as a number of other strategies. Look for this logo **LS**, which identifies strategies for different learning styles.

Meeting Individual Needs

	Description	Sources of Help/Information
Learning Disabled	All learning-disabled students have an academic problem in one or more areas, such as academic learning, language, perception, social-emotional adjustment, memory, or attention.	*Journal of Learning Disabilities* *Learning Disability Quarterly*
Behaviorally Disordered	Students with behavior disorders deviate from standards or expectations of behavior and impair the functioning of others and themselves. These children may also be gifted or learning-disabled.	*Exceptional Children* *Journal of Special Education*
Physically Challenged	Students who are physically disabled fall into two categories—those with orthopedic impairments and those with other health impairments. Orthopedically impaired children have the use of one or more limbs severely restricted, so the use of wheelchairs, crutches, or braces may be necessary. Children with other health impairments may require the use of respirators or other medical equipment.	Batshaw, M.L. and M.Y. Perset. *Children with Handicaps: A Medical Primer.* Baltimore: Paul H. Brooks, 1981. Hale, G. (Ed.). *The Source Book for the Disabled.* New York: Holt, Rinehart & Winston, 1982. *Teaching Exceptional Children*
Visually Impaired	Students who are visually disabled have partial or total loss of sight. Individuals with visual impairments are not significantly different from their sighted peers in ability range or personality. However, blindness may affect cognitive, motor, and social development, especially if early intervention is lacking.	*Journal of Visual Impairment and Blindness* *Education of Visually Handicapped* American Foundation for the Blind
Hearing Impaired	Students who are hearing impaired have partial or total loss of hearing. Individuals with hearing impairments are not significantly different from their hearing peers in ability range or personality. However, the chronic condition of deafness may affect cognitive, motor, and social development if early intervention is lacking. Speech development also is often affected.	*American Annals of the Deaf* *Journal of Speech and Hearing Research* *Sign Language Studies*
English Language Learners	Multicultural and/or bilingual students often speak English as a second language or not at all. The customs and behavior of people in the majority culture may be confusing for some of these students. Cultural values may inhibit some of these students from full participation.	*Teaching English as a Second Language Reporter* R.L. Jones (Ed.). *Mainstreaming and the Minority Child.* Reston, VA: Council for Exceptional Children, 1976.
Gifted	Although no formal definition exists, these students can be described as having above-average ability, task commitment, and creativity. Gifted students rank in the top five percent of their class. They usually finish work more quickly than other students and are capable of divergent thinking.	*Journal for the Education of the Gifted* *Gifted Child Quarterly* *Gifted Creative/Talented*

Tips for Instruction

With careful planning, the needs of all students can be met in the science classroom.

1. Provide support and structure; clearly specify rules, assignments, and duties.
2. Practice skills frequently. Use games and drills to help maintain student interest.
3. Allow students to record answers on tape and allow extra time to complete tests and assignments.
4. Provide outlines or tape lecture material.
5. Pair students with peer helpers, and provide class time for pair interaction.

---

1. Provide a clearly structured environment with regard to scheduling, rules, room arrangement, and safety.
2. Clearly outline objectives and how you will help students obtain objectives.
3. Reinforce appropriate behavior and model it for students.
4. Do not expect immediate success. Instead, work for long-term improvement.
5. Balance individual needs with group requirements.

---

1. Openly discuss with the student any uncertainties you have about when to offer aid.
2. Ask parents or therapists and students what special devices or procedures are needed, and if any special safety precautions need to be taken.
3. Allow physically disabled students to do everything their peers do, including participating in field trips, special events, and projects.
4. Help nondisabled students and adults understand physically disabled students.

---

1. Help the student become independent. Modify assignments as needed.
2. Teach classmates how to serve as guides.
3. Limit unnecessary noise in the classroom.
4. Provide tactile models whenever possible.
5. Describe people and events as they occur in the classroom.
6. Provide taped lectures and reading assignments.
7. Team the student with a sighted peer for laboratory work.

The *Student Edition* and *Teacher Wraparound Edition* contain a variety of strategies for addressing the individual learning styles of students:

- Both structured and open-ended **Activities**
- Hands-on **MiniLABS**
- **Internet Connections**
- **Visual Learning** strategies
- **Across the Curriculum** strategies
- **Learning Styles** strategies

The following Glencoe products also can help you tailor your instruction to meet the individual needs of your students:

- **Study Guide for Content Mastery**
- **Teaching Transparencies**
- **Section Focus Transparencies**
- **Science Integration Transparencies**
- **MindJogger Videoquizzes**
- **English/Spanish Audiocassettes**
- **Cooperative Learning Resource Guide**
- Various multimedia products including **The Infinite Voyage Series; Science and Technology Videodisc Series; National Geographic Society Series; Glencoe Interactive Life, Earth, and Physical Science CD-ROM products**

---

1. Seat students where they can see your lip movements easily, and avoid visual distractions.
2. Avoid standing with your back to the window or light source.
3. Using an overhead projector allows you to maintain eye contact while writing.
4. Seat students where they can see speakers.
5. Write all assignments on the board, or hand out written instructions.
6. If the student has a manual interpreter, allow both student and interpreter to select the most favorable seating arrangements.

---

1. Remember, students' ability to speak English does not reflect their academic ability.
2. Try to incorporate the student's cultural experience into your instruction. The help of a bilingual aide may be effective.
3. Include information about different cultures in your curriculum to help build students' self-image. Avoid cultural stereotypes.
4. Encourage students to share their cultures in the classroom.

---

1. Make arrangements for students to take selected subjects early and to work on independent projects.
2. Let students express themselves in art forms such as drawing, creative writing, or acting.
3. Make public services available through a catalog of resources, such as agencies providing free and inexpensive materials, community services and programs, and people in the community with specific expertise.
4. Ask "what if" questions to develop high-level thinking skills. Establish an environment safe for risk taking.
5. Emphasize concepts, theories, ideas, relationships, and generalizations.

Cultural Diversity

American classrooms reflect the rich and diverse cultural heritage of the American people. Students come from different ethnic backgrounds and different cultural experiences into a common classroom that must assist all of them in learning. The diversity itself is an important aspect of the learning experience.

Responding to diversity and approaching it as a part of every curriculum is challenging to a teacher, experienced or not. Diversity can be repressed, creating a hostile environment; ignored, creating an indifferent environment; or appreciated, creating a receptive and productive environment. The goal of multicultural education is to promote the understanding of how people from different cultures approach and solve the basic problems all humans have in living and learning.

For the 21st century, census data indicate that our society will grow more diverse as our world becomes smaller. The science students that we are educating will need to be prepared for the increasing diversity in their own society and in the technology that will bring them into contact with people all around the world. A strong grounding in science will be essential for the success of all students.

Cultural Diversity in Science

Every culture has utilized science to explore fundamental questions, meet challenges, and address human needs. The growth of knowledge in science has advanced along culturally diverse roots. No single culture has a monopoly on the development of scientific knowledge. The history of science is rich with the contributions and accomplishments of men and women from diverse cultural backgrounds. A brief look at some of the milestones in science reveals this diversity. Today, individuals from diverse cultural and ethnic backgrounds continue to make significant contributions to science. The **People and Science** feature in the **Student Edition** of *Glencoe Science* highlights some of these individuals and their contributions.

The scientific enterprise is a human enterprise that utilizes science processes (observing and inferring, classifying, and so on) to discover, explore, and explain the environment. Consequently, students must experience and understand that all people, regardless of cultural orientation, have made major contributions to science. *Glencoe Science* addresses this issue. In addition to the individuals highlighted in the **Student Edition,** the **Cultural Diversity** sections of the **Teacher Wraparound Edition** provide information about people and groups who have traditionally been misrepresented or omitted. The intent is to build awareness and appreciation for the global community in which we all live.

Goals of Multicultural Education

The *Glencoe Science* **Teacher Classroom Resources** include a **Multicultural Connections** booklet that offers additional opportunities to integrate multicultural materials into the curriculum. By providing these opportunities, *Glencoe Science* is helping to meet four major goals of multicultural education:

1. promoting the strength and value of cultural diversity
2. promoting human rights and respect for those who are different from oneself
3. promoting social justice and equal opportunity for all people
4. promoting equity in the distribution of power among groups

These goals are accomplished when students see others like themselves in positive settings in their textbooks. They also are accomplished when students realize that major contributions in science are the result of the work of hundreds of individuals from diverse backgrounds. *Glencoe Science* has been designed to enable you, the science teacher, to achieve these goals.

Books that provide additional information on multicultural education are:

Atwater, Mary, et al. *Multicultural Education: Inclusion of All.* Athens, Georgia: University of Georgia Press, 1994.

Banks, James A. (with Cherry A. McGee Banks). *Multicultural Education: Issues and Perspectives.* Boston: Allyn and Bacon, 1989.

Selin, Helaine. *Science Across Cultures: An Annotated Bibliography of Books on Non-Western Science, Technology, and Medicine.* New York: Garland, 1992.

Reflecting Diversity. New York: Macmillan/McGraw-Hill, 1993.

School-to-Work

What Is School-to-Work?

School-to-Work is a rigorous and focused program of study that aims to create a workforce in the United States that is technically literate. It is designed to prepare students enrolled in a general curriculum for the demands of further education or for employment by providing them with essential academic and technical foundations, along with problem-solving, group-process, and lifelong-learning skills. These goals can be achieved by integrating vocational study with higher-level academic study.

What are the characteristics of the School-to-Work curriculum?

In 1990, Congress passed the Carl D. Perkins Vocational and Applied Technology Act to set aside funds for the development and administration of Tech Prep programs. The criteria outlined in Title III of the Perkins Act specify that tech prep programs take place during the last two years of high school, followed by two years of post-secondary occupational education, and that this education culminate in a certificate or associate degree. The Secretary's Commission on Achieving Necessary Skills (SCANS), an arm of the U.S. Department of Labor, published a report in June 1991 that outlined several competencies that characterize successful workers. The School-to-Work curriculum seeks to address these competencies, which include the

- ability to use resources productively.
- ability to use interpersonal skills effectively, including fostering teamwork, teaching others, serving customers, leading, negotiating, and working well with individuals from culturally diverse backgrounds.
- ability to acquire, evaluate, interpret, and communicate data and information.

- ability to understand social, organizational, and technological systems.
- ability to apply technology to specific tasks.

How does School-to-Work relate to the middle school curriculum?

School-to-Work provides an alternative to the college-prep course of study that leads to a career goal. Because of this, School-to-Work is available to all students whether they are taking a college prep, vocational, or general course of study. The middle school years provide an opportunity to identify those students who might benefit from a School-to-Work curriculum once they reach high school. They also provide an opportunity to introduce unfamiliar students to technological applications leading to career opportunities, or to provide practice for students who already have some knowledge of how technology is used. Young children are fascinated by the technology that surrounds them, and they are especially quick to learn how to use and apply it. Take advantage of their enthusiasm by preparing them for a possible School-to-Work education when they are in middle school.

How does *GLENCOE SCIENCE* address School-to-Work issues?

Glencoe Science helps you develop scientific and technological literacy in your students through a variety of performance-based activities that emphasize problem solving, critical thinking skills, and teamwork. In the *Student Edition*, **Problem Solving, Using Technology,** and **Design Your Own Experiment** provide opportunities for practical applications of concepts. **Using Computers** and **Using Math** features give additional practice in computer and math applications. **Unit Internet Projects** provide ideas for activities that can be used as extended projects or for science fairs.

Assessment

Audrey B. Champagne, Ph.D.
Chair, Assessment Working Group on Science Assessment Standards, National Science Education Standards.

As a teacher, you spend a considerable portion of your professional time gathering information and making decisions based on that information. You gather information that will improve your teaching. You read professional journals, attend conferences, and participate in workshops to obtain information about strategies for improving your teaching. You make decisions based on this information. In the final analysis, the information that most influences your professional practice is the information you gather about your students.

What is assessment?

Assessment is the process of gathering information and making decisions based on that information.

How do you decide when and how to assess?

Decisions about when and how to assess should be guided by your answers to the following questions: What use will I make of the information I collect? Will the time and effort it takes to collect the information be justified by the use I make of the information? Is the information I am collecting relevant to the learning activities in which the students are engaged? Can I be confident that the information I collect will be of sufficient quality to justify the decisions I will make with it?

What information do you gather? How do you gather it? How do you use the information?

Through tests, journals, portfolios, and by direct observation, teachers gather information about their students, including their students' understanding of scientific ideas, ability to design and conduct scientific investigations, and their ability to work in groups.

You give **tests** to gather information about what your students have learned and use this information to assign grades. You read students' **journals** to gather information about how well students can communicate scientific information and report on inquiries they have conducted. You may use this information for grading. Or you may use it to plan lessons that will

improve your students' abilities to communicate and write reports. You review student **portfolios** to get information about your students' progress over time. Often, you gather information by asking questions or just by observing your students. You judge how well the lessons you are conducting are proceeding on the basis of students' responses to questions or simply by observing their body language.

How confident must I be in the validity of the information I gather?

High-stakes decisions require high-quality information.

Grades have significant influences in your students' lives. You must be able to justify your grades to students, parents, and your colleagues. Consequently, you must have confidence in the information on which you base your decisions about grades.

What do my students gain from being assessed?

Assessment is a learning process. Students learn from doing well-designed tests and assessment tasks.

Your students pay close attention to the kinds of information you collect and how you use that information. The tests you give communicate to your students the scientific facts and principles that you expect them to learn. The savvy student uses this information as he or she studies. Your attention to portfolios and journals sends the message that you value your students' ability to communicate their scientific ideas. When you gather data about students' abilities to work in groups and perform laboratory techniques, you inform your students that these are abilities you value. The feedback you provide to students about their performance gives even more direct clues about the science content you value. As students engage in well-designed assessment tasks, they often make important new connections among scientific ideas and their applications. Well-designed assessment tasks will help your students meet national and state standards for science.

Assessment Strategies

Traditional tests are useful ways of gathering information about students' knowledge of scientific terminology, their understanding of scientific principles, and their ability to apply these principles.

But science literacy implies much more than knowing and understanding science. Science literacy involves the integration of knowledge and understanding of science with the abilities to conduct scientific investigations and to communicate the results of the investigations. Assessing students' science literacy can not be accomplished with traditional tests alone. Assessing science literacy requires that students be engaged in tasks that require that they integrate knowledge with ability. When these tasks are associated with situations students are likely to encounter in their daily lives, these are called *authentic tasks.*

Performance Assessment refers to the strategies used to assess students' level of science literacy. Because success on performance tasks requires the integration of knowledge with many different abilities, scoring students' performance is a challenging and time-consuming process. Scores may be determined by observing students as they perform the task and/or by the product the student produces. If students are scored on the basis of observation, the knowledge and abilities that are required to perform the task successfully must be identified by the number of points to be awarded for knowing and performing each component set. If students are to be scored on the basis of the product, the characteristics of the expected product must be described and the number of points for a product with each of the characteristics set.

In modern assessment terminology, the specification of the knowledge and ability components or the product characteristics, as well as the point value assigned, is called a *rubric.* Rubrics define ahead of time in explicit terms teachers' expectations for students' performance and work products. Rubrics may be developed by several teachers, by individual teachers, or by teachers and their students. The latter method increases the learning that can result from engaging students in performance assessment, and discussion of the characteristics of a high-quality product helps students self-evaluate their products.

Because of their complexity, performance assessment tasks are challenging for students and are teaching-time intensive. These constraints can be ameliorated to some extent by having groups of students work together on the tasks. *Group performance assessment* is a way in which teachers can assess their students' abilities to work well in groups, as well as their ability to communicate orally about science. Group performance assessment presents its own challenges, especially with regard to scoring; that is, deciding how an individual student's contribution to the performance of the task will be rewarded. Involving students in the process of deciding how individuals should be rewarded can be an important learning experience for them.

Engagement in performance assessment provides many opportunities for student record keeping. *Science Journals* are the place where students keep records of their plans for and the results of their science investigations and performance assessment tasks. The Science Journal is where students practice written science communication and where teachers can gather information about how well students' abilities to communicate scientific ideas are developing.

Students' oral and written communications are windows on students' reasoning. Thus, teachers can use students' communication to assess their scientific reasoning.

Portfolios A student's performance on a single assessment task or scientific investigation provides a snapshot of the student's knowledge and abilities at a certain point in time. A *portfolio* containing a collection of a student's reports on tasks and investigations conducted over a period of several years provides a record of the student's progress over time.

Portfolios are useful sources of information about students' growth over time. Portfolios should contain examples of students' best work, as well as students' reflections about why the contents are their best work. Students should also have the opportunity to reflect on their progress. Portfolio reflections by students are valuable for developing the students' self-assessment skills. Students who find writing a limited form of expression may wish to communicate their understanding of science using drawing, physical models, or plays.

Dr. Champagne is Professor of Chemistry and Professor of Education at the State University of New York at Albany, Albany, New York.

Cooperative Learning

What is cooperative learning?

In cooperative learning, students work together in small groups to learn academic material and interpersonal skills. Group members each learn that they are responsible for accomplishing an assigned group task, as well as for learning the material. Cooperative learning fosters academic, personal, and social success for all students.

Recent research shows that cooperative learning results in

- development of positive attitudes toward science and toward school.
- lower dropout rates for at-risk students.
- building respect for others regardless of race, ethnic origin, or sex.
- increased sensitivity to and tolerance of diverse perspectives.

Establishing a Cooperative Classroom

Cooperative groups in the middle school usually contain from two to five students. Heterogeneous groups that represent a mixture of abilities, genders, and ethnicity expose students to ideas different from their own and help them learn to work with different people.

Initially, cooperative learning groups should work together for only a day or two. After the students are more experienced, they can work with a group for longer periods of time. It is important to keep groups together long enough for each group to experience success and to change groups often enough that students have the opportunity to work with a variety of students.

Students must understand that they are responsible for group members learning the material. Before beginning, discuss the basic rules for effective cooperative learning: (1) listen while others are speaking, (2) respect other people and their ideas, (3) stay on tasks, and (4) be responsible for your own actions.

The *Teacher Wraparound Edition* uses the code **COOP LEARN** at the end of activities and teaching ideas

where cooperative learning strategies are useful. For additional help, refer again to these pages of background information on cooperative learning.

Using Cooperative Learning Strategies

The *Cooperative Learning Resource Guide* of the *Teacher Classroom Resources* provides help for selecting cooperative learning strategies, as well as methods for troubleshooting and evaluation.

During cooperative learning activities, monitor the functioning of groups. Praise group cooperation and good use of interpersonal skills. When students are having trouble with the task, clarify the assignment, reteach, or provide background as needed. Answer questions only when no students in the group can.

Evaluating Cooperative Learning

At the close of the lesson, have groups share their products or summarize the assignment. You can evaluate group performance during a lesson by frequently asking questions to group members picked at random or having each group take a quiz together. You might have all students write papers and then choose one at random to grade. Assess individual learning by your traditional methods.

Managing Activities
In the Science Classroom

Glencoe Science engages students in a variety of experiences to provide all students with an opportunity to learn by doing. The many hands-on activities throughout *Glencoe Science* require simple, common materials, making them easy to set up and manage in the classroom.

MiniLAB

MiniLABs are intended to be short and occur many times throughout the text. The integration of these activities with the core material provides for thorough development and reinforcement of concepts.

Design Your Own Experiment

What makes science exciting to students is working in the lab, observing natural phenomena, and tackling concrete problems that challenge them to find their own answers. *Glencoe Science* provides more than the same "cookbook" activities you've seen hundreds of times before. Students work cooperatively to develop their own experimental designs in **Design Your Own Experiments.** They discover firsthand that developing procedures for studying a problem is not as hard as they thought it might be. Watch your students grow in confidence and ability as they progress through the self-directed **Design Your Own Experiments.**

Preparing Students for Open-Ended Lab Experiences

To prepare students for the **Design Your Own Experiments,** you should follow the guidelines in the *Teacher Wraparound Edition,* especially in the sections titled Possible Procedures and Teaching Strategies. In these sections, you will be given information about what demonstrations to do and what questions to ask students so that they will be able to design their experiment. Your introduction to a **Design Your Own Experiment** will be very different from traditional activity introductions in that it will be designed to focus students on the problem and stimulate their thinking without giving them directions for how to set up their experiment. Different groups of students will develop alternative hypotheses and alternative procedures. Check their procedures before they begin. In contrast to some "cookbook" activities, there may not be just one right answer. Finally, students should be encouraged to use questions that come up during the activity in designing a new experiment.

If you feel that your students are not prepared to begin designing experiments, you may want to practice some paper-and-pencil lab designs using the following format: Select a simple scenario and ask students to design an experiment that will test a hypothesis they make about the problem presented in the reading. Give them a set of questions that will lead them to an appropriate experimental design.

Laboratory Safety

Safety is of prime importance in every classroom. However, the need for safety is even greater when science is taught. The activities in *Glencoe Science* are designed to minimize dangers in the laboratory. Even so, there are no guarantees against accidents. Careful planning and preparation, as well as being aware of hazards, can keep accidents to a minimum. Numerous books and pamphlets are available on laboratory safety with detailed instructions on preventing accidents. In addition, the *Glencoe Science* program provides safety guidelines in several forms. The *Lab and Safety Skills* booklet contains detailed guidelines, in addition to masters you can use to test students' lab and safety skills. The *Student Edition* and *Teacher Wraparound Edition* provide safety precautions and symbols designed to alert students to possible dangers. Know the rules of safety and what common violations occur. Know the **Safety Symbols** used in this book. A safety symbol chart is on page 528. Know where emergency equipment is stored and how to use it. Practice good laboratory housekeeping and management to ensure the safety of your students.

It is most important to use safe laboratory techniques when handling all chemicals. Many substances may appear harmless but are, in fact, toxic, corrosive, or very reactive. Always check with the manufacturer. Chemicals should never be ingested. Be sure to use proper techniques to smell solutions or other agents. Always wear safety goggles and an apron. The following general cautions should be used.

1. Poisonous/corrosive liquid and/or vapor. Use in the fume hood. Examples: *acetic acid, hydrochloric acid, ammonia hydroxide, nitric acid.*

2. Poisonous and corrosive to eyes, lungs, and skin. Examples: *acids, limewater, iron(III) chloride, bases, silver nitrate, iodine, potassium permanganate.*

3. Poisonous if swallowed, inhaled, or absorbed through the skin. Examples: *glacial acetic acid, copper compounds, barium chloride, lead compounds, chromium compounds, lithium compounds, cobalt(II) chloride, silver compounds.*

4. Always add acids to water, never the reverse.

5. When sulfuric acid or sodium hydroxide is added to water, a large amount of thermal energy is released. Sodium metal reacts violently with water. Use extra care when handling any of these substances.

Unless otherwise specified, solutions are prepared by adding the solid to a small amount of distilled water and then diluting with water to the volume listed. For example, to make a $0.1M$ solution of aluminum sulfate, dissolve 34.2 g of $Al_2(SO_4)_3$ in a small amount of distilled water and dilute to a liter with water. If you use a hydrate that is different from the one specified in a particular preparation, you will need to adjust the amount of the hydrate to obtain the required concentration.

Chemical
Storage & Disposal

General Guidelines

Be sure to store all chemicals properly. The following are guidelines commonly used. Your school, city, county, or state may have additional requirements for handling chemicals. It is the responsibility of each teacher to become informed as to what rules or guidelines are in effect in his or her area.

1. Separate chemicals by reaction type. Strong acids should be stored together. Likewise, strong bases should be stored together and should be separated from acids. Oxidants should be stored away from easily oxidized materials, and so on.

2. Be sure all chemicals are stored in labeled containers indicating contents, concentration, source, date purchased (or prepared), any precautions for handling and storage, and expiration date.

3. Dispose of any outdated or waste chemicals properly according to accepted disposal procedures.

4. Do not store chemicals above eye level.

5. Wood shelving is preferable to metal. All shelving should be firmly attached to the wall and should have antiroll edges.

6. Store only those chemicals that you plan to use.

7. Hazardous chemicals require special storage containers and conditions. Be sure to know what those chemicals are and the accepted practices for your area. Some substances must even be stored outside the building.

8. When working with chemicals or preparing solutions, observe the same general safety precautions that you would expect from students. These include wearing an apron and goggles. Wear gloves and use the fume hood when necessary. Students will want to do as you do whether they admit it or not.

9. If you are a new teacher in a particular laboratory, it is your responsibility to survey the chemicals stored there and to be sure they are stored properly or disposed of. Consult the rules and laws in your area concerning what chemicals can be kept in your classroom. For disposal, consult up-to-date disposal information from the state and federal governments.

Disposal of Chemicals

Local, state, and federal laws regulate the proper disposal of chemicals. These laws should be consulted before chemical disposal is attempted. Although most substances encountered in high school biology can be flushed down the drain with plenty of water, it is not safe to assume that this is always true. It is recommended that teachers who use chemicals consult the following books from the National Research Council:

Prudent Practices for Handling Hazardous Chemicals in Laboratories. Washington, DC: National Academy Press, 1981.

Prudent Practices for Disposal of Chemicals from Laboratories. Washington, DC: National Academy Press, 1983.

Safety in Academic Chemistry Laboratories. Washington, DC: American Chemical Society, 1995.

Current laws in your area would, of course, supersede the information in these books.

Disclaimer

Glencoe/McGraw-Hill makes no claims to the completeness of this discussion of laboratory safety and chemical storage. The material presented is not all-inclusive, nor does it address all of the hazards associated with handling, storage, and disposal of chemicals, or with laboratory management.

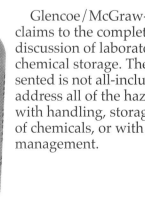

Activity Materials

Get the materials you need quickly and easily!

Glencoe and Science Kit, Inc., have teamed up to make materials selection for *Glencoe Science* easier with an activity materials folder. This folder contains two convenient ways to order materials and equipment for the program: the **Activity Plan Checklist** and the **Activity Materials List** master.

Call Science Kit at 1-800-828-7777 to get your folder.

Materials Support Provided By:

Science Kit® & Boreal®Laboratories

Your Classroom Resource

777 East Park Drive

Tonawanda, NY 14150-6784

800-828-7777; Fax 716-874-9572

Non-Consumables

It is assumed that goggles, laboratory aprons, tap water, textbooks, paper, calculator, pencil, and pen are available for all activities.

Item	Activity	MiniLAB	Explore
Advertisements, magazine	1-1		
Aquarium, 10-gal		458	
Ash			467
Balance			495
Ball, basketball		382	
Ball, golf		382	
Ball, high-bouncing	12-1		
Ball, Ping-Pong		484	
Ball, playground	12-1		
Ball, softball	14-1		
Ball, tennis	12-1		
Ball, volleyball	12-1		
Basalt sample			405
Battery, D-cell with holder	13-1, 13-2	352	
Beaker, 250-mL		477	
Beaker, 600-mL	17-1		467
Bell	13-1		
Bottle, 1-L plastic with cap		442	
Bottle, 2-L	6-1, 17-2		
Bottle, glass soft drink	18-1		
Bowl			211
Box, cardboard with lid	2-1		
Bucket	5-1	458	
Buttons, assorted	2-2		
Buzzer	13-1		
Calcite sample		411	
Canned goods		267	
Coins		110	
Compass, directional	18-2		
Container, small planting	4-1, 6-2		
Container, transparent	4-2	333	237
Cookie sheet	16-1	442	
Cork		484	
Coverslips	3-1	92	
Cup	4-2, 17-2	154, 228, 255	211
Cup, clear colorless plastic	3-2	239	
Dinosaur model, plastic with scale	5-2		
Dowel rod	11-2		
Dropper	2-1	92, 333	61
Dropper bottle	8-2		
Envelopes		46	
Flashlight	2-1	381, 397, 498	
Foam packing (peanuts)	8-1		
Fulcrum	11-1		
Funnel	4-2		
Globe		498	
Gneiss sample	15-2		
Graduated cylinder	5-2, 9-2, 17-1, 17-2	239, 333	467

Equipment List

Non-Consumables

Item	Activity	MiniLAB	Explore
Granite sample	15-2		405
Gravel	17-1, 17-2	154	
Gravel, aquarium	6-1		
Hammer	18-2		
Hand lens	4-2		89, 405
Hanger, wire coat	16-2		
Heater	6-2		
Hornblende sample		411	
Hot plate	15-1		
Jar	15-1	92, 411	
Light source	6-2		
Lightbulb, small with holder	13-1, 13-2	352	
Limestone with fossils		422	
Magnetite sample		411	
Magnets, bar	13-2		345
Marble	11-2		263
Mass, kg		267	
Measuring cup	4-2, 15-1		
Measuring tape	12-1	158	
Meterstick	1-2, 5-1, 10-1, 11-1, 12-2, 16-4		
Microscope	3-1, 4-2	75, 92	
Nail, 20-penny iron	13-2		
Obsidian sample			237
Pan, long flat	5-2	442, 477, 484	
Paper clip	8-1, 11-1, 13-2, 15-1		
Pictures of vertebrates		46	
Pie pan	8-2	215, 422	
Pin, straight	18-2		
Plastic, stiff	18-2	458	
Plate		228	
Projector, overhead	14-1		
Protractor		273	
Pulley	11-2		
Pumice sample			237
Quartz sample		411	
Ring stand with ring	16-2		
Rocks, igneous	15-1		
Rod, glass		92, 348	
Rod, wooden		484	
Rolling pin	15-2		
Ruler, metric	4-1, 6-1, 11-1, 16-2, 17-1		61, 377
Sand	6-1, 17-1, 17-2	154, 228	467
Saucepan	15-1		
Scale, spring	11-2		
Scissors	1-2, 3-2, 6-1, 10-2, 14-1, 18-2	397	61
Screw		297	

Item	Activity	MiniLAB	Explore
Shaker	13-2		
Silk		348	
Silt	17-2		
Slide, microscope of human blood cells		75	
Slide, microscope of human cheek cells	3-1		
Slide, microscope of human muscle cells		75	
Slide, microscope of human nerve cells		75	
Slide, microscope of human skin cells		75	
Slides, microscope	3-1	92	
Soil, clay	17-1, 17-2		
Soil, potting	4-1, 6-2, 17-1	154, 477	467
Spoon	4-1, 6-2, 9-2, 16-1	154, 228	377
Spoon, wooden	15-1		
Stapler			263
Strainer, metal		329	
Styrofoam		484	
Test tubes	8-2		
Test-tube holder	8-2		
Thermometer, Celsius alcohol	12-2		
Timer (stopwatch or watch with second hand)	5-1, 10-1, 16-1, 17-1, 17-2	36, 333	
Toy, spring			437
Toys, wind-up	10-1		
Washer, metal		273	
Watering can	4-1		
Weights			
Wire	13-1, 13-2	352	
Wood; wood block	8-1, 18-2; 11-1		
Wool	8-1	348	

Consumables

Item	Activity	MiniLAB	Explore
Alcohol, rubbing	4-2	239	
Aluminum foil	6-2, 8-1, 10-2		
Antacid	9-1		
Apple, fresh	2-1		
Bag, plastic sealable	9-2, 13-2	69, 411	
Bag, plastic with handles		267	
Baking soda	8-2, 9-2	442	
Balloon	8-1, 10-2, 18-1	348, 397, 504	495
Banana, fresh	2-1		
Bean seeds	6-2		
Box, round cereal (oatmeal type)		397	
Candle	8-2	329	
Candy, small pieces		69	

Equipment List

Consumables

Item	Activity	MiniLAB	Explore
Cardboard	14-1		
Cards, 3 × 5		122	61, 117
Celery, stalk with leaves	3-2		
Cereal, bran flakes	2-1		
Chalk, or white crayon			377
Clay, modeling	11-1, 15-2		
Cleaners, household	9-1		
Container, pint-sized milk		130	
Corn syrup		239	
Cornstarch	8-2, 16-1		
Craft sticks	1-2		
Cups, paper	4-1, 16-1		
Filter, coffee	4-2		467
Fish food	6-1		
Food coloring	3-2	215, 239, 333, 458	
Fruit, fresh		69	
Gelatin		69	
Glue, white	1-2, 16-1	196, 422	
Jelly beans, small assorted	5-1		
Juice, fruit	9-1		
Labels	6-2		
Lint	8-1		
Macaroni		422	
Marker, fine-tip	16-2		
Marker, permanent	4-1, 5-2, 17-1		
Markers, colored	14-2	397	122
Matches	8-2	329	
Meat tenderizer	4-2		
Milk, whole		215	
Newspaper			61
Onion	4-2		61
Paper, construction	10-2	397 (black)	263, 377
Paper, graph	13-2		345
Paper, tissue			211
Paper, waxed	10-2	273, 498	
Pasta, dry		69	
Pencils, colored	7-2, 17-2	122	
Petroleum jelly		130	
Plaster of paris		130	
Plastic wrap, colorless			61
Plate, paper		255	
Rice, uncooked	15-2		377
Rubber band, wide	10-2, 16-2		319
Salt, table	4-2, 9-1, 17-2	228, 255	467
Sandpaper		273	
Seeds			89
Soap, liquid	4-2, 9-1	215, 442	
Soft drink, colorless	9-1		

Item	Activity	MiniLAB	Explore
Steel wool, fine unoiled and unsoaped		255	
Straws, drinking	1-2, 2-1, 10-2, 18-2		
String	1-2, 10-2, 11-2, 15-1	196, 348	
Sugar	8-2, 9-1, 15-1	92	
Swabs, cotton-tipped	2-1		
Tape, clear	18-2	196, 397	
Tape, masking	1-2, 4-1, 10-1, 10-2, 11-1, 12-1, 13-1, 16-2		61, 211, 263
Toothpicks	4-2		
Twist ties	11-2		
Vegetables, fresh		69	
Vinegar, white	2-1, 8-2, 9-1	411, 442	
Water, distilled	9-1		
Yeast, dry		92	

Chemical Supplies

Item	Activity	MiniLAB	Explore
Calcium chloride	9-2		
Iodine solution, aqueous	8-2		
Iron filings	13-2	411	
pH paper, 1-14	9-1		
Phenol red solution	9-2		
Silica gel, dehydrated	16-1		

Live Organisms

Item	Activity	MiniLAB	Explore
Cuttings, from houseplant	4-1		
Elodea plants	3-1, 6-1		
Guppy	6-1		
Mealworms	2-1		

Supplier Addresses

Scientific Suppliers

Science Kit & Boreal Laboratories
777 East Park Drive
Tonawanda, NY 14150-6748

Carolina Biological Supply Co.
2700 York Road
Burlington, NC 27215

Fisher Scientific Co.
1600 W. Glenlake
Itaska, IL 60143

Flinn Scientific Co.
P.O. Box 219
770 N. Raddant Road
Batavia, IL 60510

Frey Scientific
100 Paragon Parkway
Mansfield, OH 44903

Kemtec Educational Corp.
9889 Cresent Park Drive
West Chester, OH 45069

Sargent-Welch Scientific Co.
P.O. Box 5229
911 Commerce Ct.
Buffalo Grove, IL 60089

Ward's Natural Science Establishment, Inc.
P.O. Box 92912
Rochester, NY 14692

Software Distributors

(AIT) Agency for Instructional Technology
Box A
Bloomington, IN 47402-0120

Cambridge Development Laboratory, Inc.
86 West Street
Waltham, MA 02154

COMpress
P.O. Box 102
Wentworth, NH 03282

Earthware Computer Services
P.O. Box 30039
Eugene, OR 97403

Educational Activities, Inc.
1937 Grand Avenue
Baldwin, NY 11510

Educational Materials and Equipment Company (EME)
P.O. Box 2805
Danbury, CT 06813-2805

GEMSTAR MEDIA INC
P.O. Box 50228
Staten Island, NY 10305

IBM Educational Systems
Department PC
4111 Northside Parkway
Atlanta, GA 30327

McGraw-Hill Webster Division
1221 Avenue of the Americas
New York, NY 10020

Microphys
1737 W. Second Street
Brooklyn, NY 11223

Queue, Inc.
338 Commerce Drive
Fairfield, CT 06432

School Division of the Learning Co.
6160 Summit Drive
Minneapolis, MN 55430

Texas Instruments, Data Systems Group
P.O. Box 1444
Houston, TX 77251

Ventura Educational Systems
910 Ramona, Suite E
Grover Beach, CA 93433

Audiovisual Distributors

Aims Multimedia
9710 Desoto Avenue
Chatsworth, CA 91311-4409

BFA Educational Media
468 Park Avenue S.
New York, NY 10016

CRM Films
2215 Faraday Avenue
Carlsbad, CA 92008

Diversified Educational Enterprise
725 Main Street
Lafayette, IN 47901

Encyclopaedia Britannica Educational Corp. (EBEC)
310 S. Michigan Avenue
Chicago, IL 60604

Focus Media, Inc.
485 S. Broadway, Suite 12
Hicksville, NY 11801

Hawkill Associates, Inc.
125 E. Gilman Street
Madison, WI 53703

Journal Films, Inc.
930 Pitner Avenue
Evanston, IL 60202

Lumivision
877 Federal Blvd.
Denver, CO 80204

National Earth Science Teachers Association
NESTA/MESTA Publications
C/O Lisa Bouda
1977 Broadstone
Grosse Point Woods, MI 48236

National Geographic Society Education Services
17th and "M" Streets, NW
Washington, DC 20036-4688

Phoenix Learning Group
2349 Chaffee Drive
St. Louis, MO 63146

Science Software Systems
11890 W. Pico Blvd.
Los Angeles, CA 90064

Society for Visual Education & Churchill Media
6677 N. Northwest Highway
Chicago, IL 60631-1304

Time-Life Videos
Time and Life Building
1271 Avenue of the Americas
New York, NY 10020

Universal Education & Visual Arts (UEVA)
100 Universal City Plaza
Universal City, CA 91608

Video Discovery
1515 Dexter Avenue N.
Suite 400
Seattle, WA 98109

Glencoe

SCIENCE

An Introduction to the Life, Earth, and Physical Sciences

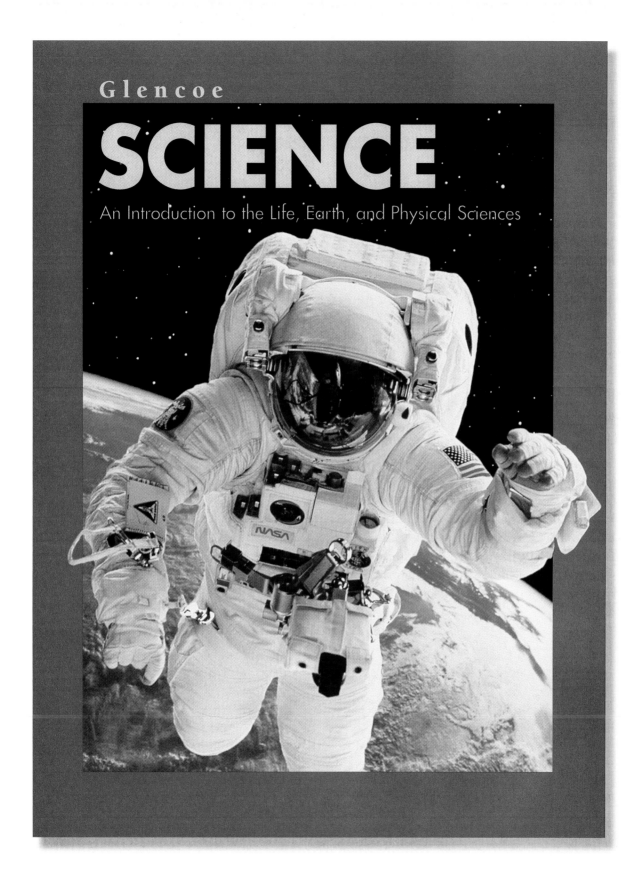

 **Glencoe McGraw-Hill**

New York, New York   Columbus, Ohio   Woodland Hills, California   Peoria, Illinois

A GLENCOE PROGRAM

Glencoe Science

Student Edition
Teacher Wraparound Edition
Study Guide for Content Mastery SE and TE
Reinforcement SE and TE
Enrichment & Critical Thinking/Problem
  Solving SE and TE
Activity Masters
Chapter Review
Chapter Review Software
Lab Manual SE and TE
Science Integration Activities

Cross-Curricular Integration
Science and Society/Technology Integration
Multicultural Connections
Performance Assessment
Assessment: Chapter and Unit Tests
Lesson Plans
Spanish Resources
MindJogger Videoquizzes and Teacher Guide
English/Spanish Audiocassettes
Computer Testbank: Windows and Macintosh
  Versions

Transparency Packages:
    Teaching Transparencies
    Section Focus Transparencies
    Science Integration Transparencies
The Glencoe Science Professional Development Series
    Performance Assessment in the Science Classroom
    Lab Safety Skills in the Science Classroom
    Cooperative Learning in the Science Classroom
    Alternate Assessment in the Science Classroom
    Exploring Environmental Issues

Cover: Science Source/Photo Researchers; inset NASA/The Stock Market

Glencoe/McGraw-Hill
A Division of The McGraw-Hill Companies

Send all inquiries to:
Glencoe/McGraw-Hill
8787 Orion Place
Columbus, OH 43240

ISBN 0-02-828315-5
Printed in the United States of America.

4 5 6 7 8 9 10   071/046   06 05 04 03 02 01 00

Authors

Dan Blaustein
Science Teacher and Author
Waukegan High School
Waukegan, Illinois

Wanda Matthias
6th Grade Science
   Teacher, retired
Plano, Texas

Louise Butler
Principal, Normandy
   School District
St. Louis, Missouri

Bryce Hixson
President of the Utah Museum of
   Science and Industry
Sandy, Utah

Contributing Authors

Carolyn Randolph, Ph.D.
Consultant for Science and
   Social Studies
Lexington School District Two
West Columbia, South Carolina

Leonard Rodriguez
Assistant Principal
First Avenue School
Arcadia, California

Contributing Writers

Mary Dylewski
Science Writer
Houston, Texas

Rebecca Johnson
Science Writer and Author
Sioux Falls, South Dakota

Ralph M. Feather, Jr., Ph.D.
Science Department Chair
Derry Area School District
Derry, Pennsylvania

Nancy Ross-Flanigan
Science Reporter
Detroit, Michigan

Consultants

Chemistry

Anne Barefoot, A.G.C.
Physics and Chemistry
    Teacher, Emeritus
Whiteville High School
Whiteville, North Carolina

Cheryl Wistrom, Ph.D.
Assistant Professor
Chemistry Department
St. Joseph's College
Rensselaer, Indiana

Physics

Albert E. Acierno
Supervisor Science Education
    K-12, retired
Columbus Public Schools
Columbus, Ohio

Safety

Jay A. Young, Ph.D.
Safety and Chemical Safety
    Consultant
Silver Spring, Maryland

Earth and Planetary Sciences

Professor Larry Lebofsky, Ph.D.
Professor
Department of Planetary
    Sciences
University of Arizona
Tucson, Arizona

James B. Phipps, Ph.D.
Instructor
Grays Harbor College
Aberdeen, Washington

Assessment

Audrey B. Champagne, Ph.D.
Professor of Chemistry
Department of Chemistry
Professor of Education
Department of Educational
    Theory and Practice
University at Albany
State University of New York
Albany, New York

Multicultural

Lorraine Cruz-Lugo
Middle School Teacher
Chicago, Illinois

Karen Muir, Ph.D.
Lead Instructor
Department of Social and
    Behavioral Sciences
Columbus State Community
    College
Columbus, Ohio

Life Science

William R. Ausich, Ph.D.
Chair and Professor
    Geological Sciences
Department of Geological
    Sciences
The Ohio State University
Columbus, Ohio

Maryanna Quon Warner
Science Teacher
Del Dios Middle School
Escondido, California

Reading

Nancy Farnan, Ph.D.
Graduate Programs
    Coordinator
School of Teacher Education
San Diego State University
San Diego, California

Elizabeth Gray, Ph.D.
Adjunct Professor
    Department of Education
Otterbein College
Westerville, Ohio

Reviewers

Linda Bodie
Pace Middle School
Milton, Florida

Teckla Dando
Troy City Schools
Troy, Ohio

Celeste Dellinger
Atlanta Middle School
Atlanta, Texas

Connie Denk
Starling Middle School
Columbus, Ohio

Lenore Gallagher
Northeast Middle School
Bristol, Connecticut

Fred George
Dr. T.F. Reszel
   Middle School
North Tonawanda,
   New York

Penny Hamisch
Greenfield School
Greenfield, California

Barbara Hartgrove
Everett Middle School
Columbus, Ohio

Michael Henson
Beery Middle School
Columbus, Ohio

Patrick J. Herak
Westerville South High
   School
Westerville, Ohio

Phyllis Herzog
Dominion Middle School
Columbus, Ohio

Mario Inchaustegui
Academy of the
   Americas
Detroit, Michigan

Kevin Kerr
Fort Lincoln School
Mandan, North Dakota

Mike Mansour
John Page Middle School
Madison Heights, Michigan

Stephanie Molesky
Davidson International
Baccalaureate
   Middle School
Davidson, North Carolina

Rebecca Morris
H. L. Harshman
   Middle School
Indianapolis, Indiana

Debbie Panebianco
South Charleston
   Middle School
Charlotte, North Carolina

Lashawn Porter
Louie Welch
   Middle School
Houston, Texas

Ramonita Torres
Thomas Giordano
   Middle School 45
Bronx, New York

Mike Walsh
Carmel Junior High
   School
Carmel, Indiana

Teacher Activity Testers

Life Science

Diana Such
Science Teacher
St. Michael's School
Worthington, Ohio

Physical Science

Tom Speece
Science Teacher
Big Walnut
   Middle School
Sunbury, Ohio

Earth Science

Debbie Huffine
Science Teacher
Noblesville
   Intermediate School
Noblesville, Indiana

Contents

UNIT 1

Contents

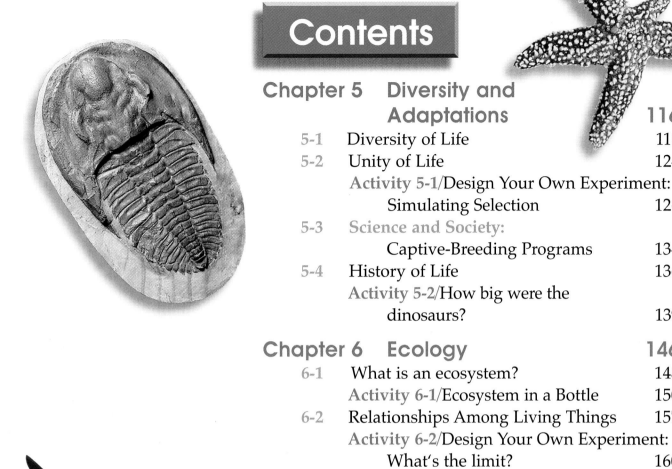

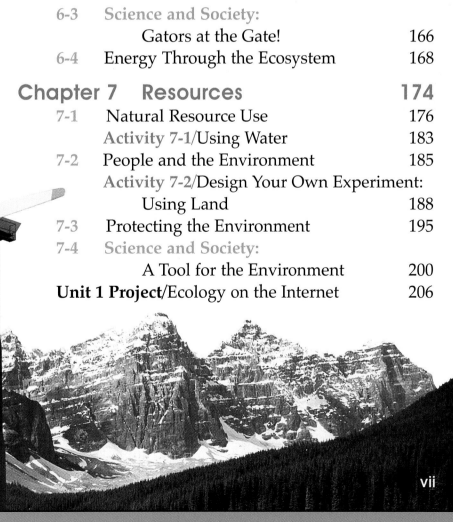

Contents

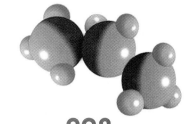

Physical Science 208

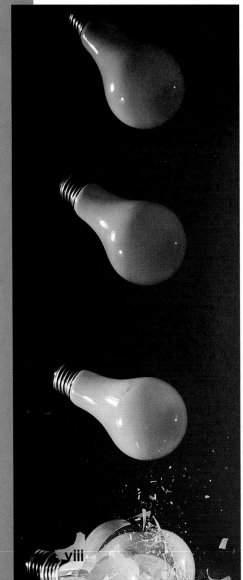

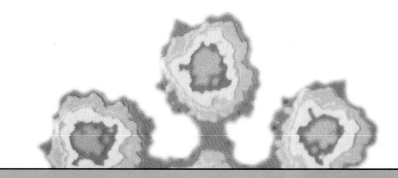

Contents

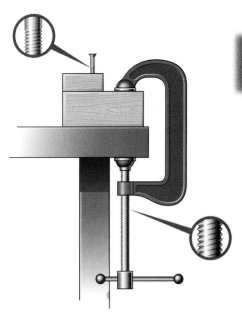

Contents

Contents

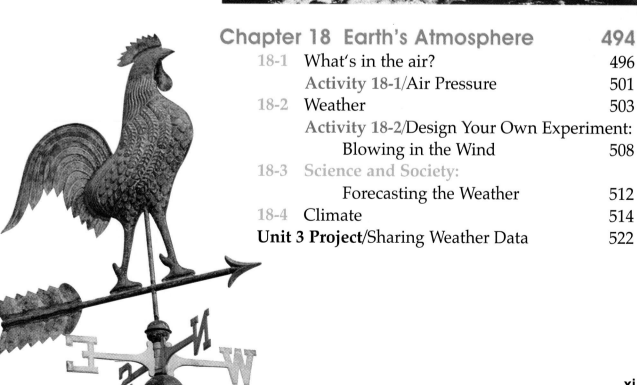

Contents

Appendices

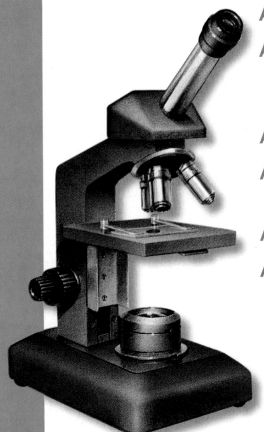

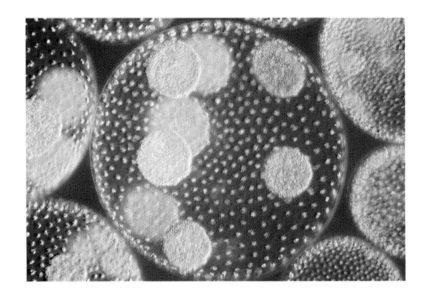

Contents

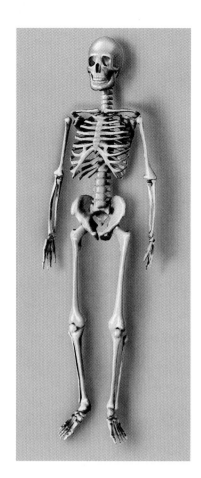

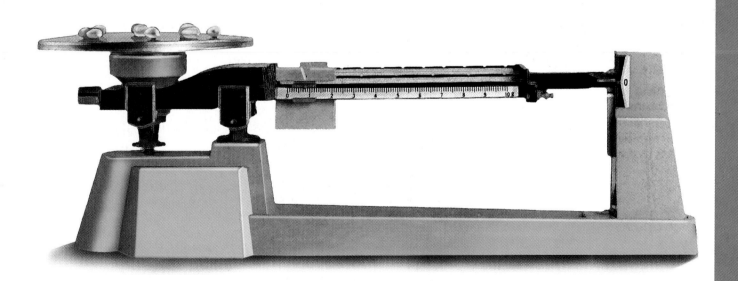

Activities

Activities

MiniLABs

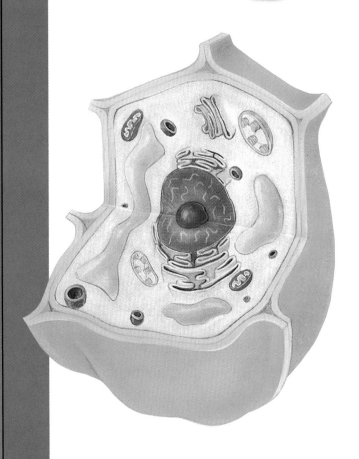

MiniLABs

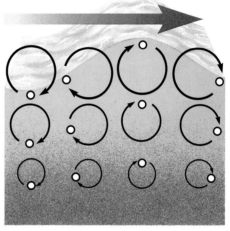

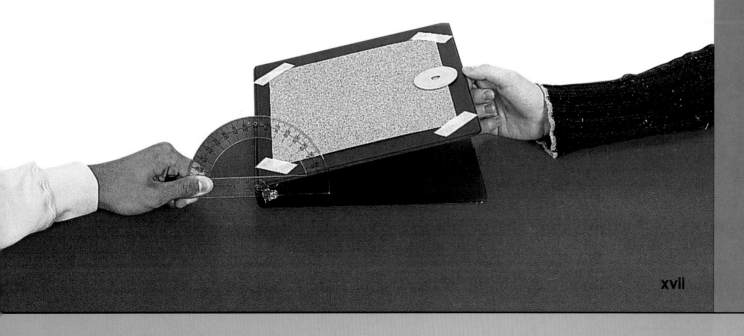

Explore Activities

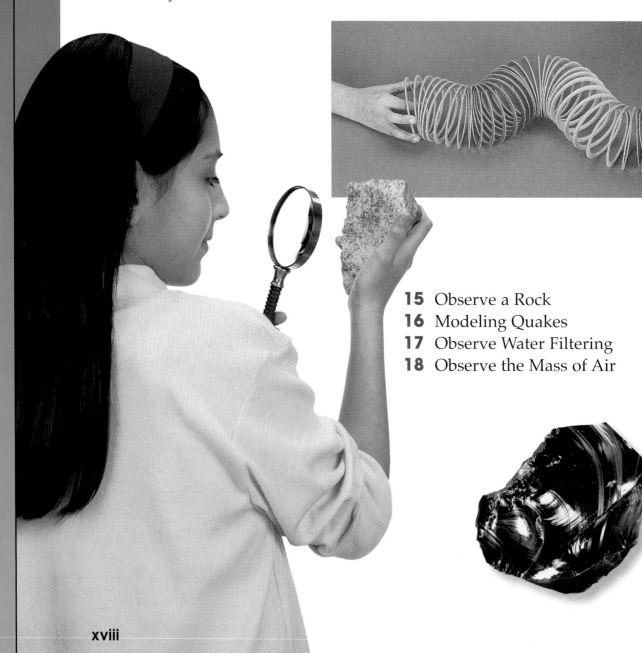

Problem Solving

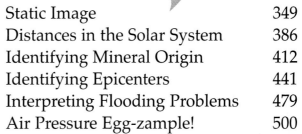

Using Technology

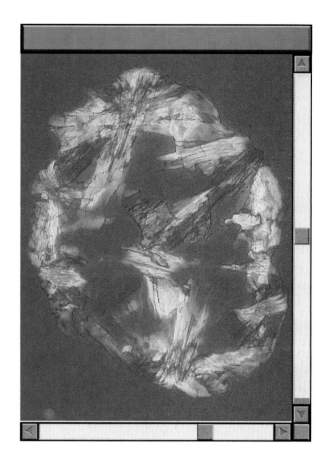

Skill Builders

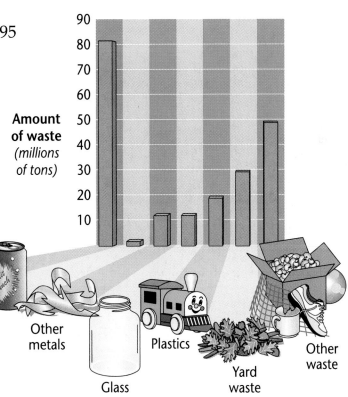

Amount of waste *(millions of tons)*

Paper products

Aluminum

Other metals

Glass

Plastics

Yard waste

Other waste

People and Science

Science Connections

Science and the Arts

Science and History

Science and Language Arts

Chapter Organizer

Section	Objectives/Standards	Activities/Features
Chapter Opener		Explore Activity: Draw a Bridge, p. 3
1-1 **What is science?** (2 sessions, 1 block)*	**1. Discuss** how science can help solve problems. **2. Describe** and **apply** the skills used in science. National Science Content Standards: (5-8) UCP2, UCP3, A1, A2, E2, F5, G1, G2	Using Technology: Science and Technology, p. 9 MiniLAB: Observing and Inferring, p. 10 Activity 1-1: Advertising Inferences, p. 11 Skill Builder: Communicating, p. 12 Science Journal, p. 12 People & Science, p. 13
1-2 **Doing Science** (3 sessions, 1½ blocks)*	**3. Demonstrate** how science can solve problems. **4. Design** an experiment. National Science Content Standards: (5-8) UCP2, A1, A2, F5	Problem Solving: Flex Your Brain, p. 17 Activity 1-2: Bridge Building, pp. 24-25 Skill Builder: Designing an Experiment, p. 26 Using Computers: p. 26

* A complete Planning Guide that includes block scheduling is provided on pages 31T-33T.

Activity Materials

Explore	Activities	MiniLABs
No materials needed.	page 11 magazine advertisements pages 24-25 meterstick, drinking straws, craft sticks, string, glue, tape, scissors	page 10 Science Journal

Need Materials? Call Science Kit (1-800-828-7777).

Teacher Classroom Resources

Reproducible Masters	Transparencies	Teaching Resources

Activity Worksheets, pp. 7-8, 11
Cross-Curricular Integration, p. 5
Enrichment, p. 7
Lab Manual 1
Reinforcement, p. 7
Science and Society/
    Technology Integration, p. 19
Study Guide, p. 7
	Science Integration Transparency 1,
	    Technology and You
	Section Focus Transparency 1, All in a
	    Day's Work
		Spanish Resources
		English/Spanish Audiocassettes
		Cooperative Learning Resource
		    Guide
		Lab Partner
		Lab and Safety Skills
		Lesson Plans

Assessment Resources

Reproducible Masters	Transparencies	Assessment Resources

Activity Worksheets, pp. 5, 9-10
Enrichment, p. 8
Lab Manual 2
Multicultural Connections, pp. 5-6
Reinforcement, p. 8
Science Integration
    Activities, pp. 39-40
Study Guide, p. 8
	Section Focus Transparency 2, Building
	    Bridges
	Teaching Transparency 1, Flex Your
	    Brain
	Teaching Transparency 2, Scientific
	    Methods
		Chapter Review, pp. 5-6
		Assessment, pp. 5-8
		Performance Assessment, p. 39
		Performance Assessment in the
		    Science Classroom (PASC)
		MindJogger Videoquiz
		Alternate Assessment in the
		    Science Classroom
		Computer Test Bank

Key to Teaching Strategies

The following designations will help you decide which activities are appropriate for your students.

L1 Level 1 activities should be appropriate for students with learning difficulties.

L2 Level 2 activities should be within the ability range of all students.

L3 Level 3 activities are designed for above-average students.

ELL ELL activities should be within the ability range of English Language Learners.

LS These activities are designed to address different learning styles.

COOP LEARN Cooperative Learning activities are designed for small group work.

P These strategies represent student products that can be placed into a best-work portfolio.

GLENCOE TECHNOLOGY

The following multimedia resources are available from Glencoe.

Science and Technology
Videodisc Series (STVS)
Animals
Chemistry
Earth & Space
Plants & Simple Organisms
Physics

The Infinite Voyage Series
To the Edge of the Earth
The Champion Within
The Geometry of Life
Miracles by Design
The Great Dinosaur Hunt
Unseen Worlds
Insects: The Ruling Class

Teacher Classroom Resources

This is a representation of key blackline masters available in the Teacher Classroom Resources.

Teaching Aids

Section Focus Transparencies

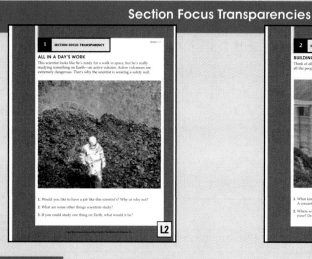

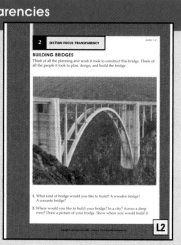

Science Integration Transparencies

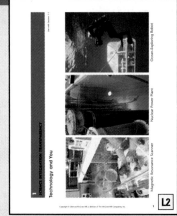

Teaching Transparencies

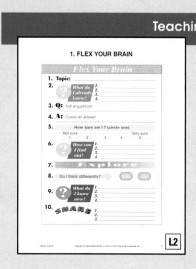

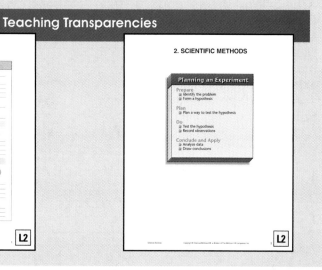

Meeting Different Ability Levels

Study Guide for Content Mastery

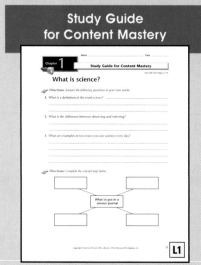

Reinforcement

Enrichment Worksheets

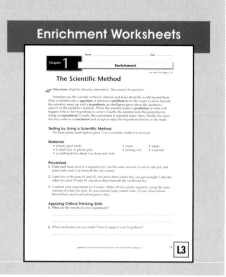

2C

Chapter 1 The Nature of Science

Hands-On Activities

Science Integration Activities

Lab Manual

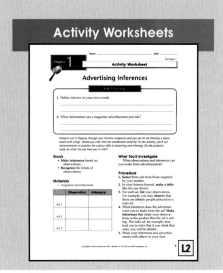

Activity Worksheets

Enrichment and Application

Cross-Curricular Integration

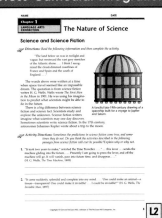

Multicultural Connections

Science and Society/ Technology Integration

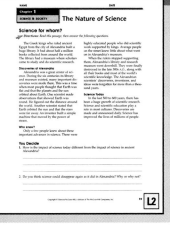

Assessment

Performance Assessment

Chapter Review

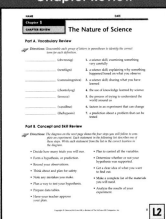

Assessment

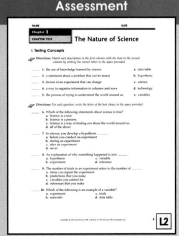

CHAPTER OVERVIEW

Section 1-1 In this section, students receive a general overview of basic science skills, such as observing, inferring, and communicating.

Section 1-2 This section demonstrates the use of basic science skills by taking students through the steps of problem solving.

Chapter Vocabulary

science
technology
hypothesis
variable

Theme Connection

Systems and Interactions A scientific method is a systematic way to solve problems. Often, this method leads to new discoveries, which, in turn, lead to new technologies that impact society.

Chapter Preview

Skills Preview

▶ **Skill Builders**
- communicate
- design an experiment

▶ **MiniLABs**
- observe
- infer

▶ **Activities**
- observe
- infer
- hypothesize
- collect data
- communicate
- compare results
- make a table

2

Learning Styles	Look for the following logo for strategies that emphasize different learning modalities.	
Kinesthetic	Activity 1-2, p. 24; Go Further, p. 25
Visual-Spatial	Visual Learning, pp. 5, 8, 19, 21, 22; Demonstration, pp. 9, 20; Activity 1-1, p. 11; Across the Curriculum, p. 20
Interpersonal	Activity, p. 7; Assessment, p. 26
Intrapersonal	Enrichment, p. 6; Inclusion Strategies, p. 17; Science at Home, p. 27
Logical-Mathematical	Explore, p. 3; Inquiry Question, p. 8; MiniLAB, p. 10; Assessment, p. 12; Problem Solving, p. 17; Activity, p. 21; Demonstration, p. 23
Linguistic	Discussion, p. 5; Science Journal, p. 16; Using Science Words, p. 19

The Nature of Science

Y ou hold your breath as you speed over the edge of the gorge. The only thing separating you from the river 200 meters below is a spidery network of steel and cables, and a whole lot of air. How can the bridge hold so much weight? Bridges are amazing achievements. They connect worlds by making travel across rivers, canyons, and bays possible. Bridges cross roads, train tracks, and even small gullies and ditches. You'll see how they are also great examples of how science affects everyday life.

EXPLORE ACTIVITY

Draw a Bridge

1. Find a bridge close to your home or school. You may find a pedestrian bridge, a highway overpass, a railroad bridge, or a bridge across a small ditch.
2. In your Science Journal, draw a detailed picture of the bridge that shows how the bridge is made. Record any other observations you make such as what the bridge is used for and what you think it is made out of.

Science Journal

In your Science Journal, write a paragraph that describes how you think the people who built the bridge knew that it would be strong enough.

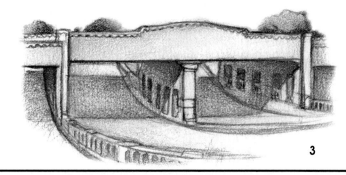

3

Assessment Planner

Portfolio
Refer to page 27 for suggested items that students might select for their portfolios.

Performance Assessment
See page 27 for additional Performance Assessment options.
Skill Builders, pp. 12, 26
MiniLAB, p. 10
Activities 1-1, p. 11; 1-2, pp. 24-25

Content Assessment
Section Wrap-ups, pp. 12, 26
Chapter Review, pp. 27-29
Mini Quiz, p. 12

Group Assessment
Opportunities for group assessment occur with Cooperative Learning Strategies and Flex Your Brain activities.

EXPLORE ACTIVITY

Purpose

LS **Logical-Mathematical** Students are to observe the details of bridge construction and to infer the purpose and composition of a bridge of their choice. **L2**

Preparation

Students will need to do this as an overnight assignment.

Materials

pencil or pen and paper

Teaching Strategies

- Use a picture of a bridge to lead students through the activity in the classroom before sending them out on their own.
- Students who have trouble drawing can take a photograph of a bridge or bring in a magazine with a picture of a bridge.

Science Journal Answers will vary, but a student could say that a railroad trestle is made of metal and has a lot of support beams to hold the weight of a train.

✓ Assessment

Process Divide students into groups and select the best drawing, complete with details and answers to questions. Working together, students can improve on the drawing until they are ready to present it to the class. Use the Performance Task Assessment List for Scientific Drawing in **PASC**, p. 55. **P**

Prepare

Section Background

This section teaches that observation leads to inferences and hypotheses that can be tested. Based on the results of the tests, hypotheses are either supported or not supported.

Preplanning

Refer to the Chapter Organizer on pages 2A-B.

1 Motivate

Bellringer

Before presenting the lesson, display **Section Focus Transparency 1** on the overhead projector. Assign the accompanying **Focus Activity** worksheet.
L2 ELL

Tying to Previous Knowledge

Have students make a list of all the decisions they made that day. Point out the past observations and inferences they used to make these decisions.

1•1 What is science?

What YOU'LL LEARN

- How science can help you solve problems
- The skills used in science

Science Words:
science
technology

Why IT'S IMPORTANT

Learning about science helps you find out more about the world around you.

FIGURE 1-1

Some bridges are designed for people to walk across. Others are designed to carry the weight of cars and trucks.

Science and Scientists

What do you think of when you think of a scientist? Do you think of someone alone in a laboratory, mixing chemicals and doing strange experiments? Maybe you imagine some person with wild hair and a lab coat. That's the image many people have of scientists. However, nothing could be further from the truth. Anyone who tries to learn something about the world is a scientist. That means anyone, including the people who designed the bridges shown in **Figures 1-1** and **1-2**, can be a scientist. Even you can be a scientist. The best way to understand science and what scientists do is to jump in and do it. Let's follow a science class as they explore the nature of science.

It was the first day of school and Gabby was unsure about what to expect in her new science class. As the rest of the students sat down and Ms. Quon took attendance, Gabby wondered what they were going to study.

"All right everyone, settle down. We need to get started," said Ms. Quon. "We've been given a special project and I want to start right away. Have any of you heard that the LEP Corporation donated some property for our school to use as an outdoor education center?"

4

Program Resources

Reproducible Masters
Activity Worksheets, pp. 7-8, 11 L2
Cross-Curricular Integration, p. 5 L2
Enrichment, p. 7 L3
Lab Manual, pp. 1-4 L2
Reinforcement, p. 7 L2
Science and Society/Technology Integration, p. 19 L2
Study Guide, p. 7 L1

 Transparencies
Science Integration Transparency 1 L2
Section Focus Transparency 1 L2

"Yes, it's great," Pablo responded, "but we'll be in high school before it's ready to be used."

"Right now, it's just an empty lot next to some old warehouses," added Sydney. "It only has a few trees and a polluted stream. Some gift that was!" He paused. "What does the lot have to do with us?"

Ms. Quon laughed. "It has a lot more to do with you than you realize. In fact, it's going to be a big part of your grade!"

Science as a Tool

Gabby raised her hand. "Ms. Quon, I thought that this was science class. What does a vacant lot have to with science?"

"Maybe we'll come up with a way to blow it up so we can start over!" joked Jim.

"No, we're not going to blow up anything. There's a lot more to science than explosions," Ms. Quon said. "We've been given a special assignment. We're going to use science to build a bridge over the stream."

"Wouldn't a hammer and nails work better?" Sydney asked, grinning.

FIGURE 1-2

Some bridges help ease the flow of traffic. *How does the highway overpass shown here differ from the "walking bridge" shown in Figure 1-1?*

Discussion

LS **Linguistic** Ask students to define science in their own words. Lead them into a discussion of the different ways science impacts their lives.

GLENCOE TECHNOLOGY

 Videodisc

The Infinite Voyage: To the Edge of the Earth
Chapter 1
Great Explorers

Teacher F.Y.I.

Bridge designs can be divided into six main types: simple girder, cantilever, arch, suspension, movable, and slab. Study the illustration on the bottom of this page for examples of each. If possible, photocopy the designs for students or draw them on the chalkboard. Have students discuss differences among the designs.

Visual Learning

Figure 1-2 How does the highway overpass shown here differ from the "walking bridge" shown in Figure 1-1? *Answers will vary but students should note that highway overpasses must support a lot of weight. These bridges are normally larger and stronger than pedestrian bridges.*
LS

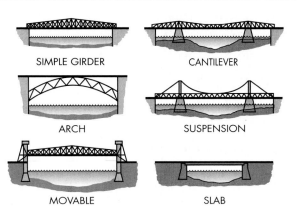

SIMPLE GIRDER

CANTILEVER

ARCH

SUSPENSION

MOVABLE

SLAB

Encourage students to discuss the stages involved in building a bridge. Ask them to put the stages in order—would they design the bridge or buy material first? Help them see that the planning stage is as crucial—if not more crucial—than the construction stage.

Enrichment

LS **Intrapersonal** Science is an ever-changing process. As new knowledge is discovered, old concepts are often discarded or modified. To help students understand how scientific knowledge and beliefs change over time, have them research and write about different views of the structure of Earth. They can compare and contrast ancient Greek, medieval, and 20th-century western views of Earth. [L3] [P]

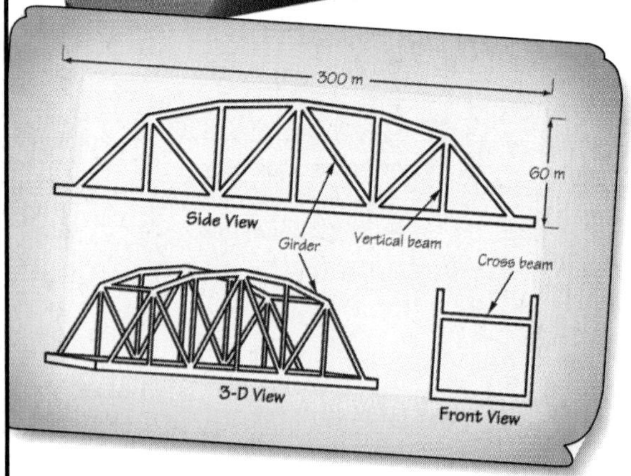

FIGURE 1-3

Planning is the first step in building a bridge. Engineers test different designs before construction begins.

"Science can be as good a tool as a hammer, Sydney," she explained.

Gabby wondered aloud, "What's so hard about building a bridge? All we need is some wood and nails, and we put them together to make a bridge. Why do we need science?"

"Well, do you know everything you need to know before you build your bridge? For example, how can we build a bridge that's strong enough to hold an entire class?" Ms. Quon pulled some books on bridges out of her backpack. "Are you sure wood is the best material to make your bridge out of? How high should your bridge be to keep it from being washed away in a flood? Where along the stream should we build the bridge?"

"I guess there's a lot we need to know before we start," replied Gabby. "But I still don't see how science can help. We haven't even had science yet. We don't know enough about engineering to answer those kinds of questions."

Using Science Every Day

"Science can help us answer those questions. Science is not just about knowing a bunch of facts and information. Science is a process. It's a way of finding out about the world around us." Ms. Quon began unrolling a poster. "It can help you solve problems by giving you a way to find out more about the problem. In fact, I'm sure you already use science every day to help you solve problems."

Using Prior Knowledge

"I don't think so, Ms. Quon," replied Tinho. "This is my first science class."

Ms. Quon thought for a moment. "Tinho, how did you decide which clothes to wear today?"

Content Background

We can divide science into three main areas of study: life science, Earth science, and physical science. Life science is the study of living things and the factors that influence them. Earth science is the study of Earth and its place in space. Physical science is the study of matter and energy.

"Well, I knew it was going to be warm, so I wore a short-sleeved shirt."

"How did you know it was going to be warm today? Did you listen to a weather report?"

"No, I know that it's always warm here in early September."

"I knew it! You did use science! You used what you've experienced for the last few years to guess what the weather would be like today. You made a prediction," said Ms. Quon. "But you didn't stop there. You made another prediction. You guessed that the best shirt to wear on a warm day is a short-sleeved shirt. How did you make that prediction?"

"I know that shirts with short sleeves are more comfortable to wear when it gets hot than shirts with long sleeves. We don't have air-conditioning in our school, so I knew that I would need to be as comfortable as possible."

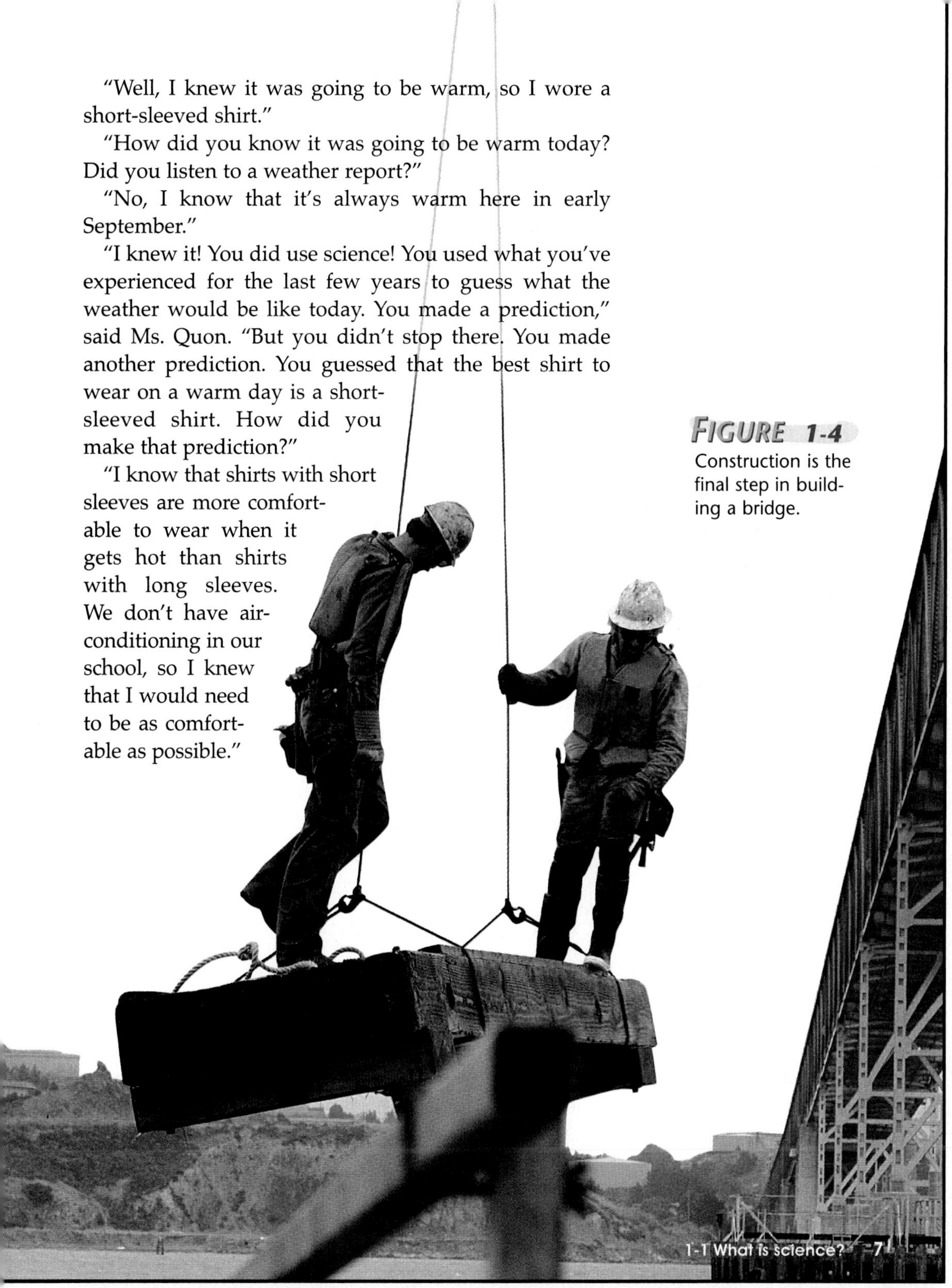

FIGURE 1-4

Construction is the final step in building a bridge.

Activity

IS **Interpersonal** Divide students into small groups and have each team choose a simple everyday problem, such as which shirt to wear. Have them pantomime the problem and the solution. Have the rest of the class guess what the problem is and what steps the group is acting out. This is similar to charades, where the student who guesses correctly wins points for his or her team. **L2** **ELL** **COOP LEARN**

GLENCOE TECHNOLOGY

Videodisc

The Infinite Voyage: The Champion Within

Chapter 1
The Chemistry of Motion

Chapter 2
Shoe Design and Diabetic Research

Content Background

Japan is made up of four main islands, all of which are linked by a series of bridges. The most impressive bridge is also the newest—it took decades to design. Opened for traffic in 1998, the Akashi Kaikyo Bridge spans the islands of Honshu and Awaji. At 3.91 km, it is the longest suspension bridge in the world. The bridge is considered an engineering masterpiece, able to withstand seasonal typhoon winds and earthquakes with magnitudes of up to 8.5 on the Richter scale. Part of its strength is due to the ten-ton pendulums located within the bridge's towers. These pendulums swing in the opposite direction of the towers' movements, helping to keep bridge stable when strong winds and earthquakes shake its foundations.

7

Using an Analogy

Tell students that playing a sport such as basketball is similar to doing science. Both athletes and scientists plan their moves ahead of time and practice their skills to get them right.

Inquiry Question

LS **Logical-Mathematical** Ask students how their lives would be different without a particular form of technology, such as automobiles. Answers will vary. Accept all reasonable answers.

GLENCOE TECHNOLOGY

Videodisc

The Infinite Voyage: The Geometry of Life

Chapter 10
Manipulating Genetic Engineering

Chapter 11
The Human Genome Project

The Infinite Voyage: Miracles by Design

Chapter 1
Auto Racing: Advancement in the Field of Metallurgy

Chapter 9
"Smart Materials" and Their Construction

FIGURE 1-5
Learning to play a musical instrument requires a lot of practice. *How is this similar to learning the skills used in science?*

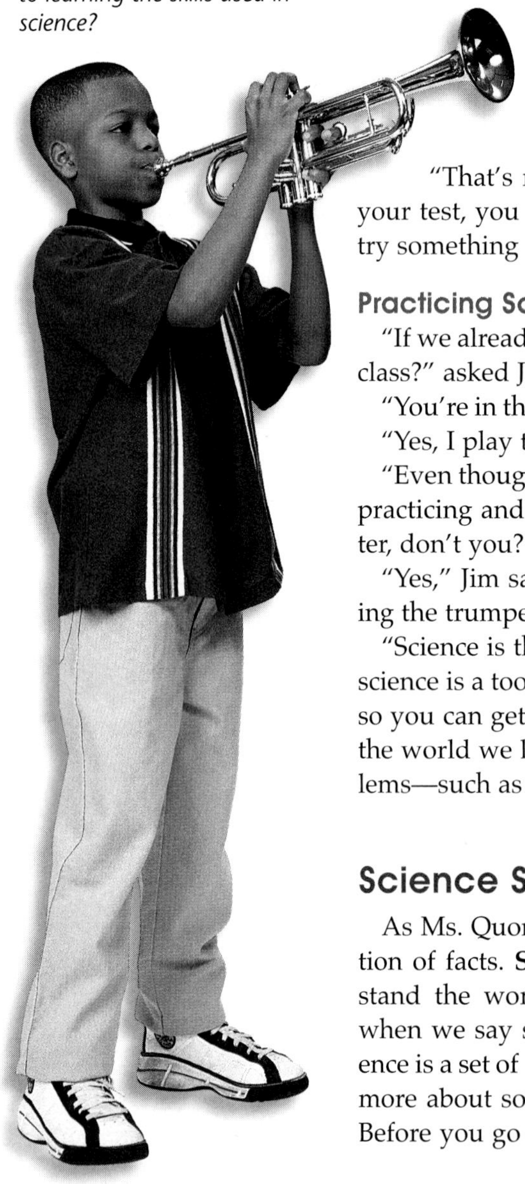

"Great! You used what you already knew about clothes to predict that the best shirt to wear today should have short sleeves. You then tested your prediction by wearing a short-sleeved shirt." Ms. Quon paused for a second to tape the poster to the bulletin board. "What will you wear tomorrow if your prediction is correct?"

"I'll wear a short-sleeved shirt, of course," Tinho said.

"What if the weather was not what you predicted and it was cold today?"

"I'd probably wear a warmer shirt tomorrow."

"That's right, Tinho. Depending on the results of your test, you either keep doing the same thing, or you try something different. That's using science!"

Practicing Science Skills

"If we already use science, why do we have to take this class?" asked Jim.

"You're in the band, aren't you, Jim?" Ms. Quon asked.

"Yes, I play the trumpet," he answered.

"Even though you can play the trumpet now, you keep practicing and taking lessons so you can play even better, don't you?"

"Yes," Jim said. "I still have a lot to learn about playing the trumpet."

"Science is the same way," said Ms. Quon. "As I said, science is a tool, and you need to practice using that tool so you can get better. That'll help you learn more about the world we live in and help you to solve harder problems—such as building a bridge."

Science Skills

As Ms. Quon's class will learn, science is not a collection of facts. **Science** is the process of trying to understand the world around you. But what do we mean when we say science is a process? That means that science is a set of skills that you can use to help you find out more about something or to solve a problem. But wait! Before you go out and start "doing science," it's a good

Visual Learning

Figure 1-5 How is this similar to learning the skills used in science? *You must practice science skills to get them right, just as you must practice playing an instrument to do it well.* LS

idea to practice the skills you need to do it well. Just as athletes and musicians practice the skills necessary to do their craft, science has skills that you can practice every day.

When many people think of science, they think of complicated machines, computers, and robots. As you are learning, however, science is a way to find out about the world around you. The gadgets and gizmos some think of as science are really technology. **Technology** is the use of knowledge learned through science. **Figure 1-6** shows an example of technology.

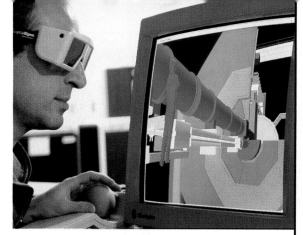

FIGURE 1-6
This virtual reality computer helps scientists test machines before the machines are constructed.

USING TECHNOLOGY
For more information on bridge designs, contact the local engineer's or planner's office in your area.

*inter*NET CONNECTION

The Glencoe Homepage at **www.glencoe.com/sec/science** provides links connecting concepts from the student edition to relevant, up-to-date Internet sites.

USING TECHNOLOGY
Science and Technology

In the late 1800s, before cars became common and traffic jams were a daily event, a New York engineer named John Roebling was given the task of building a bridge between Brooklyn and Manhattan.

To make sure that the bridge would be safe and strong, Roebling developed a new technology using a cable-weaving design. This is the design he came up with: four cables, each 38 centimeters thick, would hold up the bridge. Each cable would be made of 19 strands of wire. Each strand, in turn, would be made up of 278 long wires. Roebling designed the wires to be long, continuous strands so that the builders did not have to link wire ends together. This way, the wires would not break apart.

In 1883, the Brooklyn Bridge was completed. More than 100 years later, it is still considered a technological wonder—and a good example of how a new technology can impact millions of lives.

*inter*NET CONNECTION
Visit the Glencoe Homepage at *www.glencoe.com/sec/science* to learn more about the technology of bridges.

1-1 What is science? 9

Demonstration
LS Visual-Spatial Use a block of dry ice to demonstrate **Figure 1-7** on page 10. Dry ice (extremely cold, solid carbon dioxide) changes to its gaseous state very quickly at room temperature. It doesn't store, even in a freezer. Obtain the dry ice on the morning of the demonstration. The block of dry ice need not be larger than 5 pounds. If it is wrapped well in several layers of loosely folded towels, it should last all day. Because dry ice can quickly freeze the tissues of human hands, wear thick cotton gloves when handling the dry ice. Do not allow even small pieces to come into contact with skin. Also, make sure the room in which the dry ice is stored and displayed is well ventilated. The yellow pages of your phone book will list dry ice distributors.

Theme Connection
Systems and Interactions
Computers are a good example of the theme of Systems and Interactions. The computer was developed using a systematic scientific method. Computers are now used to help us solve different problems using scientific methods.

Mini LAB

Purpose

LS **Logical-Mathematical** This activity will reinforce students' ability to make inferences from observations. **L2**

Teaching Strategies

Discourage students from peeking at the complete illustration on page 27.

📁 **Activity Worksheets,** pages 5, 11

Analysis

1. Answers will vary.
2. Student answers will vary.

✓ Assessment

Content Have students draw their own illustrations. Afterwards, they should cover the illustration with a sheet of paper, leaving only a portion of the drawing visible. Have students show their covered drawings to the class. Have the class try to infer what is happening in the drawings. Use the Performance Task Assessment List for Making Observations and Inferences in **PASC,** p. 17. **P**

3 Assess

Check for Understanding

Revealing Preconceptions Many students believe that scientists work mainly in a lab. Tell students that many scientists work in the field. Show photos of various scientists on the job, such as geologists or oceanographers. Have students discuss why these scientists need to conduct research in the field.

FIGURE 1-7
The water in this pot is actually cold—dry ice is making it bubble.

Mini LAB

Observing and Inferring

Practice the skill of using observation to make an inference with this activity.

1. Look at the illustration on this page. It is a part of a larger illustration.
2. Record in your Science Journal everything you can observe about the illustration.
3. Use your list of observations to make inferences about what is happening in the illustration.

Analysis

1. What do you think is happening in the illustration?
2. Compare your inference with the entire illustration on page 27. How close was your inference to the illustration?

The Skill Handbook

The important skills used in science are described in the Skill Handbook on pages 541-561. The Skill Handbook can be a helpful reference as you learn and practice the skills of science.

Some of the skills may seem easy but still require practice. For example, observing is a skill that may seem simple. After all, everyone knows how to watch something, right? However, observing is more than just watching. You have to make sure you observe carefully and write down exactly what you see.

You also have to know the difference between an observation and an inference (IHN fuh runtz). An inference is an explanation of why something happened. This isn't as easy as it seems. See the pot in **Figure 1-7**? It looks like it is full of boiling water. Would you believe that the water is actually cold? Dry ice is making the water bubble as if it's boiling hot, so you infer that the water is hot. You are trying to explain what you observed—the bubbling water. However, all you can *observe* is that the water is bubbling. You *infer* that the water is hot because the water is bubbling. An observation is what you see. An inference is how you explain what you observe.

In science, it's important to make sure you know exactly what you observed. Once you know what you observed, you can begin to explain, or infer, what happened.

Advertising Inferences

Imagine you're flipping through your favorite magazine and you see an ad showing a skateboard with wings. Would you infer that the skateboard could fly? In this activity, you'll use advertisements to practice the science skills of observing and inferring. Do the products really do what the ads lead you to infer?

What You'll Investigate
What observations and inferences can you make from advertisements?

Procedure
1. **Select** three ads from those supplied by your teacher.
2. In your Science Journal, **make a table** like the one shown below.
3. For each ad, **list** your observations. For example, you may **observe** that there are athletic people pictured in a soda ad.
4. What inferences does the advertiser want you to make from the ad? **Make inferences** that relate your observations to the product that the ad is selling. The soda ad, for example, may lead you to infer that if you drink that soda, you will be athletic.
5. **Share** your inferences and advertisements with others in your class.

Conclude and Apply
1. **Analyze** the inferences you made. Are there other explanations for the observations?
2. **Create** your own ad to sell a product. Think about what people will observe in the ad and what you want them to infer from it.

	Observation	Inference
Ad 1	Observations and inferences will vary based on ads chosen	
Ad 2		
Ad 3		

Goals
- Make inferences based on observations.
- Recognize the limits of observations.

Materials
- magazine advertisements

1-1 What is science?   11

Purpose

IS **Visual-Spatial** This activity will reinforce students' ability to observe and make inferences. **L2**

Process Skills
observing and inferring, comparing and contrasting, recognizing cause and effect, hypothesizing, analyzing, collecting and organizing data, making and using tables, communicating

Time
one class period

📁 **Activity Worksheets,** pages 7-8

Teaching Strategies
Select advertisements that clearly suggest that the potential reader's life will be markedly changed should he or she buy the product.

Answers to Questions
1. Answers will vary. Generally, students will make inferences based on evidence that is not observable in the ads. Expect multiple explanations for the observations.
2. Make sure ads lead potential customers to make observations and inferences that the product or service is something they need.

✓**Assessment**

Portfolio Have each group make a magazine of their ads. See whether other groups draw the expected inferences from the ads. Use the Performance Task Assessment List for Booklet or Pamphlet in **PASC**, p. 57.
P **COOP LEARN**

Repeat the charades activity on page 7 in the teacher edition, performing the pantomime yourself. Have students guess which problem you are acting out.

Extension

For students who have mastered this section, use the **Reinforcement** and **Enrichment** masters.

4 Close

·MINI·QUIZ·

Use the Mini Quiz to check students' recall of chapter content.

1. Science is the _____ of trying to _____ the world around you. *process, understand*

2. The gadgets and gizmos some people think of as science are really _____. *technology*

Section   Wrap-up

1. Science is the process of trying to understand the world around you.

2. Science can help you solve a problem. It gives you a way to learn more information that you can then use to solve the problem.

3. **Think Critically** Communicating the results of your experiments can help others learn from what you did. It is also important to ask questions to find out what others have learned from their experiments.

Science Journal Computer probes can be used to sense temperature or light and feed the data to a computer. The computer records the data in a graph or table.

FIGURE 1-8

These scientists are sharing the results of an experiment with one another.

Communicating

Another important science skill is communication. Look at **Figure 1-8.** Scientists depend on one another to share what they have learned. One way to practice doing this is by using a Science Journal. A Science Journal is much more than a place to write down your observations. It's also a place to express your ideas about what you're investigating, make a sketch, or explore different opinions on a subject. Your Science Journal allows you to practice communicating your thoughts and ideas.

Throughout this book, you'll get many opportunities to practice observing, inferring, and many other science skills. By practicing these skills, you'll become better at solving problems and you'll learn more about the world around you. In the next section, you'll get a chance to start practicing these skills as you learn some ways to use science to solve problems.

Section   Wrap-up

1. What is science?

2. How can science help you solve a problem?

3. **Think Critically:** Why do you think communication is an important skill in science?

4. **Skill Builder**
   Communicating Choose a science skill from the Skill Handbook, such as observing or classifying. In your Science Journal, write a paragraph describing the skill. If you need help, refer to Communicating on page 542 in the **Skill Handbook.**

Science Journal

Research the Internet for information on computer probes, or if your school has computer probes, have your teacher show you how to use them. In your Science Journal, write about ways you can use this technology to help you collect data.

Skill Builder
Communicating Answers will vary. Students should describe the skill and convey its importance in science.

Assessment

Process Give students pictures of bridges and ask them to decide which one would hold the most weight and why. Use the Performance Task Assessment List for Formulating a Hypothesis in **PASC,** p. 21. [IS] [L2]

People & Science

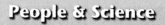

Amanda Shaw, International Science Fair Contestant

Q Ms. Shaw, tell us about competing in an international science fair.

A My project was about plants and global warming. Global warming is a rise in global temperatures that may be due to increases in certain gases in the atmosphere. I went to a regional competition where I won first place. Then I competed against hundreds of other kids in the International Science and Engineering Fair (ISEF).

Q What did you test?

A Many scientists think that increased amounts of carbon dioxide in the air may cause global warming. But plants take carbon dioxide out of the air. I predicted that plants would have a reversing effect on global warming—that is, not cause temperatures in the atmosphere to rise.

Q How did you test your prediction?

A I used glass jars to model the atmosphere. I had one jar with just air inside, one with carbon dioxide, one with a plant, and one with carbon dioxide and a plant. In sunlight, the jars containing carbon dioxide got warmer than the jar with air. But

the jar with the plant and carbon dioxide didn't get as warm. That showed that the plant reversed the carbon dioxide effect.

Q How else are you pursuing your interest in science?

A After the ISEF, I worked as a summer intern at Johns Hopkins University. I worked with scientists who study childhood diseases.

Q Do you enjoy working with scientists?

A I sure do! It seems like we learn something new every day. When we make a new discovery, the scientists in the lab get so excited! And I get excited when I think that with every discovery, we might be closer to helping somebody.

Career Connection

Scientists from many fields study global warming.

• meteorologist • geochemist • ecologist

Choose one of these careers and research the education and training required for it. Share your findings with your classmates.

• The ISEF has been called the Super Bowl of science fairs. Some 800 to 1000 finalists from more than 400 regional science fairs held in the United States and 30 international countries attend the ISEF every May. The fair is open to students in grades 9 through 12. Awards are given in 16 different categories ranging from biochemistry to zoology. Each year, two finalists are chosen to attend the Nobel prize ceremonies in Stockholm, Sweden.

Teaching Strategies

• If your school does not have a science fair, conduct a science fair for your class. Your local library should have books on winning science fair projects. Provide students with a list of simple, interesting projects that can be done in the classroom. Let students select their own projects from the list.

Career Connection

Career Path A scientist who studies global warming usually has a master's degree or Ph.D. in meteorology, ecology, environmental sciences, or a related field.

• High School: math and science
• College: chemistry, biology, and meteorology
• Post-College: Master's degree or Ph.D. in meteorology or ecology

For More Information

Science Service, Inc., coordinates ISEF. Contact this nonprofit organization at Science Service, Inc., 1719 N. St. NW, Washington, DC 20036.

1•2 Doing Science

Prepare

Section Background

Scientific methods are a series of planned steps used by scientists to solve problems. Not all scientists use the same steps or do the steps in the same order, so there are different ways to solve problems. The scientific method presented in this section involves identifying the problem, making a hypothesis, testing the hypothesis, analyzing the results, and drawing conclusions.

Preplanning

Refer to the Chapter Organizer on pages 2A-B.

1 Motivate

Bellringer

Before presenting the lesson, display **Section Focus Transparency 2** on the overhead projector. Assign the accompanying **Focus Activity** worksheet.
L2 ELL

What YOU'LL LEARN

- How to use science to solve problems
- How to design your own experiments

Science Words:
hypothesis
variable

Why IT'S IMPORTANT

Using scientific methods will help you solve many types of problems.

PREPARE
✓ Identify the Problem
✓ Form a Hypothesis

PLAN
✓ Design an Experiment to Test the Hypothesis

DO
✓ Test the Hypothesis
✓ Observe and Record

CONCLUDE & APPLY
✓ Analyze the Results
✓ Draw Conclusions

Solving Problems

Now, it's time to start practicing your science skills as you learn some ways to use science to solve problems. **Figure 1-9** shows one method to solve problems. Let's see how Ms. Quon's class uses this method to solve the bridge problem.

"All right," said Ms. Quon, "we've been given the job of designing a bridge to go across the stream at the site for the new outdoor education center. Where do we start?"

"I think we should go look at the site," said Tinho.

"Why would you start there? What would you look for?" questioned Ms. Quon.

"I don't know. Maybe we could get some ideas by seeing the place," he answered.

"You might be right, but what should we do to make the best use of our time there?" asked Ms. Quon.

"We should think about it some more and make a plan about what we want to find out by going there," offered Gabby.

Getting Organized

"That's right," said Ms. Quon. "Whenever you try to solve a problem, you need to start by making sure you know exactly what the problem is. Then you develop a plan for how to solve it. Let's start by defining the problem. What do you think it is?"

Sydney raised his hand and said, "I think the problem is: How should a bridge be built to cross the stream at the outdoor education center?"

FIGURE 1-9

This poster shows one way to solve problems using science skills.

Program Resources

 Reproducible Masters
Activity Worksheets, pp. 5, 9-10 L2
Enrichment, p. 8 L3
Lab Manual, pp. 5-8 L2
Multicultural Connections, pp. 5-6 L2
Reinforcement, p. 8 L2
Science Integration Activities, pp. 39-40 L2
Study Guide, p. 8 L1

Transparencies
Section Focus Transparency 2 L2
Teaching Transparencies 1, 2 L2

"That's a good start," said Ms. Quon. "What do we already know about the problem?"

"Well, we know that the bridge needs to be strong enough to hold people safely," suggested Tinho.

"And the bridge must be able to survive high water in the spring," said Sydney.

"Wonderful!" said Ms. Quon. "We also know that the site is safe, thanks to studies done by the city to make sure there are no harmful chemicals or wastes dumped there. Now we can ask a question based on what we already know."

"I think we have a lot of questions, like how strong it should be, or what we should make the bridge out of," said Tinho.

"That's right, sometimes we need to break a problem down into smaller problems that are easier to answer. But we have to start somewhere. What do we want to find out first?" asked Ms. Quon.

"First, let's decide where we're going to put the bridge," said Gabby. "That should be easy to figure out."

The class came up with several possible places to put the bridge. Ms. Quon smiled. "Hmm. It looks like this question isn't as easy as we thought. How can we decide which of the suggested places is best?"

Sydney suggested, "Well, we need to look at how the rest of the site is going to be used—you know, like where the trails may be. That may help us decide."

Ms. Quon agreed. "Other classes are working on different parts of the outdoor education center. Maybe we need a way of communicating what each class is doing so we can see how our project fits into the bigger picture."

interNET CONNECTION

Go to the Glencoe Homepage at **www.glencoe.com/ sec/science** to learn more about bridge design.

FIGURE 1-10
Gathering information at the library or on the Internet can make your problem-solving tasks easier.

1-2 Doing Science 15

Brainstorming

Remind students that science can be used to solve problems. Have them brainstorm ways that science might help solve the problems a city faces after being hit by an earthquake. [L2]

2 Teach

Content Background

When a scientific hypothesis has been supported by data obtained by repeated experiments over a long period of time, it is used to form a theory. An example is the Big Bang theory describing the origin of the universe. Theories differ from scientific laws, which are rules that describe the behavior of something in nature. An example is Newton's first law of motion.

interNET CONNECTION

The Glencoe Homepage at **www.glencoe.com/sec/ science** provides links connecting concepts from the student edition to relevant, up-to-date Internet sites.

? FLEX Your Brain

Use the Flex Your Brain activity to have students explore SCIENTIFIC METHODS.

📁 **Activity Worksheets,** page 5

Community Connection

Planner Contact your local civil engineer. Ask him or her to explain to students what type of planning must go into building an overpass over a heavily used road or highway. For example, he or she might explain how traffic is diverted and what safety precautions are necessary to keep debris from falling on cars passing under the construction zone.

Have students develop surveys about a school-related issue, such as whether or not the school cafeteria should have a soda-pop machine. Make sure survey questions are in simple yes or no formats. Have students gather at least 40 responses to their surveys. Students can present the results of their surveys in a table or circle graph. **L2**

COOP LEARN

Discussion

Survey results can be influenced by improper or unclear wording of the questions. Discuss with students why survey questions should not reflect the biases of the interviewer.

GLENCOE TECHNOLOGY

 Videodisc

STVS: Animals

Disc 5, Side 1
Tagging Ants (Ch. 16)

Cockroach on a Treadmill (Ch. 18)

Disc 5, Side 2
Raising Super Fish (Ch. 7)

How Bats Hear (Ch. 18)

[barcode]

FIGURE 1-11
Problem solving involves making careful observations and gathering information.

"We also need to look at the stream. Maybe there are trees, or the shape of the stream's banks may be an issue," said Tinho.

"What if we end up with two equally good choices?" asked Ms. Quon. "That is, what if both meet all of the requirements that we come up with?"

"We could make some sketches of different locations and ask people which one they prefer," suggested Gabby.

"Yeah, we could do a survey," agreed Tinho.

Carrying Out the Plan

The class made a list of requirements that the bridge location should meet. Then they went to the site and, like the student in **Figure 1-11**, took photos of the stream and made careful observations of each of the possible bridge locations. They chose the two best sites, based on the requirements that they had decided on. Then, Gabby and Tinho drew a map of the site that showed the two possible bridge locations. Using these drawings, the class surveyed people in the school and the community to find out which location they thought would be the best. The class counted the results and found a clear winner.

"Well," said Ms. Quon, "it looks like we have a location. What's the next step?"

"We need to find out how to build a bridge that's strong enough," said Sydney.

"How should we solve this new problem?" asked Ms. Quon.

"We can look at other bridges to get ideas about how we could build ours," said Tinho, pointing to the pile of books that Ms. Quon brought in.

Science Journal **Communicating Results** Tell students that scientists share information by writing articles in scientific journals or by speaking at scientific conferences. Have students write about why communicating the results of scientific experiments is so important. **L2** **P** **LS**

Using Models

Ms. Quon flipped through one of the books. "There are a lot of different bridge designs. I think many of them might work for our bridge. How will you decide which design to use?"

"We could test them to see which ones are strong enough," said Gabby.

"Won't it be hard to test different bridges? How are we going to build all of these bridges?" asked Ms. Quon.

"We can build smaller ones and test them that way," suggested Sydney.

"That's a good idea," Ms. Quon said. "That's called using models. If you need to experiment with something that takes too long, is too fast, too big, or too small, you can test it using a model. We can't build full-size bridges to test, but we can build models of them and see which designs are strong enough. After we experiment and find out which design we will use, we can then experiment with materials in the same way.

"Before we experiment, though, we need to know how to plan an experiment to make sure we find out the information we need."

Problem Solving

Flex Your Brain

Solving problems requires a plan. This plan may be a simple thing that you do in your head, or it may be something more complicated that you actually write down. Below is a process called *Flex Your Brain*, which is one way to help you organize a plan for solving a problem.

Solve the Problem:
Use the *Flex Your Brain* chart to explore different styles of bridges.

Think Critically:
Why does *Flex Your Brain* ask you to share what you learned?

Flex Your Brain

1. **Topic:** _____
2. **?** What do I already know?
   1. _____
   2. _____
   3. _____
   4. _____
3. **Q:** Ask a question _____
4. **A:** Guess an answer _____
5. **How sure am I ? (circle one)**
   Not sure   Very sure
   1   2   3   4   5
6. **?** How can I find out?
   1. _____
   2. _____
   3. _____
   4. _____
7. **Explore**
8. Do I think differently? → yes   no
9. **?** What do I know now?
   1. _____
   2. _____
   3. _____
   4. _____
10. **SHARE**
   1. _____
   2. _____
   3. _____

 Problem Solving

Solve the Problem
Answers will vary, depending on the bridges chosen by students.

Think Critically
Answers will vary, but students may note that this helps them to see how their level of sureness affects their thinking. IS

Enrichment
Have students use the Flex Your Brain chart to explore a topic of their choice.

GLENCOE TECHNOLOGY

Videodisc
STVS: Chemistry
Disc 2, Side 2
Neutron Activation Analysis of Paintings (Ch. 6)

STVS: Earth & Space
Disc 3, Side 1
Computerized Star Imaging (Ch. 10)

Disc 3, Side 2
Fibers from Rocks (Ch. 11)

Inclusion Strategies

Gifted Have students research a science-related problem in their area, such as air pollution, soil erosion, or flooding. Have students explain what is being done to address the problem. Have them propose additional strategies to solve the problem. They should back up their strategies with research. L3 IS

Planning an Experiment

Experiments are often part of a plan for solving problems. They give you a chance to test your ideas as you go through the process of solving the problem. **Figure 1-12** shows one way to test experiments. As Ms. Quon's class is finding out, experiments will help them learn more about bridges as they try to solve the problem of building the bridge across the stream.

Throughout this book, you'll have opportunities to design your own experiments. As you can see in Activity 1-2 on pages 24-25, the two-page activities in this book are divided into four major parts. First, you'll *Prepare* for the experiment. Then, you'll *Plan* the experiment. Next, you'll *Do* the experiment. And finally, you'll *Conclude and Apply.*

FIGURE 1-12
Car designers test the safety of automobiles using crash-test dummies.

PREPARE

Preparing for an experiment involves getting a clear idea of what you want to find out. You start by thinking about the problem that you are trying to solve. What do you already know that may be helpful in solving the problem? Organize this information and use it to form a hypothesis (hi POTH uh sus).

A **hypothesis** is a statement that can be tested about a problem. A hypothesis about the strongest material to make the bridge may be "Wood is the strongest of the materials we can use to build the bridge." Notice that this hypothesis can be tested. You can do an experiment that tests the strength of different materials. After doing the experiment, you may find that your hypothesis was supported. Or, you may find that there was a stronger material and your hypothesis was not supported. When this happens, you'll have to develop a new hypothesis.

18 Chapter 1 The Nature of Science

Cultural Diversity

Scientific Methods and the Inuit Scientific methods begin with observation and a search for patterns. Many traditional peoples employ informal scientific methods in their lives. Inuit carefully observed polar bears, then modeled their seal-hunting techniques after the bears' methods. In addition, Inuit use seal behavior to predict winter storms. If seals surface briefly, keeping their heads low and parallel to the water, storms are likely. Both hunting and weather forecasting are examples of the Inuit people using observation, analysis, and prediction—using scientific methods.

FIGURE 1-13

The amount of water added to the plants is the variable in this experiment. *Based on the photographs, what would you conclude about the effects of water on plants?*

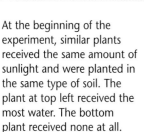

A At the beginning of the experiment, similar plants received the same amount of sunlight and were planted in the same type of soil. The plant at top left received the most water. The bottom plant received none at all.

B Three weeks later, by controlling other factors and changing only one variable—water—the results of the experiment clearly show the effect of water on plants.

PLAN

Once you've developed a hypothesis, it's time to plan a way to test the hypothesis. Planning an experiment may take longer than the experiment itself. Careful planning helps ensure that your experiment is safe and will help you get useful results.

Variables

One of the most important considerations in an experiment is controlling variables. **Variables** are factors in an experiment that can change. **Figure 1-13** gives an example of testing variables. For another example, think of bridges. If you are testing different bridge-building materials to see which is the strongest, the material you use is one variable. You may test wood, steel, or plastic. You would test them all the same way. But you test only one at a time. Each time you run the experiment, you change which material you test. Other variables in such an experiment might be the shape of the material, how you measure the material's strength, and how you test the material.

1-2 Doing Science   19

Using Science Words

Linguistic The word *hypothesis* is derived from the Greek word *hupotithenai*. Have students research the meaning of the Greek word. (*Hupotithenai* means "to suppose.") **L2**

Teacher F.Y.I.

The one factor that changes in an experiment is called the independent variable. The factor that changes as a result of the independent variable is called the dependent variable.

GLENCOE TECHNOLOGY

Videodisc

The Infinite Voyage: The Great Dinosaur Hunt
Chapter 3
"The Great American Bone Rush"

Chapter 4
The Tyrrell Museum of Paleontology

Chapter 5
Dinomation International: Building Dinosaurs

Chapter 10
New Dinosaur Discoveries and Their Link with Today

Visual Learning

Figure 1-13 Based on the photographs, what would you conclude about the effects of water on plants? *Without water, the plants in the photograph turned brown and died. These plants need water to live.* **LS**

19

Demonstration

 Visual-Spatial The strength of a material is influenced by the way it is shaped and how it is used. Obtain a wooden paint-stirring paddle from your local paint store. Place it across two desks, and hang a heavy weight from it. The weight should be heavy enough to break the paddle. Turn another paddle on edge, holding it in place as you hang the same amount of weight on it. Discuss with students why the second paddle did not break. Explain that the way a material is used can make it stronger. Drinking straws that are bent into triangles, for instance, support more weight than straight straws.

You have to make sure, though, that you change *only one* variable at a time. This is called "controlling the variables." To make sure you are testing which material is the strongest, each material you test should have the same shape, and you should test each one the same way. If you change more than one variable, you can't be sure which variable caused the results.

Number of Trials

Another thing to decide when planning an experiment is how many trials to have, or how many times you need to repeat the experiment to make sure the results are useful. Look at **Figure 1-14.** If you're testing the strength of wood, do you test just one piece of wood? What would happen if the wood you tested had a crack in it to begin with? Can you trust that the results from the cracked-wood test will show the strength of another piece of wood? What if you made a mistake measuring the strength of the wood? These kinds of errors are always going to occur. That's why it's a good idea to run more than one trial of an experiment. If there's an error, it will be balanced by the other trials.

Sometimes you may need to run ten trials; other times you may need to run 100. However, some experiments are either too costly or take too much time to run more than two or three times. When this happens, you'll have to think about the possible errors when you analyze the results.

FIGURE 1-14
To balance the possibility of errors in an experiment, it's a good idea to run several trials of the experiment. *Is this always possible? Explain why or why not.*

20 Chapter 1 The Nature of Science

Across the Curriculum

Geography The design and materials needed to build a bridge depend on what will be traveling across it, how long it will be, and how strong it needs to be. Bridge design also depends on the topography of the area in which it is built. Show students photos of a pedestrian bridge in a city, a bridge over a mountain stream, and a bridge crossing a chasm in the rain forest. Ask students to describe differences between the bridges, and to relate these differences to changes in topography.

Strength of Different Bridge Materials

Trials

Materials	1	2	3	4
Wood	broke	didn't break	didn't break	didn't break
Plastic	broke	broke	broke	didn't break

 A Most tables have a title that tells you, at a glance, what the table is about.

 B A table is divided into columns and rows. The top row lists what is being compared—in this case, the strength of different bridge materials during different test trials.

 C The first column lists items being compared—in this case, different bridge materials.

FIGURE 1-15
Data tables help you to organize your observations and test results.

Materials

After you have planned the experiment, you must make a complete list of the materials you will need to do the experiment. Try to think of alternate materials in case something you need is not available. Be resourceful. You don't have to have fancy scientific equipment to do experiments. You can learn a lot using things you have around your house and classroom.

Data Tables

A data table organizes observations in columns and rows. **Figure 1-15** shows how you might set up a data table for one particular experiment. Because you will need to record your observations during the experiment, it's a good idea to make any data tables you'll need ahead of time. This will help you make sure you make all of the observations necessary to make a conclusion about your hypothesis.

Activity

LS **Logical-Mathematical** Tell students that the science skill *classifying* is used by many scientists to organize data. A geologist, for instance, might organize rock samples by color, shape, or size. Draw at least a dozen different geometric shapes on the chalkboard, using different-colored chalk, if possible, to fill in the shapes. Have students classify the shapes into several subgroups. Students should be able to explain the characteristics they used to develop their classification schemes. Encourage students to come up with creative titles for their subgroups. **L2**

Visual Learning

Figure 1-14 Is this always possible? Explain why or why not. *It is not always possible because some experiments are either too costly or take too much time to repeat more than two or three times.* **LS**

GLENCOE TECHNOLOGY

◉ Videodisc

The Infinite Voyage: Insects: The Ruling Class
Chapter 8
Prospecting for Healing Medicine from Insects

Have students create a poster of safety rules to use in the lab. Encourage students to come up with additional safety rules that may pertain to their particular lab situation. Display the poster on a wall or bulletin board near your science lab area. **L2**
COOP LEARN

GLENCOE TECHNOLOGY

 Videodisc

STVS: Physics

Disc 1, Side 1

Detecting Flaws in Machine Parts (Ch. 7)

Laser Eye Surgery (Ch. 13)

Fiber Optics on the Farm (Ch. 16)

Disc 1, Side 2

Computer Graphics (Ch. 16)

Modular Factory (Ch. 19)

Communications Security (Ch. 21)

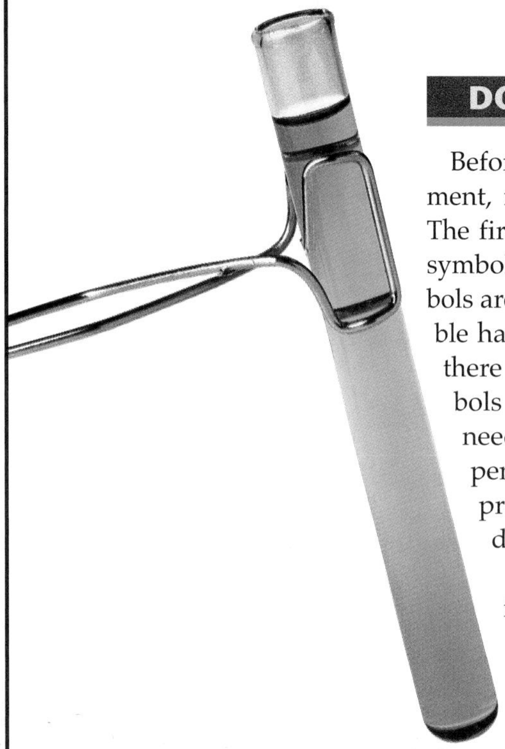

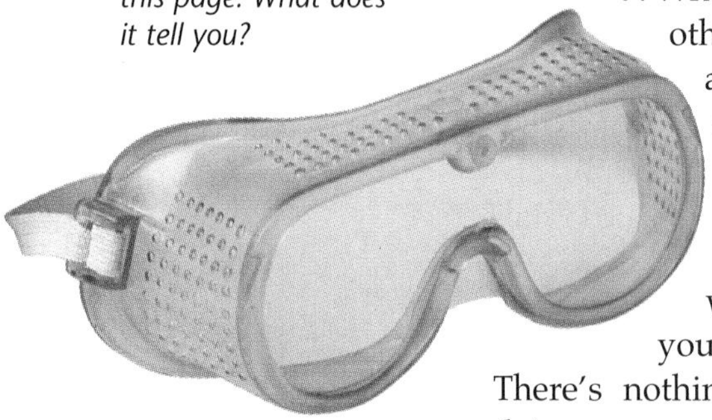

FIGURE 1-16

Lab equipment, like the test tube shown above, must be handled properly. Goggles like the ones shown below should be worn during all lab activities. *Look at the safety symbol on this page. What does it tell you?*

DO

Before you jump in and do the experiment, make sure you think about safety. The first thing to do is to look for safety symbols on the activity page. Safety symbols are pictures that tell you about possible hazards. In Appendix D on page 528, there is a table that describes all of the symbols used in this book. While you may not think you need goggles and aprons all the time, accidents happen—and when they do, you'll be glad you were protected. Follow safety rules whenever you are doing activities. **Figure 1-16** gives more safety tips.

Doing experiments can be a lot of fun, but it's important to be careful when doing them. Not paying attention to what you're doing could seriously injure you or one of your classmates. Remember, safety is no accident! Here are a few important safety rules that you should follow:

1. Before beginning any lab, understand the safety symbols shown in Appendix D.
2. Follow all safety symbols.
3. Always slant test tubes away from yourself and others when heating them.
4. Never eat or drink in the lab, and never use lab glassware as food or drink containers.
5. Never inhale chemicals, and don't taste anything.
6. Report *any* accident or injury to your teacher.
7. When cleaning up, get rid of chemicals and other materials as directed by your teacher, and always wash your hands after working in the lab.

Another thing you must do before you begin the experiment is to have your teacher approve your plan. He or she will be able to find any problems with your plan and suggest ways to improve it. There's nothing worse than spending time carefully doing an experiment only to have to start over because you forgot to include an important step in your plan.

22   Chapter 1 The Nature of Science

Visual Learning

Figure 1-16 Look at the safety symbol on this page. What does it tell you? *This symbol represents fire safety. It appears when care should be taken around open flames.* **LS**

Recording Observations

As you do the experiment, record all of your observations in your Science Journal. **Figure 1-17** shows an example of a Science Journal. Observations can include measurements and descriptions of what happened and sketches or drawings. For example, if you were testing the strength of bridge materials, you might observe that a piece of plastic held a certain amount of weight without breaking. However, you may observe that it bent a lot. You should also note any mistakes you made during the experiment.

CONCLUDE AND APPLY

After you've carried out the experiment and recorded all of your observations, it's time to look at the results and figure out what they mean. This is when you start asking *why* things happened. You'll need to look at all of the observations you wrote down and even compare your results with those of other groups that did the same experiment. Making tables and graphs can also help you see what happened in the experiment.

If you tested the strength of bridge materials, you might decide that wood is the strongest material you can use. That's called drawing a conclusion. A conclusion is a decision based on the results of the experiment. It may either support or not support your hypothesis. If you hypothesized that wood was the strongest material you could use to build the bridge, then you could say that your hypothesis was supported.

Next, you'll get a chance to design your own experiment.

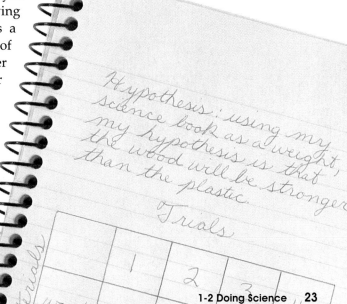

FIGURE 1-17

Record observations and conclusions in your Science Journal.

Hypothesis: using my science book as a weight, my hypothesis is that the wood will be stronger than the plastic.

Trials

Community Connection

Safety in the Workplace Invite a local construction worker, factory worker, or miner to class to discuss safety precautions in the workplace. Encourage students to ask questions about potential hazards and the procedures used to minimize these hazards. Afterwards, stress to students that safety standards are followed in most occupations—indeed, they are a part of everyday life. Lead students into a discussion about how safety issues affect them. For instance, are there warnings on the labels of certain products at home? Why do they wear seat belts while traveling in a car or plane?

Activity 1-2

Activity 1-2

PREPARE

Purpose

IS **Kinesthetic** This activity reinforces the problem-solving process. **L2** **ELL** **COOP LEARN**

Process Skills

sequencing; observing and inferring; formulating models; measuring in SI; hypothesizing; using variables, constants, and controls; experimenting; analyzing; recognizing spatial relationships

Time

1 to 2 class periods

Materials

Use two stacks of books to create the 50-cm gap.

Safety Precautions

Remind students that they should follow lab safety procedures. Have them put on safety aprons and goggles before conducting the experiment.

Possible Hypotheses

Hypotheses will vary. One possible hypothesis might be that a flat bridge of craft sticks will be strong enough to hold a science book across a 50-cm gap.

📂 **Activity Worksheets,** pages 5, 9-10

Possible Materials
- drinking straws
- craft sticks
- string
- glue
- tape
- scissors
- meterstick
- books

Design Your Own Experiment
Bridge Building

One of the problems Ms. Quon's class needs to solve is what design they should use to build their bridge. In this activity, you will plan and do an experiment to design and test a model bridge. You only get to use certain materials, your bridge must cross a gap 50 cm wide, and it must be strong enough to hold your science book. Good luck!

PREPARE

What You'll Investigate
How can you build a model bridge that will cross a 50-cm gap and hold your science book?

Form a Hypothesis

As a group, look at the materials your teacher has supplied. Think about ways to use them to build a model bridge that meets the requirements stated above. How can the materials be used to make bridge shapes you have seen? Try putting together a few pieces to see how they join together and how strong they are. Sketch some designs, and discuss the good and bad points of each. You might need to combine some of the best parts of different designs. Decide on a design that will meet the requirements.

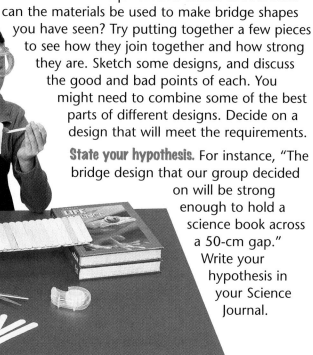

This student is working on one of many possible designs.

State your hypothesis. For instance, "The bridge design that our group decided on will be strong enough to hold a science book across a 50-cm gap." Write your hypothesis in your Science Journal.

Bridge Models

Trials

Materials		1	2	3
	Drinking straws	did not hold weight of book	did not hold weight of book	did not hold weight of book
	Craftsticks	held weight of book	did not hold weight of book	held weight of book
	Craftsticks with string and tape	held weight of book	held weight of book	held weight of book

Goals

Design a model bridge.

Construct and **test** a model bridge.

Safety Precautions

PLAN

1. **Decide** on a way to test your group's hypothesis. **List** the materials you will use in each step of your experiment. **Describe** how you will connect the materials.
2. Sketch your design. Look for any spots in the design that may be weak. How would you strengthen those spots?
3. How will you **test** the strength of your model? How many times will you repeat the test?
4. **Make a data table** to record your observations.

DO

1. Make sure your teacher approves your plan and your data table before you begin.
2. Write down the observations you make in your Science Journal.
3. **Build** your model using the materials provided by your teacher.
4. **Test** your bridge to see if it meets or exceeds the requirements.

CONCLUDE AND APPLY

1. Based on your results, did your **model** bridge meet the requirements?
2. **Compare** your model with the models of all the other groups. Which designs performed best in the tests?
3. What was different about the strongest bridge models? How did these differences make the models stronger?
4. **APPLY** Based on your observations of all the models, how would you change your design? **Draw** the new design.

1-2 Doing Science   **25**

Possible Procedures

Glue several layers of craft sticks together, with each layer going in different directions. The craft sticks can be reinforced with string and tape. Overlapping craft sticks turned on edge could be used as underpinning.

Teaching Strategies

Many of the designs will not support the book, so allow extra time for starting over.

DO

Expected Outcome

Several different types of bridges will support the weight. Refer to the bridge designs on page 5 in the teacher edition. A simple girder, slab, or cantilever design will support the weight of a book.

CONCLUDE AND APPLY

1. Answers will depend on student results.
2. Answers will vary, but students should note how well the bridges met the requirements.
3. Students should note how the materials were used to build a stronger structure and how geometric shapes such as arches and triangles were used.
4. Answers will vary but should reflect the use of information learned from the experiments to improve their designs.

Go Further

LS Kinesthetic Using what the students have learned, challenge them to build a bridge using only drinking straws. It can be done if the ends are crimped so that the straws will fit together.

 **Assessment**

Process Have students make a concept map such as an events chain showing the steps they took to build their model bridges. Use the Performance Task Assessment List for Events Chain in **PASC**, p. 91. **P**

25

4 Close

Activity

Scramble the following steps, then have students sequence them in the order they should occur: define the problem, research the problem, form a hypothesis, test the hypothesis, draw a conclusion, communicate the results. **L2**

Section Wrap-up

1. Science can help you solve problems by giving you a way to find out more about the problem.

2. Models are useful when the problem you are testing occurs too quickly or too slowly, is too dangerous, too far away, too big or too small, or is otherwise impractical to be tested directly. For example, computer models are useful for studying the movements of Earth's plates because that movement takes place slowly over hundreds of thousands of years.

3. **Think Critically** Students should suggest that she keep all of the other ingredients the same, the oven temperature the same, the time of cooking the same, and the cookware the same.

Using Computers

Posters will vary but should focus on a science skill from the Skill Handbook.

FIGURE 1-18

Sharing the results of your experiment with others is part of doing science.

Communicating the Results

Ms. Quon's class performed many experiments, testing many designs and materials for their bridge. When they were finished, they had decided on a bridge design that would be perfect for the new outdoor education center. As a class, they wrote a report describing the design and how they came to that conclusion. They included graphs and tables that summarized their observations from the various experiments. When it was all finished, they presented the report and a model of the bridge to the school board. The board was impressed with how well the class had communicated their work. The design was approved.

"You've done a fabulous job planning this bridge," said Ms. Quon. "Unfortunately, we don't have much of a budget to actually build it."

"It looks like we have another problem to solve," sighed Gabby.

"Yes, but I'm sure you can do it!" smiled Ms. Quon.

"You mean, you *hypothesize* that we can do it!" laughed Gabby.

Section Wrap-up

1. How can science help you solve problems?

2. Why do scientists sometimes use models to test the design of objects?

3. **Think Critically:** Valerie is going to do an experiment testing how much flour is needed to make the cake with the best texture. Which variables should she control?

4. **Skill Builder**
   Designing an Experiment to Test a Hypothesis Design an experiment that tests the hypothesis "Wood is a stronger material than plastic or aluminum." Describe the steps you would follow and list the variables in your experiment. If you need help, refer to Designing an Experiment to Test a Hypothesis on page 553 in the **Skill Handbook.**

Using Computers

Graphics Software Use a graphics program to design a poster that shows the importance of a particular science skill. Use pictures to help show the skill. Present your poster to your class.

Skill Builder

Designing an Experiment There are many ways to design this experiment. One way is to have samples of wood, plastic, and aluminum that are all the same length, shape, width, and thickness. The samples are placed over a gap that is the same distance. Weights are hung from the samples at the same point along each sample. Weights are added until each sample breaks. The sample that holds the most weights is the strongest. Students should note that several trials should be run.

Assessment

Portfolio Have groups of students write and perform skits showing how an everyday problem, such as finding out which bus to ride home, can be solved using scientific skills. If possible, videotape the performances. Use the Performance Task Assessment List for Skit in **PASC**, p. 75. **L2** **P** **LS**
COOP LEARN

R ead the statements below that review the most important ideas in the chapter. Using what you have learned, answer each question in your Science Journal.

1. Science is the process of learning and studying things in the world around you. *How is science like a tool?*

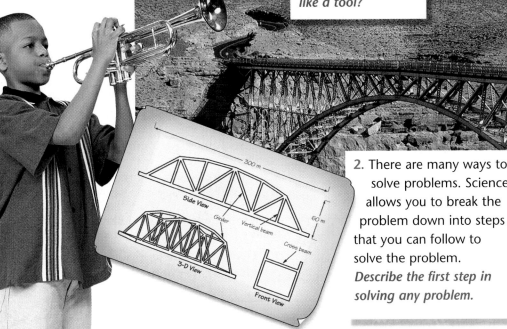

2. There are many ways to solve problems. Science allows you to break the problem down into steps that you can follow to solve the problem. *Describe the first step in solving any problem.*

3. When designing an experiment, following a set of steps such as *Prepare, Plan, Do,* and *Conclude and Apply* will help you get the answers you need to solve the problem. *Describe what a hypothesis has to do with designing an experiment.*

Review

Have students look at the illustrations on this page. Ask them to describe details that support the main ideas of the chapter found in the statement for each illustration.

Teaching Strategies

Have students explain how using a simple hand tool involves the problem-solving process. For example, when using a screwdriver to tighten a loose screw, you have to first decide which type of screwdriver to use. Then you try the screwdriver to see if it will work. If it doesn't fit the screw, you have to try again with a different screwdriver. This process is similar to defining the problem, forming a hypothesis, testing the hypothesis, and drawing a conclusion.

Answers to Questions

1. Science is a tool in that it can be used to help solve a problem. Like a tool, it takes practice and requires a certain set of skills to use effectively.

2. The first step in solving a problem is to clearly define what the problem is.

3. A hypothesis is a statement that can be tested about a problem. An experiment is performed to test this hypothesis.

✔ Assessment

Portfolio Encourage students to place in their portfolios one or two items of what they consider to be their best work. Examples include:
- MiniLab, p. 10
- Activity 1-1, p. 11
- Activity 1-2, pp. 24-25 **P**

Performance Additional performance assessments may be found in **Performance Assessment** and **Science Integration Activities.** Performance Task Assessment Lists and rubrics for evaluating these activities can be found in Glencoe's **Performance Assessment in the Science Classroom (PASC).**

Science at Home

LS **Intrapersonal** Have students watch for commercials on TV that make a claim about a product. Have them design an experiment that will test the product's claim. **L2**

Using Key Science Words

1. Before beginning an experiment, you should define the problem and organize what you already know about it. Then you form a hypothesis and develop a plan for testing it. Variables should be identified and how to control them should be included in your plan. Make a list of materials and safety concerns.

2. Science is the process of learning about the world around you, and technology is the use of knowledge gained by science.

3. A variable is something in an experiment that can change.

4. Scientists try to find out something about the world around them by using science. This includes developing hypotheses and testing them.

5. A hypothesis is a statement that can be tested about a problem.

Checking Concepts

6. d   **7.** c   **8.** a
9. b   **10.** b

Thinking Critically

11. They performed experiments testing only the strength of different bridge materials.

12. The geese are migrating to a warmer climate for the winter.

13. Answers will vary, but an example hypothesis would be: A house fern must be watered once a day.

Using Key Science Words

hypothesis
science
technology
variable

Answer the following questions in complete sentences. Use the terms in the list above in your answers.

1. What should you do before you begin an experiment?
2. What is the difference between science and technology?
3. What is a variable?
4. What does a scientist do?
5. Define *hypothesis*.

Checking Concepts

Choose the word or phrase that completes the sentence.

6. Science is _____.
   a. a way to help solve a problem
   b. a tool
   c. the process of trying to understand the world around you
   d. all of the above

7. The use of scientific discoveries is called _____.
   a. the scientific method
   b. a hypothesis
   c. technology
   d. a variable

8. Comparing and contrasting is an example of a _____.
   a. science skill
   b. hypothesis
   c. conclusion
   d. control

9. You should never _____ in a science lab.
   a. wear safety goggles
   b. eat or drink
   c. wear an apron
   d. report any accidents to your teacher

10. A _____ is a statement about a problem that can be tested.
    a. conclusion   c. variable
    b. hypothesis   d. data table

Thinking Critically

Answer the following questions in your Science Journal using complete sentences.

11. Give one example of how Ms. Quon's class controlled variables.

12. You observe a flock of geese flying south. What inference can you make?

13. Give an example of a hypothesis.

14. Describe how you have used technology today.

15. Is every problem solved using the same steps? Explain.

Assessment Resources

Reproducible Masters
Chapter Review, pp. 5-6
Assessment, pp. 5-8
Performance Assessment, p. 39

Glencoe Technology
Computer Test Bank
MindJogger Videoquiz

Skills Review

Developing Skills

If you need help, refer to the description of each skill in the Skill Handbook.

16. Forming a Hypothesis: Enrique has an aquarium. He bought a new fish and placed it in his aquarium with his other fish. Two days later, it died. Form a hypothesis about why the fish died. Design an experiment to test your hypothesis.

17. Concept Mapping: Scientists use different methods to help solve problems. Complete the events chain on this page that shows the basic steps in solving a problem.

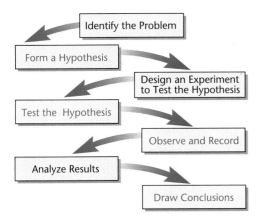

Identify the Problem

Form a Hypothesis

Design an Experiment to Test the Hypothesis

Test the Hypothesis

Observe and Record

Analyze Results

Draw Conclusions

18. Forming a Hypothesis: Your volleyball team has lost three games in a row. Form a hypothesis about what can be done to help your team play better.

19. Separating and Controlling Variables: Suppose you hypothesized that your volleyball team would play better if it practiced more. You decide to test your hypothesis. After adding one more afternoon practice to your schedule, your team wins a game! What were the variables in your test? Which variables did you control?

20. Sequencing: Maggie wants to see how many kinds of birds live in a nearby park. Sequence the following steps in the order she should follow to solve her problem.
   a. Observe birds.
   b. Analyze the results.
   c. Research the kinds of birds that typically live in her area.
   d. Hypothesize how many birds live in the park.
   e. Record her observations.

Performance Assessment

1. Bulletin Board: Think about the steps involved in solving a problem. Make a Solving Problems with Science bulletin board. Include the major steps of solving problems using science. Write a summary of what to do at each step, and include drawings or photographs that help explain the process.

2. Oral Presentation: Design an experiment that investigates the effects of sunlight on small plants. Prepare an oral presentation that describes your experimental plan and give it to your class.

Performance Assessment

1. Bulletin boards will vary. Use the Performance Task Assessment List for Bulletin Board in **PASC**, p. 59. **P**

2. Presentations will vary depending on the student's experiment. Use the Performance Task Assessment List for Oral Presentation in **PASC**, p. 71. **P**

14. Answers will vary, but students should recognize that items such as wristwatches, alarm clocks, buses, sneakers, clothing fabrics, dyes, and even their pens and pencils are examples of technology.

15. No, some problems are solved by merely observing, while others require extensive experimenting. Each problem is unique and may require an equally unique procedure to solve.

Developing Skills

16. Forming a Hypothesis Student answers will vary. Students may hypothesize that the fish was sick or that the water from the aquarium did not suit the life requirements of the fish.

17. Concept Mapping See answer on student page.

18. Forming a Hypothesis Responses will vary but may include that the volleyball team would play better if it practiced more.

19. Separating and Controlling Variables The variables include your team, the team you played against, the number of practices, the volleyball court, the weather, time of game, health of players, referees, and coaches. Most of these were probably not controlled. Most likely, your team played a different team at a different place. Enough evidence is not provided to conclude that the extra practice made a difference.

20. Sequencing The sequence should be: c, d, a, e, b.

UNIT 1

LIFE SCIENCE

In Unit 1, students are introduced to living things and find out how scientific methods are used to learn about them. The characteristics of living things are defined. Cells and how they reproduce are described. Students learn how living things respond to changes in the availability of natural resources.

Unit Contents

Unit 1 Project

The Biome Internet Project
pp. 206-207

interNET CONNECTION

The Glencoe Homepage at **www.glencoe.com/sec/science** provides links to information related to topics found in each chapter in Unit 1.

Science at Home

Natural Resources at Home Have students choose one room and list several objects found there and the natural resources that were used to make these objects. Students may list furniture, clothing, rugs, dishes, food, etc. Allow students to ask adults for help in identifying resources used in certain products. Ask students to share their lists in class. **L2** **COOP LEARN** **ELL**

Life Science

What's happening here?

Seeds can often be found some distance away from their parent plants. Most plants don't move, so how do they spread their seeds? Some seedpods explode, shooting their seeds far away. Other seeds hitch rides on air currents or passing animals. Milkweed pods burst open, releasing seeds with plumes of silky hairs. These hairs allow the seeds to ride the wind. Cocklebur seeds, as shown in the smaller photo, may travel miles on animal fur. How does traveling help seeds? By being spread far from their parents, the seeds do not have to compete for water and sunlight as they grow into new plants.

CONNECTION

Visit the Glencoe Homepage at *www.glencoe. com/sec/science* for links to information about milkweed seeds, cockleburs, and other seeds. Find out how other plants spread their seeds. In your Science Journal, write about how plants use wind, water, and other organisms for seed dispersal.

31

Chapter Organizer

Section	Objectives/Standards	Activities/Features
Chapter Opener		Explore Activity: Compare Living and Nonliving Things, p. 33
2-1 **Alive or Not** (5 sessions, 2½ blocks)*	**1. List** five characteristics of living things. **2. Describe** the main needs of living things. National Science Content Standards: (5-8) UCP1, UCP3, A1, A2, B3, C1, C2, C3	MiniLAB: Inferring Body Responses, p. 36 Activity 2-1: Mealworm Behavior, pp. 40-41 Using Technology: A New Chess Champion?, p. 42 Skill Builder: Making and Using Tables, p. 44 Science Journal, p. 44
2-2 **Classifying Life** (3 sessions, 1½ blocks)*	**3. Investigate** classification systems. **4. Compare and contrast** the kingdoms of living organisms. National Science Content Standards: (5-8) UCP1, UCP5, A1, A2, C1	MiniLAB: Classifying Animals with Bones, p. 46 Problem Solving: Classifying Animals, p. 47 Activity 2-2: Button Classification, p. 48 Skill Builder: Developing Multimedia Presentations, p. 53 Science Journal, p. 53
2-3 **Science and Society:** **Viruses** (1 session, ½ block)*	**5. Determine** how the AIDS virus and the cold virus affect humans. **6. Outline** an article as a tool for remembering the big ideas. National Science Content Standards: (5-8) UCP5, F1	Skill Builder: Outlining, p. 55 Science & Language Arts: Hatchet, p. 56 Science Journal, p. 56

* A complete Planning Guide that includes block scheduling is provided on pages 31T-33T.

Activity Materials

Explore	Activities	MiniLABs
No materials needed.	**page 40** mealworms; small cardboard box with lid, such as a shoe box; eye dropper; cotton-tipped swabs; penlight or small flashlight; dry bran flakes; chopped apple; chopped banana; vinegar; ice cubes; plastic straw; hot water; tap water **page 48** assorted buttons	**page 36** stopwatch or clock with second hand **page 46** animal photographs, envelopes

Need Materials? Call Science Kit (1-800-828-7777).

Teacher Classroom Resources

Reproducible Masters	Transparencies	Teaching Resources
Activity Worksheets, pp. 5, 13-14, 17 **Enrichment,** p. 9 **Lab Manual 3, 4** **Reinforcement,** p. 9 **Science and Society/** **Technology Integration,** p. 20 **Study Guide,** p. 9	**Science Integration Transparency 2,** Living Things and Robots **Section Focus Transparency 3,** What is a pet? **Teaching Transparency 3,** The Traits of Life **Teaching Transparency 4,** Kingdoms of Life	Spanish Resources English/Spanish Audiocassettes Cooperative Learning Resource Guide Lab Partner Lab and Safety Skills Lesson Plans
Activity Worksheets, pp. 5, 15-16, 18 **Cross-Curricular Integration,** p. 6 **Enrichment,** p. 10 **Multicultural Connections,** pp. 7-8 **Reinforcement,** p. 10 **Science Integration Activities,** pp. 41-42 **Study Guide,** p. 10	**Section Focus Transparency 4,** Losing Your Marbles	**Assessment Resources** Chapter Review, pp. 7-8 Assessment, pp. 9-12 Performance Assessment, p. 40 Performance Assessment in the Science Classroom (PASC) MindJogger Videoquiz Alternate Assessment in the Science Classroom Computer Test Bank
Enrichment, p. 11 **Reinforcement,** p. 11 **Study Guide,** p. 11	**Section Focus Transparency 5,** Viruses	

Key to Teaching Strategies

The following designations will help you decide which activities are appropriate for your students.

L1 Level 1 activities should be appropriate for students with learning difficulties.

L2 Level 2 activities should be within the ability range of all students.

L3 Level 3 activities are designed for above-average students.

ELL ELL activities should be within the ability range of English Language Learners.

LS These activities are designed to address different learning styles.

COOP LEARN Cooperative Learning activities are designed for small group work.

P These strategies represent student products that can be placed into a best-work portfolio.

GLENCOE TECHNOLOGY

The following multimedia resources are available from Glencoe.

Science and Technology Videodisc Series (STVS)
Animals
Studying Sharks
The Infinite Voyage Series
To the Edge of the Earth
National Geographic Society Series
STV: Biodiversity
STV: Plants and Simple Organisms

Glencoe Life Science CD-ROM
Glencoe Life Science Interactive Videodisc
Bacterial Action
Fungal Action

Teacher Classroom Resources

This is a representation of key blackline masters available in the Teacher Classroom Resources.

Teaching Aids

Section Focus Transparencies

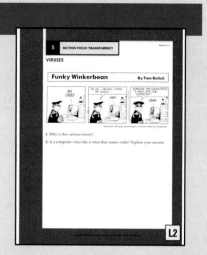

Science Integration Transparencies

Teaching Transparencies

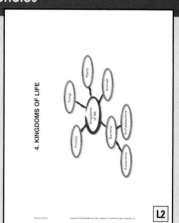

Meeting Different Ability Levels

Study Guide for Content Mastery

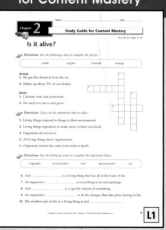

Reinforcement

Enrichment Worksheets

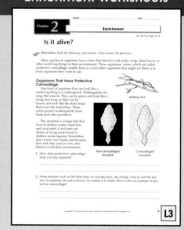

Hands-On Activities

Science Integration Activities

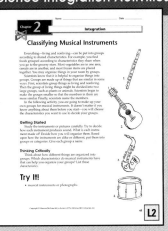
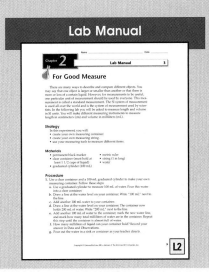

Classifying Musical Instruments

Lab Manual

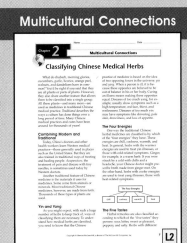

For Good Measure

Activity Worksheets

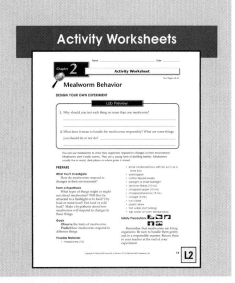

Mealworm Behavior

Enrichment and Application

Cross-Curricular Integration

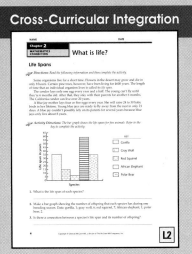

What is life?

Multicultural Connections

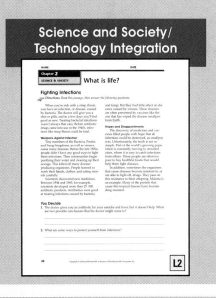

Classifying Chinese Medical Herbs

Science and Society/Technology Integration

What is life?

Fighting Infections

Assessment

Performance Assessment

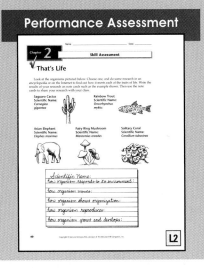

That's Life

Chapter Review

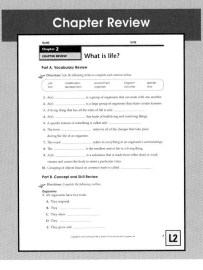

What is life?

Assessment

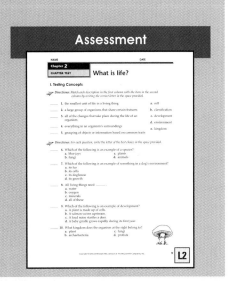

What is life?

What is life?

CHAPTER OVERVIEW

Section 2-1 The set of basic traits that characterize all living things will be discussed. The basic needs of most living things are also presented.

Section 2-2 The scientific classification of living things is presented. Students will learn about the process of classification and then be able to understand the kingdom classification system.

Section 2-3 Science and Society How the AIDS virus and the cold virus both affect humans is presented. Students will organize this information by making an outline.

Chapter Vocabulary

trait	classification
organism	species
environment	kingdom
cell	virus
development	vaccine

Theme Connection

Scale and Structure Living things range in size from the microscopic bacteria to the 200-ton blue whale. But no matter what the size, all living things share some common structures and traits.

Chapter Preview

32

Skills Preview

► **Skill Builders**
  • make tables
  • develop multimedia presentations

► **MiniLABs**
  • infer
  • classify

► **Activities**
  • hypothesize
  • design an experiment
  • classify
  • compare and contrast

Learning Styles

Look for the following logo for strategies that emphasize different learning modalities. **LS**

Kinesthetic	Explore, p. 33; MiniLAB, p. 36; Activity, p. 37; Activity 2-1, pp. 40-41; Activity 2-2, p. 48
Visual-Spatial	Demonstration, pp. 35, 37; Visual Learning, pp. 35, 36, 39, 43; MiniLAB, p. 46; Activity, p. 51; Check for Understanding, p. 52
Interpersonal	Assessment, p. 41
Intrapersonal	Enrichment, p. 38; Science Journal, p. 40
Logical-Mathematical	Problem Solving, p. 47
Linguistic	Check for Understanding, p. 43; Assessment, p. 44; Using Science Words, p. 47; Science & Language Arts, p. 56
Auditory-Musical	Science & Language Arts, p. 56

LS

Chapter 2

What is life?

First, there's just a tiny crack in the eggshell. Then, the opening becomes larger. Soon, you see a beak, then bulging eyes, and you hear a weak squeak or two. Finally, a wobbly, wet chick comes out! An egg is a protective place for the developing chick. It also stores things such as food and water that are needed by the chick for development. Just weeks after being laid, the "lifeless" egg hatches.

The baby chicken is alive. You are, too. So is the grass you walk on. But is a sidewalk alive? What about a bicycle? You probably don't think about the fact that you are surrounded by living and nonliving things. Have you ever thought about the difference between a living thing and a nonliving thing?

EXPLORE ACTIVITY

Compare Living and Nonliving Things

1. Look around your classroom. In 15 seconds, how many living things can you find? How many things can you find that are not alive?
2. Look carefully at each thing you found. Think about each one.

Science Journal

In your Science Journal, list the items you saw. Next to each object, write down why you think it's a living thing or a nonliving thing.

33

EXPLORE ACTIVITY

Purpose

LS **Kinesthetic** Use the Explore activity to help students think about what makes something living or nonliving. **L2**

Preparation

Obtain some living things such as plants, hamsters, or fish, if possible.

Teaching Strategies

Display various living things in your classroom before the students do this activity.

Safety Precautions Remind students to handle all living things gently.

Science Journal Students will list the obvious living things, but the reasoning behind their lists may differ. Accept all reasonable answers.

✓ Assessment

Content Have student groups compare and discuss the lists they made in the Explore activity. Instruct students to agree upon and put together a finalized list of organisms and prepare an operational definition of *life*. Use the Performance Task Assessment List for Group Work in **PASC,** p. 97.

COOP LEARN

Assessment Planner

Portfolio
Refer to page 57 for suggested items that students might select for their portfolios.

Performance Assessment
See page 57 for additional Performance Assessment options.
Skill Builders, pp. 44, 53, 55
MiniLABs, pp. 36, 46
Activities 2-1, pp. 40-41; 2-2, p. 48

Content Assessment
Section Wrap-ups, pp. 44, 53
Chapter Review, pp. 57-59
Mini Quizzes, pp. 43, 53

Group Assessment
Opportunities for group assessment occur with Cooperative Learning Strategies and Flex Your Brain activities.

Prepare

Section Background

What sets living organisms apart from nonliving things is the life processes that they carry out. Living things are defined on the basis of several processes including the ability to respond to stimuli in the environment, movement, organization, the ability to reproduce, and the ability to grow and develop. In other words, life is defined best by what it *does*, not by what it *is*.

Preplanning

Refer to the Chapter Organizer on pages 32A-B.

1 Motivate

Bellringer

Before presenting the lesson, display **Section Focus Transparency 3** on the overhead projector. Assign the accompanying **Focus Activity** worksheet.

L2 ELL

2•1 Alive or Not

What YOU'LL LEARN

- Five ways to decide what is alive
- The main needs of living things

Science Words:
trait
organism
environment
cell
development

Why IT'S IMPORTANT

You will understand why you are thought to be alive.

FIGURE 2-1

There are more than 50 different kinds of *Lithops* plants. Each type looks a lot like a real stone. They're small, hard, and rounded. Their colors are just like rocks. *How are living stones really alive?*

Identifying Life

Have you ever seen a living stone? Sounds like a joke, doesn't it. But there really are such things as living stones. *Lithops* (LIH thops) is the plant pictured in **Figure 2-1.** It is sometimes called a living stone or a flowering stone because, even though it is a plant, it looks like a stone. Although *Lithops* appear to be nonliving, they are just as alive as flowers, trees, and other plants with which you are familiar. Most plants and animals are easy to recognize as living things, but sometimes it's not so simple. Does something have to move to be alive? Does it have to breathe as you do?

Must certain activities take place inside plants and animals if they are to be considered living? Or must they share common physical features, or traits? A **trait** is a specific feature of something. For example, human traits include brown eyes, red hair, and the ability to walk upright. No single trait can tell you whether something is alive. Over time, however, scientists have observed common traits that are used to identify something as a living organism. An **organism** is a living thing that has all of the traits of life. **Figure 2-2** shows different organisms that have the traits of life. Look at the list of some traits of life shown below.

Some Traits of Life
- organisms respond
- organisms move
- organisms show organization
- organisms reproduce
- organisms grow and develop

Program Resources

📁 Reproducible Masters
Activity Worksheets, pp. 5, 13-14, 17 L2
Enrichment, p. 9 L3
Lab Manual, pp. 9-16 L2
Reinforcement, p. 9 L2
Science and Society/Technology
  Integration, p. 20 L2
Study Guide, p. 9 L1

🖥 Transparencies
Science Integration Transparency 2 L2
Section Focus Transparency 3 L2
Teaching Transparencies 3, 4 L2

Organisms Respond

What might you do to find out whether *Lithops* is alive? Maybe you would gently touch it and see whether anything happens. The first trait of organisms is that they respond to things in their environment. If you touch a living stone, you are checking the plant's ability to react to its environment (en VI ur munt). An organism's **environment** includes everything in its surroundings—other organisms, water, weather, temperature, soil, sound, and light—anything with which it comes in contact. Reacting to the environment is what squirrels do when they move away from loud sounds, and when plants grow toward light.

An organism can also respond to changes inside itself. For example, body temperature or disease can cause changes inside an organism. Sometimes, your body does not work the way it should. It must then make changes to return to "normal." Body temperature increases a little when it's hot or when you're working hard. In response to this internal change, you begin to sweat. Your face gets red as tiny blood vessels in your skin swell with blood. Both of these responses help cool your body so it can maintain its normal temperature.

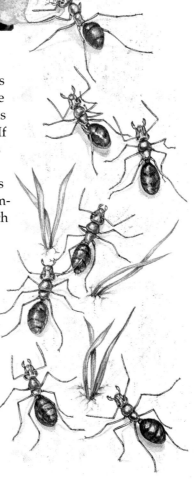

2-1 Alive or Not 35

35

Mini LAB

Inferring Body Responses

How does your body react to internal changes?

1. Find your pulse by placing the index finger of one hand on the thumb side of the wrist of your other hand, just below the base of the thumb.

2. While at rest, count how many pulse beats you feel for 10 seconds, and then multiply this number by 6. This is your pulse rate before exercise.

3. Jog quickly in place at your desk for one minute.

4. Determine your pulse rate immediately after exercise, then at 5 minutes and 10 minutes after.

Analysis

1. Did your pulse rate change while you exercised? If so, how?

2. How did your body change 5 and 10 minutes after exercising?

Organisms Move

Movement is the second trait of organisms. Birds glide through the air. Snakes slither on the ground. Dogs, cats, and people walk and run. These are all ways that animals move from place to place. As shown in **Figure 2-3**, all organisms move in some way, but they don't all move to different places. Many plants move merely by bending toward the sunlight.

Living Things Show Organization

All living things are made up of parts that are organized—the third trait of organisms. If you could look at a small piece of a *Lithops* plant under a microscope, you would see hundreds of tiny compartments set in a pattern that looks like a brick wall. These little compartments are the plant's *cells.* The **cell** is the smallest unit of life in a living thing.

Content Background

All living things are made up of cells—either a single cell or many cells. Cells carry out many functions for living things, such as generating energy and storing genetic information. Organisms, both single-celled and multicellular, that have nuclei in their cells are *eukaryotic* cells. Cells that have no nuclei are *prokaryotic* cells.

Visual Learning

Figure 2-3 How do these different organisms demonstrate the traits of life discussed—movement, showing organization, and reproduction? *Animals move and flowers move to face the sunlight; plants and animals are made up of organized cells and all have specialized body parts; young rabbits and flowering plants show reproduction.*

All organisms are made up of cells. An example of cells is pictured in **Figure 2-3.** You, like elephants and whales, are made up of trillions of cells. But as you'll learn in the next section, many other organisms are made up of only one cell, such as bacteria. In all organisms, cells do the same basic job. They take in materials and release energy and waste products.

Living things are organized in other ways, as well. For example, plants and animals have parts that perform certain jobs. Plants have leaves, stems, flowers, and roots for carrying out many different plant activities. You, as an animal, have a mouth for eating, legs and feet for moving around, and ears for hearing.

Organisms Reproduce

A fourth question you might ask when you're trying to decide whether or not something is alive is "Where did it come from?" All living things reproduce. Reproduction means that organisms make more of their own kind. Animals such as cows and deer give birth to live offspring, while some animals such as turtles and birds lay eggs. But no matter how organisms reproduce, reproduction is necessary for life to continue.

Living organisms reproduce in many different ways. For example, *Lithops* and some other plants produce flowers. Flowers contain male and female parts needed for reproduction. In Chapter 4, you'll learn more about reproduction and discover how it enables all groups of organisms to survive.

FIGURE 2-3

How do these different organisms demonstrate the traits of life discussed—movement, showing organization, and reproduction?

2-1 Alive or Not   37

Demonstration

Visual-Spatial Students will learn more about cells in Chapter 3. Set up a microscope with slides of different cells. Students can examine the cells under low power. Explain that onions, dogs, trees, people, and other living things are all made up of cells. L2 ELL

Using an Analogy

Think of cells as tiny factories. Like factories, some cells have walls, they run on energy, and they manufacture essential substances.

Activity

Kinesthetic Have students examine real flowers to learn how they carry out the important life process of reproduction. Ask florists for flowers that they are throwing away. Gladioli and lilies work best. Provide diagrams or photographs of flower anatomy. Have students learn the names of parts and their functions. Discuss flower pollination and fruit and seed production. L2

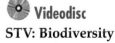 **Videodisc**

STV: Biodiversity

Newborn alligators, Everglades; Florida

Black bear cub

Content Background

Regardless of the number of offspring produced, the result of reproduction is new members of the species. Reproduction involves the transfer of genetic material from parent to offspring. This genetic material contains information that tells how the new species member will look and how it will function. In asexual reproduction, the genetic information in the offspring is exactly the same as the parent; in sexual reproduction, the information is a combination of information from two parents.

Organisms Grow and Develop

If you could watch a *Lithops* plant over a long time, you would see that it doesn't always look like a stone. *Lithops* plants grow and produce flowers for part of the year. All of the changes that take place during the life of an organism are known as its **development.** Most organisms show growth and development—the fifth trait of organisms. The activities that take place inside a chicken egg are an example of growth and development. **Figure 2-4** illustrates growth and development in frogs. You, too, are a product of growth and development. You began life as one cell. As you developed, the number of cells making up your body increased and you became larger. Development took place. Some of your cells developed into skin cells, while others changed into muscle cells. Still other cells became nerve cells and bone cells.

Development isn't easy to see in all organisms. Many single-celled organisms, such as bacteria, don't seem to develop. The single cell grows a little bit in size and then reproduces itself by splitting in half. Even though bacteria do not seem to develop, many changes are going on inside the cell as it carries out its different life activities. **Figure 2-5** illustrates the five traits of life using different organisms.

FIGURE 2-4
Frogs show major changes during their development. Fertilized eggs hatch into legless tadpoles that live in water. These tadpoles develop legs and lungs and move onto land as young adult frogs.

Theme Connection

Energy

Energy is a central concept in both physical and life science. In chemistry, for example, energy is associated with reactions between different kinds of atoms. In life science, energy gives living things the ability to carry out life processes, such as homeostasis and reproduction.

A An adult African elephant grazes on grass in an African plain. The elephant *responds* to its sense of smell to lead it to food. *How does the elephant show organization?*

FIGURE 2-5

These organisms are showing different traits of life.

B A female leopard *responds* to her cub's "play attack" by pulling out of the way. The cub shows a lot of *movement* when it leaps up, using its tail for balance. The patterns on the leopards' coats provide camouflage. *What other traits of life are shown in this picture?*

C A honeybee flies toward a clover flower and collects food from it. The shape of the flower is *organized* to ensure that pollen gets on the bee at the same time. At the next flower, the bee transfers the pollen, which fertilizes the flower. The flower produces seeds for the purpose of *reproduction*.

2-1 Alive or Not 39

Teacher F.Y.I.

You may wish to write biology "fun facts" on the chalkboard each day to peak student interest. Choose facts that relate to chapter content. Here are a few to try: *Maximum Life Spans of Various Animals:* tortoise—120-150 yr.; deep-sea clam—100 yr.; Asian elephant—78 yr.; mouse—6 yr.; housefly—17 days; *Gestation Times of Mammals:* Human—9 mo.; mouse—20-30 days; dolphin—10-12 mo.; monkey—5-6 mo.; *Heart Rates of Mammals (beats per minute):* Human—70; elephant—25; mouse—600-700; large dog—120.

Discussion

You may wish to discuss the "fun facts" listed under the Teacher F.Y.I. above. Ask: **Why do animals vary in their characteristics? Why do some animals live longer than others do? Why do they have different gestation times?** *Answers will vary. Ask students to support their opinions with facts.*

Visual Learning

Figure 2-5A How does the elephant show organization? *The elephant's trunk is organized to function as a "hand."*

Figure 2-5B What other traits of life are shown in this picture? *The cub is the result of reproduction. The cub and mother demonstrate the process of growth and development.*

PREPARE

Purpose

IS **Kinesthetic** Students will determine how mealworms respond to various stimuli in the environment. L2 ELL COOP LEARN P

Process Skills

making a hypothesis, designing an experiment, recognizing cause and effect, controlling variables

Time

50 minutes

Materials

Keep mealworms in the refrigerator at 45 to 50 degrees until ready to use. Do not keep more than a two-week supply.

Safety Precautions

Mealworms are living things and should be treated gently and responsibly. Be sure students don't squeeze the mealworms too hard when handling them.

Possible Hypotheses

Students may have some general ideas about how the mealworms will respond. Most students will predict that mealworms are repelled by vinegar due to the strong odor. Students may also predict that mealworms are repelled by hot water because of the increased temperature.

📁 **Activity Worksheets,** pages 5, 13-14

Activity 2-1

Possible Materials

- mealworms (10)
- small cardboard box with lid, such as a shoe box
- eyedropper
- cotton-tipped swabs
- penlight or small flashlight
- dry bran flakes (15 mL)
- chopped apple (15 mL)
- chopped banana (15 mL)
- vinegar (5 mL)
- ice cubes
- plastic straw
- hot water (not boiling)
- tap water at room temperature

Design Your Own Experiment
Mealworm Behavior

You can use mealworms to show how organisms respond to changes in their environment. Mealworms aren't really worms. They are a young form of darkling beetles. Mealworms usually live in moist, dark places or where grain is stored.

PREPARE

What You'll Investigate

How do mealworms respond to changes in their environment?

Form a Hypothesis

What types of things might or might not attract mealworms? Will they be attracted to a flashlight or to food? Dry food or moist food? Hot food or cold food? **Make a hypothesis** about how mealworms will respond to changes in these things.

Goals

Observe the traits of mealworms.

Predict how mealworms will respond to different things.

Safety Precautions 🤚 🤚 ✋ 🧪 🥽

Remember that mealworms are living organisms. Be sure to handle them gently and in a responsible manner. Return them to your teacher at the end of your experiment.

PLAN

1. Working in a group, **make a list** of some possible things that you can put in front of the mealworms.
2. As a group, **make hypotheses** about how the mealworms will respond to each thing.
3. As a group, **decide** how the different things will be shown to the mealworms. Will you show each item alone, or will you use more than one item at a time?

Science Journal **Larvae Research** Mealworms are not worms; they are beetle larvae. Select interested students to do some research about beetle larvae or insect larvae in general. Have students try to find out why the mealworms respond to some of the different stimuli. IS L3 P

Data Table

Items presented	Mealworm response
light source	avoidance (−)
bran flakes	attraction (+)
apples	attraction (+)
banana	attraction (+)
vinegar	avoidance (−)
ice cubes	avoidance (−)
blowing air	avoidance (−)
hot water	avoidance (−)
water at room temperature	attraction (+)

4. How many mealworms will you use?
5. **Prepare** a data table in your Science Journal.
6. With your partners, **write** a list of all the steps you will take in your experiment.
7. **Read** over your experiment to make sure that all steps are in logical order.

DO

1. **Read** over your plan to make sure that it will test your hypotheses.
2. Make sure your teacher approves your plan before you do it.
3. Carry out the experiment as planned.
4. While the experiment is going on, **record** all observations that you make and fill in the data table in your Science Journal.

CONCLUDE AND APPLY

1. Which things attracted the mealworms? Which things did the mealworms avoid?
2. Did the results of your experiment agree with your hypotheses?
3. What did the results of your experiment tell you about mealworm behavior?
4. APPLY The behaviors shown by mealworms are called instincts. Instincts are behaviors that organisms are born with. Why are instincts important to organisms? How do instincts help mealworms?

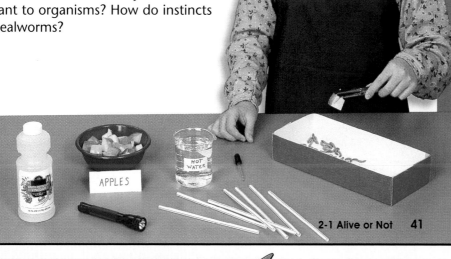

2-1 Alive or Not   41

Go Further

Students can extend this experiment by brainstorming other types of stimuli for the mealworms to respond to. Students can make hypotheses about the mealworms' responses and then do an experiment to test their hypotheses. Approve all student experimental plans before they do the experiments.

✓ Assessment

Performance Have student groups prepare poster presentations about their experiments. Students can show how mealworms respond through pictures or photographs. Encourage students to videotape their experiments. Students may edit a final version complete with narration and special effects. Use the Performance Task Assessment List for Video in **PASC**, p. 81. ⬛ P

COOP LEARN

PLAN

Possible Procedures

Place eight to ten mealworms in the cardboard box. Determine how mealworms respond to stimuli. Present each stimulus separately to the mealworms.

Teaching Strategies

Present students with basic information about mealworms before they do this activity. Mealworms are the larvae of the beetle *Tenebrio molitor*. During their life cycle, these beetles undergo complete metamorphosis—egg, larva, pupa, and adult.

Troubleshooting Active mealworms work best for this activity.

• Have students use small amounts of food in their experiments.

DO

Expected Outcome

Refer to Conclude and Apply question 1.

CONCLUDE AND APPLY

1. Food and water attract mealworms. Touching, blowing with a straw, vinegar, and hot water are usually avoided by the mealworms.
2. Answers will vary, depending upon the hypotheses that the students made.
3. Mealworms need food and water and are therefore drawn to it. They avoid vinegar, blowing air, touch, and extreme temperatures.
4. Instincts enable organisms to stay alive by making quick responses to stimuli.

- You may also want to supplement this feature with an article from a newspaper or magazine describing this event.

- Discuss the meaning of *artificial intelligence* with students. Point out that it is possible to build computers that have the ability to play near-perfect games of chess. However, scientists are far from creating computers that can duplicate actual intelligence in humans.

*inter*NET CONNECTION

The Glencoe Homepage at **www.glencoe.com/sec/science** provides links connecting concepts from the student edition to relevant, up-to-date Internet sites.

Discussion

Initiate a discussion about the needs of organisms. Using everyday examples, have students discuss the difference between a *need* and a *wish*. In biology, an organism's needs include substances for survival, such as food, water, and shelter.

FLEX Your Brain

Use the Flex Your Brain activity to have students explore ENERGY.

📁 **Activity Worksheets,** page 5

Inquiry Question

Can a person stay alive longer without food or without water? *People can actually live without food for several weeks, whereas a person can't last much longer than a week without water.*

42

Basic Needs of Living Organisms

If you did the previous activity, you may have learned that food was one thing that attracted mealworms. It shouldn't be surprising to see that mealworms are attracted to food. Aren't you sometimes willing to go out of your way for a snack when you're hungry? Mealworms need food for the same reason you need food—for energy! What other things do living things need?

Living Things Need Energy

None of life's important activities would be possible without some form of energy. Growth, movement, and reproduction could not take place without it. Energy is an important need of living things. How do organisms

GARRY KASPAROV · DEEP BLUE

The final game of their six-game championship chess tournament ended much sooner than people had expected. Garry Kasparov, the reigning world champion, quit the game after just 19 moves. His opponent showed no emotion, and no one was surprised. Kasparov's opponent was a computer. The 1997 tournament made headlines because it marked the first time ever that a computer had beaten a human champion at chess. The computer that beat Kasparov was a huge computer named Deep Blue. Kasparov had defeated Deep Blue a year earlier. But in their second matchup, Deep Blue was more prepared.

Deep Blue was programmed to allow it to choose the best move it could play. However, scientists are far from making a robot or computer that can think the same way people think. A computer can only make a choice from options that it has been given.

*inter*NET CONNECTION

What is artificial intelligence? Check the Glencoe Homepage, *www.glencoe.com./sec/science*, for a link to a site with several definitions of computer intelligence.

Content Background

When cells break down food molecules, part of the energy contained in the molecules is released as heat. Some is stored temporarily in ATP (adenosine triphosphate). ATP is a complex molecule made of a sugar, an adenine molecule, and a chain of three phosphate groups. A steady supply of ATP is necessary so that a cell can perform its life activities. When the chemical bond between two of the phosphate groups is broken, a large amount of energy is released. This energy is then used to drive the cells' many chemical reactions.

obtain the energy they need? Most living things get their energy from the sun. Plants use the sun's energy and carbon dioxide from the air to make food. This important activity takes place inside some plant cells. Plants use the food energy they make for growth and other plant activities.

Most other organisms depend on this food-making ability of plants. People obtain energy when they eat plants. They can also obtain energy when they eat animals that have eaten plants. **Figure 2-6** shows a girl obtaining energy from food.

Living Things Need Water, Oxygen, and Minerals

Have you ever forgotten to water a plant? If a plant goes too long without water, it will probably die. You can't live long without water, either. Organisms are made up mostly of water. Your body, in fact, is about 70 percent water. Cells that make up organisms need water to carry out all important life activities. So, life on Earth wouldn't be possible without water.

Whether food comes from plant or animal sources, it has energy in it. Many organisms need the chemical oxygen to release the energy stored in their food. The oxygen you use for this purpose is found in air. As shown in **Figure 2-7,** organisms take in this oxygen in different ways.

FIGURE 2-6

Plants are the source of energy for most organisms. This girl is obtaining energy by eating both plants—tomatoes, lettuce, onions, wheat—and plant-eating organisms—beef cattle.

FIGURE 2-7

Some organisms, such as earthworms, absorb oxygen through their moist skin. Fish have gills that get oxygen that is in the water around them. *How do deer obtain water, oxygen, and minerals?*

2-1 Alive or Not   43

Visual Learning

Figure 2-7 How do deer obtain water, oxygen, and minerals? *Deer drink water from streams and lakes; oxygen is taken in from the air using lungs; minerals are obtained through water and food.* LS

Across the Curriculum

History Fermentation is the simplest form of energy production in the cell. Yeast cells convert sugar (glucose) into ethyl alcohol and carbon dioxide. Fermentation occurs during wine making. Encourage students to learn more about how wines are made. You may also wish to have students gather information about the history of wine making around the world. L3

Have students brainstorm ways that living things need and use water. This could include drinking, bathing, reproduction, regulating body temperature, shelter, etc. Discuss each of the answers. Have students think about specific organisms and how they need/use water. L2

3 Assess

Check for Understanding

Discussion Write the following words on the chalkboard: *mushroom, flame, dog, glob of slime, fish, grass.* Have students study the list, then have them explain why each object is considered a living thing or a nonliving thing. LS

Reteach

As a homework assignment, have students cut out five pictures of living things and five pictures of nonliving things. Have students write a list of the photos explaining why each is a living or nonliving thing. L1

Extension

For students who have mastered this section, use the **Reinforcement** and **Enrichment** masters.

4 Close

•MINI•QUIZ•

Use the Mini Quiz to check for understanding.

1. **Eye color and hair color are examples of _____.** *traits*
2. **All of the changes that take place during the life of an organism are known as its _____.** *development*

Section Wrap-up

1. The five general traits of living things are response to things in the environment, movement, organization, reproduction, and growth and development.

2. All living things need energy, water, and minerals. Most living things also need oxygen to survive.

3. **Think Critically** Answers will vary. Although fire appears to be living because it moves and grows and it can produce more fire, it is nonliving because it is not organized, nor does it develop or reproduce.

Science Journal Answers will vary but should include tests to see if the green blob has any of the general life processes.

Assessment

Portfolio Have students write short essays that respond to the following questions. "What is biology? Why do you think biology is important to you?" Share some of the more interesting responses with the class. Use the Performance Task Assessment List for Writer's Guide to Nonfiction in **PASC**, p. 85.

P LS L2

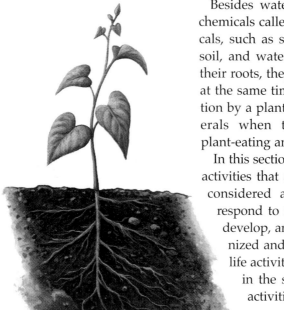

FIGURE 2-8
Water and minerals used by a plant enter by way of its roots.

Besides water and oxygen, organisms need other chemicals called minerals to live. Minerals are chemicals, such as sodium and chlorine, found in the air, soil, and water. When plants take in water through their roots, they absorb minerals that are in the water at the same time. **Figure 2-8** illustrates water absorption by a plant. Animals obtain these important minerals when they eat plants or when they eat plant-eating animals.

In this section, you've learned that there are five life activities that are common to all living things. To be considered alive, an organism must be able to respond to its surroundings, reproduce, grow and develop, and move. Organisms must also be organized and have cells to carry out these necessary life activities. But not all organisms are organized in the same way. And not all carry out these activities in the same way. Earth is filled with many different living things. In the next section, you'll discover how scientists study and learn about the different organisms on Earth.

Section Wrap-up

1. List the five traits of living things.

2. Describe the needs of organisms.

3. **Think Critically:** Is fire a living organism? Explain your answer.

4. **Skill Builder**
   Making and Using Tables Make a table of the five traits of living things. In one column, list the five traits. In the second column, give an example of each of the five traits using an organism with which you are familiar. If you need help, refer to Making and Using Tables on page 546 in the **Skill Handbook**.

Science Journal
Imagine discovering a mysterious green blob in an unexplored part of a rain forest. In your Science Journal, describe some tests that you might perform to find out whether the green blob is alive.

Skill Builder
Making and Using Tables Check students' tables for accuracy.

Sample examples

Five Traits of Life	Example
1. Organisms respond	Plants growing in the direction of the sun
2. Organisms move	Elephant running
3. Organisms show organization	Dog has ears, nose, eyes, torso, paws, tail
4. Organisms reproduce	Snakes laying eggs
5. Organisms grow and develop	Kittens growing up to be cats

Classifying Life 2•2

Classification of Living Things

Imagine walking into a video store to rent a movie and discovering that someone has rearranged all the movies on the shelves in no particular order. Would this lack of order make it hard for you to find what you're looking for? Fortunately, most stores, including grocery stores and clothing stores, are organized to help customers. Items in the store are usually grouped according to their similarities and placed together in their own sections or aisles. In a similar way, scientists organize, or classify, living things into groups. **Classification** (klas if uh KAY shun) is the grouping of objects or information based on common traits.

How are organisms classified?

When classifying organisms, scientists often use similarities in body parts to group organisms. Eastern gray squirrels and tufted-eared squirrels, shown in **Figure 2-9,** have a lot of features in common. Both squirrels are medium-sized animals with long, bushy tails and strong back legs. So, scientists have classified them in a group of organisms that share these common traits. However, the eastern gray squirrels and tufted-eared squirrels also look different. As a result, scientists call each different kind of organism a species. A **species** is a group of organisms that can mate with one another. Therefore, members of the eastern gray group can mate with one another, and members of the tufted-eared squirrels can also mate with one another.

What YOU'LL LEARN

- How to classify things
- The six kingdoms of living organisms

Science Words:
classification
species
kingdom

Why IT'S IMPORTANT

Classification is a tool that you use every day.

FIGURE 2-9

The tufted-eared squirrel and the eastern gray squirrel are two different species, even though they have many body parts in common.

2-2 Classifying Life   45

Prepare

Section Background

The biological kingdom classification system is presented. The two kingdoms, Eubacteria and Archaebacteria, are both discussed as bacteria. With an understanding of the classification process, students should be able to understand how and why living things are grouped the way they are.

Preplanning

Refer to the Chapter Organizer on pages 32A-B.

1 Motivate

Bellringer

 Before presenting the lesson, display **Section Focus Transparency 4** on the overhead projector. Assign the accompanying **Focus Activity** worksheet.
[L2] ELL

Program Resources

🗀 Reproducible Masters

Activity Worksheets, pp. 5, 15-16, 18
Cross-Curricular Integration, p. 6
Enrichment, p. 10 [L3]
Multicultural Connections, pp. 7-8
Reinforcement, p. 10 [L2]
Science Integration Activities, pp. 41-42
Study Guide, p. 10 [L1]

Transparencies

Section Focus Transparency 4 [L2]

Tying to Previous Knowledge

Students are familiar with the process of classification. Most people use the process every day. Your students probably organize clothing, videos, books, CDs, and tapes. Classification is also seen in most kinds of stores. Items are always organized and grouped according to their characteristics.

2 Teach

Mini LAB

Purpose

LS **Visual-Spatial** Students will learn about the process of biological classification by making observations of species within a vertebrate class. **L2**

Materials

photographs or illustrations of different vertebrate animals, envelopes

Teaching Strategies

• Collect pictures of different vertebrates by clipping them from newspapers and old magazines.

• "Fish envelopes" should contain pictures of four different species of fish; "bird envelopes" should contain four different birds, etc. Make up three envelopes for each of the five vertebrate groups.

📁 **Activity Worksheets,** pages 5, 18

Analysis

1. Answers will vary depending on which vertebrate group was chosen.

2. Answers will vary, but check to make sure students understand the purpose of the exercise.

Mini LAB

Classifying Animals with Bones

1. Scientists recognize five major groups of animals with bones: fish, amphibians, reptiles, birds, and mammals. Choose one of these groups to work with.

2. Obtain photographs of four different animals from the group you chose.

3. Study and compare the animals in the photographs. Make observations about the similar traits among the different species of your group.

Analysis

1. Which group of animals did you study—fish, amphibians, reptiles, birds, or mammals?

2. List the traits shared by the animals within your group.

FIGURE 2-10

Scientists once agreed that guinea pigs should be classified with hamsters because they shared similar body parts. But by studying an important chemical in their cells, they decided that guinea pigs should really be put in a group by themselves.

46 Chapter 2 What is life?

An Example of Classification

Animals are classified into groups based on shared traits. For example, some animals have bones and others don't. There are five groups of animals with bones. Fish live in water. They have gills for breathing. Most fish have fins as well for moving through the water easily. Amphibians, such as frogs, live both in water and on land. Reptiles live on land and include animals such as turtles, alligators, and snakes. They have dry skin and most lay eggs. Animals that have feathers are placed in the bird group. Except for whales, animals that have hair and give birth to live offspring are classified as mammals. Mammal mothers nurse their babies.

There is more to classification than just common body parts. Living organisms also can be classified according to similarities in the materials that make up their bodies. **Figure 2-10** shows one example of scientific classification based on chemicals found inside the organisms' bodies.

Assessment

Portfolio Have students make labeled posters about the vertebrate classes they chose in the MiniLAB activity. Have them ask, "What is a fish?" or "What is a mammal?" Then have them make posters showing the characteristics that unite species within the vertebrate group. Use the Performance Task Assessment List for Poster in **PASC**, p. 73. **P**

Community Connection

Classifying Merchandise Have students visit a store or shop in the area to learn about how the merchandise is organized or classified. Instruct students to take notes about the organization of the store and make diagrams of the classification system. **L2**

COOP LEARN

Kingdoms of Life

All organisms on Earth have been separated into six large groups called kingdoms. A **kingdom** is a large group of organisms that share certain features. In the next activity, you could say that you will classify objects within the button kingdom.

The six kingdoms of living things are Archaebacteria, Eubacteria, Protists, Fungi, Plant, and Animal. Organisms are placed into a kingdom based on the four features shown below.

- how many cells they are made up of
- what their cells look like
- whether or not they can move from place to place
- how they obtain energy

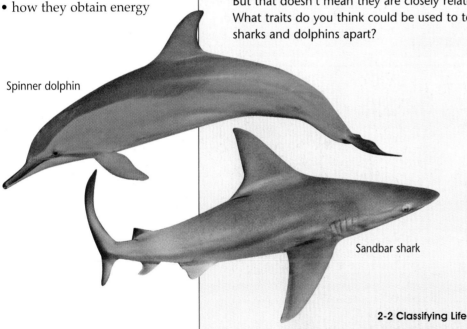

Spinner dolphin

Sandbar shark

Problem Solving

Classifying Animals

Scientists don't just classify living organisms to make things easier. They also group living things to show the relationships between the different organisms. It seems logical that if two organisms look similar, they are related. But that's not always true. Think about birds, bats, and insects. All these organisms have wings and all fly, but are they related to each other? Do you see why scientists can't classify organisms just because of the way they look or act?

Solve the Problem:
Study the illustrations of the shark and the dolphin. Observe the traits that sharks and dolphins share.

Think Critically:
Sharks and dolphins look alike in many ways. But that doesn't mean they are closely related. What traits do you think could be used to tell sharks and dolphins apart?

Content Background

The word *species* comes from the Latin, where it means "appearance" or "kind." The organisms on Earth must be described in a consistent manner. The most widely accepted species concept states that a species is a group of organisms that can interbreed freely under natural conditions, and will not interbreed with organisms that are not part of the group.

? FLEX Your Brain

Use the Flex Your Brain activity to have students explore CLASSIFICATION.

📁 **Activity Worksheets,** page 5

 Problem Solving

Solve the Problem
Students should list all of the traits they observe about the sharks and dolphins, based on the photographs and their own previous knowledge.

Think Critically
Many students will say that sharks and dolphins have different skin, teeth, and other structures. Some may say that the chemical makeup of sharks and dolphins is probably different. Accept all reasonable answers. ⅃S

GLENCOE TECHNOLOGY

 Videodisc
STVS: Animals
Disc 5, Side 2
Studying Sharks (Ch. 3)

Using Science Words

⅃S **Linguistic** A *kingdom* is also defined as "the position, rank, or power of a king." Students may be familiar with this meaning through books or movies. As you teach students about the biological meaning of the word *kingdom,* it may be useful to discuss this other definition and show how they are related.

Purpose
LS **Kinesthetic** Students will learn about the process of classification. **L2** **COOP LEARN** **P**

Process Skills
classifying, observing and inferring, comparing and contrasting

Time
45 minutes

Alternate Materials
Small objects such as birdseed mixture, candies, or marbles could be classified.

Safety Precautions
Remind students not to eat any materials used in the experiment.

📁 **Activity Worksheets,** pages 5, 15-16

Teaching Strategies
- Try to collect a variety of interesting-looking buttons for this activity. You may wish to have students bring in buttons.
- Have students compare their classification schemes. Different groups may have different schemes. Point out that this sometimes happens in science. Explain that scientists often disagree on how a particular species should be classified. The more information known about an organism, the easier it is to classify.

Activity 2-2

Goals
- Classify a group of buttons.

Materials
- assorted buttons (10 to 15 different types)
- pencil
- paper

Button Classification

Scientists classify living things so they can talk and write about organisms more easily. Classification systems also show how different species are related. To be related, the different species must share some traits in common. How do scientists make up classification systems?

What You'll Investigate
How do scientists classify organisms?

Procedure
1. **Obtain** ten to 15 different buttons.
2. **Study** the different types of buttons. **Choose** one feature that will allow you to separate the buttons into two smaller groups. Size, shape, color, number of holes, and how they feel (texture) are some possible features. **Separate** the buttons into groups based on the feature you chose.
3. Begin to make a chart of your classification system as shown on this page. **Write** down the feature you use to separate your buttons into two groups.
4. **Repeat** step 3 for each of the smaller groups you form. Keep track of your separations in your chart.
5. Continue following step 3 until there is only one button in each group.
6. **Complete** your classification chart and compare it with the results from other lab groups.

Buttons
Black   White

48   Chapter 2 What is life?

Conclude and Apply
1. Why did some groups develop different classification systems for the buttons?
2. **Compare** the way you **classified** your buttons with the way scientists classify living things.
3. Why do you end up with only one button in each group?

Answers to Questions
1. Classification schemes depend on the characteristics chosen.
2. Living things are also grouped according to similarities.
3. Every button has its own unique qualities.

✓ Assessment

Portfolio Have students make posters of their button classification schemes to share with the class. Encourage students to be creative with their diagrams and use colored pencils, markers, crayons, and other items for their posters. Use the Performance Task Assessment List for Poster in **PASC**, p. 73. **P**

Bacteria

Bacteria is a term that has been used to describe members of the eubacteria (YEW bak teer ee uh) kingdom and members of the archaebacteria (ar kee back TEER ee uh) kingdom. Bacteria in the eubacteria kingdom are known as true bacteria. Archaebacteria are bacteria that live in extreme environments. All eubacteria are single-celled organisms. They include many bacteria that cause diseases in other organisms. There are also bacteria that are helpful. Bacteria used to make yogurt and some cheeses belong to the eubacteria kingdom.

There are thousands of known species of bacteria, but some scientists predict that millions more may exist. Some bacteria species take in food for energy and move around as many animals do. Others make their own food from sunlight, just as plants do. These bacteria contain the same chemical that plants use in making food.

Archaebacteria live in some of Earth's most difficult environments, such as hot sulfur springs and extremely salty lakes. Archaebacteria don't get energy from eating food as animals do. They also don't produce food by using energy from sunlight as plants do. Instead, archaebacteria get energy by taking in chemicals from their surroundings. One group, for instance, makes energy from the chemical called carbon dioxide found in the soil where they live. Another type makes energy from another chemical—sulfur. These archaebacteria grow in sulfur springs where it is extremely hot.

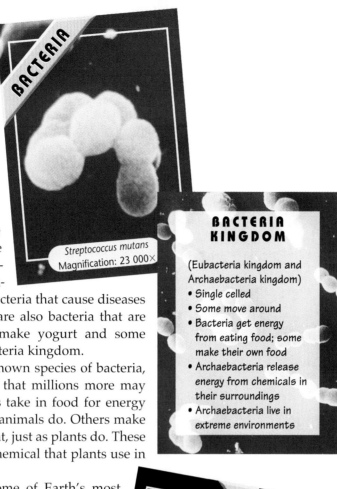
Streptococcus mutans
Magnification: 23 000×

BACTERIA KINGDOM

(Eubacteria kingdom and Archaebacteria kingdom)
- Single celled
- Some move around
- Bacteria get energy from eating food; some make their own food
- Archaebacteria release energy from chemicals in their surroundings
- Archaebacteria live in extreme environments

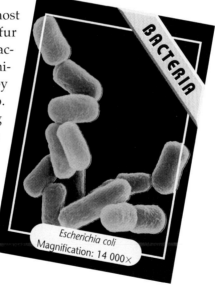
Escherichia coli
Magnification: 14 000×

2-2 Classifying Life   49

Teacher F.Y.I.

One group of Archaebacteria—the methanogens—uses carbon dioxide as a source of energy and releases methane as a waste product. Methanogens are common in the guts of animals that eat plants, such as termites and especially cows. Methanogens release approximately 2 billion tons of methane into Earth's atmosphere every year. About one-third of this methane production comes from mammalian flatulence.

Revealing Preconceptions

Students often mistakenly think that all bacteria are harmful, disease-causing organisms. Students are not always familiar with the many benefits of microorganisms to humans. Point out the importance of bacteria in the production of foods, such as yogurt, cheese, and sauerkraut or in the human digestive tract.

GLENCOE TECHNOLOGY

Videodisc
Glencoe Life Science Interactive Videodisc
Side 1, Lesson 3
Bacterial Action

33486-35028

Content Background

- Most biologists now accept the six-kingdom classification system of species. Many scientists agree that archaebacteria are different enough from other bacteria that they should be recognized as a separate kingdom. Differences between archaebacteria and the *eubacteria,* or "true bacteria," include differences in the makeup of the cell wall, ribosomes, and RNA base sequences.

- *Streptococcus mutans* is a bacteria that causes tooth decay by converting sugar to an acid that erodes tooth enamel. *Escherichia coli* (or *E. coli*) are found in the gut flora of humans and warm-blooded animals. Some strains are pathogenic. Some strains cause "traveler's diarrhea" in adults.

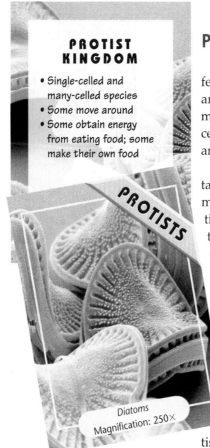

PROTIST KINGDOM

- Single-celled and many-celled species
- Some move around
- Some obtain energy from eating food; some make their own food

PROTISTS

Diatoms
Magnification: 250×

inter**NET** CONNECTION

Visit the Glencoe Homepage, *www.glencoe.com/sec/science*, to learn more about mushrooms, which are a type of fungi.

Protists

The Protist (PROH tust) kingdom includes many different single-celled and many-celled organisms. They are difficult to classify and seem to have little in common. However, all protists are classified based on their cell structure. Two types of protists include plantlike and animal-like protists.

If you have seen green algae (AL jee) growing in a fish tank, on the surface of a pond, or on the sides of a swimming pool, you're already familiar with plantlike protists. You may also be familiar with seaweeds, which are types of brown algae. Green algae and brown algae are called plantlike protists. Plantlike protists use the sun's energy to make food. Some seaweeds, such as kelp, even look like plants. Kelp is a long brown algae that can often be found on ocean beaches.

The protists also include many animal-like forms. These single-celled protists move around and take in food from the environment just like animals do. Some protists also cause diseases. One type of animal-like protist causes a disease called malaria. Malaria kills millions of people every year.

Fungi

If you like pizza with mushrooms, you are a fungus eater! Mushrooms are a type of fungi (FUN ji). You've seen other types of fungi if you've ever noticed fuzzy, black mold growing on old bread or green mold on a piece of fruit.

FUNGI KINGDOM

- Single-celled and many-celled species
- Do not move around
- Obtain energy by feeding on dead or decaying tissue

FUNGI

Scarlet waxy cap mushroom

Content Background

- Many diatom shells look like delicately carved, microscopic pillboxes with lids. Each species has its own unique shape, decorated with exquisitely fine grooves and pores.

- Ask students if they are familiar with mold growing on old bread or fruit. Explain to the class that each little black dot on the mold is a spore case that will release thousands of individual spores. Explain that each individual spore has the potential for growing into a living mold.

In some ways, fungi are like plants. In other ways, they are similar to animals. You may have thought that mushrooms were plants because they grow in the ground and don't move about. Mushrooms, however, don't make food using sunlight as plants do. Instead, mushrooms and other fungi get their nutrients from the substances they grow on. As a fungus grows, it digests the material around it. When fungus grows on bread or fruit, it's actually eating the food!

The Fungi kingdom includes single-celled and many-celled forms. People have found many uses for fungi. Besides as a food, people use fungi in making medicines.

Plants

How many different kinds of plants have you seen today? They're all around—in gardens, in lakes and streams, and even between the cracks in sidewalks. They are also found in houses and office buildings and in your salad bowl at lunchtime. Grass, trees, bushes, and ferns are all members of the Plant kingdom.

All plants are many-celled organisms. Most contain the green chemical chlorophyll (KLOR uh fihl), which enables them to make food. As you've learned, plants capture energy from sunlight and store it as food. Food made by plants is the source of energy for many organisms. Every species of animal depends in one way or another on the plants in its environment.

PLANT KINGDOM
- Many celled
- Do not move around
- Make their own food

PLANT

Viola

PLANT

Alaskan Cedar

Activity

🅛🅢 **Visual-Spatial** Students can learn about fungi by dissecting mushrooms and making spore prints. Cut whole, fresh mushrooms in half with plastic knives. Put one half to the side and study the other half. Draw what is seen. Explain that the mushroom is the reproductive part of the fungus. Remove a small piece of gill from the mushroom and place it on a microscope slide. Have students observe the gill pieces under low power. Hundreds of tiny spores will be seen. 🄻🄾 **ELL**

Discussion

Discuss the importance of fungi to people. Focus on the importance of fungi in the production of antibiotics, such as penicillin and erythromycin. Ask students if they have ever heard of these medicines. Many of them will be familiar with the drugs.

NATIONAL GEOGRAPHIC SOCIETY

 Videodisc

STV: Biodiversity
Corn and soybeans in alternating rows

46064
Waterlilies, Point Pelee National Park; Canada

46079
Evergreens, Jasper National Park; Canada

46017

Cultural Diversity

Language Development As language develops, ideas are classified or grouped. Interestingly, different cultures do this in similar ways.

For example, if a culture has only two terms used for color, they will be "black" and "white." This is true for some cultures found in Papua New Guinea. If a third color term is added, it will be "red." A fourth color will be either "yellow" or "green." (Reference: *Basic Color Terms: Their Universality and Evolution*, Berlin and Kay, California, 1992.)

Likewise, the first major division made for plants by most cultures is between "grasses" and "trees/shrubs." Students could research some different classification systems based on language.

Each kingdom contains organisms that are grouped into different phyla. For instance, in the animal kingdom there is phylum arthropoda (insects, crabs, shrimp, and lobsters), phylum mollusca (snails, clams, and oysters), and phylum chordata (animals with spinal cords). Similarly, the plant kingdom has many phyla, including mosses, ferns, and flowering plants. Phyla are further subdivided into classes, orders, and families. Vertebrates (animals with bones), for instance, are subdivided into five classes: mammals, birds, reptiles, fish, and amphibians.

3 Assess

Check for Understanding

Activity Show students slides, models, photographs, illustrations, or living examples of species from each of the six different kingdoms. Show them to students one at a time. Have students try to guess the kingdom the animal belongs to, then have students try to list the characteristics common to members of that kingdom. [LS] [L2]

Reteach

Students often organize clothing, books, CDs, etc. Have students think about and describe how their own personal items at home are organized. Have students explain how the classification of personal items is similar to the classification of living things. [L1]

Extension

For students who have mastered this section, use the **Reinforcement** and **Enrichment** masters.

ANIMAL KINGDOM
• Many celled
• All species move around
• Get energy by eating food

Red-faced macaque monkeys

Tunicates

Tropical butterfly
Euphaedra uganda

Animals

Pigeons and squirrels in the park, penguins in Antarctica, rhinoceroses in Africa—Earth is home to more than a million different species of animals. Some, such as cats and dogs in your neighborhood, are familiar to you. Others, such as the rare birds, insects, monkeys, and snakes that live in tropical rain forests and other far-away places, are less well known. Even though there is a great number of animals, they are surprisingly similar in many ways.

It's easy to recognize animals. They're the only organisms you can see moving around from place to place. Moving around in their environment helps animals find food, shelter, and mates, and allows them to escape from enemies.

The Animal kingdom is a group made up of a lot of different organisms. All animals are many-celled creatures. Some animals do not have bones. Sponges, worms, crabs, and insects are all boneless. Earlier, you learned that birds, fish, and reptiles are animals with bones.

Visual Learning

Until recently, scientists recognized five kingdoms of living organisms. When more information about archaebacteria was discovered, scientists proposed that they should be recognized as a sixth kingdom. Most scientists agree that much more is to be learned about bacteria. It is possible that more kingdoms will be recognized in the future.

Macaws

In this section, you've learned that scientists classify living things into six large groups of organisms. Later in the book, you'll learn more about the different kinds of life on Earth and how living things carry out their life activities.

Section Wrap-up

1. What is a species? Give an example of two different species.

2. What's the difference between an animal and a plant? Be specific.

3. **Think Critically:** Briefly describe how you might go about classifying a collection of CDs into two groups.

4. **Skill Builder**
   Developing Multimedia Presentations
   Give a multimedia presentation that compares and contrasts the six kingdoms of living things. Be sure to include the name of the kingdom, how its members get energy, whether or not they can move around, and whether they are made up of one or many cells. If you need help, refer to Developing Multimedia Presentations on page 564 in the **Technology Skill Handbook.**

Science Journal
In your Science Journal, write a one-page summary about an organism from one of the six kingdoms. In your summary, include where it lives, details about its appearance and behavior, and why it interests you.

4 Close

•MINI•QUIZ•

Use the Mini Quiz to check students' recall of chapter content.

1. An organism that doesn't move around but obtains its energy by eating food is called a(n)____. *fungus*

2. A many-celled organism that makes its own food is called a(n) ____. *plant*

3. The microscopic kingdom of organisms that includes animal-like and plantlike forms is known as the ____ kingdom. *protist*

Section Wrap-up

1. A species is a group of organisms that can mate with one another. Examples will vary. Examples of species are domestic cats, African elephants, etc.

2. Answers may vary but should include that plants don't move around and they make their own food. Animals move around and get energy by eating food.

3. **Think Critically** Answers will vary, but students should describe procedures in which the CDs are grouped according to certain characteristics such as type of music, performer, etc.

Skill Builder
Developing Multimedia Presentations Check students' multimedia presentations for accuracy of information.

✓ Assessment

Performance Have students make posters about the six kingdoms of living things. Have students decorate posters with illustrations and photos of organisms from each kingdom. Students should also list the characteristics that define each of the kingdoms. Use the Performance Task Assessment List for Poster in **PASC**, p. 73. L2 P

Science Journal Answers will vary. Check students' entries for completeness and accuracy.

Teaching the Content

- Review the basic characteristics of organisms.

- Have students sequence the four basic steps involved when a virus invades a cell. The four steps are attach, invade, copy, and finally, explosion of host cell and release of new viruses.

- Discuss the importance of childhood immunizations.

Content Background

- The exact origin of viruses is not known. Most scientists, however, think viruses were originally mutated genes that somehow were able to escape the cells in which they formed.

- Viruses are so small that they cannot be viewed with a light microscope. Scientists have used electron microscopes to view these microbes.

54

Science & Society

2•3 Viruses

What YOU'LL LEARN

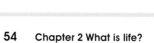

- How the AIDS virus and the cold virus both affect humans

- How to use outlining as a tool for remembering the big ideas of an article

Science Words:
virus
vaccine

Why IT'S IMPORTANT

You'll be able to use outlining to organize information.

FIGURE 2-11

This computer-enhanced image is of a virus that causes influenza in humans. Influenza is often accompanied by fever and body aches.

What is a virus?

A **virus** (VI rus) is a particle that has things in common with both living and nonliving things. Viruses are like living things in that they are able to reproduce. But they can reproduce only inside a living cell. Viruses are like nonliving things because they don't grow, eat, or respond to their environments.

Viruses need living things to survive. Once a virus attaches to a cell, it invades the cell and begins to make copies of itself. Eventually, the cell occupied by the virus explodes, releasing many new viruses. These new viruses then enter other cells.

Examples of Viruses

What do measles, AIDS, colds, and chicken pox all have in common? Viruses cause all of these diseases. While viruses cannot be classified into any of the six kingdoms, they do infect all kinds of living things. Viruses, such as the one pictured in **Figure 2-11**, infect bacteria as well as plants and animals. Some plant viruses destroy food crops such as potatoes and tomatoes. Other types can cause cancers in house cats and humans.

Viruses also cause cold sores and chicken pox in humans. Humans can be infected with either of these viruses by touching someone who already has one of these viruses. Hepatitis B and HIV are viruses that can be passed from person to person by blood and other body fluids and on needles carrying the virus. Hepatitis B damages the liver. HIV can lead to AIDS. AIDS is a disease that eventually destroys the body's ability to fight off diseases that can be life threatening.

Program Resources

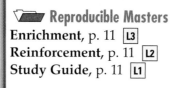 **Reproducible Masters**
Enrichment, p. 11 L3
Reinforcement, p. 11 L2
Study Guide, p. 11 L1

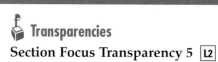 **Transparencies**
Section Focus Transparency 5 L2

Treating Viruses

Most viruses, like the one that causes the common cold, can't be treated. Other viruses that affect humans, however, can be prevented with vaccines. **Vaccines** are made from either dead or weak viruses and are given by mouth or by injection, as shown in **Figure 2-12**. Vaccines cause the body to make substances that resist particular viruses. Many states require vaccines for polio, measles, and mumps. These diseases can affect children and adults.

FIGURE 2-12

Many states require that children receive vaccinations before starting school.

 Skill Builder: Outlining

LEARNING the SKILL

1. Carefully read the material to be outlined.

2. Choose at least two or three main ideas from the passage that summarize each section or paragraph within a section. Label each main idea with a separate Roman numeral (I, II, III, etc.).

3. Now pick out key words in each section or paragraph that support each main idea. These supporting details are written separately under each main idea. Write a capital letter (A, B, C, etc.) before each.

4. Arrange the key phrases under each key word with a number before each phrase (1, 2, etc.).

5. List additional details under the numbered key phrases with a lowercase letter (a, b, c, etc.) before the details.

6. Check the accuracy of your outline by rereading the passage with your outline in hand. Make any necessary changes.

PRACTICING the SKILL

1. Make an outline using the main ideas in this feature. Use the questions that follow to help you organize your outline.

2. How many main ideas do you have?

3. What key terms and phrases summarize the definition of a *virus?*

4. List examples of viruses and the types of organisms they affect.

5. In the final part of your outline, summarize how diseases caused by viruses are prevented.

APPLYING the SKILL

Use references and what you have learned about viruses to make an outline that compares and contrasts viruses that affect plants with those that affect animals and bacteria.

Teaching the Skill

Use the first paragraph of this feature to demonstrate how to make an outline. Explain the hierarchy in an outline and how Arabic and Roman numbers and capital and lowercase letters are used. See answer 1 under Practicing the Skill for one possible outline.

Answers to Practicing the Skill

1. One possible outline is given below.

Viruses

I. **What is a virus?**
  A. General traits
    1. Has characteristics of both living and non-living things
    2. Can reproduce inside a cell
    3. Cannot grow, eat, or react to its environment
    4. Depends on other living things to survive
  B. Reproduction
    1. Attaches to cell
    2. Invades cell
    3. Makes copies of itself
    4. Causes cell to die when releasing new viruses

II. **Examples of Viruses**
  A. Viruses that infect bacteria
  B. Viruses that infect plants
  C. Viruses that infect animals
    1. Some human viruses
      a. Cold sores
        i. through touching someone with virus
      b. Chicken pox
        i. through touching someone with virus
      c. Hepatitis B
        i. through blood and needles
        ii. damages liver
      d. HIV
        i. through blood and needles
        ii. can lead to AIDS

III. **Treating Viruses**
  A. Most viruses can't be cured.
    1. colds
  B. Some viruses can be prevented with vaccines.
    1. polio
    2. mumps
    3. measles

2. three main points

3.-5. Refer to the outline given.

Source

Paulsen, Gary. *Hatchet.* Bradbury Press: New York, 1987.

Biography

Gary Paulsen is an award-winning author who writes fast-paced stories of adventure and survival. His 1986 Newberry Honor Book, *Dogsong,* was inspired by his own experiences in the Iditarod, a 1049-mile dogsled race across Alaska. Born on May 17, 1939, in Minnesota, Paulsen was raised by his grandmother and several aunts. The child of two alcoholic parents, Paulsen worked more than one job to support himself from the time he was 15 years old. Many of Paulsen's books share the themes of imperfect families, self-sufficiency, and overcoming nature and show teen protagonists living life as a challenge. In 1997, Paulsen received the Margaret A. Edwards Award for lifetime achievement in writing for young adults.

Teaching Strategies

- **Linguistic** Have students read the book *Hatchet* as an assignment. Have students write short book reports after completing the book. Or you may wish to have students write an outline of the book, asking them to give examples of how Brian uses scientific methods in each section.

- **Auditory-Musical** Listen to the audio version of *Hatchet* in class. You may wish to preview the recording and write a list of questions students can answer as they listen to the story.

Hatchet
by Gary Paulsen

Brian opened his eyes and screamed.

For seconds he did not know where he was, only that the crash was still happening and he was going to die, and he screamed until his breath was gone.

Then silence, filled with sobs as he pulled in air, half crying. How could it be so quiet? Moments ago there was nothing but noise, crashing and tearing, screaming, now quiet.

Some birds were singing.

How could birds be singing?

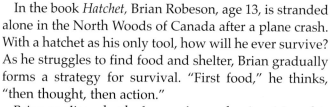

Science Journal

When do you use scientific skills and methods outside of school? Think it over. Then, in your Journal, write a short play about your experiences. That's thought, then action. What's missing? Brian would want you to have a sandwich first.

In the book *Hatchet,* Brian Robeson, age 13, is stranded alone in the North Woods of Canada after a plane crash. With a hatchet as his only tool, how will he ever survive? As he struggles to find food and shelter, Brian gradually forms a strategy for survival. "First food," he thinks, "then thought, then action."

Brian applies what he knows from school and from his parents and friends to make sense out of what his senses tell him. After he carries out a plan of action, he stops to think about it. What does he need to change to get a better result? Brian makes a guess about what kind of wood he needs to make a bow and how it should be shaped. The first result? Failure! When Brian pulls back on the shoelace he's used for a bow string, the bow snaps in half. He thinks carefully about the failure and makes a new plan, a new guess, and this time, it works. Brian thinks like a scientist, not to get an answer to a school question or to make dramatic discoveries, but to survive.

Science Journal Have students discuss the plays they wrote. Groups of students may perform select plays.

Other Works

Gary Paulsen has written and published more than 60 books for young adults since 1977, including two adventure series, as well as several books for adult readers. Interested students may wish to read *The River* (the sequel to *Hatchet*), *Dogsong, The Boy Who Owned the School, Canyons,* and *The Haymeadow,* among others.

Read the statements below that review the most important ideas in the chapter. Using what you have learned, answer each question in your Science Journal.

1. Five life traits define living things: response to the environment, organization, movement, reproduction, and growth and development. *How would you decide whether or not something is alive?*

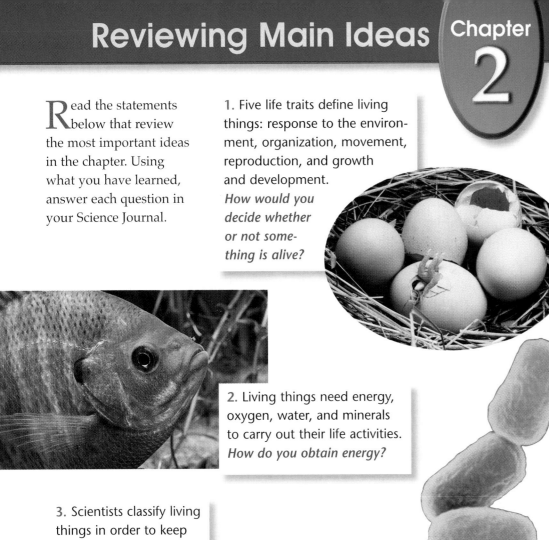

2. Living things need energy, oxygen, water, and minerals to carry out their life activities. *How do you obtain energy?*

3. Scientists classify living things in order to keep track of the millions of species on Earth. *Name three different organisms that live in your neighborhood.*

4. Organisms on Earth are classified into major groups called kingdoms. *Which kingdom of living things do you belong to?*

Chapter 2 Review **57**

Have students look at the illustrations on this page. Ask them to describe details that support the main ideas of the chapter found in each statement for each illustration.

Teaching Strategies
- Have students review illustrations on this page independently. Have them write answers to the questions in each statement on a separate sheet of paper.
- Discuss student responses to the questions together as a class.

Answers to Questions
1. Answers must include tests to see whether something has cells, whether it can respond to stimuli, whether it can move, whether it can reproduce, and whether it can grow and develop.
2. Students should recall that animals obtain energy by eating food.
3. Answers will vary, but check to make sure students understand the meaning of the word *organism*.
4. Humans are classified in the animal kingdom.

Science at Home
Students can carry out the process of classification with a variety of household items, such as kitchen items, tools, pens and pencils, and canned food items. Encourage interested students to have fun and learn by classifying objects around the house. L1

 **Assessment**

Portfolio Encourage students to place in their portfolios one or two items of what they consider to be their best work. Examples include:
- Data from MiniLAB, p. 36
- Data from Activity 2-1, pp. 40-41
- Classification schemes from Activity 2-2, p. 48 P

Performance Additional performance assessments may be found in **Performance Assessment** and **Science Integration Activities.** Performance Task Assessment Lists and rubrics for evaluating these activities can be found in Glencoe's **Performance Assessment in the Science Classroom (PASC).**

Chapter 2 Review

Using Key Science Words

1. species
2. classification
3. organism
4. vaccine
5. environment

Checking Concepts

6. c
7. c
8. d
9. d
10. b

Thinking Critically

11. Inflated balloons will respond to a sharp object, but they are still not living things because they don't possess all of the characteristics of life, such as reproduction and organization.

12. Answers will vary, but some examples might be responding to an alarm clock or responding to the smell of food.

13. Answers will vary, but common answers will be books, videotapes, magazines, silverware, etc.

14. Videotapes in a store are grouped according to similarities in themes. In the same way, living things are grouped according to their similarities.

15. Students should recall that colds are caused by viruses, and that one way to prevent the spread of virus particles is to wash hands thoroughly several times a day.

Using Key Science Words

cell	organism
classification	species
development	trait
environment	vaccine
kingdom	virus

Match each phrase with the correct term from the list of Key Science Words.

1. group of organisms that can breed with one another
2. grouping objects based on common traits
3. a living thing
4. is made from either dead or weak viruses
5. an organism's surroundings

Checking Concepts

Choose the word or phrase that completes the sentence.

6. A particle that has some things in common with both living and nonliving things is called a(n) _____.
   a. cell   c. virus
   b. vaccine   d. organism

7. Archaebacteria, Eubacteria, and Protist are names of different _____ of living things.
   a. traits   c. kingdoms
   b. species   d. environments

8. The smallest unit of life in living things is the _____.
   a. species   c. kingdom
   b. organism   d. cell

9. The changes that take place when a puppy grows into a dog are known as _____.
   a. organisms   c. homeostasis
   b. traits   d. development

10. The spots on a dalmatian dog are an example of a(n) _____.
    a. cell   c. environment
    b. trait   d. classification

Thinking Critically

Answer the following questions in your Science Journal using complete sentences.

11. Think about an inflated balloon. A sharp object can easily get a response from an inflated balloon. Are inflated balloons living things? Why or why not?

12. Give an example of when you reacted to something in your environment today.

13. People organize and classify many things in their homes. Briefly describe one example of a classification system in your home.

14. Describe how the classification of living things is similar to the way videotapes are organized in a store.

15. If you have a cold, how can you help prevent your cold from spreading to other people in your class?

Assessment Resources

Reproducible Masters
Chapter Review, pp. 7-8
Assessment, pp. 9-12
Performance Assessment, p. 40

Glencoe Technology
Computer Test Bank

MindJogger Videoquiz

Chapter 2 Skills Review

Developing Skills

If you need help, refer to the description of each skill in the Skill Handbook.

16. **Observing and Inferring:** On the way to school, you find a mysterious-looking object on the ground. The object appears to be dead. You put the object under a microscope and discover that it is organized into tiny, bricklike units. Could the object be a living thing? Explain your answer.

17. **Making and Using Tables:** The table below shows kingdoms of living things. Complete the table by filling in the missing information.

Kingdoms of Life

	How many cells does it have?	Does it move?	How does it obtain energy?
Bacteria	single celled	some move	absorb chemicals from their surroundings
Protists	single celled and many celled	some move	some get energy from eating food; some make their own food
Fungi	single celled and many celled	most do not move	eat food
Plants	many celled	do not move	make their own food
Animals	many celled	all move	eat food

18. **Outlining:** Outline the five ways to decide when something is alive. These ways were discussed in Section 2-1 of this chapter.

19. **Observing and Inferring:** When opening a can of food for lunch, you notice that your cat comes running into the kitchen. Explain why this behavior might be an important instinct for a cat in the wild.

20. **Comparing and Contrasting:** Organisms in the kingdom Protista are often called plantlike or animal-like organisms. Compare and contrast these two types of protists.

Performance Assessment

1. **Poster:** Choose an organism in the animal or plant kingdom. Research information about the organism. Make a poster showing the classification of this organism. What other organisms are closely related?

2. **Making and Using a Classification System:** Select a collection of seeds, cards, or toys. Create a classification system using these objects. Figure out a way to display the objects. Indicate the traits you used to classify the objects.

Developing Skills

16. **Observing and Inferring** Answers will vary, but many students will recognize that the object is organized and that the tiny, bricklike units might be cells. Therefore, many students will conclude that it's a living thing.

17. **Making and Using Tables** See student page for a completed table.

18. **Outlining**
   Is Something Alive?
   I. All living things must have these traits
     A. respond to their environments
     B. move
     C. show organization
     D. reproduce
     E. grow and develop

19. **Observing and Inferring** Cats recognize that cans contain food. Large cats in the wild must also be able to recognize the sounds of potential food sources.

20. **Comparing and Contrasting** Some protists are green and make food from the sun's energy, just like plants. Other protists swim around and eat food, just like animals do.

Performance Assessment

1. Check students' posters for accuracy of information and completeness. You may wish to provide students with ideas and reference materials. Use the Performance Task Assessment List for Poster in **PASC**, p. 73. P

2. Answers will vary depending on the type of object being classified. Check students' classification systems to make sure they understand the process of classification. Use the Performance Task Assessment List for Making and Using a Classification System in **PASC**, p. 49. P

Chapter Organizer

Section	Objectives/Standards	Activities/Features
Chapter Opener		**Explore Activity:** Observe Onion Cells, p. 61
3-1 **Cells—Building Blocks of Life** (5 sessions, 2½ blocks)*	1. **Compare and contrast** the parts of animal and plant cells. 2. **Describe** the purpose of different cell parts. **National Science Content Standards:** (5-8) UCP1, UCP2, UCP5, A1, C1	**Using Technology:** Microscopes, p. 64 **MiniLAB:** Modeling a Cell, p. 69 **Activity 3-1:** Comparing Cells, p. 71 **Skill Builder:** Making and Using Tables, p. 73 **Using Computers,** p. 73
3-2 **The Different Jobs of Cells** (3 sessions, 1½ blocks)*	3. **Explain** how a cell's shape sometimes relates to its function. 4. **Compare and contrast** tissues, organs, and organ systems. **National Science Content Standards:** (5-8) UCP1, UCP2, UCP5, A1, C1	**MiniLAB:** Analyzing Cells, p. 75 **Problem Solving:** Calculating Numbers of Blood Cells, p. 76 **Activity 3-2:** Where does the water go?, pp. 78-79 **Skill Builder:** Sequencing, p. 81 **Science Journal,** p. 81
3-3 **Science and Society:** **Test-Tube Tissues** (1 session, ½ block)*	5. **Investigate** how human tissues can now be grown in labs. 6. **Analyze** news sources. **National Science Content Standards:** (5-8) A2, E2, F5, G1	**Skill Builder:** Analyzing News Media, p. 83 **People & Science,** p. 84

* A complete Planning Guide that includes block scheduling is provided on pages 31T-33T.

Activity Materials

Explore	Activities	MiniLABs
page 61 index cards, plastic wrap, newspaper, onion skin, tap water, masking tape, dropper, metric ruler, scissors	page 71 microscope, slide of *Elodea* leaf cells, prepared slide of human cheek cells pages 78-79 fresh stalk of celery with leaves, clear drinking glass, scissors, red food coloring, tap water	page 69 candies, pasta, fruit and vegetable pieces or other common food items, unset gelatin, resealable plastic bags page 75 microscope; slides of human blood cells, muscle cells, skin cells, nerve cells

Need Materials? Call Science Kit (1-800-828-7777).

Teacher Classroom Resources

Reproducible Masters	Transparencies	Teaching Resources
Activity Worksheets, pp. 5, 19-20, 23 **Cross-Curricular Integration,** p. 7 **Enrichment,** p. 12 **Lab Manual 5, 6** **Reinforcement,** p. 12 **Study Guide,** p. 12	**Section Focus Transparency 6,** Building Dreams **Teaching Transparency 5,** Plant Cell **Teaching Transparency 6,** Animal Cell	**Spanish Resources** **English/Spanish Audiocassettes** **Cooperative Learning Resource Guide** **Lab Partner** **Lab and Safety Skills** **Lesson Plans**
Activity Worksheets, pp. 5, 21-22, 24 **Enrichment,** p. 13 **Reinforcement,** p. 13 **Science and Society/Technology Integration,** p. 21 **Science Integration Activities,** pp. 43-44 **Study Guide,** p. 13	**Science Integration Transparency 3,** Chemistry and You **Section Focus Transparency 7,** A Tool for Every Job	**Assessment Resources** **Chapter Review,** pp. 9-10 **Assessment,** pp. 13-16 **Performance Assessment,** p. 41 **Performance Assessment in the Science Classroom (PASC)** **MindJogger Videoquiz** **Alternate Assessment in the Science Classroom** **Computer Test Bank**
Enrichment, p. 14 **Multicultural Connections,** pp. 9-10 **Reinforcement,** p. 14 **Study Guide,** p. 14	**Section Focus Transparency 8,** Lettuce in a Test Tube	

Key to Teaching Strategies

The following designations will help you decide which activities are appropriate for your students.

L1 Level 1 activities should be appropriate for students with learning difficulties.

L2 Level 2 activities should be within the ability range of all students.

L3 Level 3 activities are designed for above-average students.

ELL ELL activities should be within the ability range of English Language Learners.

LS These activities are designed to address different learning styles.

COOP LEARN Cooperative Learning activities are designed for small group work.

P These strategies represent student products that can be placed into a best-work portfolio.

GLENCOE TECHNOLOGY

The following multimedia resources are available from Glencoe.

The Infinite Voyage Series
Miracles by Design

National Geographic Society Series
STV: The Cell

Glencoe Life Science CD-ROM

Teacher Classroom Resources

This is a representation of key blackline masters available in the Teacher Classroom Resources.

Teaching Aids

Section Focus Transparencies

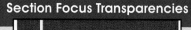

6 SECTION FOCUS TRANSPARENCY — Section 3-1

BUILDING DREAMS
Buildings, like this cathedral in Spain, are made of lots of different materials. And each material has a special job to do.

1. What are some of the materials this cathedral is made of? How are the different materials used?
2. What are organisms made of?

L2

7 SECTION FOCUS TRANSPARENCY — Section 3-2

A TOOL FOR EVERY JOB
You probably have many different tools to do many different jobs around your house. You bake with spoons and spatulas—or maybe you've helped repair an electric plug using a screwdriver and a pair of pliers. It always helps to have the right tool for the right job. You also need to have the right cell for the right job inside your body. Cells come in different shapes and sizes to do different jobs.

1. You have probably never seen a tool like this one. Does this tool remind you of a tool that you have seen before?
2. How could you use this tool?

L2

8 SECTION FOCUS TRANSPARENCY — Section 3-2

LETTUCE IN A TEST TUBE
This young lettuce plant will remain in the test tube until it is large enough to plant in soil. This plant was grown from a tiny piece of tissue from a much larger parent plant.

1. What are some other ways to start new plants?
2. What do you think plants need to grow?

L2

Science Integration Transparencies

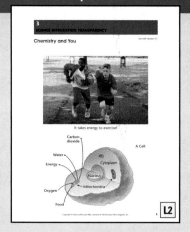

3 SCIENCE INTEGRATION TRANSPARENCY — Use with Section 3-1

Chemistry and You

It takes energy to exercise!

Carbon dioxide
Water
Energy
Oxygen
Food
Cytoplasm
Nucleus
Mitochondria
A Cell

L2

Teaching Transparencies

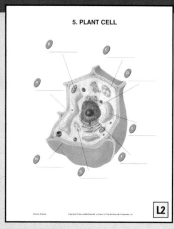

5. PLANT CELL

L2

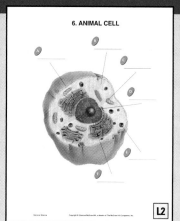

6. ANIMAL CELL

L2

Meeting Different Ability Levels

Study Guide for Content Mastery

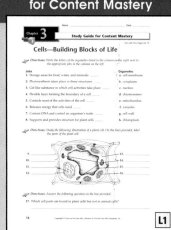

Name ____ Date ____

Chapter 3 — **Study Guide for Content Mastery**
(Use with Text Pages 62–71)

Cells—Building Blocks of Life

Directions: Write the letters of the organelles listed in the column on the right next to the appropriate jobs in the column on the left.

Jobs
1. Storage areas for food, water, and minerals ____
2. Photosynthesis takes place in these structures ____
3. Gel-like substance in which cell activities take place ____
4. Flexible layer forming the boundary of a cell ____
5. Controls most of the activities of the cell ____
6. Releases energy that cells need ____
7. Contain DNA and control an organism's traits ____
8. Supports and provides structure for plant cells ____

Organelles
a. cell membrane
b. cytoplasm
c. nucleus
d. chromosomes
e. mitochondria
f. vacuoles
g. cell wall
h. chloroplasts

Directions: Study the following illustration of a plant cell. On the lines provided, label the parts of the plant cell.

9. ____ 13. ____
10. ____ 14. ____
11. ____ 15. ____
12. ____ 16. ____

Directions: Answer the following question on the line provided.

17. Which cell parts are found in plant cells but not in animal cells?

L1

Reinforcement

Name ____ Date ____

Chapter 3 — **Reinforcement**
(Use with Text Pages 62–71)

Cells—Building Blocks of Life

Directions: Use the following words to answer the questions below.

cell   nucleus   cellular respiration   photosynthesis

1. What is the process by which food and oxygen combine to make carbon dioxide and water and release energy?
2. What word describes the way plants make their own food?
3. What is the smallest unit of life in all living things?
4. What part of the cell controls most of its activities?

Directions: Determine whether the italicized word or term makes each statement true or false. If the statement is true, write true on the line provided. If the statement is false, write the word or term that makes the statement true.

5. The cell membrane forms a boundary between the cell and its environment.
6. The chromosomes are located outside a cell's nucleus.
7. The gel-like substance in cells is called cytoplasm.
8. The cell wall is located inside the cell membrane.
9. Mitochondria release energy the cell needs to do its work.

Directions: Draw a line between each term on the left and its description on the right.

cell membrane — storage areas for food, water, and minerals
chromosomes — small green structures that contain chlorophyll
vacuoles — helps control what enters and leaves the cell
chloroplasts — contain DNA, and determine an organism's traits
nucleus — controls most of the cell's activities

L2

Enrichment Worksheets

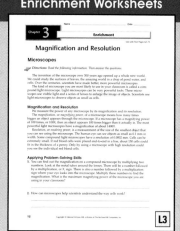

Name ____ Date ____

Chapter 3 — **Enrichment**
(Use with Text Pages 62–71)

Magnification and Resolution

Microscopes

Directions: Read the following information. Then answer the questions.

The invention of the microscope over 300 years ago opened up a whole new world. We could study the surfaces of leaves, the amazing world in a drop of pond water, and cells. Over the centuries, scientists have made better, more powerful microscopes. The kind of microscope you are most likely to use in your classroom is called a compound light microscope. Light microscopes can be very powerful tools. These microscopes use visible light and a series of lenses to enlarge the image of objects. Scientists use light microscopes to observe objects as small as cells.

Magnification and Resolution

We measure the power of any microscope by its magnification and its resolution. The magnification, or magnifying power, of a microscope means how many times bigger an object appears through the microscope. If a microscope has a magnifying power of 100 times, or 1000, then an object appears 100 times bigger than it actually is. The most powerful light microscopes have a magnification of about 1400X.

Resolution, or resolving power, is a measurement of the size of the smallest object that you can see using the microscope. The human eye can see objects as small as 0.1 mm in width. Some compound light microscopes have a resolution of 0.0002 mm. Cells can be extremely small. If red blood cells were placed end-to-end in a line, about 150 cells could fit in the thickness of a penny. Only by using a microscope with high resolution could you see the individual red blood cells.

Applying Problem-Solving Skills

1. You can find out the magnification on a compound microscope by multiplying two numbers. Look at the metal tubes around the lenses. There will be a number followed by a multiplication or X sign. There is also a number followed by a multiplication sign where your eyes look into the microscope. Multiply these numbers to find the magnification. What is the maximum magnifying power of the microscope you are using in your classroom?

2. How can microscopes help scientists understand the way cells work?

L3

Hands-On Activities

Science Integration Activities

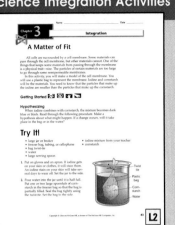

A Matter of Fit

Lab Manual

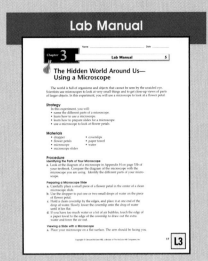

The Hidden World Around Us—Using a Microscope

Activity Worksheets

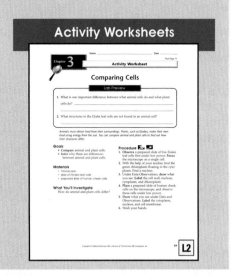

Comparing Cells

Enrichment and Application

Cross-Curricular Integration

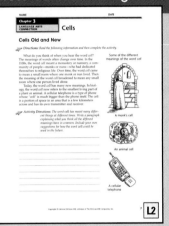

Cells Old and New

Multicultural Connections

Fighting Brain Tumors

Science and Society/Technology Integration

Growing Older—and Older

Assessment

Performance Assessment

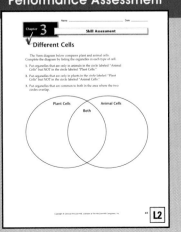

Different Cells

Chapter Review

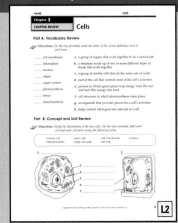

Cells

Assessment

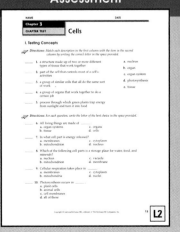

Cells

Cells

CHAPTER OVERVIEW

Section 3-1 The concept of the cell is introduced. Students learn that cells are made up of specialized parts that carry out particular functions. The section concludes with a discussion about energy and the living cell.

Section 3-2 The section discusses the relationship between cell shape and function. The section also presents a discussion about the organization of cells and tissues in living organisms.

Section 3-3 Science and Society This section discusses how some human tissues can now be grown in labs. It also helps students learn to analyze news media to help them make better decisions.

Chapter Vocabulary

cell membrane	chloroplast
nucleus	tissue
mitochondrion	organ
photosynthesis	organ system

Theme Connection

Scale and Structure On the most basic level of organization, all living things are composed of microscopic units called cells. Cells carry out important processes for organisms, such as the production of energy and important chemicals such as hormones.

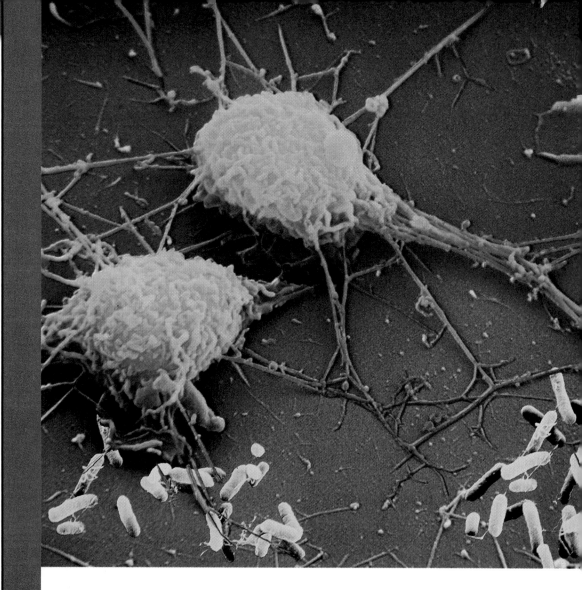

Chapter Preview

Section 3-1 Cells—Building Blocks of Life

Section 3-2 The Different Jobs of Cells

Section 3-3 Science and Society: Test-Tube Tissues

Skills Preview

▶ **Skill Builders**
- make and use a table
- sequence

▶ **MiniLABs**
- model
- compare

▶ **Activities**
- compare
- interpret
- diagram
- hypothesize
- infer

60

Chapter 3

Cells

You are about to enter an active world that exists in all of us, and in all other living things. It is a world so important that life couldn't exist without it. Yet it is a world that, most of the time, you can't see with just your eyes. Microscopes are windows into this world. The photograph on the left shows two white blood cells attacking bacteria that have invaded a human body. Welcome to the microscopic (mi kruh SKAH pihk) world of cells—the building blocks of life.

Magnification:
5400×

EXPLORE ACTIVITY

Observe Onion Cells

1. Cut a 2-cm hole in the middle of an index card.
2. Tape a piece of plastic wrap over the hole.
3. Turn down the two shorter ends of the card about 1 cm.
4. Place the hole over a piece of newspaper with the letter *e* printed on it. Draw what you see.
5. Place the hole over a piece of onion skin.
6. Put a drop of water on the plastic wrap.
7. Look through the water drop and observe the piece of onion. Draw what you see.

Science Journal

In your Science Journal, describe how the newspaper *e* and the onion skin looked when viewed with your magnifier.

61

Assessment Planner

Portfolio
Refer to page 85 for suggested items that students might select for their portfolios.

Performance Assessment
See page 85 for additional Performance Assessment options.
Skill Builders, pp. 73, 81, 83
MiniLABs, pp. 69, 75
Activities 3-1, p. 71; 3-2, pp. 78-79

Content Assessment
Section Wrap-ups, pp. 73, 81
Chapter Review, pp. 85-87
Mini Quizzes, pp. 73, 81

Group Assessment
Opportunities for group assessment occur with Cooperative Learning Strategies and Flex Your Brain activities.

EXPLORE ACTIVITY

Purpose
LS Kinesthetic Students will make and use a simple form of a microscope to examine onion skin cells. **L2 ELL**

Preparation
Cut tiny pieces from the thin, transparent skin of an onion for each student group.

Materials
dropper, water, index card, colorless plastic wrap, scissors, tape, onion skin, metric ruler, newspaper

Teaching Strategies
- Be sure adequate light to illuminate the object being observed is available.
- You may wish to have students examine other plant tissues, such as leaves or roots of common plants.

Troubleshooting If students are unable to see the cells and letter clearly, have them use microscopes to observe these items. Use low-power objective.

Science Journal Many students will describe the "e" as enlarged and the onion skin cells as bricklike.

✓ Assessment

Performance Have students use a microscope to observe the onion cells. Have them draw what they see and compare these drawings with the ones from the Explore activity. Use the Performance Task Assessment List for Scientific Drawing in **PASC**, p. 55. **LS P**

Prepare

Section Background
- Cells are the smallest functional units of living organisms.
- Cells make energy, proteins, and other important substances for organisms.
- All cells contain separate parts called organelles that perform various tasks for the cell.

Preplanning
Refer to the Chapter Organizer on pages 60A-B.

1 Motivate

Bellringer

 Before presenting the lesson, display **Section Focus Transparency 6** on the overhead projector. Assign the accompanying **Focus Activity** worksheet.

[L2] [ELL]

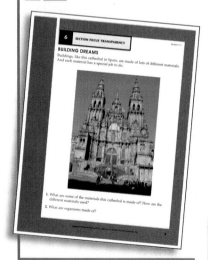

What YOU'LL LEARN
- The parts of animal and plant cells
- The purpose of different cell parts

Science Words:
cell membrane
nucleus
mitochondrion
photosynthesis
chloroplast

Why IT'S IMPORTANT
You will better understand how your cells carry out the activities of life.

FIGURE 3-1
All living things are made up of cells.

The World of Cells

Different cells have different jobs to do in living things. Cells found in onion stems help move water and other substances throughout the onion plant. White blood cells, found in humans and many other animals, help protect you from disease. White blood cells, onion cells, and all other cells are alike in many ways. A cell is the smallest unit of life in all living things. Cells are important because the activities of life take place in and around cells. They help living things carry on all of the important activities of life discussed in Chapter 2—movement, growth, and reproduction.

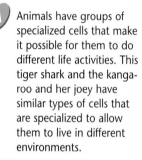 Animals have groups of specialized cells that make it possible for them to do different life activities. This tiger shark and the kangaroo and her joey have similar types of cells that are specialized to allow them to live in different environments.

62

Program Resources

📁 Reproducible Masters
Activity Worksheets, pp. 5, 19-20, 23 [L2]
Cross-Curricular Integration, p. 7 [L2]
Enrichment, p. 12 [L3]
Lab Manual, pp. 17-22 [L3] [L2]
Reinforcement, p. 12 [L2]
Study Guide, p. 12 [L1]

🔦 Transparencies
Section Focus Transparency 6 [L2]
Teaching Transparencies 5, 6 [L2]

The Microscopic Cell

All the living things pictured in **Figure 3-1** are made up of cells. The smallest organisms on Earth are bacteria. Bacteria are single-celled organisms, which means they are made up of only one cell. Larger organisms are made of many cells. These cells work together to complete all of the organisms' life activities. The living things that you see every day without using a microscope—trees, dogs, frogs, people—are many-celled organisms. Your body alone contains about 10 trillion cells. Ten trillion is written as 10 000 000 000 000.

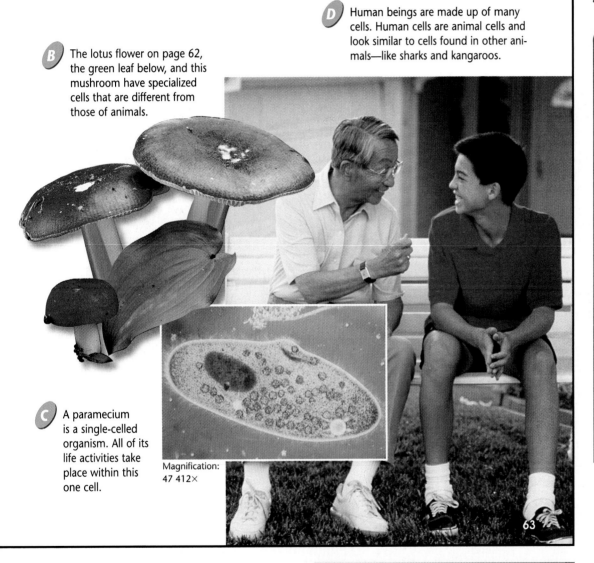

B The lotus flower on page 62, the green leaf below, and this mushroom have specialized cells that are different from those of animals.

D Human beings are made up of many cells. Human cells are animal cells and look similar to cells found in other animals—like sharks and kangaroos.

C A paramecium is a single-celled organism. All of its life activities take place within this one cell.

Magnification: 47 412×

63

Magnification: 55 000×

What makes up a cell?

As small as most cells are, as shown in **Figure 3-2**, they are made up of even smaller parts that do many different jobs. A single cell can be compared to a pizza shop, as illustrated in

FIGURE 3-2
How small are cells? If red blood cells were placed end to end in a line, about 150 cells could fit in the thickness of a penny.

USING TECHNOLOGY

Microscopes

Scientists have been able to see and study cells for only about 300 years. In that time, they have learned a lot about cells and unlocked many of their mysteries through the use of better and better microscopes. Some modern microscopes allow scientists to study the small details inside cells.

Your Microscope
The microscope used in most classrooms is called a compound light microscope. In this type of microscope, light passes through the object you are looking at and then through two or more lenses. The lenses make and enlarge an image of the object. Using the low-power lens, your microscope can let you see an object 40 times larger than its actual size. Using a high-power lens, the microscope will enlarge the image about 400 times.

Magnification: 40×

Magnification: 400×

*inter*NET
CONNECTION

Microscopes have changed throughout the years. For a history of the light microscope, visit the Glencoe Homepage. *www.glencoe.com/sec/science*

Content Background

The compound microscopes used in science classes today look much different from the earliest microscopes. In reality, they are similar in that both use two lenses to enlarge the image of an object.

Other types of microscopes are used to see tiny details inside cells, such as mitochondria and ribosomes. Cell biologists use either transmission electron microscopes (TEM) or scanning electron microscopes (SEM). Both types of microscopes use beams of electrons to help magnify objects. TEMs produce a more-or-less flat image that shows remarkable details of cell structure. SEMs enable scientists to see detailed, three-dimensional images of the surfaces of cells. Scientists use scanning tunneling electron microscopes (STM) to view living cells.

Figure 3-3. A pizza shop needs a building. The space inside is needed to perform the business activities. The different pizza ingredients—dough, cheese, tomato sauce, and toppings—must be assembled, cooked, and served. Storage for supplies and products is also needed. Electricity runs the ovens, powers the lights, and heats the shop.

A manager is in charge of the entire operation. The manager makes a plan for every employee of the shop. He or she also has a plan for every step of the process of making and selling the pizzas.

A living cell operates in a similar way. A cell has a boundary inside which its life activities take place. Smaller parts inside the cell act as storage areas and power sources. It also has parts that use ingredients—oxygen, water, minerals, and other nutrients—to release energy and make substances that are necessary for maintaining life. **Figure 3-4** on page 66 and **Figure 3-5** on page 67 show some of the important parts of the cell and the life activities that these parts perform.

FIGURE **3-3**

A cell can be compared to a pizza shop. *How is a cell like a pizza shop?*

Visual Learning

Figure 3-3 **How is a cell like a pizza shop?** *Both contain individual parts that must work together for the whole operation to run successfully. Cells contain organelles that carry out the important activities of the cell; pizza shops employ individual people to carry out essential tasks. Both cells and pizza shops operate within enclosed spaces. Cells and pizza shops both contain areas where wastes are removed, and both have sources of energy. Review this analogy with students.* **LS** **ELL**

Discussion

You may wish to discuss the magnification power of the microscopes students will be using. Most classroom microscopes are equipped with a low power, or 10×, lens. This lens magnifies an object ten times. The high-power lens is often 40×. Compound microscopes also contain an eyepiece lens, which is usually 10×. Therefore, when students examine objects under low power, they are seeing the object magnified a total of 100 times (10× times 10×) its actual size. An object under high power is magnified 400 times (10× times 40×) its actual size.

Using an Analogy

LS **Interpersonal** You may wish to have students brainstorm a list of other possible analogies for living cells. For instance, one common analogy is that cells are like factories. Both produce things (cells produce energy, proteins, and other substances), and they have specialized areas where this production takes place. In addition, production operates within enclosed spaces, and wastes are removed during the process. After student groups are finished, ask individuals to present their ideas to the class. Accept all reasonable responses. **L3** **ELL** **COOP LEARN**

Brainstorming

Have students brainstorm several different meanings of the word *cell*. Have them compare the meanings and brainstorm a list that shows how all these *cells* have similar characteristics. **L2**

NATIONAL GEOGRAPHIC SOCIETY

 Videodisc

STV: The Cell
Parts of the Cell
Unit 2
Cell Membrane

12694-15290
Nucleus

15330-19100
Other Organelles

19135-23966
Plant Cells

25786-27234

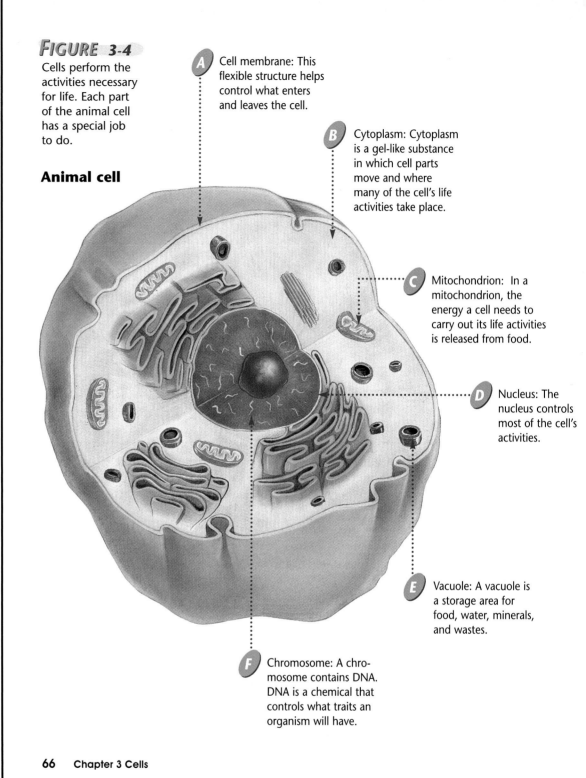

FIGURE 3-4

Cells perform the activities necessary for life. Each part of the animal cell has a special job to do.

Animal cell

A Cell membrane: This flexible structure helps control what enters and leaves the cell.

B Cytoplasm: Cytoplasm is a gel-like substance in which cell parts move and where many of the cell's life activities take place.

C Mitochondrion: In a mitochondrion, the energy a cell needs to carry out its life activities is released from food.

D Nucleus: The nucleus controls most of the cell's activities.

E Vacuole: A vacuole is a storage area for food, water, minerals, and wastes.

F Chromosome: A chromosome contains DNA. DNA is a chemical that controls what traits an organism will have.

Visual Learning

Figure 3-4 and Figure 3-5 Point out that these figures are illustrations of generalized cells. Discuss the meaning of the word *generalized.* Explain that cells don't really look like these illustrations. You also may wish to discuss the importance of scientific illustrations. Explain the value of illustrations for learning science, especially when microscopic structures, such as cells, are depicted. **LS** **ELL**

FIGURE 3-5

In addition to many of the cell parts found in an animal cell, a plant cell has two special parts—a cell wall and chloroplasts.

Plant cell

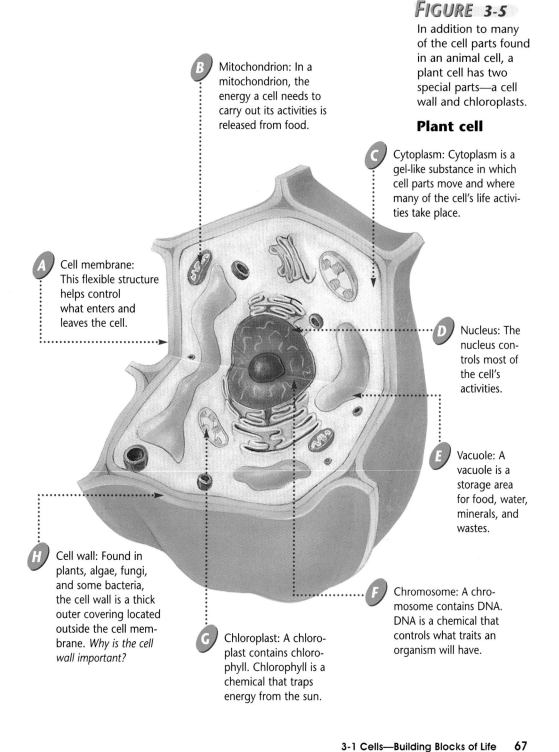

B Mitochondrion: In a mitochondrion, the energy a cell needs to carry out its activities is released from food.

C Cytoplasm: Cytoplasm is a gel-like substance in which cell parts move and where many of the cell's life activities take place.

A Cell membrane: This flexible structure helps control what enters and leaves the cell.

D Nucleus: The nucleus controls most of the cell's activities.

E Vacuole: A vacuole is a storage area for food, water, minerals, and wastes.

H Cell wall: Found in plants, algae, fungi, and some bacteria, the cell wall is a thick outer covering located outside the cell membrane. *Why is the cell wall important?*

F Chromosome: A chromosome contains DNA. DNA is a chemical that controls what traits an organism will have.

G Chloroplast: A chloroplast contains chlorophyll. Chlorophyll is a chemical that traps energy from the sun.

Teacher F.Y.I.

In all cells, mitochondria supply the energy for the cell's activities. Cells that use a lot of energy, such as muscle cells, typically have more mitochondria than other cells.

?FLEX Your Brain

Use the Flex Your Brain activity to have students explore THE CELL NUCLEUS.

📁 **Activity Worksheets,** page 5

Visual Learning

Figure 3-5H Why is the cell wall important? *The cell wall provides support for plants.* LS

Activity

LS **Kinesthetic** Give groups of students (about two to three in each group) small lumps of different-colored clay. As you teach about the parts of cells, have groups build model cells. Introduce and discuss the cell parts, and have students add clay parts to their own model cells. When you finish discussing the cell parts, have groups make small labels for each of the parts. Students can use toothpicks to place the labels on the model cells. L1 ELL COOP LEARN

Enrichment

Students who have mastered the content can construct more elaborate clay models of cells. Have them research cells and cell parts. Students can construct detailed models that display all of the important cell parts—not just the ones described in this textbook. L2 ELL COOP LEARN

Inclusion Strategies

Visually Impaired Students with visual impairments may benefit by using and constructing models of cells. Some of the large, demonstration cell models sold by science-supply houses are excellent for this purpose. Such models are typically large, detailed, and three dimensional. LS L2

Demonstration

Visual-Spatial Use a small section (6 inches) of dialysis tubing (or a plastic food-storage bag) to show how the cell membrane works. Dialysis tubing can be purchased through a science-supply company. Tie one end of the tubing with a knot, then open the other end. Fill it with a solution of cornstarch and water. Tie a knot on the end. Use this small sac to represent a cell. The dialysis tubing represents the cell membrane. Place the model cell into a beaker or jar filled with aqueous iodine solution. After a minute or so, students will see color changes within the bag. Starch turns blue-black in the presence of iodine. Point out that dialysis tubing, like the cell membrane, is filled with hundreds of pores. Both membranes are designed to allow only certain substances to pass through. Discuss with students that the dialysis tubing allowed iodine molecules to enter the model cell but prevented starch molecules from leaving the model cell. See the teacher instructions in Activity 8-2 on page 226 for disposal of the iodine solution.

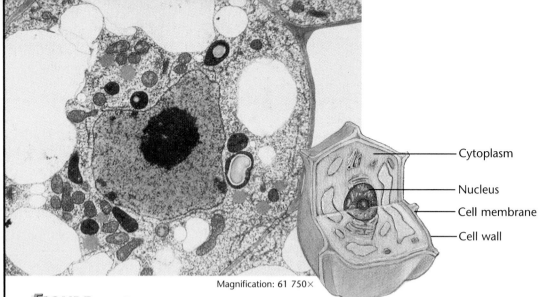

Magnification: 61 750×

FIGURE 3-6

This plant cell has a cell wall and a cell membrane. Animal cells only have a cell membrane.

Cell Structure

The cell membrane is a flexible structure that holds the cell together. The **cell membrane**, shown in **Figure 3-6**, forms a boundary between the plant or animal cell and its environment and also helps control what goes into and out of the cell. Plants, algae, fungi, and many types of bacteria have an additional tough, outer structure outside the membrane called a cell wall. The cell wall helps support the plant.

Filling the cell is a clear, gel-like substance called cytoplasm (SI toh plaz um). Cytoplasm is like the work area inside the pizza shop. It is mostly water, but it also contains chemicals the cell needs. Cytoplasm is the material in which the parts of the cell move. It provides the space and holds the materials for the cell's important activities.

Inside the Cell

Pizza shops have employees who perform the shop's activities. They also have equipment for specific purposes. Cells contain organelles (or guh NELZ) that are illustrated on pages 66 and 67. Organelles are specialized cell parts that perform the cell's activities. Examples of organelles are the nucleus, chloroplasts, mitochondria, and vacuoles. All of the cell's organelles move within the gel-like cytoplasm. You could think of these organelles as

Inclusion Strategies

Learning Disabled To help students learn about the function of the cell membrane, have students place sand and some beads into a fine mesh bag. Lift the mesh bag. The sand will pass through the bag, and the beads will remain inside the bag. Point out that the cell membrane works in a similar way by allowing the passage of only some substances. L2 ELL

the employees of the cell because each type of organelle does a different job.

The manager of the cell is the nucleus (NEW klee us). The **nucleus** controls most of the cell's activities. Inside the nucleus are the chromosomes (KROH muh zohmz), which can be likened to the business plan that the manager follows in running the shop. A business plan describes how the business operates. This could include such plans as what kinds of food and drink the business will serve. Just as the pizza shop has a business plan, chromosomes contain a plan for the cells. Chromosomes contain an important chemical called DNA that determines what kinds of traits an organism will have, such as height, hair color, and eye color. In **Figure 3-7**, you can see the nucleus and chromosomes of an animal cell.

Mini LAB

Modeling a Cell

1. Using **Figure 3-4** or **Figure 3-5** as a guide, make a model of a cell using common food items and gelatin that has not yet set firmly. Candies, pasta, and fruit and vegetable pieces are food items you could use.
2. Carefully put 125 mL of unset gelatin and pieces of food into a resealable bag. Reseal the bag.
3. Identify the cell parts by taping masking tape labels to the outside of the bag, and make a labeled sketch of your cell in your Science Journal.
4. Gently poke the center of the bag with your fingers.

Analysis

1. How is your model like a cell?
2. Can you change the shape of the cell? Does the shape stay changed? What helps keep the shape of the bag? How does the bag relate to a real cell?
3. What type of cell did you model?

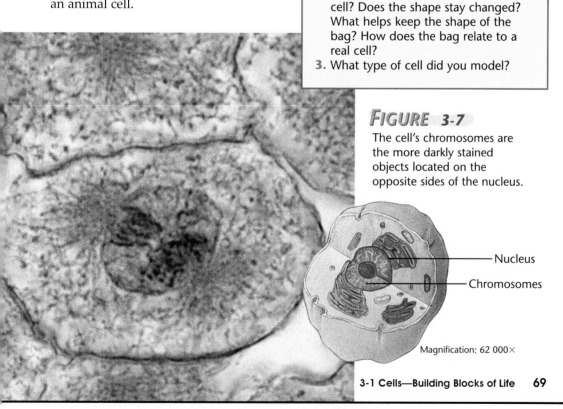

FIGURE 3-7

The cell's chromosomes are the more darkly stained objects located on the opposite sides of the nucleus.

Nucleus

Chromosomes

Magnification: 62 000×

3-1 Cells—Building Blocks of Life   69

What are some activities that you do that require energy? *Student answers might include running, walking, playing sports, or using a computer. Discuss all responses with the class.*

Demonstration

Demonstrate a by-product of cellular respiration. Explain that a waste product of cellular respiration is the invisible gas carbon dioxide. Your lungs release this waste product every time you exhale. Place 50 mL of bromothymol blue solution into a glass jar. Bromothymol blue solution changes color in the presence of carbon dioxide. Using a straw, exhale into the solution. **CAUTION:** *Make sure students do not inhale any solution.* As carbon dioxide from your lungs reacts with the bromothymol blue solution, the solution will turn bluish-green.

Pantries, closets, refrigerators, and freezers store food and other materials that a pizza business needs. Trash cans hold garbage until it can be picked up. In cells, food, water, and other substances are stored in bag-like structures in the cytoplasm called vacuoles (VAK yew ohlz). Some plant cells have large vacuoles to store the food the plant produces. Other vacuoles store wastes until the cell is ready to get rid of them.

Energy and the Cell

Power plants and gas companies supply the energy needed to turn on the lights and run the ovens at a pizza shop. Cells need energy, too. Cells contain organelles called **mitochondria** (mi tuh KAHN dree uh) that provide power for all of the cell's activities. Mitochondria are the sausage-shaped power plants of both plant and animal cells. This is where energy that the cell needs is released from food. *Mitochondrion* is the singular of *mitochondria.* As shown in **Figure 3-8,** an important chemical reaction called cellular respiration (SEL yuh lur • res puh RAY shun) takes place inside a mitochondrion. Cellular respiration is the process in which food and oxygen combine to make carbon dioxide and water, and release energy. Both animal and plant cells use the energy from cellular respiration to do all of their work.

FIGURE **3-8**

Cellular respiration takes place in the power plant of the cell—the mitochondrion.

Mitochondrion

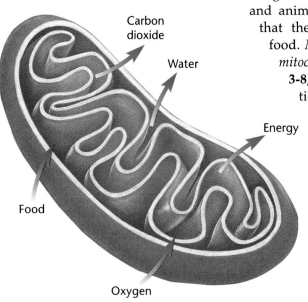

Carbon dioxide

Water

Energy

Food

Oxygen

Cellular Respiration

food and oxygen $\xrightarrow[\text{into}]{\text{is changed}}$ carbon dioxide and water and energy

Community Connection

Growing Power Plants Have students make comparisons between mitochondria and power plants. Both change stored energy to another form. Encourage students to research information and prepare a presentation about how power plants produce electricity. Students can call the local electric company or write letters requesting reference materials. **IS** **L2**

Comparing Cells

Animals must obtain food from their surroundings. Plants, such as *Elodea*, make their own food using energy from the sun. You can compare animal and plant cells to find out how their structures differ.

What You'll Investigate

How do animal and plant cells differ?

Procedure

1. Observe a slide of live *Elodea* leaf cells first under low power. **Focus** the microscope on a single cell.
2. With the help of your teacher, find the green chloroplasts floating in the cytoplasm. Find a nucleus.
3. In your Science Journal, **draw** what you see. **Label** the cell wall, nucleus, cytoplasm, and chloroplasts.
4. **Place** a prepared slide of human cheek cells on the microscope and observe these cells under low power.
5. **Draw** what you see. **Label** the cytoplasm, nucleus, and cell membrane.
6. Wash your hands.

Conclude and Apply

1. **Describe** the two types of cells that you observed.
2. **Compare** the sizes and shapes of the animal and plant cells. How are they similar? How are they different?
3. In what ways are animal and plant cells the same? Why do you think they are different in some ways?

Goals

- **Compare** animal and plant cells.
- **Infer** why there are differences between animal and plant cells.

Materials

- microscope
- slide of *Elodea* leaf cells
- prepared slide of human cheek cells

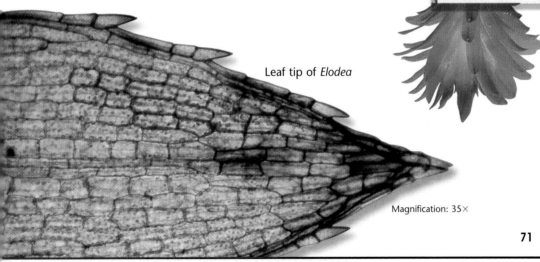

Leaf tip of *Elodea*

Magnification: 35×

71

3. All living things carry out the same basic life processes; therefore, one might expect similarities in cellular design and function. Plant and animal cells differ because they have somewhat different functions. Plants have chloroplasts for making food and cell walls to give them support as they grow upward.

✓ Assessment

Portfolio Have students make poster-sized illustrations of the observations they made using the microscope to include in their portfolios. Instruct students to label all visible structures. Use the Performance Task Assessment List for Science Portfolio in **PASC**, p. 105. [IS] [P]

Purpose

[IS] **Visual-Spatial** Students examine plant and animal cells under the microscope. They will locate and identify cell parts and make sketches of what they see. [L2] COOP LEARN

Process Skills

comparing and contrasting, observing and inferring

Time

50-60 minutes

Alternate Materials

If *Elodea* are unavailable, other plants, such as *Coleus*, also work well for this activity. Prepared slides of muscle or skin cells can be used if cheek cells are unavailable. **CAUTION:** *Do not make cheek cell slides. Using body fluids in lab activities is discouraged.*

📁 **Activity Worksheets,** pages 5, 19-20

Teaching Strategies

- Students should be taught how to use and handle a microscope properly before doing this activity.
- Encourage students to make neat and clear illustrations of the cells.
- After students have successfully focused on cells under low power and made sketches, encourage them to view the cells under high power.

Answers to Questions

1. Answers may include that *Elodea* leaf cells have many green chloroplasts and are brick-shaped. Human cheek cells are round.
2. Plant cells often display bricklike or blocklike shapes. Animal cells are more rounded and irregularly shaped. Both types of cells are similar in size.

Videodisc

STV: The Cell
Chloroplasts within plant cells

46944

3 Assess

Check for Understanding

Inquiry Question Why does every animal on the planet depend on plants? *Even if animals don't eat plants directly, they are still dependent upon plants indirectly. Plants trap the energy in sunlight and use it to make food. The food produced by plants during photosynthesis is then made available to animals and other organisms.*

Reteach

Visual-Spatial Have students get a book about cells from the library. Have them make sketches of animal and plant cells from photographs in other books and label all cell structures. **L1**

Extension

For students who have mastered this section, use the **Reinforcement** and **Enrichment** masters.

FIGURE 3-9

Photosynthesis takes place inside the chloroplasts of these heads of lettuce. Chloroplasts are located inside the cells of green leaves.

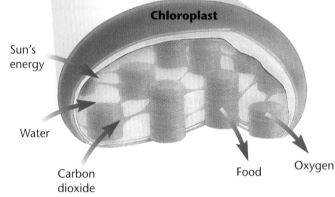

Chloroplast

Sun's energy

Water

Carbon dioxide

Food

Oxygen

Photosynthesis

sun's energy and water and carbon dioxide $\xrightarrow[\text{into}]{\text{is changed}}$ food and oxygen

Plants—Nature's Solar Energy Factories

If you've ever watched a cow grazing in a pasture, a bird pecking at worms, or a dog eating from its bowl, you know that animals get food from their surroundings. But unless you've observed a Venus's-flytrap, you've probably never seen a plant trap anything. How do plants get

72   Chapter 3 Cells

Visual Learning

Figure 3-9 Review the chemical equation for photosynthesis. Explain that a chemical equation is similar to a recipe. Both begin with a set of ingredients that come together to form one or more final products.

the food they need for energy? Plants, green algae, and many types of bacteria make their own food through a process called photosynthesis (foh toh SIHN thuh sus).

During **photosynthesis**, as shown in **Figure 3-9**, green plants trap energy from the sun. They then convert this energy into food needed by the plant. This food-making activity occurs inside the cell's chloroplasts (KLOR uh plasts). **Chloroplasts** are green organelles that trap energy from sunlight and turn it into food. This food is needed by the plant to stay alive. As the plant needs energy, its mitochondria release the food's energy. Most photosynthesis occurs in a plant's leaves. Leaves are green because leaf cells contain many of these green chloroplasts.

All cells contain organelles that perform special jobs. But all cells are not identical. In this section, you've learned about the different parts of animal and plant cells. In the next section, you'll learn that large, many-celled organisms like you are made up of many different kinds of cells that perform different functions.

interNET CONNECTION

How is photosynthesis related to leaves changing their colors in autumn? For answers to these and other questions about leaves, visit the Glencoe Homepage. *www.glencoe.com/ sec/science*

Section Wrap-up

1. How is a living cell similar to a business?

2. Why is the nucleus so important to the living cell?

3. How do cells get the energy they need to carry on their activities?

4. **Think Critically:** Suppose your teacher gave you a slide of an unknown cell. How would you tell whether the cell was from an animal or from a plant?

5. **Skill Builder**

   Making and Using Tables Make a table listing the parts of animal and plant cells and the jobs they do in the cells. If you need help, refer to Making and Using Tables on page 546 in the **Skill Handbook.**

Using Computers

Internet Visit the Glencoe Homepage at *www.glencoe.com/ sec/science* to learn more about microscopes. Draw a diagram of a microscope and label the different parts.

Skill Builder

Making and Using Tables Check tables for accuracy. Make sure students provide the correct functions for each cell part. See page 87 for a completed table.

Assessment

Content Divide the class into groups of 3-4 students. Have students use their tables from the Skill Builder to prepare "game show"-type questions from the information. When students finish, collect all the questions and quiz the groups on the material. Use the Performance Task Assessment List for Group Work in **PASC**, p. 97. **IS** **L2**

COOP LEARN

4 Close

MINI QUIZ

Use the Mini Quiz to check students' recall of chapter content.

1. **The smallest functional units of living things are called _____.** *cells*

2. **The _____ is the control center of the cell.** *nucleus*

3. **The _____ forms a boundary between a cell and its environment.** *cell membrane*

4. **The energy-making process in cells is known as _____.** *cellular respiration*

5. **Plants produce food through the process of _____.** *photosynthesis*

Section Wrap-up

1. Answers will vary but should include that, like a business, a cell is made up of different parts that perform different tasks.

2. The nucleus controls most of the cell's activities.

3. The cell's mitochondria release energy by combining food molecules and oxygen during the process of cellular respiration.

4. **Think Critically** Answers may vary, but students should indicate that most plant cells have chloroplasts and cell walls, while animal cells don't have either of these structures.

Using Computers

Check student illustrations for accuracy and make sure all parts are correctly labeled. **P**

Prepare

Section Background
- Cells are shaped according to their functions.
- Living things are organized on several levels. The cell is the most basic level.

Preplanning
Refer to the Chapter Organizer on pages 60A-B.

1 Motivate

Bellringer

 Before presenting the lesson, display **Section Focus Transparency 7** on the overhead projector. Assign the accompanying **Focus Activity** worksheet.

`L2` `ELL`

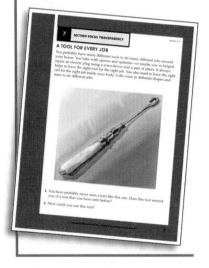

Tying to Previous Knowledge
Students should be familiar with tools, such as hammers, screwdrivers, saws, chisels, and files. Have students make observations about how the structure of each tool is related to its function. Just as tools are shaped according to their functions, so too are the cells of many-celled organisms. `L2`

74

3•2 The Different Jobs of Cells

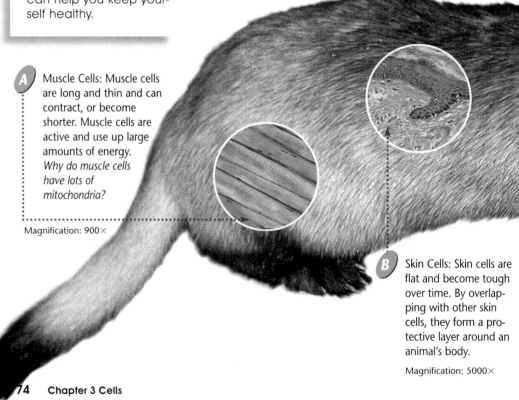

What YOU'LL LEARN
- How a cell's shape is important for doing its job
- The differences among tissues, organs, and organ systems

   Science Words:
   tissue
   organ
   organ system

Why IT'S IMPORTANT
Understanding how cells work and are organized can help you keep yourself healthy.

Special Cells for Special Jobs

Have you ever heard the phrase "choose the right tool for the right job"? It's a common expression that means that the best tool for a job is the one that has been designed for that job. For instance, you wouldn't use a hammer to saw a board in half, and you wouldn't use a saw to pound in a nail. Having the right tool applies to the world of cells, too. Inside your body, many different types of cells keep you healthy.

The large numbers of cells making up both plants and animals are classified into groups. Each group has a different job to do within the organism. This section of the chapter will give you examples of these different types of cells.

A Muscle Cells: Muscle cells are long and thin and can contract, or become shorter. Muscle cells are active and use up large amounts of energy. *Why do muscle cells have lots of mitochondria?*

Magnification: 900×

B Skin Cells: Skin cells are flat and become tough over time. By overlapping with other skin cells, they form a protective layer around an animal's body.

Magnification: 5000×

74 Chapter 3 Cells

Program Resources

📁 Reproducible Masters
Activity Worksheets, pp. 5, 21-22, 24 `L2`
Enrichment, p. 13 `L3`
Reinforcement, p. 13 `L2`
Science and Society/Technology Integration, p. 21 `L2`
Science Integration Activities, pp. 43-44 `L2`
Study Guide, p. 13 `L1`

🔖 Transparencies
Science Integration Transparency 3 `L2`
Section Focus Transparency 7 `L2`

Different Animal Cells

Your body is made up of about 200 different types of cells. The same is true for other animals. **Figure 3-10** shows some different cell types in a ferret. Notice the variety of sizes and shapes. A cell's shape and size can be related to its function in the body. The MiniLAB on this page will help you see how the shapes of human cells are related to their jobs.

C Nerve Cells: Nerve cells are long and shaped like wires, which allows them to deliver nerve messages quickly from one part of the body to another.

Magnification: 520×

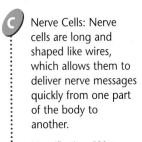
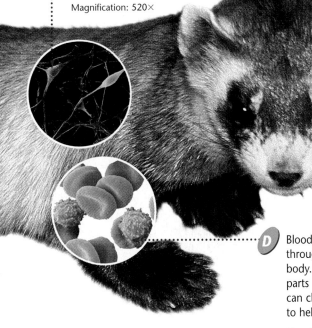

Mini LAB

Analyzing Cells

Analyze different animal cells. Determine how an animal cell's shape is related to its job.

1. Examine slides of human blood cells, muscle cells, skin cells, and nerve cells under low power of a microscope.
2. Draw each type of cell in your Science Journal. Label cell parts that you can see.

Analysis

1. In what ways were the cells that you observed similar? How were they different?
2. How are the cells' shapes related to their jobs?

FIGURE 3-10

Animals such as this ferret are made from many different types of cells. These cells come in shapes and sizes that are related to their jobs.

D Blood Cells: Red blood cells move easily through even the smallest blood vessels in the body. They deliver oxygen to different body parts and remove wastes. White blood cells can change shape and leave the blood vessels to help protect the body from disease.

Magnification: 9000×

3-2 The Different Jobs of Cells **75**

2 Teach

Mini LAB

Purpose

LS **Visual-Spatial** Students will compare and contrast a variety of different animal cells and learn how cell shape and structure are related to function. **L2** **ELL**

Materials

prepared slides of human muscle cells, nerve cells, blood cells, and skin cells; compound microscopes

Alternate Materials

If slides are not available, you may wish to use high-quality photomicrographs of different types of cells. Another alternative would be to find video programs that show cell shape and structure.

Activity Worksheets, pages 5, 24

Analysis

1. All of the cells are microscopic in size. With the exception of muscle cells, the other cells would be somewhat round in shape.
2. See **Figure 3-10** captions for answers.

Assessment

Process In the human body, structure is always closely related to function. Have students use slides or photographs of tissues (bones, joint surfaces, cross sections of blood vessels, and organs) to prepare a presentation that relates how structure is related to function. Use the Performance Task Assessment List for Slide Show or Photo Essay in **PASC**, p. 77. **L3**

Visual Learning

Figure 3-10 Review the photos of the cells with students. Discuss the functions of each of the cells. Have students describe how the shapes of the cells relate to function.

Figure 3-10A Why do muscle cells have lots of mitochondria? *Muscle cells require a lot of energy, and mitochondria release energy.* **LS** **L2**

Science Journal **Form and Function** Different types of eating utensils— forks, knives, and spoons— are also perfectly designed for their functions. Have students write brief paragraphs in their Science Journals about how the shapes of eating utensils match their functions. Have them relate this concept to cells. **LS** **L2** **P**

Problem Solving

Solve the Problem

There are 1000 milliliters in a liter.

You may wish to review SI and SI units with your students. Explain that even though SI units are not commonly used in the United States, they remain the standard units of measurement in science.

Think Critically

Five billion red blood cells are in every liter of blood. (5 million $\frac{RBC}{mL} \times 1000 \frac{mL}{L}$ = 5 billion $\frac{RBC}{L}$) Because an average person has about 3.5 L of blood, an average person's body contains about 17.5 billion red blood cells. (5 billion $\frac{RBC}{L} \times 3.5$ L = 17.5 billion RBC) 📘

Content Background

In the same way that some people specialize at particular jobs, the cells of animals are specialized to perform different functions. The shape of an animal cell provides clues about its function in the organism. Cells that perform the same function are grouped together into tissues.

The human body has four basic kinds of tissues: muscle, nerve, connective, and skin. Connective tissue provides support for structures of the body. Nerve tissue carries messages, in the form of nerve impulses, around the body. Skin tissue forms the outer surface of the body and also lines body cavities, such as the mouth. Muscle tissue is designed to produce movement of body parts.

Problem Solving

Calculating Numbers of Blood Cells

Benito wanted to solve a problem. In science class, he learned that about 200 different types of cells can be found in the human body. There are also billions of cells of each type. Benito wanted to find out how many red blood cells the human body has. After a visit to the library and a few questions to his teacher, Benito found out that every milliliter of blood contains 5 million red blood cells, and an average person has about 3.5 liters of blood.

Solve the Problem:

One piece of data Benito discovered is expressed in milliliters. The other data are expressed in liters. Find out how many milliliters there are in a liter. Refer to Appendix B on page 526 if you need help.

Think Critically:

Using the information Benito discovered, calculate the number of red blood cells in a person with 3.5 liters of blood. List the steps you would take to answer this problem.

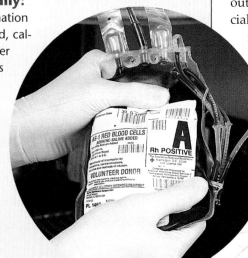

Across the Curriculum

History Have students find out about Belgian scientist Andreas Vesalius (1514-1564). Vesalius conducted some of the most accurate and thorough examinations of the human body that had ever been done. Vesalius wrote a classic book about human anatomy that has become popular because of its extremely detailed anatomical drawings. 📘

Different Plant Cells

Like animals, plants are also made of several different cell types, as shown in **Figure 3-11**. Plants, for instance, have different cells in their leaves, roots, and stems. Each type of cell has a specific job. Some cells in plant stems are long and tube shaped. Together, they form a system through which water, food, and other materials move in the plant. Other cells, like those that cover the outside of the stem, are smaller or thicker. They provide strength to the stem.

Cells that make up many-celled organisms are specialized. Unlike single-celled organisms, such as bacteria, specialized cells found in larger organisms cannot survive outside of the organism. Single-celled organisms carry out all life activities. But specialized cells work as a team to meet the life activities of every cell inside a many-celled organism. Cells are the building blocks of life.

FIGURE 3-11
Plants, like animals, must carry on different activities to survive. Their many cell types are specialized for certain activities.

A Leaves: Leaf cells are brick shaped and contain many chloroplasts. *Why do you think leaf cells contain so many chloroplasts?*

Magnification: 900×

Magnification: 1500×

B Stems: Many cells found in stems are long and tube shaped for moving water and other materials through the plant.

C Roots: Root cells are block shaped and often have small extensions called root hairs.

Magnification: 450×

3-2 The Different Jobs of Cells　**77**

Cultural Diversity

Skin Color All humans have approximately the same number of melanocytes, the cells that produce melanin. Melanin is the primary protein that gives skin its color. Although all humans have the same *number* of melanocytes, some people's melanocytes produce more pigment, making their skin darker. Some cultures use difference in skin color as a major way of deciding people's "race." In Brazil, many other factors are also involved, such as wealth, Portuguese last name, and occupation. In fact, Brazilians recognize around 40 races. Japan recognizes some people as members of the "burakumin race," although there is no difference in skin color at all. In fact, the burakumin are genetically indistinguishable from other Japanese.

Activity 3-2

Purpose

VS **Visual-Spatial** Students will relate the structure of plant tissue and its function. Students will determine which plant tissues are involved in the absorption of water from the soil by observing how the xylem tissue in a celery plant helps transport water up through the stem and into the leaves. **L2** **ELL** **COOP LEARN**

Process Skills

forming a hypothesis, designing an experiment, making and using tables, observing and inferring, separating and controlling variables

Time

30 minutes for planning and setup and 15 minutes to check and record results

Materials

Definitely use red food coloring. Blue, green, and yellow food coloring will not show up well in the celery.

Safety Precautions

• Caution students to avoid getting food coloring on clothing.
• Clean up all spills immediately.
• Clean up the work space when finished.
• Have students wash their hands after completing their experiments.

Possible Hypotheses

• Plant stems are hollow and act like drinking straws.
• Plant stems contain thin, strawlike passageways for water to flow through.

Activity Worksheets, pages 5, 21-22

Possible Materials

• fresh stalk of celery with leaves
• clear drinking glass
• scissors
• red food coloring
• water

Design Your Own Experiment
Where does the water go?

When you are thirsty, you may sip water from a glass or drink from a fountain. Plants must get their water in other ways. Their cells help with the task. In most plants, water is absorbed from the soil by cells in the roots. How does this water go into other parts of the plant?

What You'll Investigate
Where does water travel in a plant?

Form a Hypothesis

Think about what you already know about how a plant functions. In your group, **make a hypothesis** about where you think water travels in a plant. Remember that a hypothesis is a statement that can be tested. Write down your hypothesis in your Science Journal.

Goals

Demonstrate how cells in plant stems move water through a plant.

Safety Precautions

Use care when handling sharp objects such as scissors. Avoid getting red food coloring on your clothing.

78

Theme Connection
Systems and Interactions

The shapes of a plant's stem cells are directly related to function. Stem cells are shaped somewhat like straws. Through capillary action, water is drawn upward by the stem cells and carried to the plant's leaves. This is just one example of how an organism's interactions with its environment determine its adaptations.

Sample data table

Time	Observations
Start of activity	Celery is light green
End of class	Red streak moved up stalk
Next day	Celery stalk is red on the inside

PLAN

1. **Decide** on a way to test your group's hypothesis. How will you use the materials to show that water moves through the celery?
2. How will you carry out your experiment? **Make a list** of the different steps you will take during the experiment.
3. How long will your experiment take? At what time intervals will you make your observations? At the end of class? The next day?
4. **Prepare** a data table to record your observations.
5. What kinds of observations will you make? Will you use drawings?

DO

1. **Read** over your entire experimental plan to make sure that it is designed to test your hypothesis.
2. Make sure your teacher approves your plan before you proceed.
3. When you cut the celery stalk, put the celery under water and carefully use the scissors to **make a sharp cut** on the bottom of the stalk, or ask your teacher to **cut** the stalk with a sharp knife. Place your materials where they cannot be knocked over.
4. Carry out your plan. Remember to record the time you set up the experiment and the time you record your observations.
5. While doing the experiment, be sure to record all of your observations in your Science Journal.

CONCLUDE AND APPLY

1. **Compare** the color of the celery stalk before, during, and after the experiment.
2. **Make a drawing of** the stem material after the stalk is cut. Label your drawing. Why do you suppose only some of the stem material has turned red?
3. Was your hypothesis supported?

Go Further

How do environmental factors affect water transport in plants? Students can investigate this question by performing similar experiments under varying environmental conditions. Such conditions could include different temperatures and different light conditions. L3

 **Assessment**

Content Have groups of students design a similar experiment to determine whether different plants absorb water in the same way. Have students write hypotheses to answer this question. Then have them plan and carry out experiments using fresh carrots instead of celery stalks to test the hypotheses. Use the Performance Task Assessment List for Designing an Experiment in **PASC**, p. 23. P

PLAN

Possible Procedures

Place celery stalks in a glass that contains water colored with red food coloring. Leave them there overnight. The next day, remove the celery stalks, observe them, then cut them open.

Teaching Strategies

- Have students relate this activity to different plant cell types.
- To get maximum absorption, place stalks in water containing a high concentration of food coloring and cut the celery diagonally.
- The longer the celery remains in the water, the more food coloring is absorbed.

DO

Expected Outcome

Students will observe that the plants have absorbed some of the food coloring. Students should see that tubelike vessels in the stem are involved in water absorption.

CONCLUDE AND APPLY

1. The celery stalk was light green before the experiment. During the experiment, a red streak moves up the stalk. After the experiment, the celery stalk is red on the inside.
2. Only some of the tissue inside the stem carries water. These are the tissues that turned red.
3. Student answers will vary depending on student hypotheses and results.

Some students will be surprised to learn that blood is also a tissue. It's natural for students to think only of solid tissues, such as muscle tissue, nerve tissue, or skin tissue. Explain that blood is classified as a tissue because it is made up of groups of similar cells that work together to carry out a specific task.

3 Assess

Check for Understanding

Visual Learning

Figure 3-12 Have students study the different levels of organization in an alligator. Point out that all many-celled organisms show this kind of organization. You may wish to review another example with students. For instance, review the levels of organization in a tree, starting on the level of a leaf cell.

Reteach

 Logical-Mathematical Have students make a chart with two columns. In column one, have them list the different levels of organization, beginning with the level of the cell. In the other column, have students provide a real example for each different level. Have students choose an organism not discussed in the chapter.

Extension

 For students who have mastered this section, use the **Reinforcement** and **Enrichment** masters. L2

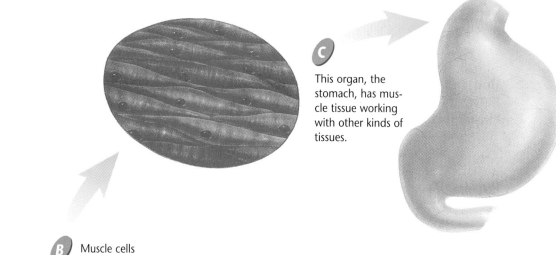

C This organ, the stomach, has muscle tissue working with other kinds of tissues.

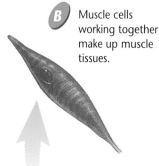

B Muscle cells working together make up muscle tissues.

A The cell is the simplest level of organization. This muscle cell is the smallest unit in the body to do "muscle work."

FIGURE 3-12
In many-celled organisms such as an alligator, the cells are arranged into different levels of organization.

Cell Organization

How well do you think your body would work if all the different cell types were just mixed together in no particular pattern? Could you walk if your leg muscle cells were scattered here and there, each doing its own thing, instead of being grouped together in your legs? How could you think if your brain cells weren't close enough together to communicate? Many-celled organisms are not just mixed-up collections of different types of cells.

Figure 3-12 shows the different levels of organization in an alligator. Cells that are alike are organized into tissues (TIH shews). **Tissues** are groups of similar cells that all do the same sort of work. For example, in animals with muscles, there is muscle tissue made up of just muscle cells. Bone tissue is made up of just bone cells, and nerve tissue is made up of just nerve cells. There is also blood, a liquid tissue, that is made up of just blood cells.

As important as individual tissues are, they do not work all by themselves. Instead, organisms are made up of many organs (OR gunz). An **organ** is a structure made up of two or more different types of tissue that work together. For example, the stomach is an organ made up of muscle tissue, nerve tissue, and blood tissue. Other organs include the heart and the kidneys.

Inclusion Strategies

Learning Disabled Students who have difficulty understanding the levels of organization may benefit by making models. First, have them make a small model cell out of clay. Next, instruct students to make a few more of the same kind of cell. Have them join the individual cells to make a tissue. The other levels of organization can be studied in similar fashion. L1

Content Background

The digestive systems of alligators and humans are not the same. Both alligators and humans have mouths, stomachs, intestines, and livers. In addition to these organs, humans have a pancreas that secretes both digestive enzymes and hormones. Human intestines are also divided into the small and large intestine, which have different jobs to do.

The stomach is an organ in the alligator's body that digests food. Other organs in its body are also involved in digesting food—a mouth, intestines, and a liver. A group of organs that work together to do a certain job is called an **organ system.** Other organ systems found in your body include the respiratory system, the nervous system, and the reproductive system.

D Organ systems are the next level of organization. Systems are made up of different organs working together. In the alligator's digestive system, the mouth, stomach, intestines, and liver are the organs working together.

Section Wrap-up

1. How is a cell's shape important to its job? Give an example.

2. What is the difference between a tissue and an organ?

3. **Think Critically:** Give an example of a human organ system, and name some of the parts that make up the organ system.

4. **Skill Builder**
   Sequencing Make a list that sequences the different levels of cell organization. Start with the cell and list all levels of organization. End with an organ system. For each level of organization, provide an example from the human body. If you need help, refer to Sequencing on page 543 of the **Skill Handbook.**

Science Journal
In terms of organization, organisms can be compared to a school band. In your Science Journal, write a short paragraph explaining how an organism is like a band.

Skill Builder
Sequencing Examples used by the students will vary, but check for accuracy. Example: Cell (muscle cell) → Tissue (muscle) → Organ (heart) → Organ System (circulatory system)

Assessment

Portfolio Have students design posters illustrating the levels of organization in living things. Posters should include (1) a type of cell in the human body; (2) groups of these cells together to show a particular tissue; (3) how this tissue is incorporated in an organ; and (4) the organ system that the organ belongs to. Use the Performance Task Assessment List for Poster in **PASC,** p. 73. [L3] [P]

4 Close

•MINI•QUIZ•

Use the Mini Quiz to check students' recall of chapter content.

1. **The most basic level of organization in living things is the _____.** *cell*

2. **A cell's _____ and size is closely related to its function.** *shape*

3. **Similar types of cells working together form a(n) _____.** *tissue*

Section Wrap-up

1. Cells come in a variety of shapes and sizes, depending on their functions. Examples might include stem cells being long and tubelike to carry water and other materials throughout the plant.

2. Tissue consists of similar cells working together. An organ consists of different tissues working together.

3. **Think Critically** Student answers might include the digestive system, nervous system, musculoskeletal system, or reproductive system, along with organs and tissues associated with the chosen system.

Science Journal
The band at school is made up of individuals who work together. Organisms are made up of different cells that must work together to keep an organism alive.

Content Background

While experts in the field of tissue engineering can currently produce skin, bone, and cartilage via cultures, laboratory-grown organs are far from reality.

Scientists at the Massachusetts Institute of Technology and the University of Massachusetts engineered a viable human ear that thrived when implanted under a mouse's skin.

Science & Society

3•3 Test-Tube Tissues

What YOU'LL LEARN

- Many human tissues can now be grown in labs
- How to analyze news sources

Why IT'S IMPORTANT

Learning to analyze news media will help you make better decisions.

Tissue Transplants

Transplanting tissues and organs from one person to another is done often in hospitals. Burn victims and patients with one or more organs that are not working properly can live normal lives with transplants. Too many times, however, donors aren't available. In other cases, tissues and organs are available, but the person's body cannot accept the new tissue or organ.

Growing Tissues

Because of these problems of tissue not being available and bodies not accepting new tissue, scientists have found ways that allow them to grow some kinds of tissue from tissue that already exists. For example, patches of skin tissue smaller in size than quarters can be used to make more skin. Special nutrients are added to the skin cells that make up the tissue. The cells then multiply to make flat sheets of new skin tissue, as shown in **Figure 3-13**. Pieces of skin as large as postcards are being made to treat some of the 13 000 burn victims who need skin grafts in the United States each year.

Parts of a magazine article that was in *Science World* on March 7, 1996, are shown on the next page. Read these paragraphs carefully and then answer the questions that follow.

FIGURE 3-13

New skin tissue is grown in laboratories from the skin cells of the patient.

Program Resources

Reproducible Masters
Enrichment, p. 14 [L3]
Multicultural Connections, pp. 9-10 [L2]
Reinforcement, p. 14 [L2]
Study Guide, p. 14 [L1]

Transparencies
Section Focus Transparency 8 [L2]

Grow Your Own Body Parts

Suddenly, fire filled the room and Emerson was engulfed in flames. When doctors saw the burns covering more than 85 percent of his body—so deep that the skin would never grow back—they thought Emerson would die.

But by the next day, Emerson's skin was growing. A miracle? Emerson's skin was growing in a laboratory 900 miles from his Kalamazoo, Michigan, hospital bed. The idea was to grow enough skin in the lab so Emerson could receive a transplant of his own skin.

Scientists experimenting with this technique, called tissue engineering, say that one day it may be used to replace other body parts that can't grow back on their own—like ears, livers, and heart valves.

"For burn patients, the most urgent need is to replace the protective, moisture-sealing layer—even if it means doing without some of the skin's other functions. . . ." says Dr. Howard Green.

Eight months after he was burned, Emerson returned to work and began playing softball again—all thanks to tissue-engineering technology. "If I'd been burned just 10 years earlier," he said, "I don't think I would have survived."

 Skill Builder: *Analyzing News Media*

LEARNING the SKILL

1. To get an accurate view of current events and technologies, you must think critically about the news items you hear or read. First, think about the audience. For example, if the item is a magazine article, is the magazine written for scientists or for everyone? If the item is on the radio or television, does the reporter state needed details or does he or she only give a brief summary?

2. Next, try to determine whether the article is true. Were any experts in the field interviewed? Was the item simply reported or was the item also analyzed and interpreted?

3. Finally, ask yourself whether the item presents both sides of the issue. Does the author or reporter of the news item have a bias?

PRACTICING the SKILL

1. From the excerpt given above, what general point is the author of the article trying to make?

2. Were any experts cited or referenced?

3. Were both sides of the issue presented? If not, where might you go to find information about the other side?

APPLYING the SKILL

Find out more about tissue engineering using many different sources. Other books, magazines, newscasts, and even the Internet can be checked for more information. Interview a doctor who does tissue transplants or someone who has had a transplant done.

ing the experience of a burn victim who received such tissue.

2. Yes, an expert named Dr. Howard Green was cited.

3. Different sides of the issue were not presented. Students could find differing opinions by reading other magazine articles, newspaper articles, books, and Internet articles and listening to news programs.

Teaching the Content

- Make a block diagram on the chalkboard that shows the three layers of human skin—the epidermis, dermis, and subcutaneous layers. The epidermis is made of both dead and living cells. The dermis contains blood vessels, nerves, nerve endings, and sweat and oil glands. The subcutaneous layer contains fat cells.

- Have students discuss how scientific advances in tissue engineering benefit medicine.

Teaching the Skill

- Have students read the article. Then, ask the questions presented under *Learning the Skill* and *Practicing the Skill.*

- Obtain the articles listed below. Have students read and analyze them. Have them explain how they differ from the article from *Science World.*

- Allen, Laura, and Susan Benesch. "Grow your own body parts." *Science World,* March 7, 1997, pp. 17+.

- Cowley, Geoffrey. "Replacement parts." *Newsweek,* January 27, 1997, pp. 66-69.

- Edelson, Edward. "The body builders." *Popular Science,* May, 1996, pp. 60+.

- Toufexis, Anastasia. "An early tail." *Time,* November 6, 1995, p. 60.

- Lipkin, Richard. "Tissue engineering: Replacing damaged organs with new tissues." *Science News,* July 8, 1995, pp. 24+.

Answers to Practicing the Skill

1. The author presents the basics of tissue growing, or engineering, while cit-

People & Science

Background

- Modern surgeons use many remarkable tools. Micro-surgery, the technique of observing through a micro-scope while operating on minute structures, involves many tiny, precision surgical instruments. Micro-surgery has made it possible for surgeons to perform delicate operations on the inner ear, eye, and brain, and to reattach limbs and digits by reconnecting severed muscles, tendons, blood vessels, and nerve fibers. Microsurgical techniques also have many applications in research on cells.

- Laser surgery involves using light, in the form of lasers, to perform various surgical procedures. Laser surgery has many advantages over more traditional techniques. There is less blood loss, reduced swelling, and minimal scarring. The chance of infection is also reduced because laser light sterilizes tissue that it passes through. Lasers are extremely precise and can be used on a target organ or cell without damaging surrounding tissues.

- The powerful photon light beams of lasers also enable researchers to "operate" on individual cells. For instance, lasers are used in genetic engineering to manipulate chromosomes and genes to better understand cell functions.

Teaching Strategies

Contact the public relations office of a local hospital and arrange for students to tour an operating room. Have students write a report about the experience. If such a trip is impossible, obtain a similar, previewed video.

84

People & Science

Dr. Khristine Lindo, Surgeon

Q Dr. Lindo, have you always been interested in science?

A I became interested in science in seventh grade when I read a book called *The Making of a Woman Surgeon* by Elizabeth Morgan. That book really inspired me. I realized I wanted to become a physician—a surgeon—and science was the way to reach that goal.

Q What areas of science did you like best?

A Well, in high school, I took every science class I could—biology, chemistry, and physics. I liked biochemistry because it focused on what was happening inside cells.

Q Why is understanding cells important for a surgeon?

A Cells truly are the building blocks of living things. And just like you can't build a house without laying one brick at a time, you also can't manage disease, illness, and especially surgical wounds without knowing a lot about cells. When you know what's going on at the cellular level, you can better understand why patients have certain symptoms, what different medicines do, and how healing occurs.

Q What steps do you take to reduce damage to cells during surgery?

A In any surgery, it's important to be precise in order to limit the damage done to surrounding tissues and cells as much as possible. And it's important to prevent infection before, during, and after surgery. Infection is destructive to cells.

Q What is the most rewarding part of your job?

A Probably when a patient doesn't ever have to see me again! Or when I see someone ten years later who remembers me and is beaming and smiling and well, and I can't remember he or she was ever a patient. Yes, when former patients have fully recovered and are enjoying life—that's the best!

Career Connection

There are many different kinds of medical doctors. Interview a doctor whose area of interest isn't surgery. Find out what areas of medicine he or she is interested in, and then write your own "People & Science" feature.

84 Chapter 3 Cells

For More Information

- Have students contact local doctors and hospitals to find out more about what specializations are available.
- Universities have information available on the educational requirements for careers in medicine.

Career Connection

Career Path Becoming a surgeon is a career choice that requires commitment and dedication, as well as a strong interest in science.

- High School: biology, chemistry (preferably advanced-placement classes)
- College: biology, chemistry, physics
- Medical School: 4 years
- Surgery Residency: 4-5 years of training in general surgery; specialization, such as plastic surgery, can take another 2 years

Read the statements below that review the most important ideas in the chapter. Using what you have learned, answer each question in your Science Journal.

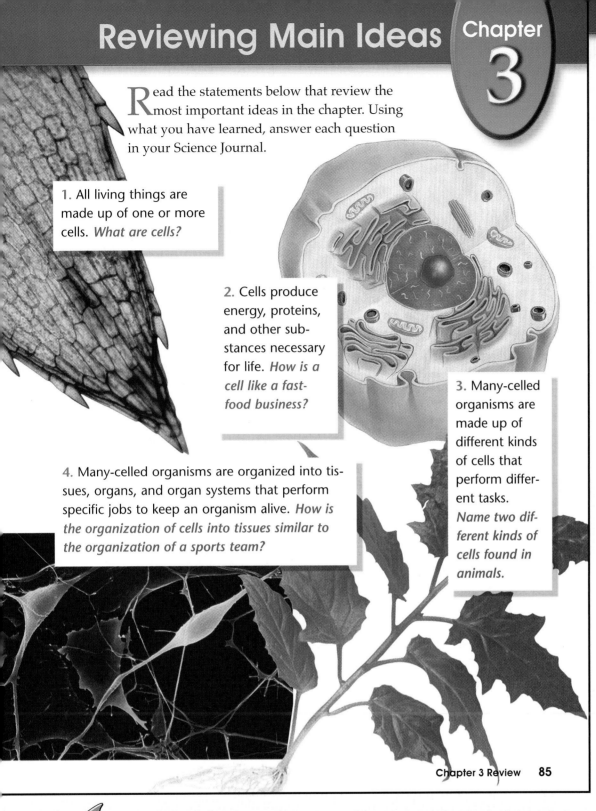

1. All living things are made up of one or more cells. *What are cells?*

2. Cells produce energy, proteins, and other substances necessary for life. *How is a cell like a fast-food business?*

3. Many-celled organisms are made up of different kinds of cells that perform different tasks. *Name two different kinds of cells found in animals.*

4. Many-celled organisms are organized into tissues, organs, and organ systems that perform specific jobs to keep an organism alive. *How is the organization of cells into tissues similar to the organization of a sports team?*

Chapter 3 Review **85**

Have students look at the four illustrations on this page. Ask them to describe details that support the main ideas of the chapter found in the statement for each illustration.

Teaching Strategies

You may wish to review each of the illustrations together as a class, discussing each statement and answering the questions as you proceed.

Answers to Questions

1. Cells are the smallest parts of living things to carry out life activities.
2. Like fast-food businesses, cells produce things. Cells are also made up of individual parts that work together and function as a whole.
3. Answers may include muscle cells, skin cells, blood cells, or nerve cells.
4. Tissues are made up of similar cells all working together. Sports teams are made up of individual players who work together as a team.

Science at Home

LS Intrapersonal During photosynthesis, plants trap energy in sunlight and convert it into chemical energy stored in food. What effect does light intensity have on plant growth? Encourage interested students to research light intensity and then design an experiment at home to answer this question. **L2**

Assessment

Portfolio Encourage students to place in their portfolios one or two items of what they consider to be their best work. Examples include:
- MiniLAB, p. 69
- Activity 3-1, p. 71
- Using Computers, p. 73 **P**

Performance Additional performance assessments may be found in **Performance Assessment** and **Science Integration Activities.** Performance Task Assessment Lists and rubrics for evaluating these activities can be found in Glencoe's **Performance Assessment in the Science Classroom (PASC).**

Using Key Science Words

1. chloroplast
2. photosynthesis
3. nucleus
4. mitochondrion
5. tissue

Checking Concepts

6. b
7. b
8. a
9. c
10. a

Thinking Critically

11. The cell would die because it couldn't get water, food, or other materials, and it couldn't release wastes.

12. The plant cell would not be able to make food for energy, and it would die.

13. All of life's activities take place within the cell of single-celled organisms. In many-celled organisms, cells work together as a team to meet all of the life activities.

14. Cells with few mitochondria present include cells with low energy requirements, such as skin cells.

15. The cell is a plant cell because it contains chloroplasts.

Chapter 3 Review

Using Key Science Words

cell membrane
chloroplast
mitochondrion
nucleus
organ
organ system
photosynthesis
tissue

Match each phrase with the correct term from the list of Key Science Words.

1. cell part that traps energy from sunlight
2. food-making process in plants
3. controls most of the cell's activities
4. provides power for the cell's activities
5. group of similar cells working together

Checking Concepts

Choose the word or phrase that completes the sentence.

6. The _____ is the cell part that controls what enters and exits the cell.
   a. mitochondrion   c. vacuole
   b. cell membrane   d. nucleus

7. _____ are found inside the nucleus of the cell.
   a. Vacuoles   c. Chloroplasts
   b. Chromosomes   d. Mitochondria

8. _____ is the gel-like substance that fills the cell.
   a. Cytoplasm   c. Chromosome
   b. Tissue   d. Mitochondrion

9. Groups of different tissues working together form an _____, such as the stomach.
   a. organelle
   b. organ system
   c. organ
   d. organism

10. Photosynthesis is a process that makes _____ for cells.
    a. food   c. energy
    b. organs   d. tissues

Thinking Critically

Answer the following questions in your Science Journal using complete sentences.

11. What would happen to a cell if the cell membrane were solid and watertight?

12. What might happen to a plant cell if all its chloroplasts were removed? Explain.

13. Why are cells called the "building blocks of life"?

14. What cells in an animal may not have a lot of mitochondria present?

15. Identify the cell shown to the right as an animal cell or a plant cell. Explain.

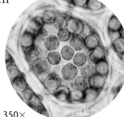

Magnification: 350×

Assessment Resources

Reproducible Masters
Chapter Review, pp. 9-10
Assessment, pp. 13-16
Performance Assessment, p. 41

Glencoe Technology
- Computer Test Bank
- MindJogger Videoquiz

Developing Skills

If you need help, refer to the description of each skill in the Skill Handbook.

16. **Comparing and Contrasting:** Compare and contrast photosynthesis and cellular respiration.

17. **Making and Using Tables:** Make a table that lists the functions of each cell part below.

Cell Part	Function
Nucleus	Controls most of the cell's activities
Cell membrane	Helps control what enters and leaves the cell
Mitochondrion	Releases the energy the cell needs to carry out its activities
Chloroplast	Traps energy from the sun

18. **Observing and Inferring:** Why is the bricklike shape of some plant cells important?

19. **Designing an Experiment:** Describe an experiment you would do to determine how different temperatures affect how water is moved through a plant.

20. **Making and Using Graphs:** Sunlight is necessary for most plants to make food. Using the graph below, determine which plant produced the most food. How much sunlight was needed every day to produce the most food?

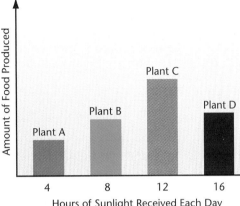

Performance Assessment

1. **Skit:** Working with three or four classmates, develop a short skit about how a living cell works. Have each group member play the role of a different cell part.

2. **Model:** Using assorted household materials such as clay, cardboard boxes, Styrofoam, pipe cleaners, yarn, buttons, dry macaroni, or other objects, make a three-dimensional model of an animal or plant cell large enough that you can walk into. On a separate sheet of paper, make a key to your model cell. On your key, indicate which materials represent the different cell parts.

Chapter 3 Review 87

Developing Skills

16. **Comparing and Contrasting** Photosynthesis and cellular respiration appear to be opposite processes. The starting materials of one process are the materials in resulting products.

17. **Making and Using Tables** See student page for answers.

18. **Observing and Inferring** Answers may vary, but students should recognize that many plants are tall and need strong, rigid cells for support.

19. **Designing an Experiment** Answers will vary, but many students will describe an experiment similar to Activity 3-2. Students will make various suggestions about how to manipulate temperature, such as the use of heaters, heating lamps, and natural sunlight. Accept all logical experimental designs.

20. **Making and Using Graphs** According to the graph, Plant C produced the most sugar. Twelve hours of sunlight were needed every day.

Performance Assessment

1. You can assess student skits by observing how detailed and informative the different characters' roles are. Use the Performance Task Assessment List for Skit in **PASC**, p. 75. **P**

2. Check models for creativity, content, and accuracy. Use the Performance Task Assessment List for Model in **PASC**, p. 51. **P**

Chapter Organizer

Section	Objectives/Standards	Activities/Features
Chapter Opener		Explore Activity: Comparing Types of Seeds, p. 89
4-1 **Continuing Life** (4 sessions, 1½ blocks)*	1. **Explain** how and why cells divide. 2. **Discuss** the importance of reproduction for living things. 3. **Compare and contrast** sexual and asexual reproduction. National Science Content Standards: (5-8) UCP5, A1, C2, C5, F5	MiniLAB: Yeast Budding, p. 92 Activity 4-1: Plant Cuttings, pp. 94-95 Using Technology: The Millennium Seed Bank, p. 98 Skill Builder: Using a CD-ROM, p. 99 Using Math, p. 99
4-2 **Science and Society:** **Cloning** (1½ sessions, 1 block)*	4. **Explain** how cloning is done and discuss possible benefits of cloning research. 5. **State** main ideas in science articles. National Science Content Standards: (5-8) A2, E2, F5, G1	Skill Builder: Identifying the Main Idea, p. 101
4-3 **Genetics—The Study** **of Inheritance** (3½ sessions, 2 blocks)*	6. **Describe** how traits are inherited. 7. **Describe** the structure and function of DNA. 8. **Identify** how mutations add variations to a population. National Science Content Standards: (5-8) A1, C1, C2, C5, E2, F5, G1	Activity 4-2: Getting DNA from Onion Cells, p. 104 Problem Solving: Analyzing a DNA Fingerprint, p. 10 MiniLAB: Probability, p. 110 Skill Builder: Concept Mapping, p. 111 Science Journal, p. 111 Science & History: A Horse for Every Job, p. 112

* A complete Planning Guide that includes block scheduling is provided on pages 31T-33T.

Activity Materials

Explore	Activities	MiniLABs
page 89 hand lens, seeds	pages 94-95 small planting container, metric ruler, potting soil, spoon, watering can, paper cups, permanent marker, masking tape, cuttings from houseplant page 104 transparent container, funnel, hand lens or microscope, measuring cup, rubbing alcohol, coffee filter, meat tenderizer, onion, table salt, liquid soap, toothpicks, cup	page 92 coverslips, jar or beaker, microscope, slides, sugar, dry yeast, dropper, very warm water, glass rod for stirring page 110 coins

Need Materials? Call Science Kit (1-800-828-7777).

Teacher Classroom Resources

Reproducible Masters	Transparencies	Teaching Resources
Activity Worksheets, pp. 5, 25-26, 29 **Enrichment**, p. 15 **Multicultural Connections**, pp. 11-12 **Reinforcement**, p. 15 **Study Guide**, p. 15	**Section Focus Transparency 9**, Babies, Babies, Babies **Teaching Transparency 7**, Mitosis	**Spanish Resources** **English/Spanish Audiocassettes** **Cooperative Learning Resource Guide** **Lab Partner** **Lab and Safety Skills** **Lesson Plans**
Enrichment, p. 16 **Reinforcement**, p. 16 **Study Guide**, p. 16	**Section Focus Transparency 10**, Leapin' Lizards	**Assessment Resources** **Chapter Review**, pp. 11-12 **Assessment**, pp. 17-20 **Performance Assessment**, p. 42 **Performance Assessment in the Science Classroom (PASC)** **MindJogger Videoquiz** **Alternate Assessment in the Science Classroom** **Computer Test Bank**
Activity Worksheets, pp. 5, 27-28, 30 **Cross-Curricular Integration**, p. 8 **Enrichment**, p. 17 **Lab Manual 7, 8** **Reinforcement**, p. 17 **Science and Society/ Technology Integration**, p. 22 **Science Integration Activities**, pp. 45-46 **Study Guide**, p. 17	**Science Integration Transparency 4**, Radiation Protection **Section Focus Transparency 11**, Generation to Generation **Teaching Transparency 8**, Passing Traits to Offspring	

Key to Teaching Strategies

The following designations will help you decide which activities are appropriate for your students.

L1 Level 1 activities should be appropriate for students with learning difficulties.

L2 Level 2 activities should be within the ability range of all students.

L3 Level 3 activities are designed for above-average students.

ELL ELL activities should be within the ability range of English Language Learners.

LS These activities are designed to address different learning styles.

COOP LEARN Cooperative Learning activities are designed for small group work.

P These strategies represent student products that can be placed into a best-work portfolio.

GLENCOE TECHNOLOGY

The following multimedia resources are available from Glencoe.

Science and Technology Videodisc Series (STVS)
Plants & Simple Organisms
  Seed Banks
  Plant Clones
  Selective Breeding in Cows
The Infinite Voyage Series
The Geometry of Life
The Dawn of Humankind

Glencoe Life Science CD-ROM
Glencoe Life Science Interactive Videodisc
Introduction
Chromosomes
Mitosis Preparation

Teacher Classroom Resources

This is a representation of key blackline masters available in the Teacher Classroom Resources.

Teaching Aids

Section Focus Transparencies

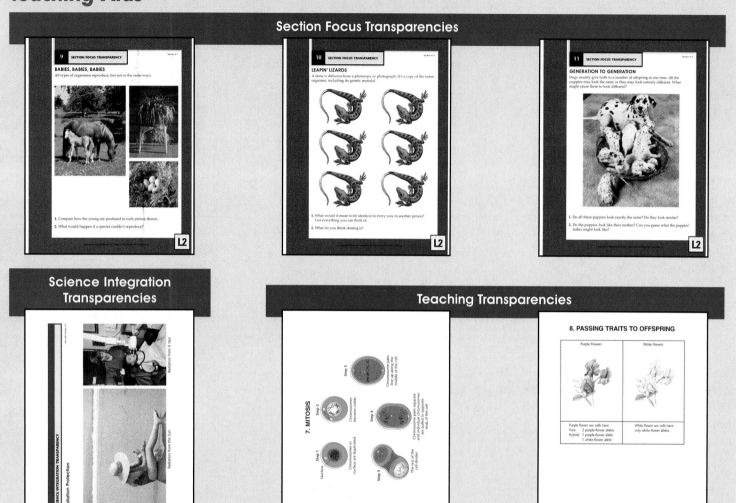

9 SECTION FOCUS TRANSPARENCY Section 4-1

BABIES, BABIES, BABIES
All types of organisms reproduce, but not in the same ways.

1. Compare how the young are produced in each picture shown.
2. What would happen if a species couldn't reproduce?

L2

10 SECTION FOCUS TRANSPARENCY Section 4-2

LEAPIN' LIZARDS
A clone is different from a photocopy or photograph. It's a copy of the entire organism, including its genetic material.

1. What would it mean to be identical in every way to another person? List everything you can think of.
2. What do you think cloning is?

L2

11 SECTION FOCUS TRANSPARENCY Section 4-3

GENERATION TO GENERATION
Dogs usually give birth to a number of offspring at one time. All the puppies may look the same or they may look entirely different. What might cause them to look different?

1. Do all these puppies look exactly the same? Do they look similar?
2. Do the puppies look like their mother? Can you guess what the puppies' father might look like?

L2

Science Integration Transparencies

SCIENCE INTEGRATION TRANSPARENCY
Radiation Protection

L2

Teaching Transparencies

7. MITOSIS

Step 1 — Nucleus — Chromosomes in nucleus are duplicated
Step 2 — Chromosomes become visible
Step 3 — Chromosome pairs line up along the middle of the cell
Step 4 — Chromosome pairs separate and individual chromosomes are pulled to opposite end of the cell
Step 5 — The rest of the cell divides

L2

8. PASSING TRAITS TO OFFSPRING

Purple Flowers	White Flowers
Purple flower sex cells have Pure: 2 purple-flower alleles Hybrid: 1 purple-flower allele 1 white-flower allele	White flower sex cells have only white-flower alleles.

L2

Meeting Different Ability Levels

Study Guide for Content Mastery

Chapter 4 Study Guide for Content Mastery

Continuing Life

Directions: Answer the following questions on the lines provided.

1. What is the method of reproduction in which two parents produce a new organism?
2. What is the division of one nucleus into two nuclei called?
3. Where are chromosomes located in the cell?
4. What is the method of reproduction in which one parent produces a new organism?
5. What do we call the process that forms sex cells?

Directions: The diagram shows the process of sexual reproduction as the cycle of life. Label each stage with one of the following words.

development   fertilization   meiosis

I. In this stage, sperm and egg cells are produced.
II. In this stage, sperm and egg cells join.
III. In this stage, the single cell divides into more cells, becoming a fully formed organism.

L1

Reinforcement

Chapter 4 Reinforcement

Continuity of Life

Directions: Determine whether the italicized term makes each statement true or false. If the statement is true, write the word true on the line provided. If the statement is false, write in the blank the term that makes the statement true.

1. In asexual reproduction, offspring receive DNA from two parents.
2. In the process of mitosis, a single cell divides into two identical cells.
3. Growing a new plant from a cutting taken from an existing plant is an example of sexual reproduction.
4. Most cells in the human body contain chromosomes. In human sex cells, however, there are half that number, or 23 chromosomes.
5. In mitosis, the number of chromosomes in a cell is divided in half.
6. During the process of mitosis, a sperm cell and an egg cell join to form a new cell.

Directions: Answer the following questions on the lines provided.

7. Explain how the offspring resulting from asexual reproduction are similar to or different from the parents.
8. Explain how the offspring resulting from sexual reproduction are similar to or different from the parents.

L2

Enrichment Worksheets

Chapter 4 Enrichment

DNA

Hereditary Information

Directions: Read the following information. Then answer the questions.

DNA, or deoxyribonucleic acid, holds the blueprint for all living organisms. Humans and other animals, plants, single-celled organisms, and even some viruses need DNA to carry the information needed to reproduce themselves and to function.

Variation in Simple Traits
DNA is responsible not only for reproduction, but also for variation. Variation is what makes organisms different from one another. For example, DNA is responsible for the color of your eyes and hair, the shape of your nose, and certain other simple physical traits. Your DNA is one of the things that makes you unique.

DNA's Shape
The DNA molecule is shaped in a long double strand. DNA is tightly packaged inside cells. If you could take a strand of DNA out of a cell and stretch it out, it would be over two meters long.
The two strands wind around each other like a spiral ladder. Four kinds of molecules, called bases, make up the rungs of the ladder. The bases are adenine, thymine, guanine, and cytosine. An adenine molecule can only attach to a thymine molecule, and guanine can only attach to cytosine. Base pairs are arranged in a unique order that codes instructions for cell function.

Applying Critical Thinking Skills
1. In what way is DNA important to all living things?
2. Suppose you find one-half of a strand of DNA with base pairs in the sequence ATGACTGCCGTA. What would be the sequence of the missing half?

L3

88C

Chapter 4 Heredity and Reproduction

Hands-On Activities

Science Integration Activities

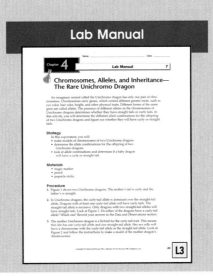

Lab Manual

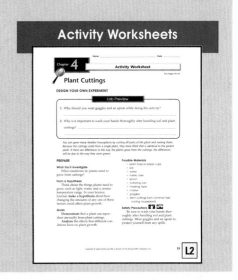

Activity Worksheets

Enrichment and Application

Cross-Curricular Integration

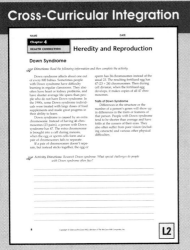

Multicultural Connections

Science and Society/Technology Integration

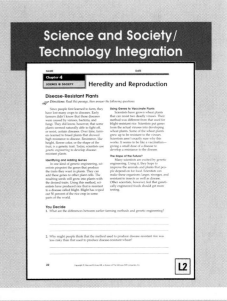

Assessment

Performance Assessment

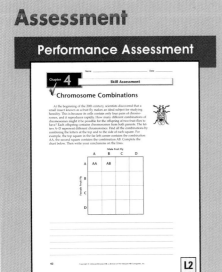

Chapter Review

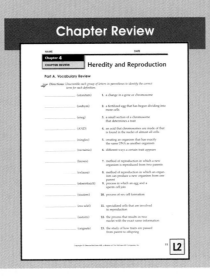

Assessment

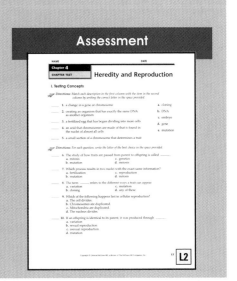

Heredity and Reproduction

CHAPTER OVERVIEW

Section 4-1 The purpose of this section is to introduce students to the process of reproduction in organisms. Reproduction is essential for the continuation of the species. Students will first learn about examples of asexual reproduction. Next, students will examine the process of sexual reproduction and learn how it differs from asexual reproduction.

Section 4-2 Science and Society In this section, students, while practicing the skill of identifying the main idea in an article, will learn about the science of cloning. This controversial topic made headlines in 1997 when scientists successfully cloned a mammal.

Section 4-3 In the final section, students are introduced to genetics. Students first learn about DNA and discover how DNA structure is related to its ability to carry genetic information. Next, students learn about heredity and discover how traits are passed from parent to offspring.

Chapter Vocabulary

mitosis	cloning
asexual	embryo
reproduction	genetics
sexual	gene
reproduction	DNA
sex cell	variation
meiosis	mutation
fertilization	

Theme Connection

Stability and Change Although Earth's millions of species show many physical and behavioral differences, all living things must reproduce if their species is to continue.

Chapter Preview

Skills Preview

▶ **Skill Builders**
- sequence
- map concepts
- use a CD-ROM

▶ **MiniLABs**
- observe
- predict

▶ **Activities**
- observe
- compare and contrast
- hypothesize
- collect data

88

Learning Styles

Look for the following logo for strategies that emphasize different learning modalities. **LS**

Kinesthetic	Explore, p. 89; Inclusion Strategies, pp. 93, 103; Reteach, pp. 98, 109; Activity 4-2, p. 104; MiniLAB, p. 110
Visual-Spatial	MiniLAB, p. 92; Activity 4-1, p. 94; Visual Learning, pp. 96, 103; Demonstration, p. 97; Assessment, p. 99
Interpersonal	Theme Connection, p. 97; Brainstorming, p. 108
Intrapersonal	Community Connection, p. 92; Using Technology, p. 98; Science at Home, p. 113
Logical-Mathematical	Inquiry Question, p. 91; Assessment, p. 104; Across the Curriculum, p. 105
Linguistic	Science Journal, pp. 92, 107; Assessment, pp. 95, 110; Using Science Words, pp. 97, 103

Chapter 4

Heredity and Reproduction

Dandelions! Those plants seem to pop out of thin air every spring, showing up in lawns, along roadsides, and even in sidewalk cracks. In just a few days, a fresh, green patch of grass can become blanketed with yellow dandelion flowers. Soon, the flowers turn to fluffy puffballs that bear the plants' seeds—thousands of future dandelions. For dandelions—and all living things—reproduction (ree proh DUK shun) is an important process. For life to continue, living things must produce more living things like themselves.

EXPLORE ACTIVITY

Comparing Types of Seeds

1. Obtain seeds from your teacher, such as pinto beans, dandelion seeds, corn, burrs, beans, and acorns.
2. Using a hand lens, study each seed. Make a list of the characteristics of each seed in your Science Journal.

Science Journal

Draw and label a picture of each type of seed in your Science Journal. Why do you think the seeds are shaped differently?

89

Assessment Planner

Portfolio
Refer to page 113 for suggested items that students might select for their portfolios.

Performance Assessment
See page 113 for additional Performance Assessment options.
Skill Builders, pp. 99, 101, 111
MiniLABs, pp. 92, 110
Activities 4-1, pp. 94-95; 4-2, p. 104

Content Assessment
Section Wrap-ups, pp. 99, 111
Chapter Review, pp. 113-115
Mini Quizzes, pp. 99, 111

Group Assessment
Opportunities for group assessment occur with Cooperative Learning Strategies and Flex Your Brain activities.

EXPLORE ACTIVITY

Purpose

LS **Kinesthetic** In this activity, students compare and contrast seeds from different plants. **L2** **ELL**

Preparation

Collect seeds with obvious dispersal mechanisms. Different types of beans can be obtained easily from the grocery store. Garden centers will have packets of many different types of seeds.

Materials

seeds of different plants, such as pinto beans, corn seeds, maple seeds, milkweed, and acorns; hand lenses

Teaching Strategies

Ask students if they have ever seen how a maple seed falls to the ground by spinning. Demonstrate this if maple seeds are available. Explain to students that seeds have different ways of being spread or different dispersal mechanisms.

Science Journal Students should recognize that the shape and structure of a seed is related to its method of dispersal. **P**

✓ Assessment

Performance Have students work alone or in teams to design seeds with unique dispersal mechanisms. Encourage students to construct seeds out of household items such as pipe cleaners, paper clips, or different types of paper. Use the Performance Task Assessment List for Invention in **PASC**, p. 45.

Prepare

Section Background

- Cells make more cells through the process of mitosis. Mitosis is cell division. During this process, two daughter cells are produced from a single parent cell. Daughter cells are identical to each other and to the parent cell.

- Asexual reproduction may occur by binary fission, stem cuttings, leaf cuttings, root divisions, cloning, budding, or by fragmentation. Each method results in offspring that are genetically identical to the parent and to one another.

Preplanning

Refer to the Chapter Organizer on pages 88A-B.

1 Motivate

Bellringer

 Before presenting the lesson, display **Section Focus Transparency 9** on the overhead projector. Assign the accompanying **Focus Activity** worksheet.

L2 ELL

What YOU'LL LEARN

- How cells divide
- The importance of reproduction for living things
- The differences between sexual and asexual reproduction

Science Words:
mitosis
asexual reproduction
sexual reproduction
sex cells
meiosis
fertilization

Why IT'S IMPORTANT

You'll understand how a living thing can inherit characteristics from its parents.

Cell Division

Have you ever seen frog or toad eggs developing in a pond? Frogs and toads lay eggs by the hundreds in gooey clumps, as in **Figure 4-1.** Some other kinds of organisms, including humans, usually produce only one offspring at a time.

Millions of different kinds of living things inhabit Earth. New organisms are the result of the process of reproduction. Reproduction is important to all living things. Without reproduction, a particular species could not continue. During reproduction, information is passed on from parent to offspring. This information is contained in a chemical called DNA found inside the nucleus of cells. DNA controls what the new individuals will look like and how their bodies work.

How did you grow from a baby to the size you are now? Most of the cells that make up organisms form new cells by dividing into two. All organisms are made up of cells, and all cells come from other cells. As you developed from an infant to a preschooler and then to the size you are now, you grew because your cells divided. Before a cell divides into two, all the information contained in the chromosomes of that cell is copied.

FIGURE 4-1

These frog eggs will hatch into free-swimming tadpoles about six days after fertilization.

90   Chapter 4 Heredity and Reproduction

Program Resources

📁 **Reproducible Masters**
Activity Worksheets, pp. 5, 25-26, 29 L2
Enrichment, p. 15 L3
Multicultural Connections, pp. 11-12 L2
Reinforcement, p. 15 L2
Study Guide, p. 15 L1

🔦 **Transparencies**
Section Focus Transparency 9 L2
Teaching Transparency 7 L2

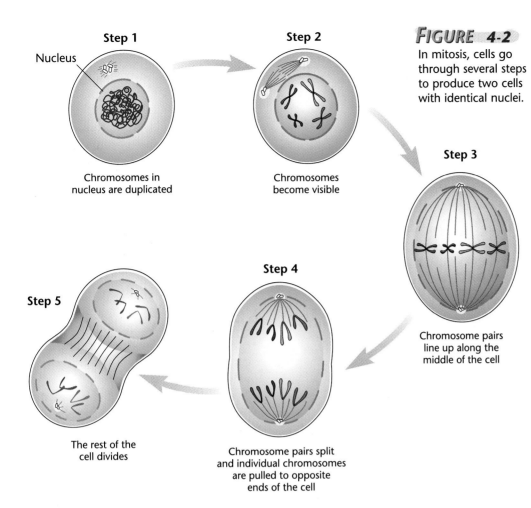

Step 1

Nucleus

Chromosomes in nucleus are duplicated

Step 2

Chromosomes become visible

FIGURE 4-2

In mitosis, cells go through several steps to produce two cells with identical nuclei.

Step 3

Chromosome pairs line up along the middle of the cell

Step 4

Chromosome pairs split and individual chromosomes are pulled to opposite ends of the cell

Step 5

The rest of the cell divides

This way, each new cell has exactly the same information as the cell it came from.

First, the chromosomes, which are located in the nucleus (NEW klee us), are duplicated. Then the nucleus divides into two. Each new nucleus receives a copy of the same chromosomes. This division of the nucleus is called mitosis (mi TOH sus). **Mitosis** is the process that results in two nuclei with the exact same information. You can see the process in **Figure 4-2**. Once the nucleus divides, the rest of the cell divides into two cells of equal size. Almost all the cells in any plant or animal divide through the process of mitosis. Cells divide for growth and to replace aging or injured cells.

4-1 Continuing Life **91**

91

Mini LAB

Yeast Budding

1. Fill a small jar or beaker halfway with very warm water.
2. Add a pinch of yeast—the amount you can pinch between your thumb and index finger—to the water. Add the same amount of sugar. Stir. Wash your hands thoroughly.
3. Use a dropper to place a drop of the yeast mixture onto a microscope slide. Add a coverslip.
4. Examine the slide under low power, then high power. Make a new slide after five minutes. Record your observations in your Science Journal.

Analysis

1. What did you observe on the first microscope slide? The second?
2. Why do you think it was necessary to add sugar?

Reproduction from One Parent

Have you ever seen shoots develop from the eyes of a potato as shown in **Figure 4-3?** The method of reproduction in which an organism can produce a new organism from one parent is called **asexual reproduction.** In asexual (ay SEK shul) reproduction, all the DNA in the new organism comes from one parent organism. In the case of the potato, the DNA of the growing eye is the same as the DNA in the cells of the parent potato.

Let's Split!

There are several different types of asexual reproduction. Some organisms, such as bacteria and other single-celled organisms, divide in half, forming two cells. The DNA in a bacterium cell is copied. When the cell divides, each new cell gets an exact copy of the parent's DNA. The two new cells are exactly alike. The parent cell no longer exists.

Budding Out and Breaking Up

Microscopic organisms, such as bacteria, are not the only species that reproduce asexually. Many species of mushrooms, plants, and even a few animals have some type of asexual reproduction. **Figure 4-4A** shows asexual reproduction in *Hydra,* a relative of jellyfish and corals. When *Hydra* reproduces asexually, a new individual develops from the parent by a process called budding. As you can see, the *Hydra* bud has the same shape and characteristics as its parent. The bud matures and will eventually break away from the parent and live on its own.

FIGURE 4-3

Many plants in addition to potatoes reproduce asexually. Some of these are cattails, strawberries, and creosote bushes.

In some species of organisms, a whole individual can grow from just a part of a parent organism, as you can also see in **Figure 4-4.** Cuttings taken from plants such as spider plants and ivy develop into whole new plants. Gardeners take advantage of this when they take cuttings from the stems, leaves, or roots of plants. They can grow many plants from one.

In a process called regeneration (ree jen uh RAY shun), some organisms are able to replace body parts that have been lost because of an injury. Lizards, such as chameleons, grow a new tail if theirs is broken off. In sea stars, regeneration is a form of asexual reproduction. If an arm is removed from a sea star, the sea star will grow a new arm. But the arm that was removed can often develop into a whole new sea star, too.

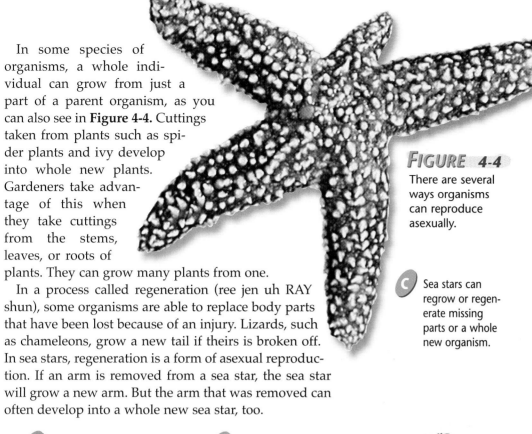

FIGURE 4-4

There are several ways organisms can reproduce asexually.

C Sea stars can regrow or regenerate missing parts or a whole new organism.

A *Hydra* reproduce asexually by budding.

B Some plants can grow from a part of a parent organism.

93

Purpose

LS **Visual-Spatial** The purpose of this activity is to teach students about the effects of the environment on living organisms. Students obtain plant cuttings from a single plant, then grow the cuttings under different environmental conditions. Because the cuttings come from the same plant, all of the cuttings are genetically identical to each other. Any differences observed in the way the plants grow must be due to environmental and not genetic differences in the plants. **L2**

COOP LEARN **ELL**

Process Skills

making and using tables, observing and inferring, recognizing cause and effect, designing an experiment to test a hypothesis, separating and controlling variables, interpreting data

Time

Allow 40 to 50 minutes for groups to set up the experiment. Students will monitor the growth of plants over a two-week period.

Materials

Paper cups can be used instead of plant trays.

Safety Precautions

If grow lights are used, be sure to instruct students to use caution when handling water near the lights.

Possible Hypotheses

Most students will hypothesize that plants will grow best when they have adequate soil or material containing nutrients, light, and water.

Activity Worksheets, pages 5, 25-26

Activity 4-1

Design Your Own Experiment
Plant Cuttings

You can grow many familiar houseplants by cutting off parts of the plant and rooting them. Because the cuttings come from a single plant, all of the cells in the plant have DNA that is identical to the parent plant. If there are differences in the way the plants grow from the cuttings, the differences will be due to the conditions under which they were grown.

Possible Materials

- plant trays or paper cups
- soil
- water
- metric ruler
- spoon
- watering can
- masking tape
- marker
- goggles
- stem cuttings from common fast-rooting houseplants

PREPARE

What You'll Investigate
What conditions do plants need to grow from cuttings?

Form a Hypothesis

Think about the things plants need to grow, such as light, water, and a certain temperature range. In your Science Journal, **make a hypothesis** about how changing the amounts of any one of these factors could affect plant growth.

Goals

Demonstrate that a plant can reproduce asexually from plant cuttings.

Analyze the effects that different conditions have on plant growth.

Safety Precautions

Be sure to wash your hands thoroughly after handling soil and plant cuttings. Wear goggles and an apron to protect yourself from any spills.

Sample Data Table for Amount of Water as a Variable

Water added per day (mL)	1	5	10
Starting height of plant (cm)			
Plant 1	4	5	6
Plant 2	5	6	5
Plant 3	4	5	4
Final height of plant (cm)			
Plant 1	6	7	7
Plant 2	6	9	5
Plant 3	5	8	5

PLAN

1. As a group, agree upon and write a hypothesis.
2. What steps will you take to **test** your hypothesis? With your group, brainstorm possible experiments.
3. **Choose** the best possible way to test your hypothesis. **Make a list** of the steps you will take for the experiment your group will perform. Be specific and describe exactly what you will do at each step.
4. Which environmental factor will you study? Be sure you choose only one factor at a time to study. You might test the number of hours a day the cutting is in light or different types of water—distilled or tap.
5. What kinds of data will you collect? Will you take measurements? How often will you collect data?
6. **Design a data table** that clearly organizes the data from your experiment.
7. Be sure all the cuttings are from the same plant.
8. As a group, review your experimental plan to make sure that all steps are in a logical order.
9. How will you present the results of your experiment? Will you use charts, graphs, photos, or drawings?

DO

1. Make sure your teacher approves your plan.
2. Carry out your plan.
3. As you do the experiment, be sure to write down all observations in your data table.

CONCLUDE AND APPLY

1. Which environmental factor did you study?
2. **Compare** any differences in the plants grown. What differences in the conditions affected the growth of the cuttings?
3. What type of reproduction did these plants have?
4. **APPLY** How would the owner of a greenhouse use information about plant cuttings and the conditions needed to grow them?

4-1 Continuing Life   95

Go Further

You may wish to have interested students extend the experiment to include more complex environmental variables, such as air temperature and humidity.

✔ Assessment

Performance Have students prepare detailed laboratory reports about this activity. Reports should include purpose, materials used, hypotheses tested, procedures, data obtained, results, and conclusions. Use the Performance Task Assessment List for Lab Report in **PASC**, p. 47. [LS] [P]

PLAN

Possible Procedures
Manipulate simple environmental variables, such as amount of light, amount of water, or differences in water quality. Students should choose only one variable to test.

Teaching Strategies
Tying to Previous Knowledge
Before doing this activity, you may wish to review with students how plants respond to changes in the environment. This discussion will help groups form useful hypotheses and may assist them with their observations.

Troubleshooting Small uniform stem cuttings will work best for this activity. Make sure the planting trays used in the activity are large enough.

DO

Expected Outcome
If the experiment is done correctly and accurately, students will see differences in plant growth, plant height, and overall plant health. Other differences may be growth of roots and leaf color. Students should examine all parts of the plant, including roots.

CONCLUDE AND APPLY

1. Answers will vary, but most groups will investigate differences in light, water, and soil.
2. Answers will vary but groups should see differences in the plants.
3. asexual reproduction
4. They could apply this information in growing plants most economically for their business.

FIGURE 4-5

A human egg cell (top) and human sperm cells (bottom) each contain 23 chromosomes.

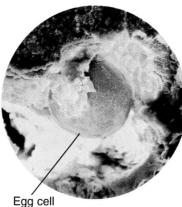

Egg cell

Magnification: 1850×

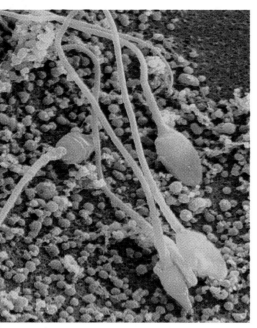

Magnification: 12 000×

Sexual Reproduction

Does a new baby look exactly like its father or its mother? Most likely, the baby has features of both of its parents, or it may even look a lot like one of its grandparents. The baby might have her dad's nose shape and hair color, and her mom's eyes and chin. But the baby probably doesn't look exactly like either of her parents. Each person is an individual. That's because humans, as well as many other organisms, are a product of sexual (SEK shul) reproduction. In **sexual reproduction**, a new organism is produced from two parents. During this process, DNA from both parents is combined to form a new individual with its own DNA.

Production of Sex Cells

In Chapter 3, you learned that your body is made up of different types of cells. Specialized cells called **sex cells** are involved in reproduction. They are shown in **Figure 4-5.** Female sex cells are called eggs and male sex cells are called sperm.

Sex cells are like most other cells in your body. They contain chromosomes, which are located inside the nucleus of the cell. Each human body cell, such as a skin cell or muscle cell, has 23 pairs of chromosomes for a total of 46 chromosomes. For each chromosome from one parent there is a matching chromosome from the other parent. This results in the total of 46 chromosomes in each cell.

When a skin cell or other body cell divides, it produces two new cells, each with this same number of chromosomes. Sex cells, however, are produced in a different way. Instead of dividing by mitosis like other cells in the body, certain cells in the reproductive organs undergo meiosis (mi OH sus). **Meiosis** is the process of sex cell formation. During meiosis, the chromosome number in each sex cell becomes half. Human eggs and sperm contain

96 Chapter 4 Heredity and Reproduction

only 23 chromosomes each, one chromosome from each pair of chromosomes. This is half as many chromosomes as body cells contain. That way, when an egg and a sperm join in a process called **fertilization,** the result is a new individual with a full set of 46 chromosomes. **Figure 4-6** shows how cells are fertilized.

FIGURE 4-6

Many animals and plants produce offspring through sexual reproduction.

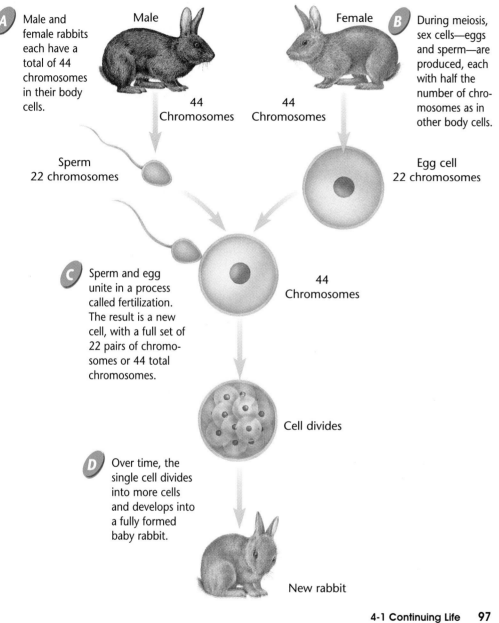

A Male and female rabbits each have a total of 44 chromosomes in their body cells.

Male

Female

B During meiosis, sex cells—eggs and sperm—are produced, each with half the number of chromosomes as in other body cells.

44 Chromosomes

44 Chromosomes

Sperm 22 chromosomes

Egg cell 22 chromosomes

C Sperm and egg unite in a process called fertilization. The result is a new cell, with a full set of 22 pairs of chromosomes or 44 total chromosomes.

44 Chromosomes

Cell divides

D Over time, the single cell divides into more cells and develops into a fully formed baby rabbit.

New rabbit

4-1 Continuing Life   97

Demonstration

Visual-Spatial Explain to students that different species of organisms have different numbers of chromosomes. List the following organisms and their chromosome numbers on the chalkboard or overhead projector as you teach about the process of meiosis: dog (78 chromosomes); rabbit (44); strawberry (56); pig (40); monkey (48); fly (12); chicken (78); horse (66); pea plant (14).

Using Science Words

Linguistic Cells having pairs of chromosomes are said to be *diploid* cells. The diploid number of chromosomes is represented by $2n$, where $n =$ the number of different pairs. Humans have 23 different pairs of chromosomes, so $2n =$ 46. Sex cells are said to be *haploid*. The haploid number is represented by n. In humans, $n = 23$. Haploid cells cannot be produced by mitosis. Rather, they are produced by the process of meiosis. Unlike mitosis, which occurs in most kinds of body tissues, meiosis occurs only in particular reproductive tissues. Meiosis ensures that gametes will contain the right number and kinds of chromosomes. As a result, the zygote formed by the union of the gametes will also contain the proper number and kinds of chromosomes.

Theme Connection

Energy Point out to students that the production of reproductive cells and the process of reproduction require the use of energy by organisms. Have students brainstorm all the activities performed by organisms each day that require energy. They will find that all activities do.

COOP LEARN

GLENCOE TECHNOLOGY

 Videodisc

STVS: Plants & Simple Organisms
Disc 4, Side 2
Seed Banks (Ch. 10)

3 Assess

Check for Understanding

Discussion Review the illustration of sexual reproduction in Figure 4-7 with your students. Carefully study each step of the process as you review the diagram. Ask students questions about how sexual reproduction compares to asexual reproduction. Ask students to list the similarities and differences. **L2** **P**

Reteach

 Kinesthetic Students who have difficulty understanding mitosis may benefit by manipulating models of chromosomes made out of pipe cleaners. Demonstrate to students how they can model the process of mitosis using pipe cleaners. **L1** **ELL**

Extension

For students who have mastered this section, use the **Reinforcement** and **Enrichment** masters.

98

USING TECHNOLOGY
The Millennium Seed Bank

In 1996, researchers at the Royal Botanic Gardens near London, England, began the Millennium Seed Bank. The Seed Bank is a project they hope will protect the future of more than 25 000 species of plants around the world. When the seed bank project is completed in the year 2000, it will be the world's largest storage and research facility for seeds. In a seed bank, large amounts of plant seed are stored in cold underground vaults, where they can last for hundreds of years.

Why save seeds?

Within 50 years, one quarter of the world's plants could disappear from the planet due to loss of habitat, pollution, or for other reasons. If this happens, other organisms that depend on the plants may disappear as well.

*inter*NET
CONNECTION

Visit the Glencoe Homepage, *www.glencoe.com/sec/science,* for a link to further information about the Millennium Seed Bank. On what areas of the world are scientists concentrating their efforts? Where will they keep all the seeds?

Seed Production

It may seem that flowers are just for decoration. But flowers contain the reproductive structures in many plants. Many flowers have male and female parts. Male flower parts produce pollen, which contains sperm cells. Female flower parts produce eggs. When a sperm and an egg join, a new cell forms. As the new cell develops, it becomes enclosed in a seed, which protects the developing cell and helps keep it alive.

Soon after fertilization, rapid changes begin to take place in the flower. The petals and most other parts fall off. The flower's ovary (OH vuh ree), where the eggs are, swells and develops into a fruit which contains seeds, as shown in **Figure 4-7.**

*inter*NET
CONNECTION

Visit the Glencoe Homepage at **www.glencoe.com/sec/science** for links to more information on seed banks.

Use the Mini Quiz to check students' recall of chapter content.

1. **The process of cell division in which two identical daughter cells are produced from a single parent cell is _____.** *mitosis*
2. **Why is reproduction important for living things?** *Reproduction allows the species to continue.*

In this section, you have learned about how cells divide. You have also learned that organisms reproduce in different ways. Asexual and sexual reproduction are different, but in both, DNA is passed on to the new organisms. Later in this chapter, you'll learn that in organisms that have sexual reproduction, the information contained in DNA is passed on in specific patterns.

FIGURE 4-7
An apple blossom, when fertilized, will develop into an apple containing seeds.

Section Wrap-up

1. Explain the process of mitosis.
2. Compare the outcome of the process of mitosis with the outcome of the process of meiosis.
3. Explain why offspring produced by asexual reproduction are identical to the parent that produced them.
4. **Think Critically:** Describe the differences between asexual reproduction and sexual reproduction.
5. **Skill Builder**
   Using a CD-ROM Use a CD-ROM encyclopedia to research information on the process of reproduction in a flowering plant. Briefly outline the steps of sexual reproduction in one type of plant. If you need help, refer to Using a CD-ROM on page 563 in the **Technology Skill Handbook.**

USING MATH

A female bullfrog produces 350 eggs. All of the eggs are fertilized and hatch in one season. Assume that half of the tadpoles are male and half are female. If all the female tadpoles survive and produce 350 eggs each one year later, how many eggs would be produced?

Section Wrap-up

1. Mitosis is cell division. During mitosis, two identical daughter cells are produced by the division of a single parent cell.
2. Mitosis produces two cells that are identical to each other and to the parent cell. Meiosis produces sex cells with half the number of chromosomes as in the parent cell.
3. All of the organism's DNA comes from this single parent.
4. **Think Critically** Asexual reproduction requires only one parent; sexual reproduction requires two. Meiosis is required in sexual reproduction. Asexual reproduction occurs in a number of ways including budding, fragmentation, and fission.

Skill Builder
Using a CD-ROM Check student steps for accuracy. Make sure students have not skipped any important steps. Make sure they have referenced the CD they used to find the information.

Assessment

Performance Have students make illustrations of the reproductive process in flowering plants to accompany their descriptions. Use the Performance Task Assessment List for Scientific Drawing in **PASC,** p. 55. **L2** **P**
LS

USING MATH
If 175 females produce 350 eggs each and all of these eggs hatch, then there would be 350×175, or 61 250 frogs.

Science
& Society

4•2 Cloning

 Before presenting the lesson, display **Section Focus Transparency 10** on the overhead projector. Assign the accompanying **Focus Activity** worksheet.

L2 ELL

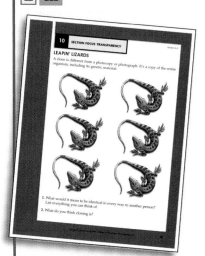

GLENCOE TECHNOLOGY

 Videodisc

STVS: Plants and Simple Organisms
Disc 4, Side 2
Plant Clones (Ch. 13)

||||||||||||

Disc 4, Side 1
Selective Breeding in Cows (Ch. 22)

||||||||||||

What YOU'LL LEARN

- How cloning is done and possible benefits of cloning research
- How to identify main ideas in science articles

  Science Words:
  cloning
  embryo

Why IT'S IMPORTANT

Knowing how to identify the main idea will make you a better reader.

Wood frog, 4-cell stage

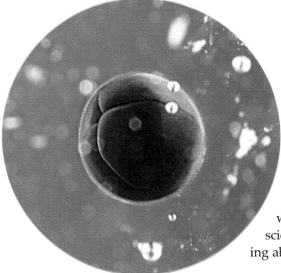

What is cloning?

Cloning means creating an organism that has exactly the same DNA as another organism. Some scientists use the word *cloning* to describe a procedure that could more accurately be called embryo splitting. An **embryo** is a fertilized egg that has begun dividing into more cells. In embryo splitting, a method that has been used for decades by animal breeders, an embryo is divided into several portions. Each portion can develop into a whole organism. For example, if frog eggs are used, each of the portions can grow into a normal frog. All of the frogs produced from one embryo are genetically identical to each other. But these frogs are not true clones. They grew from a fertilized egg cell that received half of its genetic information from the mother and half from the father. A clone receives all of its DNA from just one parent. Until now, this process was not possible except in plants.

The Science of Cloning

In 1997, a Scottish scientist, Dr. Ian Wilmut, announced that he had successfully cloned an adult Finn Dorset sheep he named Dolly. Dolly was the first successfully cloned mammal. She grew from a single cell taken from another sheep. She did not have two parents. Dr. Wilmut said the real value of cloning research is not that Dolly was created in a laboratory, but that scientists now have a better understanding about how cells work.

Program Resources

 Reproducible Masters
Enrichment, p. 16 L3
Reinforcement, p. 16 L2
Study Guide, p. 16 L1

 Transparencies
Section Focus Transparency 10 L2

Through Dolly, scientists learned more about the details of how a single cell can develop into an organism with skin cells, bone cells, muscle cells, and other body cells. Dr. Wilmut hopes other female animals can be changed genetically to produce medicines in their milk. Other scientists think that the new technology may make it possible to grow different types of cells—such as nerve and muscle cells—in the laboratory.

As you read about new developments in science, you come across many details. These details make more sense when you can connect them to one main idea. Understanding the main idea allows you to grasp the big picture. Follow the steps below to identify the main idea and the supporting details in what you just read.

Dolly

 Skill Builder: *Identifying the Main Idea*

LEARNING the SKILL

1. Before reading the material, review related background material from your text, notes, or from class. In this case, review information about cells, cell division, and DNA.

2. Study any photographs or illustrations that accompany the material.

3. What points in the passage are stated most strongly? What is the main idea in each paragraph?

4. Identify the main idea of the article.

5. Identify details that support the main idea.

PRACTICING the SKILL

1. What is the main idea of this article?

2. Briefly describe the difference between a clone and an embryo that has been split.

3. What points in the passage are stated the most strongly?

4. What details are given that support the main idea?

APPLYING the SKILL

Many people have different opinions about cloning. Bring news articles showing these differing opinions on cloning to class. Identify the main idea and supporting details. Discuss different viewpoints with your class.

Teaching the Content

Students may think that identical twins are clones. Have students brainstorm why identical twins are not clones. Explain that identical twins have the same DNA because they come from one fertilized egg that split into two. Although identical twins have the same DNA, they are not clones. A clone is defined as an individual who has the same DNA as its parent.

Teaching the Skill

- Pair students who are good readers with those who have reading problems.
- Have students choose a short newspaper or magazine article. Have them use the steps listed in Learning the Skill to understand the main idea of the article. L2

Answers to Practicing the Skill

1. A clone is an organism that is genetically identical to the parent.

2. Twins are the result of an embryo splitting. A fertilized egg is split at a stage when cell division has already begun. The DNA of these organisms has come from two parents. A clone receives all of its DNA from one parent.

3. Students' answers might include that cloning is a useful procedure to create genetically identical organisms for the purpose of producing medicines or better food products, or for understanding how our cells work.

4. Student answers might include that cloning is done in a laboratory, leads to understanding how cells work, and has useful applications.

Content Background

- The purpose of the original research on cloning was to find ways to alter the genetic makeup of farm animals. Researchers wanted to find ways to have herds of cows produce more milk, more nutritious meat, or other products consumers might want or need.

- Dolly was the only lamb to survive from 277 eggs that Dr. Wilmut treated in his experiment. Other embryos grew after his treatment, but they did not survive the complete gestation period.

Prepare

Section Background

In this section, students learn about *heredity,* the passing of traits from parent to offspring. Genetics, the science of heredity, is concerned with the questions of how and why offspring resemble their parents.

Preplanning

Refer to the Chapter Organizer on pages 88A-B.

1 Motivate

Bellringer

Before presenting the lesson, display **Section Focus Transparency 11** on the overhead projector. Assign the accompanying **Focus Activity** worksheet.

L2 ELL

Tying to Previous Knowledge

Review again the process of sexual reproduction with your class. Explain to students that in sexual reproduction, both parents contribute genetic information.

What YOU'LL LEARN

- How traits are inherited
- The structure and function of DNA
- How mutations add variation to a population

Science Words:
genetics
gene
DNA
variation
mutation

Why IT'S IMPORTANT

You will understand why you have the traits you have.

FIGURE 4-8

Family members often share similar physical features.

4•3 Genetics—The Study of Inheritance

Heredity

When you go to a family reunion or browse through family pictures, like the one in **Figure 4-8,** you can't help but notice similarities and differences among your relatives. You notice that your mother's eyes look just like your grandmother's, and one uncle is tall while his brothers are short. Scientists also think about these similarities and differences. Heredity (huh RED uh tee) is the passing on of traits from parents to offspring. Cracking the mystery of heredity has been one of the great success stories of biology.

Look around for a moment at the students in your classroom. What makes each person a unique individual? Is it hair or eye color? Is it the shape of a nose or the arch in a person's eyebrows? Eye color, hair color, skin color, nose shape, and many other features are types of traits that are inherited from a person's parents. A trait is a physical characteristic of an organism. Every organism is a collection of traits. The study of how traits are passed from parent to offspring is called **genetics** (juh NET ihks).

Genes on Chromosomes

All the traits that you have are inherited. Earlier in the chapter, you learned that you're a sort of combination of your parents. Half of the information in your chromosomes came from your father, and half came from your mother. This information was contained in the chromosomes of the sex cells that joined and eventually became you.

Program Resources

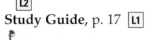

Reproducible Masters
Activity Worksheets, pp. 5, 27-28, 30 L2
Cross-Curricular Integration, p. 8 L2
Enrichment, p. 17 L3
Lab Manual, pp. 23-30 L3 L2
Reinforcement, p. 17 L2
Science and Society/Technology
 Integration, p. 22 L2

Science Integration Activities, pp. 45-46 L2
Study Guide, p. 17 L1

Transparencies
Science Integration Transparency 4 L2
Section Focus Transparency 11 L2
Teaching Transparency 8 L2

How do chromosomes work? As **Figure 4-9** shows, all chromosomes contain genes (JEENZ). A **gene** is a small section of a chromosome that determines a trait. Genes are arranged on a chromosome, one next to another. Humans have about 100 000 different genes arranged on 23 pairs of chromosomes. Genes control all of the traits of organisms, even traits that can't be seen, such as the size and shape of your stomach. Genes provide all the information needed for the growth and life of that species.

Life's Master Molecule—DNA

You've probably seen or heard about science fiction movies in which DNA is used to bring prehistoric animals back to life. Perhaps you've heard about using DNA to solve crimes. What is DNA? How does it work?

DNA is the material that chromosomes are made of. It is a chemical found in the nuclei of almost all cells. All of the information in the DNA in your chromosomes is called your genetic information. You can think of DNA as a genetic blueprint that contains all the instructions for how an organism looks and functions. Your DNA controls the texture of your hair, the shape of your ears, and even how you digest your food. The same is true for other organisms—all of an organism's traits are affected by the DNA in its cells.

Chromosome Pair

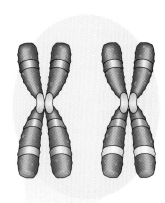

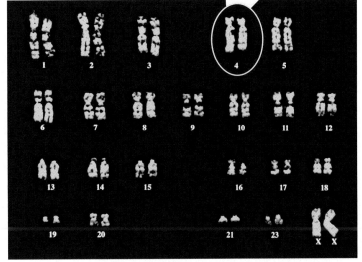

FIGURE 4-9

Each chromosome in a human body is made of DNA. All 23 pairs of chromosomes of one person are shown in this photograph. This person is a female.

4-3 Genetics—The Study of Inheritance **103**

Content Background

DNA is an example of a nucleic acid. The basic structure of a nucleic acid is a repetition of simple building blocks. In DNA, the building blocks are called *nucleotides*. A typical molecule will contain millions of nucleotides. Each nucleotide consists of a phosphate molecule, a sugar molecule, and a nitrogenous base. In DNA, there are four kinds of bases: adenine, guanine, cytosine, and thymine.

The DNA molecule is made from two chains of nucleotides. The easiest way to picture this is to think of a ladder. The bases link to each other across the chain and form the steps of the ladder. Either adenine and thymine link together or guanine and cytosine. The bases always bond in this manner. The sugar and phosphate molecules form the sides of the ladder. DNA is sometimes referred to as a double helix because the DNA ladder is twisted. The sequence of bases along the DNA ladder is the genetic code.

Using Science Words

ℕ **Linguistic** Ask students if they have ever heard the expression, "It's in the *genes*." Review what this expression means as you teach students about the words *gene* and *genetics*.

Visual Learning

Figure 4-9 As you study the illustration, discuss the Human Genome Project, which is a massive scientific project now being completed by scientists around the world. The goal of this project is to sequence and identify every gene on the human genome. **ELL** ℕ

Inclusion Strategies

Learning Disabled You may wish to have students make three-dimensional DNA models using toothpicks and assorted candies. Licorice can be used to represent the sides of the DNA ladder. Gumdrops could be the steps of the DNA ladder. Point out to students that the sequence of gumdrops or marshmallows represents the genetic code. ℕ **L1** **ELL**

Purpose
LS **Kinesthetic** In this activity, students learn how DNA can be obtained from onion cells.
L2 **COOP LEARN** **ELL**

Time
40 to 50 minutes

Materials
blender, coffee filter, small funnel or strainer, toothpicks, rubbing alcohol, warm water (1/4 cup), salt (1 teaspoon), meat tenderizer, medium onion (approximately 1 cup, chopped), liquid soap (1/4 cup), measuring cup (1 cup), large Styrofoam cup, medium glass container

Safety Precautions
Be sure students wear goggles and aprons at all times. Be sure there is no open flame in the room when alcohol is present.

Teaching Strategies
- Prepare enough onion mixture for the number of student groups you have. You may wish to have students participate in making the onion mixture.
- To prepare onion mixture: blend 1 cup chopped onion with 1/4 cup warm water and 1 teaspoon salt. Pour the mixture into a Styrofoam cup and add 1/4 cup liquid dish washing detergent. Mix gently for 5 minutes. Place a coffee filter into a small funnel. Pour the mixture into the funnel and filter out all the liquid into a small glass container. Add about 1/8 teaspoon of meat tenderizer to the liquid in the glass container.

Troubleshooting Limit the amount of alcohol in the room by having fewer but larger groups of students or have small groups perform the activity one group at a time.

104

Activity 4-2

Getting DNA from Onion Cells

Throughout the life of an organism, DNA provides instructions for the millions of cell processes that occur daily. It is found in the cells of all living things. In this activity, you will be able to see the actual material that contains all the instructions for a living thing—an onion.

Goals
- Remove DNA from onion cells.
- Practice laboratory skills.

Materials
- prepared onion mixture
- toothpicks
- rubbing alcohol
- measuring cup (1 cup)
- large glass or other glass container
- magnifying glass or microscope

What You'll Investigate
How is DNA taken out of cells?

Procedure
1. **CAUTION:** *Be sure to wear an apron and goggles throughout this activity. Avoid getting the rubbing alcohol in your eyes or on your clothing.*
2. **Obtain** a cup of prepared onion mixture from your teacher.
3. Slowly **pour** an equal amount of rubbing alcohol into the mixture. The alcohol should form a separate layer on top of the onion mixture.
4. **Observe** the gooey strings of DNA floating to the top.
5. Use a toothpick to gently **stir** the alcohol layer. Use another toothpick to remove the slimy DNA strings.
6. **Observe** DNA with a magnifying glass or a microscope. Record your observations in your Science Journal.
7. When you're finished, **pour** all liquids into containers provided by your teacher.

Conclude and Apply
1. Based on what you know about DNA, **predict** whether DNA removed from other plants would look different from the DNA you obtained.
2. **Infer** whether this method of taking DNA out of cells could be used to compare the amount of DNA between different organisms. **Describe** an experiment you might use to find out which types of plants have the most DNA.

Answers to Questions
1. Answers will vary, but most students will indicate that DNA should look the same no matter what organism it came from.
2. Answers will vary, but many students will describe a procedure that involves comparing the amounts of DNA in different plants. Procedures could suggest measuring the precise amounts of DNA in different plants and comparing them.

✓ Assessment

Performance Have students try to develop procedures for extracting DNA from other plants, such as lettuce, carrots, tomatoes, or other fruits or vegetables. Instruct students to prepare materials lists and step-by-step procedures. Use the Performance Task Assessment List for Assessing a Whole Experiment and Planning the Next Experiment in **PASC**, p. 35. **LS** **P**

Life's Code

If you could look at DNA in detail, you would see that it is shaped like a ladder. This structure, shown in **Figure 4-10,** is the key to how DNA works. The two uprights of the ladder form the backbone of the DNA molecule. The uprights support the rungs of the ladder. It is the rungs that hold all the genetic information. Each rung of the ladder is made up of a pair of chemicals called bases. The secret of DNA has to do with how these bases are put in order. A DNA ladder has billions of rungs, and the bases are arranged in thousands of different orders. The order or sequence of bases along the DNA ladder forms a sort of code. When the cell's machinery reads the code, the cell gets instructions about what to do and how to do it. The largest DNA molecule in a human is approximately 4 cm in length.

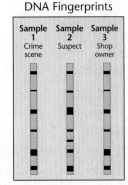

FIGURE 4-10
The sequence of bases that form the "rungs" of the DNA molecule forms a code. This code gives the instructions for running the body.

Problem Solving

Analyzing a DNA Fingerprint

A DNA fingerprint looks something like the bar codes on items you buy at the grocery store. DNA fingerprinting is a method of distinguishing individuals based on their DNA. This method produces a pattern of a person's DNA. Each person's DNA is unique.

DNA fingerprinting is an important, accurate tool for investigating crimes in which biological clues such as blood or hair samples are left behind. By comparing the DNA pattern of hair found at a crime scene to a suspect's DNA patterns, investigators can determine whether the hair is likely to have come from the suspect. DNA fingerprinting is also useful for determining whether two people are related.

DNA Fingerprints

Sample 1 Crime scene	Sample 2 Suspect	Sample 3 Shop owner

Solve the Problem:
The police wanted to find out whether a particular person was involved in the robbery of a shop. These DNA fingerprints show hair samples that were taken from the suspect compared with those left at the scene of the crime.

Think Critically:
After looking at the DNA fingerprints, is it possible that the suspect committed the crime?

4-3 Genetics—The Study of Inheritance   105

Discussion

Discuss with your class that much of our current understanding of how genetics works comes from experiments conducted by Austrian monk Gregor Mendel (1822-1884). Working by himself, Mendel performed many long experiments that helped establish the basic laws of genetics. The content of his work is now called classical or Mendelian genetics. Mendel published his findings in an obscure Austrian journal, and his work remained mostly unknown until after his death because it was ahead of its time.

GLENCOE TECHNOLOGY

 Videodisc

Glencoe Life Science Interactive Videodisc

Side 1, Lesson 1
Introduction

11-1805
Chromosomes

1807-2672

Using an Analogy

Explain that the genetic code is similar to other kinds of codes, such as Morse code and the bar codes on items at the store. Describe Morse code to the students or have them look it up in the dictionary. Point out that just as Morse code consists of a series of dots and dashes arranged in some sort of meaningful sequence, the sequence of nucleotide bases in DNA also forms a code. You may wish to point out the similarities between supermarket bar codes and the autoradiographs produced from DNA profiling.

106

FIGURE 4-11

Pea flowers can be purple or white. The chromosome pair from a pea plant shows that both have an allele for the trait of flower color. *What color will the flowers produced by a plant with this chromosome pair be?*

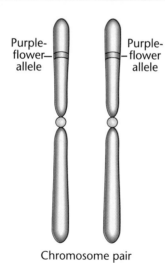

Purple-flower allele — Purple-flower allele

Chromosome pair

106   Chapter 4 Heredity and Reproduction

What determines traits?

Remember that in body cells, such as skin cells or muscle cells, chromosomes come in pairs. The genes on those chromosomes are in pairs, too. You have learned that genes control traits. A single pair of chromosomes may control many different traits.

The genes that make up a gene pair may or may not be the same—they may come in different forms. For example, the genes for the trait of flower color in pea plants may be of the purple variety or the white variety. When genes that control a trait come in different forms, those forms are called alleles (uh LEELZ), as shown in **Figure 4-11.**

The combination of alleles in a gene pair determines how a trait will be shown. If a pea plant has two copies of the purple-flower allele, it will have purple flowers. If a pea plant has two copies of the white-flower allele, it will have white flowers.

What if the plant has one purple-flower allele and one white-flower allele? Will the flowers be purple or white? That depends on something called dominance (DAW muh nunts). Dominance means that one trait covers over or masks another form of the trait. Some alleles dominate others, as **Table 4-1** shows. For instance, if a pea plant has one purple-flower allele and one white-flower allele, its flowers will all be purple, just as if the plant had two purple-flower alleles. In pea plants, purple is the dominant flower color. White is what's known as the recessive flower color. Recessive means that the trait is hidden or masked if the dominant form of the trait is present. When an organism has two identical alleles for a trait, it's called pure. If it contains different alleles for a trait, it's called a hybrid (HI brud).

Table 4-1

Purple Flowers	White Flowers
Purple-flower may have Pure: 2 purple-flower alleles or Hybrid: 1 purple-flower allele 1 white-flower allele	White-flower has only white-flower alleles.

Teacher F.Y.I.

Baldness is recessive to hairiness. Believe it or not, the gene for six fingers is dominant over the gene for five fingers.

GLENCOE TECHNOLOGY

 Videodisc

The Infinite Voyage: The Geometry of Life
Chapter 4
Master Genes and Their Influence

Revealing Preconceptions

Ask students whether they think there are more dominant or recessive alleles for a trait. Many think that dominant traits are the more numerous. This is not always the case. Dominance does not mean that there will be more of a certain allele. Often, recessive alleles show up more often in a population. For example, the color yellow is dominant to green in pea plants, but most peas commonly eaten are green.

Visual Learning

Figure 4-11 What color will the flowers produced by a plant with this chromosome pair be? *purple*

Content Background

Eye color is a polygenic trait. It is the result of more than one pair of genes interacting, thus, great variation exists in the human population. Colored blindness is recessive but is sex-linked. It is located on the X (female) chromosome. A male (XY) will be color blind if the X chromosome carries the recessive trait because there is no corresponding allele on the Y chromosome. A female will be color blind only if both recessive alleles are present.

Brainstorming

Enrichment

A Punnett square shows all the ways the alleles from two parents can combine. Have interested students research how to use Punnett squares. Have them make a Punnett square showing a cross between a hybrid purple-flowered pea plant and a white-flowered plant. They should draw the following Punnett square. L3

Purple hybrid

	W	w
w	Ww	ww
White		
w	Ww	ww

50% white flowers, 50% purple flowers

Passing Traits to Offspring

How are traits passed from parents to offspring? Let's use the trait of flower color in pea plants as an example. Suppose a hybrid purple-flowered pea plant (one with two different alleles for flower color) is mated with a white-flowered pea plant. To mate one pea plant with another, pollen from one plant is placed on the pistil of a flower on another plant. What color flowers will the offspring have?

The traits that a new pea plant will inherit depend upon which genes are carried in the parents' sex cells. Remember that sex cells are produced during meiosis. In meiosis, pairs of chromosomes separate as sex cells form. Pairs of genes therefore also separate from one another. As a result, each sex cell contains one allele for each trait. Because it is a hybrid, the purple-flowered plant in **Figure 4-13** produces half of its sex cells with the purple-flower allele and half with the white-flower allele. On the other hand, the white-flowered plant is pure. All of the sex cells that it makes contain only the white-flower allele.

FIGURE 4-12
The colors of pea flowers are a result of the alleles for flower color the plant receives from its parent plants.

In fertilization, one sperm will join with one egg. Which egg and sperm will join? We don't know. Many events, such as flipping a coin and getting either heads or tails, are a matter of chance. In the same way, chance is involved in heredity. In the case of the pea plants, there was an equal chance that the new pea plant would receive either the purple-flower allele or the white-flower allele from the sperm cell.

FIGURE 4-13

The traits an organism has depend upon the combinations of genes that were carried in the parents' sex cells. This diagram shows how the trait of flower color is passed on in pea plants.

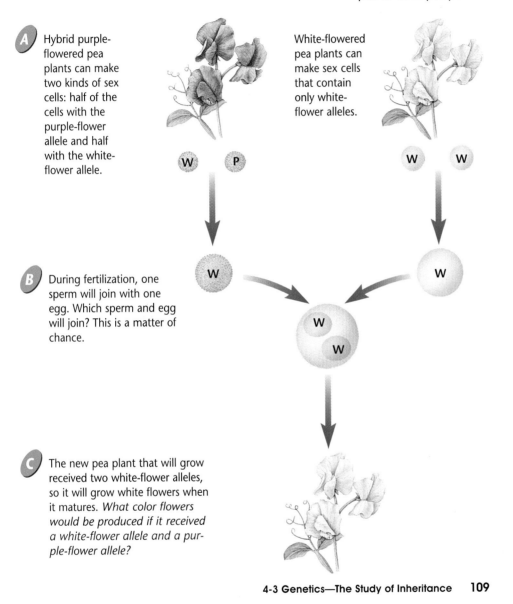

A Hybrid purple-flowered pea plants can make two kinds of sex cells: half of the cells with the purple-flower allele and half with the white-flower allele.

White-flowered pea plants can make sex cells that contain only white-flower alleles.

B During fertilization, one sperm will join with one egg. Which sperm and egg will join? This is a matter of chance.

C The new pea plant that will grow received two white-flower alleles, so it will grow white flowers when it matures. *What color flowers would be produced if it received a white-flower allele and a purple-flower allele?*

4-3 Genetics—The Study of Inheritance **109**

Visual Learning

Figure 4-13C What color flowers would be produced if it received a white-flower allele and a purple-flower allele? *The flowers would be purple.* ELL

Use the Flex Your Brain activity to have students explore GENETIC TRAITS.

Activity Worksheets, page 5

Discussion

Gregor Mendel had no knowledge at all about the mechanism of genetic transfer. At the time of his experiments, nobody knew anything about genes, chromosomes, and DNA. When his work was "discovered" 20 years after his death, scientists quickly recognized its importance. Discuss with your class how scientific discoveries don't just happen overnight. In a way, scientific knowledge builds on itself. Point out that knowledge in one discipline can be applied to other disciplines.

3 Assess

Check For Understanding

Discussion Review Figures 4-11, 4-12, and 4-13 with students. Ask them to explain dominant and recessive traits. Have students explain in their own words what is happening in the illustrations. L2

Reteach

LS Kinesthetic You may wish to have students manipulate colored beads, candies, or other small objects while they learn about single-trait inheritance. Have students use small objects to model the process shown in Figure 4-13.

109

FIGURE 4-14
Height is a trait that has many variations.

Mini LAB

Probability

Flip a coin to understand how events happen by chance.

1. Flip a coin ten times. Count the number of heads and the number of tails.
2. Now flip the coin 20 times. Count the number of heads and tails.
3. Record your data in your Science Journal.

Analysis

1. What were your results when you flipped the coin ten times? Was this what you expected?
2. What were your results when you flipped the coin 20 times? Were your observed results closer to your expected results when you flipped the coin more times?
3. How is the flipping of a coin similar to the joining of egg and sperm at fertilization?

Differences in Organisms

Now you know why a new baby can have characteristics of either of its parents. The genes he or she inherited from the parents determined hair color, skin color, eye color, and other traits. But what accounts for the differences, or variations (vayr ee AY shuns), in a family? **Variations** are the different ways a certain trait appears. An example of variations in height is shown in **Figure 4-14.**

Multiple Genes and Multiple Alleles

Earlier, you learned how the trait of flower color in pea plants is passed from parent to offspring. Flower color in pea plants shows a simple pattern of inheritance. Sometimes, though, the pattern of inheritance of a trait is not so simple. Many traits in organisms are controlled by more than two or multiple alleles. In humans, multiple alleles control blood types A, B, or O.

Traits can also be controlled by more than one gene. In humans, for example, height, weight, eye color, skin color, and hair color are traits that are controlled by several genes. This type of inheritance provides for differences or variations in a species. Humans, for instance, can be tall, short, and every height in between.

Mutations: The Source of New Variation

When you hear the words *mutation* (myew TAY shun) or *mutated,* what comes to mind? Maybe you think of a horrible creature straight out of a science fiction movie or a cartoon character with supernatural powers. But if you've ever hunted through a patch of clover until you found one with four leaves instead of three, you've come face-to-face with a real mutation. A four-leaf clover is the result of a mutation. Actually, the word *mutate* simply means "to change." In genetics, a **mutation** is a change in a gene or chromosome due to an error in meiosis or mitosis or due to an environmental factor. Many mutations happen by chance, but some mutations are caused by outside influences, such as X rays or dangerous chemicals in the environment.

What are the effects of mutations? Sometimes mutations affect the way cells grow, repair, and maintain themselves. Some mutations are harmful to organisms. Others are beneficial. But many, such as the four-leaf clover example, have no effect at all. Whether a mutation is beneficial, harmful, or neutral, all mutations add variation to the genes of a species. **Figure 4-15** shows a mutation called albinism that changes the way skin and other body cells work.

FIGURE 4-15

Mutations happen in all organisms. Some cause visible changes such as in the clover and the squirrel. Others are not seen. Some are harmful, but most are harmless.

Section ⟳ Wrap-up

1. What is heredity?

2. Describe the function of genes.

3. **Think Critically:** Why is DNA sometimes called the "blueprint" for an organism?

4. ✚▦ *Skill Builder*
   Concept Mapping Make a concept map that shows the relationships between the following concepts: genetics, genes, chromosomes, DNA, variation, and mutation. If you need help, refer to Concept Mapping on page 544 in the **Skill Handbook.**

Science Journal

Find out what a transgenic organism is, then go to the library and find books or magazine articles on these organisms. In your Science Journal, write a brief summary of your findings.

✚▦ *Skill Builder*

Concept Mapping Check students' concept maps for accuracy. Make sure students understand the definition of each key term. A possible spider map is shown here.

variations — genetics — chromosomes
caused by mutations
genes located on
made of DNA

✔ **Assessment**

Portfolio Have students do research about a particular genetic disorder. Some possible topics include color blindness, hemophilia, albinism, sickle-cell anemia, cystic fibrosis, and Down syndrome. Have them present their research in a poster project, report, or multimedia presentation. Use the Performance Task Assessment List for Poster in **PASC**, p. 73 Ⓟ

Extension

🗀 For students who have mastered this section, use the **Reinforcement** and **Enrichment** masters.

4 Close

•MINI•QUIZ•

Use the Mini Quiz to check students' recall of chapter content.

1. **What is the name of the science that studies how traits are passed from parent to offspring?** *genetics*

2. **What is the name of the complex chemical that contains the genetic code?** *DNA*

3. **What is a gene? Where are genes located?** *A gene is a section of DNA that controls a particular trait. Genes are located on chromosomes.*

Section ⟳ Wrap-up

1. Heredity is the transmission of genetic information.

2. Genes control and determine traits of organisms.

3. **Think Critically** DNA, which makes up genes, provides all of the necessary information for the growth, development, and life of an organism.

Science Journal Check students' Journal entries to make sure they understand the concept of transgenic organisms. Organisms that contain functional recombinant DNA are called transgenic organisms.

Science & History

Background

A breed is usually defined as a group of domesticated animals that are similar in appearance and, that when bred to one another, produce offspring that are similar to the original group. The great array of modern horse breeds is the result of many centuries of artificial selection.

While many horse breeds were developed to perform certain jobs, such as pulling carriages, dragging plows, or herding cattle, others have been bred for particular physical or behavioral traits, such as color and gait. Pintos, buckskins, and palominos are examples of breeds developed primarily for their distinctive coat color. The color standard for a palomino is often described as that of a "newly minted gold coin."

Several types of horses were developed specifically for the comfort of riders. These "gaited" breeds have smoother ways of walking quickly than the standard trot. The Tennessee walking horse, for example, has what is called a running gait, which is halfway between a walk and a run.

Teaching Strategies

- Have students make a bulletin board display showing different horse breeds. L1
  ELL

- Have students work in groups to research the history of the horse in North America. Some groups might focus on the role horses played in Native American cultures, while others may want to concentrate on the wild horse populations currently found in several western states and on a few coastal islands. L2
  COOP LEARN

interNET CONNECTION

Visit the Glencoe Homepage at *www. glencoe.com/sec/ science* for a link to more information on different uses of horses through history.

A Horse for Every Job

No one is sure exactly when the first wild horses were tamed, or domesticated, by people. But people in Asia were riding horses as long ago as 3000 B.C. Once horses began to play a role in human culture, people bred them to do specific jobs. Large horses were bred with strong ones to produce breeds suited for hard work. Smaller, sleeker horses were bred for riding and racing. Today, there are hundreds of different breeds, each with its own unique characteristics. However, all horses can be grouped into several basic categories.

Draft horses are large, strong breeds that were bred for pulling plows and hauling heavy loads. Draft breeds include Percherons, Clydesdales, and shires. Shires are the largest and strongest horses in the world. They stand 180 cm (72 inches) high at the withers (the ridge between a horse's shoulders) and weigh at least 1200 kg (2640 pounds). Shires can pull five times their own weight!

Today, the saddle horse breeds show the most variety. These smaller, lighter breeds were developed for such jobs as riding, herding, hunting, and racing. The high-stepping American saddlebred, the thoroughbred, and the American quarter horse are examples of saddle horses.

112   Chapter 4 Heredity and Reproduction

- Invite a veterinarian who works with large animals to visit your class to talk about working with horses.

Bibliography

Horses Through Time, edited by Sandra L. Olsen. Boulder, Colorado: Roberts Rinehart Publishers, 1996.

interNET CONNECTION

Visit the Glencoe Homepage, **www.glencoe.com/sec/science**, for a link to information on a variety of different horse breeds.

Read the statements below that review the most important ideas in the chapter. Using what you have learned, answer each question in your Science Journal.

1. Organisms can reproduce sexually or asexually. *Why does sexual reproduction provide more variety in a species than asexual reproduction?*

2. A clone is an organism that is genetically identical to another organism. *What are some possible benefits of cloning technology?*

3. Genetics is the study of how traits are passed from parent to offspring. *How is it possible that you have traits of both of your parents?*

4. Mutations are changes in a gene or chromosome. *How do mutations provide new variation in a population?*

Review

Have students look at the illustrations on this page. Ask them to describe details that support the main ideas of the chapter found in the statement for each illustration.

Teaching Strategies

Instruct students to study each of the illustrations carefully. Have them answer all questions in their Journals.

Answers to Questions

1. Sexual reproduction involves two parents. Asexual reproduction involves a single parent.
2. for making important medicines and other chemicals
3. Sex cells from both parents contain genetic information. When these cells united at fertilization, a zygote was formed.
4. Mutations are changes in a gene or chromosome. These changes add more differences, or variations, into the population.

Science at Home

LS **Intrapersonal** Have students investigate the inheritance of traits in their own families. Students can study how traits—such as earlobe shape (attached or unattached), eye color, shape of hairline (straight or widow's peak), and hair texture—are distributed in their parents and grandparents. Have students collect the information, then show them how pedigree charts can be constructed from the information. L2

Assessment

Portfolio Encourage students to place in their portfolios one or two items of what they consider to be their best work. Examples include:
- Science Journal, p. 89 and p. 92
- Activity 4-1, p. 95 **P**

Performance Additional performance assessments may be found in **Performance Assessment** and **Science Integration Activities.** Performance Task Assessment Lists and rubrics for evaluating these activities can be found in Glencoe's **Performance Assessment in the Science Classroom (PASC).**

Chapter 4 Review

Using Key Science Words

1. DNA
2. asexual reproduction or cloning
3. mutation
4. sexual reproduction
5. mitosis

Checking Concepts

6. c   7. b   8. d
9. a   10. c

Thinking Critically

11. Students should recognize that chromosomes are made up of DNA and that genes are small sections of chromosomes, which also makes them small sections of the DNA molecule. A pea plant has genes on its chromosomes that determine the flower color, as well as all the other traits of the plant.

12. Eye color is determined by several genes. Blue eyes are a recessive trait, but the parents have more alleles for eye color. It is possible for two brown-eyed parents to have a child with blue eyes.

13. Students should recognize from the photo that baby spider plants are growing from the runners that emerge from the adult spider plant. Because only one parent is involved, it is a form of asexual reproduction. Spider plants can reproduce through sexual reproduction when flowers are produced and seeds are made.

14. Meiosis is important because it's the process in which gametes are made. Each gamete will contribute chromosomes that will give the zygote a complete set of chromosomes.

114

Using Key Science Words

asexual reproduction	meiosis
cloning	mitosis
DNA	mutation
embryo	sex cells
fertilization	sexual
gene	reproduction
genetics	variation

Match each phrase with the correct term from the list of Key Science Words.

1. chemical that chromosomes are made of
2. reproduction with one parent
3. a change in a gene or chromosome
4. reproduction involving two parents
5. the process of nuclear division

Checking Concepts

Choose the word or phrase that completes the sentence.

6. The process of producing an organism that is genetically identical to another organism is called _____.
   a. fertilization
   b. sexual reproduction
   c. cloning
   d. mutation

7. Sperm and eggs are types of cells called _____.
   a. embryos   c. mutations
   b. sex cells   d. genes

8. During meiosis, _____ for a trait separate.
   a. cells   c. clones
   b. sex cells   d. genes

9. Albinism is a type of _____.
   a. mutation
   b. sexual reproduction
   c. gene
   d. embryo

10. A(n) _____ is a feature or characteristic of an organism.
    a. sex cell   c. trait
    b. embryo   d. gene

Thinking Critically

Answer the following questions in your Science Journal using complete sentences.

11. Explain the relationship among DNA, genes, and chromosomes using the pea as an example.

12. Two brown-eyed parents have a baby with blue eyes. Explain how this could have happened using what you know about heredity.

13. The photo shows a picture of a spider plant. Why is this plant an example of asexual reproduction? How could the plant reproduce through sexual reproduction?

114 Chapter 4 Heredity and Reproduction

Assessment Resources

 Reproducible Masters
Chapter Review, pp. 11-12
Assessment, pp. 17-20
Performance Assessment, p. 42

Glencoe Technology
🔘 Computer Test Bank
📼 MindJogger Videoquiz

14. How is the process of meiosis important in sexual reproduction?

15. Explain how a mutation in a gene could be harmful or beneficial to an organism.

Developing Skills

If you need help, refer to the description of each skill in the Skill Handbook.

16. Measuring in SI: Gather 50 sunflower seeds or pinto beans. Using a metric ruler, measure the total length of each seed in millimeters. Record the results in your Science Journal.

17. Making and Using Graphs: Make a bar graph from the seed data you collected in question 16. Discuss the variation in seed length for this collection.

18. Designing an Experiment: Design an experiment to test whether or not plants produced by asexual reproduction grow faster or taller than plants produced by sexual reproduction.

19. Recognizing Cause and Effect: Stomach cells produce chemicals that help digest certain foods. What effect might a mutation in a stomach cell gene have on the life of an organism?

20. Predicting: A pure, purple-flowered pea plant is crossed with a pure, white-flowered pea plant. What color of flowers will the resulting pea plant be able to produce?

Performance Assessment

1. Scientific Drawing: Use your imagination and make vocabulary illustrations for each of the following science words: *asexual reproduction, genetics,* and *mutation.*

2. Newspaper Article: Many scientists have reported that it's possible to get DNA from prehistoric creatures. Go to the library and find a newspaper article that describes the discovery of ancient DNA. Write a summary of the article in your Science Journal.

Performance Assessment

1. Check students' illustrations to make sure they understand the meaning of each term. Accept all reasonable illustrations. Use the Performance Task Assessment List for Scientific Drawing in **PASC,** p. 55. **P**

15. Answers will vary but should include the idea that if a mutation is a change in a gene or chromosome, an organism's traits will also be affected.

Developing Skills

16. Measuring in SI Give students advice on how to take the measurements. If these seeds are not available, you may wish to use pine needles.

17. Making and Using Graphs Check to make sure students know how to construct bar graphs from the data. Refer them to the Skill Handbook as needed. Make sure students clearly label and title their graphs.

18. Designing an Experiment Answers will vary, but many students will describe experiments that involve taking cuttings from a particular plant and growing them under the same environmental conditions as a different plant of the same species produced by sexual reproduction.

19. Recognizing Cause and Effect Answers may vary, but students should recognize that the mutation might impair the way the organism digests food. This may lead to a change in diet or a supplement to help digest different types of food.

20. Predicting Students should recognize that the resulting pea plant will be able to produce only purple flowers because it will have one dominant allele and one recessive allele.

Chapter Organizer

Section	Objectives/Standards	Activities/Features
Chapter Opener		**Explore Activity:** Identify Different Species, p. 117
5-1 **Diversity of Life** (3 sessions, 1 block)*	1. **Define** the term *biodiversity*. 2. **Explain** why some environments have high biodiversity. 3. **Describe** how organisms are adapted to their environments. **National Science Content Standards: (5-8)** **UCP1, UCP5, C1, C3, C5, F2, G1**	**MiniLAB:** Identifying Adaptations of Fish, p. 122 **Skill Builder:** Observing and Inferring, p. 122 **Using Math,** p. 122 **Science & Language Arts:** Using Field Guides to Explore Diversity, p. 123 **Science Journal,** p. 123
5-2 **Unity of Life** (3 sessions, 2 blocks)*	4. **Explain** how species change and new species form through natural selection. 5. **Analyze** the evidence that supports the idea that organisms change through time. **National Science Content Standards: (5-8)** **UCP2, UCP3, UCP4, A1, C5, F5**	**Activity 5-1:** Simulating Selection, pp. 126-127 **MiniLAB:** Making a Fossil, p. 130 **Problem Solving:** More Chemical Clues, p. 131 **Skill Builder:** Concept Mapping, p. 133 **Science Journal,** p. 133
5-3 **Science and Society:** **Captive-Breeding** **Programs** (1 session, ½ block)*	6. **Discuss** how captive-breeding programs help endangered species. 7. **Identify** resources. **National Science Content Standards: (5-8)** **C2, C4, F2, F5, G1**	**Skill Builder:** Using Resources, p. 135
5-4 **History of Life** (2 sessions, 1 block)*	8. **Understand** the importance of the fossil record. 9. **Describe** the importance of the geologic time scale. 10. **Relate** the possible causes of mass extinctions. **National Science Content Standards: (5-8)** **UCP2, UCP3, C4, D2, E2, F5**	**Using Technology:** Blood in the Bones, p. 137 **Activity 5-2:** How big were the dinosaurs?, p. 139 **Skill Builder:** Recognizing Cause and Effect, p. 142 **Using Math,** p. 142

* A complete Planning Guide that includes block scheduling is provided on pages 31T-33T.

Activity Materials

Explore	Activities	MiniLABs
page 117 species category cards	pages 126-127 small assorted jelly beans, meterstick, timer page 139 plastic dinosaur model, graduated cylinder, marking pen, calculator, water, small pail, utility pan	page 122 adaptation cards page 130 pint-sized milk container, petroleum jelly, plaster of paris,

Need Materials? Call Science Kit (1-800-828-7777).

Chapter 5 Diversity and Adaptations

Teacher Classroom Resources

Reproducible Masters	Transparencies	Teaching Resources

Activity Worksheets, pp. 5, 35
Cross-Curricular Integration, p. 9
Enrichment, p. 18
Lab Manual 9
Multicultural Connections, pp. 13-14
Reinforcement, p. 18
Science and Society/
  Technology Integration, p. 23
Science Integration Activities,
  pp. 47-48
Study Guide, p. 18
	Section Focus Transparency 12, An
	  Endless Variety
	Teaching Transparency 9, Life's Variety
		Spanish Resources
		English/Spanish Audiocassettes
		Cooperative Learning Resource
		  Guide
		Lab Partner
		Lab and Safety Skills
		Lesson Plans

Activity Worksheets, pp. 5, 31-32, 36 Enrichment, p. 19 Lab Manual 10 Reinforcement, p. 19 Study Guide, p. 19	Section Focus Transparency 13, Mammoths and Elephants	**Assessment Resources** Chapter Review, pp. 13-14 Assessment, pp. 21-24 Performance Assessment, p. 43

Enrichment, p. 20
Reinforcement, p. 20
Study Guide, p. 20
	Science Integration Transparency 5,
	  Fossils and Adaptations
	Section Focus Transparency 14, Save
	  the Species
		Performance Assessment in the
		  Science Classroom (PASC)
		MindJogger Videoquiz
		Alternate Assessment in the
		  Science Classroom
		Computer Test Bank

Activity Worksheets, pp. 5, 33-34
Enrichment, p. 21
Reinforcement, p. 21
Study Guide, p. 21
	Section Focus Transparency 15, A Story
	  in Bones
	Teaching Transparency 10, Geologic
	  Time Scale	

Key to Teaching Strategies

The following designations will help you decide which activities are appropriate for your students.

L1 Level 1 activities should be appropriate for students with learning difficulties.

L2 Level 2 activities should be within the ability range of all students.

L3 Level 3 activities are designed for above-average students.

ELL ELL activities should be within the ability range of English Language Learners.

LS These activities are designed to address different learning styles.

COOP LEARN Cooperative Learning activities are designed for small group work.

P These strategies represent student products that can be placed into a best-work portfolio.

GLENCOE TECHNOLOGY

The following multimedia resources are available from Glencoe.

The Infinite Voyage Series
Insects: The Ruling Class
Life in the Balance
The Great Dinosaur Hunt

National Geographic Society Series
STV: Biodiversity

Glencoe Life Science CD-ROM

Teacher Classroom Resources

This is a representation of key blackline masters available in the Teacher Classroom Resources.

Teaching Aids

Section Focus Transparencies

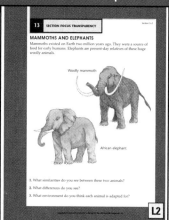

AN ENDLESS VARIETY
There are more species of insect than any other species of animal. This insect collection shows you just a small sample.

1. How many different kinds of insects can you see in the photo?
2. Why do you think it is good that there are so many different species of organisms on Earth?

L2

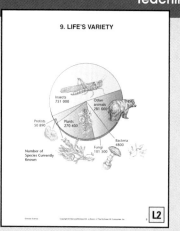

MAMMOTHS AND ELEPHANTS
Mammoths existed on Earth two million years ago. They were a source of food for early humans. Elephants are present-day relatives of these huge woolly animals.

Woolly mammoth

African elephant

1. What similarities do you see between these two animals?
2. What differences do you see?
3. What environment do you think each animal is adapted for?

L2

SAVE THE SPECIES
Some species of organisms are becoming extinct in their natural environments. For a number of reasons, including loss of living space, animals such as the puma shown here are at risk.

1. What do you think "loss of living space" might mean to the puma?
2. How might the puma be protected?

L2

Science Integration Transparencies

5
SCIENCE INTEGRATION TRANSPARENCY

Fossils and Adaptations

L2

Teaching Transparencies

9. LIFE'S VARIETY

Insects 751 000

Other animals 281 000

Protists 50 890

Plants 270 400

Bacteria 4800

Fungi 101 500

Number of Species Currently Known

L2

10. GEOLOGIC TIME SCALE

L2

Meeting Different Ability Levels

Study Guide for Content Mastery

Name _____ Date _____

Chapter 5 — Study Guide for Content Mastery

Diversity of Life

Directions: Study the following illustration, then answer the questions below. Use complete sentences.

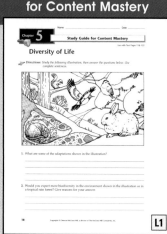

1. What are some of the adaptations shown in the illustration?

2. Would you expect more biodiversity in the environment shown in the illustration or in a tropical rain forest? Give reasons for your answer.

L1

Reinforcement

Name _____ Date _____

Chapter 5 — Reinforcement

Diversity of Life

Directions: Study the illustrations of living organisms below. On the lines provided, identify an adaptation for each organism and explain how that adaptation helps the organism survive in its environment.

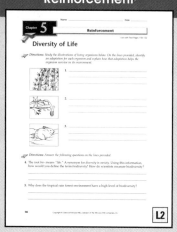

1.

2.

3.

Directions: Answer the following questions on the lines provided.

4. The root bio means "life." A synonym for diversity is variety. Using this information, how would you define the term biodiversity? How do scientists measure biodiversity?

5. Why does the tropical rain forest environment have a high level of biodiversity?

L2

Enrichment Worksheets

Name _____ Date _____

Chapter 5 — Enrichment

Size Adaptations Among Animals

Directions: Read the following information. Then answer the questions.

Is there a reason why a hippopotamus is larger than a rabbit or a whale is larger than a salmon? By studying many species, scientists have found that there is a typical range of size for every type of animal. This size is an adaptation to the animal's environment.

Light Enough to Fly
Birds are usually light in weight, which helps them fly. Although some large birds can fly, such as pelicans, there is a limit to how large a flying bird can be. If you double the weight of a bird, the amount of power needed for it to fly increases four times. A very large flying bird needs to have enormous, powerful wings.

Large Enough to Keep Warm
Animals that live in very cold climates tend to be large mammals or birds, such as polar bears, walruses, and penguins. Cold-blooded reptiles and amphibians, as well as small birds and mammals, cannot keep warm. Small mammals lose heat through the surfaces of their bodies, and these surfaces are big in relation to their size. Even in a temperate climate, a mouse has to eat one-fourth of its weight in food every day just to keep warm. In the Arctic, a mouse could not possibly eat enough food.

Too Big for Land?
Finally, consider the largest animals that have ever existed, the blue whales. These creatures can be over 30 meters long and weigh over 150 metric tons. Could such huge animals exist on land? The largest known dinosaur weighed 80 metric tons. A land animal as large as the blue whale would need such strong, bulky legs that it is doubtful it could move.

Applying Problem-Solving Skills
1. How does an animal's size affect its relationship with other animals?

2. When insects breathe, oxygen reaches many parts of their bodies by spreading fairly slowly through their cells. Why do you think very few insects are more than 1 centimeter thick?

L3

Chapter 5 Diversity and Adaptations

Hands-On Activities

Science Integration Activities

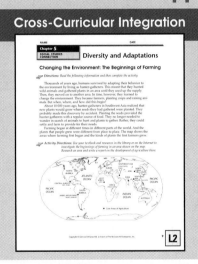

Lab Manual

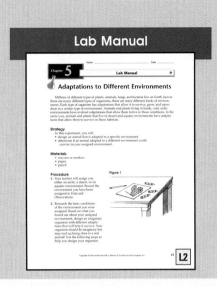

Activity Worksheets

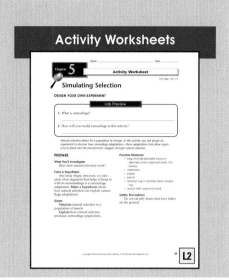

Enrichment and Application

Cross-Curricular Integration

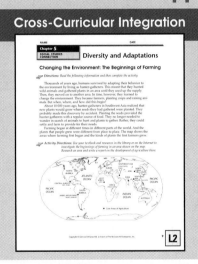

Multicultural Connections

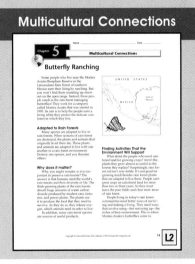

Science and Society/ Technology Integration

Assessment

Performance Assessment

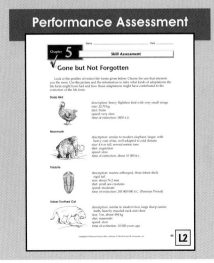

Chapter Review

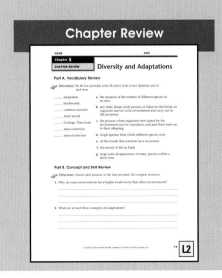

Assessment

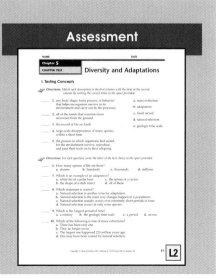

Chapter 5

Diversity and Adaptations

CHAPTER OVERVIEW

Section 5-1 The concept of bio-diversity is discussed. Students learn why tropical rain forests have high biodiversity compared to other environments. The concept of *adaptation* is also discussed.

Section 5-2 Natural selection is discussed. Students learn how the process of natural selection accounts for the diversity and adaptations.

Section 5-3 Science and Society This section encourages students to identify resource materials while finding out how captive-breeding programs are one way endangered species are being preserved.

Section 5-4 Students are introduced to the geologic time scale as they learn how the fossil record provides scientists with information about the history of life on Earth. Extinction is also discussed. Students learn how a mass extinction today is leading to the loss of biodiversity in many parts of the world.

Chapter Vocabulary

biodiversity
adaptation
natural
  selection
common
  ancestor
fossil
fossil record
captive
  breeding
geologic time
  scale
mass
  extinction

Theme Connection

Stability and Change The world's environments are under constant change. Species also change over time in order to adapt to the changing conditions around them.

Chapter Preview

Skills Preview

▶ **Skill Builders**
- observe
- infer
- make an events chain
- recognize cause and effect

▶ **MiniLABs**
- hypothesize
- infer
- compare and contrast
- predict

▶ **Activities**
- measure
- use tables
- infer
- conclude

116

Learning Styles

Look for the following logo for strategies that emphasize different learning modalities. **LS**

Kinesthetic	MiniLAB, pp. 121, 130; Activity, p. 125; Activity 5-1, p. 126; Activity 5-2, p. 139
Visual-Spatial	Visual Learning, pp. 119, 120, 125, 129, 132, 138; Assessment, pp. 121, 127; Reteach, p. 122; Activity, p. 128; Inclusion Strategies, p. 128
Interpersonal	Assessment, pp. 133, 139
Intrapersonal	Assessment, p. 117; Enrichment, p. 128; Revealing Preconceptions, p. 129; Community Connection, p. 130; Science at Home, p. 143
Logical-Mathematical	Explore, p. 117; Using Math, pp. 122, 142; Problem Solving, p. 131; Using an Analogy, p. 138; Using Science Words, p. 141; Assessment, p. 142
Linguistic	Using Science Words, p. 120; Science Journal, pp. 131, 141; Reteach, p. 132; Across the Curriculum, p. 137

LS

Diversity and Adaptations

Hiking through the rain forest, you come across the scene pictured in the photo. At first, you see nothing unusual. Then your guide whispers, "Look at that cool insect! Can you spot it?" This animal from the tropical rain forest of South America is called a stick insect. The insect blends into its surroundings. Its body shape and color make it almost invisible among the branches and leaves of the plants. Stick insects are found in many parts of the world including North America. Stick insects are just one of the many species that can be found in a tropical rain forest.

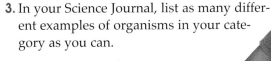

EXPLORE ACTIVITY

Identify Different Species

1. Obtain a species category card from your teacher.
2. Look at the card. How many different types of organisms can you name that fit this category?
3. In your Science Journal, list as many different examples of organisms in your category as you can.

Science Journal

Find a picture of one of the living things you listed in your Science Journal. Observe the features of this living thing. What features help it survive in its environment?

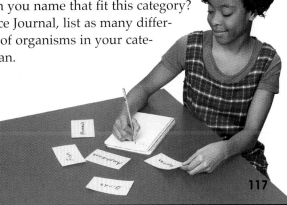

117

5•1 Diversity of Life

What YOU'LL LEARN

- The term *biodiversity*
- Why some environments have high biodiversity
- How organisms are adapted to their environments
  Science Words:
  biodiversity
  adaptation

Why IT'S IMPORTANT

Learning the connection between environments and the organisms that live there will help you make better decisions about Earth's resources.

Life's Endless Variety

Earth is filled with an enormous number of different living things. Life can be found almost everywhere on Earth—in the air, on water, on land, underground, and in the soil. Why do you think so many different forms of life exist on Earth?

How many different types of animals and plants did you name in the Explore activity? Perhaps you named people, dogs, cats, birds, squirrels, bees, and other familiar species in your environment. No matter how many species you named, they will be only a small portion of the many species known.

How do scientists keep track of all these species? All living things on the planet can be classified into categories called kingdoms. Within these kingdoms, approximately 1.4 million species have been identified and named. Scientists estimate that many more millions of species are yet to be discovered. Some scientists estimate that the total of all life on Earth may be somewhere between 10 and 100 million species. The graph in **Figure 5-1** shows that plants, insects, and other animals show the most *diversity*, or variety, of living organisms that we know. Are you surprised to learn that insects represent more than half of all known species?

FIGURE 5-1
Which group of organisms has the most members? The fewest?

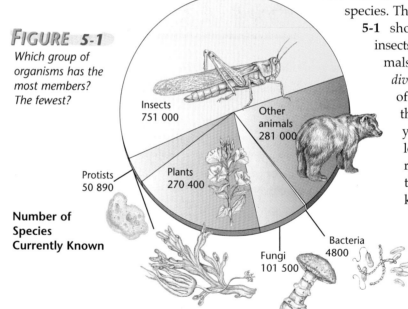

Insects
751 000

Other animals
281 000

Protists
50 890

Plants
270 400

Number of Species Currently Known

Fungi
101 500

Bacteria
4800

Diversity in the Tropical Rain Forest

You're back in the tropical rain forest now, still staring at the amazing stick insect. The forest is thick with vines, mosses, orchids, and trees, trees, trees. Insects buzz all around. Brightly colored birds flit from branch to branch. Other animals feed and rest in the treetops.

Tropical rain forests are full of life! These environments have a high number of species. Biological diversity or **biodiversity** (bi oh duh VUR suh tee) is the measure of the number of different species in an area. Tropical rain forests contain more different types of living things than any other places on the planet. They have a high biodiversity.

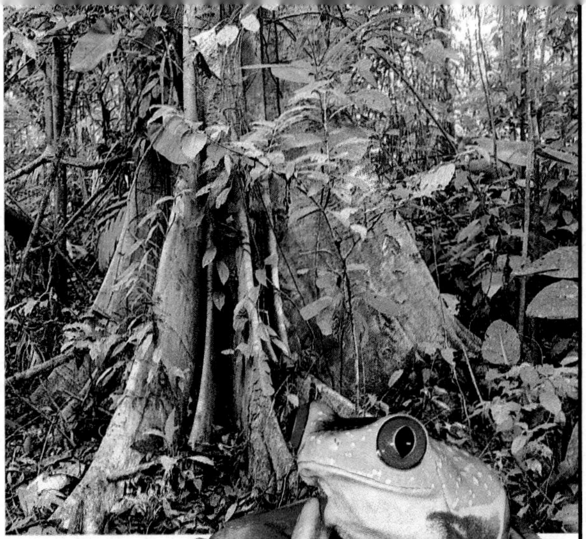

Why do tropical rain forests have such high biodiversity? To answer this question, think about an environment that contains fewer species, such as a desert. In the desert, temperatures are high and very little rain falls. Some desert landscapes have little plant life. What kinds of animals might be able to survive in this environment? Where would they live? What would they eat?

In tropical areas of the world, the climate provides more resources for organisms. Temperatures remain warm and steady all year, and rainfall is high. These conditions are perfect for the growth of plants. In the rain forest, you might find more than 200 different kinds of plants growing in an area the size of a football field. Living among these plants are many insects, birds, mammals, reptiles, and the amphibian shown in **Figure 5-2.** These animals use the plants for food and shelter. In short, tropical environments have high biodiversity because food, water, and shelter are plentiful.

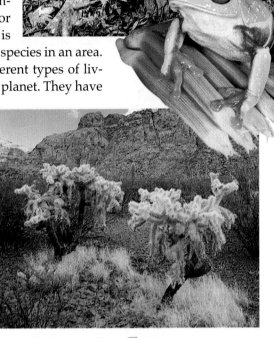

FIGURE 5-2
Animals such as the red-eyed tree frog (middle) find many places to live and many sources of food in the tropical rain forest (top). Deserts (bottom) have fewer sources of food and a harsh environment.

Tying to Previous Knowledge

Biodiversity also refers to diversity within a single species. Have students recall that sexual reproduction and mutations within the gene pool help introduce variation to a species.

2 Teach

Visual Learning

Figure 5-1 Which group of organisms has the most members? *insects* **The fewest?** *bacteria* Review and discuss each section of the circle graph with the class. Discuss why insects are so highly represented. Most reproduce in large numbers and at a rapid rate. Explain that insects can survive in many different and sometimes extreme environments. ELL LS

Teacher F.Y.I.

About three fourths of all animal species known are insect species. Within insects, about 40 percent are beetles.

FLEX Your Brain

Use the Flex Your Brain activity to have students explore BIODIVERSITY.

Activity worksheets, page 5

Inquiry Question

In addition to tropical rain forests, what other environments might have a high biodiversity? *Coral reefs have a lot of nutrients and grow in warm shallow waters. Many species live in this environment. Students might also suggest wetlands.*

Content Background

Tropical rain forests occupy only a small percentage of Earth's land surface, yet by some estimates they may contain more than half of all organisms on Earth. For example, of the known vascular plants, about 68 percent live in the tropics and subtropics. In addition, more than 1200 different species of butterflies were discovered in one Peruvian rain forest preserve, yet only 440 species live in the entire eastern United States.

Scientists have proposed a variety of explanations for why biodiversity is highest in the tropics. Many scientists think that more than one factor may be involved. Most argue that tropical ecosystems offer many opportunities for organisms, including plentiful food, water, sunlight, and shelter.

Prickly pear cactus

Namib viper

Diversity and Adaptation

From the stick insect in the rain forest, to a deep-sea fish, to a lizard in the desert, Earth is filled with living things that seem to fit into their environments. Their colors, shapes, sizes, and behaviors allow them to live in their surroundings. Any body shape, body process, or behavior that allows an organism to survive in its environment and carry out its life processes is called an **adaptation.**

As you can see in **Figure 5-3,** species of organisms have many different adaptations to help them survive. Some adaptations involve physical features, such as the body shape and color of a stick insect. These features cause the insect to blend into its environment, hidden from predators. The sharp spines of a cactus are an adaptation for protection. Some animals that have coloration that allows them to blend into the environment may also have behaviors that help protect them. The bobwhite quail and viper remain absolutely still in times of danger. As a result, these animals also stay hidden from predators.

Bobwhite quail

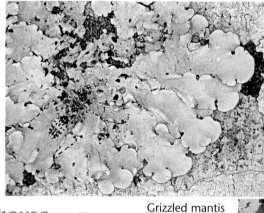

Grizzled mantis

FIGURE 5-3

Plants and animals show a variety of adaptations for survival. *What adaptations do you see in these organisms?*

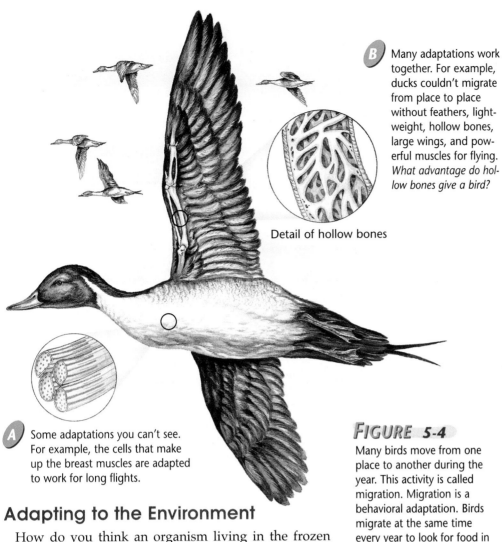

B Many adaptations work together. For example, ducks couldn't migrate from place to place without feathers, lightweight, hollow bones, large wings, and powerful muscles for flying. *What advantage do hollow bones give a bird?*

Detail of hollow bones

A Some adaptations you can't see. For example, the cells that make up the breast muscles are adapted to work for long flights.

Adapting to the Environment

How do you think an organism living in the frozen Arctic would be different from one that lives in the rain forest or the desert? Why would it have different adaptations? An organism's environment includes all of the living things in its surroundings. It also includes the nonliving things such as water, sunlight, and soil. For example, the thick white fur of a polar bear protects it from the cold and helps it blend into its icy, snowy surroundings. These are adaptations the polar bear has to its particular surroundings. Some organisms such as the bird in **Figure 5-4** can move to find a suitable environment.

FIGURE 5-4

Many birds move from one place to another during the year. This activity is called migration. Migration is a behavioral adaptation. Birds migrate at the same time every year to look for food in milder climates. Some ducks travel thousands of miles to a suitable environment. *What are some other behavioral adaptations of birds?*

Analysis

1. Answers will vary depending on the chosen adaptation cards.
2. Check answers to see that the students' choices for adaptations make their organisms suited for their particular environments.

✓ Assessment

Performance Have students design and illustrate their own fictional organisms and environments. Have students design organisms with sets of interesting and unusual adaptations. Encourage creativity. Use the Performance Task Assessment List for Cartoon/Comic Book in **PASC** p. 61.
LS **P**

3 Assess

Check for Understanding

Critical Thinking Show students a series of slides or photographs of different animal species. Have students examine the pictures to determine the animals' adaptations.

Mini LAB

Purpose

LS Kinesthetic In this activity on page 122, students explore the concept of adaptation by designing an environment for a fish with a given set of structural adaptations. **L2** **ELL**

Materials

adaptation cards, drawing paper, markers or colored pencils

Teaching Strategies

- Use blank 3 × 5 index cards to make the adaptation cards.
- Make four different categories of adaptation cards: body shape cards, coloration cards, mouth cards, and eye cards. Each category will contain different forms of the adaptation: body shape cards: snake-like body, round body, and oval-shaped body; coloration cards: blue-green, orange, red-and-white striped; mouth: large with many teeth, small with many teeth, large with a single tooth; eyes: large and round, small and round, slit-like.

📁 **Activity Worksheets,** pages 5, 35

4 Close

122

Mini LAB

Identifying Adaptations of Fish

What kinds of adaptations are useful for fish?

1. Obtain a set of four adaptation cards from your teacher.
2. Look at the cards. On a small piece of paper, draw a picture of a fish that matches the features given to you.
3. On a separate piece of paper, make a drawing of the type of environment your fish could survive in.

Analysis

1. Describe the adaptations of your fish.
2. Explain how the adaptations of your fish make it well suited to its environment.

FIGURE 5-5

A waxy outer layer and sharp spines are protective adaptations for many cacti.

Figure 5-5 shows up close an adaptation of a cactus plant. Cacti live in desert environments where water is in short supply. Thick, waxy coatings on the stems of cactus plants help prevent the cactus from drying up. This is an adaptation for conserving water. Try the MiniLab to see what adaptations a fish species has developed for its watery environment.

Section Wrap-up

1. Define the term *biodiversity.*
2. Explain why tropical rain forests have high biodiversity.
3. What are adaptations? Give an example of one.
4. **Think Critically:** What adaptations would an animal living in a desert need to survive?
5. *Skill Builder*
   Observing and Inferring Think of an organism, not a pet, that is familiar to you. Make a list of its traits. Next to each trait, describe how it helps the organism survive in its environment. If you need help, refer to Observing and Inferring on page 550 in the **Skill Handbook.**

USING MATH

Scientists have identified and named approximately 1 032 000 different species of animals. If 751 000 of these are insects, what percentage do insects make up in the total animal population? Show all your work.

RED-HEADED WOODPECKER
Melanerpes erythrocephalus

Description: — 10" (25 cm) *Whole head red*, wings and tail bluish black, with *large white patch on each wing;* white underparts; white rump, conspicuous in flight. Immature resembles adult, but has gray head, 2 dark bars on white wing patch.
Voice: — A loud *churr-churr* and *yarrow-yarrow-yarrow.*
Habitat: — Open country, farms, rural roads, open parklike woodlands, and golf courses.
Nesting: — 5 white eggs placed without nest lining in a cavity in a tree, telephone pole, or fence post.
Range: — Saskatchewan, Manitoba, and Quebec south to Florida and the Gulf Coast. Scarce in northeastern states. Winters in southern part of range.

Using Field Guides to Explore Diversity

A bird with a red head and black wings lands on a tree. It hammers into the bark with its beak. You're pretty sure it's a woodpecker, but what kind?

One place to find out is in a field guide, such as the *National Audubon Society Field Guide to North American Birds.* Field guides are handbooks for identifying living things, from mushrooms to mammals. They contain descriptions and photographs or illustrations, as well as information about where organisms live. Field guides are useful tools for making sense of the often-bewildering diversity of life on Earth.

Some field guides are organized around easily identifiable characteristics, such as shape or color. A field guide to flowers might group yellow flowers in one section, red in another, and so on. Other guides group together species that may look different, but that all belong to the same scientific family.

Using field guides is a bit like detective work. You follow clues, gradually narrowing your search until you can finally answer the question: "What's that?"

Science Journal

Spend several hours observing nature. Describe five plants or animals you saw. Make sketches and note colors, shapes, and sizes. Then, using field guides, try to identify the organisms on your list. If you have a beetle on your list, good luck—there are nearly 300 000 different kinds!

5-1 Diversity of Life   123

Source

Bull, John and John Farrand, Jr. *National Audubon Society Field Guide to North American Birds (Eastern Region).* New York: Knopf, 1994.

Background

- Most field guides describe where a given species can be found geographically, as well as the type of habitat it occupies. As students explore the diversity of life on Earth using field guides, emphasize that the adaptations exhibited by a particular organism can reveal a great deal about its survival strategies. A redheaded woodpecker, for example, has large feet with strong claws that help it cling to tree bark. It also has a strong beak for drilling into wood and stiff tail feathers for stability in a vertical position.

- While every kind of organism has its own unique set of adaptations, very different organisms that share the same habitat may have some adaptations in common. Mammals of the arctic tundra, such as arctic hares, arctic foxes, and lemmings, all have thick fur and relatively small ears and tails. These adaptations protect against heat loss to the environment.

Science Journal

Have students compile their data into a field guide for their local area.

Teaching Strategies

- As a class project, have students create a field guide to the living things that can be found near your school. They should first make detailed observations and identify species. After completing this research, they can write descriptions and create original drawings or take photographs to illustrate the guide. **L2** **COOP LEARN**

- Organize a field trip to a natural history museum or university that maintains a large reference collection of insects, such as butterflies or beetles. Afterwards, encourage students to write about their impressions of insect diversity. **L2**

- Have students put on a play in which they take the part of different animals or plants in a rain forest, along a rocky seashore, or on a dry grassland. Encourage students to use their imaginations to demonstrate the physical and behavioral adaptations of the diverse species they represent. **L2** **ELL**

Prepare

Section Background

Scientists consider evolution to be the central, unifying theme of biology. Natural selection is regarded as the main mechanism for evolutionary change.

Preplanning

Refer to the Chapter Organizer on pages 116A-B.

1 Motivate

Bellringer

 Before presenting the lesson, display **Section Focus Transparency 13** on the overhead projector. Assign the accompanying **Focus Activity** worksheet.
L2 ELL

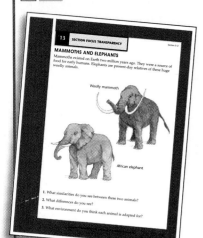

Tying to Previous Knowledge

You may wish to review some of the concepts in Chapter 4, such as genes and DNA, before you study this section.

5•2 Unity of Life

What YOU'LL LEARN

- How species change and new species form through natural selection
- About the evidence that supports the idea that organisms change through time

Science Words:
natural selection
common ancestor
fossil

Why IT'S IMPORTANT

If you know how organisms are adapted to their environments, you can better understand how the environment can affect organisms.

Why are they like that?

You probably would never confuse an eagle with a hummingbird. Both have wings, beaks, feathers, and other familiar features of birds, but they are different in size, shape, and where they live. On the other hand, you might have trouble recognizing different kinds of birds called warblers unless you were an experienced bird-watcher. Many species of these small birds look alike. Why are some species more similar than others? And how do new species form? Through the study of living and once-living organisms, scientists try to find the answers.

Adaptation Through Natural Selection

In the last section, you learned that organisms are adapted to the environments in which they live. How do these adaptations happen? Recall from Chapter 4 that individual organisms are not identical to each other, even if they belong to the same species. In every species, variations, or differences, occur in the traits of that species. In a population of gray squirrels, for example, individuals may have slightly different colors of fur. Most will have brown-gray fur color. A few will have dark fur, and some will have light fur.

FIGURE 5-6
This illustration shows how a population of squirrels can change through natural selection.

 A In all species, individual organisms have differences, or variations.

B In this population, some squirrels may be alert and hide more quickly when predators come near.

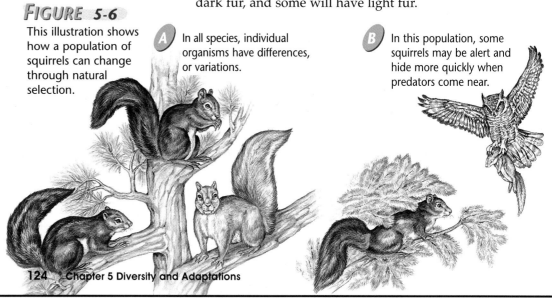

124 Chapter 5 Diversity and Adaptations

Program Resources

📁 Reproducible Masters
Activity Worksheets, pp. 5, 31-32, 36 L2
Enrichment, p. 19 L3
Lab Manual, pp. 35-38 L2
Reinforcement, p. 19 L2
Study Guide, p. 19 L1

🔦 Transparencies
Section Focus Transparency 13 L2

Now, imagine this same population of squirrels in an environment that changed. Suppose the environment changed so that squirrels with dark fur are better able to survive than squirrels with a lighter fur color because predators are not able to see squirrels with darker fur as easily. Dark squirrels would be better off in this environment. They would be more likely to survive and have offspring, as shown in **Figure 5-6**. This process, in which organisms with characteristics best suited for the environment survive, reproduce, and pass these traits to their offspring, is called **natural selection.**

Now, think about the process of natural selection happening over many, many years. With each generation, more and more dark squirrels will survive and pass their coat color on to their offspring. Squirrels with lighter fur don't survive and so produce fewer offspring. The population changes. Every generation, the population consists of more dark squirrels than the one before it. This process of natural selection is the main way changes happen in a population.

Natural selection can explain why polar bears are white, why some other bears are brown, and why some insects that live in the leaves of a tree are green. Think of how the stick insect at the beginning of the chapter came to look the way it does.

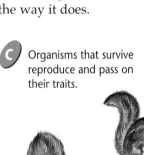

C Organisms that survive reproduce and pass on their traits.

D There is still variation, but the population has a higher percentage of individuals with beneficial variations.

2 Teach

Discussion

After you have introduced the concept of natural selection, ask students if they have ever heard the phrases "only the strong survive" or "survival of the fittest." Discuss with students the meaning of these phrases as they relate to natural selection.

Using an Analogy

Compare the way species change over time with the way cars have changed in the past 50 years. Relate the *artificial selection* of cars by car designers to the process of *natural selection.*

? FLEX Your Brain

Use the Flex Your Brain activity to have students explore NATURAL SELECTION.

📁 **Activity Worksheets,** page 5

Activity

LS **Kinesthetic** Variation in the population is essential for natural selection. Have students learn about variation in the human population. Give each student a metric ruler. Have each student measure the length of the pinky finger of his or her left hand in millimeters. Instruct students to record their measurements on the chalkboard to show how finger length varies. **L2** **ELL**

Content Background

Natural selection is considered the central mechanism of evolution. English naturalist Charles Darwin (1809–1882) is credited with developing the idea of natural selection. Nearly 150 years later, Darwin's hypothesis that natural selection is an important factor in evolutionary change is still an important one.

Visual Learning

Figure 5-6 Have students carefully read each caption. Explain that this is a hypothetical case of natural selection. Point out that random factors in the environment are involved and that the effects of natural selection cannot be predicted. **LS**

Activity 5-1

PREPARE

Purpose

IS **Kinesthetic** In this activity, students will carry out a simulation of the process of natural selection. **L2** **ELL** **COOP LEARN**

Process Skills

observing and inferring, measuring in SI, interpreting data, designing an experiment to test a hypothesis

Time

40 to 50 minutes

Safety Precautions

Make sure students do not eat the jelly beans.

Possible Hypotheses

Accept all reasonable hypotheses. Many students will say that organisms that are camouflaged are better able to avoid predators.

Activity Worksheets, pages 5, 31-32

PLAN

Possible Procedures

Start the simulation with a population of 20 to 30 insects (jelly beans). Be sure that all colors are evenly represented at the start. The data chart should indicate the starting population numbers. Spread the jelly beans evenly over a grassy patch of lawn or a solid-colored tablecloth. The bird will have 10 seconds to pick up as many jelly beans as he/she can. These jelly beans (insects) are then discarded. After this "generation," the jelly beans left on the ground are counted. All of the surviving jelly beans will reproduce. For each jelly bean color remaining, add two more jelly beans of that color to the population. Be sure

Activity 5-1

Design Your Own Experiment

Simulating Selection

Natural selection causes a population to change. In this activity, you will design an experiment to discover how camouflage adaptations—those adaptations that allow organisms to blend into the environment—happen through natural selection.

PREPARE

What You'll Investigate
How does natural selection work?

Form a Hypothesis

Any body shape, structure, or coloration of an organism that helps it blend in with its surroundings is a camouflage adaptation. Make a hypothesis about how natural selection can explain camouflage adaptations.

Goals

Model natural selection in a population of insects.

Explain how natural selection produces camouflage adaptations.

Safety Precautions

Do not eat any jelly beans.

Possible Materials
- bag of small-sized jelly beans in assorted colors (approximately 100 beans)
- meterstick
- paper
- pencil
- Astroturf rug or another solid colored rug
- watch with a second hand

PLAN

1. With a partner, **discuss** the process of natural selection. How does natural selection cause species to change?
2. In this activity, one student will play the role of the bird that eats insects (jelly beans) that live in grass. With your partner, think about how you can model natural selection using some of the materials suggested.

to indicate the changes in the data table. Carry out this procedure for at least five generations.

3. With your partner, **make a list** of the different steps you might take to model natural selection.

4. You might start the experiment with a particular insect population. How many insects (jelly beans) will be in your starting population? How will you show variation in your starting population?

5. Think about how many generations of insects you will use in the experiment.

6. What data will you collect? Will you need a data table for this experiment? If so, **design a table** in your Science Journal for recording data.

7. What happens to individual organisms if they have favorable variations for a particular environment? What happens if an organism has unfavorable variations? Make sure you think about these questions in designing your experiment.

DO

1. **Review** your list of steps to make sure that the experiment makes sense and that all of the steps are in logical order.

2. Make sure that your teacher has approved your plan before you continue.

3. While doing the experiment, **record** your observations and **complete** your data table in your Science Journal.

4. Carry out the experiment.

CONCLUDE AND APPLY

1. **Observe** which types of insects the bird in your experiment was able to locate most quickly. Why?

2. Did your population of insects change over time? **Explain** your answer.

3. **APPLY** How can your experiment be used to **explain** camouflage adaptations in organisms?

5-2 Unity of Life **127**

Go Further

As an extension to this activity, students can add different types of selection pressures to the simulation. For instance, instead of adding two new individuals for each surviving insect color, students can set up the simulation so that the different insects reproduce at different rates. Red insects, for example, might contribute more offspring every generation than other types of insects.

Visual-Spatial Show students slides or photographs of different types of cats, such as house cats, lions, tigers, and leopards. Have students make a list of all of the features these animals have in common. Explain that the many species of cats show similarities. The large number of similarities indicates that cats evolved from a common ancestor. **L2**

Revealing Preconceptions

Many students mistakenly think that *individual* organisms change over time. Point out that individuals don't evolve; populations do. Explain that natural selection does operate on individuals but that it is the *population* as a whole that changes over time.

Enrichment

 Intrapersonal Geographic isolation is one reason why island animals and plants sometimes display different types of adaptations compared to their mainland relatives. Have interested students research information about unusual island species. Encourage students to share their findings with the class in an oral presentation or with a poster project. **L3**

GLENCOE TECHNOLOGY

 Videodisc

The Infinite Voyage: Insects: The Ruling Class

Chapter 5

Caterpillars: Altering Appearances

A The members of this population of lizards all look alike and are able to interbreed.

FIGURE 5-7

One way new species form is for a population to be separated.

The Origin of New Species

If you've ever watched nature programs showing lions and tigers in the wild, you've seen the animals hunting, washing themselves, and caring for their young. You may have seen these same behaviors in a house cat. Maybe you've also noticed that all types of cats—lions, tigers, bobcats, and others—have similar features. How did the various species of cats form, and why do they share so many features?

Remember that scientists define a species as a group of organisms that look alike and reproduce. One way a new species can form is if a large population of organisms becomes separated into smaller populations that no longer breed with others of its kind. A barrier, such as a mountain range or river, can cause this separation. **Figure 5-7** shows an example.

A New Species Develops

Let's look at a population of lizards. Suppose a river divides the whole population into two smaller populations. Lizards in one group can no longer mate with lizards in the other groups. They can mate only among themselves. Over time, each small population will adapt to its own environment. Because each environment is a little bit different, each small population will develop different adaptations. Over time, the populations may become so different from each other that offspring couldn't be produced even if the lizards could again mate with one another. Each small population of lizards

---

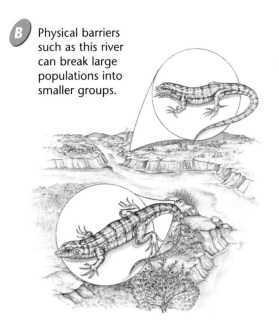

B Physical barriers such as this river can break large populations into smaller groups.

C New species can form when a population is separated and individuals in one population no longer mate with individuals on the other side of the barrier.

has become a new species. Because the different species of lizards in this example all arose from one population, we can say that they share a **common ancestor** (AN ses tur). Each of the new lizard species now has its own specific traits. But all the species share some traits that they inherited from their common ancestor.

Evidence for Change Over Time

You might say seeing is believing. But we can't always see everything we'd like to study. Objects may be too tiny or too far away for us to see. Processes may happen too quickly or too slowly for us to observe. That's the problem with natural selection. It occurs in all species, but it usually occurs so slowly that it's usually not possible to see the process in action. So how can scientists be sure that species have changed over time? They rely on several types of evidence. They compare physical traits of species that are similar. They study the DNA and proteins of species for clues to how closely they are related. They look at fossils. Let's take a closer look at each of these methods.

5-2 Unity of Life **129**

Content Background

The Hawaiian islands offer a perfect example of how geographic barriers contribute to speciation. The islands were originally colonized by birds, crickets, wasps, beetles, snails, flowering plants, and other kinds of organisms that were carried by the wind from Asia and North America. As the organisms spread into the diverse niches on the islands, they evolved in response to local environmental conditions. Eventually, the species became distinct enough from the ancestral populations left behind in Asia and North America to become new species.

Mini LAB

Purpose

LS **Kinesthetic** Students learn how mold and cast fossils are made by making their own fossils. **L2** **ELL**

COOP LEARN

Materials

petroleum jelly; pint-sized milk container; plaster of paris; water; small objects such as keys, beads, or rings

Alternate Materials

modeling clay or Playdoh instead of plaster of paris

Teaching Strategies

- You may wish to have students fill a cup with modeling clay or Playdoh and press an object into it. The plaster of paris should then be poured into the clay. The clay can be pulled from the plaster when it dries.

- The types of fossils students will make in this laboratory are mold and cast fossils.

- Show students real examples of mold and cast fossils. Have students examine the fossils to determine what kinds of organisms are preserved.

- For best results, make sure that plaster has completely dried and hardened before removing the objects.

📁 **Activity Worksheets,** pages 5, 36

Analysis

1. Answers will vary depending on how well students performed the activity. But in general, the casts of the molds usually provide the most information.

130

Mini LAB

Making a Fossil

How can you make a cast fossil from a mold?

1. A mold fossil is formed when an organism leaves an impression in clay or mud. When the mold is later filled in, a cast fossil can be formed. To make a model of a fossil, grease the inside of a pint-sized milk container with petroleum jelly.
2. Pour plaster of paris into the milk container until it is half full.
3. When the plaster begins to thicken, grease some small objects and press them into the plaster.
4. After the plaster has hardened, remove the objects. You have now made a mold.
5. Next, grease the entire layer of hard plaster. Pour another layer of colored plaster to fill in the mold.
6. After the plaster hardens, tear away the milk container and separate the two layers of plaster. The colored layer shows your cast.

Analysis

Which fossil showed the most details of the objects you used, the mold or the cast?

FIGURE 5-8
Trilobites are an example of an organism that changed through time.

Fossils Show Change over Time

Although dinosaurs have been extinct for millions of years, they come to life in the movies. Do you think those giant, roaring creatures on screen are anything like real dinosaurs were? How do scientists and the movie producers know how dinosaurs looked and acted? Scientists have learned a lot about the ancient reptiles by studying their fossils. **Fossils** are the remains or traces of ancient life.

Fossils give scientists the most direct evidence that species change over time. For example, fossils of trilobites (TRI luh bites) show that change occurred in their structure over time. Trilobites are extinct relatives of animals you know today such as lobsters, crabs, and insects. Early trilobites had generalized body structures with few segments. Fossils of later trilobites show many more specializations and differences in shape. **Figure 5-8** shows examples of these fossil organisms. Through the study of fossils, scientists have shown that trilobites changed over time.

✔ **Assessment**

Portfolio Encourage students to include their best plaster fossils in their portfolios. Use the Performance Task Assessment List for Science Portfolio in **PASC**, p. 105. **P**

Community Connection

Local Fossils Fossils provide scientists with the most direct evidence for evolution. Fossils represent physical evidence that organisms have changed over time. Have students research information about where fossils may be collected locally. Most libraries have reference books that detail geological formations in the state where fossils can be found. **L2** **LS**

FIGURE 5-9
Wolves and dogs are close relatives. *What physical similarities can you see between these two animals?*

DNA Shows Relationships

Appearances can trick you. Some species look a lot alike but aren't closely related. Scientists need something besides physical appearance to help them figure out how closely related species are. In Chapter 4, you learned that DNA determines the traits of organisms. The more similar the DNA of two species is, the more closely they are related. For example, in comparing the DNA of modern breeds of dogs, wolves, and foxes, the gray wolf, *Canis lupus,* turns out to be the closest relative of the domestic dog. These animals are shown in **Figure 5-9.**

Problem Solving

More Chemical Clues

You've learned that DNA can be used to show how organisms are related. Scientists also use other chemicals found in living things to show how closely related two species are. One type of chemical they study is protein. Proteins perform a variety of jobs in living things. Some are used in the building of living material, such as bones, muscles, and skin. Others help living things grow, digest foods, and fight diseases. Scientists can learn about the relationships between species by studying the structure of proteins. Each protein is made of building blocks put together in a specific order. In closely related species, similar proteins have a similar order of building blocks. In species that are more distantly related, the same proteins have different arrangements of building blocks.

Solve the Problem:

The illustration shows the structure of a type of protein used in digestion found in three unknown bird species (Species A, Species B, and

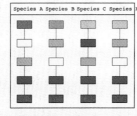

Species C) and in a known bird species (Species X). The colored blocks represent the building blocks of the digestive protein. Compare the structure of the proteins in all four species.

Think Critically:

Which bird species (A, B, C) do you think is most closely related to Species X? Why did you reach your decision?

Brainstorming

Have students examine the photographs in Figure 5-9 on page 131. Explain how DNA sequencing studies were performed to determine the relationships among dogs, wolves, and foxes. Many students will not be surprised to discover that wolves are closely related to dogs.

3 Assess

Check for Understanding

Discussion Have students make predictions about how natural selection might operate on a population of hypothetical organisms given particular environmental factors.

Reteach

LS **Linguistic** Students who have difficulty understanding the concept of natural selection may benefit by learning how different breeds of dogs arose. Explain that all of the different types of dogs belong to a single species, *Canis familiaris*. Point out that dog breeders have produced the different breeds of dogs by artificially selecting the characteristics they wanted. Have students discuss the similarities and differences in artificial selection and natural selection.

Extension

For students who have mastered this section, use the **Reinforcement** and **Enrichment** masters.

Teacher F.Y.I.

Explain that studies of the anatomy and DNA of cichlid fish in Lake Victoria indicate that the species evolved from a common ancestor that colonized Lake Victoria from nearby lakes.

Anatomy and Ancestry

If you were a biologist studying the many species of cichlid (SI klud) fish in Africa's Lake Victoria, you would see differences in the jaws and teeth of these fish. You would also see differences in their color. Some of these differences are shown in **Figure 5-10.** Some species have small mouths and sharp teeth for eating insects. Others have mouths adapted for scraping algae. Still others are adapted for eating the scales of other fish. However, despite the differences, all 170 species of cichlids in the lake have many more things in common.

Scientists compare similarities and differences in the body structures of all living things. This gives them clues to how species may have changed over time. After comparing the anatomy of many species, scientists conclude that all species may have come from a common ancestor.

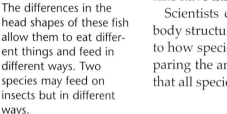

FIGURE 5-10
The differences in the head shapes of these fish allow them to eat different things and feed in different ways. Two species may feed on insects but in different ways.

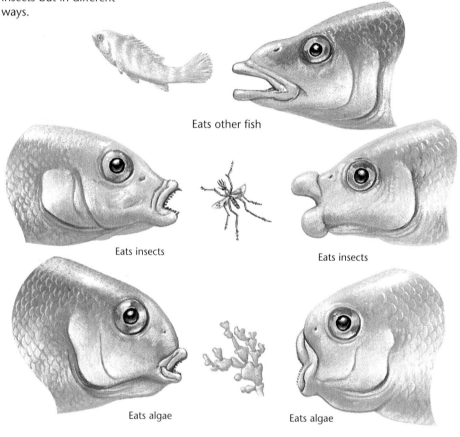

Eats other fish

Eats insects

Eats insects

Eats algae

Eats algae

Content Background

The hypothesis that dogs and wolves are closely related rests on a growing body of molecular data, as well as the striking anatomical, behavioral, and physiological similarities between the species. Most agree that dogs became domesticated when wild wolves developed close associations with hunting and gathering peoples by 12 000 years ago—possibly earlier.

Visual Learning

Figure 5-11 What do the similarities between these two organisms tell you about the probable ancestor of *Archaeopteryx*? *Scientists view these similarities as evidence of a common ancestor for these two organisms.* **LS**

Another example of similarities in anatomy between two organisms is seen in **Figure 5-11.** Do you see these similarities? If you said yes, then you made the type of observation that scientists use to find relationships among organisms. Scientists view these similarities as evidence that two organisms developed from a common ancestor.

FIGURE 5-11

Archaeopteryx (above) has many similarities to *Ornitholestes* (below), a small dinosaur. *Archaeopteryx* is considered a close relative of the ancestor of modern birds. *What do the similarities between these two organisms tell you about the probable ancestor of* Archaeopteryx?

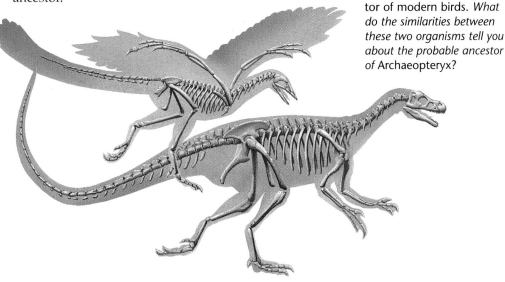

Section Wrap-up

1. Explain natural selection.

2. Briefly describe how natural selection plays a role in how species adapt to the environment.

3. **Think Critically:** Explain how fossils can show that species change over time.

4. **Skill Builder**
   Concept Mapping Make an events chain that describes how camouflage adaptations may arise in a population of birds. If you need help, refer to Concept Mapping on page 544 in the **Skill Handbook.**

Science Journal

Some scientists hypothesize that dinosaurs and birds are closely related. Imagine you are a scientist who wants to examine this hypothesis. In your Science Journal, describe methods you would use to investigate the relationship between dinosaurs and birds.

Skill Builder

Concept Mapping Birds show variation in coloration → variations give some individuals an advantage → birds with favorable variations leave more offspring than other birds → after many years, population consists mostly of birds with favorable variations.

Assessment

Performance Have students do further research on the relationship between birds and dinosaurs. Have students do research about bird ancestors, such as *Archaeopteryx* and *Protoavis.* Encourage students to present their findings in an oral presentation with visuals to the class. Use the Performance Task Assessment List for Oral Presentation in **PASC,** p. 71 [L3] [IS] **COOP LEARN**

4 Close

MINI QUIZ

Use the Mini Quiz to check students' recall of chapter content.

1. **The process by which organisms well suited for the environment survive, reproduce, and pass their traits to offspring is known as _____.** *natural selection*

2. **Barriers can split large populations into smaller ones. Eventually, two different species may form. This process is called _____.** *geographic isolation.*

Section Wrap-up

1. Natural selection is the process in which organisms with favorable variations for an environment will survive, reproduce, and pass their traits to offspring.

2. Natural selection allows a species to adapt to changes in its environment. Individuals that are not well suited to an environment leave few offspring. Individuals that are well suited for an environment will leave more offspring. Over time, the population will consist of individuals that are well adapted to the environment.

3. **Think Critically** External features of fossil species can be compared to those of modern species to show change over time.

Science Journal

Answers will vary, but most students should indicate that dinosaur and bird skeletons could be examined for similarities.

5•3 Captive-Breeding Programs

What YOU'LL LEARN

- How captive breeding programs help endangered species
- How to identify resources

Science Words: captive breeding

Why IT'S IMPORTANT

Learning how to identify and use information resources will help improve your research skills.

FIGURE 5-12

Siberian tigers once roamed Russia, China, and Korea in large numbers. It is now estimated that there are less than 200 Siberian tigers left in the wild.

Have you ever seen a Siberian tiger such as the one in **Figure 5-12** in a zoo? Zoos are great places to see and learn about animals from all over the world. But there's more to a zoo than meets the eye. In the United States, many zoos belong to the American Zoo and Aquarium Association. This group is trying to save endangered species using a variety of methods, including captive breeding programs. Endangered species are those that have only a small number of living members left.

Captive Breeding

Captive breeding is the breeding of endangered species in captivity. Usually, the species that are bred in captivity have only a few living members left in the wild. Through captive breeding, the number of individuals in the population is built up if the program is successful. Some of the animals that are bred by a zoo are released into the wild. Others are kept by the zoo or given to other zoos to breed more animals. Why do zoos trade animals for breeding? Zoos can't breed animals that are born of the same parents over and over again because this would weaken the genetic diversity and health of the species, so they trade animals with other zoos.

Breeding Programs

Let's look at one captive-breeding program that works with several states and zoos to help save black-footed ferrets from extinction.

134   Chapter 5 Diversity and Adaptations

In 1972, it was thought that the black-footed ferret was extinct. No one had seen one for years. But in 1981, a Wyoming rancher found a dead ferret on his land. Soon after, a small group of ferrets, like those shown in **Figure 5-13,** was discovered nearby. The last of these wild ferrets was captured in the early 1990s for captive breeding. This group has increased from 18 to more than 500.

Since then, captive-breeding programs for the ferrets have been started at zoos and wildlife centers from Virginia to Colorado. Some of the ferrets are being released into the wild; others are kept for breeding. By the year 2010, it is hoped that 1500 ferrets will be living in the wild.

Would you like to learn more about ferrets or other animals that are bred in captivity? You can do this if you know which resources to use. Resources are things that help you to learn more about a subject. Encyclopedias, magazines, the Internet, newspapers, people—all of these are resources.

FIGURE 5-13

Black-footed ferrets born in captivity in a breeding program (top) can be returned to their natural environment (bottom).

 Skill Builder: *Using Resources*

LEARNING the SKILL

1. To identify resources, first look for key words or events in the material you read. You'll use these things as guides to help you in your research.

2. With the help of your teacher or a librarian, list all the resources available for you to use.

3. Decide what type of information you're looking for. Are you looking for local events? A newspaper or an interview with a local resident might be the best resource for you to use. Are you looking for historical events? Then use an encyclopedia or the Internet.

PRACTICING the SKILL

1. What is captive breeding?

2. What keywords would you use to learn more about captive breeding?

3. Which resources would you use to find out more about captive breeding in general? What resources would you use to find out about captive-breeding programs in your area?

APPLYING the SKILL

Choose a topic that interests you to research. It could be anything from insects to stars and galaxies. Use at least three resources to research the topic. Share your research with the class.

5-3 Captive-Breeding Programs   **135**

Prepare

Section Background

In this section, students learn how the history of life on Earth is preserved by the fossil record. Although the fossil record is imperfect, it does give scientists a great understanding of what life was like in the past and how it has evolved over time.

Preplanning

Refer to the Chapter Organizer on pages 116A-B.

1 Motivate

Bellringer

Before presenting the lesson, display **Section Focus Transparency 15** on the overhead projector. Assign the accompanying **Focus Activity** worksheet.

L2 **ELL**

Tying to Previous Knowledge

In this section, students learn more about how fossils provide direct evidence for the fact that species change over time.

136

5●4 History of Life

What YOU'LL LEARN

- The importance of the fossil record
- About the geologic time scale
- The possible causes of mass extinctions

Science Words:
fossil record
geologic time scale
mass extinction

Why IT'S IMPORTANT

Learning about the history of life on Earth will help you understand the diversity of life that is now on Earth.

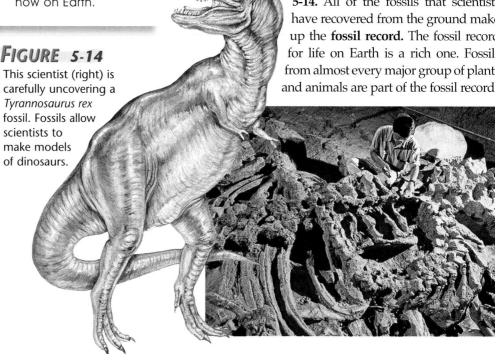

FIGURE 5-14

This scientist (right) is carefully uncovering a *Tyrannosaurus rex* fossil. Fossils allow scientists to make models of dinosaurs.

A Trip Through Geologic Time

Have you ever seen a movie or read a book about time travel? What if you could travel back in time into Earth's past? What kinds of interesting creatures would you see? Time travel will always be science fiction rather than science fact. But in a way, scientists can travel back through time by studying fossils. You learned that fossils are evidence of ancient life. Fossils help scientists form a picture of the past. They provide a history for life on Earth. Let's go along on the journey through Earth's history.

The Fossil Record

Earth's rocky crust is a vast graveyard that contains the fossil remains of species that have lived throughout Earth's history. Large fossils are carefully removed from the ground as shown in **Figure 5-14.** All of the fossils that scientists have recovered from the ground make up the **fossil record.** The fossil record for life on Earth is a rich one. Fossils from almost every major group of plants and animals are part of the fossil record.

136 Chapter 5 Diversity and Adaptations

Program Resources

📂 Reproducible Masters
Activity Worksheets, pp. 5, 33-34 **L2**
Enrichment, p. 21 **L3**
Reinforcement, p. 21 **L2**
Study Guide, p. 21 **L1**

 Transparencies
Section Focus Transparency 15 **L2**
Teaching Transparency 10 **L2**

However, the fossil record doesn't show a complete history. Although many major groups of plants and animals are represented, not every species is. It is much more common for an organism to decay without ever becoming a fossil.

USING TECHNOLOGY

Blood in the Bones

In 1991, researchers at Montana State University studying a *Tyrannosaurus rex* fossil made what may turn out to be an important discovery. "I think I've found red blood cells," one of the researchers said as she focused her microscope on a slide made from the dinosaur fossil.

The Real Jurassic Park Lab

Were the mysterious structures in the dinosaur bones really red blood cells? If so, the scientists reasoned that they would find hemoglobin (HEE muh gloh bun), the blood protein that carries oxygen to body cells. To test for hemoglobin, the dinosaur tissue was hit with laser light. The tissue produced the same chemical signature as modern hemoglobin-carrying blood cells. In the second test, an extract was made from the dinosaur tissue and compared with the blood proteins from birds, crocodiles, and humans—all organisms with hemoglobin in their red blood cells. Similarities between the blood samples gave the scientists more evidence that hemoglobin, and therefore blood cells, were present in the fossil.

If these cells are dinosaur blood cells, scientists may be able to get DNA from them. Scientists aren't interested in bringing *T. rex* back to life as the characters did in the movie *Jurassic Park*. But they could compare dinosaur DNA to DNA in living organisms. If so, the world would gain a better picture of the nature of these terrible lizards.

Scientist studying dinosaur fossils

*inter*NET
CONNECTION

To what living organisms would the scientists compare dinosaur DNA? Check the Glencoe Homepage. *www.glencoe.com/sec/science*

137

---

2 Teach

? FLEX Your Brain

Use the Flex Your Brain activity to have students explore the GEOLOGIC TIME SCALE.

📁 Activity Worksheets, page 5

USING TECHNOLOGY

- Have students read the feature. Initiate a discussion about the importance of the discovery. Ask students to discuss how the discovery of dinosaur blood cells and DNA would add to our modern understanding of dinosaurs.
- In 1997, scientists reported that DNA had been recovered from the bones of *Neanderthals,* a group of ancient humans. Have interested students research this important discovery. L3

*inter*NET
CONNECTION

The Glencoe Homepage at **www.glencoe.com/sec/science** provides links to information related to this Internet connection.

Across the Curriculum

Literature Read several passages to students from the books *Jurassic Park* and *The Lost World* by Michael Crichton. Both books deal with the subject of dinosaur paleontology. Read passages describing the environment and the animals living at that time. Have students describe how the passages made them feel and the picture the words formed for them of the time written about. L2 LS

IS Logical-Mathematical Students sometimes have difficulty grasping the incredible amount of time that species have been evolving on Earth. Tell students to imagine that the entire history of Earth is compressed into a single year. Tell them that Earth formed on January 1. On the chalkboard, write the following events and corresponding calendar dates: Earth forms—January 1; oldest fossils—March 21; primitive animals appear—November 14; dinosaurs appear—December 12; dinosaurs go extinct—December 26; people appear—December 31, 11:49 P.M.

Visual Learning

Table 5-1 Review the geologic time scale with students. Begin by discussing what the earliest forms of life were like, then show students where these organisms would fit in on the geologic time scale. Explain that there was an explosion of diversity about 590 million years ago. This is when the real fossil record of life on Earth begins. Review each of the remaining time periods with students. **IS**

Table 5-1

Geologic Time Scale

Era	Period	Million years ago	Major evolutionary events	Representative organisms
Cenozoic	Quaternary	1.6	First humans	
	Tertiary	66		
Mesozoic	Cretaceous	144	Large dinosaurs First flowering plants	
	Jurassic	208	First birds First mammals	
	Triassic	245		
Paleozoic	Permian	286		
	Pennsylvanian	320	First conifer trees; first reptiles and insects	
	Mississippian	360		
	Devonian	408	First amphibians and land plants; first bony fish	
	Silurian	438	First fish with jaws	
	Ordovician	505	First vertebrates, armored fish without jaws	
	Cambrian	544	Simple invertebrates	
	Precambrian		First fossilized animals and plants; protozoa, sponges, corals, and algae	
		4000	First fossil bacteria	

Geologic Time Scale

By studying the fossil record, scientists have put together a sort of diary for life on Earth called the **geologic time scale.** The geologic time scale helps scientists keep track of when a species appeared on Earth or when it disappeared from Earth.

You can see the geologic time scale in **Table 5-1.** As you can see, the geologic time scale is divided into four large intervals of time called eras, and each era is subdivided into periods. The beginning or end of each time period marks an important event in Earth's history, such as the appearance or disappearance of a group of organisms. The fossil record of life on Earth gives scientists strong evidence that life has changed over time.

Science Journal **Time Machine** Have students choose a time period in Earth's history they are interested in. Gather some books, magazines, and other references for students to use to find out some basic information about that time period. Students should gather basic information about the time period, such as the global climate and typical plants, animals, and other species known through fossils. After students have finished the research, have them write a short story about what it would be like if they were able to take a time machine back to visit that particular period. **L2 P IS**

How big were the dinosaurs?

Activity 5-2

Fossils can be used to estimate the size of a dinosaur. Dinosaur models also may be used.

What You'll Investigate
How can you estimate the mass of a dinosaur?

Procedure

1. **Make a data table** in your Science Journal. Include type of dinosaur, its scale, and volume of water displaced in your table.
2. Obtain a plastic dinosaur model. Fill in your data table. A good scale to use is a 1:40 scale, meaning that each dimension of the model—length, width, and height—is 1/40 of the dinosaur's actual size.
3. **Fill** the pail with water until it is almost full. Carefully place the pail in the utility pan. Fill the pail to the brim, taking care not to let any water spill over into the utility pan.
4. **Submerge** the dinosaur model into the water, allowing the water to flow out of the pail and into the utility pan.
5. Carefully remove the pail from the utility pan.
6. Pour the water from the utility pan into a graduated cylinder. The volume of the water displaced is equal to the volume of the dinosaur model.
7. **Record** your measurement in your data table.
8. To find the mass of the actual dinosaur in grams, multiply the volume of the model by the cube of the scale. If the scale is 1:40 and the model has a volume of 10 mL, you would multiply 10×40^3 ($40 \times 40 \times 40$). To find the mass in kilograms, divide by 1000.

Conclude and Apply

1. What is the estimated mass of your dinosaur in kilograms?
2. Male Indian elephants weigh about 5500 kg. What can you **infer** about the approximate size of some dinosaurs?

Goals

- Estimate the mass of a dinosaur using a model.

Materials

- plastic dinosaur model
- plastic graduated cylinder
- marking pen
- calculator
- water
- small pail
- utility pan

5-4 History of Life   **139**

Answers to Questions

1. Answers will vary, but check to make sure that the estimates are reasonable. For instance, the largest species was *Brachiosaurus* at 80 000 kg.
2. Most students will find that their dinosaurs were heavy, especially the students who used a *Tyrannosaurus rex* model or a *Brachiosaurus* model.

✔ Assessment

Performance Have students work in groups to make posters or three-dimensional models comparing the sizes of particular dinosaur species with modern animals. Use the Performance Task Assessment List for Poster in **PASC**, p. 73. **LS** **COOP LEARN** **P**

Activity 5-2

Purpose

LS **Kinesthetic** In this activity, students estimate the live mass of a dinosaur. The method used provides for integration of math and science. **L2**

Process Skills

measuring in SI, comparing and contrasting, observing and inferring

Time

30 minutes

📁 **Activity Worksheets,** pages 5, 33-34

Teaching Strategies

- Explain that some dinosaur models and toys can be used to estimate the mass of the actual dinosaur because the models are accurately proportioned and are based on actual fossil skeletons.
- The *Carnegie* dinosaur models found in museum shops, some toy stores, and in the Carolina Biological Supply catalog are model dinosaurs made to a 1:40 scale.
- Students should realize that they are assuming that the density of the dinosaur is the same as water, 1 g/mL. If the volume of the dinosaur is 40^3 mL, then the mass of the dinosaur is 40^3 g.

Troubleshooting Dinosaur models must be marked with the scale. A small dinosaur such as a *Dimetrodon* may fit in a graduated cylinder. Students would record the original water level, submerge the dinosaur, and record the new water level. The difference between the two measurements is the volume of the dinosaur.

139

140

Extinction of Species

Why don't you see dinosaurs at the zoo? These large reptiles ruled Earth for more than 100 million years, but they're all gone now. Among the dinosaurs were the largest land animals, such as the plant-eating *Brachiosaurus* and the meat-eating *Tyrannosaurus rex*. Yet, despite the great success and long history of the dinosaurs, all of them became extinct about 66 million years ago. So did many other land and sea animals.

Scientists aren't surprised that dinosaurs are now extinct. That's because about 99 percent of all species that have ever existed are now extinct. Usually, extinction is a natural event. As you have learned, Earth's environments are constantly changing. When a species can't adapt to changes in its environment, it becomes extinct forever.

FIGURE **5-15**
The second-largest mass extinction event in Earth's history occurred 66 million years ago at the end of the Cretaceous period. One-half of all living things— including many dinosaurs—became extinct at this time.

140   Chapter 5 Diversity and Adaptations

Cultural Diversity

Social Adaptations Not all adjustments to the environment are physical. Humans inhabit a wide variety of environments, primarily due to cultural and social adjustments. People adjusted to warm climates by wearing loose clothing that allows sweat to evaporate and light clothing that reflects heat, and by taking naps during the hottest hours of the day. People adjusted to cold by wearing warm furs, using sealskin to keep dry, making eye goggles out of shell to protect their eyes from bright sunlight on snow, and eating high-fat meals.

Mass Extinctions

The extinction event that killed the dinosaurs around 66 million years ago was a mass extinction. A **mass extinction** is a large-scale disappearance of many species within a short time.

The extinction of the dinosaurs was not the only mass extinction event in Earth's history. There were other large extinctions. The most severe mass extinction happened about 245 million years ago. Scientists estimate that nearly 96 percent of all animal species became extinct at this time. Most were animals without backbones, such as clams, jellyfish, sponges, and trilobites. Most fish and land species survived.

What causes mass extinctions? Scientists are not sure about the exact causes of all the mass extinctions. One idea that has been hypothesized for the extinction of the dinosaurs is that large asteroids slammed into Earth. According to this hypothesis, the collision sent huge clouds of dust into the atmosphere. Over time, the clouds blocked out sunlight. The result would have been rapid cooling of the environment. Dinosaurs and other organisms probably could not have adapted quickly enough to survive, as shown in **Figure 5-15.**

CONNECTION

Visit the Glencoe Homepage at **www. glencoe.com/sec/ science** for links to more information about dinosaurs and extinctions.

5-4 History of Life   141

interNET
CONNECTION

The Glencoe Homepage at **www.glencoe.com/sec/ science** provides links to information related to this Internet connection.

Science Journal **Persuasive Essay** Have students write an essay explaining how they feel about the plight of South America's rain forests. Students should take a position on either side of the issue and write a brief essay persuading others to their side. Encourage students to cite specific evidence to support their positions. **LS** **P**
L2

Using Science Words

LS **Logical-Mathematical** Five major mass extinctions have occurred during Earth's history. Scientists are not sure about the exact causes of every mass extinction event, but most evidence suggests that sharp environmental changes were involved. Have students use their understanding of the terms *adaptation, evolution,* and *natural selection* to explain how rapid and severe environmental changes can cause the extinction of species. **L3**

Discussion

Initiate a discussion with students about how the rapid destruction of South America's rain forests is contributing to a mass extinction of species. Discuss with students the importance of South America's rain forests and why the forests should be saved. Explain that native peoples are destroying the rain forests to make room for crops.

3 Assess

Check for Understanding

Discussion Initiate a discussion about the fossil record. Write the following question on the chalkboard: *How has the fossil record increased our understanding of the history of life on Earth?* Have students write answers to this question, then call upon individual students to share their responses with the rest of the class.

Reteach

Compare the fossil record to other types of historical records, such as sports records. Point out that the geologic time scale helps scientists organize the historical record of life on Earth.

📁 For students who have mastered this section, use the **Reinforcement** and **Enrichment** masters.

4 Close

•MINI•QUIZ•

Use the Mini Quiz to check students' recall of chapter content.

1. The _____ is a chart or time line that organizes the historical record of life on Earth. *geologic time scale*

2. The process by which many species become extinct in a short period of time is called a _____. *mass extinction*

Section Wrap-up

1. The fossil record is the total of all fossils recovered from the ground.

2. The geologic time scale is a time line that shows the appearance and disappearance of species during Earth's history.

3. A mass extinction is a large-scale extinction event in which many species became extinct in a short period of time, for example, when the last of the dinosaurs died out 65 million years ago.

4. **Think Critically** The extinction of one species can affect other species that depend upon it for survival.

USING MATH

The Precambrian makes up 89 percent of Earth's history. [LS]

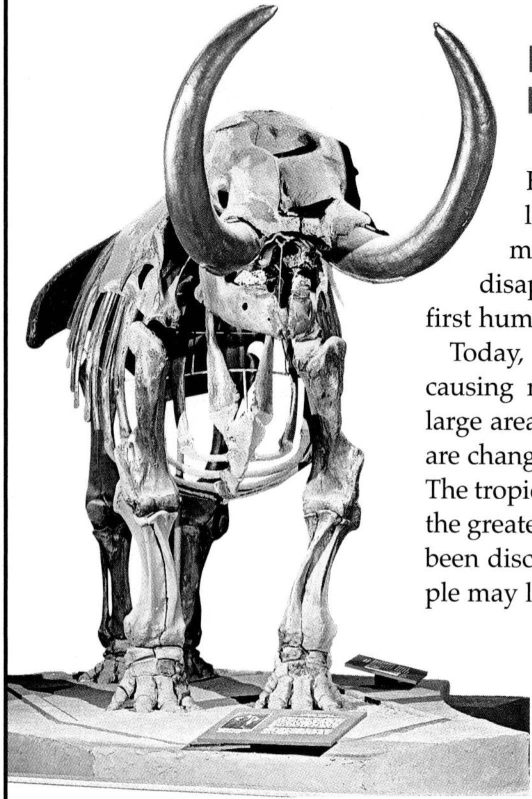

Mass Extinction and the Loss of Biodiversity

Mass extinctions are not just a part of the past. Extinctions have also been occurring within the last few thousand years. For instance, large mammals such as mammoths and mastodons disappeared from North America about the time the first humans appeared here.

Today, scientists fear that human activity is rapidly causing more species to become extinct. Humans use large areas of land for many reasons. Some of these uses are changing or limiting the habitat for many organisms. The tropical rain forests, where biodiversity is high, are in the greatest danger. Many rain forest species have not yet been discovered or named. If these species are lost, people may lose possible sources of food or new medicines.

FIGURE 5-16

Many organisms, such as the mastodon, disappear from Earth for reasons not completely known.

Section Wrap-up

1. What is the fossil record?

2. What is the geologic time scale?

3. What is a mass extinction? Give an example.

4. **Think Critically:** How might the extinction of a single plant species from a forest affect other organisms that live there?

5. 📈 *Skill Builder*
   Recognizing Cause and Effect Explain how an environmental change might cause a species to go extinct. If you need help, refer to Recognizing Cause and Effect on page 551 in the **Skill Handbook.**

USING MATH

The Precambrian represents about 4.1 billion years of Earth's 4.6-billion-year history. What percent of the total does this represent?

📈 *Skill Builder*
Recognizing Cause and Effect A species will go extinct if it can't adapt to environmental changes.

✓ Assessment

Performance Have students perform calculations similar to the one in Using Math for the other major time periods of the geologic time scale. Use the Performance Task Assessment List for Using Math in Science in **PASC**, p. 29. [LS]

R ead the statements below that review the most important ideas in the chapter. Using what you have learned, answer each question in your Science Journal.

1. Biodiversity refers to the variety of plants, animals, and other species in an area. *Why do tropical areas have high biodiversity?*

2. An adaptation is any structure or behavior that helps a species survive in its environment. *What are some adaptations of fish?*

3. Species adapt to the environment, and new species form through the process of natural selection. *What is natural selection?*

4. The geologic time scale shows when species appeared and disappeared during Earth's history. *When were the dinosaurs the dominant land animals on Earth?*

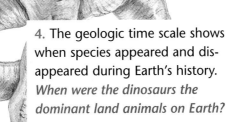

Chapter 5 Review   143

Review

Have students look at the illustrations on this page. Ask them to describe details that support the main ideas of the chapter found in the statement for each illustration.

Teaching Strategies

Have students record answers to questions on separate sheets of paper.

Answers to Questions

1. Tropical areas have high biodiversity because food, water, and shelter are abundant.
2. Fish possess fins, scales, and streamlined bodies.
3. Natural selection is the process by which species that are well-suited for an environment are better able to survive, reproduce, and pass their genes to the next generation.
4. Dinosaurs existed on Earth during the Mesozoic Era.

Science at Home

Intrapersonal Scientists have recovered many fossils; however, finding a new fossil is actually a rare event. Most living things that die never become fossilized. Design an experiment at home to see what happens to fresh chicken bones when they are left outside for a period of four weeks. Handle bones with gloves, and wash hands thoroughly after touching bones. **L2**

✓ **Assessment**

Portfolio Encourage students to place in their portfolios one or two items of what they consider to be their best work. Examples include:
- Skill Builder, p. 122
- Illustrations from MiniLab activity, p. 121
- Fossil from MiniLAB, p. 130 **P**

Performance Additional performance assessments may be found in the **Performance Assessment** and **Science Integration Activities**. Performance Task Assessment Lists and rubrics for evaluating these activities can be found in Glencoe's **Performance Assessment in the Science Classroom (PASC).**

Chapter 5 Review

Using Key Science Words

(Answer Key)

1. fossils
2. biodiversity
3. natural selection
4. adaptation
5. mass extinction

Checking Concepts

6. d   7. a   8. c
9. c   10. b

Thinking Critically

11. Many students will indicate that webbed feet would be an adaptation for living in a wetlands environment because the feet are useful when swimming in water.

12. Answers may vary, but many students will say that the DNA samples can be studied and compared. The samples showing the most similarities probably belong to the organisms that are most closely related.

13. Adaptations for desert organisms may include features that help it stay cool and conserve water.

14. Natural selection allows individuals with the most suitable characteristics to survive and reproduce. Bears with lighter fur were more successful in snowy, arctic environments. Bears with brown fur would be more successful in wooded areas and would be less successful in snowy environments.

15. The traits these organisms had were suitable for the environment. The environment did not change and neither did the animals.

Using Key Science Words

adaptation	fossil record
biodiversity	geologic time scale
captive breeding	mass extinction
common ancestor	natural selection
fossil	

Match each phrase with the correct term from the list of Key Science Words.

1. traces or remains of ancient life
2. numbers of different species in an area
3. mechanism for species' change over time
4. dolphin's flipper or a bird's wing
5. a large-scale disappearance of many species

Checking Concepts

Choose the word or phrase that completes the sentence.

6. A(n) _____ is the disappearance of many species in a short time.
   a. natural selection
   b. adaptation
   c. fossil
   d. mass extinction

7. Species adapt to the environment through the processes of mutation and _____.
   a. natural selection
   b. biodiversity
   c. captive breeding
   d. fossil formation

8. Dinosaur bones are examples of _____.
   a. species
   b. adaptations
   c. fossils
   d. biodiversity

9. New species can form when _____ split into small groups.
   a. fossils   c. populations
   b. adaptations   d. biodiversity

10. Tropical rain forests have high _____ compared with deserts.
    a. species   c. fossil records
    b. biodiversity   d. temperatures

Thinking Critically

Answer the following questions in your Science Journal using complete sentences.

11. Which of the following variations would be most beneficial to a bird living in a wetland: webbed feet, clawed feet, feet with toes for gripping branches? Explain.

12. If a scientist had DNA samples from four organisms, how could he find out which organisms were related?

13. What types of adaptations would be beneficial to an organism living in the desert?

14. Use the idea of natural selection to explain why polar bears are white and why some other bears are brown.

15. Horseshoe crabs are an example of an organism that has changed very little through time. What factors might have kept these organisms from changing?

Assessment Resources

Reproducible Masters

Chapter Review, pp. 13, 14
Assessment, pp. 21-24
Performance Assessment, p. 43

Glencoe Technology

🖳 **Computer Test Bank**

📼 **MindJogger Videoquiz**

Chapter 5 Skills Review

Developing Skills

If you need help, refer to the description of each skill in the Skill Handbook.

16. **Observing and Inferring:** Look at the chart below. Scientists hypothesize that millions more microscopic species exist. Why do you think the number of microscopic species known to science is so low?

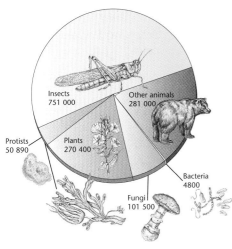

Insects 751 000

Other animals 281 000

Protists 50 890

Plants 270 400

Bacteria 4800

Fungi 101 500

17. **Hypothesizing:** How are the thorns on a rose an adaptation to the plant's environment?

18. **Recognizing Cause and Effect:** A new predator is introduced into a population of rabbits. Rabbits with longer back legs have an advantage because they can run faster. Using the concept of natural selection, explain what might happen to the rabbit population over time.

19. **Interpreting Illustrations:** Study the geologic time scale on page 138. Make a table that lists when the following groups of organisms appeared on Earth: amphibians, fish, land plants, mammals, reptiles, birds, simple animals.

20. **Hypothesizing:** Look at the fish in **Figure 5-10** on page 132. Study the mouth and head shapes and hypothesize how the shapes help the fish feed.

Performance Assessment

1. **Poster:** Make a poster that depicts the animals or plants of a particular time during Earth's history. For instance, you might make a poster about the dinosaurs of the Jurassic period, or a poster about the earliest known birds.

2. **Newspaper Article:** Write a summary of a newspaper or magazine article that discusses the discovery of a fossil species. In your summary, be sure to include information about where the fossil was found, how old it is, and its importance to science.

2. Check summaries for accuracy. You may wish to provide students with newspaper science articles or with science magazines, such as *Discover* or *Science News.* Use the Performance Task Assessment List for Writing in Science in **PASC**, p. 87.

Developing Skills

16. **Observing and Inferring** Answers may vary, but most students will mention that microscopic species are more difficult to find and study.

17. **Hypothesizing** The thorns protect the plant from animals that may feed on it.

18. **Recognizing Cause and Effect** Answers may vary, but many students will say that natural selection should produce a population of rabbits with longer back legs.

19. **Interpreting Illustrations** Amphibians: 408 million years ago; fish: 505 million years ago; land plants: 408 million years ago; mammals: 208 million years ago; reptiles: 320 million years ago; birds: 208 million years ago; simple animals: 544 million years ago.

20. **Hypothesizing** The flattened faces of the algae-eating fish allow them to scrape algae from rocks and other flat surfaces. The protruding mouths of the insect-eating fish allow them to trap and hold the insects. The long mouth of the fish-eating fish allows for more teeth and firmer grasp.

Performance Assessment

1. Check posters for accuracy of information. You may wish to give students some possible topics for their posters. Use the Performance Task Assessment List for Poster in **PASC**, p. 73.

Chapter Organizer

Section	Objectives/Standards	Activities/Features
Chapter Opener		**Explore Activity:** Observe a System, p. 147
6-1 **What is an ecosystem?** (3 sessions, 1 block)*	**1. Identify** the living and nonliving factors in an ecosystem. **2. Explain** how the parts of an ecosystem interact. National Science Content Standards: (5-8) UCP1, A1, B1, B3, B4	**Activity 6-1:** Ecosystem in a Bottle, p. 150 **MiniLAB:** Comparing Soils, p. 154 **Skill Builder:** Observing, p. 156 **Science Journal,** p. 156
6-2 **Relationships Among Living Things** (3 sessions, 2 blocks)*	**3. Give examples** of how ecologists organize living systems. **4. Examine** relationships among living things. National Science Content Standards: (5-8) UCP1, UCP3, A1, C4, G1	**MiniLAB:** Inferring, p. 158 **Using Technology:** Internet Monarch, p. 159 **Activity 6-2:** What's the limit?, pp. 160-161 **Problem Solving:** Graphing Populations, p. 163 **Skill Builder:** Developing a Multimedia Presentation, p. 164 **Using Math,** p. 164 **People & Science,** p. 165
6-3 **Science and Society: Gators at the Gate!** (1 session, ½ block)*	**5. Describe** the problems of endangered and threatened species. **6. Distinguish** fact from opinion. National Science Content Standards: (5-8) UCP4, C4, F2	**Skill Builder:** Distinguishing Fact from Opinion, p. 167
6-4 **Energy Through the Ecosystem** (2 sessions, 1 block)*	**7. Explain** how organisms get the energy they need. **8. Trace** how energy flows through ecosystems. National Science Content Standards: (5-8) B3, C4, F2	**Skill Builder:** Comparing and Contrasting, p. 170 **Using Computers,** p. 170

* A complete Planning Guide that includes block scheduling is provided on pages 31T-33T.

Activity Materials

Explore	Activities	MiniLABs
No materials needed.	**page 150** 2-L bottle, aquarium gravel, metric ruler, sand, scissors, *Elodea* plants, guppy, fish food **pages 160-161** small planting container, heater, light source, potting soil, spoon, aluminum foil, bean seeds, labels	**page 154** cup, gravel, sand, potting soil, spoon **page 158** measuring tape

Need Materials? Call Science Kit (1-800-828-7777).

Teacher Classroom Resources

Reproducible Masters	Transparencies	Teaching Resources
Activity Worksheets, pp. 5, 37-38, 41 **Enrichment**, p. 22 **Reinforcement**, p. 22 **Science Integration Activities**, pp. 49-50 **Study Guide**, p. 22	**Science Integration Transparency 6,** Space Shuttle Ecosystem: Abiotic Factors **Section Focus Transparency 16,** Parts of Ecosystems **Teaching Transparency 11,** Levels of Organization	**Spanish Resources** **English/Spanish Audiocassettes** **Cooperative Learning Resource Guide** **Lab Partner** **Lab and Safety Skills** **Lesson Plans**
Activity Worksheets, pp. 5, 39-40, 42 **Cross-Curricular Integration**, p. 10 **Enrichment**, p. 23 **Lab Manual 11** **Reinforcement**, p. 23 **Science and Society/Technology Integration**, p. 24 **Study Guide**, p. 23	**Section Focus Transparency 17,** It's Mine **Teaching Transparency 12,** Feeding Relationships	
Enrichment, p. 24 **Reinforcement**, p. 24 **Study Guide**, p. 24	**Section Focus Transparency 18,** Home Sweet Home?	**Assessment Resources** **Chapter Review**, pp. 15-16 **Assessment**, pp. 25-28 **Performance Assessment**, p. 44 **Performance Assessment in the Science Classroom (PASC)** **MindJogger Videoquiz** **Alternate Assessment in the Science Classroom** **Computer Test Bank**
Activity Worksheets, p. 5 **Enrichment**, p. 25 **Lab Manual 12** **Multicultural Connections**, pp. 15-16 **Reinforcement**, p. 25 **Study Guide**, p. 25	**Section Focus Transparency 19,** Who eats whom?	

Key to Teaching Strategies

The following designations will help you decide which activities are appropriate for your students.

L1 Level 1 activities should be appropriate for students with learning difficulties.

L2 Level 2 activities should be within the ability range of all students.

L3 Level 3 activities are designed for above-average students.

ELL ELL activities should be within the ability range of English Language Learners.

LS These activities are designed to address different learning styles.

COOP LEARN Cooperative Learning activities are designed for small group work.

P These strategies represent student products that can be placed into a best-work portfolio.

GLENCOE TECHNOLOGY

The following multimedia resources are available from Glencoe.

Science and Technology Videodisc Series (STVS)
Ecology
Preserving Duck Habitats
The Infinite Voyage Series
Secrets from a Frozen World

National Geographic Society Series
GTV: Planetary Manager
Glencoe Life Science CD-ROM

Teacher Classroom Resources

This is a representation of key blackline masters available in the Teacher Classroom Resources.

Teaching Aids

Section Focus Transparencies

Science Integration Transparencies

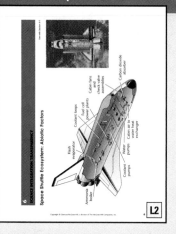

Teaching Transparencies

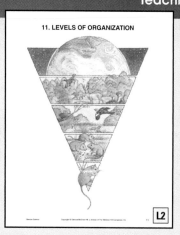

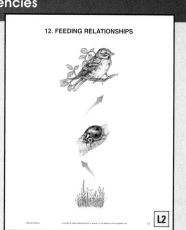

Meeting Different Ability Levels

Study Guide for Content Mastery

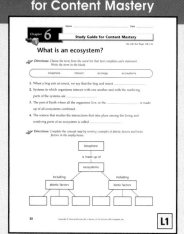

Reinforcement

Enrichment Worksheets

Hands-On Activities

Science Integration Activities

How does your garden grow?

Lab Manual

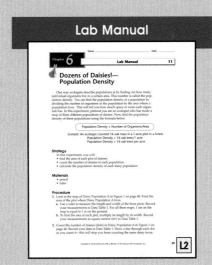

Dozens of Daisies!—Population Density

Activity Worksheets

Ecosystem in a Bottle

Enrichment and Application

Cross-Curricular Integration

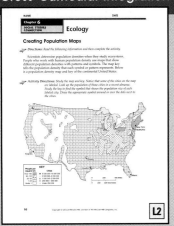

Ecology

Creating Population Maps

Multicultural Connections

A Web of Life Across Australia

Science and Society/Technology Integration

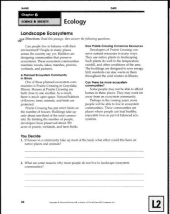

Ecology

Landscape Ecosystems

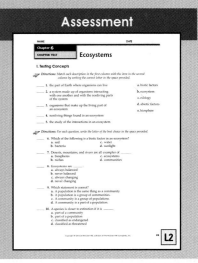

Assessment

Performance Assessment

There's No Place Like Home

Chapter Review

Ecology

Part A. Vocabulary Review

Assessment

Ecosystems

I. Testing Concepts

Chapter 6 Ecology

CHAPTER OVERVIEW

Section 6-1 This section introduces ecosystems as dynamic systems composed of interacting biotic and abiotic components.

Section 6-2 This section discusses the characteristics of populations and how populations interact within a community.

Section 6-3 Science and Society This section distinguishes fact from opinion in the context of discussing the impact of humans on the environment.

Section 6-4 This section explores the flow of energy through an ecosystem and the cycling of matter through the biosphere.

Chapter Vocabulary

ecosystem	habitat
ecology	endangered
biosphere	species
biotic factor	threatened
abiotic factor	species
population	producer
community	consumer
limiting factor	decomposer
niche	

Theme Connection

Systems and Interactions Learning about ecosystem processes and the interactions among populations, communities, and their environment will enable students to better understand the interconnectedness of all living things. After examining natural systems, students will better understand their role in the environment.

Chapter Preview

Section 6-1   What is an ecosystem?

Section 6-2   Relationships Among Living Things

Section 6-3   Science and Society: Gators at the Gate!

Section 6-4   Energy Through the Ecosystem

146

Skills Preview

▶ **Skill Builders**
• observe
• compare and contrast

▶ **MiniLABs**
• compare
• infer

▶ **Activities**
• observe
• collect data
• graph
• hypothesize
• compare results

Learning Styles Look for the following logo for strategies that emphasize different learning modalities. **LS**

Kinesthetic	Activity 6-1, p. 150; Assessment, p. 150; Activity, p. 153; MiniLAB, p. 154; Activity 6-3, p. 160
Visual-Spatial	Explore, p. 147; Visual Learning, pp. 149, 152, 169; Across the Curriculum, p. 158; Science Journal, p. 162
Interpersonal	Community Connection, p. 152; Inclusion Strategies, p. 155; Assessment, p. 170
Intrapersonal	Enrichment, p. 152; Across the Curriculum, p. 153
Logical-Mathematical	Activity, p. 149; Science Journal, pp. 152, 163; MiniLAB, p. 158; Problem Solving, p. 163; Using Math, p. 164
Linguistic	Using Science Words, p. 149; Reteach, p. 155; Science Journal, p. 159
Auditory-Musical	Across the Curriculum, p. 149

LS

Chapter 6

Ecology

The day is warm and bright, perfect for a field trip to a local stream. Carefully, quietly, you push aside the cattails and kneel in the marshy grass to view an insect's progress down a water-soaked log. You lean in for a closer look when—WHAM! A sticky tongue latches onto the insect and flings it into the waiting mouth of a frog. Startled, you jump back. Your sudden movement sends the frog leaping into the water. SPLASH! You're all wet. You have just observed a system in action.

EXPLORE ACTIVITY

Observe a System

1. Choose a small area near your school to observe, such as a plot of grass or weeds. Identify the boundaries of your plot.
2. Carefully observe and record everything in your plot. Be sure to include all the parts of the plot, including air and soil.
3. Classify what you observe into two groups—things that are living and things that are not living.

Science Journal

A system is a group of things that interact with one another. In your Science Journal, describe how you think the parts of the plot you observed form a system.

147

Assessment Planner

Portfolio
Refer to page 171 for suggested items that students might select for their portfolios.

Performance Assessment
See page 171 for additional Performance Assessment options.
Skill Builders, pp. 156, 164, 167, 170
MiniLABs, pp. 154, 158
Activities 6-1, p. 150; 6-2, pp. 160-161

Content Assessment
Section Wrap-ups, pp. 156, 164, 170
Chapter Review, pp. 171-173
Mini Quizzes, pp. 156, 164, 170

Group Assessment
Opportunities for group assessment occur with Cooperative Learning Strategies and Flex Your Brain activities.

EXPLORE ACTIVITY

Purpose

Visual-Spatial Use the Explore activity to give students experience in observing and identifying the different parts of an ecosystem. **L1** **ELL**

Preparation

Scout out an area ahead of time where students can observe. Check carefully to ensure that there are no pieces of broken glass or other dangerous items in the area.

Alternate Materials

If students are unable to go outside, have them examine an indoor terrarium or aquarium.

Teaching Strategies

Ask students to consider organisms that are not visible to the naked eye. Ask them to predict whether their plots contain bacteria and fungi.

Science Journal The students should recognize that they are observing a system composed of interacting living and nonliving components. The interactions between these two factors create an ecosystem.

✓ Assessment

Oral Ask students to discuss what their plots would contain at different times of the year. How does the environment change with the seasons? How would seasonal changes influence organisms living in an area? Use the Performance Task Assessment List for Oral Presentation in **PASC**, p. 71.

Prepare

Section Background
- An ecosystem is a system of interactions between biotic and abiotic factors in the environment.
- Biotic factors include all living organisms in an ecosystem. Abiotic factors include all nonliving parts of an ecosystem such as weather, soil, water, and sunlight.

Preplanning
Refer to the Chapter Organizer on pages 146A-B.

1 Motivate

Bellringer

 Before presenting the lesson, display **Section Focus Transparency 16** on the overhead projector. Assign the accompanying **Focus Activity** worksheet.
L2 ELL

Tying to Previous Knowledge
Ask students to consider how adaptations have allowed different organisms to cope with different environmental conditions. Refer to Chapter 5.

148

What YOU'LL LEARN

- About the living and nonliving factors in an ecosystem
- How the parts of an ecosystem interact

Science Words:
ecosystem
ecology
biosphere
biotic factor
abiotic factor

Why IT'S IMPORTANT

Understanding interactions of an ecosystem will help you understand your role in your ecosystem.

FIGURE 6-1
Let's identify the living and nonliving parts of this stream ecosystem.

 Rocks and water are nonliving. Pond skaters are insects that skim the surface of the water.

6•1 What is an ecosystem?

Ecosystems

Take a walk outside and look around. What do you see? Woods? A street? A patch of weeds growing in a sidewalk crack? If you observe one of these areas closely, you may see many different organisms (OR guh nih zumz) living there. In a forest, for instance, there are birds, deer, insects, plants, mushrooms, and trees. In your backyard, you might see squirrels, birds, insects, grass, and shrubs. These organisms, along with the nonliving things in the woods or yard, such as soil, air, and light, make an ecosystem (EE koh sihs tum). An **ecosystem** is made up of organisms interacting with one another and with nonliving factors to form a working unit. **Figure 6-1** shows an example of a stream ecosystem.

148 Chapter 6 Ecology

Program Resources

📁 **Reproducible Masters**
Activity Worksheets, pp. 5, 37-38, 41 L2
Enrichment, p. 22 L2
Reinforcement, p. 22 L2
Science Integration Activities, pp. 49-50 L2
Study Guide, p. 22 L1

🖥 **Transparencies**
Science Integration Transparency 6 L2
Section Focus Transparency 16 L2
Teaching Transparency 11 L2

What does it mean to say that an organism interacts with another organism? Think back to the field trip to the stream on page 147. When the frog ate the insect, an interaction occurred between two organisms living in the same ecosystem.

What does it mean to say that an organism interacts with the nonliving parts of an ecosystem? Think about the field trip again. What did the frog do when it spotted your movement? It dove into the stream, probably for safety. The frog uses the stream for shelter. This is an example of an interaction between a living organism and a nonliving part of an ecosystem.

B Algae, fish, crayfish, and mosses covering rocks are living parts of this ecosystem. *How do these organisms interact with nonliving parts of the ecosystem?*

6-1 What is an ecosystem?   149

2 Teach

Activity

LS **Logical-Mathematical** Lead students in analyzing the "classroom ecosystem." Ask students to identify different living and nonliving factors in their classroom and discuss how they interact. Ask students how they interact with the living and nonliving parts of the classroom ecosystem. **L2**

Visual Learning

Figure 6-1 What other nonliving things can be found in this ecosystem? *Air, soil, and sunlight are other nonliving parts of this ecosystem.* **How do these organisms interact with nonliving parts of the ecosystem?** *The fish, algae, and mosquitoes depend on the nonliving parts of the ecosystem for protection, a place to live, and food.* **ELL** **LS**

Using Science Words

LS **Linguistic** Write the word *ecology* on the chalkboard. Explain that the word *ecology* comes from the Greek words *oikos*, which means "place to live," and *ology*, which means "study of." Ask students why the word *ecology*, meaning "the study of place to live," is an appropriate name for the study of our environment.

Across the Curriculum

Language Ask students to find or write a poem that describes animals or plants interacting with their environment. Have students present their selection to the class and discuss why they chose their particular poem. **L2** **P** **LS**

149

Activity 6-1

Ecosystem in a Bottle

You may think of ecosystems as large areas. But ecosystems can be any size. You can even make an ecosystem that fits in a plastic bottle.

What You'll Investigate
What are the parts of an ecosystem?

Goals
• Make an ecosystem
• Observe an ecosystem

Materials
• 2-L bottle
• scissors
• sand
• aquarium gravel
• metric ruler
• water
• *Elodea* plants
• guppy
• fish food

Procedure 🥽 🧤 🐾
1. **Rinse** out a 2-L plastic bottle with water. Using scissors, carefully **cut** off the top of the bottle.
2. **Pour** a layer of sand 5 to 10 cm deep in the bottom of the bottle.
3. **Fill** the bottle to within 5 cm of the top with water that has stood in an open container for about two days. Keep a supply of this aged water on hand to replace the water that evaporates from the bottle. The level should always be about the same.
4. **Plant** the *Elodea* and add a 2-cm layer of gravel.
5. When the water clears, **add** a guppy.
6. **Feed** the fish one or two small flakes of food every day.
7. Now you've made an ecosystem. **Observe** your ecosystem every day and **record** your observations in your Science Journal. Be sure to make observations about the living and nonliving parts of your ecosystem.

Conclude and Apply
1. **Describe** how the parts in the bottle work together to form an ecosystem.
2. What is needed to keep the ecosystem healthy?

150

Answers to Questions
1. The soil, sand, and gravel provided anchorage and nutrients for the plant. The plant photosynthesized and added oxygen to the water. The guppy used oxygen and ate the fish food. The guppy also used the water and rocks as its habitat.
2. The fish food and additional water were necessary to maintain a balanced ecosystem.

✔ **Assessment**

Performance Have students create a model or diagram of the relationships in their ecosystem. Ask students to demonstrate how their ecosystem functioned. Use the Performance Task Assessment List for Model in **PASC**, p. 51. P IS

The Study of Ecosystems

When you study the interactions in an ecosystem, you are studying the science of ecology (ee KAH luh jee). **Ecology** is the study of the interactions that take place among the living organisms and nonliving parts of an ecosystem. Ecologists spend a lot of time outdoors, observing their subject matter up close. Just as you knelt quietly in the cattails on your field trip, an ecologist might spend hours by a stream, watching, recording, and analyzing what goes on there. In addition, like other scientists, ecologists also conduct experiments in laboratories. For instance, they might need to analyze samples of stream water. But, most of the ecologist's work is done in the field.

The Largest Ecosystem

Ecosystems come in all sizes. Some are small, like a pile of leaves. Others are big, like a forest or the ocean. **Figure 6-2** shows the biosphere (BI oh sfeer), the largest ecosystem on Earth. The **biosphere** is the part of Earth where organisms can live. It includes the topmost layer of Earth's crust; all the oceans, rivers, and lakes; and the surrounding atmosphere. The biosphere is made up of all the ecosystems on Earth combined.

How many different ecosystems are part of the biosphere? Let's list a few. There are deserts, mountains, rivers, prairies, wetlands, forests, plains, oceans—the list can go on and on, and we haven't even gotten to smaller ecosystems yet, such as a vacant lot or a rotting tree trunk. The number of ecosystems that make up the biosphere is almost too many to count. How would you describe your ecosystem?

FIGURE 6-2

The biosphere is the part of Earth that contains all the living things on the planet. Each ecosystem that you study is part of the biosphere.

Inquiry Question

Why would most of an ecologist's work be done in the field? *Ecologists spend most of their time in the field because they study the interactions of organisms with their outdoor environment.*

Using an Analogy

Many ecologists compare the biosphere to a living organism. Just as an animal is composed of different interacting parts, the biosphere is also composed of different interacting parts such as oceans, atmosphere, and crust.

Cultural Diversity

Natural Balance The Balinese government tried to increase rice production by using accepted methods such as pesticides. The chemicals failed to kill the pests, but did kill other food sources such as the eels living in irrigation ditches. Rice crops actually decreased. When scientists studied the ecological system, they found that timing of irrigation was key. All fields received the water needed and all were barren at the same time. This starved the pests that damaged crops. When scientists developed a computer model to explain this system, it also showed that it was the most productive one.

Figure 6-3 Lead a discussion of how biotic interactions are important in a desert ecosystem. Ask students to identify as many biotic interactions as possible. **LS**

How is the snake interacting with abiotic factors in this ecosystem? *The snake is interacting with the ground as a habitat. The rocks absorb heat, and the organism can absorb the heat from the rocks. The snake could also be protected from the sun and predators by hiding under the rocks.* **What abiotic factors are important in this ecosystem?** *The high temperatures and lack of water are important abiotic factors in this ecosystem.* **ELL** **LS**

Teacher F.Y.I.

Desert ecosystems are extremely harsh for organisms. Deserts receive little rain and frequently have searing summer heat and chilly nights. Most deserts receive less than 20 cm of water per year and have daytime temperatures that reach more than 30°C.

Enrichment

LS **Intrapersonal** Ask students to research how organisms interact with their environment at different stages of their life cycles. Do younger plants and animals have different requirements than older ones? If possible, have students visit a zoo or botanical garden to interview staff members about unique requirements some young organisms might have. **L3**

Living Parts of Ecosystems

Each of the many ecosystems in the biosphere contains many different living organisms. Think about a rotting tree trunk. It's a small ecosystem compared to a forest, but the tree trunk may be home to bacteria, bees, beetles, mosses, mushrooms, slugs, snails, snakes, wildflowers, woodpeckers, and worms. The organisms that make up the living part of an ecosystem are called **biotic factors.** An organism depends on other biotic (bi AH tihk) factors for food, shelter, protection, and reproduction. **Figure 6-3** shows some of the biotic factors in a desert ecosystem.

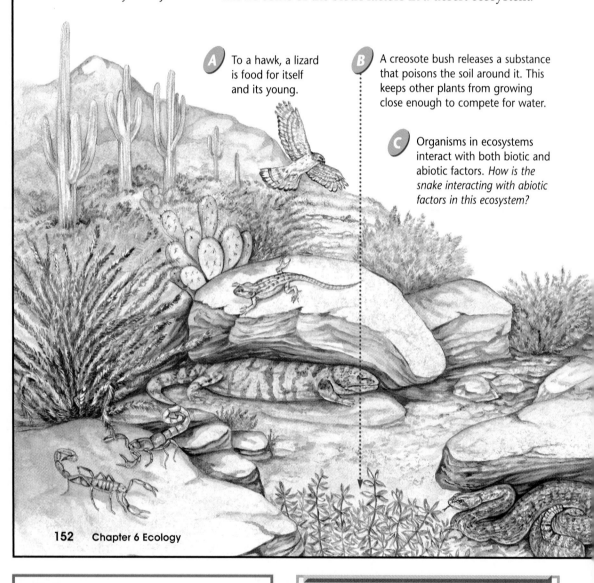

FIGURE 6-3
The living and nonliving components in this desert ecosystem interact in a variety of ways.

A To a hawk, a lizard is food for itself and its young.

B A creosote bush releases a substance that poisons the soil around it. This keeps other plants from growing close enough to compete for water.

C Organisms in ecosystems interact with both biotic and abiotic factors. *How is the snake interacting with abiotic factors in this ecosystem?*

Science Journal **Ecosystems** Have students describe the ecosystem they live in. They should choose home, school, or their city or town. Be sure they define their boundaries. Have students list both biotic and abiotic factors that occur in their ecosystem, and predict what would happen if the environment changed dramatically. **L2** **P** **LS**

Community Connection

Ecologists Invite a local ecologist such as a park ranger, an environmental scientist, or a university ecology or environmental science professor to visit the class and explain how ecologists study the environment. Have students ask what types of organisms and interactions ecologists study. **LS** **COOP LEARN**

Nonliving Parts of Ecosystems

Earlier, you listed the parts that make up an ecosystem near your school. Was your list limited to the living or biotic factors only? No. You included nonliving factors, too, such as air and soil. The nonliving things found in an ecosystem are called **abiotic factors.** Look for some abiotic (AY bi AH tihk) factors in the desert shown in **Figure 6-3.** Abiotic factors have roles in ecosystems, too. They affect the type and number of organisms living there. Let's take a closer look at some abiotic factors.

Soil

Soil is an abiotic factor that can affect which plants and other organisms are found in an ecosystem. Soil is made up of ingredients much like a recipe. It is made up of a combination of minerals, water, air, and organic matter—the decaying parts of plants and animals. You know that salt, flour, and sugar are found in many recipes. But not all foods made from these same ingredients taste or look the same. Cakes and cookies look and taste different because different amounts of salt, flour, and sugar are used to make them. It's the same with soil. Different amounts of minerals, organic matter, water, and air make different types of soil.

 D Abiotic factors are important in all ecosystems. *What abiotic factors are important in this ecosystem?*

 **E** A bird seeks shelter among the spines of a cactus.

6-1 What is an ecosystem? 153

interNET
CONNECTION

Visit the Glencoe Homepage, *www.glencoe.com/sec/science,* for a link to more information about desert ecosystems.

? FLEX Your Brain

Use the Flex Your Brain activity to have students explore ECOSYSTEM INTER-ACTIONS.

Activity Worksheets, page 5

interNET
CONNECTION

Visit the Glencoe Home-page, **www.glencoe.com/sec/science,** for links to desert information.

Activity

LS **Kinesthetic** Have students research soils in different areas of the country. They should compare the south-east to southwest or prairie to deciduous forest. They should make a table or chart comparing and contrasting soils from different areas. They could take it further by including information on the types of vegetation found in each area. **L3**

 NATIONAL GEOGRAPHIC SOCIETY

 Videodisc
GTV: Planetary Manager
Soil

48168

48212

Across the Curriculum

History During the mid-1800s, many Americans migrated westward across North America to the west coast in wagon trains. Research some of the experiences of these travelers and report on the environmental conditions they lived with. The *Little House* books by Laura Ingalls Wilder or the CD-ROM game *The Oregon Trail* could be sources for students. **L2** **LS**

153

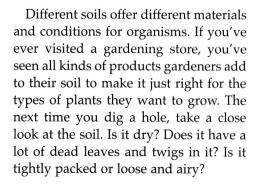

Mini LAB

Purpose

LS **Kinesthetic** Students will observe how soil composition influences the ability of soil to hold water. **L1** **ELL**

Materials

cups, different soil components (e.g., sand, clay, peat, topsoil, rocky soil), trowel or spoon

Teaching Strategies

Avoid oversaturating the soil. Make sure students do not add too much water to their containers. You can obtain soil components at a garden-supply store.

Activity Worksheets, pages 5, 41

Analysis

1. Answers will vary according to the amounts and types of materials included in the samples.

2. Results will vary by student soil recipe. Soil samples composed of large particles will not hold water as well as soils composed of small particles. Plants and animals need special adaptations to live in soils that are very wet or very dry.

Assessment

Performance Ask students to design an experiment that would examine how plants respond to different soil conditions such as dryness or mineral content. Students should try to grow the same plant in different soil compositions and compare the results. Use the Performance Task Assessment List for Designing an Experiment in **PASC,** p. 23.

Mini LAB

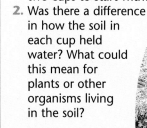

Comparing Soils

Observe soil characteristics.

1. Fill two cups with soil. Use different amounts of the materials available to create two different "soil recipes." Pack the soil equally into each cup.
2. Pour equal amounts of water into each cup.
3. Tip the cups over to see if any water pours out.
4. Observe the characteristics of the soils you made. Record your observations in your Science Journal.

Analysis

1. What was the difference between the soil in the two cups to start with?
2. Was there a difference in how the soil in each cup held water? What could this mean for plants or other organisms living in the soil?

Different soils offer different materials and conditions for organisms. If you've ever visited a gardening store, you've seen all kinds of products gardeners add to their soil to make it just right for the types of plants they want to grow. The next time you dig a hole, take a close look at the soil. Is it dry? Does it have a lot of dead leaves and twigs in it? Is it tightly packed or loose and airy?

Temperature

Soil is only one of the factors that affect the organisms that live in an ecosystem. Temperature also determines which organisms live in a particular place. How do the tropical plants shown in **Figure 6-4** compare with the mountainside plants? Predict what would happen if the organisms on the mountainside were moved to a hot climate such as a tropical rain forest.

FIGURE 6-4

Plants have adaptations for their environments. The mountainside wildflowers (above) grow in clusters close to the ground, which protects them from strong winds. The tropical plants (left) have large leaves to absorb as much light as possible in the dim understory.

Water

Another important abiotic factor is water. In the field trip to the stream at the beginning of the chapter, maybe you saw a sleek trout dart through the water. Some organisms, such as fish, whales, and algae (AL jee), are adapted for life in water, not on land. But these organisms depend upon water for more than just a home. Water helps all living things carry out important life processes such as digestion and waste removal. In fact, the bodies of most organisms are made up mostly of water. Scientists estimate that two-thirds of the weight of the human body is water, as shown in **Figure 6-5.** Do you know how much you weigh? Calculate how much of your weight is made up of water.

Because water is so important to living things, it is also important to an ecosystem. The amount of water available in an ecosystem can determine how many organisms can live in a particular area. It can also serve as shelter and as a way to move from place to place.

Sunlight

The sun is the main source of energy for most organisms on Earth. Energy from the sun is used by green plants to produce food. Humans and other animals then obtain their energy by eating these plants and other organisms that have fed on the plants. When you eat food produced by a plant, you are consuming energy that started out as sunlight. You'll learn more about the transfer of energy in an ecosystem in Section 6-4.

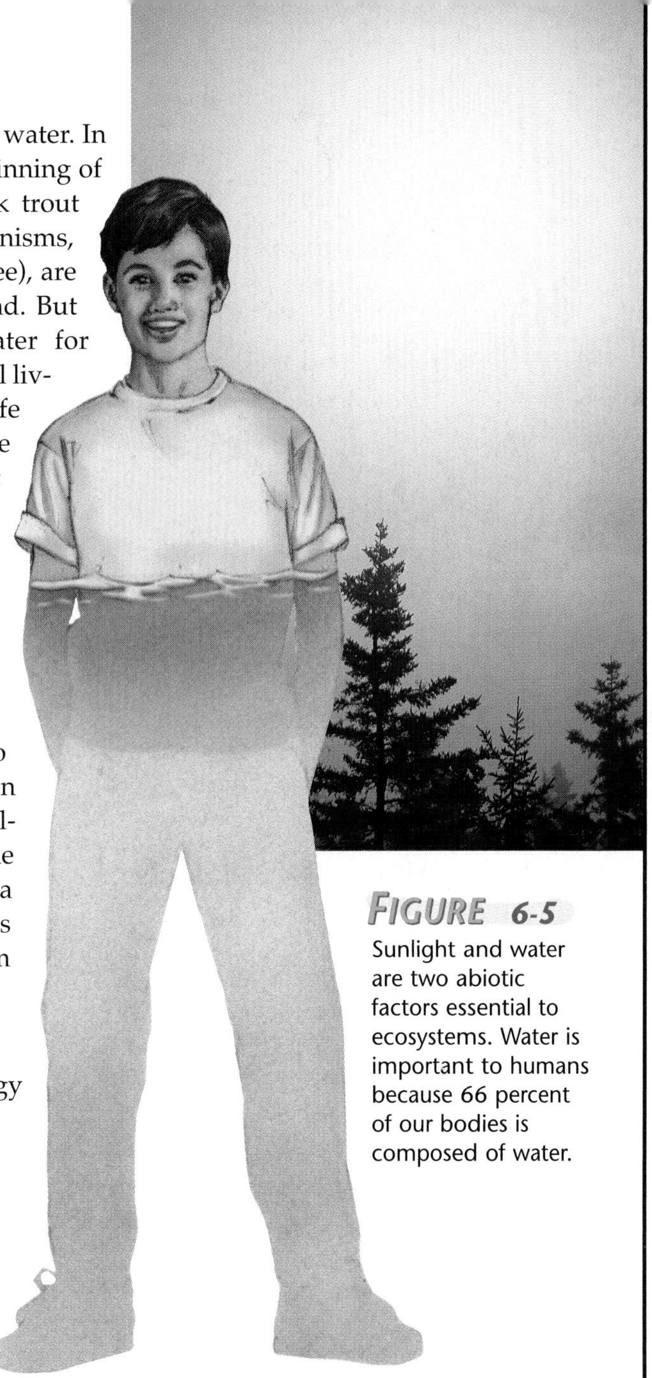

FIGURE 6-5
Sunlight and water are two abiotic factors essential to ecosystems. Water is important to humans because 66 percent of our bodies is composed of water.

Student Text Question

How do tropical plants compare with plants you might see on a cool mountainside? *Tropical plants are larger in size and have large leaves. Plants on a cool mountainside have small, thicker, waxy leaves to protect against cold temperatures and are more compact for protection from winds.* LS

Brainstorming

Have students brainstorm all of the different ways they use water in their daily activities. They should then brainstorm the actions students can take to conserve water. L1

Inquiry Question

Why are most aquatic plants found near the surface of the water? *Plants require sunlight to grow. Light does not penetrate deeply into the water, so most plants are found near the surface.*

3 Assess

Check for Understanding

Brainstorming Have students brainstorm the biotic and abiotic factors they might encounter in a polar region. They should then discuss the adaptations organisms would need to have to live in that environment. L1

Reteach

LS **Linguistic** On the chalkboard, write a list of different ecosystem factors. Ask students to identify each as biotic or abiotic. Ask for examples of the possible role the factors may play in the ecosystem. L1

Extension

For students who have mastered this section, use the **Reinforcement** and **Enrichment** masters.

Inclusion Strategies

Gifted Have students research sunlight-deficient ecosystems such as the ocean depths and caves. How have organisms adapted to their special conditions? Students should present their findings to the class. L3 COOP LEARN LS

4 Close

•MINI•QUIZ•

Use the Mini Quiz to check students' recall of chapter content.

1. A(n) _____ is a system of organisms interacting with the living and the nonliving parts of their environment. *ecosystem*
2. The study of interactions in an ecosystem is called _____ . *ecology*
3. The nonliving parts of an ecosystem are called _____ . *abiotic factors*

Section 3 Wrap-up

1. An ecosystem is a system of organisms interacting with the nonliving parts of their environment.
2. They determine the number and type of organisms that can live there.
3. **Think Critically** Answers will vary. Students may respond that humans traveling in a space station will not be able to bring all of the food, water, oxygen, and other resources they would need. If the space station were set up as a self-supporting ecosystem, they would be able to support themselves without outside resources.

Science Journal Journal entries should reflect that fire would change the ecosystem's biotic and abiotic factors. A fire would destroy vegetation, causing many animals to leave the area. Different plants and animals would move into the new conditions created by the fire and form a new ecosystem.

A Balanced System

Every ecosystem is made up of many different factors working together. When these factors are in balance, the system is in balance, too. Does an ecosystem ever get out of balance? Ecosystems are always changing. Many events can affect the balance of a system. A good example would be a long period of time without rain (called a drought). Predict what would happen if a drought occurred at the stream where you took your field trip. You can see a possible result in **Figure 6-6.** Some organisms, like a fish, would not survive for long periods of time without water. Some might have to find new homes. And, organisms that couldn't survive in a stream environment before might find the dried-up stream to be just the ecosystem they need.

FIGURE 6-6

Ecosystems are always changing. Some changes are small. Others, such as a stream drying up, are much larger with many more effects.

Section 3 Wrap-up

1. What is an ecosystem?
2. How are abiotic factors important to an ecosystem?
3. **Think Critically:** If you were designing a space station on the moon, how would you use your knowledge of ecosystems to help you design it?
4. **Skill Builder**

   Observing Like all living things, people are part of an ecosystem. Describe the ecosystem you are a part of. Include both biotic and abiotic factors. If you need help, refer to Observing and Inferring on page 550 in the **Skill Handbook.**

Science Journal

Write a paragraph in your Science Journal describing the changes that you think would occur in a forest ecosystem following a fire. How might the interactions between biotic and abiotic factors change?

Skill Builder

Observing Student answers will vary depending on their surroundings. However, most students should include biotic factors such as family, friends, pets, and food and abiotic factors such as housing, air, ground, water, and other nonliving parts of the environment. **P**

Assessment

Performance Fires are not the only catastrophic event that can occur in an ecosystem. Ask students how other catastrophes such as tornadoes, floods, earthquakes, and droughts may influence an ecosystem. Would the ecosystems remain unchanged? Use the Performance Task Assessment List for Oral Presentation in **PASC**, p. 71. **L2**

Relationships Among Living Things

Organizing Ecosystems

Imagine trying to study all of the living things on Earth at once! When ecologists study living things, they usually don't start by studying the entire biosphere. Remember, the biosphere consists of all the parts of Earth where organisms can live. It's much easier to begin by studying smaller parts of the biosphere.

To separate the biosphere into smaller systems that are easier to study, ecologists find it helpful to organize living things into groups. They then study how members of a group interact with each other and their environments.

Groups of Organisms

Look at the sea horse in **Figure 6-7.** This particular sea horse lives in a coral reef in the warm, shallow waters near Australia. The sea horse uses energy, grows, reproduces, and eventually dies. The coral reef is the ecosystem the sea horse lives in. All of the sea horses that live in this particular coral reef make up a population. A **population** is a group of the same type of organisms living in the same place at the same time. Some other populations that you might find in a coral reef ecosystem are sponges, algae, sharks, and coral. What are some populations of organisms that live around your school?

FIGURE 6-7

This sea horse belongs to the population of sea horses living in a coral reef ecosystem.

What YOU'LL LEARN

- How ecologists organize living systems
- To describe relationships among living things

Science Words:
population
community
limiting factor
niche
habitat

Why IT'S IMPORTANT

Organizing living things will help you to study ecosystems and understand your role in them.

Prepare

Section Background

- Interacting populations of organisms make up communities. The size of a population is influenced by biological and environmental factors. Populations cannot expand forever. At some point, a limiting factor will impede population growth.

Preplanning

Refer to the Chapter Organizer on pages 146A-B.

1 Motivate

Bellringer

 Before presenting the lesson, display **Section Focus Transparency 17** on the overhead projector. Assign the accompanying **Focus Activity** worksheet. L2 ELL

Program Resources

 Reproducible Masters
Activity Worksheets, pp. 5, 39-40, 42 L2
Cross-Curricular Integration, p. 10 L2
Enrichment, p. 23 L3
Lab Manual, pp. 39-42 L2
Reinforcement, p. 23 L2
Science and Society/Technology
   Integration, p. 24 L2
Study Guide, p. 23 L1

Transparencies
Section Focus Transparency 17 L2
Teaching Transparency 12 L2

Tying to Previous Knowledge

Ask students to review the resources necessary for life. In reviewing these resources, students may better understand how limiting factors influence populations. L2

2 Teach

Mini LAB

Purpose

LS **Logical-Mathematical**
Students will calculate the population density of their classroom. **L1** **ELL**

Materials
measuring tape

Teaching Strategies
- Have students work in teams to obtain measurements.
- If the class is too large, provide students with several room measurements and have them calculate the square footage.

Troubleshooting If the classroom is oddly shaped, have students determine the area of different parts of the room and add the areas together to determine the total area of the room.

Activity Worksheets, pages 5, 42

Analysis

1. Population density would increase if the number of students in the classroom doubled and the amount of space remained the same.

Assessment

Performance Ask students what would happen to the classroom population density if the classroom area were doubled. The population density would decrease, and students would have twice as much space. Use the Performance Task Assessment List for Oral Presentation in **PASC,** p. 71.

158

FIGURE 6-8
Many populations make up a coral reef ecosystem.

Mini LAB

Inferring

Calculate the population density of your classroom.

1. Calculate the area of your classroom in square meters by multiplying the length by the width.
2. Count the number of students in the class.
3. Divide the area of the room by the number of students to determine how much space is available for each student.

Analysis

1. What would happen to the population density if the number of students in your classroom doubled?

158   Chapter 6 Ecology

Groups of Populations

Many populations live in an ecosystem like the coral reef in **Figure 6-8.** All of the populations that live in an area make up a **community** (kuh MYEW nuh tee). The members of a community depend on each other for food, shelter, and other needs. For example, a shark depends on the fish populations for food. The fish populations, on the other hand, depend on coral animals to build the reef that they use to hide from the sharks.

No matter where you live, you live in and are part of a community. Make a list of as many of the populations that make up your community as you can. Compare your list with the lists of your classmates. How many populations did the class come up with?

Characteristics of Populations

Look around your classroom. Is the room big or small? How many students are in your class? Are there enough books and supplies for everyone? Ecologists ask questions like these to describe populations. They want to know the size of the population, where its members live, and how it is able to stay alive.

Population Density

Think about your classroom. A population of 25 students in a large room has plenty of space. How would the same 25 students fit into a smaller room? Ecologists determine population density (DEN suh tee) by comparing the size of a population with its area. For instance, if 100 dandelions are growing in a field that is 1 square kilometer in size, then the population density is 100 dandelions per square kilometer.

Content Background

Populations can be described by population size, spacing arrangement (whether they are spaced evenly or randomly), and by population density. Population density is the number of organisms contained in a given amount of space. It can vary over time in response to biological and environmental factors.

Across the Curriculum

Geography Have students look at a global map of human population densities. Have students identify the most densely populated areas and the least densely populated areas. Ask students what types of environments support high and low population densities. Provide students with a climate map to answer this last question. **L2** **LS**

Limits to Populations

Populations cannot grow larger and larger forever. There wouldn't be enough food, water, living space, and other resources to go around. The things that limit the size of a population, such as the amount of rainfall or food, are called **limiting factors.** Think back to the stream. One biotic limiting factor in the stream ecosystem is the mosquito population. How can a mosquito population be a limiting factor? If you were a frog, you might know the reason. Frogs eat mosquitoes. If lack of rain caused the mosquito population to go down, then the frog population might not have enough food and its population size might go down. What are some other limiting factors in a stream ecosystem?

USING TECHNOLOGY

Internet Monarch

A common sign of summer is the appearance of butterflies fluttering over flower beds and fields. Some butterflies live only a short time. Others, such as monarchs, can live for years. Monarchs travel to warm climates during the winter and return to the same location year after year. This seasonal travel is called migration.

To study the monarch migration, a "monarch watcher"—often a student—carefully catches a monarch and attaches a tag to one of its wings. When someone else catches that butterfly and reads that tag, he or she can figure out how far the butterfly has flown.

Monarch watchers can use the Internet to collect and organize their data. A butterfly may be tagged in Illinois and caught several days later in South Carolina. This information is combined with similar information from other places to build a picture of the migration.

*inter*NET CONNECTION

Visit the Glencoe Homepage, *www.glencoe.com/sec/science*, for a link to information about butterfly migration. Find out how data is collected from monarch watchers and how these data are used.

159

Activity 6-2

Design Your Own Experiment
What's the limit?

How many blades of grass are in a park? It may seem to you like there's no limit to the number of blades of grass that can grow there. However, as you've discovered, there are many factors that organisms like the plants in the park need to live and grow. By experimenting with these factors, you can see how they limit the size of the population.

PREPARE

Purpose

Kinesthetic Students will determine how light, water, temperature, or space limit plant populations. L1 ELL

Process Skills

observing and inferring, making and using tables, making and using graphs

Time

50 minutes for setup; 10 minutes every 3 to 4 days for 2 weeks for data collection

Materials

Peas or corn may be substituted for bean seeds.

Safety Precautions

Students should wash their hands carefully after handling seeds and soil.

Possible Hypotheses

Student hypotheses will reflect that space, light, water, or temperature may limit the number of plants that can grow in a container.

📁 **Activity Worksheets,** pages 5, 39-40

PLAN

Possible Procedures

- Plant the same number of seeds in each pot and manipulate the variable you are interested in observing while keeping all other variables the same. Observe the results.

- Plant an increasing number of seeds in a series of pots to determine how space influences plant growth when all other variables are the same. Choose one variable to test at a time. Test light, water, temperature, or space.

Possible Materials
- bean seeds
- small planting containers
- soil
- water
- labels
- spoons
- aluminum foil
- sunny window or other light source
- refrigerator or heater

PREPARE

What You'll Investigate
How do space, light, water, and temperature limit plant populations?

Form a Hypothesis

Think about what you already know about the needs of plants. In your group, **make a hypothesis** about how one abiotic factor may limit the number of bean plants that can grow in a single pot.

Goals

Observe how space, light, water, or temperature affect how many bean plants are able to grow in a pot.

Design an experiment that shows whether a certain abiotic factor limits a plant population using the materials listed.

Safety Precautions

Wash your hands after you handle soil and seeds.

Sample Data Table for Temperature as a Variable

Temperature (°C)	Number of Plants		
	Trial 1	Trial 2	Trial 3
10	2	0	1
20 (Control)	5	7	7
30	4	3	5

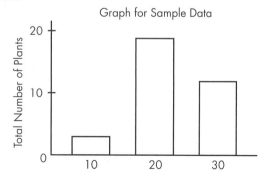

Graph for Sample Data

PLAN

1. **Decide** on a way to test your group's **hypothesis**. Make a complete materials list as you plan the steps of your experiment.
2. What is the one abiotic factor you will be testing? How will you test it? What factors will you need to control? Be specific in describing how you will handle the other abiotic factors.
3. How long will you run your experiment? How many trials of your experiment will you run?
4. Decide what data you will need to collect. **Prepare a data table** in your Science Journal.
5. **Read** over your entire experiment and imagine yourself doing it. Make sure the steps are in logical order.

1. Make sure your teacher has approved your plan and your data table before you proceed.
2. Carry out your plan.
3. While doing the experiment, **record** your observations and **complete** your data table in your Science Journal.

CONCLUDE AND APPLY

1. **Compare** your results with those of other groups. How did different factors affect the plants?
2. **Graph** your results using a bar graph. **Compare** the number of bean seeds that grew in the experimental containers with the number of bean seeds that grew in the control containers.
3. How did the abiotic factor you tested affect the bean plant population?
4. **APPLY Predict** what would happen to your plant population if you added another kind of plant or animal to the containers.

6-2 Relationships Among Living Things   161

Teaching Strategies

Troubleshooting Students should be careful not to plant seeds too deeply in the soil and to avoid overwatering the containers. Be sure students have a control.

DO

Expected Outcome

The control plants will do better than the experimental plants. Most students' results will show that seeds need water, light, space, and moderate temperatures in order to grow.

CONCLUDE AND APPLY

1. Student results should be similar to those of other students who observed the same variable. Limiting any of the variables will inhibit plant growth.
2. Graphs should show that more bean seeds germinate and grow in the control pots.
3. Student answers should state what factor limited growth and how it did.
4. Adding another organism to the container would either benefit or harm the plants, depending on how the two organisms interacted and how the introduced organism reacted to the limiting factor.

Go Further

Different types of organisms have different resource requirements. Students should select two or three different types of organisms and test how the absence of one abiotic factor will influence the different organisms. This will enable students to observe that environmental changes may influence populations of organisms in different ways.

Assessment

Performance Have students make posters of their experimental designs, results, and conclusions to share with the class. Use the Performance Task Assessment List for Poster in **PASC**, p. 73.

FIGURE 6-9
One of the most common ways organisms interact in a community is by being food for another organism.

Interactions in Communities

Are frogs the only organisms in the stream community that eat mosquitoes? No, there are many animals that eat them, including some birds and spiders. That means that frogs must compete with birds and spiders for the same food. Feeding interactions such as those in **Figure 6-9** are the most common interactions among organisms in a community.

Imagine a large bowl of popcorn in your classroom. As long as there is enough popcorn to go around, you don't have to worry that you'll get your share. But if the bowl of popcorn were small, you would have to compete with your classmates to get some. The greater the population size of an area, the greater the competition for resources such as food. Food isn't the only resource that organisms compete for. Organisms will compete for any resource that is in limited supply. Space, water, sunlight, and shelter are all resources that may be limited in a particular ecosystem.

Eat or Be Eaten

Have you ever heard the phrase "birds of prey"? A falcon is a bird of prey, which means it captures and eats other animals. A falcon, with its razor-sharp talons, will swoop down from the sky to snatch up a field mouse. The falcon is a predator (PRED uh tur). Predation (pruh DAY shun) is the act of one organism feeding on another.

Organisms That Live Together

Predation doesn't sound like a good deal for the field mouse, does it? The falcon population, however, is limited by the size of the mouse population. There are other types of relationships among organisms. In one type of interaction, both organisms in the relationship benefit. The African tickbird, for instance, gets its food by eating insects off the skin of zebras. The tickbird gets food, while the zebra gets rid of harmful insects. In another type of relationship, only one organism benefits. The other is not harmed. A bird building a nest in a tree is an example of this. The bird gets protection from the tree, but the tree isn't harmed. In still another relationship, one organism is helped while the other is harmed. The insects on the zebra's skin, for example, benefit from the zebra. However, these insects can cause harm to the zebra. Have you ever been bitten by a mosquito? That's a firsthand experience of this type of relationship.

Where and How Organisms Live

How can a small ecosystem such as a classroom aquarium support a variety of different organisms? It's possible because each type of organism has a different role to play in the ecosystem. A typical classroom aquarium may contain snails, fish, algae, and bacteria. The role of snails is to feed on algae, which keeps the glass of the aquarium clear for light to get in. The role of the algae is to provide oxygen for the system through photosynthesis. What do you think the role of the fish might be? The role of an organism in an ecosystem is called the organism's **niche** (NIHCH). The niche of the fish might be to feed on algae to keep

Problem Solving

Graphing Populations

One way to understand more about relationships among organisms in an ecosystem is to keep track of, or monitor, and graph populations. Use the information below to make a graph of population size over time for barn owls and field mice. Then, answer the questions that follow.

Table 6-1

Monthly Population Size per Hectare (in 100s)									
Month	J	F	M	A	M	J	J	A	S
Field Mice	6	5	4	3	3	4	5	4	6
Barn Owls	2	3	4	4	2	1	4	3	4

Solve the Problem:

Set up your graph with months on the *x*-axis and numbers of organisms on the *y*-axis. Use two colors to plot your data. For more help, refer to Graphing in the Skill Handbook. Use your graph to infer how the population of field mice affects the population of barn owls. Predict how the next two months of the graph will look.

Think Critically:

Field mice eat green plants and grains. What do you think would happen to the population of barn owls if there were no rain in the area for a long time?

6-2 Relationships Among Living Things

Science Journal **Critical Thinking** Have students examine the niche humans occupy in their ecosystem. What conditions are required for survival? What habitats are humans able to occupy? What habitats are not suitable for human habitation and why?

L3 P LS

Problem Solving

Solve the Problem

When mice are plentiful, the number of barn owls increases until the mouse population is not large enough to feed the barn owl population. The barn owl population then decreases. The next two months of the graph should show a decrease in the number of field mice and an increase in the number of barn owls.

Think Critically

A lack of rain would cause plant populations to decrease, causing the field mouse population to decline. As the field mouse population continued to decline, the barn owl population would also decline.
LS

Student Text Question

Page 164 How would you define your niche in your ecosystem? *Students go to school, eat certain foods, create a certain amount of waste, do certain chores, etc.*

3 Assess

Check for Understanding

Brainstorming Have students brainstorm the interactions between biotic and abiotic factors in a garden community.

Reteach

Review the different interactions that occur between organisms in an ecosystem.

Extension

For students who have mastered this section, use the **Reinforcement** and **Enrichment** masters.

163

4 Close

·MINI·QUIZ·

Use the Mini Quiz to check students' recall of chapter content.

1. A(n) _____ is a group of organisms living in the same place at the same time. *population*

2. Things that limit the size of populations are called _____. *limiting factors*

3. The place where an organism lives is its _____. *habitat*

Section Wrap-up

1. Limiting factors such as food, water, predation, space, and other resources can limit a population's growth and prevent it from increasing.

2. A community and a population are both parts of an ecosystem. A population is a group of similar organisms living in the same place at the same time. In contrast, a community is a group of different populations of organisms living in the same place at the same time.

3. **Think Critically** Student answers will vary. Organisms interact for shelter, food, protection, and many other resources.

USING MATH

L·S **Logical-Mathematical** Population variation and stability will vary depending on the population and organism selected.

FIGURE 6-10
Each organism in this ecosystem has its own job or niche in the ecosystem.

its population in check. All the conditions an organism needs to survive are part of its niche. How would you define your niche in your ecosystem?

The place where an organism lives out its life is called its **habitat** (HAB uh tat). Different species share a habitat. Describe the habitat of the organisms in **Figure 6-10.** Resources such as food, space, and shelter are all shared among the organisms in the habitat. This is possible because each organism has its own niche. The different ways each organism feeds and uses other resources allow all the organisms to live together.

Section Wrap-up

1. Identify three reasons a population can't continue to grow larger and larger forever.

2. List similarities and differences in a community and a population.

3. **Think Critically:** Identify several interactions between organisms in an ecosystem. Give an example of each interaction.

4. **Skill Builder**
   Developing a Multimedia Presentation
   Choose a community such as a coral reef, grassland, wetland, tundra, or forest. Develop a multimedia presentation that describes the community and some interactions that occur within it. If you need help, refer to Developing a Multimedia Presentation on page 564 in the **Technology Skill Handbook.**

USING MATH

Find data about the population size of an organism such as bison, mountain lions, whales, or an endangered species over a period of ten years. Encyclopedias or the Internet can provide this information. Make a graph that shows how the size of the population varied over those years.

Skill Builder
Developing a Multimedia Presentation
Students could use a program such as *Hyperstudio* to build their multimedia presentations. They could also use a combination of photographs, slides, videos, and/or audio in their presentations. They should show the different environmental conditions in their ecosystem and the interactions the organisms in the community have with their environment and other organisms.

✓Assessment

Performance Have students research in groups the niche requirements of a plant or animal of their choice and present their findings to the class. Use the Performance Task Assessment List for Group Work in **PASC,** p. 97.

People & Science

Laura Collins, Bird Curator

Q Ms. Collins, tell us about your job.

A I'm the curator of birds at Sea World Ohio. We have about 400 birds on exhibit here, representing roughly 90 species.

Q What goes into building an exhibit that recreates a bird's natural habitat?

A It's a team effort, involving many people and a lot of research. We determine what the birds' needs are and then design an exhibit that's appropriate for them. For example, a flamingo needs a shallow pond so it can wade and filter feed, while a diving duck needs a deeper pond. Our Penguin Encounter exhibit has probably been one of our most challenging projects as far as designing a habitat for a bird.

Q Tell us about that exhibit. How do you make the penguins feel at home?

A We have Macaroni, Adélie, and emperor penguins. In the exhibit, there are mountains, rocks for the penguins that like to climb, a 45 000-gallon saltwater pool, and snow on the ground. Two snow

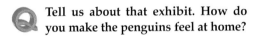

machines on top of the exhibit make 5000 pounds of snow a day. All the water in the pool is filtered every 25 minutes, and it's chilled to around 40°F (4°C). The air is filtered and chilled, too, down to about 27°F (−3°C). Even on a 90° day, it's still below freezing in there!

Q That sounds like a huge project!

A The engineering and life-support systems, and the number of people that it takes to monitor those systems and keep them running, are phenomenal. But the real success story is that the penguins breed in the exhibit. They build nests, lay eggs, and raise chicks—it's a self-sustaining population.

Career Connection

Many careers involve working with animals. Interview a person in your community who has a career working with animals. Some examples include a keeper in a zoo, a veterinarian, a dog groomer, or an animal breeder. Present a summary of your interview to your class.

6-2 Relationships Among Living Things   **165**

- Have students select an animal and research its needs. Then have each student create a shoe-box diorama that depicts an appropriate setting for his or her chosen species. **L1**

For More Information

Have students research penguins or penguin breeding on the Internet. Students may also want to read *The Seaworld Book of Penguins* by Frank S. Todd, Harcourt Brace Jovanovich, New York, 1984.

Career Connection

Career Path Keepers and curators at zoos and animal parks generally have a background in animal biology.
- High School: biology and chemistry
- College (often necessary for higher-level positions): biology or zoology; veterinary science
- Hands-on experience working with animals is also important.

Background

- The breeding season for Adélie and Macaroni penguins begins in September (which is spring for the birds because they are kept on an Antarctic, or austral, light cycle). At that time, the keepers bring in loads of small rocks for the penguins to use in building nests. The birds pair up with their mates. After staking out nest sites, pairs build nests and lay eggs. Both Macaroni and Adélie penguins usually lay two eggs.
- Emperor penguins breed in April and May. Females lay a single egg that is incubated—usually on the feet of the male—for two months during the dark days of the austral winter.
- Laura and members of her staff are able to monitor the penguins' activities in the exhibit using closed-circuit TV cameras. All of the birds wear identification bracelets on one of their flippers. So without even entering the exhibit, keepers can monitor the penguins closely and record on videotape their behaviors as they go about incubating eggs and rearing chicks. That video record is a useful tool for spotting changes in behavioral patterns that could indicate a problem with a particular bird or pair of birds. It can also be an important source of information for the keepers in the event that a chick must be hand-reared.

Teaching Strategies

- Discuss with students the meaning of the term *curator*. A curator is a person in charge of a museum, zoo, or other place of exhibit.

165

6•3 Gators at the Gate!

Science & Society

Bellringer

 Before presenting the lesson, display **Section Focus Transparency 18** on the overhead projector. Assign the accompanying **Focus Activity** worksheet.

L2 **ELL**

GLENCOE TECHNOLOGY

🔘 **Videodisc**

STVS: Ecology
Disc 6, Side 1
Preserving Duck Habitats
(Ch. 7)

What YOU'LL LEARN

- How to recognize the problems of endangered and threatened species
- How to distinguish fact from opinion

   Science Words:
   endangered species
   threatened species

Why IT'S IMPORTANT

The needs of wildlife and the needs of humans are in conflict all over the world.

FIGURE 6-11

Alligators are a common sight along the banks of rivers and canals in Florida neighborhoods.

When you think about Florida, you probably picture sandy beaches and palm trees. But do you think about alligators? Alligators are among the best-known animals that live in Florida. They can grow to be 13 feet long and weigh more than 600 pounds.

Endangered Alligators

By the 1960s, numbers of alligators were greatly reduced in Florida due to hunting and habitat loss. The numbers became so low that alligators were placed on the endangered species list. A species is considered an **endangered species** when so few of its members are living that the entire species may become extinct. In the United States, it became illegal to hunt alligators. The number of alligators went up. By 1977, they were renamed as a threatened species. A **threatened species** still needs to be protected but is not in immediate danger of becoming extinct. Now, more than a million alligators live on farms and in the wild. Good news, right? Think again. There are problems—big problems from big alligators.

As you read the following quotes from newspaper articles about alligators, try to distinguish facts from opinions. Facts can be proven. Opinions cannot; they reflect feelings and beliefs.

Alligator Problems

More and more people live, work, and play in areas where alligators live. Some people, such as Sandy Bonilla, think the alligators have to go. "I found an alligator in my swimming pool," Bonilla says. "My neighbor saw one on a golf course. I believe this whole situation is out of control. With people and gators living in the

---

Program Resources

 Reproducible Masters
Enrichment, p. 24 **L3**
Reinforcement, p. 24 **L2**
Study Guide, p. 24 **L1**

 Transparencies
Section Focus Transparency 18 **L2**

same areas, attacks become more likely. These animals are the greatest danger we face."

But Juana Aravjo disagrees. "It seems to me that the problem is not the alligators," she says. "The problem is that there's no room for the alligators." Aravjo explains that before, when an alligator was caught in someone's swimming pool, a professional trapper released it back into a wilderness area. "Today, there are fewer wilderness areas, so when an alligator is considered a nuisance to people, it is removed by a trapper and killed," Aravjo says. "If we don't act quickly, alligators will probably become extinct."

FIGURE 6-12

Alligators eat frogs, fish, snakes, birds, and small mammals. But they've been known to eat sticks, rocks, and even aluminum cans.

 Skill Builder: *Distinguishing Fact from Opinion*

LEARNING the SKILL

1. To tell the difference between facts and opinions, study the statements to see if they can be checked for accuracy and if they can be proven. Facts can be proven; opinions cannot.

2. Facts answer specific questions such as: What happened? Who did it? When and where did it happen? Why did it happen? They give specific information about an event.

3. Opinions tell someone's feelings or what they believe. They often begin with phrases such as "I think," " I believe," or "it seems to me." They often include words such as *probably, greatest, best,* or *worst.*

PRACTICING the SKILL

1. Compare and contrast an endangered species with a threatened species.

2. Examine the quotes from the newspaper. List one fact and one opinion from each speaker. Explain why the statements are facts or opinions.

3. Based on what you have read, should alligator habitats be protected? If so, how?

APPLYING the SKILL

Other species besides alligators are threatened. Bring a newspaper or magazine article to class about a threatened or endangered species. Distinguish fact from opinion in the article.

6-3 Gators at the Gate!   **167**

Teaching the Content

• Have students discuss the following issues. Should the number of alligators be reduced to decrease the possibility of attacks on people? If so, how? Who would decide how many alligators should be kept? Should alligator habitats be protected?

Teaching the Skill

Bring in several articles from the newspaper. Have students work in groups to find the main idea in each article. Have two groups get together to compare their work.

COOP LEARN

Answers to Practicing the Skill

1. A species is endangered when the number of its living members is so small that it is in danger of becoming extinct. A threatened species still needs to be protected but is not in danger of becoming extinct.

2. *Example of a fact* 1. I found an alligator in my swimming pool. *Example of a fact* 2. Today, there are fewer wilderness areas, so when an alligator is considered a nuisance to people, it is removed by a trapper and killed. Both facts give specific information that can be checked for accuracy and proven. *Example of an opinion* 1. These animals are the greatest danger we face. *Example of an opinion* 2. It seems to me that the problem is not the alligators.

3. Answers will vary.

Content Background

• The American alligator is one of 20 species of crocodilians (lizard-like, egg-laying meat eaters) found worldwide. The largest of the modern reptiles, crocodilians are a living link to prehistoric, dinosaur-like reptiles.

• Crocodilians in the wild may live up to 100 years; in captivity, they generally live about 40 years.

• American alligators inhabit freshwater locations in the southeastern United States from North Carolina to Florida and west to the lower Rio Grande. Large adults can stay under water for more than an hour. They are powerful swimmers, using their tails to propel them. They also use their tails as weapons or to knock land prey into the water.

Prepare

Section Background

- Organisms are classified by how they obtain their food. Producers manufacture their own food, consumers eat other organisms, and decomposers eat dead organisms.

- Energy travels through an ecosystem in the form of food. A food chain is a series of interactions between organisms that make up this transfer of energy. A food web is a model that reflects all feeding relationships in an ecosystem.

- Matter is constantly recycled through an ecosystem. Water and carbon are some of the materials recycled.

Preplanning

Refer to the Chapter Organizer on pages 146A-B.

1 Motivate

Bellringer

Before presenting the lesson, display **Section Focus Transparency 19** on the overhead projector. Assign the **Focus Activity** worksheet. [L2] [ELL]

6•4 Energy Through the Ecosystem

What YOU'LL LEARN

- How organisms get the energy they need
- How energy flows through an ecosystem

Science Words:
producer
consumer
decomposer

Why IT'S IMPORTANT

Knowing about food webs and material cycles will help you understand how living things depend on each other.

It's All About Food

Think about the interactions that we've talked about so far. The frog and the mosquito, the falcon and the mouse—most of the interactions involve food. Energy moves through an ecosystem in the form of food.

Many different populations interact in a backyard ecosystem, including plants, birds, insects, squirrels, and rabbits, as shown in **Figure 6-13.** The plants in the ecosystem produce food through photosynthesis. An organism that makes its own food, like a plant, is called a **producer.** The grasshopper that nibbles on the plants is a **consumer.** A consumer eats other organisms.

Some of the consumers in an ecosystem are so small that you might not notice them. But they have an important role to play. They are the decomposers, such as bacteria and fungi. **Decomposers** are organisms that use dead organisms and the waste material of other organisms for food.

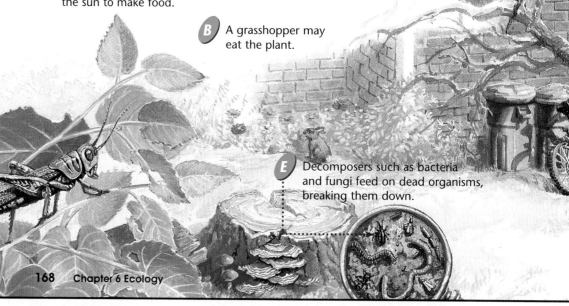

A The plants in this backyard are producers. They use energy from the sun to make food.

B A grasshopper may eat the plant.

E Decomposers such as bacteria and fungi feed on dead organisms, breaking them down.

Program Resources

 Reproducible Masters
Activity Worksheets, p. 5 [L2]
Enrichment, p. 25 [L3]
Lab Manual, pp. 43-46 [L2]
Multicultural Connections, pp. 15-16 [L2]
Reinforcement, p. 25 [L2]
Study Guide, p. 25 [L1]

Transparencies
Section Focus Transparency 19 [L2]

Modeling the Flow of Energy

The food chain in **Figure 6-9** on page 162 is a simple model that shows how energy from food passes from one organism to another. Each organism is linked by an arrow. The arrows show that energy moves from one organism to another in the form of food.

Can you spot any problems with the model? How do the other organisms in the community fit into the food chain? We need a more complex model to show the true feeding interactions in the backyard ecosystem.

In an ecosystem, food chains often overlap. For instance, a bird may eat seeds and in turn be eaten by a cat. But, a cat could also eat the rabbit or the mouse. One food chain cannot model all these overlapping relationships. Scientists use a more complicated model, called a food web, to show the transfer of energy from food in an ecosystem. A food web is a series of overlapping food chains that shows all the possible feeding relationships in an ecosystem. Draw a food web for this backyard ecosystem.

FIGURE 6-13
In any community, energy flows from producers to consumers.

C A bird may eat the grasshopper.

D A cat may prey on a bird, rabbit, or chipmunk.

6-4 Energy Through the Ecosystem 169

169

Extension

📁 For students who have mastered this section, use the **Reinforcement** and **Enrichment** masters.

4 Close

Use the Mini Quiz to test students' recall of chapter content.

1. **An organism that makes its own food is called a(n) _____.** *producer*

2. **A simple model that shows how energy is passed from one organism to another is called a(n) _____.** *food chain*

3. **All matter is _____ through the ecosystem.** *cycled*

Section Wrap-up

1. Answers will vary from student to student. Any organism that consumes another organism—plant or animal—is a consumer.

2. Energy is lost as it moves up the food chain as organisms use energy to search for food and for other activities.

3. **Think Critically** A plant converted sunlight into energy. The chicken obtained energy by eating the plant, and the student obtained energy by eating the chicken.

Using Computers

Diagrams will vary based on the ecosystem chosen. All diagrams will start with plants and move up the food chain to consumers.

FIGURE 6-14

Horses get the materials they need to grow and maintain their bodies by eating food such as grass. *Where do you think the grass gets the materials it needs?*

Cycling of Materials

What happens when you recycle a soda can? The can is taken to a place that melts it so that the aluminum can be used again. This is an example of a simple cycle. The same aluminum can be used over and over again. Cycles are important to ecosystems. Instead of aluminum cans, however, it's the materials that make up organisms that get recycled.

The bodies of living things are made up of matter. They are composed of materials like water, and chemicals like nitrogen and carbon. To get the matter needed to build bones, muscles, and skin, you need to eat food made of the right kinds of materials, as the horse in **Figure 6-14** is doing. In an ecosystem, matter cycles through food chains. The amount of matter on Earth never changes. So matter in ecosystems is recycled, or used again and again.

Living things depend on these cycles for survival. Living things also depend on one another for food, shelter, and other needs. All the different things that make up the biosphere—from a tiny insect to a raging river—have a unique role to play.

Section Wrap-up

1. Name some organisms that are consumers. Give an example of the type of food each eats.

2. Why is less energy available in the fourth link of a food chain than at the first?

3. **Think Critically:** Imagine that you just ate a chicken sandwich. Starting with the sun, trace the flow of energy in the food chain from the beginning to you.

4. *Skill Builder*
   Comparing and Contrasting Compare and contrast a food chain and a food web. If you need help, refer to Comparing and Contrasting on page 550 in the **Skill Handbook.**

Using Computers

Model Using graphics software, make a diagram that shows how energy flows through your local ecosystem in a food web.

Skill Builder

Comparing and Contrasting Both food webs and food chains are models of energy transfer through an ecosystem. A food chain explains only one path of energy transfer in an ecosystem. In contrast, a food web shows all possible transfers of energy in an ecosystem.

✓ Assessment

Performance Ask students to form groups to research one of the following matter cycles: water, nitrogen, or carbon. Have each group create a poster and present their cycle to their classmates. Use the Performance Task Assessment List for Poster in **PASC**, p. 73.

IS   COOP LEARN

Chapter 6

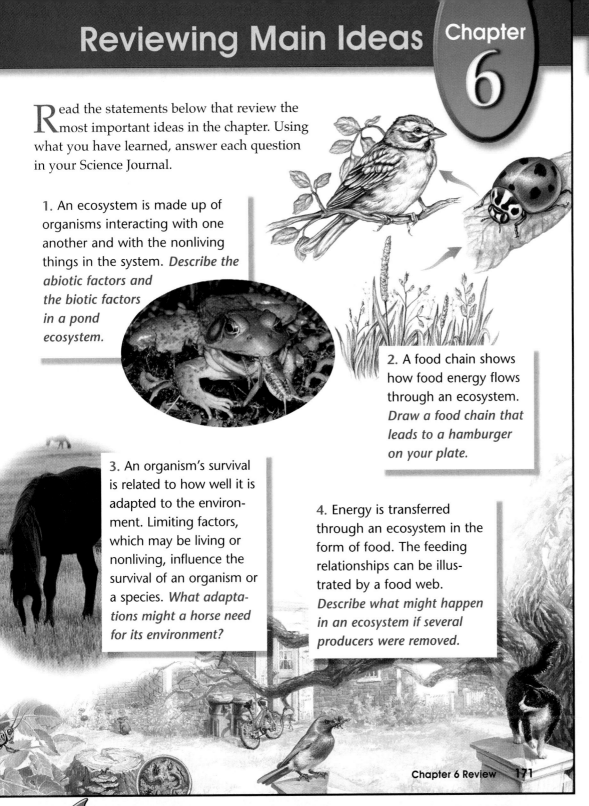

Read the statements below that review the most important ideas in the chapter. Using what you have learned, answer each question in your Science Journal.

1. An ecosystem is made up of organisms interacting with one another and with the nonliving things in the system. *Describe the abiotic factors and the biotic factors in a pond ecosystem.*

2. A food chain shows how food energy flows through an ecosystem. *Draw a food chain that leads to a hamburger on your plate.*

3. An organism's survival is related to how well it is adapted to the environment. Limiting factors, which may be living or nonliving, influence the survival of an organism or a species. *What adaptations might a horse need for its environment?*

4. Energy is transferred through an ecosystem in the form of food. The feeding relationships can be illustrated by a food web. *Describe what might happen in an ecosystem if several producers were removed.*

Review

Have students look at the illustrations on this page. Ask them to describe details that support the main ideas of the chapter found in the statement for each illustration.

Teaching Strategies

Use the pictures and diagrams to reinforce the idea that ecosystems are communities of populations of organisms interacting with each other and with their environment.

Answers to Questions

1. Student answers will vary. Abiotic factors: air, soil, light, water; Biotic factors: grass, trees, birds, insects, bacteria

2. A plant converts sunlight into energy. Next, a cow obtains energy by eating the plant. In the final step, the student obtains energy by eating the hamburger.

3. Horses have hooves for running at great speeds and dealing with uneven terrain. Their teeth are adapted for eating grasses.

4. Producers are the basis of every food chain. If several were removed, there would be less food for those higher on the food chain.

Science at Home

Based on what students have learned in this chapter, students can build a terrarium and observe the interactions occurring in their small ecosystem. They can then determine what resources are necessary to maintain a working system. L1

Assessment

Portfolio Encourage students to place in their portfolios one or two items of what they consider to be their best work. Examples include:

- Across the Curriculum, p. 149
- Skill Builder, p. 156
- Science Journal, p. 163

Performance Additional performance assessments may be found in **Performance Assessment** and **Science Integration Activities.** Performance Task Assessment Lists and rubrics for evaluating these activities can be found in Glencoe's **Performance Assessment in the Science Classroom (PASC).**

Using Key Science Words

1. ecology
2. biosphere
3. biotic factor
4. habitat
5. population

Checking Concepts

6. b   **7.** d   **8.** b
9. d   **10.** a

Thinking Critically

11. Decomposers obtain their food by breaking down dead organisms. They do not manufacture their own food so they are consumers.

12. An organism that eats low on the food chain would consume large amounts of plants.

13. Answers will vary. Accept all reasonable responses. Be sure students define their ecosystem. A pond, puddle, garden or forest are all possibilities.

14. Student answers may vary. Three possible limiting factors are light, temperature, and food. Too little light could limit the amount of plant life in the aquarium. Temperatures that are too warm or too cold can affect the living things in the aquarium. Too little food can reduce the populations of organisms in the aquarium.

15. Answers will vary. The student's habitat is his home. The student's niche includes all of the resources necessary for survival such as food, shelter, water, and relationships with other organisms. It also includes tasks the student may do such as going to school, etc.

Chapter 6 Review

Using Key Science Words

abiotic factor	endangered species
biosphere	habitat
biotic factor	limiting factor
community	niche
consumer	population
decomposer	producer
ecology	threatened species
ecosystem	

Match each phrase with the correct term from the list of Key Science Words.

1. the study of all the interactions between living and nonliving parts of an ecosystem
2. the part of Earth where an organism can live
3. any living thing in the environment
4. the place where an organism lives
5. individuals of the same species living in the same place at the same time

Checking Concepts

Choose the word or phrase that completes the sentence.

6. All are biotic factors except
   _____.
   a. raccoons   c. pine trees
   b. sunlight   d. mushrooms

7. Ponds, streams, and prairies are examples of _____.
   a. niches   c. populations
   b. producers   d. ecosystems

8. A group of the same type of organism living in the same place at the same time is a _____.
   a. habitat
   b. population
   c. community
   d. ecosystem

9. A food chain is a model that shows the transfer of _____.
   a. producers   c. plants
   b. consumers   d. energy

10. An example of a producer is a
    _____.
    a. grass   c. fungus
    b. horse   d. fish

Thinking Critically

Answer the following questions in your Science Journal using complete sentences.

11. Why is it correct to say that decomposers are also consumers?

12. Give examples of foods you would eat if you were eating low on the food chain.

13. Draw an ecosystem. Label biotic and abiotic factors. Describe three interactions among organisms in the ecosystem.

14. Identify three possible limiting factors for an aquarium ecosystem. Describe how each factor can limit population growth.

15. Describe your own habitat and niche.

Assessment Resources

Reproducible Masters
Chapter Review, pp. 15-16
Assessment, pp. 25-28
Performance Assessment, p. 44

Glencoe Technology
Computer Test Bank
 MindJogger Videoquiz

Developing Skills

If you need help, refer to the description of each skill in the Skill Handbook.

16. **Classifying:** Make a list of ten of your favorite foods. Classify each as coming from a producer, consumer, or decomposer. Write a short explanation of your classification of each item.

17. **Observing and Inferring:** Observe the photo of the aquarium on page 164. Identify at least one food chain you might find in the aquarium ecosystem.

18. **Making and Using Graphs:** The following graph shows the changes in the size of a population of insects living on roses over the course of a year. During what month is the insect population the smallest? During what month is the population the largest?

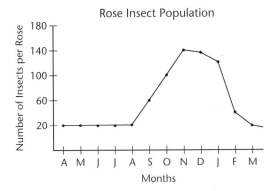

Rose Insect Population

19. **Designing an Experiment:** Design an experiment to test the hypothesis that water is a limiting factor for plants. Include a list of procedures.

20. **Predicting:** Predict what would happen to an ecosystem if its decomposers were removed.

Performance Assessment

1. **Slide Show or Photo Essay:** Find slides or photographs that show a variety of different ecosystems. Arrange a slide presentation or photo display of these images. Use titles or captions to identify each photograph or slide.

2. **Poster:** Choose an ecosystem to research. Find out what plant and animal species are found there and how they interact. Make a poster illustrating a food web in this ecosystem.

Developing Skills

16. **Classifying** Student answers will vary. Students should classify plant materials as producers, animal materials as consumers, and products from fungi and bacteria as decomposers.

17. **Observing and Inferring** Algae convert sunlight to energy, and snails eat the algae.

18. **Making and Using Graphs** The population of insects is smallest from April to August. It is highest in November.

19. **Designing an Experiment** Question: Is water a limiting factor for plant growth? State the hypothesis: Water is a limiting factor for plant growth. Design the Experiment: Grow one plant with water and another plant without water. Observe and conclude.

20. **Predicting** Dead organisms would not decompose, and matter would not be recycled through the ecosystem.

Performance Assessment

1. Slides and photographs should depict different ecosystems and illustrate that ecosystems are composed of communities of organisms. Use the Performance Task Assessment List for Slide Show or Photo Essay in **PASC**, p. 77. **P**

2. The poster should illustrate biotic and abiotic relationships between organisms and also depict a food web. Use the Performance Task Assessment List for Poster, **PASC**, p. 73. **P**

Chapter Organizer

Section	Objectives/Standards	Activities/Features
Chapter Opener		Explore Activity: Classify Resources, p. 175
7-1 **Natural Resource Use** (4 sessions, 2 blocks)*	**1. Give examples** of how resources are used. **2. Classify** resources as natural, renewable, or nonrenewable. National Science Content Standards: (5-8) UCP1, UCP3, A1, A2, C4, E1, E2, F2, F4, F5	MiniLAB: Analyzing Gift Wrap, p. 180 Using Technology: From Plastic Bottles to Play Towers, p. 181 Activity 7-1: Using Water, p. 183 Skill Builder: Sequencing, p. 184 Using Math, p. 184
7-2 **People and the Environment** (4 sessions, 2 blocks)*	**3. Describe** how people affect the environment. **4. Compare and contrast** the different types of pollution. National Science Content Standards: (5-8) UCP2, C4, E1, E2, F2	Activity 7-2: Using Land, pp. 188-189 Skill Builder: Developing Multimedia Presentations, p. 194 Science Journal, p. 194
7-3 **Protecting the Environment** (2 sessions, 1 block)*	**5. Analyze** the problems of solid waste. **6. Describe** how to reduce, reuse, and recycle resources. National Science Content Standards: (5-8) UCP2, C4, F2	MiniLAB: Making Models, p. 196 Problem Solving: Uses for Plastic Things, p. 197 Skill Builder: Recognizing Cause and Effect, p. 199 Science Journal, p. 199
7-4 **Science and Society: A Tool for the Environment** (1 session, ½ block)*	**7. Describe** a life-cycle analysis. **8. Develop** note-taking skills. National Science Content Standards: (5-8) UCP1, A1, A2, E2, F2, F4, F5, G1	Skill Builder: Taking Notes, p. 201 Science & the Arts: Xeriscaping, p. 202

* A complete Planning Guide that includes block scheduling is provided on pages 31T-33T.

Activity Materials

Explore	Activities	MiniLABs
No materials needed.	page 183 calculator pages 188-189 grid paper, colored pencils	page 180 various wrapping materials, objects to wrap page 196 discarded nonfood items, such as newspapers, clean cans, or glass; glue, string, or tape

Need Materials? Call Science Kit (1-800-828-7777).

Teacher Classroom Resources

Reproducible Masters	Transparencies	Teaching Resources
Activity Worksheets, pp. 5, 43-44, 47 **Enrichment**, p. 26 **Lab Manual 13** **Reinforcement**, p. 26 **Study Guide**, p. 26	**Science Integration Transparency 7**, Renewable and Nonrenewable Resources **Section Focus Transparency 20**, Skiing with a Splash **Teaching Transparency 13**, Meeting World Energy Needs	**Spanish Resources** **English/Spanish Audiocassettes** **Cooperative Learning Resource Guide** **Lab Partner** **Lab and Safety Skills** **Lesson Plans**
Activity Worksheets, pp. 5, 45-46 **Enrichment**, p. 27 **Lab Manual 14** **Reinforcement**, p. 27 **Science and Society/Technology Integration**, p. 25 **Science Integration Activities**, pp. 51-52 **Study Guide**, p. 27	**Section Focus Transparency 21**, Toxic Waste Scare	
Activity Worksheets, pp. 5, 48 **Cross-Curricular Integration**, p. 11 **Enrichment**, p. 28 **Multicultural Connections**, pp. 17-18 **Reinforcement**, p. 28 **Study Guide**, p. 28	**Section Focus Transparency 22**, Second-Hand Jackets? **Teaching Transparency 14**, Solid Waste	**Assessment Resources** **Chapter Review**, pp. 17-18 **Assessment**, pp. 29-32, 33-34 **Performance Assessment**, pp. 33-34, 45 **Performance Assessment in the Science Classroom (PASC)** **MindJogger Videoquiz** **Alternate Assessment in the Science Classroom** **Computer Test Bank**
Enrichment, p. 29 **Reinforcement**, p. 29 **Study Guide**, p. 29	**Section Focus Transparency 23**, Give It to Me Straight, Doc	

Key to Teaching Strategies

The following designations will help you decide which activities are appropriate for your students.

L1 Level 1 activities should be appropriate for students with learning difficulties.

L2 Level 2 activities should be within the ability range of all students.

L3 Level 3 activities are designed for above-average students.

ELL ELL activities should be within the ability range of English Language Learners.

LS These activities are designed to address different learning styles.

COOP LEARN Cooperative Learning activities are designed for small group work.

P These strategies represent student products that can be placed into a best-work portfolio.

GLENCOE TECHNOLOGY

The following multimedia resources are available from Glencoe.

Science and Technology Videodisc Series (STVS)
Chemistry
Wind Power
Hydroelectric Power
Mini-Hydroelectric Power Plants

The Infinite Voyage Series
National Geographic Society Series
STV: Water
Glencoe Life Science CD-ROM

Teacher Classroom Resources

This is a representation of key blackline masters available in the Teacher Classroom Resources.

Teaching Aids

Section Focus Transparencies

Science Integration Transparencies

Teaching Transparencies

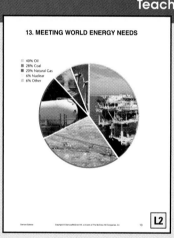

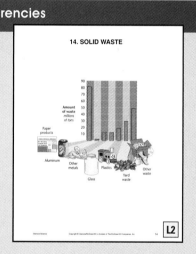

Meeting Different Ability Levels

Study Guide for Content Mastery

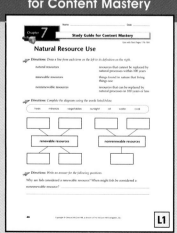

Reinforcement

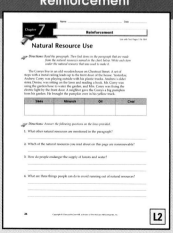

Enrichment Worksheets

174C

Hands-On Activities

Science Integration Activities

Chapter 7   Integration

Oil and Water Don't Mix

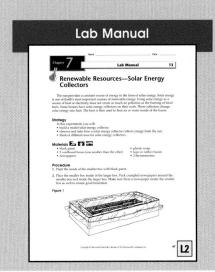

Try It!

Lab Manual

Chapter 7   Lab Manual   13

Renewable Resources—Solar Energy Collectors

Strategy

Materials

Procedure

Figure 1

Activity Worksheets

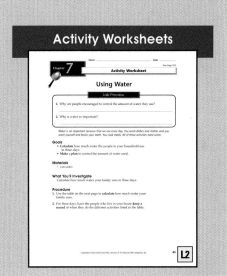

Chapter 7   Activity Worksheet

Using Water

Lab Preview

Goals

Materials

What You'll Investigate

Procedure

Enrichment and Application

Cross-Curricular Integration

Chapter 7 MATHEMATICS CONNECTION   Resources

Animals in Danger

Multicultural Connections

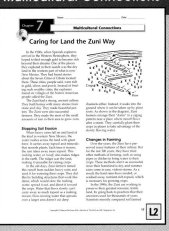

Chapter 7   Multicultural Connections

Caring for Land the Zuni Way

Stopping Soil Erosion

Changes in Farming

Science and Society/ Technology Integration

Chapter 7 SCIENCE & SOCIETY   Resources

An American Dream?

The Trouble with Cars

Mass Transit

You Decide

Assessment

Performance Assessment

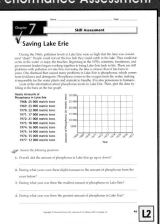

Chapter 7   Skill Assessment

Saving Lake Erie

Chapter Review

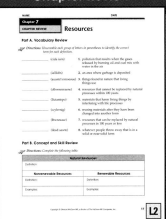

Chapter 7 CHAPTER REVIEW   Resources

Part A. Vocabulary Review

Part B. Concept and Skill Review

Assessment

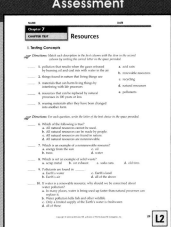

Chapter 7 CHAPTER TEST   Resources

I. Testing Concepts

Resources

CHAPTER OVERVIEW

Section 7-1 How people use Earth resources is discussed. Resources are classified as renewable or nonrenewable.

Section 7-2 The impact of humans on land use, on water, and on air is discussed. The problems caused by the pollution of land, water, and air are presented.

Section 7-3 Ways to reduce solid waste are discussed. Reducing, reusing, and recycling resources are also introduced.

Section 7-4 Science and Society Life-cycle analysis is introduced as a way of figuring out the environmental impact of a product through its entire life. The importance of note-taking skills is addressed.

Chapter Vocabulary

natural resource
renewable resource
nonrenewable resource
landfill
pollutant
acid rain
solid waste
recycling
life-cycle analysis

Theme Connection

Systems and Interactions The interaction of human use of resources with the environment supports this theme.

Chapter Preview

174

Skills Preview

▶ **Skill Builders**
- sequence
- develop multimedia presentations
- recognize cause and effect

▶ **MiniLABs**
- analyze
- model

▶ **Activities**
- make and use a table
- calculate
- diagram
- think critically

Learning Styles

Look for the following logo for strategies that emphasize different learning modalities.

Kinesthetic	MiniLAB, pp. 180, 196; Making a Model, p. 186
Visual-Spatial	Explore, p. 175; Visual Learning, pp. 177, 182, 186, 187; Demonstration, p. 177; Reteach, 193; Problem Solving, p. 197
Interpersonal	Inclusion Strategies, p. 188
Intrapersonal	Community Connection, p. 177; Science at Home, p. 203
Logical-Mathematical	Using an Analogy, p. 178; Inquiry Question, p. 181; Activity 7-1, p. 183; Using Math, p. 184; Activity 7-2, pp. 188-189
Linguistic	Using Science Words, p. 178; Science Journal, pp. 179, 181
Auditory-Musical	Activity, p. 192
Naturalist	Activity, p. 179; Demonstration, p. 186; Enrichment, p. 192; Teaching Strategies, p. 202

Chapter 7

Resources

Imagine sailing across the ocean on the HMS *Rose*—the largest active wooden ship in the world. Do you see anything about it that looks unusual? Its sails are made of materials that were once plastic soda bottles and car fenders. These plastic materials were made into new things, instead of being thrown away. Using old things to make new things is one way to help the environment.

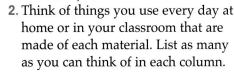

EXPLORE ACTIVITY

Classify Resources

1. In your Science Journal, make a table with five columns: *Plastic, Paper, Metal, Glass,* and *Wood.*
2. Think of things you use every day at home or in your classroom that are made of each material. List as many as you can think of in each column.

Science Journal

In your Science Journal, record how many items are listed for each category. Which column had the longest list? Where do you think these materials come from? Which category do you think you depend on most? Why?

175

EXPLORE ACTIVITY

Purpose
LS Visual-Spatial Use the Explore activity to help students become more aware of their dependence on natural resources. **L2**

Preparation
Obtain objects made from plastic, paper, metal, glass, and wood. Place these objects around the classroom.

Materials
objects made from plastic, paper, metal, glass, and wood

Teaching Strategies
Have students look around the room for objects they believe have been made from natural resources.

Science Journal Student answers will vary. The longest column may be the column of plastic items. Students may know that paper and wood come from trees, plastic items are made from oil, glass is made from sand, and metal objects are made from ore. Students may indicate that they depend on plastic most because they use so many different plastic products.

✓ Assessment
Oral As students read their lists, tally the results on the chalkboard. See which category is most popular. Graph and discuss the results. Use the Performance Task Assessment List for Conducting a Survey and Graphing the Results in **PASC,** p. 35.

Assessment Planner

Portfolio
Refer to page 203 for suggested items that students might select for their portfolios.

Performance Assessment
See page 203 for additional Performance Assessment options.
Skill Builders, pp. 184, 194, 199, 201
MiniLABs, pp. 180, 196
Activities 7-1, p. 183; 7-2, pp. 188-189

Content Assessment
Section Wrap-ups, pp. 184, 194, 199
Chapter Review, pp. 203-205
Mini Quizzes, pp. 182, 194, 199

Group Assessment
Opportunities for group assessment occur with Cooperative Learning Strategies and Flex Your Brain activities.

Prepare

Section Background
- Renewable resources can be replaced in 100 years. Some are in danger of being used up before that can happen.
- The first coal mines in the United States began operating in the 1740s. Deep seams of coal are mined through shafts. When coal is near the surface, strip mining is used.

Preplanning
Refer to the Chapter Organizer on pages 174A-B.

1 Motivate

Bellringer
Before presenting the lesson, display **Section Focus Transparency 20** on the overhead projector. Assign the accompanying **Focus Activity** worksheet.

L2 ELL

Tying to Previous Knowledge
Help students recall any products they or their friends have purchased lately that were overpackaged.

176

What YOU'LL LEARN
- How resources are used
- How resources are classified

Science Words:
natural resource
renewable resource
nonrenewable resource

Why IT'S IMPORTANT
If you understand where resources come from and how they are used, you can make better decisions about the things you buy or use.

FIGURE 7-1
On average, an area of rain forest about the size of the state of Pennsylvania is cut down every year. Pennsylvania has an area of 116 083 square kilometers.

7●1 Natural Resource Use
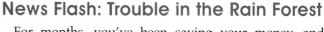

News Flash: Trouble in the Rain Forest

For months, you've been saving your money, and today you're going shopping for a CD player. On the way to the mall, you turn on the car radio. The news reporter is saying that rain forests are being cut down at a rate of about 117 000 square kilometers per year. Look at **Figure 7-1.** That's a lot of land!

The news report goes on to explain that the rain forest problem is not a simple one. People cut down trees in the rain forests to graze animals or grow crops on the cleared land. People also use rain forest trees for fuel or to make things such as furniture or paper. Large companies from foreign countries may pay local peoples to cut down the trees to make wood products. Many of the local families need this money to buy food and supplies for their families. The news report says that everyone needs to act now to save the rain forests because these activities have a negative effect on the soil, the climate, and the organisms that normally live there. The report goes on to say that if enough people show concern and start to take positive actions, change will occur.

Program Resources

📁 **Reproducible Masters**
Activity Worksheets, pp. 5, 43-44, 47 L2
Enrichment, p. 26 L3
Lab Manual, pp. 47-49 L2
Reinforcement, p. 26 L2
Study Guide, p. 26 L1

🔦 **Transparencies**
Science Integration Transparency 7 L2
Section Focus Transparency 20 L2
Teaching Transparency 13 L2

Things You Use Affect the Environment

You think about this news report as you reach the shopping mall. When you walk into the different stores, you can't help but notice that many products and the packages they come in are made of cardboard, a wood product. Even though this wood may not have come from a rain forest, it did come from a forest somewhere. Could these products have been packaged in a different way?

Let's take a look at the CD player you want to buy, shown in **Figure 7-2.** It is made of plastic and often comes in a package made of cardboard. Its wires, screws, and some inside parts are made of metal. Metal and plastic aren't made from trees. So where did they come from?

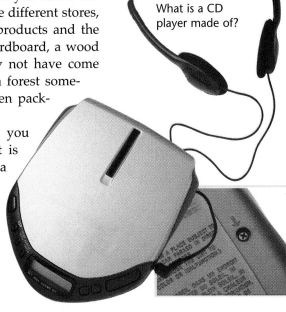

FIGURE 7-2
What is a CD player made of?

Natural Resources

Most of the things that you buy or use are made of materials that come from natural resources. **Natural resources** are things found in nature that living things use. **Figure 7-3** shows some other examples of natural resources. Living things use natural resources to meet their needs. Vegetables that you eat are natural resources. They fill your need for nutrients. The trees and the minerals that were used to make the lumber, plastic, and metal in your house are natural resources. They fill your need for a place to live. Natural resources are also used to make other things in our lives, such as CD players.

FIGURE 7-3
Cotton, gold, trees, and water are examples of natural resources.

7-1 Natural Resource Use   **177**

---

2 Teach

Student Text Question

Could these products have been packaged in a different way? *Yes, they could have been packaged in recycled material.*

Visual Learning

Figure 7-2 What is a CD player made of? *plastic case and earphones; metal wires, screws, and plates* **LS**

Demonstration

LS **Visual-Spatial** Display various items in the front of the class that are classified as or made from natural resources. Items to show include container of oil, wooden log, stereo speaker wire, glass of water, apples and oranges, and a rubber car tire. Demonstrate that all items you are showing either are natural resources or are made from them.

Community Connection

Earth-Sheltered Housing Have students research the concept of earth-sheltered housing. Have them check at the courthouse, or any location with records of housing in the area, to determine whether any local earth-sheltered houses exist. If so, invite the builders to class for a discussion of alternate housing options.
L3 COOP LEARN LS

Inclusion Strategies

Visually Impaired Help your visually impaired students understand and participate in the demonstration of natural resources by allowing them to study the displayed samples with their other senses of touch, smell, and hearing. **L2**

Using Science Words

Linguistic Crude oil (petroleum) is a thick, flammable mixture of substances made from hydrogen and carbon. It forms when marine organisms decompose in an oxygen-free (anaerobic) environment. The word *petroleum* comes from the Latin words *petra* meaning "rock" and *oleum* meaning "oil."

Discussion

Discuss with students that people in the housing industry are trying to find inexpensive, renewable resources to replace the more traditional wood, brick, and stone. In some areas of the country, bales of hay are being used to build houses. They are treated to resist fire and insects. The bales are stacked on each other and then covered inside and out with stucco. The advantages are that the house is built quickly for very little cost. The hay is great insulation and makes a good sound barrier. Bales can be arranged into any floor plan, and windows can be shaped with a chain saw.

Using an Analogy

Logical-Mathematical Relate the process of making screws to the process of knitting a scarf for a friend. The "raw material"—yarn—must first be purchased. Imperfect lengths of yarn must be removed. Then, the yarn must be processed, or knitted, to make the finished product. Finally, the scarf is delivered to your friend, ready to be put to use.

What goes into making a CD player?

You already know that the cardboard box the CD player came in was made from trees, and that trees are a natural resource. What about the plastic used to make the CD player? Where did it come from? Plastic is made from oil, a resource that's found underground. Oil is a thick, dark liquid. Deep holes have to be drilled in the ground to reach oil. Products made from this oil are then taken to a factory to be made into plastic and other products, such as gasoline or plastic soda bottles. What about the metal? Where did the metal come from to make the screws that hold the CD player together? **Figure 7-4** answers this question.

FIGURE 7-4

The screws used in your CD player are usually made of steel. Iron ore is the natural resource used to make this steel.

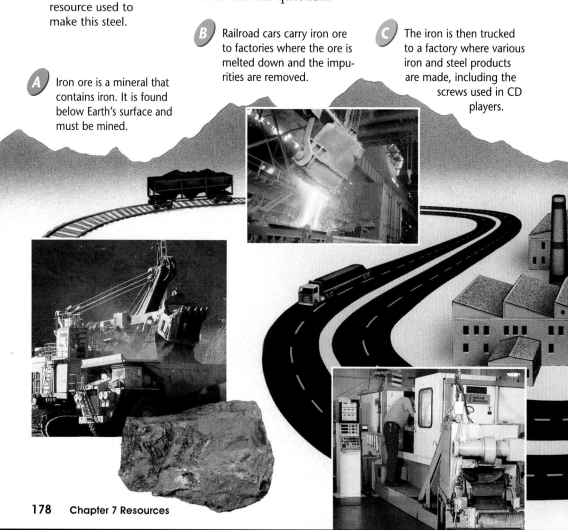

A Iron ore is a mineral that contains iron. It is found below Earth's surface and must be mined.

B Railroad cars carry iron ore to factories where the ore is melted down and the impurities are removed.

C The iron is then trucked to a factory where various iron and steel products are made, including the screws used in CD players.

Theme Connection

Systems and Interactions

Many interactions occur in the manufacture of products from natural resources. The theme of Systems and Interactions is supported by the complex, step-by-step interaction needed for the successful production of steel from iron ore as illustrated in Figure 7-4.

More Natural Resources

Trees, oil, minerals—were any other natural resources used to make the CD player? Cutting down trees, drilling for oil, mining, and getting natural resources to factories all require energy. Once the natural resources are at the factories, it takes energy to make them into plastic, cardboard packaging, and metal wires and screws. Where does all this energy come from?

If you guessed natural resources, you're right. Trucks that take the natural resources to the factories use gasoline, and gasoline is made from oil. In many parts of the country, the electricity used to power machines that make natural resources into materials for CD player parts may come from burning coal. Like oil, coal is a natural resource that forms underground. It takes energy to mine the coal and this energy comes from natural resources.

D Different parts used in making electronic equipment are transported to a factory where the parts are assembled into products.

E The final product—the CD player—is packaged and then trucked to a store, where it is sold.

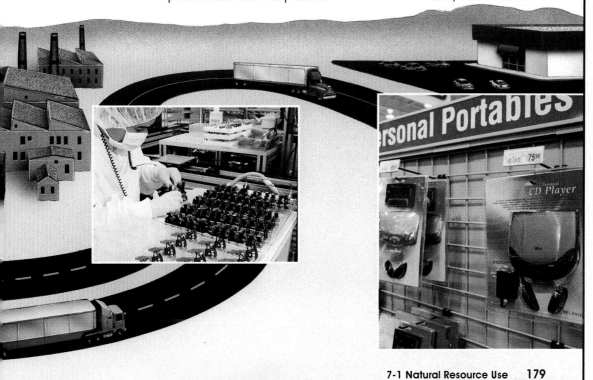

Content Background

Coal forms when pieces of dead plants are buried under other sediments. Plants are made up of molecules that contain atoms of carbon, hydrogen, and oxygen. These plant materials are chemically changed by microorganisms. The resulting sediments are compacted over millions of years to form coal. When we burn coal, it is the hydrogen that burns and is the source of heat. Bituminous coal, or soft coal, is the coal used most often. It provides lots of heat energy when burned, but it also pollutes the air.

Activity

IS **Naturalist** Obtain a sample of iron ore (magnetite) and other minerals that are ores of other metals (bauxite—aluminum ore; malachite—copper ore). Inform students that the iron ore called magnetite is magnetic. Give students the four samples described above and a magnet. **Determine which of the minerals is the iron ore.** *Magnetite will attract a magnet. Students should discover these two iron ores.* **L2** **COOP LEARN** **ELL**

Science Journal **From There to Here** Ask students to select a product they use often. Ask them to write one or two paragraphs in their Science Journal describing the step-by-step process that is followed from the time the natural resource from which the product is made is obtained, through the time when the final product is placed in the store. Students should not use the example of the CD player. **L2** **P** **IS**

179

Mini LAB

Purpose

LS **Kinesthetic** Students will use different methods and materials for wrapping objects in order to better understand the importance of natural resources. **L2** **ELL** **P**

Materials

objects to wrap; wrappings such as paper bags, plastic bags, and newspaper; scissors, string, or transparent tape

Teaching Strategies

- Be sure that at least one type of wrapping is reusable.
- Encourage students to bring in unusual wrapping materials.
- Transparent tape will not stick well to all materials. Provide various types of tape and some string.
- Students may conclude that no wrapping is a valid option. Encourage students to defend all positions taken.

Safety Precautions Caution students to use scissors with care.

Activity Worksheets, pages 5, 47

Analysis

1. Student answers will vary but could include the size, the cost, the availability, and the attractiveness of the wrap.

2. Student answers will vary but could include reference to the most colorful or the most efficient wrapping. Some students will select reusable or recyclable materials.

Mini LAB

Analyzing Gift Wrap

Can you think of a better way to wrap a gift?

1. Your teacher will give you an object to wrap. Discuss different methods and unusual materials for wrapping this object with your group.

2. Think about how different ways of wrapping could make the best use of the materials you have. Does the wrap waste material? Could it be used again? Is it easy to dispose of?

3. Wrap the object. In your Science Journal, write down all the resources that would have been used to wrap it in gift wrap.

Analysis

1. What problems did you have with your method of wrapping?

2. Why do you think your choice of wrapping material is a good one?

FIGURE 7-5

Natural resources are found everywhere—in national parks and in the middle of large cities like Boston.

180 Chapter 7 Resources

Assessment

Content Encourage students to list the properties of the "perfect" environmental gift wrap and give suggestions as to how it could be made. Encourage students to invent samples of the "perfect" environmental gift wrap. Use the Performance Task Assessment List for Invention in **PASC**, p. 45.

You're beginning to see that it takes a variety of natural resources to make one CD player, doesn't it? Think of all the natural resources that are used to make something large like a house or a school building. Now think of all the houses and apartments in the world. Are there enough natural resources to meet everybody's need for a place to live?

Maybe. But people use natural resources to meet other needs, too. In fact, all living things on Earth use natural resources. Plants and animals use natural resources for food and shelter. Will we ever run out of natural resources? That may depend on the particular resource.

Availability of Resources

Imagine that you're riding your bike on a warm spring day. Your destination: the city park, such as the one shown in **Figure 7-5.** When you get to the park, you head straight for the pond. You hop off your bike, take an apple out of your backpack, then lean against a tree. Later, you might take a hike around the pond. But for now, you're content just to watch the sunlight sparkling on the water.

Sunlight, water, trees, apples . . . These are all natural resources. They have something else in common, too. They will likely be around for a long time. Why? Because they are all renewable (ree NEW uh bul).

Renewable Resources

Resources that can be replaced by natural processes in 100 years or less are called **renewable resources.** Look at **Figure 7-6.** Energy from the sun is a renewable resource because the sun gives off light energy every day, and it will continue to do so every day for millions of years. Trees are renewable resources because most trees will grow back and be cut again in less than 100 years. Water is another renewable resource because we use the same water over and over again.

FIGURE 7-6
Sunlight and trees are both renewable resources.

USING TECHNOLOGY

From Plastic Bottles to Play Towers

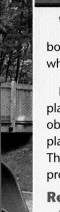

What happens to all of those clear, 2-L plastic soft-drink bottles people put into recycling bins? What is recycling, and what objects can be made from recycled plastic?

Recyclable plastic objects are first taken to a recycling plant where the different types of plastic are sorted. The objects are chopped into small pieces and washed. After the plastic is dried, it is melted and pushed through a screen. This step forms plastic pellets that will be used to make new products.

Recycled Products

Many recycled plastic pellets are re-formed into plastic packaging for different products. Pellets are also used to make stuffing for toys and beanbag chairs. Trash cans, pet food bowls, flowerpots, park benches, and plastic lumber are other products made from recycled plastic pellets. The play tower pictured here is made of recycled materials. It is made from 15 765 recycled plastic containers, 20 232 recycled aluminum cans, and 6180 recycled soup cans.

interNET
CONNECTION

Check out the link at the Glencoe Homepage at *www.glencoe.com/sec/ science* to find out about other products that are made from recycled materials.

interNET
CONNECTION

The Glencoe Homepage at **www.glencoe.com/sec/science** lists sites dealing with recycling plastics and other recycled materials.

Inquiry Question

LS **Logical-Mathematical** **The percentage of loose plastic waste is much greater than the percentage of landfill waste made up by plastic. Why is this true?** *Today, plastic bottles are much thinner and can be crushed easily. Thinner plastic is easier to make and less costly.*

Science Journal **Garage Sales** With today's emphasis on recycling, garage sales have taken on a new role. Have students explain in their Science Journals how recycling could make "one person's trash another person's treasure." **L2** **P** **LS**

3 Assess

Check for Understanding

Discussion Lead students in a discussion of how natural resources are used in students' recreation.

Reteach

Visual Learning

Figure 7-7 What energy sources do you use in your home? *Answers will vary.* LS

Extension

📁 For students who have mastered this section, use the **Reinforcement** and **Enrichment** masters.

4 Close

·MINI·QUIZ·

Use the Mini Quiz to check students' recall of chapter content.

1. **Things found in nature that living things use are called _____.** *natural resources*
2. **Resources that can be replaced by natural processes in 100 years or less are called _____.** *renewable*
3. **Gold is an example of _____.** *a nonrenewable resource*

☐ 40% Oil
☐ 28% Coal
☐ 20% Natural Gas
☐ 6% Nuclear
☐ 6% Other

FIGURE 7-7

As you can see from this circle graph, 40 percent of the world's energy needs are met by oil and 28 percent are met by coal. Scientists estimate that we have enough coal to last another 200 years. But if we continue using oil at present rates, some scientists estimate that we will run out of this natural resource in 30 or 40 years. *What energy sources do you use in your home?*

Nonrenewable Resources

Do you see coal or oil among Earth's energy sources in **Figure 7-7**? Coal, natural gas, and oil take millions of years to form inside Earth. They are examples of non-renewable (NAHN ree new uh bul) resources. **Nonrenewable resources** are resources that cannot be replaced by natural processes within 100 years. If all the coal and oil that we can recover is used up during our lifetimes, there won't be any more available for use for millions of years. So, because nonrenewable resources form slowly over long periods of time, they need to be used carefully. Also, scientists are finding other energy sources. **Figure 7-7** shows how the world's energy needs are being met. Sections 7-2 and 7-3 will discuss ways to reduce the amount of nonrenewable resources used.

182 Chapter 7 Resources

Content Background

Due to robotic submersibles, new sources of oil are being found beneath the ocean's floor. These new reserves of oil appear to be large but will be difficult to develop. Discuss with students the types of environmental problems that might be caused by developing these resources.

Using Water

Water is an important resource that we use every day. You wash dishes and clothes and you wash yourself and brush your teeth. You cook meals. All of these activities need water.

What You'll Investigate
Calculate how much water your family uses in three days.

Procedure
1. Use the table below to **calculate** how much water your family uses.
2. For three days, have the people who live in your house **keep a record** of when they do the activities listed in the table.

Conclude and Apply
1. The numbers in the table describe approximately how many liters one person uses in a single day for the activity listed. **Multiply** these numbers by the number of people who did these activities.
2. **Add** up the totals for each day and the final sum will be the total amount of water used for these activities in three days.
3. **List** ways in which your family could control the amount of water used.

Activity	Conditions	Amount of Water Used
Washing dishes by hand	Water is running all the time	113 L/person/day
Washing dishes by hand	Sink is filled with water	19 L/person/day
Washing clothes in machine	Small load with high water setting	68 L/person/day
Washing clothes in machine	Full load with high water setting	45 L/person/day
Taking a shower	10 minutes long	150 L/person/day
Taking a bath	Bathtub is full of water	113 L/person/day
Flushing the toilet	Water-saving toilet	23 L/person/day
Brushing teeth	Water is running all the time	17 L/person/day

7-1 Natural Resource Use   **183**

Activity 7-1

Goals
- Calculate how much water the people in your household use in three days.
- Make a plan to control the amount of water used.

Materials
- calculator

Activity 7-1

Purpose
Logical-Mathematical Students will become aware of the amount of clean water used each week in their homes. **L2**

Process Skills
observing and inferring, measuring in SI, collecting and organizing data, interpreting data, making and using tables, using numbers, recognizing cause and effect

Time
three days for observations, 35 minutes to complete report

Materials
calculator

Activity Worksheets, pages 5, 43-44

Teaching Strategies
- Have students meet with their families to describe the nature of the experiment and to ask for their help and cooperation.
- If students' families forget to mark down water usage, allow these students to just list their own water use.

Answers to Questions
1. **and 2.** Student numbers will vary depending on the size of the family and how many people participate in the study.
3. Ways could include filling the sink when washing dishes, adjusting the water level to the amount of clothes in the washing machine, taking a bath instead of a shower, taking shorter showers, and turning off water while brushing teeth.

Inclusion Strategies

Learning Disabled Help your students better understand water conservation by having them prepare the table in which they will record data. Help students begin by filling out an example of the data they will obtain at home. Contact parents or guardians so the project becomes a family effort. **L1**

Assessment

Performance Have students propose a schedule of water usage for members of their families that will conserve water yet allow for needed use. Write the schedule in the form of a letter to their families. Use the Performance Task Assessment List for Letter in **PASC**, p.67. **P**

Section Wrap-up

1. A natural resource is something found in nature that living things use. A squirrel stores nuts for winter in hollowed-out trees.

2. Items that cannot be replaced in less than 100 years are considered nonrenewable. Coal, silver, and oil are nonrenewable resources. Renewable resources are resources that can be replaced by natural processes in 100 years or less. Trees, water, and corn plants are renewable resources.

3. **Think Critically** Land is a renewable resource if it is used for different things over a 100-year span. However, land on which a city is built would be a nonrenewable resource.

USING MATH

During a five-minute shower, the 15-L-per-minute showerhead will put out 75 L of water per day or 525 L per week. During the same time, the 9.5-L-per-minute showerhead will put out 47.5 L of water per day or 332.5 L per week. The savings for the week would be 192.5 L of water.
(15 L/min × 5 min/day) × 7 days − (9.5 L/min × 5 min/day) × 7 days = 192.5 L

Natural resources make up the environment that we live in. These resources are classified into nonrenewable and renewable resources. Examples of renewable and nonrenewable resources are listed in **Table 7-1**. In Section 7-2, you will learn how people affect these natural resources.

Table 7-1

Natural Resources	
Renewable Resources	**Nonrenewable Resources**
Plants	Coal, oil, and gas
Sunlight	Some ores and metals
Water	Land

Section Wrap-up

1. What is a natural resource? Give some examples of how a squirrel uses natural resources.

2. List three examples of nonrenewable resources and three examples of renewable resources. Explain why your choices fit into each category.

3. **Think Critically:** Is land a renewable or a nonrenewable resource? Explain your answer.

4. *Skill Builder*
   Sequencing Make an events chain showing, in sequence, the steps that are taken to make a tree into a baseball bat displayed in a store. If you need help, refer to Sequencing on page 543 in the **Skill Handbook.**

USING MATH

A regular showerhead puts out 15 L of water per minute. A water-saving showerhead puts out only 9.5 L of water per minute. If you take a five-minute shower every day, how much water would you save in one week by using a water-saving showerhead?

 Skill Builder
Sequencing tree grows; tree is cut down; tree is cut into lumber; lumber is cut to the size and shape of a baseball bat; the bat is finished, stained, and varnished; the bat is shipped to a store and displayed

Assessment

Process Have each student in class estimate how much water he or she uses for showers each week, assuming showers are taken daily and showerheads use 12 L of water per minute. Use the Performance Task Assessment List for Using Math in Science in **PASC**, p. 29. L2

People and the Environment

Exploring Environmental Problems

Look at **Figure 7-8.** Have you ever seen a construction site for a new highway? Sometimes, hillsides have to be dynamited to make room for the highway. The trees and plants that grew on the hillside are destroyed. The animals that lived on the hillside depended on the trees and plants for food and shelter. Some might die if their food source is destroyed, but most will survive and find new habitats to live in. Construction companies now have to restore land that they dynamite so that plants and animals will be able to continue to live there.

What if there isn't another place to live? Many plants and animals lose their natural habitats because people use land for growing crops, grazing animals, or building homes. That's what's happening in the rain forest. Because large areas of rain forest are being destroyed at a rapid rate, certain species in the rain forest have no habitat to fill their needs for food, shelter, and other things. They are threatened with extinction.

Frequently, human activities affect some of our most precious natural resources: land, water, and air. Let's see how this happens.

What YOU'LL LEARN

- How people affect the environment
- The different types of pollution

Science Words:
landfill
pollutant
acid rain

Why IT'S IMPORTANT

Knowing how your actions affect the environment will help you to make choices that could help reduce environmental problems.

FIGURE 7-8

Construction destroys some parts of the environment. Laws are now in place to reduce the amount of destruction that takes place.

Prepare

Section Background

- Solid wastes found in landfills are mostly paper and things made from paper. Other solid wastes found in landfills are yard waste, plastic containers, metal products, and glass.
- A sanitary landfill is one in which each day's deposit is covered with dirt.

Preplanning

Refer to the Chapter Organizer on pages 174A-B.

1 Motivate

Bellringer

Before presenting the lesson, display **Section Focus Transparency 21** on the overhead projector. Assign the accompanying **Focus Activity** worksheet.
L2 **ELL**

7-2 People and the Environment 185

Program Resources

Reproducible Masters
Activity Worksheets, pp. 5, 45-46 **L2**
Enrichment, p. 27 **L3**
Lab Manual, pp. 51-53 **L2**
Reinforcement, p. 27 **L2**
Science and Society/Technology
  Integration, p. 25 **L2**
Science Integration Activities, pp. 51-52 **L2**
Study Guide, p. 27 **L1**

Transparencies
Section Focus Transparency 21 **L2**

Help students recall the last time they were asked to take out the garbage. Ask them whether they have ever considered where all of the garbage goes. Much of the solid waste is collected and placed in a landfill.

2 Teach

Demonstration

LS **Naturalist** Make popcorn for the class. Have the students list the ingredients and tell where each comes from, what type of resource it is, and its impact on the environment. Discuss how much land and water each requires. Don't forget about the electricity and the appliance you are using to pop the corn.

For example, corn oil comes from corn, a renewable resource. The corn needed land, water, fertilizer, insecticide (land and water pollutants), harvesters, trucks to take it to market (gasoline), and refineries to extract the corn oil (possible air pollutant), bottles to put the oil in, paper to use for labels, cardboard boxes, trucks to ship it to market, and grocery stores to sell it. The salt is a nonrenewable resource that is mined.

The point of the demonstration is to emphasize our dependence on resources.

Making a Model

LS **Kinesthetic** Have students research the use of land for a landfill. Ask students to make a model of a sanitary landfill. Students might wish to use a large plastic container in which to build their model landfill. **L3** **COOP LEARN** **ELL**

Our Impact on Land

How much space do you need? You will need to think about more than just your home. Think about where your food comes from, your school, and other space you use. If you start adding it all up, the amount of space you use is much larger than you may think. A simple peanut butter-and-jelly sandwich requires land to grow the wheat needed to make bread, land to grow the peanuts for the peanut butter, and land to grow the sugarcane and fruit for the jelly. A hamburger? Land is needed to raise cattle and to grow the grain that the cattle eat.

Using Land Wisely

All of the things we use in our everyday lives take some amount of land, or space, to produce. That means that every time we build a house, a mall, or a factory in a city as illustrated in **Figure 7-9,** we use a little more land. All you have to do is look at a globe, however, to see that the amount of land available for us to use is limited.

People need food, clothing, jobs, and a place to live, and each of these things takes space. But preserving natural habitats is also important. Remember, a habitat is the place where an organism lives. Once a wetland is filled in to build an apartment building, the wetland is lost.

FIGURE **7-9**

Land is used for many different things other than growing food. *What takes up space in the city of Pittsburgh pictured below?*

186   Chapter 7 Resources

Community Connection

Land Use Obtain a map of the local town or city and surrounding country. Ask students to study the use of land in and around your school. Encourage them to speculate as to whether they think the land has been used wisely. Ask them to explain how they might better use the land for all people living in or around your local town or city. **L2**

Visual Learning

Figure 7-9 **What takes up space in the city of Pittsburgh pictured below?** *Answers should include housing, office buildings, roads, and green space (trees and fields).* **LS**

More and more, there are laws to help protect against habitat loss and to help us use land wisely. Before major construction can take place in a new area, the land must be studied to determine what impact the construction will have on the natural habitat, the living things, the soil, and water in the area. If there are endangered organisms living there, or if the impact will be too great, construction may not be allowed. These are important studies. At stake are jobs, homes, and habitats.

Landfills

Each day, every person in the United States produces about 1.8 kg of garbage. Where does it go? About 80 percent of our garbage goes to landfills. A **landfill,** shown in **Figure 7-10**, is an area where garbage is deposited.

Most of the things we throw into a landfill are not dangerous to the environment. However, sometimes items such as batteries, paints, and household cleaners end up in landfills. These things contain potentially harmful chemicals that can leak into the soil, and eventually into rivers and oceans. Any material that can harm living things by interfering with life processes is called a **pollutant** (puh LEW tunt). Modern landfills are lined with plastic or clay to keep these chemical pollutants from leaking. However, some chemicals still find their way into the environment. If these pollutants get into the food that we eat or the water that we drink, they can cause health problems.

FIGURE 7-10

Each day, trash is put in a sanitary landfill. This trash is later covered with a thin layer of dirt and then watered down to keep the trash from blowing away.

7-2 People and the Environment   **187**

Purpose

LS **Logical-Mathematical** Students will plan a small town and emphasize that land is a resource that needs to be used carefully, just like any other nonrenewable resource. **L2** **COOP LEARN** **P**

Process Skills

classifying, observing and inferring, comparing and contrasting, formulating hypotheses, designing an experiment to test a hypothesis, making models, recognizing cause and effect, measuring in SI, interpreting scientific illustrations, making and using tables

Time

40 minutes

Think Critically

- Land resources are to be shared among living areas, recreational areas, commercial areas, schools, and landfills.
- Commercial areas will be planned first, followed by living areas. After land use for commercial and living areas is determined, other uses of land will be planned.

📂 **Activity Worksheets,** pages 5, 45-46

Possible Procedures

First, zone the plan into areas for living quarters, then office buildings and commercial areas, and finally other parts of the town.

188

Activity 7-2

Design Your Own Experiment
Using Land

Imagine planning a small town. Your job in this activity is to draw up a master plan to decide how 100 square units of land can be turned into a town.

Possible Materials
- grid paper (10 squares × 10 squares)
- colored pencils

PREPARE

What You'll Investigate
How should land resources be used?

Think Critically

People need homes in which to live, places to work, and stores from which to buy things. Children need to attend schools and have parks in which to play. How can all of these needs be met when planning a small town?

Goals

Design a plan in which 100 square units of land can be turned into a town.

PLAN

1. A 100-square-unit piece of land can be represented as a square divided into 100 blocks. One way to represent this is to make a square graph 10 blocks across and 10 blocks down.
2. The table on page 189 shows the different parts of a town that need to be included in your plan. The office buildings and industrial plant are places where the people of the town will work. They are each 6 blocks in size. These blocks cannot be divided but must be treated as one group. The landfill is 4 blocks in size and cannot be broken up.
3. All the other town parts can be broken up as needed. Stores and businesses are areas in which shops are located as well as medical offices, restaurants, churches, and cemeteries.

188 Chapter 7 Resources

Inclusion Strategies

Gifted Ask your gifted students to use computer programs on city planning to produce their town plans. Have these students present their work to the class, and have them work with others in the class to modify the plans in order to produce the most efficient use of land. **L3** **LS**
COOP LEARN

4. As a group, **discuss** how the different parts of the town might be put together. Should the park be in the center of town or near the edge of the town? Should the school be near the offices or near the houses? Where should the landfill go?

5. How will you show the different town parts on your grid paper?

Parts of Your Town	Number of Blocks Needed
Office buildings	6 blocks in one group
Industrial plant	6 blocks in one group
School	1 block
Landfill for garbage	4 blocks in one group
Houses and apartments	44 blocks—can be broken up
Stores and businesses	19 blocks—can be broken up
Park	20 blocks—can be broken up

DO

1. As a group, **plan** your town. Check over your plan to make sure that all of the town parts are accounted for.

CONCLUDE AND APPLY

1. Where did you place the office buildings and the industrial plant? Why were they placed there? Where did you place the houses, school, and the stores and businesses? Explain why you placed each one as you did.

2. Did you make one park or many parks with the land designated for park use? What are the advantages of your park(s) plan?

3. Where did you place your landfill? Will any of the townspeople be upset by its location? What direction does the wind usually blow from in your town?

4. **APPLY** Where would you put an airport in this town? Keep in mind safety issues, noise levels, and transportation needs.

189

Go Further

Encourage students to include traffic patterns, mass transit, railroads, and airports in their plans. Discuss the environmental (noise, air pollution) and commercial impacts of each.

Teaching Strategies

- Have students decide in which direction the wind blows. This may affect where factories and landfills are located.
- Instead of using a 10×10 square grid, different shapes of the 100 squares could be used—e.g., a 4×25 grid or an irregular shape.

Tying to Previous Knowledge Discuss what students like and dislike about where they live, where their school is, and where their parents work.

Troubleshooting No allowances are made for roads, highways, or railroads although these could be included.

DO

Expected Outcome

Students' town plans will show areas for office buildings and commerce, houses, apartments, parks, schools, roads, and landfills.

CONCLUDE AND APPLY

1. Student answers should include reasonable explanations for each part of the plan.

2. Student answers should indicate that the locations of the parks should be in areas easily accessible by all residents.

3. Landfills should not be placed near houses and apartments or upwind from most stores, businesses, and homes.

4. An airport should be placed far away from schools but located close to businesses, shopping malls, and hotels.

190

Our Impact on Water

Did you know that you cannot live long without water? You need clean water for drinking, as well as dozens of other things, as you found out in Activity 7-1. The average person in the United States uses about 397 L of water each day. Though water is a renewable resource, in some places it is being used up faster than natural processes can replace it.

Only a small amount of Earth's water, as shown in **Figure 7-11**, is freshwater that people can drink or use for other needs. Many places around the world are running out of usable water. How do you think your life would change if your area were running out of water?

Water Pollution

People have always used water, but everyday activities can pollute water. How? When you scrub a floor with a mixture of water and a household cleaner, what do you do with the mixture afterwards? You pour it down the drain. The polluted water usually goes to a water-treatment plant, where it is cleaned before being used again.

FIGURE 7-11
Although 70 percent of Earth's surface is covered by water, less than one percent is freshwater that can be used for drinking.

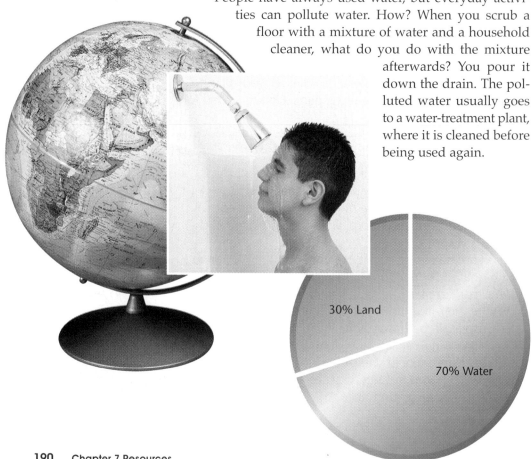

30% Land

70% Water

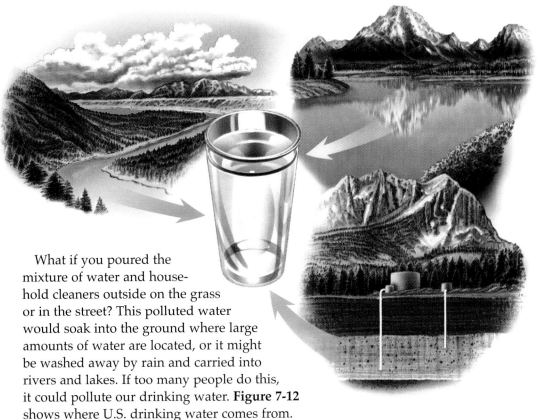

What if you poured the mixture of water and household cleaners outside on the grass or in the street? This polluted water would soak into the ground where large amounts of water are located, or it might be washed away by rain and carried into rivers and lakes. If too many people do this, it could pollute our drinking water. **Figure 7-12** shows where U.S. drinking water comes from.

Cleaning Up the Water

Countries are working together to reduce the amount of water pollution. For example, the United States and Canada have made agreements to clean up the pollution in Lake Erie, a lake that borders both countries. The U.S. government has also passed several laws to keep water supplies clean. The Safe Drinking Water Act is a set of government standards that makes sure that our drinking water is safe. The Clean Water Act gives money to the states for building water-treatment plants. Wastewater is cleaned at such plants.

Remember, Earth has a lot of water, but only a small amount of Earth's water is freshwater that people can drink or use. The best way people can protect Earth's water is by being aware of how they use it.

FIGURE 7-12
Much of our drinking water comes from rivers, lakes, and underground sources of water. This water is treated to remove impurities before it is used by people in towns and cities.

Community Connection

Water Treatment Contact a local water-treatment plant. Invite a water-treatment chemist to visit the class and discuss the ways that water is treated before being used by people. If such a visit is not possible, ask about a videotaped version of a field trip of the plant for the class to view.

Brainstorming

Have students brainstorm different ways to conserve water. Ask students to decide which ideas would be the most practical to use. One idea might be to keep absorbent material in the bottom of sinks to absorb water from daily activities. This water could then be used to water plants. L2

Content Background

Water-treatment plants use a colloid—a heterogeneous mixture that never settles out—to remove impurities in water. To make this colloid, measured amounts of lime and alum are added to the water. These compounds form colloids of the compound aluminum hydroxide. Later, the gel can be filtered out through beds of fine sand. From this point, the water may be sprayed through air to add oxygen to the water. It can then be treated with chlorine for further purification.

Teacher F.Y.I.

The 1986 Safe Drinking Water Act is a law to ensure that drinking water in the U.S. is safe. Eighty-seven percent of the 58 000 public water systems met the government standards in 1996. However, in 1990, about 30 million U.S. citizens still drank from potentially unsafe water supplies.

The 1987 Clean Water Act also controls runoff from streets, mines, and farms. Runoff caused up to half of the water pollution in the U.S. before 1987.

The U.S. Environmental Protection Agency (EPA) makes sure that cities comply with both of these laws.

Teacher F.Y.I.

In some parts of the United States, the sunsets are ablaze with magnificent colors. It has been said that the Hawaiian Islands have some of the most beautiful sunsets in the world, yet when the sun sets, there are no colors at all. The sun simply moves down below the rim of the horizon. The colors in the sunset are produced by the pollution in the air. The reason the Hawaiian Islands have no color in their sunset is because they are in the middle of pollution-free air and are surrounded by ocean.

Activity

Auditory-Musical Have students write a song about how the world has become polluted and how they will clean it up and improve the environment. **L2 COOP LEARN P**

Enrichment

Air Pollution Have students place flat paper plates coated with petroleum jelly in various locations around the school grounds. Place rocks on the plates to hold them in position. After a day, collect the plates and have students write in their Science Journals about the type of air pollution found in and around their community. **L2 P LS**

Visual Learning

Figure 7-13 What activities illustrated in this picture cause air pollution? *Forest fires, wood-burning in fireplaces, and exhaust from vehicles, factories, and planes all contribute to air pollution.*

Our Impact on Air

If you live in a city, you may have noticed that on some days, the air looks hazy. Pollutants such as dust and gases in the air cause this haziness. Air pollution can be caused by natural events such as a volcano eruption that releases smoke and ash into the air. But people cause most air pollution.

Figure 7-13 shows some sources of air pollution. The two biggest sources of air pollution are cars and factories, including power plants that produce electricity. One source of pollution is the fumes that come from cars. Cars need gasoline to run. When gasoline is burned, pollutants are released into the air. The polluting fumes of cars and other vehicles cause more than 30 percent of all air pollution.

Many factories and power plants burn coal or oil for the energy they need. This activity also releases pollutants into the air. Pollutants can interfere with life processes. Air pollution can make your throat feel dry or your eyes sting. Many people have trouble breathing when air pollution levels are high. For people with lung or heart problems, air pollution can be deadly. In the United States, about 60 000 deaths each year are linked to air pollution.

FIGURE 7-13

Air pollution is caused by many different activities. *What activities illustrated in this picture cause air pollution?*

Content Background

El Niño refers to a warming of the ocean surface off the western coast of South America. When this occurs—about every seven years—dead air falls across South America. In 1997, the landowners burned the weeds off their fields at the worst possible time—just before El Niño. The smoke was caught in the nonmoving air and hung like a cloud over several countries for weeks. The smoke was thick in city streets, causing the populations to wear face masks. Many became sick from breathing the smoke-filled air for such a long time. Especially strong El Niño events in 1982-83 and 1997-98 were responsible for heavy precipitation, flooding, severe storms, and damaging waves along the west coast of North America.

People aren't the only living things that are harmed by air pollution. Acid rain causes a lot of damage to other organisms. **Acid rain** happens when the gases released by burning oil and coal mix with water in the air to form acidic rain or snow. Some scientists hypothesize that when acid rain falls on the ground, plants and trees die. When acid rain falls into rivers and lakes, it can kill the fish.

Spare the Air

The best solution for all types of pollution, including air pollution, is prevention. Reducing the number of pollutants is easier to do than cleaning up the pollution that results. Automobiles being produced today release fewer harmful gases and use less fuel than did vehicles in years past. Governments around the world are also meeting to find ways to reduce the amount of air pollutants being released into the atmosphere by factories.

It may seem at first that you have no control over sources of pollution, but think again. Think about what power plants produce. They produce electricity. When power plants burn oil or coal to make electricity, harmful pollutants enter the atmosphere and cause smog, acid rain, and other problems.

You can help protect the atmosphere by limiting the amount of energy you use at home. Conserve electricity by turning off lamps, radios, fans, and other appliances that you aren't using. Keep doors and windows closed to save heat energy in the winter or to keep a cool environment in the summer. And encourage your families to buy energy-efficient lightbulbs, like those illustrated in **Figure 7-14.**

FIGURE 7-14
Turning down your home's thermostat is one easy way to reduce the amount of energy used. Energy-efficient lightbulbs use one quarter of the energy of standard incandescent lightbulbs, and they last up to ten times longer.

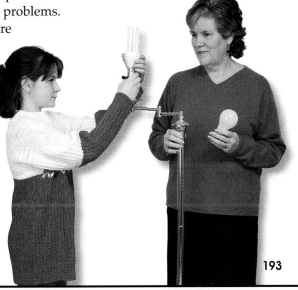

193

Across the Curriculum

Architecture Recent breakthroughs in photovoltaic technologies have made the cells available to be used as part of the building materials to make the walls and roofs of houses. This would make the house able to produce its own energy. Discuss how this new technology could change our future.

For example, what would happen to air pollution, how would this affect our dependence on fossil fuels, what would happen if the sun went behind a cloud, and how could this technology be expanded to power things other than a house?

GLENCOE TECHNOLOGY

 Videodisc

The Infinite Voyage: Crisis in the Atmosphere
Chapter 10
Los Angeles Smog and Air Quality Plan

3 Assess

Check for Understanding

Visual Learning

Figure 7-14 Ask students: **Can you think of other ways to limit the amount of energy used in your home?** *turning off lights and appliances when not needed, washing and drying full loads of clothing, etc.*

Reteach

LS **Visual-Spatial** Have the students explain the water cycle as you draw it on the chalkboard. Make it in the form of a big flow chart. Now, introduce a pollutant, such as exhaust fumes, and carry it through the water cycle to see if it has an effect. The rain absorbs the pollutant, the clouds turn black, and the acid rain falls on the ground.

Extension

For students who have mastered this section, use the **Reinforcement** and **Enrichment** masters.

4 Close

Use the Mini Quiz to check students' recall of chapter content.

1. **An area where garbage is deposited is called a(n)** _____. *landfill*

2. **Materials that can harm living things by interfering with life processes are called** _____.
pollutants

3. **When gases released by burning oil and coal mix with water in the air, they can form** _____. *acid rain*

Section Wrap-up

1. a material that can harm living things by interfering with life processes, such as batteries, paints, and household cleaners

2. land—cleared for highways, factories, or homes, making landfills; water—drinking and boating; air—driving cars and working in factories

3. **Think Critically** If gasoline leaked from underground storage tanks, it could enter and pollute the groundwater. Gasoline that was poured on the ground could flow into streams and rivers.

Science Journal Student paragraphs will vary depending on student experiences. They should include how their own actions have affected the environment. **P**

FIGURE 7-15

The people who live in Anchorage, Alaska, are surrounded by many natural resources—mountains, water, and abundant wildlife.

Land, water, and air are all parts of our natural environment, as illustrated in **Figure 7-15**. To maintain the health of the environment, people need to think how their actions will affect the land, water, and air that make up Earth. Pollution is easier to prevent than to clean up.

Section Wrap-up

1. What is a pollutant? Give examples.

2. Describe some ways that humans affect land, water, and air.

3. **Think Critically:** You know that gasoline fumes cause air pollution. Describe one way that gasoline itself can cause water pollution.

4. *Skill Builder*
   Developing Multimedia Presentations
   Based on what you have learned in this section, choose one type of pollution—land, water, or air. Develop a multimedia presentation showing how human activities have affected the specific part of the environment you have chosen and what is being done to correct it. If you need help, refer to Developing Multimedia Presentations on page 564 in the **Technology Skill Handbook.**

Science Journal

In your Science Journal, list three things you did today. Write a paragraph describing how these actions might affect the environment.

Skill Builder
Developing Multimedia Presentations Student multimedia presentations will vary but should show how some type of pollution affects the land, water, or air. Presentations should also show how humans are attempting to correct the problem.

Assessment

Oral Have students explain the environmental impact of using classroom materials. For example, does using the overhead projector reduce chalk waste? Or would the extra use of electricity be worse for the environment? Use the Performance Task Assessment List for Making Observations and Inferences in **PASC**, p. 17. **L2**

Protecting the Environment

Cutting Down on Waste

The United States faces a huge waste problem. Litter gathers along highways. Landfills leak and overflow with garbage. Five billion tons is the estimated amount of solid waste thrown away each year in the United States. **Solid waste** is whatever people throw away that is in a solid or near-solid form. Look at **Figure 7-16** for examples.

Most waste is produced when coal, oil, and other natural resources are taken from the ground. Households and businesses produce only about four percent of this country's waste. However, household and business waste is still a lot of waste—nearly 200 million tons each year.

Most of the waste from our homes, schools, and businesses is paper and cardboard products. In the cafeteria at school, it's easy to see why this is so. School lunch programs all over the country depend upon paper plates, straw wrappers, milk cartons, paper bags, drink boxes, and napkins. What if individuals just tried to reduce the amount of trash they throw away each day?

Solid-waste management for individuals can be summed up by the 3 *Rs—reduce, reuse,* and *recycle* waste.

FIGURE 7-16
Solid waste includes everything from old newspapers and pickle jars to old plastic toys and scrap metal from manufacturing processes.

Amount of waste *(millions of tons)*

90
80
70
60
50
40
30
20
10

Paper products

Aluminum

Other metals

Glass

Plastics

Yard waste

Other waste

7-3 Protecting the Environment   195

What YOU'LL LEARN

- About the problems of solid waste
- How to reduce, reuse, and recycle resources

Science Words:
solid waste
recycling

Why IT'S IMPORTANT

You can do many simple things to help protect the environment.

Prepare

Section Background

We have been called the "disposable generation." If something breaks or gets old, we throw it away. Many products are designed to become obsolete every few years so that we will replace them. Products produced by competing companies are not always compatible, leading to redundant products and causing a waste of our resources.

Preplanning

Refer to the Chapter Organizer on pages 174A-B.

1 Motivate

Bellringer

Before presenting the lesson, display **Section Focus Transparency 22** on the overhead projector. Assign the accompanying **Focus Activity** worksheet.
L2 ELL

Tying to Previous Knowledge

Help students recall a time when they threw something away and then needed it the next day.

---

2 Teach

Mini LAB

Purpose
LS Kinesthetic Students will explore the possibilities of reducing, reusing, and recycling resources. **L2**

Materials
clean items that are normally thrown out; glue; string; tape; cardboard, posterboard, or foam board

Teaching Strategies
- Use an old cardboard box to mount the artwork. This will emphasize reuse.
- Have students arrange the items before they start to fasten them together. The artwork could be a 3-D sculpture that can stand on its own.
- Students could work together to make a mascot for their school or some other theme.

Safety Precautions Caution students to use only items that do not pose safety problems. Take care that students are not burned if using hot glue. Do not have students remove items from the trash cans.

Troubleshooting Glue that does not dry quickly will cause problems during this activity.

📁 **Activity Worksheets,** pages 5, 48

Analysis
1. Student answers will vary, but students should list each item individually.
2. All three; the amount of trash thrown out was reduced, the items were reused, and the items were recycled into a work of art.

Mini LAB

Making Models
Use things that are generally thrown out to make a piece of artwork.

1. Collect different items that would normally be thrown out. Such items could include newspapers, clean cans or glass, packaging, etc. Do not collect any food items or items that could be harmful. Do not take any items out of the garbage.
2. Using glue, string, or tape, create an item of artwork.
3. Give your piece of art a name.

Analysis
1. What items did you use to make your piece of artwork?
2. Is this activity an example of reducing, reusing, or recycling? Explain.

FIGURE 7-17
School cafeterias are busy places at lunchtime. *How could these students help reduce waste?*

196   Chapter 7 Resources

✓ Assessment

Process Help students to think critically by holding up a discarded item. Challenge the class to find a use for it. Students may illustrate their thoughts in the form of a cartoon. Use the Performance Task Assessment List for Cartoon/Comic Book in **PASC,** p. 61. **P**

Reducing Waste

Most people would agree that there are no simple solutions to the solid-waste problem. But trying to *reduce* the total amount of waste produced is the simplest and most effective way that an individual can help. In **Figure 7-17,** you can see how you could reduce waste.

Reducing includes cutting down on the amount of trash you throw out. Reducing waste could mean buying a model car in smaller packaging. Such an activity helps protect the environment. If you buy a product that has no packaging, it means that no paper from trees or plastic from oil has been used to pack it. No energy has been used to make this packaging and no landfill will be needed for its disposal. But it may also mean that the product could break during shipping.

Visual Learning

Figure 7-17 How could these students help reduce waste? *Students could use reusable lunch bags, plates, and utensils. They could bring food and drinks in reusable plastic containers instead of individually-wrapped items and single-serving containers.*

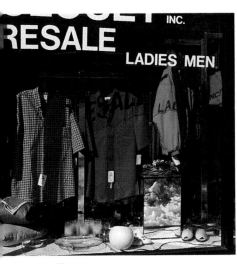

FIGURE 7-18

Secondhand stores are great places to find bargains. Like-new clothing and other items are being reused—a good way to help protect the environment.

Reusing Things

Reusing items is another way of reducing trash. Old clothes can be used as cleaning rags. Old newspapers can be used to line pet cages, to wrap gifts, or to cover the floor when painting. Some stores, as shown in **Figure 7-18**, sell used items. Books, magazines, clothes, computer disks, video games, glass jars, and cardboard boxes can also be reused. When you no longer need such items, you can give them away to somebody else who may want or need them.

Problem Solving

Uses for Plastic Things

Have you ever seen aluminum cans of beverages sold in packs of six in grocery stores? The six cans are usually held together with a plastic collar. Once the cans are released from this plastic collar, what do you do with it? Thrown away in one piece, it is dangerous to wildlife. Fish and ducks can get caught in the plastic collars, while other animals have choked on them.

Solve the Problem:
Are there other ways to deal with these plastic collars? How many ways can you think of to reuse the plastic collars in a useful way?

Think Critically:
Can you think of some other ways to package six aluminum cans of beverage together that won't hurt the environment?

Problem Solving

Solve the Problem
Student answers will vary. Possible uses could include organizing items of clothing such as socks and scarves, making bracelets, used as packing material, or part of 3-D artwork.

Think Critically
Student answers will vary. One possible answer is that they could be packaged in sheets of reusable plastic or in cardboard cartons. ⬛

Content Background

Old tires can be shredded and used for fuel, new rubber, plastic products, and as substitutes for concrete in road pavement.

Community Connection

Secondhand Store Invite the owner of a secondhand shop to class or take your class on a trip to the shop. Allow students to look through the articles for sale in the shop. Allow them to realize that many secondhand materials are as good as new.

Check for Understanding

Flex Your Brain Use the Flex Your Brain activity to have students explore RECYCLING.

📁 Activity Worksheets, page 5

Reteach

Have students tell you what to write on the chalkboard as you ask questions leading them through the production of glass. Students might state the following. Sand is collected; it is refined into glass from which bottles are made. They are then used and thrown away. Now the cycle must be repeated. Explain to students that instead of throwing away the glass, they could recycle it. The glass could be washed, melted, and formed into new bottles. L2

Extension

📁 For students who have mastered this section, use the **Reinforcement** and **Enrichment** masters.

FIGURE 7-19
Recycling products reduces the amount of energy used to make products.

A It takes 95 percent less energy to produce aluminum from recycled aluminum than from ore.

Recycling Things

Did you know that plastic soda bottles might have been used to make the carpeting in your home? It's true. Look at **Table 7-2.** Many items that people normally throw out—glass jars, newspapers, magazines, soda cans, and plastic milk containers—are now being recycled into other useful products. **Recycling** (ree SIKE ling) means reusing materials after they have been changed into another form. Recycling is done by bringing bottles, cans, and newspapers to a recycling center or leaving them on the curb in special recycling containers.

Buying products made from recycled materials helps protect the environment. Recycling usually saves energy, water, and other resources. **Figure 7-19** shows some examples of resources.

Table 7-2

Items to Be Recycled	Resulting Products
Newspapers, telephone books, magazines, and catalogs	Newsprint, cardboard, egg cartons, and building materials
Aluminum beverage cans	Beverage cans, lawn chairs, siding, cookware
Glass bottles and jars	Glass bottles and jars
Plastic beverage containers	Insulation, carpet yarn, textiles

Community Connection

Local Recycling Invite the owner of a local recycling plant to come and talk to the class about the best types of materials to recycle and which material would bring the most money. Ask what they do to the materials after they get them.

B It takes 75 percent less energy to make steel from scrap than from iron ore.

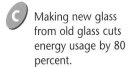

C Making new glass from old glass cuts energy usage by 80 percent.

Habits for a Healthier Environment

Practicing the 3 *Rs* makes you part of the solution to the solid-waste problem. Can you see how changing your everyday habits—the way you pack your lunch, the types of items you buy, the way your throw away your trash—can have a good effect on the environment? The best way to protect the environment is to develop habits that promote a healthy environment.

Section Wrap-up

1. What does the term *reducing waste* mean? Give an example.

2. Describe how recycling a glass bottle helps protect the environment.

3. **Think Critically:** How might a person getting food from a fast-food restaurant practice reducing waste?

4. **Skill Builder**
   Recognizing Cause and Effect
   Explain how keeping a television on has an impact on the environment. If you need help, refer to Recognizing Cause and Effect on page 551 in the **Skill Handbook.**

Science Journal

In your Science Journal, make a list of specific ways you and your family can help protect the environment.

Skill Builder
Recognizing Cause and Effect If a family always has the television on whether or not anyone is in the room, they are wasting electricity. When electricity is wasted, more coal than necessary was burned to produce the wasted electricity. As more coal is burned, more pollution enters the air, thus affecting the environment.

✓ Assessment

Performance Using plastic gloves, remove the solid waste from the trash can in the classroom. List the items contained in the can on the chalkboard. Ask students to discuss and come up with a list of how the waste could be reduced, recycled, or reused. Use the Performance Task Assessment List for Analyzing the Data in **PASC,** p. 27. L2

---

4 Close

Use the Mini Quiz to check students' recall of chapter content.

1. **America's huge waste problem is that there is too much solid _____ and there aren't enough places to put it.** *waste*

2. **Turning plastic bottles into play towers is an example of _____.** *recycling*

3. **Using a smaller box with less padding is an example of _____ the amount of resource needed.** *reducing*

4. **Washing out a mayonnaise jar and using it to store split peas is an example of _____.** *reusing*

Section Wrap-up

1. Reducing waste means that less material is used in the production or packaging of a product. Reducing the size of the package that holds a new game is an example.

2. Recycling a glass bottle reduces the need to extract resources from Earth. When lesser amounts of resources are extracted, the environment is also affected less.

3. **Think Critically** The person could request only one napkin and only as many packets of ketchup as are really needed.

Science Journal Student answers will vary but should include conserving water, turning off lights and televisions in rooms not in use, and combining short trips to reduce the use of the family car.

199

Science & Society

7●4 A Tool for the Environment

Visual Learning

Figure 7-20 What can you do with old clothes you no longer want or fit into? *Clothes could be repaired if necessary and given to someone else to wear. Clothing fabric could be re-used to make other products such as quilts and clothing. Old fabric could be used as rags.*

What YOU'LL LEARN

- About life-cycle analysis
- How to improve your note-taking skills

Science Words:
life-cycle analysis

Why IT'S IMPORTANT

Good note-taking skills will help you remember important things.

FIGURE 7-20

What can you do with old clothes you no longer want or fit into?

Are you an environmentally friendly shopper? When you buy things, do you think about how they affect the environment? Scientists wonder, too, about how the things we make and use impact the environment. That's why they've developed a tool to help them figure out the environmental impacts of products. The tool is called life-cycle analysis. **Life-cycle analysis** is a way of figuring out the environmental impact of a product through its entire life. Most scientists break down the life cycle of a product into six stages.

Life Stages of a Product

To do a life-cycle analysis, scientists start at the very beginning when natural resources are obtained to make a product. This is the first stage in the product's life. The six stages are:

1. getting the natural resources to make the product;
2. manufacturing in a factory or plant;
3. transportation to a home, store, or business;
4. use and reuse;
5. recycling;
6. disposal in a landfill or by burning.

Natural resources and energy are used during each stage of a product's life. In addition, each stage has an impact on the environment. The environmental impact might be air pollution, human health problems, use of a nonrenewable resource, or habitat loss, among others.

200 Chapter 7 Resources

Program Resources

 Reproducible Masters
Enrichment, p. 29 `L3`
Reinforcement, p. 29 `L2`
Study Guide, p. 29 `L1`

Transparencies
Section Focus Transparency 23 `L2`

When scientists do a life-cycle analysis, they add up all the natural resources and energy that are used to get the product from stage one to stage six. Then they figure out the environmental impact of each stage in the product's life. They analyze these factors to come up with a complete picture—a life-cycle analysis—of a product.

Environmentally Friendly Shopping

Once a life-cycle analysis has been completed, scientists can compare the product to similar products to see which one is better for the environment. Companies use life-cycle analyses, too, to help them figure out how to reduce the environmental impact of the products they make. Shoppers can also use life-cycle analyses. Can you figure out how? A life-cycle analysis can help you choose which products to buy. In this way, you can be an environmentally friendly shopper.

FIGURE 7-21
Turning fabric scraps into a colorful quilt is a good way to reuse different materials.

✦ Skill Builder: *Taking Notes*

LEARNING the SKILL

1. To take good notes, first read the material to identify the main ideas. The heads and subheads in the material are clues to main ideas.

2. Look for italicized or boldfaced words in the material. These are also clues to important ideas.

3. Identify details or sentences in the material that support the main ideas.

4. Using the main ideas and the details or sentences that support them, take notes about the material.

PRACTICING the SKILL

1. List three main ideas in this passage.

2. What is a life-cycle analysis? What are the six stages in a product's life?

3. What are some ways that life-cycle analyses are used?

APPLYING the SKILL

Think of a product that you would like to buy. Research the life cycle of the product. Take careful notes, and share what you've learned about the product with the class.

7-4 A Tool for the Environment **201**

Content Background

Some life-cycle analyses have shown conflicting results. Makers of plastic grocery bags find that their product is less harmful to the environment than paper grocery bags. Makers of paper grocery bags find the opposite to be true.

Given the lack of standardization, the main benefit thus far has been to help companies target areas of potential energy reduction. Manufacturers of household liquid cleaners find that roughly one-half of the total energy use associated with their products occurred during stage four (product-use stage) when people mix the cleaners with warm water. Manufacturers are currently working to develop liquid cleaners that can be used in cold water.

Science & the Arts

Source

Ellefson, Connie, Tom Stephens, and Doug Welsh. *Xeriscape Gardening: Water Conservation for the American Landscape.* New York: Macmillan, 1992.

Background

In 1981, the Front Range Xeriscape Task Force of the Denver Water Department coined the word *xeriscape.* The group was charged with developing guidelines for creating visually attractive landscapes using plants chosen for their water efficiency. Interest in xeriscaping is growing, not only as a strategy for water conservation, but because xeriscapes require little maintenance once established.

Teaching Strategies

- Invite a landscape architect or horticulturist to your classroom to talk to students about xeriscapes.

- Arrange a field trip to a university horticulture department or a local nursery so students can see water-conserving plants, including native varieties.

- Have students work in groups to research water-conserving plants and then to design a small xeriscape that could be grown in your area. **L2** **COOP LEARN** **LS**

*inter*NET CONNECTION

Visit the Glencoe Homepage, *www.glencoe.com/sec/science,* to find out how one city with an average rainfall of nine inches per year helps its citizens plan their xeriscaping.

Xeriscaping

Shoosh, shoosh, shoosh—it's summertime and the sound of lawn sprinklers fills the air. Green grass is pretty, but it guzzles water. Half of the water used in American homes goes to keep thirsty lawns and gardens green.

Fresh, clean, drinkable water is a limited resource. As the world's population grows, the demand for water increases. Communities everywhere face shrinking water supplies and are looking for creative ways to save water.

Xeriscaping (ZEER uh skay ping) is one solution. The word *xeriscape* comes from the Greek word *xeros,* meaning "dry." But a xeriscape is not a dusty brown patch of gravel with a few cactus plants placed in the middle. A xeriscape is a cool, colorful landscape made up of many different types of water-saving plants that are native to the particular area.

Creating a xeriscape takes the right plants and careful planning. Native plants—those that grow naturally where you live—are especially good choices. Native grasses, flowers, and shrubs can often survive on rainfall alone, even during a drought.

Using specific native plants certainly limits which plants you can choose, but there are still many beautiful plants that can be used. Many of these plants that have survived dry weather conditions over the years have produced tough, colorful flowers. Such flowers are often fragrant and attract birds and bees that help in pollination.

Xeriscaping helps homeowners to reduce their water use by up to 75 percent over traditional landscaping. In fact, a successful xeriscape hardly needs to be watered at all!

*inter*NET CONNECTION

The Glencoe Homepage at **www.glencoe.com.sec/science** provides links to information related to xeriscaping.

For More Information

The National Xeriscape Council, Inc. (P.O. Box 767936, Roswell, GA 30076) is a clearinghouse for information about water-conserving landscape practices. The NXCI also supports xeriscape demonstration gardens throughout the United States.

Read the statements below that review the most important ideas in the chapter. Using what you have learned, answer each question in your Science Journal.

1. Natural resources are things found in nature that living things use. *What is the difference between a renewable resource and a nonrenewable resource?*

3. Reducing, reusing, and recycling are three things that people can do to help the environment. *When you buy a product that has no packaging, how have you helped the environment?*

4. A life-cycle analysis is a way of figuring out the environmental impact of a product through its entire life. *How does knowing this information help you to make better choices about the environment?*

2. Human activities affect the environment. *List some ways that human activities impact land, air, and water.*

Chapter 7 Review **203**

Have students look at the illustrations on this page. Ask them to describe details that support the main ideas of the chapter found in the statement for each illustration.

Teaching Strategies

Photocopy these pictures. Put them on a transparency and ask students to match them to the correct captions.

Answers to Questions

1. A renewable resource is one that can be replaced by natural processes in 100 years or less. A nonrenewable resource cannot be replaced by natural resources within 100 years.
2. land—farming, construction, and landfills; water—drink, cleaning, and recreation; air—by burning materials that release pollutants.
3. There is no waste to be disposed of.
4. You can shop for products with favorable life-cycle analyses.

Science at Home

Intrapersonal Ask the students to count the number of trash bags that they carry out in one week. The next week, have them use different bags for recyclable trash and for general trash. Ask students to compare the number of general trash bags sent to the landfill in week two to the total bags sent in week one. **L2**

✔**Assessment**

Portfolio Encourage students to place in their portfolios one or two items of what they consider to be their best work. Examples include:

- Activity 7-2, pp. 188-189
- MiniLAB, p. 180
- MiniLAB, p. 196 **P**

Performance Additional performance assessments may be found in **Performance Assessment** and **Science Integration Activities.** Performance Task Assessment Lists and rubrics for evaluating these activities can be found in Glencoe's **Performance Assessment in the Science Classroom (PASC).**

Chapter 7 Review

Review

Using Key Science Words

1. pollutant
2. recycling
3. landfill
4. solid waste
5. acid rain

Checking Concepts

6. c
7. a
8. b
9. b
10. c

Thinking Critically

11. We do not run out of trees because trees are constantly being replanted by lumber companies.
12. Most of the water that covers Earth's surface is salt water that cannot be used by people unless the salt is removed.
13. Pollutants from the landfill can leak out and enter the groundwater and other parts of the environment.
14. Use of the plant must be limited so that natural processes can replenish the supply of the plant within a 100-year time period.
15. The number of new bags needed is reduced and the amount of natural resources needed to make new bags is also reduced.

Using Key Science Words

acid rain	pollutant
landfill	recycling
life-cycle analysis	renewable resource
natural resource	solid waste
nonrenewable resource	

Match each phrase with the correct term from the list of Key Science Words.

1. a material that harms living things by interfering with life processes
2. reusing materials after they have been changed into another form
3. an area where garbage is deposited
4. things people throw away that are in solid or near-solid forms
5. when gases released by burning oil and coal mix with water in the air

Checking Concepts

Choose the word or phrase that completes the sentence.

6. An example of a nonrenewable resource is _____.
   a. sunlight   c. oil
   b. water   d. a tree

7. Using an old newspaper to line a pet cage is an example of _____ the newspaper.
   a. reusing   c. reducing
   b. recycling   d. buying

8. Collecting used paper and sending it to a factory to be made into new paper is an example of _____.
   a. reusing   c. reducing
   b. recycling   d. buying

9. Breathing polluted air can cause _____.
   a. acid rain
   b. health problems
   c. solid waste
   d. water pollution

10. A life-cycle analysis of a product will indicate its environmental impact during its _____.
    a. daily use
    b. production time
    c. entire life
    d. decay time

Thinking Critically

Answer the following questions in your Science Journal using complete sentences.

11. If people use so many paper products, why don't we run out of trees?

12. Almost 70 percent of Earth's surface is covered by water, but less than one percent can be used by people. Why?

13. When a landfill can't hold any more solid waste, it is closed down. How can a landfill be an environmental problem even though people are no longer depositing trash there?

14. Three thousand new plants grow naturally in a field every year. How must these plants be treated to make sure this type of plant is a renewable resource?

Assessment Resources

Reproducible Masters
Chapter Review, pp. 17-18
Assessment, pp. 29-32, 33-34
Performance Assessment, pp. 33-34, 45

Glencoe Technology
Computer Test Bank
MindJogger Videoquiz

Chapter 7 Skills Review

15. Some people take their own bags with them when shopping. How might this affect natural resources?

Developing Skills

If you need help, refer to the description of each skill in the Skill Handbook.

16. **Concept Mapping:** Use the following phrases to make a concept map showing the complete life-cycle analysis of an aluminum can: *refine the aluminum; mine aluminum ore; use the aluminum can; shape the aluminum into cans; recycle the can; melt the ore in a factory; transport the aluminum can to where it will be used.*

17. **Recognizing Cause and Effect:** Acid rain is caused by air pollution. But acid rain often falls on places that are far from sources of air pollution—such as factories and cars. Research acid rain in the library or on the Internet and then write a short paragraph in your Science Journal explaining why acid rain can be a problem for places that are far from sources of pollution.

18. **Classifying:** Classify the following resources as renewable or non-renewable: sunlight, water, oil, trees, air, coal, soil.

19. **Making and Using Tables:** Record the things that your family throws away for one week. Make a table listing which of these items could be recycled. In another column, list the resulting products that can be made from the recycled items. Refer to **Table 7-2** for help.

Items for Recycling	Products made from recycled material

20. **Using Numbers:** If everyone in the United States recycled their newspapers, 500 000 trees would be saved each week. How many trees would be saved in one year?

Performance Assessment

1. **Design an Experiment:** Lemon juice is an acidic liquid. Design an experiment showing the effects of lemon juice on a plant. In your Science Journal, relate the results of your experiment to what you've learned about acid rain.

2. **Newspaper Article:** Write a newspaper article describing an environmental problem in your own community and possible solutions.

Chapter 7 Review **205**

Performance Assessment

1. Student answers will vary, but one idea is to spray the leaves of a plant with lemon juice and to add the lemon juice to the soil around the plant, just as acid rain would. Students should design the experiment, conduct it, and collect data. The data should be placed in a table. Use the Performance Task Assessment List for Designing an Experiment in **PASC**, p. 23. P

2. Student articles will vary based on what each student considers an important environmental problem in or near the community or school. Be sure students include a description of the problem and a possible solution. Use the Performance Task Assessment List for Newspaper Article in **PASC**, p. 69. P

Developing Skills

16. **Concept Mapping** Concept map should have the following steps: mine aluminum ore → melt the ore in a factory → refine the aluminum → shape the aluminum into cans → transport the aluminum can to where it will be used → use the aluminum can → recycle the can.

17. **Recognizing Cause and Effect** The pollution that causes acid rain may originate in one location, but wind will blow the pollution to other locations. As the pollution moves through the air, it can react with water to form acid rain that can fall far from the source of pollution.

18. **Classifying** renewable—sunlight, water, trees, air, and soil; nonrenewable—oil and coal

19. **Making and Using Tables** Student answers will vary but should include items similar to the following: newspapers, aluminum cans, glass and plastic bottles, and yard waste. The newspapers, aluminum cans, glass bottles, and plastic bottles can be recycled. Products that can be made from the recycled material are recycled paper; aluminum cans and other aluminum products; glass for windows and containers; and plastic for bathtubs, containers, and insulation.

20. **Using Numbers** 500 000 trees saved per week × 52 weeks per year = 26 000 000 trees saved per year.

205

Unit 1 Project

Objectives

LS **Intrapersonal** Students will search for, gather, and analyze information from the Internet, newspaper, books, television, and other reference sources about North American biomes. They will identify the different types of biomes in North America and learn about the many types of animals and plants that live in them. With this information, students will construct a biomes poster. **L1**

Summary

Internet Students will visit the Glencoe Homepage at **www.glencoe.com/sec/science** to find links to biome websites. Students can access the websites to learn more about biomes and to identify the type of biome that they live in. Students can also use the websites to find out about the different types of biomes in North America, as well as the diversity of plants and animals in these biomes. At the Glencoe Homepage, students will also be able to print out a blank map of North America. Students will mount the map on posterboard and, using crayons, markers, or colored pencils, indicate the geographical locations of various biomes. The Glencoe website will also provide students with links to websites where they can gather climate data.

Non-Internet Sources If you do not have access to the Internet, have students use encyclopedias, newspapers, and other reference sources to find data about biomes.

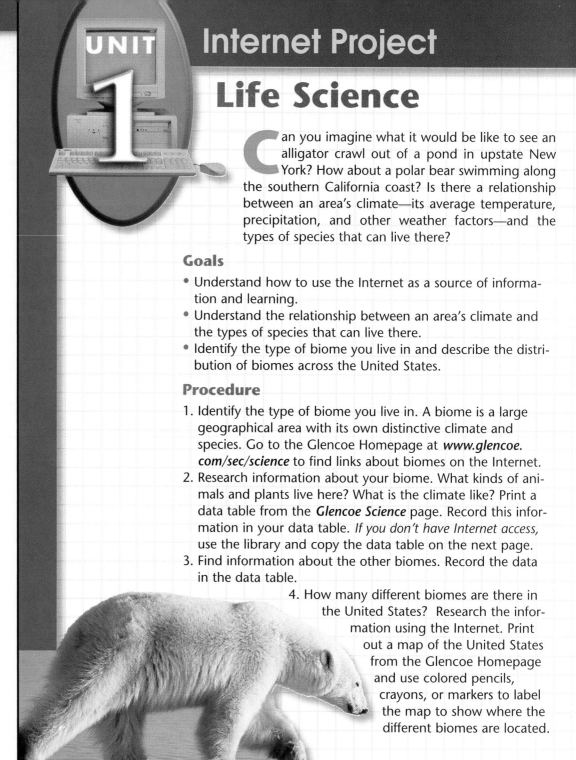

UNIT 1

Internet Project
Life Science

Can you imagine what it would be like to see an alligator crawl out of a pond in upstate New York? How about a polar bear swimming along the southern California coast? Is there a relationship between an area's climate—its average temperature, precipitation, and other weather factors—and the types of species that can live there?

Goals

- Understand how to use the Internet as a source of information and learning.
- Understand the relationship between an area's climate and the types of species that can live there.
- Identify the type of biome you live in and describe the distribution of biomes across the United States.

Procedure

1. Identify the type of biome you live in. A biome is a large geographical area with its own distinctive climate and species. Go to the Glencoe Homepage at **www.glencoe.com/sec/science** to find links about biomes on the Internet.
2. Research information about your biome. What kinds of animals and plants live here? What is the climate like? Print a data table from the *Glencoe Science* page. Record this information in your data table. *If you don't have Internet access,* use the library and copy the data table on the next page.
3. Find information about the other biomes. Record the data in the data table.
4. How many different biomes are there in the United States? Research the information using the Internet. Print out a map of the United States from the Glencoe Homepage and use colored pencils, crayons, or markers to label the map to show where the different biomes are located.

206

Time Required

Students will need three or four days to complete this project.

Preparation

- **Internet** Access the Glencoe Homepage at **www.glencoe.com/sec/science** to run through the steps the students will follow.
- **Non-Internet Sources** Use the library to find books and reference sources that contain data about biomes and climate in North America.

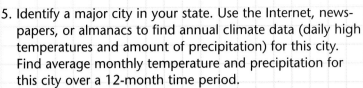

5. Identify a major city in your state. Use the Internet, newspapers, or almanacs to find annual climate data (daily high temperatures and amount of precipitation) for this city. Find average monthly temperature and precipitation for this city over a 12-month time period.

6. Use the Internet, newspapers, or the local library to find out the names of common plants and animals in your state. Post the list of species and your climate data to the Glencoe Homepage to share with students from all over the world!

7. Go to the Glencoe Homepage. Find species lists and climate data from cities within each of the remaining biomes on your map. Complete your biomes map by mounting species lists on the different biomes.

Conclude and Apply

1. Compare and contrast two different biomes in North America.

2. What kinds of plants and animals live in a desert environment? What characteristics do these organisms have to survive in a desert climate?

3. Explain the relationship between an area's climate and the types of species that can live there.

Go Further

Look up data for another country. Find out annual climate data and the species that live in different areas. Determine the biome.

Data Table

Biome	Common Animals	Common Plants	Climate	Temperature	Rainfall
Tundra					
Taiga					
Tropical Rain Forest					
Grasslands					
Desert					
Temperate Forest					

3. Students should answer that the species that are found in a particular environment are adapted to survive in the special conditions of that environment.

Assessment

To assess the results of the project, use the Performance Task Assessment List for Poster in **PASC**, p. 73.

Go Further

Encourage interested students to prepare biomes posters covering all of the world's biomes. Show students how various search engines (Yahoo!, Excite, Webcrawler, Alta Vista) can be used to find links to many different websites. Students can use the links to find data about climate and species in the biomes not found in North America.

Teaching Strategies

- Before students begin the project, initiate a discussion about biomes by asking students about weather and climate in the United States. **Is weather and climate uniform throughout the United States?** Most students will know that different parts of the country have different climates. Display a large map of North America. Have students begin thinking about the weather and climate in different parts of North America.

- Ask students about the different types of plants and animals in the United States. Can species live anywhere, or are they restricted to areas with particular climates?

- If your school does not have Internet access, many libraries now provide access to the Internet. Additionally, some students may have Internet access at home. Encourage interested students to use the Internet to find additional links that might be useful for the project.

Answers to Conclude and Apply

1. Make sure students list clear similarities and differences between the two biomes they choose. For example, both tropical rain forests and temperate forests have large trees. Many trees in temperate forests are deciduous. Trees in tropical forests are not deciduous.

2. Answers will vary. Students will most likely name a variety of different cacti, reptiles, and birds. The organisms they name will have adaptations to conserve water and, if an animal, to regulate body temperature.

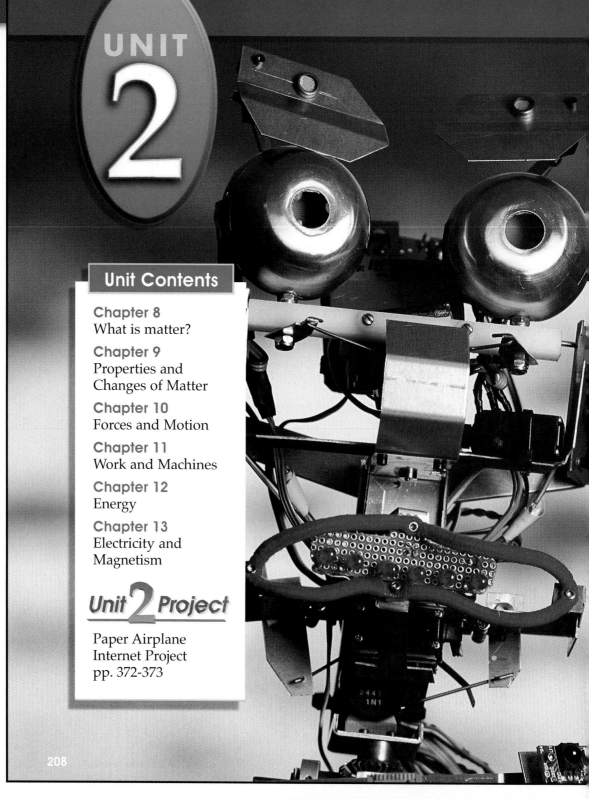

UNIT 2

Physical Science

In Unit 2, students are introduced to the fundamentals of physical science, including matter and energy. Concepts such as force, motion, and work are defined. Students discover how machines enable humans to work more efficiently. The unit concludes with a discussion of magnetism and electricity.

CONTENTS

Unit Contents

Unit 2 Project

Paper Airplane
Internet Project
pp. 372-373

interNET CONNECTION

The Glencoe Homepage at **www.glencoe.com/sec/science** provides links to information related to topics found in each chapter in Unit 2.

Science at Home

Checking for Energy Leaks Have each student test areas in his or her home where energy may be escaping. In the kitchen, students can test seals on refrigerator, oven, or freezer doors by holding a strip of paper between the door and the seal as they close the door. If the paper pulls out easily, the seal is not tight. Students can also check temperature differences in different parts of a room. **L2** **ELL**

Physical Science

What's happening here?

Robots are machines that may be powered by electricity, magnetism, or even by the sun. Researchers have found that people are more likely to work with robots if they have "faces." Scientists plan to use robots with faces in hospitals and nursing homes, where they will interact with patients. The robot in the larger photograph is named IT and was designed to respond to human emotions. The robot in the smaller photograph helps people pick up heavy loads. When a person puts the robot on, it measures the force of the wearer's arms, then multiplies it to control a pair of stronger robotic arms. Robots often are programmed to do work that humans find too messy, dangerous, or boring.

interNET CONNECTION

You can visit the Glencoe Homepage at *www.glencoe.com/sec/science* for links to information about robots. Find out what kinds of jobs robots do and why. In your Science Journal, write about what you found out.

Cultural Diversity

Automata Automata are mechanized objects that are relatively self-operating once they are set in motion. Early automata were operated by jets of air or water and coiled springs. Mechanized landscapes and flowers that opened metal petals are common types of automata built by Chinese and Muslim inventors as early as the third century.

Unit Internet Project

The Project for this unit guides students through using the Internet to investigate flight. In preparation for this project, found on pages 372 and 373, follow the suggestions indicated on page 372. Students can begin working on the project after Chapter 10.

INTRODUCING THE UNIT

What's Happening Here?

Have students look at the photos and read the text. Ask them to give some examples of work humans do that may be messy, dangerous, or boring. Explain that in this unit, students will learn how matter and energy can be harnessed to build machines such as robots that can work for humans. In this unit, students also will discover how electricity is generated.

Background

- A robot is defined as any automatically operating machine that replaces human effort. The term *robot* is derived from a Czech word, *robota*, meaning "forced labor."
- Robots in use today include automated teller machines (ATMs); the Sojourner robot that explored the surface of Mars; solar-powered, self-steering lawn mowers; and robotic welders in automobile assembly plants.

Previewing the Chapters

Have students look at the art and photos in the chapters in this unit. Ask them to list the art and photos that show machines that perform work for humans. Have students also identify the possible sources of energy depicted.

Tying to Previous Knowledge

Encourage students to identify other "robotic" machines that they depend upon every day. Some students may include the family car, which does fit the definition of robot.

Chapter Organizer

Section	Objectives/Standards	Activities/Features
Chapter Opener		**Explore Activity:** Observe That Matter Occupies Space, p. 211
8-1 **Matter and Atoms** (5 sessions, 3 blocks)*	1. **Define** what matter is, and **describe** what makes up matter. 2. **Describe** the parts of an atom and their charges. National Science Content Standards: (5-8) UCP2, A1, B1, B2, E2, G1, G3	**MiniLAB:** Using a Model, p. 215 **Problem Solving:** Comparing Gravel and Iron, p. 21 **Activity 8-1:** The Electron Magnet, pp. 220-221 **Skill Builder:** Making Models, p. 222 **Using Computers,** p. 222
8-2 **Types of Matter** (3 sessions, 1 block)*	3. **Define** and **give examples** of elements, compounds, and mixtures. 4. **Identify** elements by their names and symbols, and identify compounds by their formulas. National Science Content Standards: (5-8) A1, A2, B1, F5, G3	**Using Technology:** Goo to the Rescue, p. 225 **Activity 8-2:** The Mystery Mixture Investigation, p. 22 **MiniLAB:** Observing, p. 228 **Skill Builder:** Using a Computerized Card Catalog, p. 229 **Using Math,** p. 229
8-3 **Science and Society:** **Synthetic Elements** (1 session, ½ block)*	5. **Recognize** the elements that do not occur naturally on Earth. 6. **Take notes** to provide a summary of important information. National Science Content Standards: (5-8) B1, E2, F5, G1	**Skill Builder:** Taking Notes, p. 231 **Science & the Arts:** Maya Lin's Civil Rights Memoria p. 232

* A complete Planning Guide that includes block scheduling is provided on pages 31T-33T.

Activity Materials

Explore	Activities	MiniLABs
page 211 cup, tissue paper, bowl, water, tape	page 220 foam packing peanuts, paper clip, wood, wool, balloon, lint, aluminum foil page 226 dropper bottle, pie pan, test tubes, test-tube holder, baking soda, candle, cornstarch, matches, sugar, white vinegar, aqueous iodine solution	page 215 pie pan, food coloring, whole milk, liquid soap page 228 cup, plate, sand, table salt, spoon

Need Materials? Call Science Kit (1-800-828-7777).

Teacher Classroom Resources

Reproducible Masters	Transparencies	Teaching Resources
Activity Worksheets, pp. 5, 49-50, 53 **Enrichment**, p. 30 **Lab Manual 15** **Reinforcement**, p. 30 **Study Guide**, p. 30	**Section Focus Transparency 24**, Mass and Space **Teaching Transparency 15**, Atomic Structure	**Spanish Resources** **English/Spanish Audiocassettes** **Cooperative Learning Resource Guide** **Lab Partner** **Lab and Safety Skills** **Lesson Plans**
Activity Worksheets, pp. 5, 51-52, 54 **Cross-Curricular Integration**, p. 12 **Enrichment**, p. 31 **Lab Manual 16** **Multicultural Connections**, pp. 19-20 **Reinforcement**, p. 31 **Science and Society/Technology Integration**, p. 26 **Science Integration Activities**, pp. 53-54 **Study Guide**, p. 31	**Science Integration Transparency 8**, Add Minerals (And Mix) **Section Focus Transparency 25**, A Matter of Makeup **Teaching Transparency 16**, Chemical Structures	**Assessment Resources** **Chapter Review**, pp. 19-20 **Assessment**, pp. 35-38 **Performance Assessment**, p. 46 **Performance Assessment in the Science Classroom (PASC)** **MindJogger Videoquiz** **Alternate Assessment in the Science Classroom** **Computer Test Bank**
Enrichment, p. 32 **Reinforcement**, p. 32 **Study Guide**, p. 32	**Section Focus Transparency 26**, Bear It in Mind	

Key to Teaching Strategies

The following designations will help you decide which activities are appropriate for your students.

L1 Level 1 activities should be appropriate for students with learning difficulties.

L2 Level 2 activities should be within the ability range of all students.

L3 Level 3 activities are designed for above-average students.

ELL ELL activities should be within the ability range of English Language Learners.

LS These activities are designed to address different learning styles.

COOP LEARN Cooperative Learning activities are designed for small group work.

P These strategies represent student products that can be placed into a best-work portfolio.

GLENCOE TECHNOLOGY

The following multimedia resources are available from Glencoe.

Science and Technology Videodisc Series (STVS)
Chemistry
   Images of Atoms
The Infinite Voyage Series
Unseen Worlds
Glencoe Physical Science Interactive Videodisc
Periodicity
Alkali Metals

Electron Structure
Stability and Reactivity
Noble Gases
Conclusion
Glencoe Physical Science CD-ROM

Teacher Classroom Resources

This is a representation of key blackline masters available in the Teacher Classroom Resources.

Teaching Aids

Section Focus Transparencies

24 SECTION FOCUS TRANSPARENCY Section 8-1

MASS AND SPACE

If you walk through a shallow puddle of water, your feet hardly know the water's there. The water has mass and may take up space, but not much. But if this surfer takes a spill, he'll know right away that water has mass and takes up the space where he wants to be.

1. How will the surfer know that water has mass and takes up space?
2. What other things in the photo have mass?
3. What other things in the photo take up space?

L2

25 SECTION FOCUS TRANSPARENCY Section 8-2

A MATTER OF MAKEUP

You use pencils and paper almost every day, but do you ever think about what is used to make them? The pencil "lead" is actually graphite, a form of carbon that rubs off easily on paper. Other parts of a pencil and paper are made from other things.

1. Look closely at a pencil. What materials do you think make up a pencil?
2. Look closely at a piece of paper. Can you see small pieces of the materials that make up paper?
3. Trade pieces of paper with someone else. Do you see any differences?

L2

26 SECTION FOCUS TRANSPARENCY Section 8-3

BEAR IT IN MIND

The animal on the right might let out a roar, and you'd be wise to stay away from it. The one on the left stays quiet and won't object to being held. Both are bears. What makes them so different?

1. Which of the bears shown is human-made?
2. Name other pairs of artificial and natural things. How do they differ?

L2

Science Integration Transparencies

8 SCIENCE INTEGRATION TRANSPARENCIES Use with Section 8-2

Add Minerals (and Mix)

Granite
A Mixture of Minerals

Quartz (a compound of the elements silicon and oxygen)

Feldspar (compounds of the elements silicon, oxygen, and aluminum with potassium, sodium, or calcium)

Mica (contains the elements silicon, oxygen, aluminum, potassium, and others)

L2

Teaching Transparencies

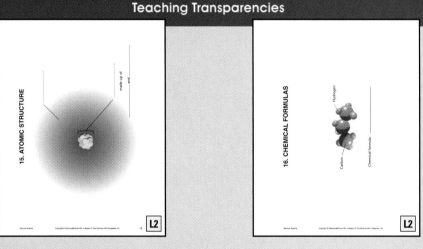

15. ATOMIC STRUCTURE

made up of ___ and ___

L2

16. CHEMICAL FORMULAS

Hydrogen

Carbon

Chemical Formula

L2

Meeting Different Ability Levels

Study Guide for Content Mastery

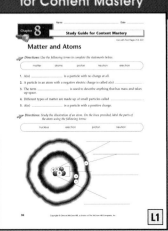

Chapter **8** Study Guide for Content Mastery

Matter and Atoms

Directions: Use the following terms to complete the statements below.

| matter | atoms | proton | neutron | electron |

1. A(n) ___ is a particle with no charge at all.
2. A particle in an atom with a negative electric charge is called a(n) ___.
3. The term ___ is used to describe anything that has mass and takes up space.
4. Different types of matter are made up of small particles called ___.
5. A(n) ___ is a particle with a positive charge.

Directions: Study the illustration of an atom. On the lines provided, label the parts of the atom using the following terms:

| nucleus | electron | proton | neutron |

L1

Reinforcement

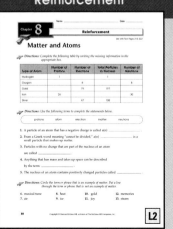

Chapter **8** Reinforcement

Matter and Atoms

Directions: Complete the following table by writing the missing information in the appropriate box.

Type of Atom	Number of Protons	Number of Electrons	Total Particles in Nucleus	Number of Neutrons
Hydrogen	1			
Oxygen		8		8
Gold		79	197	
Ion	26			30
Silver		47	108	

Directions: Use the following terms to complete the statements below.

| protons | atom | electron | matter | neutrons |

1. A particle of an atom that has a negative charge is called a(n) ___.
2. From a Greek word meaning "cannot be divided," a(n) ___ is a small particle that makes up matter.
3. Particles with no charge that are part of the nucleus of an atom are called ___.
4. Anything that has mass and takes up space can be described by the term ___.
5. The nucleus of an atom contains positively charged particles called ___.

Directions: Circle the term or phrase that is an example of matter. Put a line through the term or phrase that is not an example of matter.

6. musical tune
7. air
8. heat
9. ice
10. gold
11. joy
12. memories
13. steam

L2

Enrichment Worksheets

Chapter **8** Enrichment

Atoms

The Periodic Table and Atomic Structure

Directions: Read the following information. Then answer the questions.

The periodic table is an organized list of all the elements. In the table, there is a number above the abbreviation for each element. This number is that element's atomic number. It is the number of protons in the nucleus of one atom of the element. For example, you'll see that carbon has an atomic number of 6. This means that there are six protons in the nucleus of one carbon atom.

Below the abbreviation for each element on the periodic table is another number, the atomic mass. The atomic mass is approximately equal to the total number of protons and neutrons in an atom.

The table shows the numbers of protons, neutrons, and electrons in each of the six most common elements found in living things.

Element	Protons	Neutrons	Electrons
Carbon	6	6	6
Hydrogen	1	0	1
Nitrogen	7	7	7
Oxygen	8	8	8
Phosphorus	15	16	15
Sulfur	16	16	16

Applying Problem-Solving Skills

1. What is the atomic number of each of the six elements?

2. What is the atomic mass number of each of the six elements?

L3

Chapter 8 What is matter?

Hands-On Activities

Science Integration Activities

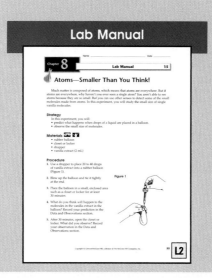

Chapter 8 — Integration

Soda, Stalactites, and Stalagmites

Have you ever made a rock using water? Of course, you can't really make rocks out of plain water. There must be something dissolved in the water that can make the rocks. A similar process happens in nature. When certain solid substances are mixed in water, you can't see them anymore. When the water evaporates, these substances become solids again.

If you have ever been in a cave, you may have noticed large rocks hanging from the ceiling of the cave or built up on the floor of the cave. The rocks hanging from the ceiling are called stalactites. The rocks built up on the floor of the cave are called stalagmites. They are made from compounds dissolved in the water found in caves.

Getting Started

You can use a mixture of water and washing soda to make rock crystals. In this activity, you will make a stalactite and a stalagmite out of water containing this compound of soda.

Thinking Critically

Read through the following procedure. When scientists perform an experiment, they often perform a second similar experiment as a control. In a control, only one thing is different from the original experiment. For example, if you thought that salt water evaporates, leaving salt behind, you would set up an experiment with salt water. As a control, you would do the same experiment using plain water. That way, you would know that results you observe are a result of adding salt to the water. Think of how you could do a control for this experiment.

Try It!

Lab Manual

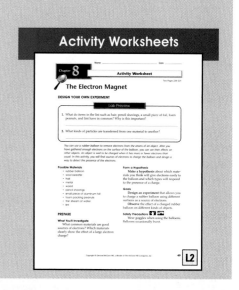

Chapter 8 — Lab Manual 15

Atoms—Smaller Than You Think!

Much matter is composed of atoms, which means that atoms are everywhere. But if atoms are everywhere, why haven't you ever seen a single atom? You aren't able to see atoms because they are so small. But you can use other senses to detect some of the small molecules made from atoms. In this experiment, you will study the small size of single vanilla molecules.

Strategy
In this experiment, you will.
- predict what happens when drops of a liquid are placed in a balloon.
- observe the small size of molecules.

Materials
- rubber balloon
- closet or locker
- dropper
- vanilla extract (2 mL)

Procedure

Figure 1

Activity Worksheets

Chapter 8 — Activity Worksheet

The Electron Magnet

DESIGN YOUR OWN EXPERIMENT

Lab Preview

Enrichment and Application

Cross-Curricular Integration

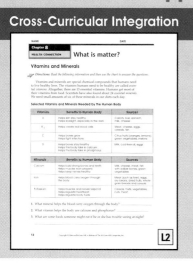

Chapter 8 HEALTH CONNECTION — What is matter?

Vitamins and Minerals

Multicultural Connections

Chapter 8 Multicultural Connections

From Bathroom Sink to Research Lab

Science and Society/Technology Integration

Chapter 8 SCIENCE & SOCIETY — What is matter?

Nuclear Power

Assessment

Performance Assessment

Chapter 8 — Skill Assessment

Separating Mixtures: Comparing Methods

Chapter Review

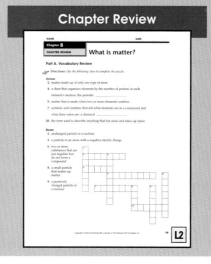

Chapter 8 CHAPTER REVIEW — What is matter?

Part A. Vocabulary Review

Assessment

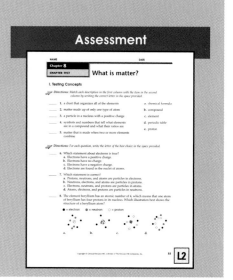

Chapter 8 CHAPTER TEST — What is matter?

I. Testing Concepts

What is matter?

CHAPTER OVERVIEW

Section 8-1 This section presents the atomic model of matter, including a brief history of its development.

Section 8-2 Elements, compounds, and mixtures are defined and described. The periodic table and chemical formulas are introduced.

Section 8-3 Science and Society In the context of exploring the manufacture and uses of synthetic elements, students examine the importance of taking accurate notes.

Chapter Vocabulary

matter	element
atom	periodic table
proton	compound
neutron	chemical formula
electron	mixture

Theme Connection

Scale and Structure Chapter 8 introduces the atomic model of matter. This model is related to the structure of elements and more complex matter—compounds and mixtures—and to radioactive decay.

Chapter Preview

210

Skills Preview

Skill Builders
- make models
- use a computerized card catalog
- take notes

MiniLABs
- infer
- observe

Activities
- hypothesize
- design your own experiment
- observe
- compare
- conclude

Learning Styles	Look for the following logo for strategies that emphasize different learning modalities. **LS**
Kinesthetic	Explore, p. 211; Activity, p. 217; Activity 8-1, p. 220; Assessment, p. 222; Activity 8-2, p. 226; Check for Understanding, p. 227; MiniLAB, p. 228
Visual-Spatial	Demonstration, p. 214; MiniLAB, p. 215; Visual Learning, pp. 219, 222, 227; Reteach, p. 219
Interpersonal	Across the Curriculum, p. 215
Intrapersonal	Enrichment, p. 219; Science at Home, p. 233
Logical-Mathematical	Science Journal, p. 213; Across the Curriculum, p. 214; Problem Solving, p. 216; Go Further, p. 220; Inquiry Question, p. 227; Using Math, p. 229; Teaching Strategies, p. 232
Linguistic	Using Science Words, p. 218; Using an Analogy, p. 225; Assessment, p. 226

What is matter?

Perched on an enormous rock overhanging a river, you feel the gritty texture of granite under your feet and smell the pine trees along the shore. Suddenly, you realize how hot you are. Taking your empty squirt bottle down to the river, you hold the bottle under water. A stream of bubbles rises from the mouth of your bottle to the water's surface. That's puzzling, you think. If the bottle were empty, where did the bubbles come from? The Explore activity below will help you solve the puzzle.

EXPLORE ACTIVITY

Observe That Matter Occupies Space

1. Take a square of tissue paper, wad it into a ball, and tape it to the bottom of the inside of a plastic or paper cup. Turn the cup upside down. The tissue should stay inside the cup.
2. Hold the cup upside down over a bowl of water. Slowly push the cup into the water as far as you can.
3. Raise the cup out of the water, and remove the tissue.

Science Journal

In your Science Journal, describe the tissue paper before and after you put the cup in the water. Explain what you think happened. What do you think was in the cup besides the tissue paper?

211

211

Prepare

Section Background

Many ancient philosophers tried to define matter. Democritus's model is the closest to modern ideas.

Preplanning

Refer to the Chapter Organizer on pages 210A-B.

1 Motivate

Bellringer

 Before presenting the lesson, display **Section Focus Transparency 24** on the overhead projector. Assign the accompanying **Focus Activity** worksheet.

`L2` `ELL`

Tying to Previous Knowledge

Ask students how their ideas have changed over time. For example, at one time, students may have thought the sun moved across the sky. Now students might know that the sun's apparent motion results from Earth's rotation. Emphasize that as students' experiences increased, so did their knowledge.

212

8•1 Matter and Atoms

What YOU'LL LEARN

- What matter is and what makes up matter
- The parts of an atom and their charges

Science Words:
matter
atom
proton
neutron
electron

Why IT'S IMPORTANT

Everything in and around you is made up of matter.

FIGURE 8-1

Democritus was a Greek philosopher who lived from about 460 B.C. to about 370 B.C.

What is matter?

When you pulled the tissue paper out of the cup in the Explore activity, it wasn't wet. You had placed the cup in the bowl of water, so why didn't water rush into the cup and get the tissue wet? More than tissue paper was in the cup. Air was trapped inside, too.

The air took up space and kept the water from touching the tissue paper. If you could put it on a balance, you would find that air also has mass. **Matter** is the term used to describe anything that has mass and takes up space. Therefore, air is matter. Anything you can see, touch, taste, or smell is matter. Doesn't that include just about everything you can imagine? What isn't matter? Light has no mass and takes up no space. Heat has no mass and takes up no space. Emotions, thoughts, and ideas aren't matter, either. Can you think of anything else?

What makes up matter?

People have always been interested in matter. What is it made of? Is all matter alike? As the tools of science developed, so did people's ideas about matter.

An Early Idea

An early idea about matter came from Democritus (dih MAW kruh tus), shown in **Figure 8-1.** He thought that the universe was made of empty space and tiny bits of stuff (atoms). Democritus didn't really know what the tiny bits of stuff were, but he thought that they must be incredibly small—so small that they couldn't be divided into smaller pieces. In fact, the word *atom* comes from a Greek word that means "cannot be divided." In science today, an **atom** is defined as a small particle that makes up most types of matter. Democritus's idea that matter is made up of small particles and empty space was a start, but many questions still needed to be answered.

Program Resources

📁 **Reproducible Masters**
Activity Worksheets, pp. 5, 49-50, 53 `L2`
Enrichment, p. 30 `L3`
Lab Manual, pp. 55-56 `L2`
Reinforcement, p. 30 `L2`
Study Guide, p. 30 `L1`

🔦 **Transparencies**
Section Focus Transparency 24 `L2`
Teaching Transparency 15 `L2`

Oxygen

+

FIGURE 8-2

When a fire burns, the total mass of the fuel (the wood) and the oxygen it reacts with equals the total mass of what is produced by the fire (water vapor, carbon dioxide, and ash). No matter is gained or lost.

Water vapor and carbon dioxide

+

Later Developments

Almost 2000 years passed without much change in Democritus's idea of what makes up matter. By that time, people thought that matter could appear from nothing, or it could completely disappear. Another common belief was that the "stuff" part of matter could change its form. The French chemist Antoine Lavoisier (luh vwah zee AY) challenged these thoughts in the late 1700s.

Lavoisier showed that there was no change in mass during common chemical reactions such as burning and rusting. Have you ever sat by a fire, such as the one shown in **Figure 8-2,** watching the wood blaze brightly, then turn to embers, and finally to ashes? What starts out as a mountain of logs is just a little heap of ashes by the end of the evening. Surely, some mass has been lost in the process—some matter has disappeared. But Lavoisier showed that the total mass of the wood and the oxygen used when it burns equals the total mass of the water, carbon dioxide, and ash produced—the amount of matter stayed the same.

8-1 Matter and Atoms   **213**

2 Teach

Content Background

The idea that matter cannot be created or destroyed is called the law of conservation of mass. A closed system that starts with a certain mass will have the same mass after various chemical and physical changes occur. However, during nuclear reactions, mass can be converted to energy and vice versa, which is stated in the law of conservation of mass-energy.

Revealing Preconceptions

Students may think that matter is created, as when a plant grows, or that matter vanishes, as when the plant dies and decomposes. Emphasize that matter is conserved in a closed system in which nothing enters or leaves. Closed systems do not interact with things outside the system. For example, astronauts in the space shuttle are in a nearly closed system—they bring the air they will breathe, and they recycle their water.

Science Journal
Elements One group of Greek philosophers thought that all matter was made of air, water, fire, or earth. Have students list in their Science Journals things around them, then try to classify each item according to the four categories. **P**
LS L2

FLEX Your Brain

Use the Flex Your Brain activity to have students explore RATIOS.

📁 **Activity Worksheets,** page 5

Demonstration

LS Visual-Spatial Show students that the apparent disappearance of matter doesn't mean it isn't there. Fill a glass with warm water. Stir in a small amount of salt or sugar. Do not allow students to taste the solution. Ask students whether the salt or sugar is still there, even though it seems not to be. If students need confirmation, the water can be evaporated from a small portion of the solution, leaving the salt or sugar behind.

Visual Learning

Figure 8-3 Have students sketch what they think a molecule of sulfur trioxide would look like if the ratio of sulfur atoms to oxygen atoms were 1 to 3. **L3** **P** **ELL**

FIGURE 8-3
A ratio is a simple fraction that compares the amounts of two items.

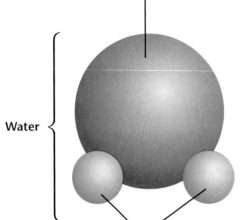

Oxygen atom

Water

Hydrogen atoms

Ratio: $\dfrac{2 \text{ Hydrogen atoms}}{1 \text{ Oxygen atom}}$

Rusting is another example of a chemical reaction. Rusting takes place between iron and the oxygen that is in the air. The iron gradually turns to rust. But, there is no change in the amount of matter. The total mass of the iron plus the oxygen equals the total mass of the rust when the reaction is complete. Observations like these changed the way people thought about matter. They realized that mass is never created or destroyed when things react with one another. This principle is called the law of conservation of matter.

Proust's Work

Another French chemist, Joseph Proust (PREWST), explored the "stuff" part of Democritus's space-and-stuff idea. He stated that the basic building blocks of any substance are always the same and are always put together in the same way. For example, a single unit, or molecule, of water contains two atoms of hydrogen and one atom of oxygen. If you counted all the atoms in a cup of water, there would always be a ratio (RAY shee oh) of two hydrogen atoms to every one oxygen atom. Look at **Figure 8-3.** Whether you have a drop of water or a whole ocean, the ratio of hydrogen atoms to oxygen atoms is always two to one.

The ideas of these and other scientists suggested that matter is made of small particles that behave in ways that can be predicted and measured. They began searching for a model to explain what an atom is made of and how it behaves.

Models of the Atom

To find out more about how an atom might look and act, scientists started to make models. There are many different ways to make models. Making a mental picture of something that you can't easily see is one way to make a model.

Across the Curriculum

Mathematics Ask students to make a list of ratios they encounter during a day. Their list might include speeds, such as 55 mi/h; parts of a pizza, such as 1/4; or the size of a container of milk, 1/2 gallon. **LS** **L3**

What is a model?

Another way to make a model is to make a small working version of something larger. For example, if you wanted to design a new kind of sailboat, would you just come up with a design, build a full-sized boat, and hope it would float? It would be smarter—and safer—to first build and test a small model of your design, as in **Figure 8-4.** Then, if it doesn't float, you can change your design, build another model, and keep trying until the model works. As you could tell from the model sailboat, models can be changed as new information is gained.

A model can be a paragraph, a math equation, a drawing, a computer program, or an actual structure, like the small sailboat. A model lets you create an image in your mind of something that is too big, too small, or too complicated to understand otherwise. For example, a model of the solar system can help you see the sizes of planets in relation to each other.

A model is also used when an actual event would occur too quickly or too slowly or is too dangerous to watch directly. For example, scientists use models to study what earthquakes and fires might do to a building. Weather maps are models that help predict weather and climate changes. Scientists sometimes use models to compare the results of experiments with their ideas of what they predicted would happen.

Mini LAB

Using a Model
Use a model to infer how atoms behave.

1. Fill a pie pan with warm, whole milk.
2. After the milk has stopped moving around, add one drop each of red, blue, and yellow food coloring. Make a triangle pattern with the dots, putting them about 2 cm apart.
3. In the middle of the triangle pattern, add one drop of liquid soap.

Analysis
1. What happened to the colors in the milk?
2. Use this model to describe how freely the atoms that make up the milk move.

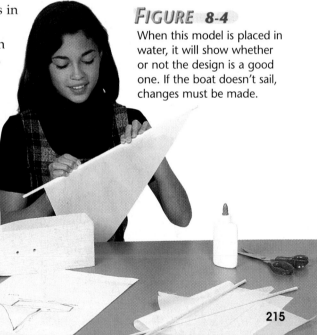

FIGURE 8-4

When this model is placed in water, it will show whether or not the design is a good one. If the boat doesn't sail, changes must be made.

215

Problem Solving

Comparing Gravel and Iron

The atoms that make up different materials determine what those materials are like and how they behave. Suppose that an artist who makes sculptures from pieces of metal lives on your block. You like to ride your bike to her house to watch her work. Because small pieces of iron may be hot as they fly off the metal she is grinding, she takes her grinder outside and uses it in the gravel driveway. But, there's a problem. You park your bike in the driveway and the small slivers of metal might give you a flat tire—or two. How can she get the small iron pieces out of the gravel?

Solve the Problem:
Think about what iron is like and how it behaves. What are some common features of iron?

Think Critically:
1. Why might a paper towel or old rag be helpful in separating the gravel from the iron? Where would you put it?
2. Explain why stirring the iron and gravel into water would not separate them.

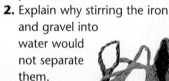

216

Dalton's Model of the Atom
In 1803, English chemist John Dalton introduced a model to describe matter and atoms. He called his model "the atomic theory of matter" and defined an atom much as scientists do today—as the smallest particle that makes up most types of matter. Dalton's model said that all of the atoms in something such as gold are identical. In other words, a chunk of gold is made up of gold atoms only, and each gold atom is like every other gold atom. Dalton's model also said that atoms of different elements are different from one another. For example, gold atoms are not the same as iron atoms.

Behavior Patterns
In his studies, Dalton also noticed that reactions between different chemicals always occur in patterns that can be predicted. For example, if 2 L of hydrogen gas react with 1 L of oxygen, it can be predicted that 20 L of hydrogen gas will react with 10 L of oxygen. This verified, or supported, what Joseph Proust had predicted. Dalton's model allowed scientists to understand the findings of both Lavoisier and Proust.

Community Connection

Manufacturing Recycling centers have to separate types of plastic. Foundries separate ore and waste products called slag. Ask someone from a local industry to explain how his or her company separates substances.

Inclusion Strategies

Learning Disabled Have students make a time line showing the development of the atomic theory. [L2]

What makes up an atom?

Dalton's ideas are close to the model scientists use today to describe matter and atoms. However, as scientists continued to study atoms, using better and better tools, they changed Dalton's model to make it more like what they observed. One of their conclusions is that atoms are mostly empty space. If you bump into a brick wall, you probably don't think that the atoms making up those bricks are mostly empty space. Can a model help you understand this?

Sizes of Atoms

Atoms are extremely small. It wasn't until the 1980s that improvements in electron microscopes let scientists see an atom, as in **Figure 8-5.** To give you a better idea of how small atoms are, look at **Figure 8-6.** Imagine you are holding an orange in your hand. If you wanted to use only your eyes to see the individual atoms on the surface of the orange, you would need to increase the size of the orange to the size of Earth. Then, cover it with billions and billions of marbles. Each marble would represent one of the atoms that make up the skin of the orange.

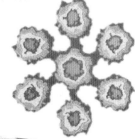

FIGURE 8-5

This photo shows six uranium atoms arranged around a central uranium atom. The atoms are enlarged about 20 million times.

FIGURE 8-6

An atom is too small to visualize without using a model, such as this one.

217

Activity

Kinesthetic Scientists discovered the structure of the atom by shooting alpha particles (helium nuclei) at atoms and observing what happened. Let several students represent a regular array of atoms in an extremely thin sheet of gold. Leave at least 2 m between students. Have other students, who represent test particles, walk in a straight line toward the array of students. If a test particle bumps into an atom, it bounces off to the side. If not, it continues straight through. Most should continue straight through, indicating that most of each atom in the array is space, with concentrations of mass. Ask students the following questions. **What results would you expect if you spread your arms?** *More test particles would be deflected.* **If you stood 10 m apart?** *Fewer test particles would be deflected.*

L2 **ELL** **COOP LEARN**

GLENCOE TECHNOLOGY

 Videodisc

The Infinite Voyage: Unseen Worlds
Chapter 2
Supercomputer Models: Photosynthesis, Space Exploration, Weather

---

Content Background

The electron microscope was developed because a light microscope cannot be used to see things that are close to the size of the wavelength of visible light. The effective wavelength of an electron is considerably smaller than that of visible light. An electron microscope directs a tightly focused beam of high-energy electrons back and forth across a tiny object. The high-energy electrons knock other electrons called secondary electrons out of the object being viewed. In a transmission electron microscope, the secondary electrons are spread out and hit a phosphorous plate, making a picture. In a scanning electron microscope, the electrons that are scattered off the front of the object make a picture.

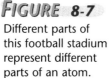
What makes up each of these individual atoms? To be able to peek inside, you would have to expand your model even more. Pick one of the marbles on the surface of your Earth-sized orange, and make it as big as a football stadium. This model of an atom, shown in **Figure 8-7**, is now large enough that you can imagine the center of the atom and what surrounds this center.

The Center of the Atom

Pretend you are standing on the edge of the football field. As you look toward the center of the football field, you see what looks like a tiny cluster of grapes. This cluster represents the center of the atom, which is called the nucleus (NEW klee us). Between you and the nucleus is nothing but empty space. The same relationship of empty space to stuff exists inside an atom.

The nucleus of an atom is made of two kinds of particles—protons (PROH tahns) and neutrons (NEW trahns). A **proton** is a particle with a positive charge. A **neutron** is a particle with no charge at all. Almost all of the mass of an atom is in its nucleus. It probably looks like a bunch of grapes, with the grapes representing the protons and neutrons snuggled up close to one another.

FIGURE 8-7

Different parts of this football stadium represent different parts of an atom.

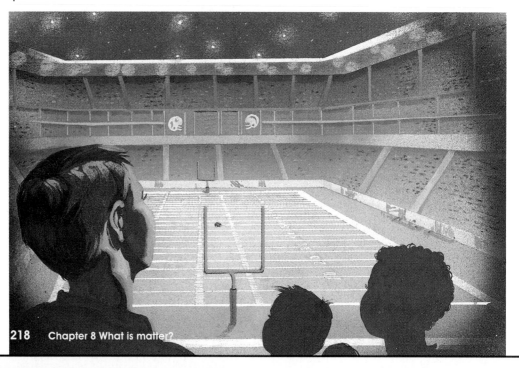

Theme Connection

Scale and Structure

The structure of the atom can be difficult to describe because the physics at such a small scale is hard to visualize. Models are used to emphasize certain important features of atoms.

FIGURE 8-8

This oxygen atom has eight protons and eight neutrons in its nucleus. Eight electrons, traveling at great speeds, form a cloud around the nucleus.

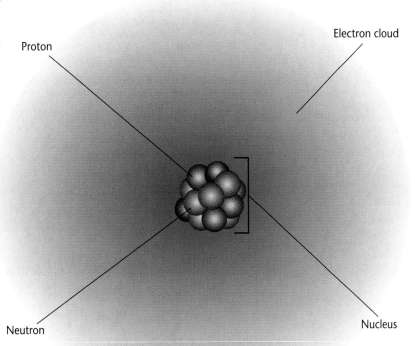

Proton

Electron cloud

Neutron

Nucleus

Around the Nucleus

You are still standing on the edge of the football field, just inside the atom. If you look up toward the top of the stadium, you'll see fireflies flying around. They stay inside the stadium, but they don't fly in a set pattern. These fireflies represent particles around the nucleus of the atom that are called electrons (ee LEK trahns).

An **electron** is a particle in an atom with a negative electric charge. An atom has equal numbers of electrons and protons, so the negative charge of the electrons equals the positive charge of the protons. Electrons, neutrons, and protons are similar in size, but a proton or neutron is about 1800 times heavier than an electron.

Examine a model of an atom, shown in **Figure 8-8.** Because they are located toward the outside of the atom, electrons sometimes are easily removed from the atom. You'll see signs that electrons are removed from atoms in the next activity.

8-1 Matter and Atoms   **219**

Enrichment

LS **Intrapersonal** Ask students to research the idea of electron orbits and electron clouds. Have students explain why the electron cloud model more accurately shows the location of electrons in an atom. **L3**

3 Assess

Check for Understanding

Brainstorming Ask students to describe other models they have encountered in or out of science class. Are there any similarities to or differences from the atomic model of matter? **L2**

Reteach

LS **Visual-Spatial** Use a double-pan balance to explain that atoms have no overall charge. Label one pan "+" and the other pan "–." Balance weights labeled "+" on one pan with an equal number of weights labeled "–" on the other pan. You must have the same amount of positive charge (protons) and negative charge (electrons) for an atom to be in balance. The number of neutrons does not affect this.

Extension

For students who have mastered this section, use the **Reinforcement** and **Enrichment** masters.

Activity 8-1

PREPARE

Purpose
LS **Kinesthetic** Students will give a balloon a static electric charge and determine what materials can be used to show this charge. **L2** **ELL** **COOP LEARN**

Process Skills
observing and inferring, predicting, comparing and contrasting, recognizing cause and effect, experimenting, formulating hypotheses, making and using tables

Time
45 minutes

Materials
Make sure to have plenty of rubber balloons on hand. The water needs to be falling from a tap or cup; a thin stream on a plate won't be affected. You might add a piece of silk to the materials. Silk will not work well to charge the balloon because it also tends to gain electrons.

Safety Precautions
Goggles must be worn; balloons may pop.

Possible Hypotheses
Some students will know that a balloon rubbed on hair or wool clothing becomes charged. Students might expect lint to become charged, like socks from the drier. Water will be affected because electricity can travel through water. Foil will be affected—a metal doorknob has shocked most students.

📁 **Activity Worksheets,** pages 5, 49-50

Possible Materials
- rubber balloon
- wool sweater
- hair
- metal
- wood
- pencil shavings
- small piece of aluminum foil
- foam packing peanuts
- thin stream of water
- lint

Design Your Own Experiment
The Electron Magnet

You can use a rubber balloon to remove electrons from the atoms of an object. After you have gathered enough electrons on the surface of the balloon, you can see their effects on other objects. An object is said to be charged when it has more or fewer electrons than usual. In this activity, you will find sources of electrons to charge the balloon and design a way to detect the presence of the electrons.

PREPARE

What You'll Investigate
What common materials are good sources of electrons? Which materials clearly show the effect of a large electron charge?

Form a Hypothesis
Make a hypothesis about which materials you think will give electrons easily to the balloon and which types will respond to the presence of a charge.

Goals
Design an experiment that allows you to charge a rubber balloon using different surfaces as a source of electrons.

Observe the effect of a charged rubber balloon on different kinds of objects.

220

Sample Data Table

Materials to be tested	Results
foam peanuts	attracted to balloon
wood	not attracted to balloon
stream of water	attracted to balloon
pencil shavings	attracted to balloon
metal	not attracted to balloon
aluminum foil	attracted to balloon
lint	attracted to balloon

Go Further
LS **Logical-Mathematical** Ask students to hypothesize why you can get a shock from a metal doorknob but not a wooden door. Metal conducts electricity; wood does not.

Safety Precautions

Wear goggles when using the balloons. Balloons occasionally burst.

PLAN

1. As a group, **list** the steps that you need to take to test your hypothesis. **Describe** exactly which surfaces and materials you will use to charge the balloon and which materials you will **test** once the balloon is charged.
2. Before you begin, make in your Science Journal a data table that allows room to list each material you are going to test and the results of the test.
3. **Read** over your entire experiment to make sure that all the steps are in logical order.
4. Will you **repeat** any portion of the experiment?

DO

1. Make sure that your teacher approves your plan and data table and that you have included any changes.
2. Carry out the experiment as planned and approved.
3. Be sure to **record** your observations in the data table as you complete each test in your plan.

1. **Compare** your findings with those of other groups.
2. **List** the materials that are good sources of electrons and those that are not. Write a general rule that will let you predict whether a material would be a good source of electrons.
3. Using your data table, **conclude** which materials are attracted to a supply of electrons.
4. **APPLY** Explain why you could be shocked when you touch a doorknob after you walk across a carpet on a cold day.

8-1 Matter and Atoms **221**

4. Walking across the carpet is like rubbing a balloon on a wool sweater—you pick up extra electrons. The electrons then jump from your body to the metal doorknob.

 Assessment

Process Rub a balloon on your hair or a student's (fine, straight, light hair will work best), and hold the balloon so the hair goes up from the head to touch the balloon. Ask students to explain what they observe. Use the Performance Task Assessment List for Making Observations and Inferences in **PASC**, p. 17.

Visual Learning

Figure 8-9 What do you see here that is made up of atoms? *Student answers might include anything pictured, such as the student, the book, or the milk.* IS

4 Close

•MINI•QUIZ•

Use the Mini Quiz to check students' recall of chapter content.

1. **What happens to the wood burned in a fire?** *It is converted with oxygen to ash, water, and carbon dioxide.*

2. **What particles are in the center of an atom?** *protons and neutrons*

3. **What type of charge does an electron have?** *negative*

Section Wrap-up

1. Student answers might include anything in the room, including air.

2. Drawings should include a nucleus containing two circles with plus signs and two circles with no markings. Two circles with minus signs are outside the nucleus.

3. **Think Critically** A globe is a small version of an object too large to observe directly.

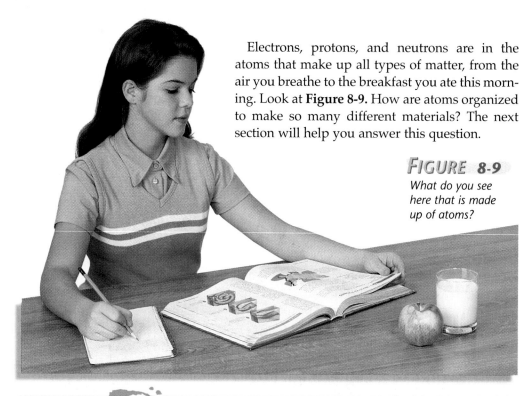

Electrons, protons, and neutrons are in the atoms that make up all types of matter, from the air you breathe to the breakfast you ate this morning. Look at **Figure 8-9.** How are atoms organized to make so many different materials? The next section will help you answer this question.

FIGURE 8-9
What do you see here that is made up of atoms?

Section Wrap-up

1. List ten things in your classroom that are made of matter.

2. An atom of the element helium contains two protons, two neutrons, and two electrons. Diagram a picture of a helium atom, using a circle to show each particle. Identify the protons with a plus sign (+), include the two neutrons with no markings, and identify the electrons with a minus sign (–).

3. **Think Critically:** Explain why a globe of Earth is a model.

4. **Skill Builder**
   Making Models Using the materials of your choice, create a model of a carbon atom. Include carbon's six protons, six neutrons, and six electrons. If you need help, refer to Making Models on page 557 in the **Skill Handbook.**

Using Computers

Spreadsheet Create a spreadsheet that will fill in the number of electrons in an atom when the number of protons and the number of neutrons are given.

Using Computers

A spreadsheet could have column A, number of protons, and column B, number of neutrons. Column C, number of electrons, would copy the number from Column A on that row. (Example: C2 = A2, C3 = A3, etc.)

Skill Builder
Making Models Student models should show a nucleus with six protons and six neutrons. Around the atom are six electrons. Models can be made in clay, with a plastic atom-model set, with markers and paper, with foam balls, or using other materials.

Assessment

Performance Give students the names of several elements and the number of protons and neutrons in each. Have students use modeling clay to make a model of an atom of one of these elements. Have them trade models with other students and identify what element is represented. Use the Performance Task Assessment List for Model in **PASC,** p. 51. IS L2 **ELL COOP LEARN**

Types of Matter 8•2

Elements

Most matter is made up of combinations of a basic group of building blocks. These building blocks are called elements. An **element** is matter made up of only one type of atom. For example, gold is an element made of only gold atoms, and iron is an element made of only iron atoms. At this time, 112 elements have been found.

Natural and Synthetic Elements

Ninety-one of the 112 elements can be found in nature. These elements make up gases in the air, ores in rocks, and liquids such as water. Examples of elements that you are probably familiar with include nitrogen and oxygen in the air you breathe and the metals gold, copper, aluminum, iron, and silver.

Scientists have also discovered another 21 elements that are known as *synthetic* (sihn THET ihk) elements. Synthetic elements are not found in nature. They are made in machines called particle accelerators (ak SEL uh ray turs), shown in **Figure 8-10.** You'll learn more about these elements and their uses in Section 8-3.

What YOU'LL LEARN

- Definitions and examples of elements, compounds, and mixtures
- Names, symbols, and formulas for elements and compounds

Science Words:
element
periodic table
compound
chemical formula
mixture

Why IT'S IMPORTANT

If you know what types of materials make up things around you, you will be able to use these materials better.

FIGURE 8-10

This particle accelerator is at Fermilab, which is near Chicago, Illinois.

Prepare

Section Background

After developing the periodic table, Mendeleev predicted the existence and properties of six undiscovered elements—scandium, gallium, germanium, technetium, rhenium, and polonium. The first three were found a few years later.

Preplanning

Refer to the Chapter Organizer on pages 210A-B.

1 Motivate

Bellringer

Before presenting the lesson, display **Section Focus Transparency 25** on the overhead projector. Assign the accompanying **Focus Activity** worksheet.

L2 ELL

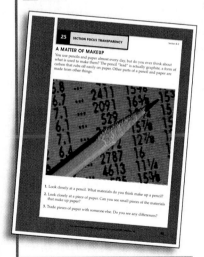

Tying to Previous Knowledge

Write H_2O on the board. Ask students if they have ever heard water called *H-two-O*. Ask them what they think it means. Students should be able to justify their answers.

Program Resources

Reproducible Masters

Activity Worksheets, pp. 5, 51-52, 54 L2
Cross-Curricular Integration, p. 12 L2
Enrichment, p. 31 L3
Lab Manual, pp. 57-58 L2
Multicultural Connections, pp. 19-20 L2
Reinforcement, p. 31 L2
Science Integration Activities, pp. 53-54 L2

Science and Society/Technology Integration, p. 26 L2
Study Guide, p. 31 L1

Transparencies

Science Integration Transparency 8 L2
Section Focus Transparency 25 L2
Teaching Transparency 16 L2

223

2 Teach

Inquiry Question

How many elements are in the top row of the periodic table? *2* The second row? *8*

Revealing Preconceptions

Students may think that an element or compound has only one form. Show them that forms can differ as a result of change of state or structure. For example, water, ice, and steam are all the same compound. Charcoal, graphite, and diamond are all composed of carbon.

224

Table 8-1

Some Elements and Their Symbols	
Name	**Symbol**
Aluminum	Al
Arsenic	As
Calcium	Ca
Chlorine	Cl
Gold	Au
Helium	He
Manganese	Mn
Neon	Ne
Nitrogen	N
Oxygen	O
Potassium	K
Silver	Ag
Sodium	Na
Tungsten	W
Uranium	U

The Periodic Table

With more than 100 natural and synthetic elements to keep track of, scientists need a system of organization. Scientists use a chart based on the work of Russian chemist Dimitri Mendeleev (men duh LAY uv). This chart is called the periodic (peer ee AW dihk) table of the elements. The **periodic table** is a chart that organizes elements by the number of protons in each element's nucleus. A periodic table is found inside the back cover of your textbook. Refer to it as you examine the information included on the table. In addition to listing the elements by their numbers of protons and their symbols, the periodic table provides other useful information. If you know how to use the table, you can find out the mass of an atom, the pattern of electrons, and whether an element is a solid, a liquid, or a gas at room temperature.

Chemical Symbols

Each element is identified on the chart with a one-, two-, or three-letter abbreviation. These symbols are a chemical shorthand that allows chemists to represent elements without writing out their full names. Some of the abbreviations are initials, but others are more complicated.

Sometimes the abbreviation is a letter or two from the name of the element. For example, the element hydrogen is abbreviated H, helium is He, lithium is Li, and sulfur is S. Other times, the abbreviations come from Latin words. For example, the symbol for iron is Fe, which comes from the Latin word for iron, *ferrum*. **Table 8-1** gives more examples of elements and their symbols.

Temporary Names

Why do some elements have three-letter symbols? The three-letter symbols are based on the number of protons in the element. These elements have been discovered so recently that they do not yet have permanent names. When scientists around the world agree on a name, the three-letter symbol is replaced by a one- or two-letter symbol representing the name of the element.

Content Background

- Lavoisier established many of the names and symbols used for elements. Consistent names for elements enabled scientists working in different places to understand each other's work. Establishing a common language is an important step in the development of a science.

- The first column of the periodic table has elements that have one electron free to form bonds, the second column has two free electrons, and so on. Column 17 has elements that have one space for an electron, and the last column has nonreactive elements, with electrons free to form bonds only under special circumstances.

FIGURE 8-11

Your favorite fruit contains sugar, which is made up of the elements carbon, hydrogen, and oxygen.

Compounds

A **compound** (KAHM pownd) is a form of matter that is made when two or more elements combine. Millions of compounds can be made from combinations of elements. For example, the elements oxygen and hydrogen combine to make the compound water. Another example is shown in **Figure 8-11.** Carbon and hydrogen also make up most compounds found in living things.

USING TECHNOLOGY

Goo to the Rescue

Why would you want a compound that holds water? Can you imagine any uses for such a compound? One recently developed family of compounds is called *dehydrated gels* (dee HI dray tud • jelz). They can capture and hold more than 100 times their volume in water and other liquids.

Uses of Gels

Small packages containing dehydrated gels are packed with camera and video equipment to absorb water that might get inside the packing case and ruin the equipment. Magicians use the gels to make water seem to disappear. If you ever helped care for a baby, you can probably think of another good use for dehydrated gels—disposable diapers!

*inter***NET**
CONNECTION

Silica gel is one of the dehydrated gels being used today. Visit the Glencoe Homepage at *www.glencoe.com/sec/science* for a link to a site that explains a use of this gel.

In areas with limited rainfall, gels are added to soil to increase the amount of water available to the soil. The gel holds onto water and releases it gradually into the soil, providing more useful water to crops. Can you think of uses for oil-absorbing gels?

8-2 Types of Matter **225**

USING TECHNOLOGY

- Oil-absorbing gels might be used to clean up oil spills.
- Using a disposable diaper or anti-humidity pack, experiment to find out how much water it can absorb without leaking.
- Students may not be familiar with the words *dehydrated* and *gel*. A substance is dehydrated if the water it normally contains has been removed. Food designed for camping trips is sometimes dehydrated. A gel is a semi-solid material, like jelly. Some shaving creams and hair-styling products are gels.

*inter***NET**
CONNECTION

The Glencoe Homepage at **www.glencoe.com/sec/science** provides links connecting concepts from the student edition to relevant, up-to-date Internet sites.

? FLEX Your Brain

Use the Flex Your Brain activity to have students explore COMPOUNDS.

📁 **Activity Worksheets,** page 5

Using an Analogy

LS **Linguistic** Students probably will be unfamiliar with finding the identity of an unknown material. To prepare them for Activity 8-2 on the next page, explain that finding the identity of an unknown is like solving a mystery in a book. From a list of possible suspects, the investigator devises ways to solve the mystery.

Theme Connection

Scale and Structure

Atoms are the building blocks of elements. Elements are the building blocks of compounds. Individual elements and compounds are the building blocks of mixtures.

Cultural Diversity

Attractive Compounds A natural attractant for insect pests is found in compounds derived from the neem tree, which is native to India and Burma. Compounds taken from its bitter seeds and leaves attract unwanted insects, providing nontoxic and long-lived substitutes for synthetic pesticides.

225

Activity 8-2

Purpose

LS **Kinesthetic** Students will perform tests and use their observations to determine the composition of an unknown sample. **L2** **ELL**
COOP LEARN

Process Skills

observing, classifying, inferring, comparing and contrasting, interpreting data

Time

45 minutes

Alternate Materials

Powdered sugar sold in grocery stores usually contains cornstarch. Use powdered sugar that contains no cornstarch or table sugar that has been ground into a powder.

Safety Precautions

Use caution when heating the samples. Use only aqueous iodine solution, not tincture of iodine, which is iodine dissolved in alcohol. Iodine will temporarily stain skin and permanently stain clothing.

Disposal

Collect in a permanently labeled container all samples that used iodine and any iodine solution that is not to be saved. Add a quart of water. Give the container and its contents to a high school chemistry teacher for treatment with sodium thiosulfate solution to reduce the iodine to iodide as required by EPA regulations before disposal.

📁 **Activity Worksheets,** pages 5, 51-52

Teaching Strategies

• If time is limited, have certain groups each do one of the three powder tests and share results.
• Have students label the compounds in the pie pan

The Mystery Mixture Investigation

Cornstarch, baking powder, and powdered sugar are compounds that look alike. To avoid mistaking one for another, you may need to identify each one. You can learn chemical tests that identify different compounds.

What You'll Investigate

Which compounds are present in a mixture that contains from one to three common compounds?

Procedure 🧤 🦺 🥽

1. **Copy** the data table into your Science Journal. **Record** your results for each of the following steps.
2. **Place** a small amount of cornstarch, sugar, and baking soda on the pie pan. **Add** a drop of vinegar to each.
3. **Place** another small amount of each compound on the pie pan. **Add** a drop of iodine solution to each.
4. **Place** a small amount of each compound in a test tube. Hold the test tube with the test-tube holder. Gently **heat** the bottom of each test tube with the candle.
5. Now, **test** your mystery mixture and find out which of these compounds are in the mixture.

Data and Observations

Sample Data for Mystery Mixture

To be tested	Vinegar fizzes	Iodine turns blue	Compound melts
Cornstarch	No	Yes	No
Sugar	No	No	Yes
Baking Soda	Yes	No	No
Mystery Mixture	Yes	No	Yes

Conclude and Apply

1. **Conclude** which compounds are in your mystery mixture. Describe how you arrived at your conclusion.
2. How would you be able to tell if all three compounds were absent from your sample?

Goals

• Test for the presence of certain compounds.
• Decide which of these compounds are present in an unknown mixture.

Materials

• test tubes (3)
• compound samples
• dropper bottles (2)
• iodine solution
• white vinegar
• candle
• test-tube holder
• small pie pan
• matches
• mystery mixture

so they know the identity of each compound.

Answers to Questions

1. Students' results should match the composition of the mystery mixture.
2. If the powder didn't fizz with vinegar, turn blue with iodine, or melt with heat, results would indicate that it didn't contain any of the three compounds.

✓ **Assessment**

Portfolio Have students write up their experiment with special care and explain how similar tests might be used by scientists studying soil samples. Use the Performance Task Assessment List for Lab Report in **PASC**, p. 47. **P** **LS**

A compound is usually much different from the elements that make it up. For example, table salt is very different from the gray sodium solid and the yellow-green chlorine (KLOR een) gas that it is made from. You may remember from Section 8-1 that a compound is always made of the same elements in the same ratio. Table salt is always made of one atom of sodium for each atom of chlorine. The chemical name for table salt is sodium chloride (KLOR ide).

It Takes a Formula

Just as scientists use symbols to represent elements, they write chemical formulas to represent compounds. A **chemical formula** is made up of symbols and numbers that tell what elements are in a compound and what their ratios are. How would you write a chemical formula for table salt? Remember that it contains sodium (whose symbol is Na) and chlorine (whose symbol is Cl) in a one-to-one ratio. That makes the chemical formula for table salt NaCl. Now try another familiar compound, water. Water is made of two atoms of hydrogen (H) and one atom of oxygen (O). Chemists write the formula for water as H_2O. The little 2 after the H is called a subscript. It tells you that there are two hydrogen atoms for every one oxygen atom. No subscript is used when only one atom is present. **Figure 8-12** shows the chemical formulas for some common compounds.

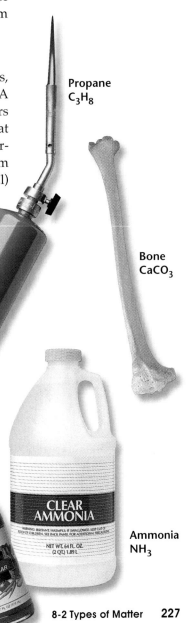

FIGURE 8-12

These common compounds have chemical formulas. *How many of each type of atom is in each formula?*

Propane
C_3H_8

Bone
$CaCO_3$

Vinegar
$HC_2H_3O_2$

CLEAR AMMONIA

Ammonia
NH_3

Battery acid
H_2SO_4

8-2 Types of Matter 227

Mini LAB

Observing

Record your observations as you separate a mixture of two compounds.

1. Pour some warm water into a large drinking glass.

2. Add a mixture of salt and sand to the glass, and stir with a spoon until it appears that all the salt has dissolved. Let the sand settle.

3. Pour about 5 mL of the salt water onto a clean, dry plate. Set the plate in a warm place, and let the water evaporate.

4. Pour the sand onto a paper towel to dry.

Analysis

1. What was it about these three compounds that allowed you to separate them?

2. Explain why this method would not work for a mixture of water, salt, and sugar.

FIGURE 8-13

This brass instrument is made of a mixture that is the same throughout.

Mixtures

If two or more substances are put together but do not combine to make a compound, the result is called a **mixture.** Air is a mixture of nitrogen, oxygen, and other gases. A rock is a mixture of different minerals. What other mixtures do you see around you?

There are two kinds of mixtures. One type of mixture is the same throughout, such as air or a soft drink. You can't see the different parts in this type of mixture. An example of this type of mixture is shown in **Figure 8-13.** No matter how closely you look, you can't see the individual parts of copper and zinc that make up brass.

In the other kind of mixture, the material is made of several different items that you can see. The granite rock by the side of the river on the first page of the chapter probably contained pieces of the minerals mica, quartz, and feldspar. If you look closely at

228   Chapter 8 What is matter?

the rock, you can see the different parts. Examine another example by looking at the salad in **Figure 8-14.** Other mixtures in this category would include things like pizza, stew, a toy box full of toys, or your laundry basket at the end of the week.

FIGURE 8-14

It's easy to see that this mixture is made of different parts. A salad taken from one side of the bowl will differ from a salad taken from a different part of the bowl.

Section Wrap-up

1. Using the periodic table of elements on the inside back cover of this book for reference, write the names of each of these elements: Be, Al, P, Ar, K, Ca, V, Cu, Pb, and Au.

2. The chemical formula for table sugar is $C_{12}H_{22}O_{11}$. What does the formula tell about the makeup of this compound?

3. **Think Critically:** Explain two ways that you could separate a mixture of iron filings and table salt.

4. **Skill Builder**
   Using a Computerized Card Catalog
   Use a computerized card catalog to investigate the names of the following elements: tungsten, curium, oxygen, and another element of your choice. Find out why each element was given its name. If the symbol is not from this name, find out the origin of the symbol. If you need help, refer to Using a Computerized Card Catalog on page 562 of the **Technology Skill Handbook.**

USING MATH

Using the periodic table inside the back cover for reference, make a line graph of the number of protons in an atom versus the mass of an atom for the first 30 elements. Write down two inferences that can be supported with your graph.

Skill Builder

Using a Computerized Card Catalog *Tungsten* is Old Norse for "heavy stone"; *W* is from "wolfram," another name for tungsten. *Curium* is named for Pierre and Marie Curie. *Oxygen* is from *oxys,* the Greek word meaning "sharp" and *genes,* the Greek word for "forming." Student answers for the other elements will vary.

Assessment

Process Provide students with an entry from the periodic table. Have them round off the atomic mass values to the nearest whole number. Ask them the number of protons, neutrons, and electrons for their element. Use the Performance Task Assessment List for Using Math in Science in **PASC**, p. 29. L2

---

4 Close

•MINI•QUIZ•

Use the Mini Quiz to check students' recall of chapter content.

1. **An element is made up of how many types of atoms?** *one*

2. **The periodic table is a chart that arranges _____ in order.** *elements*

3. **A chocolate cake is an example of a(n) _____.** *mixture*

Section Wrap-up

1. beryllium, aluminum, phosphorus, argon, potassium, calcium, vanadium, copper, lead, gold

2. It contains 12 carbon atoms, 22 hydrogen atoms, and 11 oxygen atoms in one unit of the compound.

3. **Think Critically** Use a magnet to attract the iron filings. Pour the mixture into water to dissolve the salt, remove the iron filings, then evaporate the water to leave the salt.

USING MATH

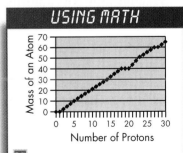

Number of Protons

Logical-Mathematical Inferences might include that as the number of protons increases, the atomic mass increases, and that the atomic mass is roughly twice the number of protons.

Teaching the Content

• Ask students if they know anyone who has had medical tests that involved radioactive tracers.

• Bring in a portable smoke detector from home. Open the unit and point out the location of the americium.

Teaching the Skill

• To confirm organization, have students try rewriting their notes as a flow chart or concept map.

• Have students take notes on a short science video, then explain how this is different from and similar to taking notes from an article.

Science & Society

8●3 Synthetic Elements

What YOU'LL LEARN

• That some elements do not occur naturally on Earth

• That taking notes provides a summary of important information

Why IT'S IMPORTANT

Taking notes on material that is unfamiliar will help you better understand what you read.

FIGURE 8-15

Americium (am uh REE shee um) is a synthetic element used in some home smoke detectors.

A Smashing Success

In this chapter, you've learned that an element is matter that is made of the same kind of atoms. Copper, hydrogen, sodium, iodine, and technetium (tek NEESH um) are among the 112 elements. Copper, hydrogen, sodium, and iodine are found in nature. However, technetium and 21 other elements are synthetic elements. A synthetic element is an element that is made in a machine called a particle accelerator, which was shown in **Figure 8-10**. Except for technetium and promethium, all of the synthetic elements have more than 92 protons.

The particle accelerators used to create synthetic elements speed up atoms and cause them to smash into, or collide with, other particles of matter at extremely high speeds. The tremendous force of these collisions causes atoms to join, making new elements.

Falling Apart at the Seams

Synthetic elements are unstable and break down, or decay, into other elements. Some decay very fast—in millionths of a second. Others stick around for as little as a few days or as long as millions of years. Most synthetic elements can be made only in small amounts. It may seem like the great expense of making synthetic elements is not worth the effort, but some of them have important uses, as in **Figure 8-15**.

Technetium, for example, is a synthetic element that decays about every six and a half hours. This time period is short enough to cause no damage to human tissue but long enough to be tracked and detected in the human body. This element can be used to diagnose disorders of many organs of the human body—the thyroid, brain, lungs, kidneys,

---

Program Resources

Reproducible Masters
Enrichment, p. 32 [L3]
Reinforcement, p. 32 [L2]
Study Guide, p. 32 [L1]

Transparencies
Section Focus Transparency 26 [L2]

FIGURE 8-16

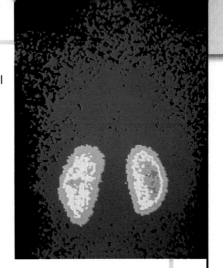

The person in this picture has been given a small amount of technetium, which gathered in the kidneys. Certain medical machines detect this element. A doctor can look at this picture and tell whether or not this person has a kidney problem.

heart, and liver, to name just a few, as shown in **Figure 8-16.** When the synthetic element neptunium decays, it forms another synthetic element named plutonium (plew TOH nee um). Some forms of plutonium are used in nuclear reactors and bombs. Small amounts of another form of this element power some pacemaker batteries.

When reading about topics such as synthetic elements, it is helpful to take notes on what you read. These notes will help you organize your thoughts on the topic.

Skill Builder: *Taking Notes*

LEARNING the SKILL

1. Read the entire text you wish to take notes on.

2. Identify the main topic in each paragraph. Write down each main idea in your own words. Leave room after each one.

3. After each main idea, jot down words and phrases that support the main idea. You may use letters or numbers to help organize your notes.

4. Read your notes aloud, using the words and phrases to make complete sentences. This reading checks your understanding of the topic.

5. If your notes don't make sense or seem incomplete, reread the passage and make any necessary changes in your notes.

PRACTICING the SKILL

1. In your own words, state the main topic of each paragraph on these two pages. List supporting details for each main idea.

2. List at least five words or phrases that describe synthetic elements.

3. Use words and phrases to state some of the advantages and disadvantages of making elements.

APPLYING the SKILL

Recall what you've learned in this chapter about elements. Use note-taking skills to briefly compare and contrast naturally occurring elements and synthetic elements.

231

Biography

Maya Lin grew up in the rural Midwest, the child of Chinese immigrant parents. As a 20-year-old student at Yale in 1980, she submitted the winning entry for the design of the Vietnam Veterans' Memorial. After completing the memorial, she returned to Yale to complete her Masters of Architecture degree.

Teaching Strategies

- **LS** **Logical-Mathematical** Ask students how softer rocks, such as limestone, might be used to convey an idea in a sculpture or a memorial.

- Ask students the following questions. **What ideas might be represented well by using steel? By using cedar? What do designers need to consider besides appearance in choosing a material for a sculpture or a memorial?**

Other Works

- The Vietnam Veterans' Memorial in Washington, D.C.
- The Peace Chapel at Juniata College in Pennsylvania

*inter*NET CONNECTION

The Glencoe Homepage at **www.glencoe.com/sec/ science** has a link with more information about architecture.

*inter*NET CONNECTION

Find out more of the story of the Civil Rights Memorial at the Glencoe Homepage, *www.glencoe.com/sec/ science*

Maya Lin's Civil Rights Memorial

Chinese-American architect Maya Lin became famous when she designed the Vietnam Veterans' Memorial in Washington, DC. Since then, she has created other monuments. One of those is the Civil Rights Memorial in Montgomery, Alabama.

This memorial honors 40 people—including Dr. Martin Luther King, Jr.—who died during the Civil Rights Movement, which started during the mid 1950s. Maya Lin was inspired by Dr. King's famous 1963 speech called "I Have a Dream." In it, King quoted: ". . . until justice rolls down like water and righteousness like a mighty stream." Lin used this quotation in the memorial.

The Right Materials

Lin used flowing water and dark granite in the memorial. The black granite wall stands in back, inscribed with the quotation, with water flowing over it. The granite, which is a mixture of the minerals feldspar, quartz, and mica, is a rock that formed when melted rock cooled. Granite is hard because the molecules that form it are held tightly together. It won't be worn down by the water that rushes over it.

In front is a round granite table on which are carved the names of those who died in the movement for racial equality. Water, which is a compound, gently flows off the edges all around the tabletop. "The water remains very still until you touch it," Lin has said. "Your hand causes ripples, which transform and alter the piece, just as reading the words completes the piece."

R ead the statements below that review the most important ideas in the chapter. Using what you have learned, answer each question in your Science Journal.

1. Matter is anything that has mass and takes up space. The basic unit of matter is the atom. The modern atomic theory was developed by many different scientists over hundreds of years. *What was the main idea of Dalton's atomic theory?*

2. Atoms are made up of smaller particles. The two heavier particles are found in the nucleus. The lighter particles are found in the space surrounding the nucleus. *What are the three particles that make up an atom? What is the charge of each?*

3. There are several types of matter. Most elements occur naturally, but some are synthetic. A compound is made up of more than one type of element. *Describe each of the two types of mixtures.*

233

Using Key Science Words

1. neutron
2. compound
3. chemical formula
4. matter
5. electron

Checking Concepts

6. d
7. c
8. d
9. b
10. a

Thinking Critically

11. Oxygen can be present in compounds with many other elements. The compounds can be solids or liquids.
12. No, it means it hasn't been mixed with other types of juice.
13. It is a model, which is a representation of the real thing.
14. 9; 1 Ca, 2 N, 6 O
15. It is a compound. The aluminum combined with the oxygen to make the solid. The solid is different from both the metal and the oxygen.

Using Key Science Words

atom	matter
chemical formula	mixture
compound	neutron
electron	periodic table
element	proton

Match each phrase with the correct term from the list of Key Science Words.

1. particle in the nucleus of the atom that does not carry a charge
2. made by combining two or more different kinds of elements
3. an abbreviation used by chemists to identify compounds quickly
4. anything that has mass and occupies space
5. the lightest of the three main particles found in an atom

Checking Concepts

Choose the word or phrase that completes the sentence.

6. A mixture can contain which of the following kinds of matter?
   a. elements   c. compounds
   b. atoms   d. all of the above

7. All known elements are organized on the periodic table _____.
   a. by the date they were discovered
   b. in alphabetical order
   c. according to the number of protons in an atom of the element
   d. as solids, liquids, or gases

8. An atom is composed of _____.
   a. electrons, nuclei, and protons
   b. compounds, mixtures, and nuclei
   c. protons, compounds, and neutrons
   d. protons, neutrons, and electrons

9. A chemical formula is composed of _____.
   a. a variety of atoms and their weights
   b. chemical symbols and numbers of atoms
   c. various compounds and mixtures
   d. protons, electrons, and neutrons

10. The particle that is negatively charged in an atom is the _____.
    a. electron   c. neutron
    b. nucleus   d. proton

Thinking Critically

Answer the following questions in your Science Journal using complete sentences.

11. As an element, oxygen is a gas. Explain how oxygen can be the most common element in Earth's crust, yet Earth's crust is not a gas.

12. When apple juice is advertised as being *pure*, does that mean that apple juice is an element? Explain.

13. Is the art shown here a molecule of propane or a model of a propane molecule? Explain.

Assessment Resources

Reproducible Masters
Chapter Review, pp. 19-20
Assessment, pp. 35-38
Performance Assessment, p. 46

Glencoe Technology
Computer Test Bank
MindJogger Videoquiz

14. Parentheses show that a group of atoms is a unit. How many atoms are in calcium nitrate, Ca(NO₃)₂?

15. Oxygen from the air comes in contact with the metal in an aluminum chair. The result is a white powder coating the aluminum. Is this powder an element, a compound, or a mixture? How do you know?

Developing Skills

If you need help, refer to the description of each skill in the Skill Handbook.

16. Interpreting Data: Fritz had a sample of sand from the beach. He added some of it to water and stirred, then dried it. It weighed the same as it did originally. He then heated the sample and weighed it again. It was still the same. Finally, he ran a magnet around in the sand and again weighed the sand. This time, its mass was less than the original mass. What was removed from the sand? Explain.

17. Making and Using Tables: The name of the element gold starts with the letter *G*, but the symbol is Au. Using the periodic table of the elements, make a table listing five other elements whose symbols don't start with their first letter and give their symbols.

18. Comparing and Contrasting: Compare and contrast the compound sugar with the mixture chicken soup.

19. Observing and Inferring: Infer the effect of a negatively charged balloon, loaded with electrons, on a pile of hair that you brushed off your dog. Explain.

20. Interpreting Scientific Illustrations: Using the pictures of the compounds below, write the chemical formula for each one.

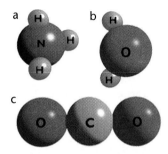

Performance Assessment

1. Display: Create a display of increasingly more complex kinds of matter. You may use models. Show the parts of an atom, a complete atom, a compound, and a mixture. Label the appropriate parts of each.

2. Designing an Experiment: You are able to separate a mixture of sand and sugar and then a mixture of iron filings and gravel. Design an experiment that allows you to separate a mixture of iron filings, gravel, and sugar.

2. Student answers might include the following. Use a magnet to remove the iron filings. Add water to dissolve the sugar, then evaporate the water. If students actually perform the experiment, be sure their procedure is approved beforehand. Use the Performance Task Assessment List for Designing an Experiment in **PASC**, p. 23. P

Developing Skills

16. Interpreting Data Small pieces of iron, which were removed by the magnet, were in the sand.

17. Making and Using Tables Tables might include sodium (Na), potassium (K), iron (Fe), silver (Ag), and antimony (Sb).

18. Comparing and Contrasting The compound sugar contains identical grains that can be broken down into identical molecules. Chicken soup has parts mixed together but still distinctly separate. A piece of chicken from the soup cannot be broken down into smaller chicken soup particles.

19. Observing and Inferring Excess electrons attract dog hair to the balloon because the balloon is negatively charged and the hair is not negatively charged.

20. Interpreting Scientific Illustrations Because of students' inexperience with formulas, accept the elements in any order.

a. NH₃

b. H₂O

c. CO₂

Performance Assessment

1. A complete atom includes protons, neutrons, and electrons. A compound is two or more kinds of atoms combined. A mixture has different kinds of atoms and/or molecules mixed together but not combined. Use the Performance Task Assessment List for Display in **PASC**, p. 63. P

Section	Objectives/Standards	Activities/Features
Chapter Opener		Explore Activity: Compare Characteristics, p. 237
9-1 **Physical and Chemical Properties** (4½ sessions, 2 blocks)*	1. **Define** and **give examples** of physical and chemical properties of matter. 2. **Compare and contrast** the properties of acids and bases. National Science Content Standards: (5-8) UCP3, A1, A2, B1, E2, F1, F3, F4, F5, G1, G2	MiniLAB: Compare Densities, p. 239 Using Math, p. 240 Using Technology: Properties and Pools, p. 244 Activity 9-1: Homemade pH Scale, pp. 246-247 Skill Builder: Classifying, p. 248 Science Journal, p. 248 Science & History: The *Hindenburg* Disaster, p. 249
9-2 **Science and Society:** **Road Salt** (1 session, ½ block)*	3. **Describe** how road salt can make travel safer but damage the environment. 4. **Identify** the main ideas in scientific writings. National Science Content Standards: (5-8) B1, C4, E2, F2, F4, F5	Skill Builder: Identifying the Main Idea, p. 251
9-3 **Physical and Chemical Changes** (3½ sessions, 2 blocks)*	5. **Define** and **identify** physical and chemical changes. 6. **Discuss** how physical and chemical changes affect the world we live in. National Science Content Standards: (5-8) UCP2, UCP3, A1, B1, B2, D1, F2	MiniLAB: Compare Chemical Changes, p. 255 Activity 9-2: Sunset in a Bag, p. 257 Skill Builder: Developing Multimedia Presentations, p. 258 Using Math, p. 258

* A complete Planning Guide that includes block scheduling is provided on pages 31T-33T.

Activity Materials

Explore	Activities	MiniLABs
page 237 samples of obsidian and pumice, container, water	pages 246-247 pH paper, pH color chart, distilled water, fruit juices, vinegar, salt, sugar, soft drinks, household cleaners, soaps, detergents, antacids page 257 baking soda, calcium chloride, phenol red solution, measuring spoons, resealable plastic bag, graduated cylinder	page 239 food coloring; corn syrup; clear, colorless cup; rubbing alcohol, graduated cylinder page 255 steel wool, salt, paper plate, cup

Need Materials? Call Science Kit (1-800-828-7777).

Teacher Classroom Resources

Reproducible Masters	Transparencies	Teaching Resources
Activity Worksheets, pp. 5, 55-56, 59 **Cross-Curricular Integration**, p. 13 **Enrichment**, p. 33 **Lab Manual 17** **Reinforcement**, p. 33 **Study Guide**, p. 33	**Section Focus Transparency 27**, How alike are they? **Teaching Transparency 17**, Calculating Density **Teaching Transparency 18**, The pH Scale	**Spanish Resources** **English/Spanish Audiocassettes** **Cooperative Learning Resource Guide** **Lab Partner** **Lab and Safety Skills** **Lesson Plans**
Enrichment, p. 34 **Lab Manual 18** **Reinforcement**, p. 34 **Study Guide**, p. 34	**Section Focus Transparency 28**, Salt of the Earth	**Assessment Resources** **Chapter Review**, pp. 21-22 **Assessment**, pp. 39-42 **Performance Assessment**, p. 47 **Performance Assessment in the Science Classroom (PASC)** **MindJogger Videoquiz** **Alternate Assessment in the Science Classroom**
Activity Worksheets, pp. 5, 57-58, 60 **Enrichment**, p. 35 **Multicultural Connections**, pp. 21-22 **Reinforcement**, p. 35 **Science and Society/Technology Integration**, p. 27 **Science Integration Activities**, pp. 55-56 **Study Guide**, p. 35	**Science Integration Transparency 9**, What happens in a forest fire? **Section Focus Transparency 29**, Chomp Change	**Computer Test Bank**

Key to Teaching Strategies

The following designations will help you decide which activities are appropriate for your students.

L1 Level 1 activities should be appropriate for students with learning difficulties.

L2 Level 2 activities should be within the ability range of all students.

L3 Level 3 activities are designed for above-average students.

ELL ELL activities should be within the ability range of English Language Learners.

LS These activities are designed to address different learning styles.

COOP LEARN Cooperative Learning activities are designed for small group work.

P These strategies represent student products that can be placed into a best-work portfolio.

GLENCOE TECHNOLOGY

The following multimedia resources are available from Glencoe.

Science and Technology Videodisc Series (STVS)
Plants and Simple Organisms
Salt-Resistant Crops

The Infinite Voyage Series
Living with Disaster

Glencoe Physical Science Interactive Videodisc
Gases Expand to Fill Container

Gases Exert Pressure
Carbonic Acid
Limestone
Cave Formation

Glencoe Physical Science CD-ROM

Teacher Classroom Resources

This is a representation of key blackline masters available in the Teacher Classroom Resources.

Teaching Aids

Section Focus Transparencies

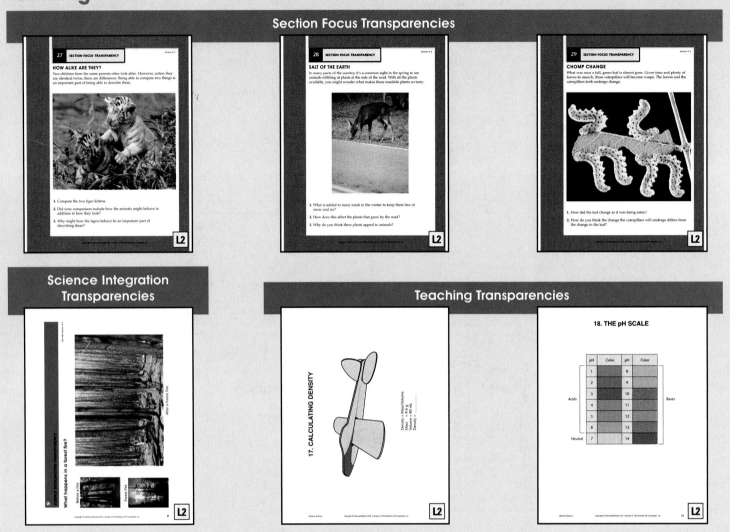

27 SECTION FOCUS TRANSPARENCY Section 9-1

HOW ALIKE ARE THEY?
Two children from the same parents often look alike. However, unless they are identical twins, there are differences. Being able to compare two things is an important part of being able to describe them.

1. Compare the two tiger kittens.
2. Did your comparison include how the animals might behave in addition to how they look?
3. Why might how the tigers behave be an important part of describing them?

L2

28 SECTION FOCUS TRANSPARENCY Section 9-2

SALT OF THE EARTH
In many parts of the country, it's a common sight in the spring to see animals nibbling at plants at the side of the road. With all the plants available, you might wonder what makes these roadside plants so tasty.

1. What is added to many roads in the winter to keep them free of snow and ice?
2. How does this affect the plants that grow by the road?
3. Why do you think these plants appeal to animals?

L2

29 SECTION FOCUS TRANSPARENCY Section 9-3

CHOMP CHANGE
What was once a full, green leaf is almost gone. Given time and plenty of leaves to munch, these caterpillars will become wasps. The leaves and the caterpillars both undergo change.

1. How did the leaf change as it was being eaten?
2. How do you think the change the caterpillars will undergo differs from the change in the leaf?

L2

Science Integration Transparencies

What happens in a forest fire?
After a Forest Fire
Before a Fire
Forest Fire

L2

Teaching Transparencies

17. CALCULATING DENSITY
Density = Mass/Volume
Mass = 9.6 g
Volume = 80 mL
Density =

L2

18. THE pH SCALE

pH	Color	pH	Color
1		8	
2		9	
3		10	
4		11	
5		12	
6		13	
7		14	

Acids   Bases

Neutral

L2

Meeting Different Ability Levels

Study Guide for Content Mastery

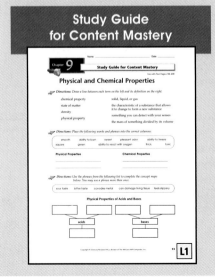

Name _____ Date _____

Chapter 9 **Study Guide for Content Mastery**

Physical and Chemical Properties

Directions: Draw a line between each term on the left and its definition on the right.

chemical property — solid, liquid, or gas
state of matter — the characteristic of a substance that allows it to change to form a new substance
density — something you can detect with your senses
physical property — the mass of something divided by its volume

Directions: Place the following words and phrases into the correct columns.

smooth | ability to burn | sweet | pleasant odor | ability to freeze
square | green | ability to react with oxygen | thick | toxic

Physical Properties   **Chemical Properties**

Directions: Use the phrases from the following list to complete the concept maps below. You may use a phrase more than once.

sour taste | bitter taste | corrodes metal | can damage living tissue | feels slippery

Physical Properties of Acids and Bases

acids   bases

L1

Reinforcement

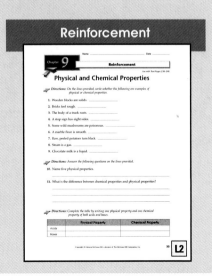

Name _____ Date _____

Chapter 9 **Reinforcement** Use with Text Pages 238-248

Physical and Chemical Properties

Directions: On the lines provided, write whether the following are examples of physical or chemical properties.

1. Wooden blocks are solids. _____
2. Bricks feel rough. _____
3. The body of a truck rusts. _____
4. A stop sign has eight sides. _____
5. Some wild mushrooms are poisonous. _____
6. A marble floor is smooth. _____
7. Raw, peeled potatoes turn black. _____
8. Steam is a gas. _____
9. Chocolate milk is a liquid. _____

Directions: Answer the following questions on the lines provided.

10. Name five physical properties.

11. What is the difference between chemical properties and physical properties?

Directions: Complete the table by writing one physical property and one chemical property of both acids and bases.

	Physical Property	Chemical Property
Acids		
Bases		

L2

Enrichment Worksheets

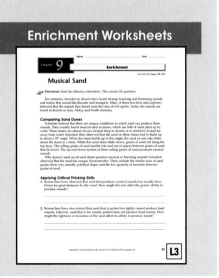

Name _____ Date _____

Chapter 9 **Enrichment** Use with Text Pages 238-248

Musical Sand

Directions: Read the following information. Then answer the questions.

For centuries, travelers in deserts have heard strange booming and humming sounds that sound like thunder and trumpets. Many of these travelers and explorers believed that the sounds they heard were the cries of evil spirits. Today, the sounds are heard in deserts in Asia, Africa, and North America.

Comparing Sand Dunes
Scientists noticed that there are unique conditions in which sand can produce these sounds. They usually found musical sand in dunes, which are hills of sand piled up by wind. These dunes are almost always located deep in deserts or in stretches of sand far away from water. Scientists then observed that the sand on these dunes had to build up to about a 35° angle. When the dune builds up to this angle, the sand on one side slides down the dune in a sheet. While this sand sheet slides down, grains of sand roll along the top layer. The rolling grains of sand tumble into and out of spaces between grains of sand that lie below. The up-and-down motion of these rolling grains of sand produce musical sounds.
Why doesn't sand on all sand dunes produce musical or booming sounds? Scientists observed that the sand has unique characteristics. These include the similar sizes of sand grains; their very smooth, polished shape; and the low quantity of moisture between grains of sand.

Applying Critical Thinking Skills
1. Researchers have observed that sand that produces musical sounds has usually been blown for great distances by the wind. How might this fact affect the grains' ability to produce sounds?

2. Researchers have also noticed that sand that is packed too tightly cannot produce loud sounds. Likewise, sand that is too loosely packed does not produce loud sounds. How might the tightness or looseness of the sand affect its ability to produce sound?

L3

Chapter 9 Properties and Changes

Hands-On Activities

Science Integration Activities

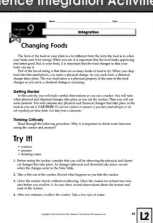

Lab Manual

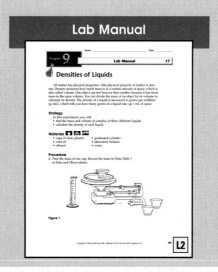

Activity Worksheets

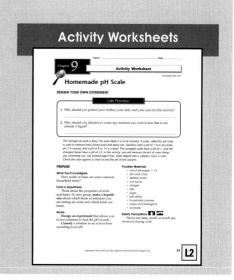

Enrichment and Application

Cross-Curricular Integration

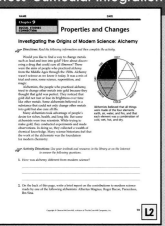

Multicultural Connections

Science and Society/Technology Integration

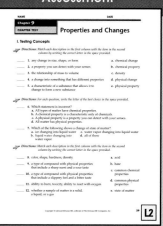

Assessment

Performance Assessment

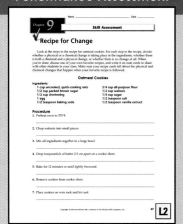

Chapter Review

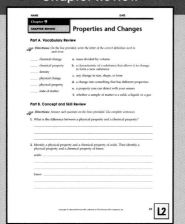

Assessment

Properties and Changes

CHAPTER OVERVIEW

Section 9-1 Physical and chemical properties are defined and applied in this section.

Section 9-2 Science and Society This section discusses the benefits and risks of road salt. Students are given practice in identifying the main idea of a selection.

Section 9-3 Physical and chemical changes are defined and explored. Weathering is used as an example.

Chapter Vocabulary

physical property
density
state of matter
chemical property
physical change
chemical change

Theme Connection

Stability and Change Students will learn to describe and classify matter using physical and chemical properties and changes. Several properties and changes involving acids and bases, road salt, and erosion are used to illustrate the concepts.

Chapter Preview

Section 9-1   Physical and Chemical Properties

Section 9-2   Science and Society: Road Salt

Section 9-3   Physical and Chemical Changes

236

Skills Preview

▶ **Skill Builders**
- classify
- develop multimedia presentations

▶ **MiniLABs**
- observe
- compare

▶ **Activities**
- design an experiment
- classify
- hypothesize
- observe
- compare
- conclude

Learning Styles	Look for the following logo for strategies that emphasize different learning modalities.
Kinesthetic	Explore, p. 237; MiniLAB, pp. 239, 255; Enrichment, p. 245; Activity 9-1, p. 246; Teaching the Content, p. 250
Visual-Spatial	Visual Learning, pp. 239, 241, 243, 244, 253; Inclusion Strategies, p. 243; Activity 9-2, p. 257
Interpersonal	Activity, p. 241; Community Connection, p. 245; Assessment, p. 258
Intrapersonal	Go Further, p. 247; Science at Home, p. 259
Logical-Mathematical	Inquiry Question, p. 240; Using an Analogy, p. 242; Check for Understanding, p. 245; Across the Curriculum, p. 254; Using Math, p. 258
Linguistic	Using Science Words, p. 242; Assessment, p. 257

9

Properties and Changes

Y ou've just received your favorite magazine. You open it up and see a picture of a place where volcanic ash once thundered down the sides of exploding volcanoes. Sounds as if you're looking at a Hollywood movie set. No, it's the Cascade Mountains of Oregon and Washington. What else could you see there? Cinder cones formed by bubbly lava. Lava tubes once filled with oozing melted rock, or magma. Before you look any further, try the activity below to compare some of the kinds of rocks you'll see.

EXPLORE ACTIVITY

Compare Characteristics

1. Obtain samples of the volcanic rocks obsidian (ahb SIH dee un) and pumice (PUH mus). The samples should be about the same size.
2. Which sample is heavier?
3. Compare the colors of the two rocks.
4. Look at the surfaces of the two rocks. How are the surfaces different?
5. Place each rock in water and observe whether or not it floats.

Science Journal

In your Science Journal, make a table that compares the observations you made about the rocks.

Obsidian

Pumice

237

EXPLORE ACTIVITY

Purpose

LS **Kinesthetic** Use the Explore activity to introduce the physical properties of matter. **L2** **ELL**

Materials

obsidian, pumice, water, container

Alternate Materials

If rock samples are not available, have students use the photos of them to answer the questions.

Teaching Strategies

Bubbles in the pumice should be obvious to students, and they can use this information to infer that the pumice is lighter. You may have to tell students that pumice floats.

Troubleshooting Advise students to be cautious not to crumble the pumice when handling it.

Safety Precautions Caution students to watch out for sharp edges on the obsidian.

Science Journal **Sample Table**

	Pumice	Obsidian
Color	brown	black
Floats	yes	no
Weight	light	heavier
Surface	rough	smooth

✓Assessment

Performance Have students draw a concept map that summarizes their observations about pumice and obsidian. Use the Performance Task Assessment List for Concept Map in **PASC**, p. 89. **P**

Assessment Planner

Portfolio
Refer to page 259 for suggested items that students might select for their portfolios.

Performance Assessment
See page 259 for additional Performance Assessment options.
Skill Builders, pp. 248, 251, 258
MiniLABs, pp. 239, 255
Activities 9-1, pp. 246-247; 9-2, p. 257

Content Assessment
Section Wrap-ups, pp. 248, 258
Chapter Review, pp. 259-261
Mini Quizzes, pp. 248, 258

Group Assessment
Opportunities for group assessment occur with Cooperative Learning Strategies and Flex Your Brain activities.

Section 9•1

Prepare

Section Background

Volume must be known to find the physical property of density. The volume of an irregularly shaped, insoluble object can be measured by finding the volume of water the object will displace.

Preplanning

Refer to the Chapter Organizer on pages 236A-B.

1 Motivate

Bellringer

Before presenting the lesson, display **Section Focus Transparency 27** on the overhead projector. Assign the accompanying **Focus Activity** worksheet.
`L2` `ELL`

Tying to Previous Knowledge

Although they may not have used the terms *physical properties* and *chemical properties*, students frequently use characteristics of items to describe them.

9•1 Physical and Chemical Properties

What YOU'LL LEARN

- Physical properties and chemical properties of matter
- The properties of acids and bases and the differences between them

Science Words:
physical property
density
state of matter
chemical property

Why IT'S IMPORTANT

When you learn about physical and chemical properties, you can better describe the world around you.

Physical Properties

Would it surprise you to know that both of the rocks that you examined in the Explore activity started out as the exact same kind of lava? So why are they so different? As you will find out later, they were produced by two different kinds of volcanic activity. Characteristics that you observe about matter are often related to how that matter was formed.

In the Explore activity, you used your senses to classify different types of matter. This classification helps you better understand what the types of matter are and how they were formed.

Common Physical Properties

Do you have a favorite souvenir—an unusual rock or seashell, a funny hat, or a special cup? Take a minute to describe that souvenir in as much detail as you can. What features did you use in your description—color, shape, and hardness? These features are all properties, or characteristics, of the souvenir. Scientists use the term **physical properties** to describe characteristics you can detect with your senses. All matter, such as the hat in **Figure 9-1**, has physical properties. Practice identifying physical properties by listing some physical properties of your science book.

For most of your life, you have been aware of some physical properties, such as color, shape, smell, and taste. You may not be as familiar with others, such as texture and density. Texture is how rough or smooth something is.

FIGURE 9-1

What are some of the physical properties of this hat?

Program Resources

Reproducible Masters
Activity Worksheets, pp. 5, 55-56, 59 `L2`
Cross-Curricular Integration, p. 13 `L2`
Enrichment, p. 33 `L3`
Lab Manual, pp. 59-62 `L2`
Reinforcement, p. 33 `L2`
Study Guide, p. 33 `L1`

Transparencies
Section Focus Transparency 27 `L2`
Teaching Transparencies 17, 18 `L2`

Density relates the mass of something to how much space it takes up, which is its volume.

Density

In the Explore activity, you compared the densities of the rock samples when you decided which equal-sized rock was heavier. Which rock was more dense—pumice or obsidian? If someone asked you which is heavier, the grocery bag in **Figure 9-2** or a grocery bag full of cans of soda, you'd probably say that the bag of canned soda is heavier. It is heavier because the density of the cans of soda is greater than the density of the snacks. If you found the mass of each full bag and divided each mass by its volume, you could find the density of what was in each bag.

Whether you realize it or not, you see examples of density every day. Ice floats in a soft drink because the density of ice is less than that of the soft drink. You shake a bottle of oil-and-vinegar salad dressing before using it. Oil is less dense than vinegar, so the oil rises to the top.

Density also can be used to identify compounds and elements because each has its own density. For example, suppose you have 20 mL (volume) of wood alcohol. It has a mass of 15.8 g. You can find the density of this alcohol by dividing its mass by its volume.

$$\text{Density} = \text{mass}/\text{volume}$$

$$D = m/v = 15.8 \text{ g}/20 \text{ mL} = 0.79 \text{ g/mL}$$

Mini LAB

Compare Densities

Compare the densities of some common liquids.

1. Lightly color about 20 mL of water with red food coloring and about 20 mL of white corn syrup with blue food coloring.
2. Pour the water into a clear, colorless, plastic cup.
3. Slowly add the corn syrup to the cup by pouring it down the side.
4. Next, in the same way, add about 10 mL of rubbing alcohol to the cup.

Analysis
1. What did you observe?
2. Explain what happened in terms of the densities of each liquid that you added.

FIGURE 9-2

This bag is easy to carry because its contents have a low density.

9-1 Physical and Chemical Properties   **239**

---

2 Teach

Mini LAB

Purpose

[LS] **Kinesthetic** Students will observe what happens when liquids with different densities are combined.
[L2] **ELL**

Materials

water, white corn syrup, rubbing alcohol, red and blue food coloring, clear plastic cup

Alternate Materials

Molasses or dark corn syrup without food coloring can replace the corn syrup.

Teaching Strategies

Safety Precautions Caution students not to drink the corn syrup. Have no more than 200 mL of rubbing alcohol in the room at any time, and have no more than 100 mL in a single container. When alcohol is in the room, be sure there are no open flames.

Troubleshooting If students pour too quickly, the layers may mix.

Disposal

Dilute liquid with water to ten times its original volume, and pour down the drain.

 **Activity Worksheets,** pages 5, 59

Analysis

1. The water floated on the corn syrup, and the alcohol floated on the water.
2. Corn syrup is the densest of the three, so it sinks. Alcohol is the least dense, so it floats.

Visual Learning

Figure 9-1 What are some of the physical properties of this hat? *Student answers may include the size or shape of the hat, the different textures of different parts of the hat, or that the hat is flexible.* **[LS]**

✓ Assessment

Process Provide students with an illustration of the layers in a lake bed. Ask them to explain the layering in terms of densities. Use the Performance Task Assessment List for Making Observations and Inferences in **PASC**, p. 17.

Practice Problem Answer

Practice Problem Answer

Find the density of the bracelet by dividing its mass by its volume.

$d = 42\ \text{g} / 5\ \text{mL} = 8.4\ \text{g/mL}$

Compare this density to the density of silver. The density of silver is 10.5 g/mL, so the metal is not pure silver.

Inquiry Question

LS **Logical-Mathematical** A gram is defined as the mass of 1 mL of water. What is the density of water? *1 g/mL*

Brainstorming

Have students brainstorm examples of things that float in other things. In each case, have them state which density is less. Examples might include that wood floats in water because wood is less dense than water, and helium balloons float in air because helium is less dense than air. **L2**

Teacher F.Y.I.

Unit analysis is a good way to check that you have set up a problem correctly. Leave the units in, cancel appropriately, and check that the answer will be in the desired units. For example, if students have difficulty remembering how to find density, the density units—g/mL, for example— will tell them to divide mass by volume. Remind students that a slash line in such a unit means "are divided by."

Student Text Question

from page 239 **Which rock was more dense—pumice or obsidian?** *obsidian*

Calculating Density

Example Problem:

An astronomer found a small meteorite. The astronomer knew that nickel and iron are often found in meteorites. Nickel has a density of 8.9 g/mL, and iron has a density of 7.9 g/mL. The meteorite had a mass of 176 g and a volume of 20 mL. What is the density of the meteorite? Is the meteorite mostly nickel or mostly iron?

Problem-Solving Steps

1. Find the density of the meteorite by dividing its mass by its volume.

$$176\ \text{g} \div 20\ \text{mL} = 8.8\ \text{g/mL}$$

2. Compare this density to the densities of iron and nickel. The density is closer to the density of nickel.

Solution: The density of the meteorite is 8.8 g/mL, which is close to the density of nickel. Therefore, the meteorite is mostly nickel.

Practice Problem

On a trip to the Southwest, you bought a bracelet that was labeled as pure silver. Is it really pure silver? The density of silver is 10.5 g/mL. The bracelet has a mass of 42 g and a volume of 5 mL. What is the density of the metal in the bracelet? Is the metal pure silver? How do you know?

Formation

The physical properties of many materials are related to the way they form. Remember the two rocks you described in the Explore activity. The two rocks are made of the same materials, but their physical properties are different. One is dark; one is light in color. One sinks; one floats. One is smooth; the other is rough and jagged. Their physical properties are different because they formed in different ways. Take a closer look at what happened.

Pumice formed when melted, or molten, rock came into contact with water while it was still under the ground. This extremely hot mixture of water and molten

240   Chapter 9 Properties and Changes

Science Journal Density Have students compare and contrast mass, volume, and density, with examples. **L3**

Content Background

Have students think of submerging themselves in a deep pool of water that's been in the sun. Or have them think about smoke rising over a fire. Warm water is forced up by cold water sinking because warm water is less dense. Warm air is forced up by cold air sinking because warm air is less dense. These two processes are the basis of weather patterns.

rock exploded out the top of the volcano. The water then boiled, forming lots of little holes in the rock. The lava cooled quickly, preserving the little holes as it hardened. The result is a brown-colored rock that is so light it floats on water. Obsidian, on the other hand, is a dark rock that sinks. It came from the same kind of lava, and it also cooled quickly. But it didn't contact water as it formed, so no bubbles resulted.

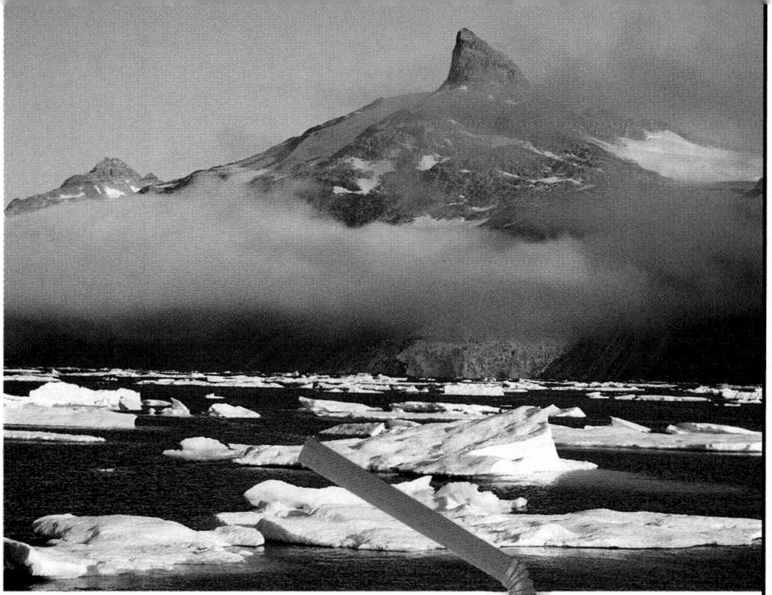

FIGURE 9-3

All three states of water are present here—solid, liquid, and gas.

State of Matter

You know that solid pumice is different from liquid magma. State of matter is another physical property. The **state of matter** tells you whether a sample of matter is a solid, a liquid, or a gas. For example, if you hold an ice cube in your hand, you have water in its solid state. Water exists in the liquid state in oceans, in rivers, and in your bathtub. It also exists as a gas in the air. In each case, each molecule of water is the same—two hydrogen atoms and one oxygen atom. But water appears different because it exists in different states, as shown in **Figure 9-3.**

Physical Properties of Acids and Bases

What do you think of when you hear the word *acid?* Do you picture a dangerous chemical that can burn your skin, make holes in your clothes, and even destroy metal? Some acids, such as hydrochloric acid, are like that. But not all acids are possibly harmful. One example is shown in **Figure 9-4.** Every time you eat a citrus fruit such as an orange or a grapefruit, you eat citric and

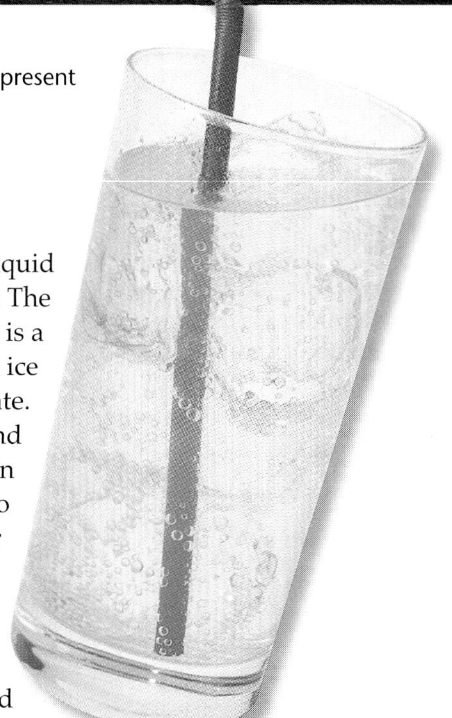

FIGURE 9-4

When you sip a carbonated soft drink, you drink carbonic and phosphoric (faws FOR ihk) acids.

GLENCOE TECHNOLOGY

Videodisc

Glencoe Physical Science Interactive Videodisc
Side 1, Lesson 3
Gases Expand to Fill Container

23416

23418-23508
Gases Exert Pressure

23602

Activity

[LS] **Interpersonal** Let a 2 m × 2 m space represent a container. Ask for four student volunteers. To represent a gas, the students can move freely all over the container. To represent a liquid, they are spread across the bottom. To represent a solid, they should hold elbows in a ring and lock their arms. [L1] **COOP LEARN**

? FLEX Your Brain

Use the Flex Your Brain activity to have students explore ACIDS.

Activity Worksheets, page 5

Content Background

• A solid has a fixed volume and shape. A liquid has a fixed volume, but it has the shape of its container. A gas has the shape and volume of its container.

• A fourth state of matter, plasma, is the least common state of matter on Earth but the most common state of matter in the universe. A plasma consists of charged particles formed when matter absorbs energy and breaks apart.

Using Science Words

LS **Linguistic** A base is also called an alkali. *Alkali* comes from the Arabic term for "the ashes" (*al qalíy*). *Acid* comes from the Latin word for "sour" (*acidus*).

Using an Analogy

LS **Logical-Mathematical** Physical and chemical properties are used to classify matter. For example, elements can be grouped based on whether each is a solid, a liquid, or a gas at room temperature. Students will be familiar with using properties to classify objects. For example, band instruments can be classified as brass, woodwind, or percussion based on their appearance and how they make sound.

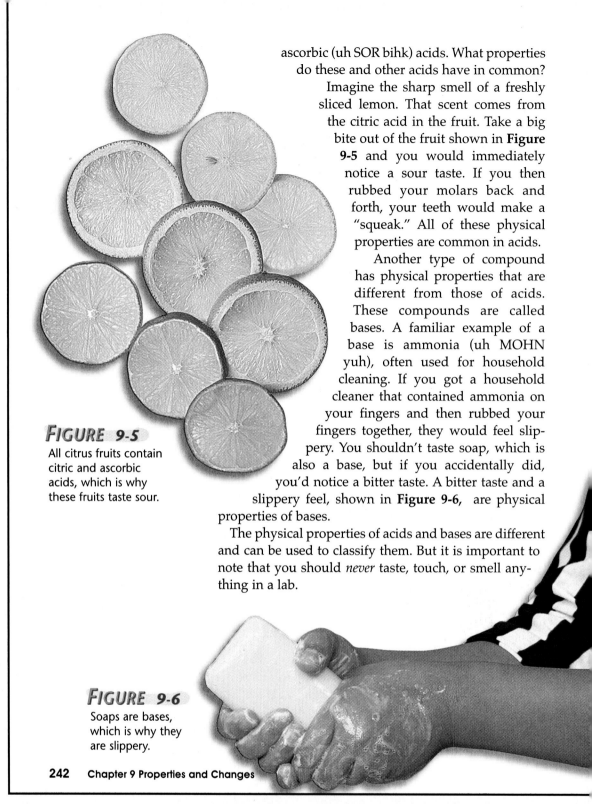

FIGURE 9-5

All citrus fruits contain citric and ascorbic acids, which is why these fruits taste sour.

ascorbic (uh SOR bihk) acids. What properties do these and other acids have in common? Imagine the sharp smell of a freshly sliced lemon. That scent comes from the citric acid in the fruit. Take a big bite out of the fruit shown in **Figure 9-5** and you would immediately notice a sour taste. If you then rubbed your molars back and forth, your teeth would make a "squeak." All of these physical properties are common in acids.

Another type of compound has physical properties that are different from those of acids. These compounds are called bases. A familiar example of a base is ammonia (uh MOHN yuh), often used for household cleaning. If you got a household cleaner that contained ammonia on your fingers and then rubbed your fingers together, they would feel slippery. You shouldn't taste soap, which is also a base, but if you accidentally did, you'd notice a bitter taste. A bitter taste and a slippery feel, shown in **Figure 9-6,** are physical properties of bases.

The physical properties of acids and bases are different and can be used to classify them. But it is important to note that you should *never* taste, touch, or smell anything in a lab.

FIGURE 9-6

Soaps are bases, which is why they are slippery.

242   Chapter 9 Properties and Changes

Cultural Diversity

Tasty Properties Some people taste certain substances that other people do not taste because of their genes. A dominant gene controls PTC (phenylthiocarbamide) tasting. If you are a "taster," PTC is bitter. PTC-like substances are found in foods such as buttermilk, spinach, and turnips, so tasters and nontasters have different responses to these foods. Some poisonous plants also contain PTCs, so being a taster would be useful to a food gatherer living where these plants grow. The ability to taste PTC varies from population to population. In Sub-Saharan Africa, only 5 percent of the alleles are for nontasting; in Denmark, 57 percent are.

Chemical Properties

You've noticed the density and the state of an ice cube. You've described the color and texture of rocks. You've noticed the taste of acid in a lemon and the slippery feel of a base such as ammonia. However, a description of something using only physical properties is not complete. What type of property describes how matter behaves?

Common Chemical Properties

If you strike a match on a hard, rough surface, the match will probably start to burn. The element phosphorus (FAWS for us) and the wood in the match combine with oxygen in the air to form new materials. Why does that happen? The phosphorus and the wood both have the ability to burn. The ability to burn is a chemical property. A **chemical property** is a characteristic of a substance that allows it to change to a new substance.

You see an example of a chemical property when you leave a half-eaten apple on your desk, and it turns brown. The property you observe is the ability to react with oxygen. Two other chemical properties are shown in **Figure 9-7**.

FIGURE 9-7

The chemical properties of a material often require a warning about its careful use.

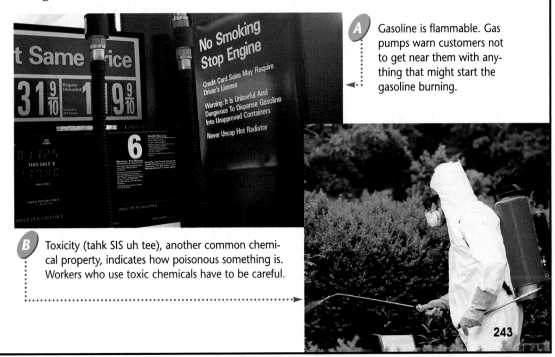

A Gasoline is flammable. Gas pumps warn customers not to get near them with anything that might start the gasoline burning.

B Toxicity (tahk SIS uh tee), another common chemical property, indicates how poisonous something is. Workers who use toxic chemicals have to be careful.

243

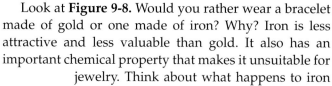

FIGURE 9-8

Gold and iron have different chemical properties that make them suitable for different uses.

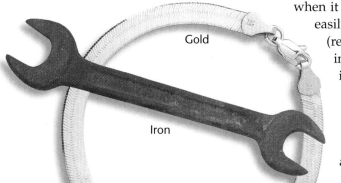

Gold

Iron

Look at **Figure 9-8.** Would you rather wear a bracelet made of gold or one made of iron? Why? Iron is less attractive and less valuable than gold. It also has an important chemical property that makes it unsuitable for jewelry. Think about what happens to iron when it is left out in the air. Iron rusts easily because of its high reactivity (ree ak TIV uh tee) with oxygen in the air. Reactivity is how easily one thing reacts with something else. The low reactivity of silver and gold, in addition to their desirable physical properties, makes those metals good choices for jewelry.

USING TECHNOLOGY

Properties and Pools

Did you ever wonder why chlorine is added to swimming pools? Chlorine compounds change the properties of the pool water. Chlorine reacts with cellular substances, killing bacteria, insects, algae, and plants.

Any time you have standing water, mosquitoes and other insects can lay eggs in it. Various plants and algae can also turn a sparkling blue pool into a slimy green mess. Bacteria are another problem. When you go swimming, you bring along millions of uninvited guests—the normal bacteria that live on your skin. The chlorine compounds kill the bacteria—as well as insects, algae, and plants that may be in the pool.

Chlorine makes the water more acidic, but it can cause other problems as well. Have your eyes ever burned after swimming in a pool? The chlorinated water can irritate the skin and eyes of swimmers.

*inter*NET
CONNECTION

Pool owners need to keep track of several different chemicals as they care for their swimming pools. Visit the Glencoe Homepage at **www.glencoe. com/sec/science** for a link about pool maintenance.

Chemical Properties of Acids and Bases

You have learned that acids and bases have physical properties that make acids taste sour and bases taste bitter and feel slippery. But the chemical properties of acids and bases are what make them both useful and sometimes harmful.

Many acids react with, or *corrode*, certain metals. Have you ever used aluminum foil to cover leftover spaghetti and tomato sauce? The next day, you might find small holes in the foil where it has come into contact with the tomatoes in the sauce, as shown in **Figure 9-9.**

The acids in tomato sauce, oranges, carbonated soft drinks, and other foods won't hurt you. However, many acids can damage plant and animal tissue. Nitric (NITE rihk) acid and sulfuric (sulf YER ihk) acid are found in small amounts in rain in areas that have a lot of pollution in the air. This rain, called acid rain, harms plant and animal life in areas such as the one shown in **Figure 9-10,** where acid rain falls. Sulfuric acid that has no water mixed with it is useful in many industries because it removes water from certain materials. However, that same property causes burns on skin that touches sulfuric acid.

A strong base also can damage living tissue. It is not uncommon for someone who smells strong ammonia to get a bloody nose from the fumes. And the reason that ammonia feels slippery to the touch is that the base actually reacts with fat that lies under the top layer of skin cells in your fingertips.

FIGURE 9-9

Aluminum reacts easily with acids, which is why acidic food, such as tomatoes, should not be cooked or stored in aluminum.

FIGURE 9-10

These trees show damage caused by acid rain.

245

Enrichment

Kinesthetic When acids are added to water, the molecules lose hydrogen. Bases form a hydroxide group when they are added to water. A hydroxide group is made up of one hydrogen atom and one oxygen atom. Have students make models of hydrogen and hydroxide using plastic molecular models, balls, or drawings on the chalkboard. Let students show that the atoms can be recombined to make water, H_2O. **L3**

3 Assess

Check for Understanding

Inquiry Question An antacid is advertised as "neutralizing stomach acid." Do you think the antacid is an acid, a base, or neutral? *A base; when it combines with stomach acid, it brings the stomach acidity closer to neutral.*

Reteach

Ask each student to write the name of something on a slip of paper. As a class, draw slips of paper and name a physical and a chemical property for each item. **L1**

Extension

For students who have mastered this section, use the **Reinforcement** and **Enrichment** masters.

Content Background

When an acid or base is placed in water, some of its units form positive and negative ions. An ion is a charged atom or group of atoms. The higher the percentage of ions formed, the stronger the acid or base.

Community Connection

Soils Ask students to research what types of crops grow well in acidic or basic soils and present their findings to the class. They may find information from a local horticulturist or lawn and garden store. Ask students to infer what kind of soil is in your area. **L3** **COOP LEARN**

PREPARE

Purpose

LS **Kinesthetic** Students will determine the pH of various solutions and classify the solutions as acidic, basic, or neutral. **L2** **COOP LEARN**

Process Skills

observing, classifying, using numbers, communicating, measuring, predicting, interpreting data, experimenting

Time

45 minutes

Materials

Clear fruit juices with little color, such as apple juice or white grape juice, work best. Use colorless soft drinks.

Safety Precautions

Do not use household cleaners that are dangerous to inhale or can irritate skin, such as strong ammonia solutions and drain and oven cleaners. Be sure to read and follow all caution statements on the labels. Be sure no cleaners are mixed together.

Possible Hypotheses

Fruit juices, vinegar, and soft drinks are acidic. Soaps and antacids are basic.

📁 **Activity Worksheets,** pages 5, 55-56

Possible Materials

- vial of pH paper, 1-14
- pH color chart
- distilled water
- fruit juices
- vinegar
- salt
- sugar
- soft drinks
- household cleaners
- soaps and detergents
- antacids

pH	Color	pH	Color
1		8	
2		9	
3		10	
4		11	
5		12	
6		13	
7		14	

Design Your Own Experiment

Homemade pH Scale

The stronger an acid or base, the more likely it is to be harmful. A scale, called the pH scale, is used to measure how strong acids and bases are. Solutions with pH of 1 to 6 are acids; pH 7 is neutral; and a pH of 8 to 14 is a base. The strongest acids have a pH of 1, and the strongest bases have a pH of 14. In this activity, you will measure the pH of some things you commonly use. Use treated paper that, when dipped into a solution, turns a color. Check the color against the chart below to find the pH of the solution.

PREPARE

What You'll Investigate

How acidic or basic are some common household items?

Form a Hypothesis

Think about the properties of acids and bases. In your group, **make a hypothesis** about which kinds of solutions you are testing are acids and which kinds are bases.

Goals

Design an experiment that allows you to test solutions to find the pH of each.

Classify a solution as an acid or base according to its pH.

Safety Precautions 🧤 🥽 🧹

Never eat, taste, smell, or touch any chemical during a lab.

Data and Observations

Sample Data

Solution to be tested	pH	Acid, base, or neutral
dish-washing detergent	8	base
salt water	7	neutral
apple juice	5	acid

Content Background

The term *pH* means "potential of hydrogen." In water, hydrogen ions combine with water to form hydronium ions (H_3O^+). pH is a measure of the concentration of hydronium ion in a solution. A pH scale is not linear, but logarithmic. A solution with a pH of 8 has ten times as many hydronium ions as a solution with a pH of 9 contains.

PLAN

1. As a group, **decide** what materials you will test. If a material is not a liquid, **dissolve** it in water so you can test the solution.
2. **List** the steps and materials that you need to **test** your hypothesis. Be specific. Will you **repeat** any part of the experiment?
3. Before you begin, **copy** a data table like the one shown into your Science Journal. Be sure to leave room to **record** results for each solution tested. If there is to be more than one trial for each solution, include room for the additional trials, too.
4. **Reread** the entire experiment to make sure that all the steps are in logical order.

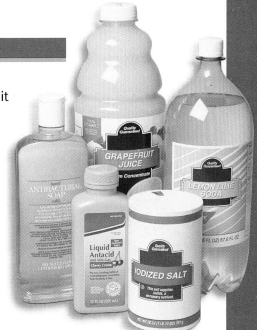

DO

1. Make sure that your teacher approves your plan and data table. Be sure that you have included any suggested changes.
2. Carry out the experiment as planned and approved. Wash your hands when done.
3. Be sure to **record** your observations in the data table as you **complete** each test.

CONCLUDE AND APPLY

1. **Compare** your findings with those of other student groups.
2. Were any materials neither acids nor bases? How do you know?
3. Using your data table, **conclude** which types of materials are usually acidic and which are usually basic.
4. **APPLY** Perhaps you have been told that you can use vinegar to dissolve hard-water deposits because vinegar is an acid. If you run out of vinegar, which of the following—lemon juice, ammonia, or water—could you most likely use instead of vinegar for this purpose?

9-1 Physical and Chemical Properties   **247**

Go Further

LS **Intrapersonal** Have students research some practical applications of pH and present a multimedia presentation of their findings. Applications might include testing swimming pool water, soil, or acid rain or various industrial applications. If any tests are actually done by students, the teacher should approve their plans in advance.

247

4 Close

Use the Mini Quiz to check students' recall of chapter content.

1. **What is a physical property?** *a characteristic of matter you can detect with your senses*

2. **Which would you expect to have greater density, a piece of balsa wood or a piece of iron?** *iron*

3. **What forms when you mix an acid and a base?** *water and a salt*

Section Wrap-up

1. Answers will vary but might include physical properties of milk, a desk, a computer disk, or air. Students will probably list relative hardness instead of a scientific definition of hardness.

2. Answers will vary, but each should relate clearly the warnings.

3. **Think Critically** The wax is still wax; the state has changed, but not the identity of the substance.

Science Journal Chemical; usually you are concerned about a dangerous reaction.

FIGURE 9-11
These everyday items contain salts.

What happens in reactions between acids and bases? Acids and bases are often studied together because they react with each other to form water and other useful compounds called *salts*. Look at **Figure 9-11.** That familiar stuff in your salt shaker—table salt—is the most common salt. Table salt, sodium chloride, is formed by the reaction between the base sodium hydroxide and hydrochloric acid. Other useful salts are calcium carbonate, which is chalk, and ammonium chloride, which is used in certain batteries.

Section Wrap-up

1. Make a chart and list ten things in your home. On the chart, include the following physical properties for each item: color, state of matter, texture, and hardness.

2. Design three separate symbols that could be put on containers to warn chemists that a certain chemical could burn easily and be toxic and corrosive.

3. **Think Critically:** Explain why the temperature at which wax melts is a physical property and not a chemical property of wax.

4. **Skill Builder**
   Classifying Classify each of the following properties as being either physical or chemical.
   a. Iron will rust when left out in the air.
   b. Lye feels slippery.
   c. Iodine is poisonous.
   d. Solid sulfur shatters when struck.

   If you need help, refer to Classifying on page 542 of the **Skill Handbook.**

Science Journal

Think about safety factors you should check around your home. Are they based upon physical properties or upon chemical properties? Explain.

✔ Check battery in smoke alarm

✔ Plan an escape route if there is a fire

Skill Builder
Classifying a. chemical, b. physical, c. chemical, d. physical

✔ Assessment

Portfolio Choose an item or material in the classroom. Ask students to write a fiction story about it, describing all the chemical and physical properties of it that they can. Examples of items or materials include water, fabric, wood, plastic, and lab chemicals. Use the Performance Task Assessment List for Writer's Guide to Fiction in **PASC,** p. 83. **P** **L2**

The *Hindenburg* Disaster

On May 6, 1937, the *Hindenburg* drifted down to land at Lakehurst, New Jersey. The huge German airship had just flown across the Atlantic Ocean with 97 people on board. The *Hindenburg* was one of the largest airships ever built. It was 245 m long and 41 m wide.

Airships were the first flying machines that had a crew and could be steered. Most had a wooden or metal framework inside the balloon covering.

Airships were filled with lighter-than-air gas. Hydrogen, which has the physical property of being the least dense of all gases and is therefore the lightest element, provided the greatest amount of lift. But hydrogen gas has chemical properties that can make it dangerous to use. Hydrogen is flammable, and it burns explosively.

Up In Flames

As the *Hindenburg* prepared to land, onlookers saw a small burst of flame near the ship's tail. Suddenly, the entire tail section was on fire. The fire roared forward along the ship. In seconds, the *Hindenburg* was completely engulfed in flames. When it crashed to the ground, 36 people were killed.

To this day, no one is exactly sure what caused the *Hindenburg* to catch on fire. One hypothesis is that a spark of electricity in the air near a leak in the ship's hull ignited the gas. Another hypothesis is that the skin of the airship first caught on fire, which then set the hydrogen burning. Whatever its cause, the fiery *Hindenburg* disaster marked the end of airships as a way for people to travel.

*inter*NET CONNECTION

In the Netherlands, university students have been studying the environmental benefits of bringing back the use of airships. Visit the Glencoe Homepage at ***www.glencoe.com/sec/science*** for a link about their findings.

Source

Stein, R. Conrad. *The Hindenburg Disaster.* Children's Press: Chicago, 1993.

Background

The German company that built the *Hindenburg* wanted to inflate the airship with nonflammable helium. The United States was the world's major source of helium at that time, and government officials were suspicious of the company's political ties. They feared that the airships might be used for military purposes. When the Helium Control Act was passed in Congress, it became impossible for the manufacturer to obtain enough helium to inflate the *Hindenburg*. Hydrogen was used instead.

Bibliography

DiChristina, Mariette. "What Really Downed the *Hindenburg*." *Popular Science,* Nov. 1997, pp. 70-76.

Ransom, Candice F. *Fire in the Sky.* Carolrhoda Books: Minneapolis, 1997.

*inter*NET CONNECTION

The Glencoe Homepage at **www.glencoe.com/sec/science** provides links connecting concepts from the student edition to relevant, up-to-date Internet sites.

Teaching Strategies

- Discuss the differences between hydrogen and helium. Provide students with information about the flammability of the two gases.
- Obtain several helium-filled balloons and demonstrate how such a lighter-than-air gas can be used to lift objects. Be sure balloons are purchased from a dealer who uses appropriate safety measures while filling the balloons.
- Tell students that the photo shown was colorized by engineer and historian Addison Bain. Bain's investigations have led him to the conclusion that the cause of the fire was ignition of a flammable compound used to treat the skin of the airship, not ignition of the hydrogen. Whatever the cause, the chemical property of flammability was the basis for the disaster.

Science & Society

9•2 Road Salt

Science & Society

Bellringer

Before presenting the lesson, display **Section Focus Transparency 28** on the overhead projector. Assign the accompanying **Focus Activity** worksheet.

[L2] [ELL]

Teaching the Content

• Prior to their reading this feature, have students sprinkle a thin layer of salt on half of a clean paper plate. Have them place one ice cube on the salt and another ice cube directly on the plate where there is no salt. The cubes must be as identical in size and shape as possible. Have students observe and note which cube melts faster. [IS] [L2]

• Have students brainstorm other uses of deicing compounds, such as deicing the wings of a plane. [L3]

What YOU'LL LEARN

• How road salt can both make travel safer and damage the environment

• Why main ideas must be identified to understand scientific writings

Why IT'S IMPORTANT

Knowing the properties of materials helps you decide how to use them.

Salt and Ice

In many climates, weather changes with the seasons. Summers are hot and winters are cold. Areas that have such climates can receive a lot of snow in the winter. Changes in temperature can cause the snow to melt, then refreeze, turning it to ice. Ice is the solid form of water. It can be dangerous to cars and trucks traveling on icy roads and to people walking on ice-covered sidewalks.

In some areas, such as the one shown in **Figure 9-12,** people spread rock salt on slick sidewalks and roads. Why? Salt changes one physical property of water—the temperature at which it freezes. When salt is added to water, the salt causes the water to freeze at a temperature lower than the freezing temperature of pure water, which is 0°C. Applying salt to slick areas melts any ice present, unless the ground temperature is colder than the normal freezing temperature of water.

Pass (on) the Salt, Please

Salt can make winter travel safer. But it also can cause problems. If they have not been protected from salt, cars driven on salted roads can begin to rust after only a few winters. Another rust problem is shown in **Figure 9-13.**

FIGURE 9-12
This salt will help melt the snow and ice on the highway, making it safe for travelers.

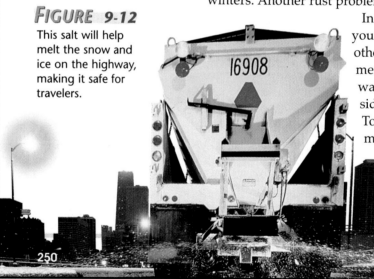

In addition to damage that you can easily see, salt causes other problems in the environment. As snow melts, salty water runs off the roads and sidewalks onto nearby land. Too much salt in the soil makes it difficult for many types of plants to grow. Salty water also flows into waterways, polluting nearby rivers,

250

Program Resources

 Reproducible Masters
Enrichment, p. 34 [L3]
Lab Manual, pp. 63-66 [L2]
Reinforcement, p. 34 [L2]
Study Guide, p. 34 [L1]

Transparencies
Section Focus Transparency 28 [L2]

lakes, and streams that provide much of the water needed by towns, cities, and farmlands.

Salt Substitutes

Because of the damage done to the environment, many places now use sand, cinders, or crushed gravel to provide the traction needed to prevent slipping on snowy roads and walkways. These materials do not harm the environment as salt can.

Chemicals called deicers also can be used rather than salt. A deicer keeps snow or ice from attaching to the pavement. A chemical called CMA is a deicer that causes little or no damage to steel. CMA also breaks down into harmless materials and does not hurt plants or drinking-water supplies.

FIGURE 9-13
Salt also speeds the rusting of steel structures, such as bridges, guardrails, and lampposts.

 Skill Builder: *Identifying the Main Idea*

LEARNING the SKILL

1. Read carefully. Use the title and headings to get an idea of what the whole article is about.

2. Reread each paragraph, one at a time. Find the main idea of each paragraph by choosing a single sentence that best states the topic of the paragraph.

3. If a single sentence does not clearly state the main idea of a paragraph, reword what is in the paragraph to state the main idea.

4. Pick out details that support the main idea. These details can be phrases or words from the paragraph, pictures, or captions.

PRACTICING the SKILL

1. What is the main idea of the first paragraph in the section called "Salt and Ice"?

2. **a.** What are deicers? How do they compare with salt?

   b. Which question in part **a** refers to the main idea? Which question refers to supporting details?

3. Reread "Road Salt." Study your answers to the questions. In your own words, what is the main idea?

APPLYING the SKILL

Cloud seeding is another process that can help humans but could harm the environment. Find the main ideas in readings about cloud seeding. From these main ideas, find out how cloud seeding relates to weather and how it affects the environment.

Answers to Practicing the Skill

1. Ice on roads and sidewalks is dangerous to pedestrian and vehicle traffic.

2. **a.** A deicer keeps snow or ice from attaching to pavement. Salt melts ice that is already attached.

   b. the first question; the second question

3. People need to get rid of ice on sidewalks and roads. Salt does this well, but it damages cars and the environment. Other materials can provide traction or deicing without harming the environment.

Teaching the Skill

- Pair students with reading problems with those who are proficient readers.

- Read aloud a paragraph from an exciting story. First paragraphs are a good choice because they must set the stage in an interesting way. Have students use the steps listed in Learning the Skill to identify the main idea. **L2**

Content Background

- Road salt is used in both anti-icing and deicing techniques. Anti-icing techniques prevent snow and ice from bonding to the pavement and are more effective than deicing techniques, which destroy an existing bond between the pavement and the ice.

- Nearly 11 million tons of rock salt are dumped on winter roads each year in the United States to prevent ice from forming or to cause the ice to melt.

- Deicers can be mixed with rock salt to improve the rate at which ice melts to temperatures of about 0°F. Deicers, alone or in combination with road salt, reduce corrosion of steel and concrete by nearly 80 percent over rock salt alone.

GLENCOE TECHNOLOGY

 Videodisc
STVS: Plants & Simple Organisms
Disc 4, Side 2
Salt-Resistant Crops (Ch. 12)

9•3 Physical and Chemical Changes

Prepare

Section Background

The type of change matter undergoes is based on the forces involved. *Intraparticle* forces (forces within the particle) are responsible for chemical properties and chemical changes. *Interparticle* forces (forces that particles have for each other) are responsible for physical properties and physical changes.

Preplanning

Refer to the Chapter Organizer on pages 236A-B.

1 Motivate

Bellringer

Before presenting the lesson, display **Section Focus Transparency 29** on the overhead projector. Assign the accompanying **Focus Activity** worksheet.

L2 **ELL**

Tying to Previous Knowledge

Physical and chemical changes of matter are based on physical and chemical properties.

252

What YOU'LL LEARN

- How to identify physical and chemical changes
- Examples of how physical and chemical changes affect the world we live in

Science Words:
physical change
chemical change

Why IT'S IMPORTANT

If you understand the changes in the things around you, you will better be able to use these changes.

9•3 Physical and Chemical Changes

Physical Change

It's picnic time. You offer to help carry the food to the picnic table—grilled chicken, potato salad, baked beans, and a big watermelon. On the last trip—with the watermelon—you suddenly lose your hold on the large, slippery fruit. It goes crashing to the ground and splits open, becoming an example of a principle that you will learn in this section—physical change.

What is physical change?

Most matter can undergo **physical change**, which is any change in size, shape, or form. The identity of the matter stays the same. Only the physical properties change. Look at **Figure 9-14.** The watermelon underwent a physical change. It went from being one large, round melon to being many smaller pieces splattered all over the floor. It is still watermelon; it just looks different.

FIGURE 9-14

The physical change this watermelon underwent changed its form.

Examples of Physical Changes

How can you recognize a physical change? Just look to see whether or not the matter has changed size, shape, or form. If you cut a watermelon into chunks, the watermelon has changed both size and shape. That's definitely a physical change. If you pop one of those chunks into your mouth and bite it, you have changed the watermelon's size and shape again.

252   Chapter 9 Properties and Changes

Program Resources

Reproducible Masters
Activity Worksheets, pp. 5, 57-58, 60 **L2**
Enrichment, p. 35 **L3**
Multicultural Connections, pp. 21-22 **L2**
Reinforcement, p. 35 **L2**
Science and Society/Technology Integration, p. 27 **L2**
Science Integration Activities, pp. 55-56 **L2**
Study Guide, p. 35 **L1**

Transparencies
Science Integration Transparency 9 **L2**
Section Focus Transparency 29 **L2**

There's another way that matter can undergo a physical change. It can change from one state to another. After dropping the watermelon, suppose you save the picnic by bringing out Popsicles for dessert. It's a hot day, and the Popsicles start to melt and drip onto the patio as you and your friends eat them. What kind of physical change is happening? A Popsicle is a frozen, solid mixture of water, sugar, food coloring, and flavoring. In the hot sun, the water in the Popsicle changes state and melts, turning into a drippy liquid. As the drops of melted Popsicle on the patio sit in the sun, the water changes state again, evaporating to become a gas. In each case, the Popsicle is composed of the same ingredients. But the water it contains is in different states as it changes from a solid to a liquid to a gas. Other examples of change of state are shown in **Figure 9-15.**

FIGURE 9-15
The four most common changes of state are shown here.

C A solid will melt, becoming a liquid.

B As it cools, this liquid metal will become solid steel.

A Water vapor in the air changes to liquid water when dew forms.

D Liquid water in perspiration changes to a gas when it evaporates from your skin.

253

Discussion

Discuss with students the physical changes involved in making paper from a tree. Simply speaking, the wood from the tree is cut into small pieces and mixed with water, forming wood pulp. Most of the water is then pressed out. The pulp is spread into a sheet and then dried. Elicit as many changes as possible from the students. L2

Revealing Preconceptions

Students may think that when any properties change, matter has undergone a chemical change. A type of matter can change to a different form and still be the same type of matter. For example, adding salt to water changes the boiling and freezing points, but it's still water and salt. Boiling water changes the state, but it's still H_2O.

Demonstration

Bring in bread and a toaster. Toast a piece of bread, and ask students to explain how they know that a chemical change has taken place. A change of color and odor indicates that a new substance has been formed. Be sure students don't eat the bread or toast.

FIGURE 9-16

You can see dramatic examples of physical weathering caused by water and wind on rocky coastlines and in deep canyons.

254

Some physical changes, such as the melting Popsicle, happen quickly. Others take place over a long time. Physical weathering is a physical change that is responsible for much of the shape of Earth's surface. Examples are shown in **Figure 9-16**. Examples also can be found in your own schoolyard. All of the soil that you see comes from physical weathering. Wind and water erode rocks, breaking them into small bits. Water fills cracks in rocks. When it freezes, the ice splits the rock into smaller pieces. No matter how small the pieces of rock are, they are still made up of the same things that made up the bigger pieces of rock. The rock has simply undergone a physical change. Gravity, plants, animals, and the movement of land during earthquakes also help cause physical changes on Earth.

Content Background

The law of conservation of mass states that no mass is gained or lost in a chemical change. The existing mass is simply rearranged into one or more new substances.

Across the Curriculum

Geography The Grand Canyon was formed over time by physical changes to the rock, as were the stone pillars in Bryce National Park. Have students look through several magazines that show remarkable landscapes. Have them explain how physical weathering could have formed some of these structures. LS L2

Chemical Changes

Your bicycle gets a chip in the paint. Soon, the bike has a spot of rust. A shiny copper penny becomes dull and dark. An apple left out too long begins to rot. What do all these changes have in common? In each of these examples, the original materials that make up the bike, the penny, and the apple change into other materials that have different properties; such a change is called a **chemical change.**

Examples of Chemical Change

Chemical changes are going on around you—and inside you—every day. When you eat, food undergoes chemical changes so that your body can use it. When plants use water and carbon dioxide to make sugar and oxygen, a chemical change occurs. Many industries make use of chemical changes to manufacture useful products from raw materials. Most products—from the toothpaste you use every morning to the clothes you wear—are produced by chemical changes.

The surface of Earth is a product of chemical changes. Remember that physical weathering breaks down rocks. Chemical changes occur as chemical weathering. Caves, as shown in **Figure 9-17,** are formed by chemical weathering. Acid rain also brings about chemical weathering. Look at a statue that has been outside for a long time. How do you think it has changed over the years? The acid in acid rain is responsible for damaging marble statues and building materials, as well as for making some lakes toxic to wildlife.

FIGURE 9-17

Over many years, rainwater slowly reacts with layers of limestone rock. It forms caves and collects minerals that it later deposits as cave formations.

Mini LAB

Compare Chemical Changes

Show how salt speeds up the chemical change between oxygen and iron.

1. Separate a piece of fine steel wool into two halves.
2. Dip one half in tap water and the other half in the same amount of salt water.
3. Place both pieces of steel wool on a paper plate and let them sit overnight. Observe any changes.

Analysis

1. What happened to the steel wool that was dipped in the salt water?
2. What might be a common problem with machinery that is operated near an ocean?

Across the Curriculum

Life Science When your body releases or uses energy and necessary nutrients from food, it uses chemical changes. It has been found that some combinations of food facilitate or hinder chemical changes. For example, rice and beans together provide a complete protein, and adequate magnesium makes it easier for your body to absorb calcium.

Mini LAB

Purpose

Ⓚ **Kinesthetic** Students will observe the different reaction rates when wet steel and salty wet steel react with air. Ⓛ2

Materials

fine, unsoaped steel wool; water; salt; paper plate

Teaching Strategies

Ask students to predict the outcome based on the information in the Science and Society section.

Troubleshooting Some steel wool is lightly coated with oil. Use steel wool that has neither soap nor oil.

📁 **Activity Worksheets,** pages 5, 60

Analysis

1. It showed much more rust than the other piece.
2. If not properly protected, salt from the ocean would speed rusting of the steel.

✔ Assessment

Performance Ask students to infer the benefits and problems of using plastic for car bodies. Be sure they include whether any problems are with physical properties of plastic or with its chemical properties. Have them present their findings in a paragraph. Use the Performance Task Assessment List for Consumer Decision-Making Study in **PASC,** p. 43. Ⓛ3 Ⓟ

Students may have heard that *hard water* and soap form soap scum. Explain to them that hard water contains minerals—usually compounds containing calcium or magnesium—that undergo a chemical change to form solids when mixed with soap. *Soft water* does not contain these compounds.

GLENCOE TECHNOLOGY

 **Videodisc**

Glencoe Earth Science Interactive Videodisc

Side 1, Lesson 2
15881-20194
Carbonic Acid

21218-23684

Limestone

23686-25474

Cave Formation

25476-28140

3 Assess

Check for Understanding

Discussion Have students discuss the physical and chemical changes that occur when someone eats an apple.

Reteach

A chemical change is a chemical *reaction* in which atoms in substances are rearranged. If students can't describe a change as a reaction in which a new substance with different properties is formed, then the change is probably physical.

256

Signs of Change

Ice melts, paper is cut, metal is hammered into sheets, and clay is molded into a vase. Seeing signs of these physical changes is easy—something changes shape, size, form, or state.

Sometimes, it's just as easy to tell that a chemical change has occurred. When wood burns, you see it change to ash and other products. When you combine soap and some types of tap water, soap scum forms a ring in your sink. Another example is shown in **Figure 9-18**. When this cake is baked, changes occur that make the cake become solid. The chemical change that occurs when baking powder mixes with water forms bubbles that make the cake rise. Raw egg in the batter undergoes changes that make the egg solid.

In all of these examples, you know that a chemical change occurs because you can see that a new substance forms. It's not always so easy to tell when new substances are formed. What are other signs of chemical change? Do Activity 9-2 to find out.

FIGURE 9-18
Chemical changes are common when food, such as cake, is cooked.

256   Chapter 9 Properties and Changes

Across the Curriculum

Art Leonardo da Vinci was one of the world's great artists, but only a few of his artworks remain. He experimented with using oil paint instead of watercolors to paint frescoes. Because of chemical changes caused by the interaction of the oil paint, air, and plaster, *The Battle of Anghiari* was lost within ten years, and *The Last Supper* must be restored continuously.

Sunset in a Bag

How do you know when a chemical change occurs? You'll see some evidence of chemical change in this activity.

What You'll Investigate
What is evidence of a chemical change?

Procedure 🌡️ 🥽

1. **Add** 20 mL of warm water and 5 mL of phenol red solution to the plastic bag. **Seal** the bag, and gently **slosh** the solution around to mix it.
2. Now, **add** a teaspoon of calcium chloride to the solution in the bag. Again, **seal** the bag and **slosh** the contents to mix the solution. **Record** any change in temperature.
3. **Open** the bag and quickly **add** a teaspoon of baking soda. Again, **seal** the bag and **slosh** the contents to mix the ingredients together. **Observe** what happens.

Conclude and Apply

1. What evidences of chemical change did you **observe**?
2. Is it always easy to tell whether or not energy is released? **Give an example** of a chemical change that does not show an obvious energy change.

Goals
- Observe a chemical change.
- Identify some signs of chemical change.

Materials
- baking soda
- calcium chloride
- phenol red solution
- warm water
- teaspoons, 2
- resealable plastic bag
- graduated cylinder

257

Activity 9-2

Purpose
LS **Visual-Spatial** Observe some of the signs of a chemical change. **L2** **ELL**

Process Skills
observing, inferring, recognizing cause and effect, experimenting, formulating hypotheses

Time
45 minutes

Alternate Materials
Many products sold to spread on icy pavement contain calcium chloride. The percentage of calcium chloride in these products varies. If you choose to use one of these products instead of laboratory calcium chloride, try the activity ahead of time to make sure the energy change can be noticed. If phenol red is not available, a purple solution made from cooked red cabbage leaves can be used, but the color change is less dramatic.

📁 **Activity Worksheets,** pages 5, 57-58

Teaching Strategies
Troubleshooting Caution students to be sure the bags are sealed before sloshing the contents.

Answers to Questions

1. an energy change (warmth) when adding the calcium chloride, gas released from adding the baking soda, color changes in the phenol red
2. No; examples might include steel rusting or an apple turning brown.

Process Ask students to write a lab report, summarizing their procedures and results and specifying their variables and controls. Use the Performance Task Assessment List for Lab Report in **PASC**, p. 47. **LS**

For students who have mastered this section, use the **Reinforcement** and **Enrichment** masters.

4 Close

·MINI·QUIZ·

Use the Mini Quiz to check students' recall of chapter content.

1. **Classify each of the following as either a physical change or a chemical change.**

a. **snow melting** *physical*

b. **baking muffins** *chemical*

c. **burning gas in a car** *chemical*

d. **weaving a blanket** *physical*

Section Wrap-up

1. Student answers will vary but might include breaking an egg (change in shape), freezing ice cubes (change of state), or blowing up a balloon (change in size).

2. Chemical; the structure of the egg changes during cooking; new substances form.

3. **Think Critically** Making ice cream changes liquid ingredients to a solid.

USING MATH

$(1000 \text{ g} + 16 \text{ g}) - 976 \text{ g} = 40 \text{ g}$ ⓛⓢ

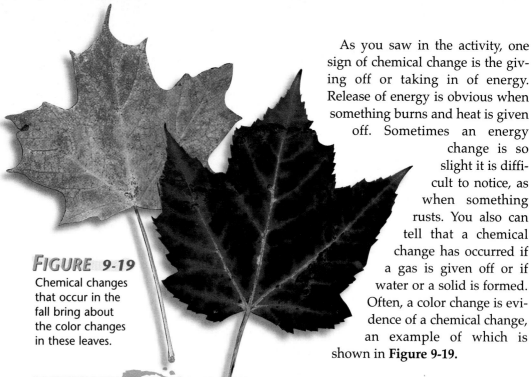

FIGURE 9-19
Chemical changes that occur in the fall bring about the color changes in these leaves.

As you saw in the activity, one sign of chemical change is the giving off or taking in of energy. Release of energy is obvious when something burns and heat is given off. Sometimes an energy change is so slight it is difficult to notice, as when something rusts. You also can tell that a chemical change has occurred if a gas is given off or if water or a solid is formed. Often, a color change is evidence of a chemical change, an example of which is shown in **Figure 9-19.**

Section Wrap-up

1. List five physical changes you can observe in your home. Explain how you decided that each change is physical.

2. When you cook an egg, what kind of change occurs—physical or chemical? Explain.

3. **Think Critically:** Which of the following involves a change of state: grinding beef into hamburger, pouring milk into a glass, making ice cream, or allowing soup to cool in a bowl? Explain.

4. *Skill Builder*

   Developing Multimedia Presentations
   Prepare a multimedia presentation that shows the steps in preparing, lighting, and extinguishing a wood fire. Identify each step as a physical or a chemical change. If you need help, refer to Developing Multimedia Presentations on page 564 of the **Technology Skill Handbook.**

USING MATH

No mass is lost during a chemical change, even if it seems like there might be. If 1000 g of wood burn in 16 g of oxygen, 976 g of ash are produced. How many grams of gases are also produced?

Skill Builder

Developing Multimedia Presentations Presentations might include the following steps: gather wood, physical; chop wood, physical; assemble kindling and wood, physical; light match, chemical; burn wood, chemical; cover fire with dirt, physical; add water to ashes, physical. Ⓟ

✔Assessment

Content Have students work in groups to create a work of art using both physical and chemical changes. Explain what each change is and how it contributes to the artwork. Use the Performance Task Assessment List for Group Work in **PASC**, p. 97. Ⓟ

Ⓛ2 ⓛⓢ **COOP LEARN**

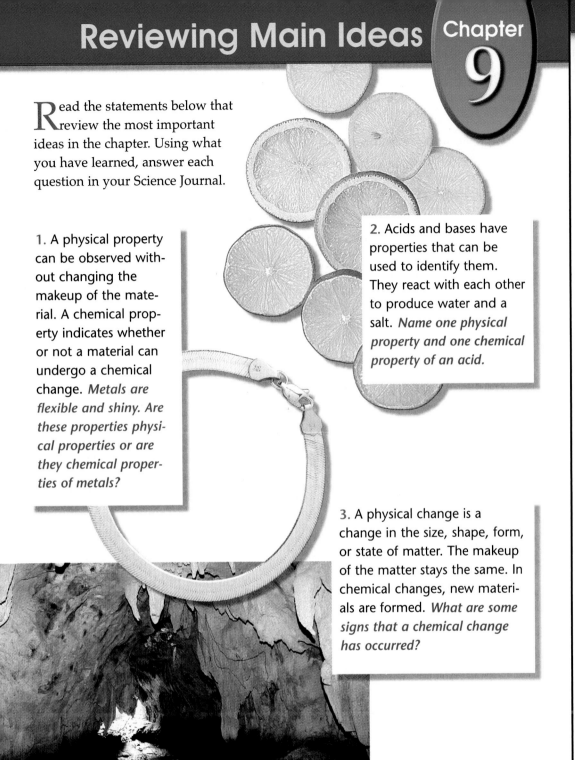

Read the statements below that review the most important ideas in the chapter. Using what you have learned, answer each question in your Science Journal.

1. A physical property can be observed without changing the makeup of the material. A chemical property indicates whether or not a material can undergo a chemical change. *Metals are flexible and shiny. Are these properties physical properties or are they chemical properties of metals?*

2. Acids and bases have properties that can be used to identify them. They react with each other to produce water and a salt. *Name one physical property and one chemical property of an acid.*

3. A physical change is a change in the size, shape, form, or state of matter. The makeup of the matter stays the same. In chemical changes, new materials are formed. *What are some signs that a chemical change has occurred?*

Chapter 9 Review **259**

Review

Have students look at the illustrations on this page. Ask them to describe details that support the main ideas of the chapter found in the statement for each illustration.

Teaching Strategies

Be sure students can distinguish between properties and changes, as well as between physical and chemical. Give them several examples and have them identify each as either a property or a change and then as either physical or chemical.

Answers to Questions

1. physical properties
2. Answers will vary but may include "a sour taste" as a physical property and "reacts with metals" as a chemical property.
3. release of a gas, change in energy, formation of a solid, color change

Science at Home

IS **Intrapersonal** Fill one small freezable container with water and one with salt water containing a couple of teaspoons of salt. The end cubes on an ice-cube tray work well as containers. Place the containers in the freezer and check regularly to see which one freezes first. Relate the results to the effect of salt on icy surfaces. **L2** **ELL**

✔ Assessment

Portfolio Encourage students to place in their portfolios one or two items of what they consider to be their best work. Examples include:
- Assessment, p. 247
- Skill Builder, p. 258
- Assessment, p. 258 **P**

Performance Additional performance assessments may be found in **Performance Assessment** and **Science Integration Activities**. Performance Task Assessment Lists and rubrics for evaluating these activities can be found in Glencoe's **Performance Assessment in the Science Classroom (PASC)**.

Using Key Science Words

1. density
2. physical property
3. physical change
4. chemical change
5. state of matter

Checking Concepts

6. d
7. b
8. a
9. d
10. d

Thinking Critically

11. yes, if the bag of feathers has a much greater volume than the bag of rocks
12. It is a physical property because the compounds water and sugar remain unchanged in the solution. No chemical change occurred.
13. carbon and hydrogen
14. **a.** physical
    b. chemical
    c. physical
15. **a.** physical
    b. chemical
    c. chemical

Chapter 9 Review

Using Key Science Words

chemical change   physical change
chemical property   physical property
density   state of matter

Match each phrase with the correct term from the list of Key Science Words.

1. mass divided by volume
2. describing an object using color, shape, size, texture, odor, or form
3. snowballs melting in the sun
4. acid rain damaging marble statues
5. matter as a solid, a liquid, or a gas

Checking Concepts

Choose the word or phrase that completes the sentence.

6. The temperature at which something boils is a _____.
   a. chemical change
   b. chemical property
   c. physical change
   d. physical property

7. A chemical change might be identified by release of a gas or _____.
   a. change of state
   b. energy
   c. liquids
   d. physical properties

8. New compounds are formed during a _____ change.
   a. chemical   c. seasonal
   b. physical   d. state

9. Solid, liquid, and gas are _____.
   a. physical changes
   b. physical properties of soil
   c. chemical changes
   d. three states of matter

10. A broken window would be considered a _____ change.
    a. chemical   c. neutral
    b. weathering   d. physical

Thinking Critically

Answer the following questions in your Science Journal using complete sentences.

11. Think about what you know about density. Could a bag of feathers be heavier than a bag of rocks? Explain.

12. Sugar dissolves in water. Is this a physical property or a chemical property of sugar?

13. When butane burns, it combines with oxygen in the air to form carbon dioxide and water. What two elements must be present in butane?

14. Identify each of the following as either a physical property or a chemical property.
    a. Sulfur shatters when hit.
    b. Gasoline burns explosively.
    c. Baking soda is a white powder.

15. Identify each of the following as either a physical change or a chemical change.
    a. Metal is drawn out into a wire.
    b. Sulfur in eggs tarnishes silver.
    c. Baking powder bubbles when water is added to it.

Assessment Resources

Reproducible Masters
Chapter Review, pp. 21-22
Assessment, pp. 39-42
Performance Assessment, p. 47

Glencoe Technology
◉ **Computer Test Bank**
▥ **MindJogger Videoquiz**

Developing Skills

If you need help, refer to the description of each skill in the Skill Handbook.

16. **Comparing and Contrasting:** Compare and contrast physical and chemical changes.

17. **Interpreting Scientific Illustrations:** Review the pictures below and determine whether each is a chemical change or a physical change.

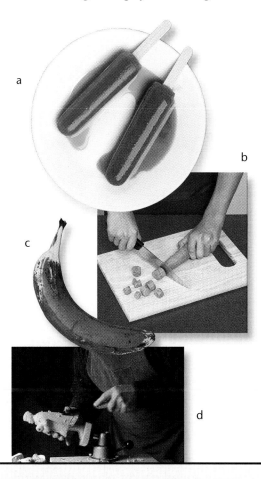

a

b

c

d

18. **Measuring in SI:** Suppose you have a rock that has a mass of 92 g. You pour water into a graduated cylinder to the 37-mL mark. After adding the rock, the water level reads 83 mL. What is the density of the rock?

19. **Observing and Inferring:** When you mixed two substances together, you observed that heat, gas, and some light were produced. Is this change chemical or physical? Explain.

20. **Concept Mapping:** Make a concept map that uses the terms *matter, physical properties, chemical properties, physical changes,* and *chemical changes.* Include an example for each term. Use the connecting terms *has* and *undergoes.*

Performance Assessment

1. **Display:** Create a display that demonstrates the characteristics of a chemical change. Be sure your display shows release of energy, change of color, and the formation of a solid.

2. **Designing an Experiment:** You know that some acids react with metal to produce a chemical change. Design an experiment that allows you to decide whether acids react with all metals or whether only a few combinations produce chemical changes.

Developing Skills

16. **Comparing and Contrasting** Both change matter. A physical change results in a change in size, shape, or form, but not identity. A chemical change results in a change in the identity of the matter.

17. **Interpreting Scientific Illustrations**
    a. physical
    b. physical
    c. chemical
    d. physical

18. **Measuring in SI**
    Volume = 83 mL − 37 mL = 46 mL;
    Density = mass / volume = 92 g / 46 mL = 2 g/mL

19. **Observing and Inferring** Producing energy (heat and light) is one sign of a chemical change, as is the formation of a new substance (the gas).

20. **Concept Mapping** Concept maps should show that matter has physical properties and chemical properties and undergoes physical changes and chemical changes. At least one example should be provided for each property and change.

Performance Assessment

1. Displays should make it clear that all of these characteristics do not happen with each chemical change. If students make a display that actually involves a chemical reaction, the teacher first must check the plans for safety. Use the Performance Task Assessment List for Display in **PASC**, p. 63. **P**

2. Collect several metals. Guide students to choose at least one nonreactive metal, such as gold, and at least one relatively reactive metal, such as aluminum. Expose the metals to an acid and observe any changes. If a student actually performs this experiment, the procedure must be approved ahead of time, and the teacher should supervise the use of acids. Students can use a saturated washing soda (sodium carbonate) solution instead of an acid. Use the Performance Task Assessment List for Designing an Experiment in **PASC**, p. 23. **P**

Chapter Organizer

Section	Objectives/Standards	Activities/Features
Chapter Opener		Explore Activity: Marble Skateboard Model, p. 263
10-1 **What is gravity?** (2 sessions, 1 block)*	1. **Investigate** the effects of gravity. 2. **Compare and contrast** weight and mass. 3. **Distinguish** methods of measuring weight and mass. National Science Content Standards: (5-8) UCP2, UCP3, A1, A2, B2, D3, E1, E2, G3	Problem Solving: Weighty Matters, p. 266 MiniLAB: Estimating Mass, p. 267 Skill Builder: Inferring, p. 268 Using Math, p. 268
10-2 **How fast is "fast"?** (2½ sessions, 1½ blocks)*	4. **Describe** movement of objects. 5. **Determine** factors that affect movement. National Science Content Standards: (5-8) UCP2, UCP3, A1, A2, B2, E2, F5	Activity 10-1: Time Trials, p. 271 Using Technology: Making Skateboards, p. 272 MiniLAB: Analyzing Friction Data, p. 273 Skill Builder: Recognizing Cause and Effect, p. 274 Using Math, p. 274
10-3 **How do things move?** (3½ sessions, 1½ blocks)*	6. **Apply** Newton's three laws to everyday life. National Science Content Standards: (5-8) UCP1, UCP2, UCP3, A1, A2, B2, E2, F5, G1	Activity 10-2: Rocket Races, pp. 280-281 Technology Skill Builder: Using E-Mail, p. 283 Science Journal, p. 283
10-4 **Science and Society:** **Air-Bag Safety** (1 session, ½ block)*	7. **Investigate** air-bag design and safety issues. 8. **Analyze** information from news media. National Science Content Standards: (5-8) B2, E2, F1, F5, G1	Skill Builder: Analyzing News Media, p. 285 People & Science, p. 286

* A complete Planning Guide that includes block scheduling is provided on pages 31T-33T.

Activity Materials

Explore	Activities	MiniLABs
page 263 construction paper, stapler, marble, masking tape	page 271 meterstick, timer, wind-up toys, masking tape pages 280-281 scissors, aluminum foil, balloon, construction paper, waxed paper, wide rubber band, drinking straws, string, masking tape	page 267 canned goods, plastic bags with handles, 1-kg mass page 273 metal washer, sandpaper, waxed paper, protractor

Need Materials? Call Science Kit (1-800-828-7777).

Teacher Classroom Resources

Reproducible Masters	Transparencies	Teaching Resources
Activity Worksheets, pp. 5, 65 **Enrichment,** p. 36 **Multicultural Connections,** pp. 23-24 **Reinforcement,** p. 36 **Study Guide,** p. 36	**Science Integration Transparency 10,** On the Moon **Section Focus Transparency 30,** Weighty Matters	**Spanish Resources** **English/Spanish Audiocassettes** **Cooperative Learning Resource Guide** **Lab Partner** **Lab and Safety Skills** **Lesson Plans**
Activity Worksheets, pp. 5, 61-62, 66 **Cross-Curricular Integration,** p. 14 **Enrichment,** p. 37 **Lab Manual 20** **Reinforcement,** p. 37 **Science Integration Activities,** pp. 57-58 **Study Guide,** p. 37	**Section Focus Transparency 31,** And the Winner Is… **Teaching Transparency 19,** Speed vs. Time	
Activity Worksheets, pp. 5, 63-64 **Enrichment,** p. 38 **Lab Manual 19** **Reinforcement,** p. 38 **Science and Society/Technology Integration,** p. 28 **Study Guide,** p. 38	**Section Focus Transparency 32,** Lazy Days **Teaching Transparency 20,** Forces	**Assessment Resources** **Chapter Review,** pp. 23-24 **Assessment,** pp. 43-46 **Performance Assessment,** p. 48 **Performance Assessment in the Science Classroom (PASC)** **MindJogger Videoquiz** **Alternate Assessment in the Science Classroom** **Computer Test Bank**
Enrichment, p. 39 **Reinforcement,** p. 39 **Study Guide,** p. 39	**Section Focus Transparency 33,** Crash Test/Pop Quiz	

Key to Teaching Strategies

The following designations will help you decide which activities are appropriate for your students.

L1 Level 1 activities should be appropriate for students with learning difficulties.

L2 Level 2 activities should be within the ability range of all students.

L3 Level 3 activities are designed for above-average students.

ELL ELL activities should be within the ability range of English Language Learners.

LS These activities are designed to address different learning styles.

COOP LEARN Cooperative Learning activities are designed for small group work.

P These strategies represent student products that can be placed into a best-work portfolio.

GLENCOE TECHNOLOGY

The following multimedia resources are available from Glencoe.

Science and Technology Videodisc Series (STVS)
Physics
   Science of Bowling
   Safer Roads
   New Skid Control
   Laminar Flow over
      Airplane Wings
   Wind Engineering
The Infinite Voyage Series
Living with Disaster

Glencoe Physical Science Interactive Videodisc
Introduction
Distance and Displacement
Speed
Velocity
Acceleration as a Change in Velocity
Free-fall
Conclusion
Friction
Mass and Acceleration
Glencoe Physical Science CD-ROM

Teacher Classroom Resources

This is a representation of key blackline masters available in the Teacher Classroom Resources.

Teaching Aids

Section Focus Transparencies

30 SECTION FOCUS TRANSPARENCY

WEIGHTY MATTERS

Machines like the one below were common in the past. They often appeared at drugstores, carnivals, and arcades.

1. What does the machine do? How do you know?
2. What similar machines have you seen?

L2

31 SECTION FOCUS TRANSPARENCY

AND THE WINNER IS . . .

With a final burst of speed, the runner crosses the finish line. It was a close race, and the winner may have just broken a track record.

1. What would you need to know to judge how impressive the runner's achievement was?
2. What do the judges have to measure to determine who won the race?

L2

32 SECTION FOCUS TRANSPARENCY

LAZY DAYS

Garfield ®   by Jim Davis

1. What first rule of physics is Garfield thinking about?
2. How does the rule relate to what Garfield is doing?

L2

Science Integration Transparencies

10 SCIENCE INTEGRATION TRANSPARENCY

On the Moon

L2

Teaching Transparencies

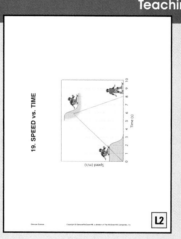

19. SPEED vs. TIME

L2

20. FORCES

L2

Meeting Different Ability Levels

Study Guide for Content Mastery

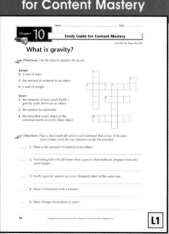

Chapter 10   Study Guide for Content Mastery

What is gravity?

Directions: Use the clues to complete the puzzle.

Across
3. a unit of mass
5. the amount of material in an object
6. a unit of weight

Down
1. the measure of how much Earth's gravity pulls down on an object
2. the symbol for kilometer
4. the force that every object in the universe exerts on every other object

Directions: Place a check mark (✓) next to each statement that is true. If the statement is false, write the true statement on the line provided.

1. Mass is the amount of material in an object.
2. A bowling ball will fall faster than a pencil when both are dropped from the same height.
3. Earth's gravity speeds up every dropped object at the same rate.
4. Mass is measured with a balance.
5. Mass changes from place to place.

L1

Reinforcement

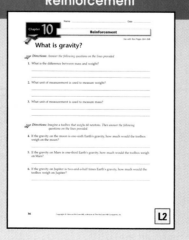

Chapter 10   Reinforcement

What is gravity?

Directions: Answer the following questions on the lines provided.

1. What is the difference between mass and weight?
2. What unit of measurement is used to measure weight?
3. What unit of measurement is used to measure mass?

Directions: Imagine a toolbox that weighs 60 newtons. Then answer the following questions on the lines provided.

4. If the gravity on the moon is one-sixth Earth's gravity, how much would the toolbox weigh on the moon?
5. If the gravity on Mars is one-third Earth's gravity, how much would the toolbox weigh on Mars?
6. If the gravity on Jupiter is two-and-a-half times Earth's gravity, how much would the toolbox weigh on Jupiter?

L2

Enrichment Worksheets

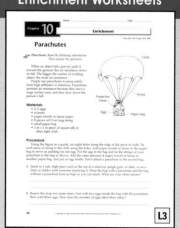

Chapter 10   Enrichment

Parachutes

Directions: Read the following information. Then answer the questions.

When an object falls, gravity pulls it toward the ground. But air resistance slows its fall. The bigger the surface of a falling object, the more air resistance.

People use parachutes to jump safely from high altitudes or distances. Parachutes increase air resistance because they have a large surface area, and they slow down the person's fall.

Materials
• 2–3 eggs
• scissors
• paper towels or tissue paper
• 8 pieces of 0.5-m long string
• small paper bag
• 1 m x 1 m piece of square silk or other light cloth

Procedure

Using the figure as a guide, cut eight holes along the edge of the piece of cloth. Tie each piece of string to the cloth using the holes. Add paper towels or tissue to the paper bag to serve as padding for one egg. Put the egg in the bag and tie the strings of your parachute to the bag as shown. Add the same amount of paper towels or tissue to another paper bag, and put an egg inside. Don't attach a parachute to the second bag.

1. Stand in a safe, high place such as the top of a staircase, jungle gym, or slide, or on a chair or ladder with someone steadying it. Drop the bag with a parachute and the bag without a parachute from as high as you can reach. What are your observations?

2. Repeat the drop two more times, first with two eggs inside the bag with the parachute, then with three eggs. How does the number of eggs affect their safety?

L3

262C

Hands-On Activities

Science Integration Activities

Going my way?

You know that friction stops motion. But do you know that friction can sometimes help things go? When friction causes one object to stick to another, the second object may catch a ride with the first.

Seeds are moved in many different ways. Some seeds are blown by the wind so that they are far from the parent plant. Other seeds are eaten by animals and then dropped far from where they came from. Still other types of seeds are held to the bodies of animals by friction. These types of seeds are sometimes called burs. Burs have little hooks, or spines, on them. The spines get tangled in animal fur or on people's clothing, which helps them be carried far away.

One inventor, George de Mestral, looked carefully at seeds. He took an idea from nature and created a hook-and-loop fastener, often called Velcro. His invention modeled burs and the way they were carried. Today, Velcro is used on everything from athletic shoes to the space shuttle.

Getting Started

In this activity, you will try to find the best material to help move spiny seeds. You will study Velcro and then make a comparison between the material that moves seeds and animal fur.

Hypothesizing

Read through the following procedure. Look at the fuzzy side of the Velcro. Which material do you think is most similar to this fuzzy side? Make a hypothesis about which kind of material will transport the most seeds the farthest distance.

Try It!

- Velcro
- shoe box or other small container
- 5 numbered material samples
- 6 craft sticks
- glue
- scissors

1. First make some models of seeds from pieces of Velcro. Take the side of Velcro that has many small hooks on it, like little stiff spines. Cut it into 1-cm squares. Take two of the 1-cm squares. Put glue on their flat side and glue them together so that the spiny side of each faces out. Continue making seeds until all the squares have been used. Put your seeds in the container, or "seed box."

Seed models

57 · L2

Lab Manual

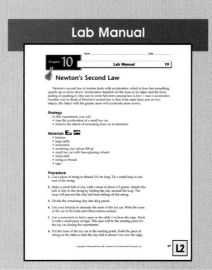

Lab Manual 19

Newton's Second Law

Newton's second law of motion deals with acceleration, which is how fast something speeds up or slows down. Acceleration depends on the mass of an object and the force pulling or pushing it. One way to write Newton's second law is *force = mass × acceleration*. Another way to think of Newton's second law is that if the same force acts on two objects, the object with the greater mass will accelerate more slowly.

Strategy

In this experiment, you will:
- time the acceleration of a toy car.
- observe the effects of increasing mass on acceleration.

Materials

- balance
- large table
- meterstick
- modeling clay (about 300 g)
- small toy car with free-spinning wheels
- stopwatch
- string or thread
- tape

Procedure

1. Cut a piece of string or thread 110 cm long. Tie a small loop in one end of the string.
2. Make a small ball of clay with a mass of about 2.5 grams. Attach this ball of clay to the string by folding the clay around the loop. The loop will prevent the clay ball from falling off the string.
3. Divide the remaining clay into 40-g pieces.
4. Use your balance to measure the mass of the toy car. Write the mass of the car in the Data and Observations section.
5. Use a meterstick to find a spot on the table 1 m from the edge. Mark it with a small piece of tape. This spot will be the starting point for the toy car during the experiment.
6. Put the front of the toy car at the starting point. Hold the piece of string on the table so that the clay ball is about 3 cm over the edge.

47 · L2

Activity Worksheets

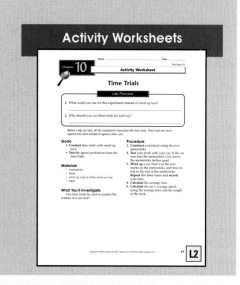

Activity Worksheet

Time Trials

Lab Preview

1. What could you use for this experiment instead of wind-up toys?

2. Why should you run three trials for each toy?

Before a big car race, all the contestants must pass the time trials. Time trials are races against the clock instead of against other cars.

Goals
- **Conduct** time trials with wind-up toys.
- **Test** the speed predictions from the time trials.

Materials
- metersticks
- timer
- wind-up cars or other wind-up toys
- tape

What You'll Investigate
Can time trials be used to predict the winner of a car race?

Procedure
1. **Construct** a racetrack using the two metersticks.
2. **Test** your track with your car if the car runs into the metersticks a lot, move the metersticks farther apart.
3. **Wind up** a car. Start it at the zero marks on the metersticks, and time its trip to the end of the metersticks. **Repeat** this three times and **record** your data.
4. **Calculate** the average time.
5. **Calculate** the car's average speed using the average time and the length of the track.

61 · L2

Enrichment and Application

Cross-Curricular Integration

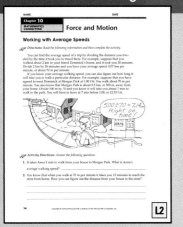

MATHEMATICS CONNECTION Force and Motion

Working with Average Speeds

Directions: Read the following information and then complete the activity.

You can find the average speed of a trip by dividing the distance you traveled by the time it took you to travel there. For example, suppose that you walked about 2 km to your friend Dominick's house, and it took you 30 minutes. Divide 2 km by 30 minutes and you have your average speed: 0.07 km per minute, or about 70 m per minute.

If you know your average walking speed, you can also figure out how long it will take you to walk a particular distance. For example, suppose that you have agreed to meet Dominick at Morgan Park at 1:00 P.M. You walk about 70 m per minute. You also know that Morgan Park is about 0.5 km, or 500 m, away from your home. Divide 500 m by 70 and you know it will take you about 7 min to walk to the park. You will have to leave at 7 min before 1:00, or 12:53 P.M.

500 ÷ 70 = 7.14, Right on time!

Activity Directions: Answer the following questions.

1. It takes Anna 5 min to walk from your house to Morgan Park. What is Anna's average walking speed?

2. You know that when you walk at 70 m per minute it takes you 15 minutes to reach the store from home. How can you figure out the distance from your house to the store?

14 · L2

Multicultural Connections

Multicultural Connections

Weighed in the Balance

Have you ever compared the weight of two objects by holding one in each hand and deciding which was heavier? If so, you were using what may have been humans' earliest way of measuring weight.

Egyptian Scales

As long as 7000 years ago, the Egyptians improved on this friction method of weighing things. They made a simple balancing scale to compare weights. Although the ancient Egyptians wouldn't have described it in these words, their scale depended on the principle that gravity pulls harder on heavier objects.

To make their scale, they tied a string around a straight stick at about its center. When they held the free end of the string, the stick hung in balance; each end of it would be the same distance from the ground. To compare the weight of two objects, they would hang them from opposite ends of the stick. If the stick still balanced, the objects were about the same weight. If one object pulled its end of the stick down toward Earth, then it was heavier. Later, people suspended a pan or dish from each end of the pole. Placing things to be weighed in pans was easier than tying them to the ends of the stick.

Standard Weights

Around 3000 B.C., the Egyptians started making weights from stone. Later they used copper. Objects to be weighed were placed on one pan of the scale, and these weights were added until the scale was balanced. Each stone or copper weight weighed a certain fixed amount or a multiple of that amount. We call such an

amount a standard unit of weight. In ancient Egypt, the basic unit was called a kite. It was roughly the same as the unit of weight we call an ounce. If nine one-kite weights in one pan balanced some grain in the other, then that grain weighed nine kites. A farmer would set a price of so much per kite of grain. By weighing what a buyer needed, the farmer could tell exactly how much the buyer should pay. Because a kite was a standard weight, the buyer knew just how much grain he or she was getting.

Not an Easy Task

It was one thing to decide that a kite would be the unit of weight and another would be the unit of a moving car to make sure all kites were the same. The ancient Egyptians had a civilization that lasted nearly 3000 years. Archaeologists, scientists who study ancient cultures, have found that the kite varied in weight from about 4.8 g (0.16 ounce) to about 32 g (1.06 ounce), depending on the period of history.

23 · L2

Science and Society/ Technology Integration

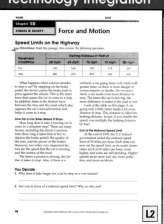

SCIENCE & SOCIETY Force and Motion

Speed Limits on the Highway

Directions: Read this passage, then answer the following questions.

Braking Distance in Feet at:

Pavement Conditions	60 mph	65 mph	70 mph	75 mph	80 mph
Dry	140	164	190	218	247
Wet	280	410	475	545	618

What happens when a driver decides to stop a car? By stepping on the brake pedal, the driver causes the brake pads to press against the wheels. This is the main force that causes the car to come to a stop. In addition, there is the friction force between the tires and the road, which also opposes the car's forward motion and helps it come to a stop.

How Far a Car Goes Before It Stops

How long does it take a moving car to come to a complete stop? There are many factors, including the driver's reaction time (how long it takes him or her to depress the brake pedal), the quality of the tires, and the power of the brakes. However, two other very important factors are the speed that the car is moving and the condition of the road.

The faster a person is driving, the farther it takes to stop. Also, if there is a

collision, a car going faster will crash with greater force, so there is more danger of serious injuries or deaths. On wet pavement, a car needs even more distance to stop. The farther the car is traveling, the more difference it makes if the road is wet.

Look at the table on this page. A car going only a little faster needs a lot more distance to stop. This distance is called the braking distance. In fact, if you double the speed, you multiply the braking distance by four.

End of the National Speed Limit

At the end of 1995, the U.S. federal government ended the national speed limit of 55 miles per hour. Each state can now set the speed limit on its roads. Some states set it at 65 miles per hour; some higher, and some are still deciding. Higher speeds mean more fuel use, more pollution, and more accidents.

You Decide
1. Why does it take longer for a car to stop on a wet surface?

2. Are you in favor of a national speed limit? Why or why not?

28 · L2

Assessment

Performance Assessment

Skill Assessment

A Day at the Races

You can see for yourself how friction and gravity affect acceleration by using just a cotton ball, a marble, and a simple ramp. Make a ramp with a slight incline by using a book to prop up one end of a board or piece of stiff cardboard. Hold the cotton ball and the marble at the top of the ramp, then release them at the same time and see which one crosses the finish line at the bottom first. Hold several races, increasing the angle of the ramp by adding another book for each race. Describe what happens on the lines below.

Race #1:

Race #2:

Race #3:

Race #4:

48 · L2

Chapter Review

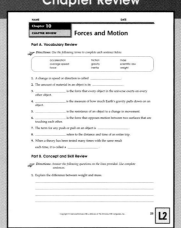

CHAPTER REVIEW Forces and Motion

Part A. Vocabulary Review

Directions: Use the following terms to complete each sentence below.

acceleration	friction	mass
average speed	gravity	scientific law
force	inertia	weight

1. A change in speed or direction is called _____.

2. The amount of material in an object is its _____.

3. _____ is the force that every object in the universe exerts on every other object.

4. _____ is the measure of how much Earth's gravity pulls down on an object.

5. _____ is the resistance of an object to a change in movement.

6. _____ is the force that opposes motion between two surfaces that are touching each other.

7. The term for any push or pull on an object is _____.

8. _____ refers to the distance and time of an entire trip.

9. When a theory has been tested many times with the same result each time, it is called a _____.

Part B. Concept and Skill Review

Directions: Answer the following questions on the lines provided. Use complete sentences.

1. Explain the difference between weight and mass.

23 · L2

Assessment

CHAPTER TEST Forces and Motion

I. Testing Concepts

Directions: Match each description in the first column with the item in the second column by writing the correct letter in the space provided.

___ 1. a change in speed or direction
___ 2. force that every object in the universe exerts on every other object
___ 3. resistance of an object to a change in movement
___ 4. any push or pull on an object
___ 5. the speed of an object over an entire trip

a. acceleration
b. force
c. gravity
d. inertia
e. average speed

Directions: For each question, write the letter of the best choice in the space provided.

___ 6. Which statement is true?
a. The weight of an object always stays the same, but its weight can change.
b. The weight of an object always stays the same, but its mass can change.
c. The mass and weight of an object both stay the same.
d. The mass and weight of an object both can change.

___ 7. Newton's second law states that
a. things don't stay or start by themselves
b. for every action force there is an equal and opposite reaction force
c. acceleration depends on the mass of an object and the force pushing or pulling the object
d. the push or pull that opposes motion between two surfaces that are touching each other is called friction

___ 8. To calculate average speed, you need to know
a. the distance something traveled
b. the time it took for something to travel
c. both the distance and the time it took for something to travel
d. neither the distance nor the time it took for something to travel

45 · L2

Chapter 10

Forces and Motion

CHAPTER OVERVIEW

Section 10-1 The relationships among mass, weight, and gravity are introduced. How each is measured is explained.

Section 10-2 Velocity and acceleration are explained. Friction is introduced as a force that opposes movement.

Section 10-3 Newton's three laws are presented, explained, and applied.

Section 10-4 Science and Society Air-bag safety for all passengers is explored. Students also learn to analyze news media.

Chapter Vocabulary

gravity	friction
weight	scientific law
mass	force
average speed	inertia
acceleration	

Theme Connection

Stability and Change Newton's three laws of motion are used to describe the stability of an object, as well as any change in motion that the object might undergo.

Chapter Preview

Section 10-1 What is gravity?

Section 10-2 How fast is "fast"?

Section 10-3 How do things move?

Section 10-4 Science and Society: Air Bag Safety

Skills Preview

▶ **Skill Builders**
- infer
- recognize cause and effect
- use E-mail

▶ **MiniLABs**
- estimate
- analyze data

▶ **Activities**
- observe
- collect data
- calculate
- predict
- hypothesize

262

Learning Styles

Look for the following logo for strategies that emphasize different learning modalities. **LS**

Kinesthetic	Explore, p. 263; MiniLAB, p. 267; Inclusion Strategies, p. 273; Activity 10-2, p. 280
Visual-Spatial	Activity, p. 265; Reteach, p. 268; Visual Learning, pp. 270, 277; Community Connection, p. 276; Demonstration, p. 278
Interpersonal	Check for Understanding, p. 282; Career Connection, p. 286
Intrapersonal	Science at Home, p. 287
Logical-Mathematical	Assessment, pp. 267, 268; Check for Understanding, p. 268; Activity 10-1, p. 271; Science Journal, p. 273; Using Math, p. 274; Enrichment, p. 277
Linguistic	Content Background, p. 265; Using Science Words, p. 266

Chapter 10

Forces and Motion

Skateboarders fly up into the air, turn, and come down onto the ramp as if skateboarding were simple. Studying such motion helps you understand forces at work. After you've studied this chapter, you'll understand a lot more about the forces behind the motions of skateboarding. You can start learning about forces in the activity below.

EXPLORE ACTIVITY

Marble Skateboard Model

1. Using the picture as a guide, staple two pieces of construction paper together at one end to make a ramp.
2. Use the smallest ring binder you have to make a slope. Tape one edge of the ramp to the binder. Set a book on the other side to hold the ramp in place.
3. Roll a marble down the slope so that it travels up the ramp. Use the pen to mark the marble's highest point on the ramp.
4. Put a book under the ring binder, and roll the marble again.

Science Journal

In your Science Journal, tell which slope made the marble roll higher up the ramp. What property of the slopes made the marble roll to different heights?

263

Prepare

Section Background

- Gravity is directly related to the masses of two objects and inversely related to the distance between them.
- Because of Earth's gravity, all free-falling objects near Earth accelerate downward at the same rate unless drag (air resistance) slows the descent.

Preplanning

Refer to the Chapter Organizer on pages 262A-B.

1 Motivate

Bellringer

Before presenting the lesson, display **Section Focus Transparency 30** on the overhead projector. Assign the accompanying **Focus Activity** worksheet.

`L2` `ELL`

10•1 What is gravity?

What YOU'LL LEARN

- How gravity affects everything
- The difference between weight and mass
- How weight and mass are measured

Science Words:
gravity
weight
mass

Why IT'S IMPORTANT

On Earth, gravity affects you every moment of every day.

Why Things Fall

To move an object, you usually have to do something to it. You have to push or pump a skateboard, open doors, or turn the pages of this book. But, sometimes all you have to do is let go of the object. It moves without being pushed, turned, or pumped. In the Explore activity on the previous page, the marble rolled down the ramp by itself. When a skateboarder jumps into the air, he or she always comes back to the ground. There is a reason why things fall or roll without being thrown or pushed.

Throughout history, people have tried to explain why things fall. The modern explanation for why things fall was first proposed in the 1600s. A scientist named Sir Isaac Newton said that everything—you, a marble, your desk, the moon—pulls on everything else. He called this pull gravity (GRA vuh tee). **Gravity** is a pull that every object exerts on every other object. **Figure 10-1** shows the pull of gravity in action.

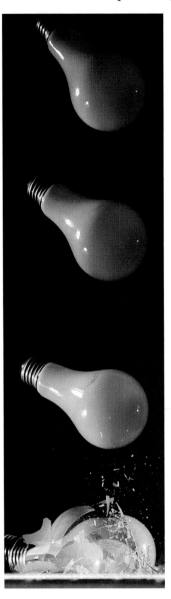

FIGURE 10-1
This lightbulb falls because of gravity between the lightbulb and Earth.

264 Chapter 10 Forces and Motion

Program Resources

📁 Reproducible Masters
Activity Worksheets, pp. 5, 65 `L2`
Enrichment, p. 36 `L3`
Multicultural Connections, pp. 23-24 `L2`
Reinforcement, p. 36 `L2`
Study Guide, p. 36 `L1`

Transparencies
Science Integration Transparency 10 `L2`
Section Focus Transparency 30 `L2`

Gravity Treats Everything the Same

Newton said that gravity acts on everything, large or small, in the same way. For example, if a tennis ball and a bowling ball are dropped at the same time, which will hit the ground first? The answer is that they will both hit the ground at the same time. When the astronauts walked on the moon, they demonstrated this fact with a feather and a lead weight. On Earth, air slows a feather's fall, just as it slows the fall of the skydiver in **Figure 10-2.** On the moon, there isn't any air to get in the way. Puffs of dust on the moon let everyone know exactly when the feather and the lead weight hit the ground. Newton was right! The two objects hit the moon at exactly the same time.

You Have Gravity!

Remember, Newton's explanation said that everyone and everything has its own gravity. If that's true, why aren't you pulled toward a desk at the other side of the room? The answer has to do with matter. The more stuff, or matter, something has, the more pull it has. Earth is huge compared to a desk. The pull of Earth's gravity is so great that you must notice it. It keeps your feet on the ground. That desk across the room is pulling, but the pull is so small that you never notice it.

The More Matter, the More Gravity

Our moon and Jupiter are good examples of how matter affects gravity. The moon's gravity is about one-sixth as strong as Earth's. Jupiter, on the other hand, is much larger than Earth. Jupiter's gravity is about two and one-half times as strong as Earth's. Imagine how different skateboarding would be on the moon or on Jupiter.

Have you seen movies of the astronauts walking on the moon? Lower gravity and bulky space suits make it hard to walk like they do on Earth, so they hop. It seems dangerous because they jump so high. How high do you think astronauts could jump if they were on solid ground on Jupiter?

FIGURE 10-2

Air slows the fall of this sky-diver. Without the parachute, he or she would fall much faster.

10-1 What is gravity?   **265**

Tying to Previous Knowledge

Even though they may not be able to verbalize a formal definition of *gravity*, all students have observed the effects of *gravity*. Have them relate examples of times they have experienced gravity at work, such as when they have fallen or dropped something breakable.

2 Teach

Activity

LS **Visual-Spatial** Have students drop identical pieces of paper from identical heights; one piece of paper should be wadded up and the other one flat. They can see that air resistance slows the flat paper.
L1 ELL

Teacher F.Y.I.

NASA has videotapes available of the astronauts on the moon. Search for the NASA site on the Internet to find out how to contact NASA.

? FLEX Your Brain

Use the Flex Your Brain activity to have students explore PARACHUTES.

📁 **Activity Worksheets,** page 5

Inquiry Question

Does your book lay on your desk because of your desk's gravity or because of Earth's gravity? Explain. *Earth's gravity; the mass of Earth is so much greater than the mass of the desk that Earth's gravity is much greater than that of the desk.*

Student Text Question

How high do you think astronauts could jump if they were on Jupiter? *They could not jump as high as they could jump on Earth because of increased gravity on Jupiter.*

Content Background

LS **Linguistic** Students frequently think that gravity is related to the amount of air present. Emphasize that gravity is related to mass, not air. For example, gravity is less on the moon because the moon has less mass than Earth, not because there is less air.

IS Linguistic Emphasize that a balance is used to find the mass of objects, and a scale, such as a bathroom scale, finds weight.

Problem Solving

Solve the Problem

1. The proportions of the ramps would be the same. No matter how strong or weak gravity is, it accelerates you down the ramp and slows you as you go up the ramp. The moon's gravity is less than Earth's gravity. On the moon, you would be going slower at the bottom of the ramp, but decreased gravity would slow you less as you go up the other side.

2. See the answer to question 1. On Jupiter, you would be going faster by the time you reached the bottom of the ramp, but that same greater gravity would slow you down just as fast as you go up the other side.

3. The main problem would be using materials that would stand up under Jupiter's high gravity. The ramp on the moon could use weaker materials.

Think Critically

The ramps would be the same proportions, and you would probably go just as high on each. The differences could be in the strength of the design and the construction materials.

Problem Solving

Weighty Matters

You've done an activity with slopes and ramps. Now it's time to design two ramps of your own. One will be on the moon and one will be on Jupiter.

You'll need to design a jump ramp for the moon, where there is less gravity than on Earth, so things weigh less than on Earth. It won't take much force to shoot you high up into the dark sky.

Next, design one for Jupiter, where gravity is two and one-half times what it is here. You'll be heavy, but gravity will provide great acceleration. On Jupiter, it will be hard to launch the skateboard high.

Solve the Problem:

1. **How much steeper will the jump ramps need to be for the moon? Remember, the moon has only one-sixth of the gravity of Earth.**
2. **How much less steep could the ramps be for Jupiter?**
3. **Will you have to use different materials to support the different weights?**

Think Critically:

In what ways could both ramps be just like one on Earth? In what ways will they need to be different?

How do you measure gravity?

How do you suppose scientists know that the moon's gravity is one-sixth of Earth's gravity? How can you measure Earth's gravity? You measure the pull of gravity every time you use a bathroom scale.

Weight is a measure of how much Earth's gravity pulls down on an object. Every time you weigh yourself on a scale, you are measuring the pull of Earth's gravity. If you weigh 100 pounds on Earth, your weight on the moon would be 17 pounds. How much would you weigh on Jupiter? The answer is 250 pounds. Take your own weight and multiply it by 2.5. Imagine what it would be like to jump, run, or even walk on Jupiter.

Astronomers learned that Jupiter is larger than the moon by observing their sizes using telescopes. But you know that gravity doesn't depend on size—it depends on the amount of matter present. **Mass** is the amount of matter in an object. Mass doesn't change from place to place. Your weight would be less on the moon, but your mass wouldn't change. You have to take away or add something to change an object's mass. For example, you could shrink the mass of a candy bar by taking a bite from it.

How do you measure mass?

Mass is measured with a tool called a balance. See **Figure 10-3.** A balance is like a small, precise seesaw. To measure your mass, you could sit on one end of a seesaw while someone else stacked objects on the other end until the seesaw balanced on its pivot point. On Jupiter or the moon, it would still take the same mass of objects to balance your mass. Mass doesn't change from place to place.

It's important to remember that a balance measures mass, but a scale measures the effect of gravity on an object by measuring weight. A balance and a scale are not the same thing.

FIGURE 10-3

Both a balance and a seesaw can be used to find the mass of an object. When they balance, masses are equal on both sides.

Mini LAB

Estimating Mass
Your body is going to be a balance.

1. Get two plastic grocery bags—one with a kilogram of mass in it and one that is empty.
2. Hold one bag in each hand.
3. Have someone put cans of food into the empty bag until you think the two bags have the same mass and balance each other.
4. Add the masses of the cans together.

Analysis
1. A kilogram equals 1000 grams. How close did you come to balancing the kilogram of mass?
2. Would this bag balance be accurate enough for a science lab? Explain.

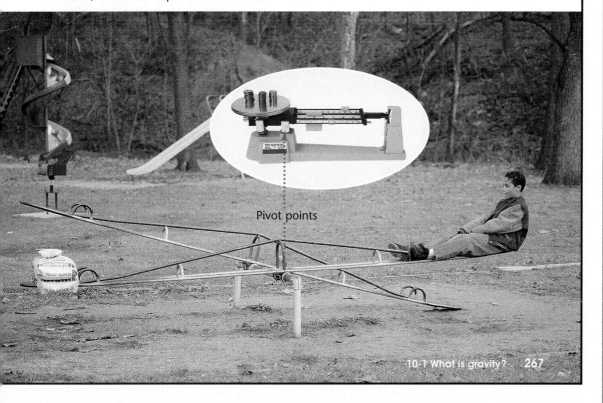

Pivot points

Mini LAB

Purpose
Kinesthetic Students will experience the scientific principles involved in using a balance. **L2 ELL**
COOP LEARN

Materials
two plastic grocery bags with handles, kilogram mass, different sizes of canned goods

Alternate Materials
Canned goods can be replaced with any nonbreakable product that lists mass and weight on the label. The kilogram mass can be 1 kg of salt or sand in a sealed plastic bag.

Teaching Strategies
• Before starting, demonstrate use of a balance.
• Use canned goods, such as tomato sauce, that have the labels painted on or make sure the labels are taped on securely.

Troubleshooting Avoid heavy cans. They can dent, leak, or hurt a foot if dropped.

Activity Worksheets, pages 5, 65

Analysis
1. Answers will vary.
2. No; good scientific results are repeatable.

✓Assessment

Performance Have students use the canned goods to calculate the approximate weight in pounds of a kilogram mass (about 2 pounds, 3 ounces). Use the Performance Task Assessment List for Using Math in Science in **PASC**, p. 29. **LS**
L3 ELL

Check for Understanding

Inquiry Question One ball is thrown horizontally from the same height at the same time that another ball is dropped. Which ball hits the ground first? *They hit at the same time.* `LS`

Reteach

`LS` **Visual-Spatial** Have students draw a diagram that shows a skydiver's body in a position in which it would move downward fastest and slowest because of increased air resistance. `L1` `ELL` `P`

Extension

For students who have mastered this section, use the **Reinforcement** and **Enrichment** masters.

4 Close

·MINI·QUIZ·

Use the Mini Quiz to check for understanding.
1. **What is gravity?** *the force that objects exert on each other because of mass*
2. **The more mass a planet has, the stronger its _____ is.** *gravity*
3. **What instrument do you use to measure mass?** *a balance*

Section Wrap-up

1. 100 cm on Earth; 600 cm on the moon
2. **Think Critically** Earth's gravity would be less.

USING MATH

mass, 60 kg; weight, 200 N

FIGURE 10-4
The mass of each of these things always stays the same. Their weights, however, can be different in different places.

You may have heard someone say he or she "weighs" 50 kilograms. This isn't really correct. Mass and weight are two different things, as explained in **Figure 10-4**. It's important to use different units for them. Mass is measured in grams (g) or kilograms (kg). Weight is measured in newtons (N), named in honor of Sir Isaac Newton. On Earth, a medium-sized apple weighs about 1 N. A kilogram weighs about 10 N on Earth. So if your mass is 50 kg, your weight is about 500 N on Earth.

Section Wrap-up

1. If an astronaut could hop 40 cm high on Jupiter, how high could he or she jump on Earth? On the moon?

2. **Think Critically:** If Earth's mass were smaller, how would Earth's gravity be affected?

3. **Skill Builder**
   Inferring Because gravity on Mars is about one-third of that on Earth, could you infer that Mars has more mass than Earth? Why? If you need help, refer to Inferring on page 550 in the **Skill Handbook.**

USING MATH

Imagine that you had a mass of 60 kg and a weight of 600 N. What would your weight and mass be on Mars if the gravity there were one-third as strong as Earth's?

Skill Builder
Inferring No, it would have less mass. Less gravity implies less mass because gravity depends on the mass of each of two objects and the distance between them.

Assessment

Performance Have students calculate their weight on Earth, Jupiter, Mars, and the moon and anonymously submit it. Then have the class compile and put the group results on a poster. Use the Performance Task Assessment List for Poster in **PASC**, p. 73. `LS` `L2`

How fast is "fast"? 10•2

Faster and Faster

Think of skating down Dead Man's Hill. Your heart starts to pound as you go faster and faster. You feel every bump in the pavement through your feet and into your knees as the curving path seems to move faster and faster past you. It's the pull of gravity that makes the trip down so much fun and the climb up such hard work. Keep thinking about this trip as you read on.

How fast were you going?

What was your speed near the bottom of the hill? Thirty kilometers per hour? Five kilometers per hour? Speed is a measure of how far you move in a given time. To find your speed, you need to know both time and distance. The path down Dead Man's Hill is 100 m long. Suppose a friend timed your trip down and found that it took you 20 s. Divide 100 m by 20 s and you've found your average speed: 5 meters per second, or about 18 kilometers per hour. "Meters per second" is abbreviated *m/s*. "Kilometers per hour" is abbreviated *km/h*. Speed can be read directly from a speedometer, such as the one shown in **Figure 10-5.**

What YOU'LL LEARN

- How to talk about the movement of objects
- What things affect movement

Science Words:
average speed
acceleration
friction

Why IT'S IMPORTANT

You can use speed, acceleration, and friction every time you move or cause something to move.

FIGURE **10-5**
The speedometer in this car uses units of mi/h, or mph, and units of km/h.

269

Prepare

Section Background
- Gravity causes falling objects near Earth to accelerate at 9.8 m/s^2.
- Friction changes some of an object's kinetic energy into thermal energy. The loss of kinetic energy slows down the object.

Preplanning
Refer to the Chapter Organizer on pages 262A-B.

1 Motivate

Bellringer

Before presenting the lesson, display **Section Focus Transparency 31** on the overhead projector. Assign the accompanying **Focus Activity** worksheet.

L2 **ELL**

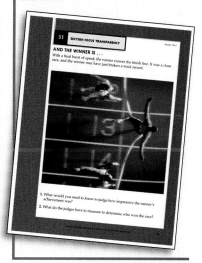

Tying to Previous Knowledge

LS **Linguistic** Although they may not know an official definition, students are familiar with the concept of speed. Ask students what the word *speed* brings to mind.

Program Resources

Reproducible Masters
Activity Worksheets, pp. 5, 61-62, 66 **L2**
Cross-Curricular Integration, p. 14 **L2**
Enrichment, p. 37 **L3**
Lab Manual, pp. 71-74 **L2**
Reinforcement, p. 37 **L2**
Science Integration Activities, pp. 57-58 **L2**
Study Guide, p. 37 **L1**

Transparencies
Section Focus Transparency 31 **L2**
Teaching Transparency 19 **L2**

2 Teach

Remember the Explore activity at the beginning of the chapter? At the top of the ramp, the marble wasn't moving at all. It gradually moved faster as it rolled down the ramp. Your trip down the hill was similar. You weren't moving at 5 m/s at the beginning, and you were moving faster than 5 m/s at the end. But your average speed was 5 m/s.

Average speed describes the movement for an entire trip. If you traveled at exactly 5 m/s from top to bottom of the hill, the 100-m trip would take 20 s. Look at **Figure 10-6** for another example of average speed.

FIGURE 10-6

Look at the graph. If the path of the skateboarder from start to finish is 52 m, what is his average speed?

Content Background

Units used for speed vary depending on the distance and time units used. Most students are used to seeing mi/h (mph) as a unit for speed. A common derived unit in SI for speed is m/s. Most speedometers are calibrated in both mi/h and km/h.

Visual Learning

Figure 10-6 If the path of the skateboarder from start to finish is 52 m, what is his average speed? *6.5 m/s* LS

Time Trials

Before a big car race, all the contestants must pass the time trials. Time trials are races against the clock instead of against other cars.

What You'll Investigate
Can time trials be used to predict the winner of a car race?

Procedure
1. **Construct** a racetrack using the two metersticks. Use the picture to help you.
2. **Test** your track with your car. If the car runs into the metersticks a lot, move the metersticks farther apart.
3. **Wind up** a car. Start it at the zero marks on the metersticks, and time its trip to the end of the metersticks. **Repeat** this three times and **record** your data.
4. **Calculate** the average time.
5. **Calculate** the car's average speed using the average time and the length of the track.

Conclude and Apply
1. **Compare** the average speed of your car to those of the other cars in your class.
2. **Predict** which car would win a race based on the results of your time trials. Test your prediction.
3. **Explain** whether time trials accurately predict which car will win a race.

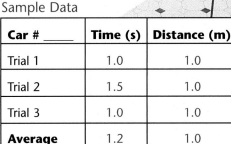

Sample Data

Car # _____	Time (s)	Distance (m)
Trial 1	1.0	1.0
Trial 2	1.5	1.0
Trial 3	1.0	1.0
Average	1.2	1.0

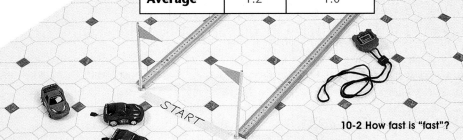

10-2 How fast is "fast"? **271**

Answers to Questions
1. Answers will vary according to student results.
2. Students will probably predict that the toy with the lowest average time would win.
3. Time trials are a good, but not perfect, predictor of which car will win a race.

Purpose
LS **Logical-Mathematical** Students calculate average speeds, then use them to make predictions. **L2** **ELL** **COOP LEARN**

Process Skills
observing and inferring, measuring in SI, collecting and organizing data, interpreting data, experimenting, analyzing, using numbers

Time
40 minutes

Alternate Materials
If sufficient metersticks are unavailable, students can calibrate and use any straight pieces of wood or plastic.

Safety Precautions
Be sure toys are not left where they will be stepped on.

📁 **Activity Worksheets,** pages 5, 61-62

Teaching Strategies
- Use masking tape on the floor to mark the start and finish lines.
- This activity lends itself to students working in groups of four or more. Each group can then have a student be a timer, a recorder, a starter, and a toy releaser.
- Review averaging with students before doing the activity.

Troubleshooting Use a 2-m track for larger toys. Caution students about overwinding the toys.

Goals
- Conduct time trials with wind-up toys.
- Test the speed predictions from the time trials.

Materials
- metersticks (2)
- timer
- wind-up cars or other wind-up toys
- tape

✔ **Assessment**

Process Ask a track coach to provide the results for all schools participating in a specific event at a track meet. Use these results to start a discussion of the many variables involved. Use the Performance Task Assessment List for Analyzing the Data in **PASC**, p. 27.

271

Relate these concepts to a bicycle tire. When bicycle tires are underinflated, more of the tires touch and stick to the road, which makes riding more work. Properly inflated tires touch less of the road and make riding easier.

*inter*NET CONNECTION

Be sure students recognize the unbalanced forces present and the effect of friction.

Mini LAB

Purpose
Students will analyze how different surfaces affect friction. **L2** **ELL**

Materials
ring binder; metal washer; different types of surfaces such as plain paper, waxed paper, or sandpaper; protractor

Alternate Materials
A wide board can be used instead of the binder. Other surfaces that can be used include different grades of sandpaper, construction paper, or aluminum foil.

Teaching Strategies
Students may need to tape surfaces to the ring binder.

Troubleshooting Be prepared to demonstrate how to use the protractor to measure the angle formed by the binder cover and the table-top.

Activity Worksheets, pages 5, 66

How do you measure "speeding up"?
Think back to your skateboarding trip down Dead Man's Hill. Instead of timing your trip like you timed the wind-up cars in the activity, you could use a speedometer or radar gun to measure your speed on the hill. These tools would give your speed at one particular moment in time. At the top of the hill, your speed was 0 m/s. Near the bottom of the hill, it was close to 10 m/s, the fastest speed you went.

As you were going downhill, you were going faster and faster. As you went down the hill, you followed the curving path. At the bottom, you slowed down and stopped. All these changes in speed and direction are examples of acceleration (ak sel uh RAY shun). **Acceleration** is a change in speed or direction. An acceleration can speed you up, slow you down, or change

USING TECHNOLOGY
Making Skateboards

As you might guess, a skateboard's wheels play an important part in its design. What the board is best used for is determined by the size and hardness of the wheels.

When Speed Is King
Skateboard wheels are made out of urethane, a kind of plastic. Some urethanes are softer than others. Softer wheels give a better grip. They are used for "skating" hills where the skater likes to carve his or her way down a hill—where the skater likes to pick up as much speed as possible. Softer wheels are also wider and taller. The increased height makes them faster because they travel farther than a short wheel each time they go around.

When Style's the Thing
Harder, smaller wheels are used by skaters who want to do tricks, such as tail slides, sliding the back of the board on curbs, and cutbacks, which are quick turns. The smaller size helps keep the board under control.

*inter*NET CONNECTION

In both in-line skating and skateboarding, forces and motion come into play. Visit the Glencoe Homepage found at *www.glencoe.com/sec/science* for a link to a site discussing the physics of in-line skating.

Analysis
1. Answers will vary. The smoothest surface usually requires the smallest angle because the surface creates less friction.
2. You could wax or oil the surfaces.

✓ Assessment

Process Discuss which moving parts of bicycles, skateboards, and skates have the most friction. Pass out several ball-bearing sets. Ask students to write an explanation of how these devices help reduce friction. Use the Performance Task Assessment List for Writing in Science in **PASC**, p. 87.

your direction of travel. From top to bottom of the hill, you accelerated from 0 m/s to about 10 m/s. When you stopped, you accelerated from 10 m/s to 0 m/s.

Use your trip down Dead Man's Hill as an example to help explain acceleration. As you were going down the hill, you went a little faster each second—first 0 m/s, then 1 m/s, then 2 m/s, and so on. Your speed was increasing 1 m/s each second. Your acceleration was 1 m/s per second.

Stopping Is Important

At the bottom of the hill, it would be wise to stop before you run into a tree! Stopping is a change in speed, so it is also a kind of acceleration. Traveling down the hill, you kept moving faster. That was adding a bit of speed each second. Braking is moving slower and slower. That is like subtracting speed each second.

Gravity speeds you up as you come down the hill, but what makes you stop? When you are on a skateboard, you drag a foot or slide the board sideways to stop. If you are using in-line skates, you might use a heel stop. Each of these ways involves rubbing something against the ground to slow down and stop. This is using friction (FRIHK shun) to help you stop. **Friction** is the push or pull that opposes motion between two surfaces that are touching each other. Think of rubbing together two pieces of coarse sandpaper. Friction makes it hard to slide the two pieces back and forth.

Mini LAB

Analyzing Friction Data

Skateboards, bicycles, and in-line skates are no fun when there's too much friction.

1. Use your ring binder for a slope as you did in the Explore activity.
2. Make the slope so steep that a flat metal washer will slide all the way to the bottom.
3. Put different types of surfaces on your slope such as plain paper, sandpaper, and waxed paper. Find the angle of the slope for each surface that is just steep enough to make the metal washer slide all the way to the bottom. For each material, measure the angle of the slope with a protractor, as shown below, and record your data in a data table.

Analysis

1. Of the surfaces, which one required the smallest angle of slope for the ring to slide to the bottom of the ramp? What makes that surface different from the others?
2. How could you change the surfaces to make the angle of incline needed to move the washer smaller?

- Many students may incorrectly believe that friction depends on surface area. Friction actually depends on the surface materials and the force pushing the surfaces together.
- Students might think that acceleration always has a positive value. Some of this preconception could result from a vehicle's accelerator being the pedal used to speed up the vehicle.

Content Background

- Be sure students understand that acceleration can be either positive or negative and that what is commonly called *deceleration* is acceleration with a negative value.
- A rolling object has very little friction between itself and the ground because no sliding occurs between the two surfaces.

? FLEX Your Brain

Use the Flex Your Brain activity to have students explore LUBRICANTS.

📁 **Activity Worksheets,** page 5

3 Assess

Check for Understanding

Using an Analogy Have students rub the palms of their hands together and notice how easily they slide. Now ask them to make a fist, rub their knuckles together, and notice how different this feels. Ask students to relate this analogy to friction as the result of texture of a surface. When the surface is smooth, there is little friction. L1 ELL

Science Journal **Texture Change** After doing the MiniLAB, use the same ramp to roll marbles over the same surfaces. See what effect the roughness has on the rolling marbles. Have students explain why results differ when comparing the washers and the marbles. LS L2 P

Inclusion Strategies

Visually Challenged Have students feel the surfaces used in the MiniLAB and rate them from 1 to 10, 10 being the roughest. The rest of the class can use these ratings to describe their results. A student also could put a hand at the bottom of the slope so that the rolling marbles hit the hand. Have them rate each surface according to how hard the marble hits. LS

Reteach

Have students calculate the speed for each time and distance listed below. L2

Time (s)	Distance (m)	Speed (m/s)
0	0	0
1	0.5	0.5
2	2.0	1.0
3	4.5	1.5
4	8.0	2.0

Extension

For students who have mastered this section, use the **Reinforcement** and **Enrichment** masters.

4 Close

•MINI•QUIZ•

Use the Mini Quiz to check for understanding.

1. **What tells you how fast something is going?** *speed, which is distance/time*

2. **What is acceleration?** *a change in speed or direction*

3. **Which would have the greater friction, a freshly waxed kitchen floor or a concrete sidewalk?** *the sidewalk*

Section Wrap-up

1. 30 km/h; 8.3 m/s

2. **Think Critically** Brakes press against the wheel rims. The wheels are slowed because of friction between the brakes and the wheels.

USING MATH

IS Logical-Mathematical The race car accelerates at 7.1 m/s². The sports car accelerates at 8.8 m/s². The sports car accelerates faster.

274

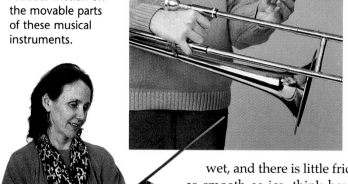

FIGURE 10-7

The rosin increases friction, and the oil reduces friction on the movable parts of these musical instruments.

Uses of Friction

In the MiniLAB, you explored friction with several materials. When you stop a skateboard, you notice the friction between your foot and the ground, but there is also friction between your other foot and the skateboard. Think of rubbing two ice cubes together. The surfaces of the cubes are smooth and wet, and there is little friction. If the skateboard were as smooth as ice, think how hard it would be to stop without slipping off the board and falling. Friction can be useful, or it can be a problem, as shown in **Figure 10-7**. It's useful when you are trying to stop or stand up, but it slows down the wheels. You want the wheels to turn as easily as possible. Otherwise, skating will be more work and you won't be able to go as fast.

Section Wrap-up

1. If you rode with your cousin for a two-hour trip to a town 60 km away, what was your average speed in km/h? In m/s?

2. **Think Critically:** Why is friction an important part of stopping a car?

3. **Skill Builder**
Recognizing Cause and Effect On a skateboard, where do you think friction is a problem? What will it do to the skateboard after a lot of use? What could you do to reduce the amount of friction? If you need help, refer to Recognizing Cause and Effect on page 551 in the **Skill Handbook.**

USING MATH

A race car moves from standing still to 50 m/s in 7 s. Is it accelerating faster or slower than a sports car that can accelerate from 0 m/s to 70 m/s in 8 s? Explain.

Skill Builder

Recognizing Cause and Effect On a skateboard, the friction is between the wheel and the axle. Over a long time, friction wears away the metal. Using a lubricant such as machine oil reduces the friction and wear, which causes the skateboard to go faster and last longer.

Assessment

Process Have students look at the bottoms of their shoes and vote on which shoe has the most friction and which has the least friction. Ask students to write why they think one shoe would be better than another and plan a way to test their hypotheses. Use the Performance Task Assessment List for Making Observations and Inferences in **PASC**, p. 17. L2 P

How do things move?

It's a Law

Remember Sir Isaac Newton's explanation of gravity from earlier in this chapter? During the last 400 years, Newton's theories have been tested again and again here on Earth and in space. When a theory has been tested many times and the results are the same each time, it may be called a **scientific law.** A scientific law isn't like the laws enforced by a government. A scientific law is an accurate description of some important part of the natural world. Three important laws about how things move are named after Sir Isaac Newton. These three laws apply to anything that moves—spaceships, bikes, skates, or people.

Before we begin talking in detail about Newton's laws, you need to learn an important word: *force*. A **force** is a push or a pull. Gravity is a force. Friction is a force. Every time you walk, open a door, or turn a page, you are using force to move something. Look at **Figure 10-8.** Now, let's explore how Newton's laws relate to forces and to movement.

What YOU'LL LEARN

- How Newton's three laws of motion apply to everyday life
- **Science Words:**
  scientific law
  force
  inertia

Why IT'S IMPORTANT

Newton's laws of motion explain how most sports happen, including skateboarding, biking, in-line skating, and wheelchair racing.

FIGURE 10-8
The people who hold this hide provide an upward force greater than the downward force of gravity on the girl.

275

Prepare

Section Background
- Newton's first law is about inertia. The second law relates force, mass, and acceleration. The third law explains balanced forces.
- Newton's laws hold true for typical speeds. As objects move closer to the speed of light, their motions are explained by Einstein's theory of relativity.

Preplanning
Refer to the Chapter Organizer on pages 262A-B.

1 Motivate

Bellringer
Before presenting the lesson, display **Section Focus Transparency 32** on the overhead projector. Assign the accompanying **Focus Activity** worksheet.
L2 ELL

Program Resources

📂 Reproducible Masters
Activity Worksheets, pp. 5, 63-64 L2
Enrichment, p. 38 L3
Lab Manual, pp. 67-70 L2
Reinforcement, p. 38 L2
Science and Society/Technology
  Integration, p. 28 L2
Study Guide, p. 38 L1

📦 Transparencies
Section Focus Transparency 32 L2
Teaching Transparency 20 L2

Tying to Previous Knowledge

Ask students how many have played air hockey. This game is a good example of inertia. Because there is very little friction, a puck can move at a slow, constant speed down the table until it is hit by a player or deflected by the walls.

2 Teach

Revealing Preconceptions

Weightlessness in orbit is not due to the lack of gravity. Instead, it is similar to the free fall simulated in an aircraft when it dives.

Discussion

Have students discuss how they feel when they go over the top of a hill on a roller coaster. Tell students that this reduced weight is similar to the apparent weightlessness that orbiting astronauts experience.

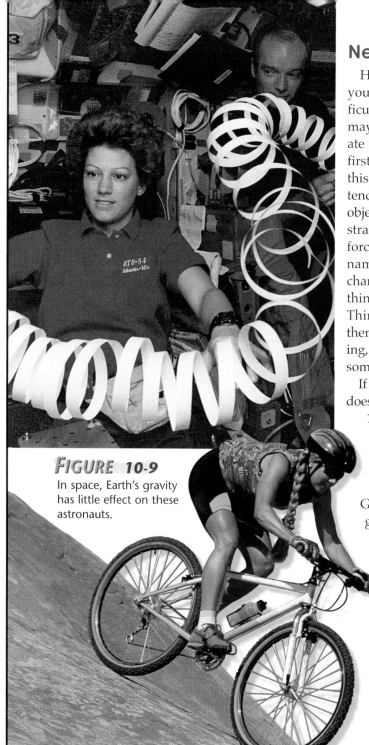

FIGURE 10-9
In space, Earth's gravity has little effect on these astronauts.

Newton's First Law

Have you ever noticed that once you get your bike stopped, it's difficult to get it moving again? You may have to pedal hard to accelerate it back up to speed. Newton's first law of motion explains why this happens: An object at rest tends to stay at rest, and a moving object tends to keep moving in a straight line until it is affected by a force. **Inertia** (ih NUR shuh) is the name for this resistance to a change in motion. Another way to think of Newton's first law is: Things don't stop or start by themselves. If a bike is not moving, it will stay that way until some force makes it move.

If Newton's laws are true, why doesn't the bike shown in **Figure 10-10** coast forever in a straight line? The reason is friction. On Earth, everything eventually slows down and stops because of friction and gravity. Gravity pulls things down to the ground. On the ground, they eventually roll or drag to a stop because of friction between them and the ground.

FIGURE 10-10
Inertia keeps this bicycle rider going until another force—usually friction—causes her to stop.

Across the Curriculum

Language Arts Have students pretend they're astronauts in space for the first time, and have them write in their Journals about a day of weightlessness. They may be surprised at how things float and follow Newton's laws, especially inertia. An object will float in midair unless someone moves it or the air-circulation fans slowly blow it away. LS L2 P

Community Connection

Recreation Safety Have a local law enforcement officer or members from a qualified scout troop speak to the class about bicycle safety. Emphasize to students that proper safety equipment usually either reduces the force on certain body parts or increases or decreases friction. LS

In space, it's easy to see Newton's first law at work, as in **Figure 10-9.** You may have seen astronauts moving equipment outside a spaceship. Even a gentle push starts an object moving, and it doesn't stop until it hits something or the astronaut retrieves it.

Balanced and Unbalanced Forces

It's possible for many forces to act on the same thing at once. If all the forces cancel out, or balance, nothing will change. Think about stopping a bike on a hillside, **Figure 10-11.** You use the brakes. After you've stopped, you put your feet down and let up the brakes a little. Gravity is still pulling the bike down the hill. Both your feet and the bike's brakes are using friction to balance that pull down the hill. If you release the brakes completely or lift your feet, the forces become unbalanced and the bike begins to move again unless you hold it back.

Newton's Second Law

Newton's second law involves force, mass, and acceleration. You know that it's harder to move a large, massive bike than to move a smaller, less massive one. Newton's second law says: Acceleration depends on the mass of an object and the force pushing or pulling the object. Here are two mathematical ways to write Newton's second law:

$$\text{Force} = \text{Mass} \times \text{Acceleration}$$
$$F = m \times a$$

or

$$\text{Acceleration} = \text{Force}/\text{Mass}$$
$$a = F/m$$

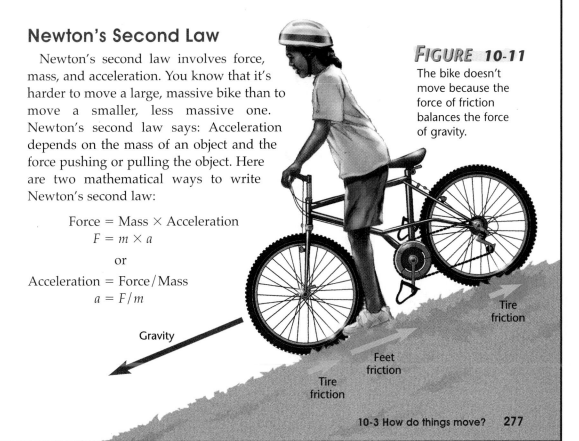

Gravity

Tire friction

Feet friction

Tire friction

FIGURE **10-11**

The bike doesn't move because the force of friction balances the force of gravity.

10-3 How do things move? 277

*inter*NET
CONNECTION

Isaac Newton helped explain many natural laws. Visit the Glencoe Homepage at **www.glencoe.com/ sec/science** to find out more about Newton.

Content Background

Most astronauts are at least mildly space sick when they first experience weightlessness. One way to determine how space sick a particular astronaut will be is to have astronauts ride in a large plane that flies high and then repeatedly dives in a parabolic curve. The astronauts fall inside the empty padded cabin and, during the dive, are weightless.

Enrichment

LS **Logical-Mathematical** Have students use a bicycle and the equation $F = m \times a$ to illustrate Newton's second law. Show that if you increase anything on one side of the equation, the other side must increase. If a student pushes the pedals of a bicycle with more force, the bicycle will accelerate faster (mass stays constant). You must pedal harder (provide more force) to accelerate a heavier bicycle. **L3**

GLENCOE TECHNOLOGY

○ Videodisc
STVS: Physics
Disc 1, Side 1
Science of Bowling (Ch. 2)

*inter*NET
CONNECTION

The Glencoe Homepage at **www.glencoe.com/sec/ science** provides links connecting concepts from the student edition to relevant Internet sites.

Visual Learning

Figure 10-10 Have students study the rider in the photo. **How do bicycle riders reduce friction with the air?** *Student answers may include that their clothes are tight and that they bend over rather than ride straight up.* **LS**

ELL

Demonstration

LS **Visual-Spatial** Newton's second law explains how force, mass, and acceleration are related. Use two identical roller skates; masses; and two identical, medium-sized rubber bands to show how different masses are accelerated differently by the same force.

Load one roller skate with marbles or other small, heavy objects. Use a rubber band to pull each roller skate across a table with the same force. Be sure that the rubber band is stretched the same for each skate so that the force is the same.

The loaded skate will be moving slower than the unloaded skate at the end of a set time period. Thus, the larger mass was accelerated less than the smaller mass.

Here are two simpler ways to think of Newton's second law. Apply them to **Figure 10-12.**

1. It takes less force to accelerate something with a small mass as fast as something with a large mass. A large, massive truck would need a powerful motor to start moving from a complete stop and keep up with a small sports car.

2. If you want to go faster, you have to use more force. You have to pedal harder to accelerate a bike to top speed in 5 s than to do the same in 2 minutes.

FIGURE 10-12

If both riders pedal with the same force, the rider moving less mass accelerates faster.

Theme Connection

Stability and Change

Although gravity is a constant force, its effect on acceleration depends on whether it acts in the same direction or in the opposite direction of the motion. When a bullet is shot straight up into the air, gravity slows it down until it stops. Then the bullet begins to fall as it is accelerated downward by gravity.

Newton's Third Law

If you are walking down the hall and give your friend a playful shove, your friend may react by pushing you back. What happens if you walk over to a wall and give it a push? Does the wall push you back? Believe it or not, it does. Remember, if forces aren't balanced, then something has to change speed or direction. If the force of your push were not balanced by the force of the wall pushing back, you would make the wall move. Newton's third law explains such balanced forces. The third law says: For every action force, there is an equal and opposite reaction force. Another way to think of this law is: Forces always come in pairs. When you push against a wall, as in **Figure 10-13,** it pushes back just as hard.

Does the floor push up on you?

Sure it does, just like the wall! You know your weight is pushing down on the floor, so the floor must be pushing up on you. Think about quicksand. It looks like it will hold you up, but when you step on it, you sink. The sand does push up with an opposite force, but that force is smaller than your weight, and you sink. Any time there is movement, when you sink or fall, the forces acting on you are out of balance. When there is no change in movement, the forces are equal and balanced, such as your force on the floor and the floor's force on you. When you step into quicksand, the forces are not equal and there is definite movement: you sink. In the activity on the next pages, you can find out how action-reaction forces affect rocket movement.

FIGURE 10-13

As this girl pushes on the wall, the wall pushes back with equal force.

10-3 How do things move?   279

FLEX Your Brain

Use the Flex Your Brain activity to have students explore WEIGHTLESSNESS.

Activity Worksheets, page 5

Content Background

If you could try to walk on the planet Jupiter, increased gravity would not be the only problem. Jupiter is a gaseous planet. As you tried to walk, forces would be unbalanced. Although high gravity and low temperatures keep the surface of Jupiter from staying gaseous, the surface is not solid, either. Walking on Jupiter would be like trying to walk on a milkshake.

GLENCOE TECHNOLOGY

Videodisc

STVS: Physics
Disc 1, Side 1
Safer Roads (Ch. 3)

New Skid Control (Ch. 4)

Disc 1, Side 2
Laminar Flow over Airplane Wings (Ch. 6)

Wind Engineering (Ch. 7)

Cultural Diversity

Motion and Speed Some people are interested in pushing the limits of speed in vehicles with high-power engines. But not everyone has joined the race.

In the United States, some conservative Amish may use engines to power farm equipment, but they rely strictly on human and animal energy for transport. Hay balers may be run by engines, but they still have to be pulled on platforms by the trusty family horses.

Sometimes wheeled vehicles simply are not useful. Incas in Peru built a system of roadways and rest houses, but they used no wheeled vehicles. The mountainous terrain made human runners faster and more efficient as messengers and for transport.

Activity 10-2

PREPARE

Purpose
L⁣S **Kinesthetic** Students will design and carry out an experiment that demonstrates Newton's third law. **L2** **ELL**

Process Skills
observing and inferring; comparing and contrasting; recognizing cause and effect; formulating models; using variables, constants, and controls; analyzing; collecting and organizing data; hypothesizing; interpreting data; experimenting

Time
2 class periods

Materials
So that they slide easily on the string, drinking straws should be smooth, not the bendable type. Balloon punching bags are slower but will travel further; they're fun to include.

For string, 20-pound test or stronger nylon monofilament fishing line works best. If you use another kind of string, make sure it won't break or ravel during use.

Possible Hypotheses
Students may hypothesize that the large, long balloons will have the least air resistance and the most air to provide force. Some students will suggest that adding fins or nose cones will reduce drag.

📁 **Activity Worksheets,** pages 5, 63-64

Design Your Own Experiment
Rocket Races

Going into space for the first time was an exciting moment for humanity. Scientists used rockets to test new technology outside our atmosphere for the first time. Rocketry uses Newton's third law. In this experiment, you will have a chance to design your own rocket and race it against others.

PREPARE

Possible Materials
- balloons of various shapes and sizes
- drinking straws
- tape
- construction paper
- waxed paper
- foil
- rubber bands
- scissors
- string or fishing line

What You'll Investigate
Design a balloon rocket that travels fast enough and has enough fuel to cross a finish line first.

Form a Hypothesis
Develop a hypothesis about the design and size of your rocket.

Goals
Design and **build** an efficient rocket that uses Newton's third law for its propulsion.

Compare results.

Safety Precautions
🥽 🧤 Be sure you review and understand all safety precautions before beginning the activity.

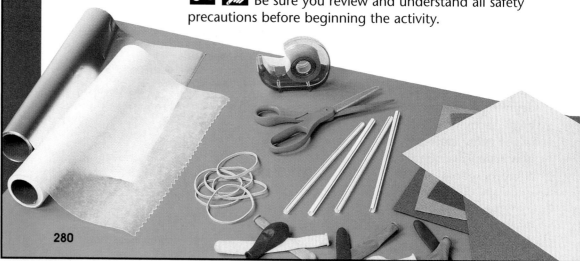

280

✔ Assessment

Portfolio Have students draw and label a new balloon rocket design and tell why they think this would produce the best results. Use the Performance Task Assessment List for Scientific Drawing in **PASC**, p. 55. **L2** **P**

PLAN

1. Make sure your group has agreed on a **hypothesis** statement.
2. **Sketch** a design of your rocket. See the photo and the list of possible materials for ideas.
3. **List** detailed, orderly steps that you will follow to build your rocket.
4. **Choose** your balloon carefully. Which balloon has the best mass? Remember, $F = m \times a$. Will it have enough fuel to finish the race? Will the neck size be important? Which shape do you think will fly the fastest?
5. How long a straw will you attach to your balloon, and how many pieces will you need to guide it along the string?
6. What shape will the rocket have? How will its shape affect its flight?
7. Make sure your teacher approves your plan and that you have included any suggested changes in the plan.

DO

1. **Gather** materials and **build** the rocket as planned.
2. **Race** it against your classmates' rockets.

CONCLUDE AND APPLY

1. **Describe** the winning rocket. What gave it an advantage over the other rockets?
2. What parts of your design were successful? Why?
3. What would you change on your design next time? Why?
4. **APPLY** Which of Newton's laws played an important part in this experiment? **Explain.**

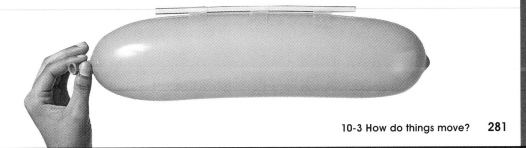

4. Applying Newton's third law, the air expanding against the nozzle is the action force. The nozzle pushing back against the expanding air is the reaction force. When this force pair is greater than the friction between the straw and the string, the balloon moves forward.

Go Further

Challenge students to design and build a two-stage balloon rocket. One solution is to use a paper cylinder around the first balloon. The nozzle of the second-stage balloon can be pinched closed between the cylinder and the first-stage balloon. The second stage is then released when the first-stage balloon deflates. ⬛L3

PLAN

Possible Procedures

Use a rocket made from an inflated balloon taped to a straw. Have string suspended tightly, with the string running through the straw. Release the balloon at the same time as others release their balloons. Other balloons should be set up the same but with different shapes and sizes of balloons. See which balloon crosses a finish line first. Things, such as nose cones, that will reduce air resistance can be added to the balloons. Rockets can be modified and raced again.

Teaching Strategies

The day before doing the activity, compare and contrast various sizes and shapes of balloons—more fuel and more mass versus less fuel, smaller nozzle, less mass, etc.

Troubleshooting Hold the line tight. Some balloons may start to make large, circular motions if the string is too loose.

DO

Expected Outcome

Some rockets will not make it to the finish line. The winner is usually a moderately large, round balloon with one straw attachment and no additional nozzles or fins.

CONCLUDE AND APPLY

1. Student answers depend on results. See Expected Outcome.
2. Student answers depend on results.
3. Student answers depend on results.

281

Check for Understanding

Brainstorming Have two students lean back-to-back in order to balance and support each other. Have students brainstorm as many reasons as possible why this is an example of Newton's laws. Answers might include that the bodies stay at rest (inertia) and that as long as the bodies don't move, the forces are balanced. **IS** **COOP LEARN** **L2**

Reteach

Keeping Newton's laws in mind, have students discuss the qualities needed for a good racing bicycle and bicycle rider. Qualities might include that the rider's clothing and the bicycle itself should minimize friction with the road and with the air.

Extension

📁 For students who have mastered this section, use the **Reinforcement** and **Enrichment** masters.

4 Close

·MINI·QUIZ·

Use the Mini Quiz to check for understanding.

1. **Which of Newton's laws is based on acceleration?** *the second*

2. **Which of Newton's laws is related to action and reaction forces?** *the third*

3. **Which of Newton's laws describes inertia?** *the first*

4. **When an object's motion changes, you know that the forces on it must be _____.** *unbalanced*

5. **Does the floor push up on you? How do you know?** *Yes. I'm not sinking.*

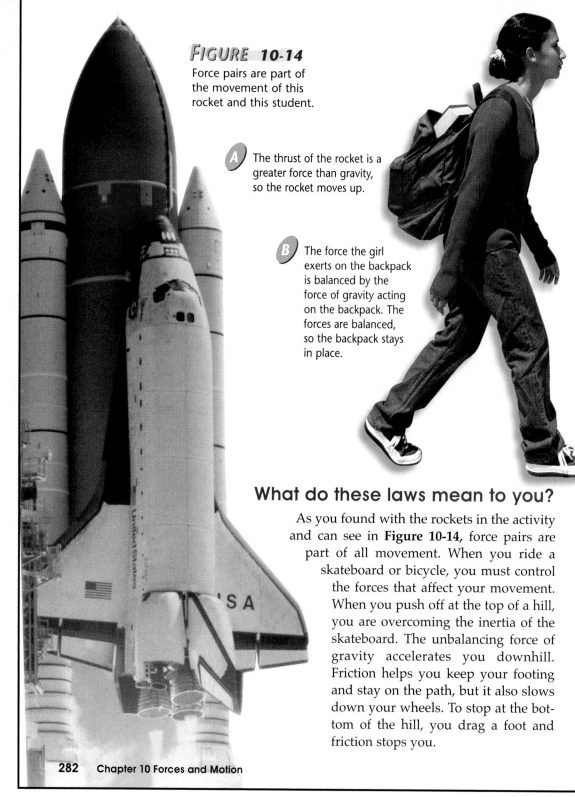

FIGURE 10-14
Force pairs are part of the movement of this rocket and this student.

A The thrust of the rocket is a greater force than gravity, so the rocket moves up.

B The force the girl exerts on the backpack is balanced by the force of gravity acting on the backpack. The forces are balanced, so the backpack stays in place.

What do these laws mean to you?

As you found with the rockets in the activity and can see in **Figure 10-14**, force pairs are part of all movement. When you ride a skateboard or bicycle, you must control the forces that affect your movement. When you push off at the top of a hill, you are overcoming the inertia of the skateboard. The unbalancing force of gravity accelerates you downhill. Friction helps you keep your footing and stay on the path, but it also slows down your wheels. To stop at the bottom of the hill, you drag a foot and friction stops you.

282 Chapter 10 Forces and Motion

What do Newton's laws mean to someone in a wheelchair?

To some people, dealing with gravity is a challenge every day. Imagine being in a wheelchair, such as the boy in **Figure 10-15,** and having to go up and down ramps all day. The friction of the wheels plus gravity make it hard to get up a ramp.

Later, keeping control as the wheelchair accelerates back down the same ramp takes effort. Inertia makes the chair difficult to start and stop. People in wheelchairs understand Newton's laws and how to use each one to their advantage.

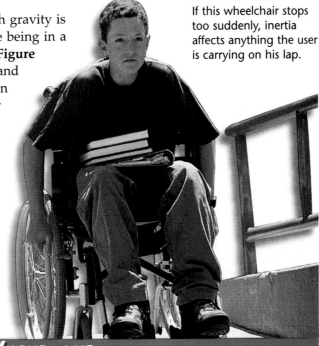

FIGURE 10-15
If this wheelchair stops too suddenly, inertia affects anything the user is carrying on his lap.

Section Wrap-up

1. When a person in a wheelchair stops suddenly, why might the books in his or her lap sometimes fly onto the floor?

2. Tell which of Newton's three laws of motion was the most difficult for you to understand and why. Then, tell how you'd explain one of the laws to a younger student.

3. **Think Critically:** Use a diagram to explain the balanced and unbalanced forces that allow you to sit at your desk and write.

4. **Technology Skill Builder**
   Using E-mail Use E-mail to contact your state or local government. Find out the requirements for building ramps for public buildings. If you need help, refer to Using E-mail on page 565 in the **Technology Skill Handbook.**

Science Journal

In this chapter, a bicycle was used to help explain $F = m \times a$. In your Science Journal, use a skateboard to explain this same law.

Section Wrap-up

1. Because of inertia, the books have a tendency to keep moving in a straight line.

2. The first part of the answer should name one of Newton's laws. The second part should state that law in simple terms, preferably with common examples.

3. **Think Critically** Drawings should show student's weight pushing against the seat and the seat pushing back; the chair pushing against the floor and the floor pushing back; the pen pushing against the paper and the paper pushing back, etc.

Science Journal The more mass the rider has, the more force he or she will need to reach the same speed and go the same height up the ramp as a lighter rider.

Skill Builder

Using E-Mail Students probably can acquire the needed information from a local contractor, also.

Assessment

Performance The effect of gravity is an application of Newton's second law. Have students design a space station that turns and shows where the floor would be. To demonstrate, put water in a plastic bucket and whirl the bucket in a large, vertical circle. The water will not pour out. Use the Performance Task Assessment List for Model in **PASC,** p. 51. L2 P

Science & Society

10•4 Air Bag Safety

Bellringer

Before presenting the lesson, display **Section Focus Transparency 33** on the overhead projector. Assign the accompanying **Focus Activity** worksheet.

[L2] [ELL]

What YOU'LL LEARN

- Air bag design and safety issues
- How to analyze news media

Why IT'S IMPORTANT

You'll be able to think critically about the news you read or hear.

Why the concern about air bags?

To help reduce traffic deaths, the U.S. government required air bags in all new cars by 1998. Air bags are bags that inflate inside a car immediately on impact. They prevent the driver and the front-seat passenger from smashing against the steering wheel, front panel, or windshield of the car because of their inertia.

While air bags have reduced traffic deaths, they have also caused some deaths, particularly among children. Read the following fictional newspaper article to learn more about the safety issues surrounding air bags. After you've finished reading, you'll learn how to analyze news media for fairness and accuracy. News media—radio, television, news magazines, and newspapers—are the sources from which we get our news. You need to stay informed to be a good citizen, but you also need to think critically about the news you read or hear.

Teaching the Content

- Have students find out whether their family vehicles have no air bags, an air bag on one side only, or air bags on both sides. Make them aware of the importance of being properly informed about how their vehicles are equipped regarding air bags.

- Ask students whether they know of anyone who was saved from serious injury by an air bag.

- Emphasize to students the importance of wearing seat belts whether or not they have air bags in their family vehicles.

- Be sure students can relate air-bag use to Newton's first two laws.

Air Bag Alarm

WASHINGTON, DC — Since the law was passed requiring air bags in all new cars, parents have expressed alarm over their children's safety. In one six-month period alone, the force of inflating air bags caused the deaths of 15 children.

But Fred Jones, government spokesperson, claimed that air bags have saved more lives than they've taken. He admitted that air bags can hurt children, but countered that children should not be riding in the front seat.

"Children under 12 should always ride in the back seat, securely buckled in a seat belt," he said.

Jones added that in cars that don't have enough rear space for children, the government will allow car companies to install cut-off switches that can turn off air bags on the passenger sides of cars.

One problem is that many people forget to turn off the switch when they enter their cars. Because of this, car companies are developing "sensors" that can detect the presence of a child's car seat. These sensors would deactivate the air bags auto-

284 Chapter 10 Forces and Motion

Program Resources

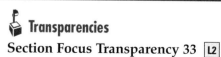

📁 **Reproducible Masters**
Enrichment, p. 39 [L3]
Reinforcement, p. 39 [L2]
Study Guide, p. 39 [L1]

🗒 **Transparencies**
Section Focus Transparency 33 [L2]

matically. The problem with this solution is that all car seat manufacturers would have to agree to install the

sensors in their car seats. In the meantime, some car companies are "depowering" air bags so that they inflate

with less force. It remains to be seen whether this will reduce injuries to children.

 ## Skill Builder: *Analyzing News Media*

LEARNING the SKILL

1. To gather news, reporters talk to people who are involved in an issue. These people are called sources. Reporters should identify their sources. If you know the sources, you can look up the facts to see whether they are true.

2. News reporters are supposed to be unbiased. That is, they should not include their own views in a news report unless the report is meant to be an analysis of an issue. Still, a reporter's biases sometimes come through. Look for words like *admits* and *claims,* or look for instances where the reporter stated something without backing it up with a source.

3. Ask yourself whether the report is balanced. Does it fairly represent both sides of the issue? Are some sources quoted while others are not?

PRACTICING the SKILL

1. Why are parents concerned about air bags? What are some of the things being done to address these concerns?

2. Did the reporter identify all sources by name?

3. Did you see any instances of bias in the article? If so, list them.

4. Were all sides of the issue fairly represented? Why or why not?

APPLYING the SKILL

If you follow the news, you know that people disagree on many issues. Choose an issue that interests you and bring news articles about the topic to class. Examine the different views presented by sources in the articles. Analyze the news articles for examples of biases.

Content Background

According to the National Highway Traffic Safety Administration (NHTSA), air bags have saved nearly 2000 lives. Beginning in model year 1998, all new cars are required to have air bags. Air bags commonly inflate with a force of 100 to 150 miles per hour, although some newer ones inflate with less force. They are particularly dangerous to infants in rear-facing safety seats or to children moving about in the front seat. NHTSA is undertaking an educational campaign to alert parents to the dangers of letting children ride in the front seat.

For additional information, have students contact the National Highway Traffic Safety Administration at 400 Seventh St., SW, Washington, DC 20590.

People & Science

People & Science

Background

- Aikido is a relatively modern Japanese martial art founded by Morihei Ueshiba (1883–1969).

- Ueshiba mastered several different martial arts, philosophy, and religion. He then developed what is now called *aikido: ai* means "harmony"; *ki*, "spirit"; *do*, "the way." Thus, *Aikido* is "the way of the spirit of harmony."

- Ueshiba practiced and taught Aikido until a few months before his death at age 86. The Japanese government has since declared him a Sacred National Treasure of Japan.

Teaching Strategies

- If there is an Aikido *dojo* (training center) in your area, invite an instructor to your classroom to demonstrate Aikido techniques. Ask students to identify examples of Newton's laws in the techniques.

- Have students research several martial arts and compare and contrast how force and motion are used in each.

Josh Pankratz, Student of Aikido

Q What is Aikido, and what does it emphasize?

A Aikido is a Japanese martial art that focuses on self-defense, on fending off attackers without injuring them. Aikido uses only self-defensive moves and skills.

Q Could you describe some of those moves and how they use forces?

A In Aikido, you learn to redirect, rather than oppose, the force of an attacker's motions. Imagine that someone is coming at you with an arm raised to hit you. In Aikido, you could step forward quickly and deflect the person's arm as it comes down. You also could step backwards at an angle and simply avoid the blow. Or you could step in from the side and grab the person's hand. Then, by redirecting the force he or she is exerting, you can turn, unbalance, and throw your attacker to the ground.

Q Does a person need to be really strong to practice Aikido?

A No, not at all. The whole idea is not to use force—to use as little muscle power as possible. It's mostly ways of moving, really, that allow you to turn and throw someone or to use his or her motion to deflect an attack. Physical size and strength don't matter very much.

Q Is Aikido considered a competitive sport?

A No, Aikido isn't competitive at all. There are no tournaments or contests. In Aikido classes, partners take turns throwing and being thrown by each other. It's really a partnership between two people. The idea of "trying to win" simply isn't there.

Career Connection

Think about careers that deal with the use of motion. Interview someone in your community who has such a career. Make a list of all the careers that were considered by other students in the class.

For More Information

Aikido is practiced by millions of men, women, and children worldwide. Search the Internet for Aikido sites or visit a local dojo.

Career Connection

Career Path Because so many careers involve motion to some degree, students will elicit a wide range of careers that involve motion. They will range from those that require little formal education, such as the operator of a ride at an amusement park, to those that require advanced degrees in math and sciences such as rocket scientists.  **COOP LEARN**

Read the statements below that review the most important ideas in the chapter. Using what you have learned, answer each question in your Science Journal.

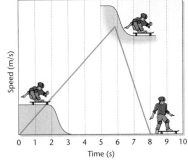

1. Gravity pulls all things toward Earth in the same way. *What are two examples that show this?*

2. Average speed during a trip is the total distance traveled divided by the total time of the trip. *How would stopping at a gas station and playing an arcade game affect your average speed during a car trip?*

3. According to Newton's first law, objects at rest tend to stay at rest, and objects that are moving tend to keep moving in a straight line. *How does this law relate to inertia?*

4. Air bags have saved lives but can be dangerous under special conditions. *What are two ways to make air bags safer?*

Chapter 10 Review 287

Review

Have students look at the illustrations on this page. Ask them to describe details that support the main ideas of the chapter found in the statement for each illustration.

Teaching Strategies

Put a steel ball bearing on an overhead projector. While discussing inertia, bring a strong magnet near the ball bearing, but out of the view of students. When the marble starts to roll, apparently by itself, lead students into a discussion of forces and inertia.

Answers to Questions

1. the bowling ball and the tennis ball in the chapter

2. The time to make the trip would be increased, and the average speed would be lower.

3. Inertia is the tendency of an object to resist change, and Newton's first law explains inertia.

4. to adjust the seat as far back as possible, to wear the shoulder harness so the person will not be thrown forward into the air bag, to put children in the back seat, to reduce the speed of the air bag or the force with which air bags inflate

Science at Home

Intrapersonal Roll a skate or a skateboard down a small slope. Measure how far it will roll across level ground. Lubricate the wheels and repeat the experiment. How far did it roll this time? Why did using a lubricant make a difference? L1

Assessment

Portfolio Encourage students to place in their portfolios one or two items of what they consider to be their best work. Examples include:
- Assessment, p. 272
- Science Journal, p. 273
- Assessment, p. 280 P

Performance Additional performance assessments may be found in **Performance Assessment** and **Science Integration Activities.** Performance Task Assessment Lists and rubrics for evaluating these activities can be found in Glencoe's **Performance Assessment in the Science Classroom (PASC).**

Using Key Science Words

1. force
2. acceleration
3. gravity
4. friction
5. average speed

Checking Concepts

6. b
7. a
8. d
9. a
10. d

Thinking Critically

11. The stopping distance and the time; the more quickly you stop, the more effective the brakes.
12. A scientific law is a former theory that was tested and proved correct many times. They would need to show repeated testing with similar results.
13. At least one time during the drive, the car was going 90 km/h. During most of the trip, it was traveling slower.
14. The car with less mass. According to Newton's second law, it will accelerate faster.
15. The force pair between your hand and the cart is greater than the friction force pair opposing the cart's forward movement.

Using Key Science Words

acceleration	inertia
average speed	mass
force	scientific law
friction	weight
gravity	

Use one of the Key Science Words above to answer each of the questions below.

1. What is the word for a push or a pull?
2. What word means that something is speeding up or slowing down?
3. What is the force that pulls all objects toward each other?
4. What is the force that resists motion when two surfaces are sliding on each other?
5. What is the result of dividing measured distance by measured time?

Checking Concepts

Choose the word or phrase that answers the question.

6. When you are sinking in quicksand, which of the following is true?
   a. The forces on you are balanced.
   b. The forces on you are unbalanced.
   c. The sand is giving an equal and opposite reaction force.
   d. Inertia is causing you to sink.

7. When you are accelerating from the top of a hill, what helps you stop?
   a. friction   c. gravity
   b. inertia   d. mass

8. On Jupiter, what would you expect to experience?
   a. Gravity would decrease.
   b. Your weight would stay the same.
   c. Your mass would change.
   d. Gravity would be stronger.

9. On the moon, which would be true?
   a. Your weight would be less.
   b. Your weight would stay the same.
   c. Your mass would change.
   d. Gravity would be stronger.

10. To go faster on a skateboard ramp, what would you do?
    a. Put a rough surface on the ramp for a better grip.
    b. Leave your safety equipment behind to lighten the load.
    c. Add more weight to the board so you will have more acceleration at the bottom of the slope.
    d. Make the slope steeper to increase the acceleration.

Thinking Critically

Answer the following questions in your Science Journal using complete sentences.

11. What would you need to measure to test how good brakes are? Explain.

12. If someone told you he or she had discovered a new scientific law, what would that person need to show you to prove it?

288 Chapter 10 Forces and Motion

Assessment Resources

Reproducible Masters
Chapter Review, pp. 23-24
Assessment, pp. 43-46
Performance Assessment, p. 48

Glencoe Technology
Computer Test Bank
MindJogger Videoquiz

13. For a long drive, a car's average speed was 60 km/h, but its speed halfway there was measured at 90 km/h. How is this possible?

14. Two cars are identical, but one carries an extra 100 kg of equipment. Which car will probably win a race? Explain.

15. You're pulling a cart. The forces between your hand and the handle are equal and opposite. The cart is accelerating. What must be true?

Developing Skills

If you need help, refer to the description of each skill in the Skill Handbook.

16. **Concept Mapping:** Use the following words to make a network tree: *Newton's laws, action and reaction forces, gravity, Newton's first law, Newton's second law, Newton's third law, inertia, F = m × a.*

17. **Making and Using Graphs:** Make a bar graph showing your weight on Earth, the moon, Mars, and Jupiter.

18. **Comparing and Contrasting:** You did one experiment with an accelerating marble. Then you did another lab with a sliding, flat, metal washer. What are the similarities and differences between the marble's movement and the washer's?

19. **Recognizing Cause and Effect:** In terms of forces, list causes for each of the following effects: (a) a baseball flies out of the ballpark; (b) a car brakes and slides to a stop; (c) a person in a wheelchair stops suddenly and her books fly off her lap; (d) a marble falls and rolls downhill; (e) you lean against a wall and the wall doesn't fall down; and (f) you sink into quicksand.

20. **Interpreting Data:** Use the following data to answer the questions.

Marble Motion

Time	Speed	Distance
Start	0	0
1 second	1 m/s	0.5 m
2 seconds	2 m/s	2 m
3 seconds	3 m/s	4.5 m

Was this marble speeding up? When was it going its average speed? What was its speed at 2 s?

Performance Assessment

1. **Predicting:** Use the data presented in question 20 above to predict the speed and distance of the marble after 6 s.

2. **Designing an Experiment:** Design an experiment that determines what shoe has the correct tread for an activity such as running.

Developing Skills

16. **Concept Mapping**

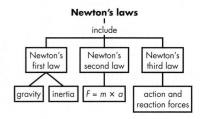

Newton's laws

include

Newton's first law | Newton's second law | Newton's third law

gravity | inertia | F = m × a | action and reaction forces

17. **Making Graphs** Student graphs should show moon weight one-sixth the height of the Earth bar; Mars, one-third; and Jupiter, two and one-half times.

18. **Comparing and Contrasting** The marble was affected less by friction with the surfaces. Both move farther and faster on the steeper slopes.

19. **Recognizing Cause and Effect** Sample answers:
   a. A bat exerted an unbalanced force on the ball.
   b. Friction between the brakes and the wheels stops the wheels from turning. As the wheels slide, friction between the wheels and the road causes the car to stop.
   c. Inertia causes the books to continue forward after the wheelchair stops.
   d. Gravity accelerates the marble.
   e. The wall is exerting an equal and opposite force to your push.
   f. The forces on you are unbalanced, and you sink.

20. **Interpreting Data** yes; sometime between 1 s and 2 s; 2 m/s

Performance Assessment

1. Speed is increasing 1 m/s/s, so speed will be 6 m/s at 6 seconds. Distance for each time period is $\frac{1}{2}$ × time × speed. Distance for 6 s is $\frac{1}{2}$ × 6 s × 6 m/s, or 18 m. Use the Performance Task Assessment List for Using Math in Science in **PASC**, p. 29. **P**

2. Experiments will vary. Students could compare tread on different shoes and match their results to the need for friction for a particular activity. If students actually do the experiment, be sure their procedure is approved by the teacher and the activity is supervised. Use the Performance Task Assessment List for Designing an Experiment in **PASC**, p. 23. **P**

Chapter Organizer

Section	Objectives/Standards	Activities/Features
Chapter Opener		**Explore Activity:** Inferring, p. 291
11-1 **What is work?** (3 sessions, 1½ blocks)*	**1. Explain** what work really means. **2. Determine** how to **calculate** the amount of work. National Science Content Standards: (5-8) UCP2, A2, B2	**MiniLAB:** Calculating Work, p. 294 **Skill Builder:** Recognizing Cause and Effect, p. 295 **Science Journal,** p. 295
11-2 **Simple Machines** (3 sessions, 1½ blocks)*	**3. Investigate** how simple machines make work easier. **4. Infer** how to use different simple machines. National Science Content Standards: (5-8) UCP2, B2, E1, E2, F5, G1	**MiniLAB:** Comparing Ramps and Screws, p. 297 **Activity 11-1:** The First Wonder of the World, pp. 302-303 **Problem Solving:** The Force of Flowing Water, p. 305 **Skill Builder:** Comparing and Contrasting, p. 306 **Science Journal,** p. 306
11-3 **What is a compound machine?** (2 sessions, 1 block)*	**5. Describe** how simple machines are combined to do work. National Science Content Standards: (5-8) UCP2, E1, E2, F5	**Using Technology:** Exercise Equipment, p. 308 **Activity 11-2:** Food out of Reach, p. 309 **Skill Builder:** Sequencing, p. 311 **Science Journal,** p. 311
11-4 **Science and Society:** **Access for All** (1 session, ½ block)*	**6. Investigate** how simple machines make sports and other activities accessible. **7. Outline** a written summary of a text. National Science Content Standards: (5-8) E2, F5	**Skill Builder:** Outlining, p. 313 **Science & the Arts:** Rube Goldberg's Wonderful, Wacky Machines, p. 314

* A complete Planning Guide that includes block scheduling is provided on pages 31T-33T.

Activity Materials

Explore	Activities	MiniLABs
No materials needed.	pages 302-303 fulcrum, meterstick, spring scale, wood block, paper clip, tape, clay, ruler page 309 dowel rod, marble, pulley, string, twist ties	page 297 screw

Need Materials? Call Science Kit (1-800-828-7777).

Teacher Classroom Resources

Reproducible Masters	Transparencies	Teaching Resources
Activity Worksheets, pp. 5, 71 **Enrichment,** p. 40 **Reinforcement,** p. 40 **Study Guide,** p. 40	**Section Focus Transparency 34,** Climbing the Walls	**Spanish Resources** **English/Spanish Audiocassettes** **Cooperative Learning Resource Guide** **Lab Partner** **Lab and Safety Skills** **Lesson Plans**
Activity Worksheets, pp. 5, 67-68, 72 **Enrichment,** p. 41 **Lab Manual 21** **Reinforcement,** p. 41 **Science Integration Activities,** pp. 59-60 **Study Guide,** p. 41	**Science Integration Transparency 11,** Leaping Kangaroos **Section Focus Transparency 35,** Clamping Down **Teaching Transparency 21,** Simple Machines	
Activity Worksheets, pp. 5, 69-70 **Cross-Curricular Integration,** p. 15 **Enrichment,** p. 42 **Lab Manual 22** **Multicultural Connections,** pp. 25-26 **Reinforcement,** p. 42 **Science and Society/Technology Integration,** p. 29 **Study Guide,** p. 42	**Section Focus Transparency 36,** Machine Teams **Teaching Transparency 22,** Compound Machines	**Assessment Resources** **Chapter Review,** pp. 25-26 **Assessment,** pp. 47-50 **Performance Assessment,** p. 49 **Performance Assessment in the Science Classroom (PASC)** **MindJogger Videoquiz** **Alternate Assessment in the Science Classroom** **Computer Test Bank**
Enrichment, p. 43 **Reinforcement,** p. 43 **Study Guide,** p. 43	**Section Focus Transparency 37,** Up and ADA'Em	

Key to Teaching Strategies

The following designations will help you decide which activities are appropriate for your students.

[L1] Level 1 activities should be appropriate for students with learning difficulties.

[L2] Level 2 activities should be within the ability range of all students.

[L3] Level 3 activities are designed for above-average students.

ELL ELL activities should be within the ability range of English Language Learners.

[LS] These activities are designed to address different learning styles.

COOP LEARN Cooperative Learning activities are designed for small group work.

[P] These strategies represent student products that can be placed into a best-work portfolio.

GLENCOE TECHNOLOGY

The following multimedia resources are available from Glencoe.

Science and Technology Videodisc Series (STVS)
Physics
  Bipedal Robot
  Micro-Machine Shop
Glencoe Physical Science Interactive Videodisc
Simple Machines and Mechanical Advantage

Mechanical Advantage Comparisons
Introduction to Bicycles
National Geographic Society Series
Human Body Vol. 2
Glencoe Physical Science CD-ROM

This is a representation of key blackline masters available in the Teacher Classroom Resources.

Teaching Aids

Section Focus Transparencies

34 SECTION FOCUS TRANSPARENCY Section 11-1

CLIMBING THE WALLS
This climber is working hard—and enjoying it. One handhold at a time, she slowly and carefully climbs toward the top of the cliff.

1. Why is climbing work? How do you know?
2. What are some other examples of fun activities that are also work?

L2

35 SECTION FOCUS TRANSPARENCY Section 11-2

CLAMPING DOWN
This toy chair is held together by glue. A clamp is used to hold the pieces together until the glue dries. Clamps are one type of machine a furniture maker uses.

1. What are the advantages of using clamps while making or repairing furniture?
2. What other kinds of machines might a furniture maker use?

L2

36 SECTION FOCUS TRANSPARENCY Section 11-3

MACHINE TEAMS
Factories aren't the only places where many machines work together. This rod and reel is made of many machines working together, too.

1. What machines are in the rod and reel?
2. Where else do many machines work together?

L2

Science Integration Transparencies

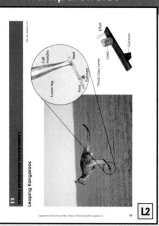

Leaping Kangaroos

L2

Teaching Transparencies

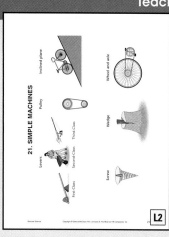

21. SIMPLE MACHINES

L2

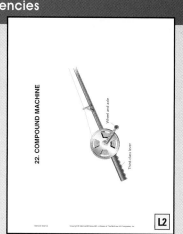

22. COMPOUND MACHINE

L2

Meeting Different Ability Levels

Study Guide for Content Mastery

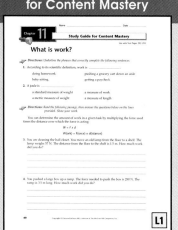

Chapter 11 Study Guide for Content Mastery

What is work?

L1

Reinforcement

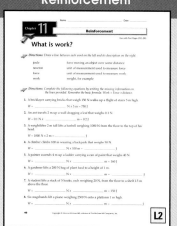

Chapter 11 Reinforcement

What is work?

L2

Enrichment Worksheets

Chapter 11 Enrichment

Tameshiwara—Boardbreaking!

L3

Hands-On Activities

Science Integration Activities

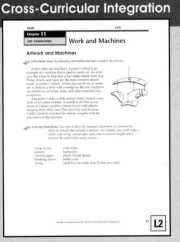

Why not wheels?

Lab Manual

Experiments with an Inclined Plane

Activity Worksheets

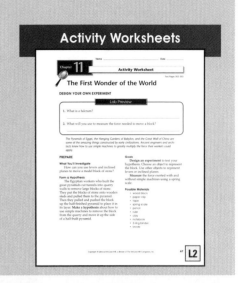

The First Wonder of the World

Enrichment and Application

Cross-Curricular Integration

Work and Machines

Artwork and Machines

Multicultural Connections

Chinese Machines

Science and Society/Technology Integration

Work and Machines

Molecular Manufacturing

Assessment

Performance Assessment

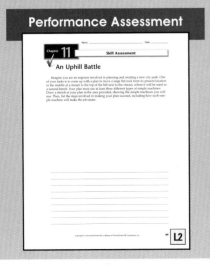

An Uphill Battle

Chapter Review

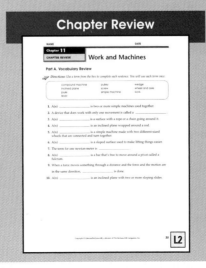

Work and Machines

Part A. Vocabulary Review

Assessment

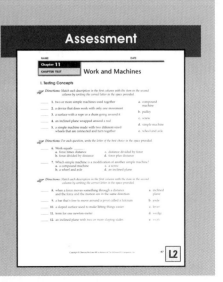

Work and Machines

Chapter 11

Work and Machines

CHAPTER OVERVIEW

Section 11-1 In this section, the scientific meaning of work and how to measure it is introduced.

Section 11-2 Simple machines are defined and demonstrated. How they make work easier is also discussed.

Section 11-3 This section explains how simple machines are combined to make compound machines and demonstrates how compound machines are used daily to make tasks easier.

Section 11-4 Science and Society The use of simple machines in recreational activities for the handicapped is discussed. It also helps students to organize information in an outline.

Chapter Vocabulary

work	wedge
joule	lever
simple machine	wheel and axle
inclined plane	pulley
screw	compound machine

Theme Connection

Systems and Interactions A simple machine is a system that makes work easier. Simple machines can be combined and interact to form compound machines.

Chapter Preview

290

Skills Preview

▶ **Skill Builders**
- recognize cause and effect
- compare and contrast
- sequence

▶ **MiniLABs**
- calculate
- compare

▶ **Activities**
- hypothesize
- design an experiment
- draw scientifically
- make a model

Learning Styles

Look for the following logo for strategies that emphasize different learning modalities.

Kinesthetic	Explore, p. 291; MiniLAB, p. 294; Reteach, p. 294; Activity 11-1, p. 302; Enrichment, p. 304; Activity 11-2, p. 309; Inclusion Strategies, p. 310
Visual-Spatial	Visual Learning, p. 293; MiniLAB, p. 297; Reteach, p. 305; Assessment, p. 306; Using Technology, p. 308
Interpersonal	Check for Understanding, p. 305; Teaching Strategies, p. 314
Intrapersonal	Science at Home, p. 315
Logical-Mathematical	Assessment, pp. 295, 309; Inquiry Question, p. 298; Science Journal, p. 301; Problem Solving, p. 305
Linguistic	Using Science Words, pp. 293, 308

Work and Machines

An outdoor vacation can be a lot of fun, but getting ready for it isn't always easy. There is so much to do—meals to plan and supplies, tents, and equipment to organize. Don't forget the fun stuff like canoes, paddles, and fishing poles. Everything needs to be lifted, moved, and packed. This involves a great deal of work. That's what this chapter is about—work and how to make it easier. You can start learning about work in the Explore activity.

EXPLORE ACTIVITY

Inferring

1. On this vacation, you will carry a canoe. Practice by using two or three books to represent the weight of the canoe. Stand up and hold the books with both hands.
2. Lift the books up over your head three times.
3. Walk around the room twice with the books over your head. Don't let the books rest on the top of your head.
4. Which do you infer was more work, lifting the canoe or carrying it? Why do you think so?

Science Journal

In your Journal, discuss the easiest way to load everything into a van for a class camping and canoeing trip.

291

Prepare

Section Background

Work is defined as $F \times d$. If there is a lot of force but no movement, then no work is done. Also, the force and the movement must be in the same direction.

Preplanning

Refer to the Chapter Organizer on pages 290A-B.

1 Motivate

Bellringer

Before presenting the lesson, display **Section Focus Transparency 34** on the overhead projector. Assign the accompanying **Focus Activity** worksheet.

L2 ELL

What YOU'LL LEARN

- What *work* really means
- How to figure out how much work you've done

Science Words:
work
joule

Why IT'S IMPORTANT

If you know what work is, you can learn how to make it easier.

FIGURE 11-1

The student is "working" on a school project. *Is he doing work in the scientific sense? Explain.*

Doing Work

What do you think of when you hear the word *work*? You may think that reading this chapter is work. On a vacation, carrying a canoe probably seems like work, too. Most adults talk about work every day. Would you be surprised to learn that none of these examples is work in the scientific sense?

When are you working?

Work. The very word makes some people tired. That's because their idea of work is anything that takes effort. That can mean homework, as shown in **Figure 11-1**, or running a marathon. But effort does not always equal work in the scientific sense. If you push against a wall until you are tired but the wall doesn't move, a scientist would say you haven't done any work on the wall. **Work** is done when a force moves something through a distance and the force and the motion are in

This judo throw looks like work. *Is the girl throwing the other girl doing work? Explain why or why not.*

Tying to Previous Knowledge

Only a few students may have been canoeing or mountain climbing, but many of them have gone up stairs carrying something heavy, such as a book bag or groceries or pushing a bike, and can use that experience.

Program Resources

Reproducible Masters
Activity Worksheets, pp. 5, 71 L2
Enrichment, p. 40 L3
Reinforcement, p. 40 L2
Study Guide, p. 40 L1

 Transparencies
Section Focus Transparency 34 L2

the same direction. An example of this definition of work is lifting a canoe, as shown in **Figure 11-2.** The force is in an upward direction and the canoe moves up, so it agrees with both parts of the definition of work. If you carry the canoe along a level path, you are still holding up the canoe, but the canoe is moving forward. The force and the movement are not in the same direction. Even though carrying the canoe feels like work, it doesn't fit the scientific definition.

How much work?

Work depends on force and distance. You can find the amount of work by multiplying the force used by the distance over which the force is acting.

If the canoe weighed 300 newtons (N) and you lifted it 1.5 meters (m), then you did 450 N•m of work on the canoe. This is read as 450 newton-meters.

$$W = F \times d$$
$$W = 300 \text{ N} \times 1.5 \text{ m}$$
$$W = 450 \text{ N•m}$$

To make it easier to talk about work, there's a unit of measure with a shorter name. A **joule** (J) is one newton-meter (N•m), so you did 450 J of work on the canoe when you lifted it. The joule (JEWL) was named in honor of James Joule, a scientist who studied work, energy, and heat.

FIGURE 11-2

Work is done when the canoe is lifted up because the canoe is moved, and the force and motion are in the same direction.

Visual Learning

Figure 11-1 Is he doing work in the scientific sense? Explain. *The student is putting in a lot of effort, but it's not work in the scientific sense. Nothing is moved. The person making the judo throw is doing work because she is lifting the other person.* **Is the girl throwing the other girl doing work? Explain why or why not.** *Yes. She is using a force to throw the other girl, and the force and motion are in the same direction.* Ⓛ

---

2 Teach

❓ FLEX Your Brain

Use the Flex Your Brain activity to have students explore WORK.

📁 **Activity Worksheets,** page 5

Using Science Words

Ⓛ **Linguistic** In daily life, the word *work* means the same as *effort*, but in science, the effort must move something and the object must move in the direction of the effort to be classified as work.

Content Background

The term *newton* is named for Sir Isaac Newton (1642-1727), the famous scientist. Newton is best known for his three laws of motion and his theory of gravity.

Teacher F.Y.I.

To help understand what a joule is, students can be told that a joule is approximately the amount of energy needed to lift a glass of milk from the table to their mouths.

GLENCOE TECHNOLOGY

 Videodisc

STVS: Physics
Disc 1, Side 2
Bipedal Robot (Ch. 2)

Micro-Machine Shop (Ch. 12)

3 Assess

Content Review Review the concept of scientific work with the students.

Reteach

LS **Kinesthetic** Have the students pull a weighted wagon. Discuss the amount of pull needed and the direction of the pull. Observe the distance traveled and its direction. Turn the tongue of the wagon as far as it will go to one side and pull again. The wagon may turn over. The pull is now in a different direction, and work is done in the direction of the force.

Mini LAB

Purpose

LS **Kinesthetic** The MiniLAB reinforces the concept that real work is the force required times the distance, traveled, not the amount of effort involved. **L1**

Materials
ten books and a desk

Teaching Strategies
Make the stacks of books heavy enough that it is difficult but not impossible for a student to lift.

Safety Precautions Warn students not to hurt their backs trying to lift stacks of books all at once. Caution students to lift with their legs, not their backs. Make sure no students have any physical limitations, such as heart problems, severe asthma, etc.

📁 **Activity Worksheets,** pages 5, 71

Mini LAB

Calculating Work

Suppose your teacher asked you to move ten books from the floor to the desk. What would be the easiest way to do the work?

1. Lift all ten books to the desk.
2. Lift two books at a time to the desk.

Analysis
1. Which way of lifting was easier?
2. Assume the books weighed about 10 N each and you lifted them 1 m. How much work did you do when you lifted the whole stack at once? How much work did you do when you lifted two books at a time?
3. Which was more work?

FIGURE 11-3

Who does more work getting to the top, the person hiking up the trail or the person climbing up the cliff? If they start at the same place and both reach the top, they will do the same amount of work even though it is much harder to climb up the cliff.

294

Is going uphill work?

Now instead of picking up and carrying a canoe, imagine carrying your own weight up a mountain. When you hike up the mountain, you do work by moving your weight up the trail just as you did work lifting the canoe. This time, you won't be walking on a level path. You are going to hike up a hill moving your weight. You will be going up, so you will be doing work. How do you calculate the amount of work done? Remember, $W = F \times d$. You are lifting yourself up the hill with every step you take. The distance equals how high you went up, and the force is your weight.

Analysis
1. Lifting two books at a time was easier.
2. 100 J of work were done lifting the whole stack. The total amount of work done moving all the books two at a time is 20 J per trip × 5 trips = 100 J.
3. The total work for both tasks is the same, 100 J.

✔ Assessment

Process Give the students examples of several tasks. Have them hypothesize how they could be done a difficult way and an easy way, but the same amount of work must be done either way. Examples of tasks are tearing a phone book in half or breaking a bundle of sticks. Use the Performance Task Assessment List for Formulating a Hypothesis in **PASC**, p. 21.

Effort Doesn't Always Equal Work

Which seems like more work, hiking or mountain climbing as shown in **Figure 11-3**? You can compare the amount of work and effort involved in the two activities. Suppose you climb straight up the side of a mountain, climbing a rock cliff using just your fingers and toes. Meanwhile, a friend takes the easy way up to the mountaintop, strolling along a hiking trail. Which of you did more work? Before you answer, think back to the way you learned to calculate the amount of work done. It doesn't seem possible, but you and your hiking friend did exactly the same amount of work, assuming you both weigh the same. You both lifted the same amount of weight (force) and you both traveled up the same distance. Even though the trip up the hiking trail was easier than climbing, the same force and the same distance make the same work.

Section Wrap-up

1. You picked up a 10-kg (98-N) barbell and lifted it 1.5 meters. How much work did you do on the barbell? What is the force? What is the distance?

2. Which of the following is work: pushing your little brother on his tricycle, lifting a box, holding books over your head, climbing up stairs, doing homework, or carrying a box uphill?

3. **Think Critically:** When an astronaut is weightless and lifts an object over her head, is work being done? Explain.

4. **Skill Builder**
   Recognizing Cause and Effect You do the same amount of work on two identical lunch boxes by moving them. However, one box traveled farther than the other. Are the forces you used to lift the two boxes different? How do you know? If you need help, refer to Recognizing Cause and Effect on page 551 in the **Skill Handbook.**

Science Journal

You are helping tear down an old garden shed, but one wall refuses to fall. You have pushed it and pulled it until you are tired. Did you do any work? Explain.

Skill Builder
Recognizing Cause and Effect Yes, the force must be different because the distance is different. $W = F \times d$; if the work stays the same, then the force must change if the distance changes.

 **Assessment**

Performance Have the students draw the two lunch boxes in the Skill Builder and calculate distances and forces that would make the work equal. Use the Performance Task Assessment List for Using Math in Science in **PASC**, p. 29.

For students who have mastered this section, use the **Reinforcement** and **Enrichment** masters.

4 Close

•MINI•QUIZ•

Use the Mini Quiz to check students' recall of chapter content.

1. Work is done when a ____ moves something through a distance, and the force and the motion are in the ____ direction. *force, same*

2. If a force of 20 N were necessary to drag a piece of furniture across the room for a distance of 3 m, how much work was done? *60 J*

3. If you tried to move the furniture and couldn't, but you put a lot of effort into it, how much work was done? *none*

Section Wrap-up

1. 147 J, 98 N, 1.5 m
2. pushing tricycle, lifting, climbing up stairs, carrying something uphill
3. **Think Critically** Yes, a force is applied over a distance, and the force and motion are in the same direction.

Science Journal No, if the wall didn't move, no work was done.

Prepare

Section Background

The screw and wedge are both types of inclined planes. The three types of levers—the fixed and movable pulleys, and the wheel and axle—are presented. Each is defined and explained; examples are given for each.

Preplanning

Refer to the Chapter Organizer on pages 290A-B.

1 Motivate

Bellringer

Before presenting the lesson, display **Section Focus Transparency 35** on the overhead projector. Assign the accompanying **Focus Activity** worksheet.

L2 ELL

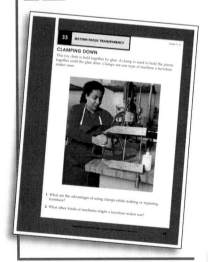

Tying to Previous Knowledge

Review the concept of work covered in Section 11-1 before discussing how simple machines make work easier.

What YOU'LL LEARN

- That simple machines make work easier
- How to use different simple machines

Science Words:
simple machine
inclined plane
screw
wedge
lever
wheel and axle
pulley

Why IT'S IMPORTANT

Simple machines make work easier.

FIGURE 11-4

What simple machine is being used here to help load the boxes into the truck? The ramp is an inclined plane. An inclined plane makes it easier to load heavy boxes up into the truck.

11•2 Simple Machines

Simpler Than You Think

Remember the hiker and climber from the last section? You probably didn't notice the machine that made work easier for the hiker. The word *machine* probably makes you think of a complicated gadget with metal, plastic, gears, and motors. But the hiking trail was a kind of machine—a simple machine. A **simple machine** is a device that does work with only one movement. Simple machines are the most basic form of useful tool. In this section, you'll learn to recognize the six types of simple machines that are all around you. The six simple machines are the inclined plane, screw, wedge, levers, the wheel and axle, and the pulley. As you read on, try to figure out which type of simple machine the hiker used.

What is an inclined plane?

On moving day, the first thing movers do is put a ramp up to the back of the truck, as shown in **Figure 11-4**. A ramp makes it easier to move things up and into the truck. This type of ramp is called an inclined plane. An **inclined plane** is a sloped surface used to make lifting things easier.

Imagine that you're canoeing and you need a place to lift the canoe out of the water and onto the riverbank. What's the first thing you'll look for? You'll follow the mover's example and search for a riverbank that looks like a ramp. If the bank has straight sides, you might not even be able to walk up it, much less drag a canoe up it. If the bank has a gentle slope, it will be easier to pull the canoe out of the water and up the bank. The riverbank with a gentle slope is an inclined plane.

296

Program Resources

Reproducible Masters
Activity Worksheets, pp. 5, 67-68, 72
Enrichment, p. 41 L3
Lab Manual, pp. 75-77 L2
Reinforcement, p. 41 L2
Science Integration Activities, pp. 59-60
L2
Study Guide, p. 41 L1

Transparencies
Science Integration Transparency 11 L2
Section Focus Transparency 35 L2
Teaching Transparency 21 L2

What is a screw?

The inclined plane made it easier to unload the moving truck and get the canoe out of the water. But if you change the shape of an inclined plane, you can do even more jobs with it. One way to change its shape is to make it into a screw. You've probably seen screws in a hardware store or used them to hold things together, but have you ever looked at one closely? If you do, you'll notice that a **screw** is an inclined plane wrapped around a rod as shown in **Figure 11-5**. In the MiniLAB, you'll learn more about how screws and inclined planes are related.

Just as ramps can have different slopes and lengths, screws come in different sizes and shapes. In the MiniLAB, some screws had small, gently sloped, closely spaced threads and some had large, steep, widely spaced threads. Screws are designed for what they do—holding wood or metal, going through thick or thin pieces of material.

Screws aren't found just in toolboxes—screws are found all around you. When you unpack the tent, you may find a new type of tent stake. It looks like a big spiral. The directions say it's easier to use and holds better than the old, straight kind. You try it and sure enough, it is easy to put into the ground. Then, you give it a good, hard tug and it doesn't move at all.

Mini LAB

Comparing Ramps and Screws

1. Make an inclined plane out of a piece of paper. Cut a piece of paper in half diagonally from corner to corner.
2. Use markers or a pencil to draw a dark line along the cut edge.
3. Start with the biggest end and wrap the paper around a pencil so that you can see the dark line. Observe the dark line along the edge of the inclined plane.
4. Now, observe the screws your teacher gives you.

Analysis

1. What geometric shape does the line around the pencil form?
2. How are the screws similar to the pencil and paper?
3. What would you have to change about the paper and pencil to make them more closely match each of the screws?

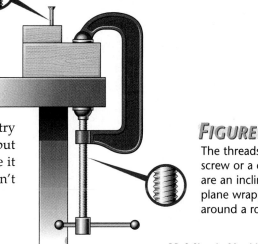

FIGURE 11-5

The threads in a screw or a clamp are an inclined plane wrapped around a rod.

2 Teach

Inquiry Question

LS **Logical-Mathematical** If a canoe and equipment weigh 1000 N, how much work is done going up a 45° slope? Up a 10° slope? (Assume the height of both slopes is 2 m.) *2000 J, 2000 J*

Teacher F.Y.I.

Inclined planes were used to move the giant blocks of the pyramids into place in ancient Egypt.

Discussion

Ask students about the inclined planes used in amusement park slides, such as water slides.

? FLEX Your Brain

Use the Flex Your Brain activity to have students explore INCLINED PLANES.

📁 **Activity Worksheets,** page 5

Inquiry Questions

How much work is done if a force of 10 N moves something 2 m? 1 m? *20 J, 10 J*

A force of 80 N does 400 J of work on an 8-kg object. How far did the object move? *5 m*

What is a wedge?

After the tent is up, it's time to clear a space for the campfire. You need to cut up some dead tree branches lying on the ground for firewood, so you get out a hatchet to cut the branches into small enough pieces to carry. When you use a hatchet, you are using another version of an inclined plane called a wedge. A **wedge** is an inclined plane with one or two sloping sides. Pushing the hatchet into the branch forces the wood apart. Can you think of any wedges that you use every day? A kitchen utensil drawer and a carpenter's toolbox are good places to look. Anything with a sharp edge is a wedge—how about scissors, a knife, a can opener, a chisel, and a saw?

What is a lever?

Your campsite is almost finished. You have pitched the tent and cleared the area except for one large rock. It would make a great seat by the campfire, but it's too far away, and it's buried too deep to move by hand. Another

FIGURE 11-6
You may not realize that you use simple machines every day.

A This camper is using a simple machine to help cut firewood. The hatchet is a wedge.

298   Chapter 11 Work and Machines

type of simple machine can help you—a lever. A **lever** is a bar that's free to move around a pivot point. The pivot point is called a fulcrum (FUL krum). In a lever, the work you do on your end (force you exert times the distance your end moves) equals the work on the other end (weight of the rock times the distance that end moves). There are three kinds, or classes, of levers. Moving the rock is a job for a first-class lever, as shown in **Figure 11-6.**

First-Class Levers

Find a thick branch and put one end under the rock. Put a smaller rock between the branch and the ground to be the fulcrum. Now, push down on the branch. You can easily pry the big rock up and out of the ground. A first-class lever allows you to move things that you could not move without it.

A seesaw is another example of a first-class lever. The fulcrum is in the middle, and each end has a force pushing down. Other types of levers have the fulcrum in different places. Let's look at some other types of levers.

B This camper is using another simple machine to move the heavy rock. The branch is a lever, and the small rock is a fulcrum. By using the lever and fulcrum, the rock can be moved more easily.

Content Background

Introductory physics classes normally ignore the mass of the lever by considering forces much larger than the lever weight. Ignoring the mass of the lever can cause problems when using a ruler to balance light objects, such as paper clips, that weigh much less than a ruler balance. Under these circumstances the ruler must pivot at its *center of mass.*

? FLEX Your Brain

Use the Flex Your Brain activity to have students explore LEVERS.

📁 **Activity Worksheets,** page 5

Demonstration

Take a pair of scissors apart to show students that the blades are wedges. Intact scissors are more efficient than if you use one of the blades to cut.

GLENCOE TECHNOLOGY

💿 Videodisc

Glencoe Physical Science Interactive Videodisc
Side 1, Lesson 2
Simple Machines and Mechanical Advantage

10627-13348
Mechanical Advantage Comparisons

13381-13399

Content Background

The wheelbarrow was invented in Europe in the Middle Ages, during the construction of the great cathedrals. It was developed from the *barrow.* The barrow required two people to carry the load between them.

Kinesthetic Have students try to move a door by pushing at different spots: near the knob, near the hinge, and in between.

Ask the following questions. **What is the fulcrum? The resistance force? The effort force?** *hinge, weight of door, hand pushing* **What class of lever is a door?** *second* **Why is the doorknob located where it is?** *to make it easier to push the door* L2

Revealing Preconceptions

Students may think that machines make work easier by reducing the amount of work that has to be done. However, machines often require more work, but they allow you to multiply force, speed, or distance.

Second-Class Levers

A second-class lever also increases your force. Suppose you need to move a canoe closer to camp. The canoe is still loaded with supplies and is too heavy for one person to lift and carry. Instead, you pick up one end and drag the canoe closer to camp.

It's easier to lift one end of the canoe because the canoe is a second-class lever as shown in **Figure 11-7**. The ground where the tip of the canoe touches is the fulcrum. The lever is the whole canoe plus the fulcrum. A second-class lever transfers some of the weight to the fulcrum, making it easier for you to lift and move the canoe, supplies and all. Another example of a second-class lever is a wheelbarrow. The wheel at the front is the fulcrum and carries some of the weight while you lift the handles at the back.

FIGURE 11-7

The forces in and out are in different places in second-class and third-class levers.

The person dragging the canoe up to the campsite is using a second-class lever.

Cultural Diversity

Work In the past, people used basic principles of machinery to move large objects. Students can research examples like the following.

Easter Island—huge stone heads were carved inland. They were moved miles to the coast of the island, then raised upright using sleds and pulleys. **Stonehenge (and other megaliths)**—huge stones were arranged in patterns. In some cases, a crosspiece stone was placed on top. **Pyramids**—rather than lifting materials to build the taller parts of structures, sand and mud were used to form ramps or inclined planes. Large stones were then rolled or pulled up the ramp and placed on top of existing structures.

Third-Class Levers

The third and last class of lever doesn't increase your force, but it can make some kinds of work easier. It also can help you have fun.

Now that the hard job of setting up camp is done, it's time to go fishing. You cast a baited hook into the river, and soon you've hooked a fish. The fishing rod you use is a third-class lever as shown in **Figure 11-7B.** This class of lever does not make your force larger, it makes things move faster. The far end of the rod makes large, fast movements. Your wrist is the fulcrum. Moving the bottom of the rod a little makes the top of the rod move a lot. That fast motion is just what you need to cast your baited hook far out into the water. To see how much this type of lever helps you, imagine fishing with only a baited hook and some line. Can you think of other examples of third-class levers? A canoe paddle, a baseball bat, and a broom are three examples.

It's smart to use a simple machine any time it will make your work faster and easier. In the activity on the next two pages, you can explore other ways to use simple machines.

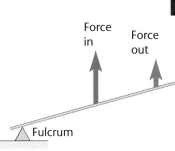

 A fishing rod is an example of a third-class lever. By moving the bottom of the fishing rod, the top moves a lot.

11-2 Simple Machines   **301**

Discussion
Have students discuss how other third-class levers such as a broom, a baseball bat, or a canoe paddle make an activity easier.

Content Background
Not all scientific work is done by people. Forces such as gravity and electromagnetism can also do work on objects.

GLENCOE TECHNOLOGY

Videodisc

Glencoe Physical Science Interactive Videodisc
Side 1, Lesson 2

13369

13372

13375

13378
Introduction to Bicycles

13404-14388

Across the Curriculum

Language Arts Have students look up the words *work, effort,* and *lever* in a dictionary. Students will find several definitions of work and effort. Have them compare these definitions to the scientific definitions. The word *lever* comes from the Latin word *levare* (to lift). Have a volunteer explain how the meaning of the word shows its origin.

Science Journal **Levers** Have the students use their knowledge of levers to develop a seesaw that needs only one person. Have them draw and label its parts and explain how it works. They could put a heavy weight on the side without the person, or they could make that side long and move the fulcrum close to the student. **[LS] [P] [L2]**

301

Activity 11-1

PREPARE

Purpose

[K] **Kinesthetic** Students will model some of the steps in building a pyramid. **L1** **ELL**

Process Skills

observing, using numbers, measuring, predicting, comparing and contrasting, experimenting, formulating hypotheses, formulating models

Time

half period to plan, full period to do and analyze

Materials

The object representing the block should be heavy enough to register a noticeable weight on the spring scale. Possibilities include a can of cat food, a weight, a rock, or a wood block. Attach a handle (e.g., tape on a bent paper clip) for hooking the scale.

Possible Hypotheses

- Levers were used to move the blocks out of the quarry.
- Inclined planes were used to move the blocks up the pyramid.

📁 **Activity Worksheets,** pages 5, 67-68

Design Your Own Experiment

The First Wonder of the World

The Pyramids of Egypt, the Hanging Gardens of Babylon, and the Great Wall of China are some of the amazing things constructed by early civilizations. Ancient engineers and architects knew how to use simple machines to greatly multiply the force their workers could apply.

Possible Materials

- wood block
- paper clip
- tape
- spring scale
- pencil
- ruler
- clay
- notebook
- 3-ring binder
- books

PREPARE

What You'll Investigate

How can you use levers and inclined planes to move a model block of stone?

Form a Hypothesis

The Egyptian workers who built the great pyramids cut tunnels into quarry walls to remove large blocks of stone. They put the blocks of stone onto wooden sleds and pulled them to the pyramid. Then they pulled and pushed the block up the half-finished pyramid to place it in its layer. **Make a hypothesis** about how to use simple machines to remove the block from the quarry and move it up the side of a half-built pyramid.

Content Background

The architect of the first Egyptian pyramid was Imhotep, who lived in the 27th century B.C. and worked for the Pharaoh Zoser. Around the 16th century, pharaohs started to use hidden tombs in the Valley of the Kings in an attempt to foil tomb robbers. The early Egyptians did not have the use of pulleys or wheeled carts, although they may have used log rollers. The Pyramids of Giza are the only one of the Seven Wonders of the Ancient World still standing.

Goals

Design an experiment to test your hypothesis. Choose an object to represent the block. Use other objects to represent levers or inclined planes.

Measure the force exerted with and without simple machines using a spring scale.

PLAN

1. **Discuss** what materials will represent the stone block, a lever and fulcrum, an inclined plane, and a half-built pyramid.
2. **Decide** how you will **measure** the force exerted in each task with and without a simple machine.
3. **Draw** a diagram of the experiment.
4. **Plan** how you will **record** the design and results of your experiment.

DO

1. Make sure your teacher has approved your design.
2. Carry out the experiment.
3. **Record** your observations.

CONCLUDE AND APPLY

1. What was the force exerted without a simple machine to move the block onto a sled? Up the pyramid? What was the force with a simple machine?
2. **Compare and contrast** your results for each case.
3. **APPLY** Predict a way to modify your simple machines and exert even less force.

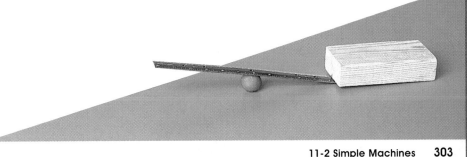

3. Make the lever arm from fulcrum to effort force longer. Make the ramp longer and less steep.

Assessment

Portfolio Write up the design and results of the experiment with illustrations. Provide background about what each item in the experiment would represent. Use the Performance Task Assessment List for Lab Report in **PASC**, p. 47. **P**

Go Further

One method of quarrying stone involved inserting a wooden wedge into a crack and wetting it so the wood swelled. Ask students to indicate the forces and explain how this simple tool worked.

PLAN

Possible Procedures

Use a rock to represent the block. Use a pencil lever and clay fulcrum to pry it onto the edge of a notebook sled. Use a 3-ring binder as an inclined plane to pull it up a pyramid of three textbooks.

Teaching Strategies

- Ask a student to research the building of the pyramids and present a brief report before students make their plans.
- Students might use inclined planes for both tasks. Encourage them to use a lever for one. (The Egyptians used levers and ropes to move the blocks onto sleds.)

Troubleshooting Students may need to orient the lever so that they can pull the spring scale down over the edge of the desk.

- Make sure students include a fulcrum in their lever models. Clay or a finger could work.
- To measure the force used without a simple machine, lift the block straight up with the spring scale.

DO

Expected Outcome

Effort should be reduced with each simple machine.

CONCLUDE AND APPLY

1. answers will vary, without a simple machine, the block's weight; with, less
2. The force without a simple machine was the same in both cases. Force with a simple machine will depend on design and should be less.

Enrichment

Teacher F.Y.I.

A zipper uses inclined planes to join and separate two rows of interlocking teeth. The zipper slide contains an upper wedge (two inclined planes) to force the teeth apart and two lower wedges to force the teeth back together.

Videodisc

STV: Human Body Vol. 2

Elbow joint (art)

48679

Knee joint within leg (art)

48677

Knee joint with tendons (art)

48681

Knee joint with lateral muscles (art)

48682

FIGURE 11-8

Doorknobs and fishing reels are two examples of a wheel and axle.

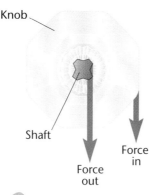

Knob

Shaft

Force in

Force out

A A small force moving a large distance (around the wheel) turns into a large force moving a small distance (around the axle).

B In a doorknob, without the knob (wheel), the shaft (axle) is hard to turn.

Working Smart

When you use levers to lift something, as you did in the activity, is the amount of work greater or less than if you didn't use a machine at all? It's the same. Simple machines usually let you use less force but cause you to go a longer distance. For example, on page 294, the shorter path to get to the top of the mountain was straight up the steep side, but the longer route on the gentle slope of the hiking trail was easier. The purpose of simple machines is to make work easier—to allow you to use less force, or move a smaller distance, or change the direction to get the same amount of work done. Some people call this working smarter, not harder.

What is a wheel and axle?

As you stand on the riverbank, casting your baited hook and reeling in fish, the lever isn't the only simple machine you're using. The reel on your fishing rod is a simple machine called a wheel and axle. A **wheel and axle** is a simple machine made with two different-sized wheels that are connected and turn together.

Usually, a wheel and axle makes your force greater. A doorknob is a wheel and axle. If you took the knob off a door, you'd see the small, square axle, as shown in **Figure 11-8.** When you turn the knob, the axle turns, too, and opens the latch or lock. Then you can pull the door open. The large knob on a smaller axle makes your force larger. Without the knob, turning the small axle with your fingers is difficult.

On the reel of a fishing rod, the handle turns a large wheel. The large wheel turns a smaller wheel, and it pulls in the fishing line. Because it's a big wheel turning a smaller wheel, it also makes your force greater. Can you imagine trying to catch a fighting fish by hand without the help of a wheel and axle?

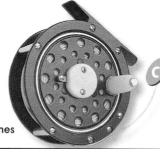

C A fishing reel is an example of a wheel and axle. The crank handle turns the axle.

Content Background

- Skateboards and skates use wheels that turn *on* an axle, but skateboards and skates are not wheels and axles because the two wheels don't turn each other.

- The ancient Romans had an advanced water-supply system, but even they had to work hard to keep up with demand. Part of the reason so much water was used was that many households didn't have plugs or faucets; water ran continually.

What is a pulley?

The fishing was good, and you've just finished the last bites of a fine fish dinner. Now it's time to put the supplies away so that raccoons, skunks, and bears can't get into them overnight. You throw a rope over a tree limb, tie it to the supplies, pull them up off the ground, and leave them hanging in midair. You just used another simple machine called a fixed pulley. A **pulley** is a surface with a rope or chain going around it. Pulleys are used for many purposes today, but the first pulley ever used was probably a tree limb with a rope over it, just like yours.

Fixed Pulleys

The limb is a fixed pulley because when you pull down on the rope, the limb doesn't move, but the supplies do. This type of pulley changes the direction of your force, but not its size, as shown in **Figure 11-9** on the next page. When you pull down, the supplies go up. Think how much more effort it would take if you had to climb the tree to pull up the supplies.

What could you do if the supplies were too heavy to pull up with a pulley? You could add another type of pulley to the fixed one.

Problem Solving

The Force of Flowing Water

Flowing water can have a lot of force. Imagine trying to hold back the water by covering the faucet with your hand. But you can use two simple machines to stop the flow. Look at the diagram of a faucet below.

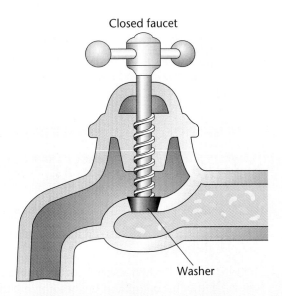

Closed faucet

Washer

Solve the Problem:
1. What part of the faucet is a wheel and axle?
2. What part of the faucet is an inclined plane?

Think Critically:
Describe what would happen if each simple machine were replaced. Would it be easier or harder to turn the faucet on and off?

Problem Solving

Solve the Problem
1. the handle and the shaft connecting it to the washer
2. the screw on the shaft

Think Critically
If you removed the screw ridges, you would have to yank the washer up and down with the handle. More effort would be needed to turn the faucet on and off. If you removed the wheel from the axle, you would have to exert enough force to turn the screw holding only the rod. It would be harder to turn the faucet on and off. IS

Inquiry Question
If the force and the movement of the object must be in the same direction before work is done, does a fixed pulley do work? *Yes, the force travels along the rope and is pulling up while the object is moving up.*

3 Assess

Check for Understanding
Brainstorming Have the students brainstorm all the simple machines they use on a daily basis. IS **COOP LEARN**

Reteach
IS **Visual-Spatial** Bring in a fishing rod and reel to demonstrate a third-class lever, a fixed pulley, and a wheel and axle.

Extension
For students who have mastered this section, use the **Reinforcement** and **Enrichment** masters.

Theme Connection
Systems and Interactions

Two pulleys working together make a pulley system. Four pulleys combined make the block and tackle that mechanics use to lift motors out of cars. Two pulleys are fixed and two are movable.

Content Background

Pulleys do work even though the force and movement appear to be in opposite directions. By definition, a pulley changes the direction of the force; therefore, consider just the force acting on the object. The force on the object is upward, and because its movement is upward, work is being done.

4 Close

Use the Mini Quiz to check students' recall of chapter content.

1. **A device that does work with only one movement is called a(n) _____.** *simple machine*

2. **A(n) _____ is two inclined planes put together, and a(n) _____ has an inclined plane wrapped around it.** *wedge, screw*

3. **Which lever is best for lifting a rock? Which lever is like a wheelbarrow? Which lever is your arm like?** *first-class, second-class, third-class*

4. **In a wheel and axle, both the wheel and the axle _____ together.** *turn*

5. **The top of a flagpole has a(n) _____ pulley.** *fixed*

Section Wrap-up

1. See Figures 11-6 and 11-7 on pages 298-301.

2. Answers might include that you can screw two pieces of wood together.

3. **Think Critically** This is a first-class lever. See page 299.

Science Journal Student answers may include using a doorknob (wheel and axle), unscrewing the lid (screw) off the peanut butter, using the knife like a lever to get the peanut butter, and using a knife to pry up the lid (lever).

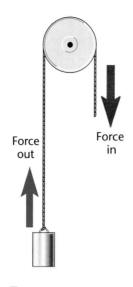

FIGURE 11-9
A pulley makes it easier to lift heavy things. *Why do most people find it easier to pull down than lift up?*

Movable Pulleys

Because the supplies are too heavy, you will need to make your force greater. One way is by using two pulleys. First, attach a pulley to the supplies. Then, tie one end of a rope to the tree limb and run the other end through the pulley attached to the supplies. Now, throw the rope up and over the limb just like last time. This time the new pulley moves with the supplies. Because it doesn't stay in one place, it is called a movable pulley.

Having two ropes supporting the supplies instead of one made pulling them up easier. You had to pull about half as hard. Each rope is supporting half of the weight of the supplies.

Helpful Machines

Are you surprised that machines made your camping trip easier? Think about how many simple machines you were able to use during this trip. It's a lot of simple machines for one trip, but each one provided a different way to make the trip easier and more fun.

Section Wrap-up

1. Draw and label the three types of levers.

2. Explain how you would use an inclined plane tool to hold two objects together.

3. **Think Critically:** When a skateboarder puts a foot on one end of his board and pushes down hard, the other end of the board pops up. Draw and label a diagram that shows which type of lever the skateboarder is using when he does this.

4.  **Skill Builder**
   Comparing and Contrasting Compare a wheel and axle to a wheel *on* an axle. Tell how they are different and how they are alike. If you need help, refer to Comparing and Contrasting on page 550 in the **Skill Handbook.**

Science Journal
In your Journal, list at least five simple machines that you have used today and tell how you used each one.

Skill Builder
Comparing and Contrasting They are shaped the same and both wheels turn, but the wheel and axle turns as a single unit, while the wheel on an axle doesn't.

Assessment

Process Provide students with a variety of kitchen gadgets, and have them identify the simple machines used. Use the Performance Task Assessment List for Making Observations and Inferences in **PASC**, p. 17. ℕ L2
ELL

What is a compound machine?

Putting It Together

On your camping trip, you used pulleys, levers, and inclined planes. You also use them every day. Now that you know what they are, you'll notice simple machines in many places. Often, however, more than one machine is used at a time. A **compound machine** is two or more simple machines used together. Look at **Figure 11-10** for some examples of compound machines.

What YOU'LL LEARN

- That simple machines can be put together to do work
  Science Words:
  compound machine

Why IT'S IMPORTANT

Combinations of simple machines make work easier to do.

FIGURE 11-10
A compound machine is two or more simple machines used together.

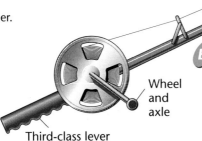

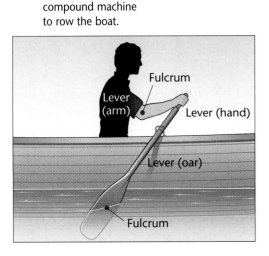

A The illustration shows how the person is using two levers as a compound machine to row the boat.

Third-class lever

Wheel and axle

Fulcrum
Lever (arm)
Lever (hand)
Lever (oar)
Fulcrum

B A fishing rod and reel is a compound machine because it combines two machines—a wheel and axle and a lever. The entire rod is a third-class lever.

C Scissors are two levers, a fulcrum, and two wedges.

Prepare

Section Background

Two or more simple machines make up a compound machine. Sometimes a part of the human body is involved as a simple machine, such as when the arm acts as a lever with the hatchet.

Preplanning

Refer to the Chapter Organizer on pages 290A-B.

1 Motivate

Bellringer

Before presenting the lesson, display **Section Focus Transparency 36** on the overhead projector. Assign the accompanying **Focus Activity** worksheet. L2 ELL

Tying to Previous Knowledge

If students have used a can opener or a pair of scissors, they have used a compound machine. Show students where the levers and the wedges are located.

Program Resources

📁 Reproducible Masters
Activity Worksheets, pp. 5, 69-70 L1
Cross-Curricular Integration, p. 15 L2
Enrichment, p. 42 L3
Lab Manual, pp. 79-80 L2
Multicultural Connections, pp. 25-26
Reinforcement, p. 42 L2
Science and Society/Technology
  Integration, p. 29 L1

Study Guide, p. 42 L1
📽 Transparencies
Section Focus Transparency 36 L2
Teaching Transparency 22 L2

2 Teach

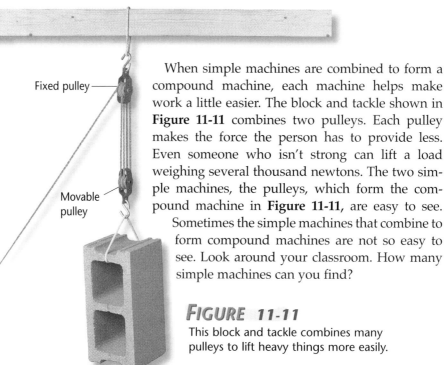

Fixed pulley

Movable pulley

When simple machines are combined to form a compound machine, each machine helps make work a little easier. The block and tackle shown in **Figure 11-11** combines two pulleys. Each pulley makes the force the person has to provide less. Even someone who isn't strong can lift a load weighing several thousand newtons. The two simple machines, the pulleys, which form the compound machine in **Figure 11-11**, are easy to see. Sometimes the simple machines that combine to form compound machines are not so easy to see. Look around your classroom. How many simple machines can you find?

FIGURE 11-11

This block and tackle combines many pulleys to lift heavy things more easily.

USING TECHNOLOGY

Exercise Equipment

Many apartment complexes, hotels, recreation centers, and sports clubs have fitness centers. In these centers, you'll find every possible combination of levers, pulleys, and weights to work out almost every muscle. There are different ideas about what the best exercise plan is.

Fitness technology is changing so fast that fitness centers quickly become outdated. As fast as the fitness industry is changing, one thing stays the same: regular, moderate exercise is still necessary for good health, and many ways to get exercise involve simple machines.

*inter*NET CONNECTION

Inventors have fun with simple and compound machines. Learn more about them on the Internet when you visit the link at the Glencoe Homepage. *www.glencoe.com/sec/science*

◄ **308**

Food out of Reach

A problem on any camping trip is keeping the animals out of the food. This is where a compound pulley machine can help.

What You'll Investigate

How can you make a model of a way to keep food out of the reach of animals?

Procedure

1. **Place** two chairs so that the chair backs can support the dowel rod.
2. Use a twist tie to **hang** a pulley from the dowel rod.
3. **Set up** the string as shown. It needs to pass around a pulley and then be tied to the dowel rod.
4. **Tie** one bag of marbles to the second pulley as shown.
5. Slowly let go of the bag and observe.
6. **Tie** on a second bag of marbles, slowly let go, and observe what happens.

Conclude and Apply

1. What happened when the marbles and pull were the same? What happened when you doubled the amount of marbles?
2. Do the pulleys allow you to pick up the most marbles with less effort? Explain your answer.
3. **Draw a diagram** of the setup. Add labels indicating the forces on the strings. Explain why it is easier to lift the marbles with two pulleys.

Goals

- Construct a pulley machine.
- Compare two uses of pulleys.

Materials

- string
- twist ties
- bags of marbles or other weights
- 2 pulleys
- dowel rod or broomstick

309

Assessment

Process Have the students determine how many of their classmates they think they could pick up if they were using the setup pulley. Use the Performance Task Assessment List for Making Observations and Inferences in **PASC**, p. 17.

Inclusion Strategies

Physically Challenged For Activity 11-2, move the dowel rod to a higher location for physically challenged students. The dowel rod could go between the tops of two cabinet doors and the pulleys could be suspended by long pieces of string to place them at the most convenient height.

Purpose

Kinesthetic Students will determine through experimentation how pulleys work and which type of system maximizes the force expended. **L2** **ELL**

Process Skills

observing and inferring, comparing and contrasting, recognizing cause and effect, formulating models, experimenting, analyzing, interpreting scientific illustrations, recognizing spatial relationships

Time

one class period

Alternate Materials

Any type of equal weights will do. Small spools of plastic fishing line that are found in smaller sizes in craft centers make great inexpensive pulleys.

Activity Worksheets, pages 5, 69-70

Teaching Strategies

Check often to see that the students are using the string and pulleys correctly.

Troubleshooting When the students disconnect the bags, the marbles may hit the floor. Make sure the bags are well sealed.

Answers to Questions

1. The pull bag fell, and when the second bag of marbles was added, the two sides balanced each other.
2. Yes, the pulleys require less effort to pick up the supplies, because one bag was able to pick up two. The force was doubled.
3. There are two ropes holding the supplies. Half of the weight is being held up by the dowel rod.

Discussion Lead students in a group discussion of how simple and compound machines impact their daily lives.

Reteach

Make a table with the list of simple machines down one side and across the top. Match up the two simple machines in the squares, and think of a real compound machine that has these two simple machines together. As an alternative, have students invent something new.

Extension

For students who have mastered this section, use the **Reinforcement** and **Enrichment** masters.

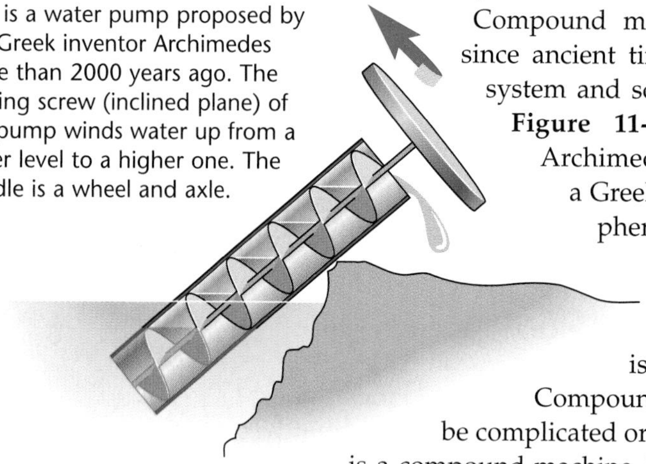

FIGURE 11-12

This is a water pump proposed by the Greek inventor Archimedes more than 2000 years ago. The turning screw (inclined plane) of the pump winds water up from a lower level to a higher one. The handle is a wheel and axle.

Compound Machines

Compound machines have been used since ancient times. The water-pumping system and screw conveyors shown in **Figure 11-12** were proposed by Archimedes (ar kih MEE deez), a Greek mathematician, philosopher, and inventor.

Any machine or device that is made up of two or more simple machines is a compound machine. Compound machines don't have to be complicated or fancy. A hand can opener is a compound machine. The crank is a wheel and axle, the handles are levers, and it has a circular wedge that cuts down into the can. Look at a bicycle and see how many simple machines you can find. A bicycle is another example of a compound machine.

The two-pulley system was only one of the compound machines you used on the camping trip. You may not have noticed several others. The biggest was the car that took you and your canoe to the river. The fishing rod and reel and the hatchet are two more examples.

Where else would you look for examples of compound machines, like the ones shown in **Figures 11-13** and **11-14?** Try the nearest gym, farm, kitchen, grocery store, or recreation center. Combinations of pulleys, inclined planes, and levers help people in many ways to make their lives easier.

FIGURE 11-13

This grain conveyor is another example of a compound machine. The corkscrew is a screw and a wheel and axle.

Content Background

Archimedes invented a device called the water screw to bail the water out of the bottom of the king's boats. Only one person was needed while the rest could row. The ships were lighter and faster, which helped to win more battles at sea. The water screw is still used in some countries as a cheap, portable water pump.

Inclusion Strategies

Visually Challenged Form copper tubing into a large spiral. Use a small metal ball bearing or several BBs to represent water. Let the student hear and feel the BBs as they move through the tubing and out the other end. **LS**

A The lever system is at rest before the key is played.

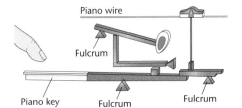

Piano wire

Fulcrum

Piano key   Fulcrum   Fulcrum

B When the key is played, the levers cause the hammer to strike the wire.

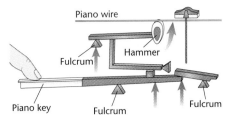

Piano wire

Fulcrum   Hammer

Piano key   Fulcrum   Fulcrum

FIGURE 11-14

A piano is a compound machine. When a key is played, a system of levers causes a felt-tipped hammer to strike the piano wire. This causes the note that you hear.

Section Wrap-up

1. What is a compound machine?

2. Why are scissors a compound machine?

3. **Think Critically:** Explain why a hatchet is a compound machine, and list all of the simple machines that are part of it.

4. **Skill Builder**

   Sequencing Two pulleys form a compound machine. A movable and fixed pulley were used to lift the supplies on the camping trip. Starting at the tree, describe the path of the rope all the way from the pulleys to the hands of the camper. If you need help, refer to Sequencing on page 543 in the **Skill Handbook.**

Science Journal

Pick any compound machine you use at home or in the classroom. Draw it in your Science Journal and tell which simple machines it uses.

4 Close

•MINI•QUIZ•

Use the Mini Quiz to check students' recall of chapter content.

1. **Two or more simple machines used together is a(n) _____.** *compound machine*

2. **What simple machines make a can opener?** *lever, wheel and axle, and wedge*

3. **When a fixed and movable pulley are both used, only _____ the force will do the same amount of work.** *half*

Section Wrap-up

1. two or more simple machines used together

2. They combine two levers (each handle-blade piece) and two wedges (the sharp edges).

3. **Think Critically** It has two simple machines—wedge and lever.

Science Journal
Student answers will vary.

P L2 ELL

Skill Builder

Sequencing The rope is tied to a tree limb, goes down and around the pulley on the supplies, then up around the fixed pulley and down into the campers' hands.

Assessment

Oral See how many simple and compound machines the students can identify in one of Rube Goldberg's machines. Use the Performance Task Assessment List for Oral Presentation in **PASC**, p. 71. L1 ELL

Science & Society

11•4 Access for All

Before presenting the lesson, display **Section Focus Transparency 37** on the overhead projector. Assign the accompanying **Focus Activity** worksheet.

L2 ELL

What YOU'LL LEARN

- That simple machines make many sports and recreation activities accessible to all
- That an outline is an orderly, written summary of a text

Why IT'S IMPORTANT

Outlining helps you organize information in a meaningful way.

FIGURE 11-15

These athletes are playing bank-shot basketball.

312

Content Background

It is estimated that between 40 and 50 million Americans have some sort of physical or mental disability. The Americans with Disabilities Act requires that employers make reasonable accommodations to facilitate the jobs and duties of their disabled employees. The act also requires that public facilities and services be made accessible to everyone.

The Americans with Disabilities Act

Have you ever been on crutches? If you have, you know how hard it is to go up and down stairs and get around at home. Some people have to deal with those problems every day. The Americans with Disabilities Act (ADA) is designed to stop discrimination against people with disabilities (dihs uh BIHL uh teez). A disability is a physical or mental limitation. The ADA became a law in 1990. The ADA requires that public buildings and services be easy for disabled persons to use. Constructing new equipment and buildings or changing older ones has made many places and activities available to all.

Happy Campers

The use of simple machines in day and summer camps over the past few years has helped many campers. Portable ramps make hiking nature trails possible for people who use wheelchairs. In some camps, levers replace doorknobs. Levers also make it easier to turn light switches on and off.

Be a Good Sport!

Simple and compound machines allow persons with disabilities to play along. Bank-shot basketball, for example, is a non-running sport that requires strategy, concentration, and hoop-shooting skill. Bank-shot basketball is similar to traditional basketball, as shown in **Figure 11-15.** But, bank shot features 18 stations whose degree of difficulty depends on levers, ramps, and other simple machines.

On the golf course, golf clubs, which are third-class levers, are used to increase the speed of the golf ball. Golfers usually stand

Program Resources

📁 **Reproducible Masters**
Enrichment, p. 43 L3
Reinforcement, p. 43 L2
Study Guide, p. 43 L1

🔲 **Transparencies**
Section Focus Transparency 37 L2

while hitting the ball. Some disabled golfers must hit from a sitting position. A golf cart with a turning seat that uses a screw may be the answer for these golfers.

Skiing is another sport in which simple machines allow persons with a variety of disabilities to hit the slopes, as shown in **Figure 11-16.** Some ski boots have wedges attached to their bottoms that aid stability and assist people with poor balance. A ski slant board does the same thing. The slant board is a metal base attached to a ski binding. The board can be tilted forward, backward, or side to side. This clever device allows a disabled skier to position himself or herself in a balanced position. The slant board can also be used by skiers whose legs are different lengths.

FIGURE 11-16

This skier is competing in the Winter Paralympics in Lillehammer, Norway.

 ## Skill Builder: Outlining

LEARNING the SKILL

1. Carefully read the material to be outlined.

2. Choose at least two or three ideas from the passage that summarize each section. Depending on the length of the information to be summarized, your outline may have more than two or three main ideas.

3. Now, pick out key words and phrases that are important to the passage. These supporting details are written under the main idea headings.

4. Arrange the key words and phrases under the correct titles. Use numbers and letters to sequence the words and phrases.

5. Check the accuracy of your outline by comparing the outline to the passage. Is the order correct? Does the outline make sense? Make any necessary changes.

PRACTICING the SKILL

1. Make an outline using the bold-faced titles in the article as your main ideas.

2. Use what you have learned to make another outline of the information on pages 312 and 313. Use these main ideas: levers, pulleys, and inclined planes.

APPLYING the SKILL

Find an article in the library about how simple machines make many playgrounds accessible to all. Make an outline of the main points in the article.

- If you have a student with a physical disability, privately ask that student if he or she feels comfortable talking to the class about the disability and what he or she does to overcome it. If this is not possible, invite an adult who is disabled to speak to the class. Ask students to imagine that they have limited use of their arms and hands. Have them brainstorm a list of simple changes that could be made to assist them in performing everyday tasks. Responses might include replacing zippers (which are compound machines) with Velcro, using modified computer systems to enter and retrieve data, using mouth-operated joysticks, and so on. L2

- Have students propose and, if time allows, invent devices using simple machines that might make life easier for someone with a disability. L2

Teaching the Skill

Use the first paragraph of this feature to demonstrate how to make an outline. Explain the hierarchy of outlines and how they use different numbers and letters.

Answers to Practicing the Skill

1. Outlines should have clear structure, use multiple levels, and accurately reflect the order and content of the section.

2. Outlines will vary. Students should include all the section content but in a different organization.

Sources

- Mazio, Peter C. *Rube Goldberg: His Life and Work.* New York: Harper and Row, 1973.
- The Internet has several good Rube Goldberg pages.

Biography

Reuben Lucius Goldberg (1883-1970) worked as a newspaper cartoonist in San Francisco, CA. In 1970, he was the first cartoonist honored by an exhibition of his work at the Smithsonian Institution in Washington, DC.

Teaching Strategies

- Have students work in small groups to draw a Rube Goldberg machine that will accomplish a simple task such as closing a window or turning a page in a book. Have each team explain to the class how their machine works. Display the finished designs on a bulletin board. L2 ELL LS

  COOP LEARN

- Have students research famous inventors and the devices they created. As part of a class discussion, have students explain what problem each inventor was trying to solve and how the machines worked. L1

Other Works

Stewart, Tabori, and Chang, *Rube Goldberg Inventions*, under license from Rube Goldberg, Inc.

*inter*NET
CONNECTION

The Glencoe Homepage at **www.glencoe.com/sec/science** provides links to relevant, up-to-date Internet sites.

314

Science & the Arts

Rube Goldberg's Wonderful, Wacky Machines

Do you like gadgets? Many little devices we use every day make it easier to do jobs. Think of a pencil sharpener or a can opener. But if you are like most people, you've probably used at least one gadget that was so hard to operate that it would have been simpler to do the job without the gadget.

Rube Goldberg (1883-1970) understood that problem. Goldberg was an American cartoonist who poked fun at gadgets and machines. He thought that technology often made life more complicated than it should be. From the 1920s until the early 1960s, Goldberg drew hundreds of cartoons that featured funny and complicated inventions. These enormously complex machines were designed to do simple things like lick a stamp, crack open an egg, or scratch a mosquito bite.

Each machine had many steps and combined tools, household objects, and animals in odd ways. For example, Goldberg's "Automatic Screen Door Closer" used flies, a spider, a potato bug, a mechanical soldier, and a circus monkey (who was an expert bowler) just to close a door!

Rube Goldberg entertained newspaper audiences for many years with his imaginative cartoons. They were considered humorous and weird, but they also made people think. Even today, his cartoon machines remind us that the best approach to solving a problem is often the simplest one.

*inter*NET
CONNECTION

How many steps does it take to scratch your back? Visit a link through the Glencoe Homepage at *www.glencoe.com/sec/science* to see how many steps Rube Goldberg used in one of his cartoons.

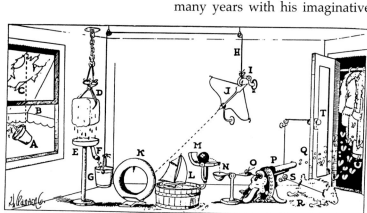

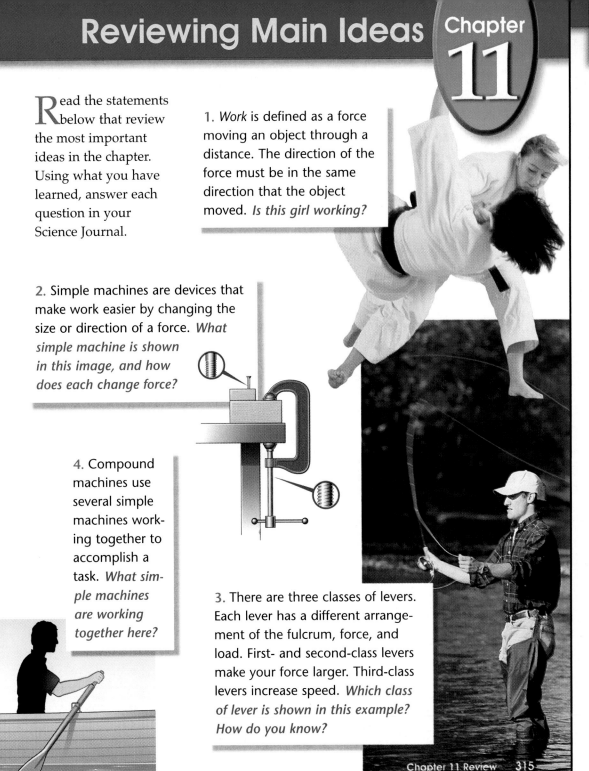

Read the statements below that review the most important ideas in the chapter. Using what you have learned, answer each question in your Science Journal.

1. *Work* is defined as a force moving an object through a distance. The direction of the force must be in the same direction that the object moved. *Is this girl working?*

2. Simple machines are devices that make work easier by changing the size or direction of a force. *What simple machine is shown in this image, and how does each change force?*

4. Compound machines use several simple machines working together to accomplish a task. *What simple machines are working together here?*

3. There are three classes of levers. Each lever has a different arrangement of the fulcrum, force, and load. First- and second-class levers make your force larger. Third-class levers increase speed. *Which class of lever is shown in this example? How do you know?*

Chapter 11 Review   315

Review

Have students look at the illustrations on this page. Ask them to describe details that support the main ideas of the chapter found in the statement for each illustration.

Teaching Strategies

If students are having difficulty understanding simple machines, have them draw each machine to help them visualize how each one makes work easier.

Answers to Questions

1. Yes, the directions of the force and movement are the same.
2. An inclined plane or screw reduces force by increasing the distance over which the force is applied.
3. A fishing rod is a third-class lever. The fulcrum is at the end of the lever arm, and by moving the bottom of the rod, the top moved a lot.
4. The paddle, arm, and hand make a first-class lever. The muscles in the arm are a third-class lever. The paddle in the water acts as a fulcrum to move the canoe forward. The force is from the arms as they pull back, causing the entire unit to function as a third-class lever.

Science at Home

LS **Intrapersonal** Have the students look around their homes and describe as many simple and compound machines as they find. See who has the most complete list. Discuss. **L1**

✓ **Assessment**

Portfolio Encourage students to place in their portfolios one or two items of what they consider to be their best work. Examples include:
- Assessment, p. 295
- MiniLAB, p. 297
- Activity 11-1, p. 303 **P**

Performance Additional performance assessments may be found in **Performance Assessment** and **Science Integration Activities**. Performance Task Assessment Lists and rubrics for evaluating these activities can be found in Glencoe's **Performance Assessment in the Science Classroom (PASC)**.

Chapter 11 Review

Using Key Science Words

1. A lever
2. simple machines
3. a wedge
4. A screw
5. work

Checking Concepts

6. a 7. b 8. d. 9. c 10. a

Thinking Critically

11. the same, but a wheelbarrow lets you use less force than simply lifting the mulch because it is a second-class lever, and the fulcrum takes some of the weight

12. It is easier to use the long-handled broom because it is a third-class lever, and it allows you to move the broom over a larger distance per movement.

13. A wrench is a second-class lever because the fulcrum is at one end and the force is at the other. The longer the wrench, the more force is applied to the object.

14. They are both combinations of simple inclined planes.

15. The screwdriver (wheel and axle) and a screw (inclined plane) are both simple machines, so together they form a compound machine. The screwdriver handle is a wheel that turns the shaft or axle, which then turns the screw.

Using Key Science Words

compound machine
inclined plane
joule
lever
pulley
screw
simple machine
wedge
wheel and axle
work

Using the list above, replace the underlined words with terms that include the correct key science word.

1. <u>A bar or rod free to rotate around a single point</u> can be used to lift heavy objects.
2. Pulleys, inclined planes, and levers are all examples of <u>a device that does work with one movement</u>.
3. A hatchet is an example of <u>two inclined planes combined</u>.
4. <u>An inclined plane wrapped around a rod</u> is often used to hold objects together.
5. Lifting a barbell is <u>exerting a force over a distance in the same direction as the object's motion</u>.

Checking Concepts

Choose the word or phrase that completes the sentence.

6. The _____ is the unit used for work.
   - a. joule
   - b. newton
   - c. pound
   - d. meter

7. A _____ is a compound machine.
   - a. screw
   - b. fishing rod and reel
   - c. ramp
   - d. wheel and axle

8. An example of an inclined plane is a(n) _____.
   - a. mountain climber
   - b. elevator
   - c. seesaw
   - d. moving van ramp

9. A _____ uses a lever.
   - a. vise
   - b. stairway
   - c. seesaw
   - d. moving van ramp

10. _____ is not work.
    - a. Carrying a canoe on level ground
    - b. Lifting a canoe
    - c. Carrying a canoe uphill
    - d. Paddling a canoe across a lake

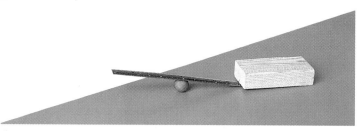

Assessment Resources

Reproducible Masters
Chapter Review, pp. 25-26
Assessment, pp. 47-50
Performance Assessment, p. 49

Glencoe Technology
Computer Test Bank
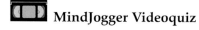 MindJogger Videoquiz

Thinking Critically

Answer the following questions in your Science Journal using complete sentences.

11. Does using a wheelbarrow to lift a load of mulch make you do more, less, or the same amount of work? Explain.

12. Which is easier to use to sweep a large area, a broom with a long handle or a broom with a short handle? Why?

13. What kind of simple machine is a wrench? How does the length of a wrench affect your force?

14. How are a needle and an ax related?

15. Explain how a screwdriver and a screw form a compound machine.

Developing Skills

If you need help, refer to the description of each skill in the Skill Handbook.

16. **Observing:** Make a list of the simple machines that make up a manual can opener. Refer to **Table 11-1** for help.

17. **Using Numbers:** Elena exerted a force of 200 N over a distance of 4 m in the direction of the force. How much work did she do?

18. **Concept Mapping:** Make a network tree concept map that includes the six types of simple machines and two examples of compound machines.

19. **Comparing and Contrasting:** Compare and contrast a first-class lever and a wheel and axle.

20. **Predicting:** Raul ties a bag weighing 400 N to one end of a rope. He tosses the rope over a tree branch and pulls down with 200 N of force. What happens?

Performance Assessment

1. **Invention:** Design a compound machine that will give a dog a treat any time the dog wants one.

2. **Scientific Drawing:** Draw each of the classes of levers, identify all the parts, and explain how they make tasks easier.

Table 11-1

Simple machine	Example
Inclined plane	the hiking trail up the mountain
	the gentle slope to pull the canoe onshore
	the screw anchors to hold the tent in place
	the wedge on a hatchet
Lever	First-class—crowbar to pry up the rock
	Second-class—canoe used to haul supplies
	Third-class—fishing rod
Wheel and axle	reel on the fishing rod
Pulley	fixed—rope over a limb
	movable—pulley on the supplies
	combination—using the tree limb and the movable pulley to lift the supplies

Chapter 11 Review   317

Developing Skills

16. **Observing** levers, wedges, and wheel and axle

17. **Using Numbers** 800 J

18. **Concept Mapping** See below for sample map.

19. **Comparing and Contrasting** Both are used to turn a small force over a large distance into a large force over a small distance. A lever is a straight rod; a wheel and axle is usually round and is rotated.

20. **Predicting** Nothing will happen because the resistive force of the bag is greater than the effort force of Raul. The bag continues to sit on the ground.

Performance Assessment

1. Student answers might include that the dog steps on a lever that opens a flap on the bottom side of a box of doggie treats and a treat falls out. If a student decides to build the machine, the design should be checked with the teacher before the student begins. Use the Performance Task Assessment List for Invention in **PASC**, p. 45. P

2. Check students' work with Figures 11-6 and 11-7 on pages 298-301. Use the Performance Task Assessment List for Scientific Drawing in **PASC**, p. 55. P

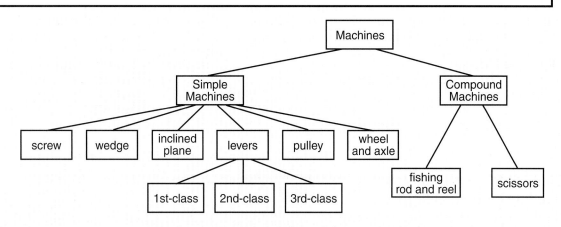

Section	Objectives/Standards	Activities/Features
Chapter Opener		Explore Activity: Observe Energy in Action, p. 319
12-1 **How does energy change?** (4 sessions, 2 blocks)*	1. **Define** *energy* and **describe** the forms that it takes. 2. **Compare and contrast** potential energy and kinetic energy. National Science Content Standards: (5-8) UCP2, UCP3, A1, A2, B2, B3, C3, E2	Problem Solving: Observing Kinetic Energy, p. 325 Activity 12-1: Colliding Objects, p. 326 Skill Builder: Making and Using Graphs, p. 327 Science Journal, p. 327
12-2 **How do you use thermal energy?** (4 sessions, 2 blocks)*	3. **Differentiate** among thermal energy, heat, and temperature. 4. **Determine** important uses of thermal energy and **describe** how thermal energy moves. National Science Content Standards: (5-8) UCP2, UCP3, A1, A2, B3, E2, F5	MiniLAB: Observe a Strained Flame, p. 329 Activity 12-2: Does thermal energy affect kinetic energy?, pp. 330-331 MiniLAB: Comparing Energy Content, p. 333 Using Technology: The Microwave Oven, p. 335 Skill Builder: Making Models, p. 337 Using Math, p. 337
12-3 **Science and Society: Sun Power** (1 session, ½ block)*	5. **Explain** how using solar energy helps conserve Earth's limited resources. 6. **Make generalizations.** National Science Content Standards: (5-8) UCP2, UCP3, A1, A2, B3, D1, E2, F5, G1	Skill Builder: Making Generalizations, p. 339 Science & Language Arts: To Build a Fire, p. 340 Science Journal, p. 340

* A complete Planning Guide that includes block scheduling is provided on pages 31T-33T.

Activity Materials

Explore	Activities	MiniLABs
page 319 wide rubber band	page 326 high-bouncing ball, playground ball, volleyball, tennis ball, measuring tape, masking tape pages 330-331 freezer, 3 identical balls, meterstick, Celsius alcohol thermometer	page 329 candle, matches, metal strainer page 333 transparent container, dropper, graduated cylinder, timer, food coloring

Need Materials? Call Science Kit (1-800-828-7777).

Teacher Classroom Resources

Reproducible Masters	Transparencies	Teaching Resources
Activity Worksheets, pp. 5, 73-74 **Enrichment**, p. 44 **Lab Manual 24** **Multicultural Connections**, pp. 27-28 **Reinforcement**, p. 44 **Science and Society/Technology Integration**, p. 30 **Study Guide**, p. 44	**Section Focus Transparency 38**, What is energy?	**Spanish Resources** **English/Spanish Audiocassettes** **Cooperative Learning Resource Guide** **Lab Partner** **Lab and Safety Skills** **Lesson Plans**
Activity Worksheets, pp. 5, 75-78 **Enrichment**, p. 45 **Lab Manual 23** **Reinforcement**, p. 45 **Science Integration Activities**, pp. 61-62 **Study Guide**, p. 45	**Science Integration Transparency 12**, Warm- and Cold-Blooded Animals **Section Focus Transparency 39**, Energy Takes the Cake **Teaching Transparency 23**, Temperature Scales **Teaching Transparency 24**, Convection Currents	**Assessment Resources** **Chapter Review**, pp. 27-28 **Assessment**, pp. 51-54 **Performance Assessment**, p. 50 **Performance Assessment in the Science Classroom (PASC)** **MindJogger Videoquiz** **Alternate Assessment in the Science Classroom** **Computer Test Bank**
Cross-Curricular Integration, p. 16 **Enrichment**, p. 46 **Reinforcement**, p. 46 **Study Guide**, p. 46	**Section Focus Transparency 40**, Run on Sun	

Key to Teaching Strategies

The following designations will help you decide which activities are appropriate for your students.

L1 Level 1 activities should be appropriate for students with learning difficulties.

L2 Level 2 activities should be within the ability range of all students.

L3 Level 3 activities are designed for above-average students.

ELL ELL activities should be within the ability range of English Language Learners.

LS These activities are designed to address different learning styles.

COOP LEARN Cooperative Learning activities are designed for small group work.

P These strategies represent student products that can be placed into a best-work portfolio.

GLENCOE TECHNOLOGY

The following multimedia resources are available from Glencoe.

Science and Technology Videodisc Series (STVS)
Earth & Space
   Greenhouse Effect
Physics
   Images of Heat
   Solar Food Dryer
Chemistry
   Solar House
   Spiral Solar Concentrator
   Solar Tower

The Infinite Voyage Series
The Champion Within
Glencoe Physical Science Interactive Videodisc
Warm Area
Cool Area
National Geographic Society Series
GTV: Planetary Manager
Glencoe Physical Science CD-ROM

Teacher Classroom Resources

This is a representation of key blackline masters available in the Teacher Classroom Resources.

Teaching Aids

Section Focus Transparencies

38 SECTION FOCUS TRANSPARENCY

WHAT IS ENERGY?
You have used the term *energy* for years. You may have used it to describe how you feel when you run a race or what makes a ball bounce. But there's more to energy than meets the eye.

1. What in the photo reminds you of the term *energy*?

2. Do you think that any of the energy in the photo changes from one form of energy to another? Explain.

L2

39 SECTION FOCUS TRANSPARENCY

ENERGY TAKES THE CAKE
It's your birthday. Your cake may not be the most beautiful one on Earth, but you can't wait to blow out the candles and cut a piece.

1. What did the cake look like before it was baked?

2. What made the cake change from batter to cooked cake?

L2

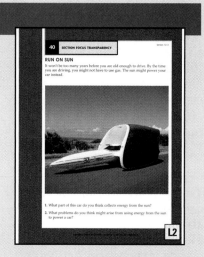

40 SECTION FOCUS TRANSPARENCY

RUN ON SUN
It won't be too many years before you are old enough to drive. By the time you are driving, you might not have to use gas. The sun might power your car instead.

1. What part of this car do you think collects energy from the sun?

2. What problems do you think might arise from using energy from the sun to power a car?

L2

Science Integration Transparencies

Warm- and Cold-Blooded Animals

L2

Teaching Transparencies

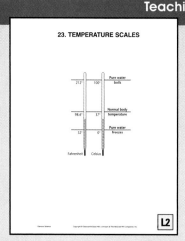

23. TEMPERATURE SCALES

Pure water boils — 212° / 100°
Normal body temperature — 98.6° / 37°
Pure water freezes — 32° / 0°

Fahrenheit   Celsius

L2

24. CONVECTION CURRENTS

L2

Meeting Different Ability Levels

Study Guide for Content Mastery

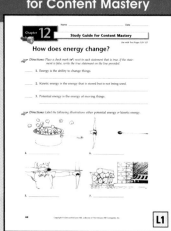

Chapter **12**   Study Guide for Content Mastery

How does energy change?

Directions: Place a check mark (✓) next to each statement that is true. If the statement is false, write the true statement on the line provided.

1. Energy is the ability to change things.

2. Kinetic energy is the energy that is stored but is not being used.

3. Potential energy is the energy of moving things.

Directions: Label the following illustrations either potential energy or kinetic energy.

L1

Reinforcement

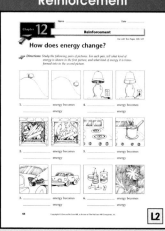

Chapter **12**   Reinforcement

How does energy change?

Directions: Study the following pairs of pictures. For each pair, tell what kind of energy is shown in the first picture, and what kind of energy it is transformed into in the second picture.

L2

Enrichment Worksheets

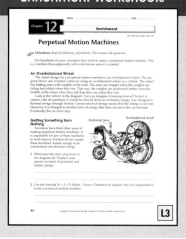

Chapter **12**   Enrichment

Perpetual Motion Machines

Directions: Read the following information. Then answer the questions.

For hundreds of years, inventors have tried to make a perpetual motion machine. This is a machine that supposedly will work forever once it is started.

An Overbalanced Wheel
The oldest design for a perpetual motion machine is an overbalanced wheel. The diagram shows one inventor's plan for using an overbalanced wheel on a vehicle. The wheel has folding arms with weights at the ends. The arms are straight when the weights are falling but folded when they rise. That way, the weights are positioned farther from the middle of the wheel when they fall than they are when they rise.

Look at the vehicle in the diagram. Can you imagine it running forever? In fact, it cannot. Like all machines, it would be slowed down as its kinetic energy was changed to thermal energy through friction. Conservation of energy means that the energy is not lost. However, it is changed to another form of energy that does not move the car forward. Eventually, the car must stop.

Getting Something from Nothing
Inventors have tried other ways of making perpetual motion machines. It is impossible for any of these machines to work forever. Friction always causes these machines' kinetic energy to be transformed into thermal energy.

1. What does the start/stop lever in the diagram do? Explain your answer in terms of potential and kinetic energy.

2. Use the formula $W = F \times D$ (Work = Force × Distance) to explain why it is impossible to build a perpetual motion machine.

L3

318C

Hands-On Activities

Science Integration Activities

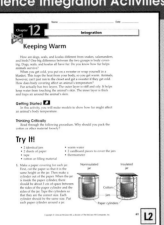

Lab Manual

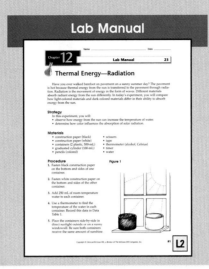

Activity Worksheets

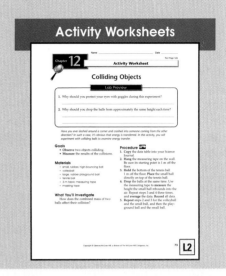

Enrichment and Application

Cross-Curricular Integration

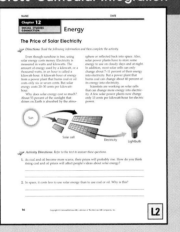

Multicultural Connections

Science and Society/Technology Integration

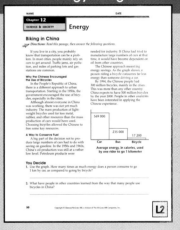

Assessment

Performance Assessment

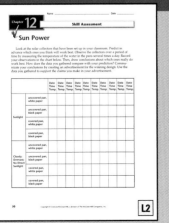

Chapter Review

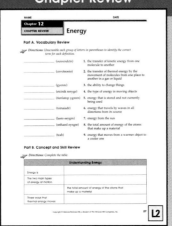

Assessment

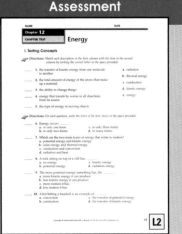

Energy

CHAPTER OVERVIEW

Section 12-1 Energy is defined as the ability to change things. Changes in temperature, shape, speed, position, and direction are examples of energy. Energy can be defined as potential (stored) or kinetic (in motion).

Section 12-2 This section distinguishes among thermal energy, heat, and temperature. The ways that thermal energy can be transferred—conduction, convection, and radiation—are explored.

Section 12-3 Science and Society This section explores solar power as an alternative energy source to nonrenewable energy sources. It also guides students in how to make generalizations.

Chapter Vocabulary

energy	heat
kinetic energy	radiation
potential energy	conduction
thermal energy	convection
	solar energy

Theme Connection

Energy/Stability and Change Students will discover that energy comes in many forms. It changes in ways that can both help and hurt humans but always remains the same in total amount available.

Chapter Preview

318

Skills Preview

▶ **Skill Builders**
- make a graph
- make a model
- make generalizations

▶ **MiniLABs**
- observe
- compare

▶ **Activities**
- observe
- measure
- sequence
- design an experiment
- graph
- compare
- predict

Learning Styles   Look for the following logo for strategies that emphasize different learning modalities. **LS**

Kinesthetic	Explore, p. 319; Activity 12-1, p. 326; Activity 12-2, p. 330
Visual-Spatial	MiniLAB, pp. 329, 333; Visual Learning, pp. 323, 324, 329
Interpersonal	Brainstorming, p. 322; Assessment, p. 337
Intrapersonal	Go Further, p. 331; Science at Home, p. 341
Logical-Mathematical	Problem Solving, p. 325; Theme Connection, p. 336
Linguistic	Using Science Words, p. 322; Across the Curriculum, p. 333; Science Journal, p. 340
Auditory-Musical	Science Journal, p. 321

Energy

With the wind rushing past her and the snow flying up around her, this snowboarder feels the thrill of moving at high speed. If you could watch her in action, you would notice that she can change the direction and speed of her board by changing her body position. A surfboard rider feels the same sense of connection between his or her body position and the board's motion. So does a skier, a cyclist, and even a young child on a swing. All are aware of something changing—position, speed, or direction. Do you realize that energy plays a part in these changes?

EXPLORE ACTIVITY

Observe Energy in Action

1. Obtain a new wide rubber band.
2. Hold the rubber band in both hands and touch it to your lower lip.
3. Quickly stretch the rubber band several times, and touch it to your lip again.
4. What difference do you notice?

Science Journal

Changes such as the one you noticed in the Explore activity occur all around you. In your Science Journal, list other such changes that you have observed today.

319

Prepare

Section Background
- One specific definition of energy is the ability to do work.
- Energy can change the temperature, shape, speed, position, or direction of an object.
- Energy is neither created nor destroyed, but it can change form. Humans use energy's ability to change form to do useful work.

Preplanning
Refer to the Chapter Organizer on pages 318A-B.

1 Motivate

Bellringer
Before presenting the lesson, display **Section Focus Transparency 38** on the overhead projector. Assign the accompanying **Focus Activity** worksheet.
`L2` `ELL`

12•1 How does energy change?

What YOU'LL LEARN
- What energy is and the forms that it takes
- The difference between potential energy and kinetic energy

Science Words:
energy
kinetic energy
potential energy

Why IT'S IMPORTANT
The more you know about energy and how it changes, the better you can use energy.

Energy

Energy (EN ur jee) is a term you probably have used many times. You may have said that eating a plate of spaghetti gives you energy or that a peppy Olympic gymnast has a lot of energy. But do you know that a burning fire, a bouncing ball, and a gallon of gasoline also have energy? Exactly what is energy?

What is energy?

Energy is the ability to change things. Energy can change the temperature, shape, speed, position, or direction of an object. Soccer players use energy to change the direction of soccer balls by hitting them with their feet or with their heads, as shown in **Figure 12-1.** Energy can change the shape of modeling clay or the temperature of a cup of water. You can use the energy of your muscles to change the speed of a bicycle, and a snowboarder uses energy to change position on the path the snowboard takes.

Energy Transformations

If you were to ask your friends to give examples of energy, you would probably get many different answers. Some of them might mention the energy in a flame. Others might suggest the energy needed to run a race. Energy occurs in several different forms. The flame has heat and light energy. Fat stored in your body contains chemical energy.

FIGURE 12-1

These soccer players use energy to change the speed, the position, and the direction of the soccer ball.

320

Program Resources

Reproducible Masters
Activity Worksheets, pp. 5, 73-74 `L2`
Enrichment, p. 44 `L3`
Lab Manual, pp. 85-88 `L2`
Multicultural Connections, pp. 27-28 `L2`
Reinforcement, p. 44 `L2`
Science and Society/Technology
  Integration, p. 30 `L2`

Study Guide, p. 44 `L1`

Transparencies
Section Focus Transparency 38 `L2`

Change Can Cause Change

In causing other things to change, energy itself often changes from one form to another. Any change of energy from one form to another is called an energy transformation. Energy transformations take place all around you every day. When a car sits in the sun all day, the energy of light waves changes to a form of energy that warms the inside of the car. In fireworks, chemical energy in the ingredients changes to light, sound, and heat energy.

During a transformation, such as when light changes to heat, the total amount of energy stays the same—no energy is lost or gained. Only the form of energy changes, not the amount.

Energy Changes in Your Body

You may not realize it, but you transform energy every time you eat. Like the logs, twigs, and leaves on the burning campfire shown in **Figure 12-2**, the corn on the cob, tossed salad, and baked potato you eat for dinner contain chemical energy. The energy from any meat you eat originally came from plants that had trapped sunlight in their leaves. What happens to the chemical energy in the food you eat? Energy contained in your food is changed into forms of energy that keep your body warm and move your muscles. Your body eventually releases the energy as heat when your muscles contract. Have you ever noticed how warm you become while playing, walking, running, or working hard? How many times have you thought it was cool outside until you started moving around? You soon found yourself taking off your jacket or sweater because your body was releasing energy in the form of heat when your muscles contracted.

Fat stores chemical energy that all of your muscles need, including your heart and those that help cause air to move into

FIGURE 12-2

In a campfire, the chemical energy in the wood is changed, or transformed, into heat and light energy.

Tying to Previous Knowledge

Have students recall from Chapter 8 that matter is made of atoms and from Chapter 9 that matter has properties that can change. These changes involve energy.

2 Teach

Teacher F.Y.I.

Pioneers crossing the prairie relied on the transformation of chemical energy to thermal energy. They burned the plant material in buffalo "chips" for fuel for their fires on the treeless plain.

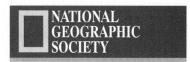
NATIONAL GEOGRAPHIC SOCIETY

 Videodisc

GTV: Planetary Manager
Energy

47903

47976

47999

Science Journal **Sing a Song of Energy** Have each student write a short story that involves both music and energy transformations. Have them use highlighters to mark the transformations. Transformations might include changing kinetic energy to sound when a drum is hit, vocal cords vibrating, or electrical energy changed to sound in a stereo. LS L2 P

Brainstorming

IS **Interpersonal** The transformation of other forms of energy to light is a useful energy change that humans have engaged in for a long time. Have students work in groups to make a brief pictorial history of people's efforts to provide light to darkened areas. **COOP LEARN** **L2**

Using Science Words

IS **Linguistic** Elicit from students that climbing to the top of a water slide uses kinetic energy, but the climber acquires potential energy. Have them explain how this energy relationship occurs.

GLENCOE TECHNOLOGY

 Videodisc

The Infinite Voyage: The Champion Within

Chapter 6

Glycogen: Fuel for Muscles

FIGURE 12-3

Low-fat items provide a small amount of fat to your diet.

FIGURE 12-4

A sled can have either potential energy or kinetic energy.

322

your lungs. Look at the low-fat items in **Figure 12-3**. All healthy bodies need a certain amount of fat—not too much, but not too little. Grocery store shelves are full of products labeled "fat free" that claim to be good for your health. But you would not be as healthy as you could be if you were on a completely fat-free diet.

Useful Changes

Since the earliest times, human beings have changed energy into forms that are useful to them. Early humans learned to change chemical energy to heat and light energy when they built a fire. Today, electrical energy changes into heat and light energy in a lightbulb. Chemical energy in gasoline changes to heat that runs the engine in your family car. The water heater in your home transforms chemical or electrical energy to thermal energy. Electrical energy is changed to energy of motion in an air conditioner. Hydroelectric and wind power plants transform the energy of moving water and wind into electrical energy.

Kinetic and Potential Energy

You've seen that energy can take many forms, such as light, heat, and motion. Two main types of energy that relate to motion are called kinetic (kih NET ihk) energy and potential (poh TEN shul) energy.

Content Background

Energy transformations are necessary for heating and cooling systems. In water heaters, for example, either electrical energy or the chemical energy in natural gas or propane is transformed into the thermal energy that is used to heat the water. Cooling systems use electrical energy in processes that remove thermal energy from a system by condensing a gas.

Energy transformations are also essential for production of electricity. Electricity is produced by changing another form of energy into electrical energy. For example, hydroelectric power plants use generators to transform the kinetic energy of moving water into electrical energy. Wind power plants transform the kinetic energy of moving air into electrical energy.

Kinetic Energy

If you were asked whether a baseball flying toward center field has energy, you might say that it does because it is moving. Moving objects have a type of energy called **kinetic energy.** As shown in **Figure 12-4**, a sled moving down a hill has kinetic energy. Molecules in air that is blown through a musical instrument also have kinetic energy.

Not all moving objects have the same amount of kinetic energy. Which would have more kinetic energy, a sled or a molecule of oxygen in the air? The amount of kinetic energy an object has depends on the mass and speed of the object. If the two objects were traveling at the same speed, the sled would have more kinetic energy than the oxygen molecule because the sled has more mass. Look at another example in **Figure 12-5.**

Potential Energy

If you were asked whether the sled sitting at the top of the hill in **Figure 12-4** has energy, what would you say? An object does not have to be moving to have energy—it may have what is called potential energy. **Potential energy** is energy that is stored. This is not energy that comes from motion; it is energy that comes from position or condition. The sled at the top of the hill has potential energy because even though it is not moving, it has potential for movement because of its position. As you can see, potential and kinetic energy are related to each other. Just as you can fill a water bottle to be used later at the basketball court, you can store potential energy to be used later when you need it.

FIGURE 12-5

If both have the same mass and each is running at its top speed, which would have more kinetic energy, a cheetah—one of the fastest land animals— or a dog?

Visual Learning

Figure 12-5 If both have the same mass and each is running at its top speed, which would have more kinetic energy, a cheetah—one of the fastest land animals— or a dog? *a cheetah* LS ELL

? FLEX Your Brain

Use the Flex Your Brain activity to have students explore KINETIC ENERGY. P

📁 **Activity Worksheets,** page 5

Demonstration

Show students that kinetic energy depends on mass. Lay a ruler on a smooth table. Place several quarters against the edge of the ruler. Push the coins at the same speed across the table and then stop the ruler. Choose two quarters that slide the same distance before stopping. Now tape another coin on top of one quarter to increase its mass. Again push the two coins with the ruler. Students should be able to see that the stopping distance, and thus the kinetic energy, depends on the mass.

Inclusion Strategies

Physically Challenged Because of the common association of energy with activity, students who are physically challenged may have difficulty relating to that type of transformation. Be sure to mention mostly transformations with which all students can identify.

Students may believe that more weight will make a pendulum move faster. In truth, only the length of the pendulum affects the period, or the time for one complete swing, of the pendulum. This is best illustrated by having students construct and work with pendulums. A pencil, string, and washers make a convenient desk-side pendulum. L1 ELL

FIGURE 12-6

This clock contains a pendulum. *When does the pendulum have the greatest potential energy?*

Pendulum

Arc

Potential to Kinetic

Potential energy becomes kinetic energy when something acts to release the stored energy. For example, it takes a lot of energy to tightly coil the wire in a coiled spring toy. Energy is stored in the coiled wire. When the toy walks down a flight of steps, the stored energy of the spring allows the movement to continue. It needs only the slightest help from gravity to keep it going.

One of the easiest ways to see the difference between potential and kinetic energy is to work with a pendulum (PEN juh lum), as shown in **Figure 12-6.** A pendulum is a weight that swings back and forth from a single point. A child's swing is an example of a pendulum. You can add potential energy to a swing by pulling back on it. Similarly, you can add potential energy to a pendulum by moving the weighted end to one side. When you let go of a swing or any other pendulum, gravity releases the potential energy, and the pendulum swings down in a curved path, called an arc. The instant the pendulum moves, potential energy is transformed into kinetic energy.

The more potential energy something has, the more kinetic energy results from the transformation of potential energy to kinetic energy. When a ski lift takes a skier to the top of a ski run, the skier's potential energy increases.

FIGURE 12-7

A bowling ball ready to be rolled down the alley has potential energy.

 When you roll it down the alley, your muscles give the bowling ball kinetic energy.

Across the Curriculum

History James Prescott Joule (1819-1889) suspected that energy was conserved. Through experimenting with energy transformations, he determined that a measured amount of kinetic energy could be transformed into a specific amount of thermal energy. He submerged a paddle into a water-filled bucket. The water was a certain temperature. He then attached one end of a thin rope to a known mass and the other end to the handle of the paddle. He arranged the system so that the paddle would turn in the water as the mass fell. The repeated release of the mass produced a measurable temperature change in the water. The unit for energy, the joule, is named in his honor.

That energy becomes kinetic energy when the skier moves down the mountain. The higher the ski run, the greater the potential energy the skier has at the top and the more kinetic energy she has skiing down the slope.

Transfer of Kinetic Energy

Kinetic energy can be transferred from one object to another when those objects interact. Look at the changes and transfer of energy shown in **Figure 12-7.** Energy can even be transferred from one bowling pin to another, knocking them all down with one roll of the ball, even though the ball does not touch all of the pins.

It is important to know the difference between potential and kinetic energy. It is this difference that allows us to store energy for later use.

  When it hits the pins, the ball transfers its energy to them.

Problem Solving

Observing Kinetic Energy

Baseball players both love and hate a fastball. A fastball is a pitch that comes straight across the plate at the fastest rate of speed the pitcher can deliver. Speeds of 90 mph are not unheard of when a good pitcher delivers a fastball. A team values a pitcher who can throw good fastballs. A good batter has mixed feelings about a fastball. The ball is harder to hit as it whizzes by, but some of the most dramatic home runs hit by batters are hit off fastballs.

Solve the Problem:
It may be difficult for a batter to connect with a fastball. But when it does happen, why is there a good chance that the ball will be hit a great distance? The answer lies in the presence of potential and kinetic energy in the ball and the bat.

Think Critically:
If you were a major league ballplayer, would you want to face a fastball pitcher? Explain.

Problem Solving

Solve the Problem
The potential energy of the bat held by the batter becomes kinetic energy when the batter swings the bat. Upon impact, the kinetic energy of the ball is increased by the kinetic energy of the bat. The more the initial kinetic energy of the ball and bat, the more kinetic energy will be in the hit ball because the kinetic energy is conserved.

Think Critically
Student answers will reflect whether each student prefers quick reflexes and the fastball or the greater strength needed to hit a different pitch as far. **IS** **P**

3 Assess

Check for Understanding

Demonstration Hold a fried corn chip with tongs. Use a match to light the chip. Have students observe that the chip burns for a long time and produces a lot of energy. Have students equate the fat and plant material's burning to energy from the food that they burn in their bodies. **ELL**

Reteach

Compare potential and kinetic energy to storing supplies for use later. Have students plan a hypothetical canoe trip and relate examples of activities—from the retrieval of the canoes up-river to the storage of water—to potential and kinetic energy. **L1**

Extension

For students who have mastered this section, use the **Reinforcement** and **Enrichment** masters.

325

Purpose

IS **Kinesthetic** Students will infer a cause-and-effect relationship between mass and kinetic energy. **L1** **ELL**

Process Skills

predicting, observing, recording data, inferring, making and using tables, measuring in SI

Time

40 minutes

Alternate Materials

The use of the hard, rubber, high-bouncing ball is necessary. Balls of different sizes can be substituted for the other balls.

Activity Worksheets, pages 5, 73-74

Teaching Strategies

- Students should be allowed to explore the characteristics of each ball before they begin to address the question of mass and energy.

- Averaging should be emphasized in this activity because measurements cannot be exact.

Troubleshooting If necessary, have students practice dropping the balls at the same time. Students may have difficulty steadying the smaller ball on top of the larger ball without adding energy to it. If the small ball does not bounce fairly straight up, do not count the trial. Be sure all balls are at the same temperature.

326

Activity 12-1

Goals

- Observe two objects colliding.
- Measure the results of the collisions.

Materials

- small, rubber, high-bouncing ball
- volleyball
- large, rubber playground ball
- tennis ball
- 3-m fabric measuring tape
- masking tape

Colliding Objects

Have you ever dashed around a corner and crashed into someone coming from the other direction? In such a case, it's obvious that energy is transferred. In this activity, you will experiment with colliding balls to examine energy transfer.

What You'll Investigate

How does the combined mass of two balls affect their collision?

Procedure

1. **Copy** the data table into your Science Journal.
2. **Hang** the measuring tape on the wall. Be sure its starting point is 1 m off the floor.
3. **Hold** the bottom of the tennis ball 1 m off the floor. **Place** the small ball directly on top of the tennis ball.
4. **Drop** the balls at the same time. Use the measuring tape to **measure** the height the small ball rebounds into the air. **Repeat** steps 3 and 4 three times, and **average** the data. **Record** all data.
5. **Repeat** steps 2 and 3 for the volleyball and the small ball, and then the playground ball and the small ball.

Conclude and Apply

1. **Rank** the balls from lightest to heaviest. What is the relationship between the mass of the two balls dropped and the height the small ball rebounds into the air?
2. Based on your observations, is the mass of an object related to the amount of potential energy it has?

Data and Observations
Sample Data

Small ball and . . .	Trial 1	Trial 2	Trial 3	Average
tennis ball	1.4 m	1.5 m	1.9 m	1.6 m
volleyball	2.5 m	2.0 m	2.4 m	2.3 m
playground ball	3.8 m	3.5 m	3.2 m	3.5 m

326 Chapter 12 Energy

Answers to Questions

1. The greater the mass of the other ball, the higher the bounce will be.
2. Yes; assuming other conditions are identical, the greater the mass, the greater the potential energy.

Content Have students use what they have learned about mass and energy to write a persuasive argument for using seat belts in cars. Arguments should include that seat belts decrease the amount of kinetic energy of a human body that is transferred to the car. Use the Performance Task Assessment List for Investigating an Issue Controversy in **PASC**, p. 65. **P** **L2**

FIGURE 12-8

Both the cars and the people have kinetic energy and potential energy several times during a roller coaster ride.

Section Wrap-up

1. Think of a roller coaster like the one shown in **Figure 12-8** climbing to the top of the steepest hill on its track. When does the first car have the greatest potential energy? When does it have the greatest kinetic energy?

2. Use the relationship between weight and kinetic energy to explain why some amusement parks have weight limits for their rides.

3. **Think Critically:** You get up in the morning, get dressed, eat breakfast, walk to the bus stop, and ride to school. List three different energy transformations that have taken place.

4. **Skill Builder**

   Making and Using Graphs A pendulum was found to swing seven times per minute. If the string were half as long, the pendulum would swing ten times per minute. If the original length were twice as long, the pendulum would swing five times per minute. Make a bar graph that shows these data. Draw a conclusion from the results. If you need help, refer to Making and Using Graphs on page 547 in the **Skill Handbook.**

Science Journal

In your Science Journal, write a paragraph about what energy transformations took place when last night's dinner was prepared.

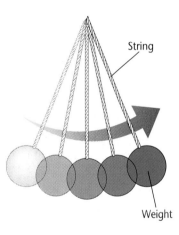

String

Weight

12-1 How does energy change? **327**

Skill Builder

Making and Using Graphs The longer the string, the slower the pendulum moves.

(Bar graph: Swings per Minute (y-axis, 0 to 10) vs. Length (x-axis: ½, 1, 2). Bars at heights 10, 7, and 5 respectively.)

✓ Assessment

Performance Have students use their graphs to predict the number of swings per minute if the pendulum string were one-fourth the length of the original string. Use the Performance Task Assessment List for Written Summary of a Graph in **PASC**, p. 41. [L3]

327

Prepare

Section Background

- Thermal energy is the total amount of kinetic and potential energy in the atoms that make up any material.
- Heat always moves from warmer to colder areas—that is, from areas of greater molecular activity to areas of less molecular activity—thereby increasing molecular activity.
- Thermal energy moves by radiation, convection, and conduction.

Preplanning

Refer to the Chapter Organizer on pages 318A-B.

1 Motivate

Bellringer

 Before presenting the lesson, display **Section Focus Transparency 39** on the overhead projector. Assign the accompanying **Focus Activity** worksheet.

L2 ELL

What YOU'LL LEARN

- The difference among thermal energy, heat, and temperature
- Important uses of thermal energy and how thermal energy moves

Science Words:
thermal energy
heat
radiation
conduction
convection

Why IT'S IMPORTANT

Life would not be possible if thermal energy did not exist and did not move.

12•2 How do you use thermal energy?

Thermal Energy

Imagine that you have just come inside from playing outside in freezing weather. What do you think about? A cup of hot cocoa? A nice, warm fire in the fireplace or a warm stove? What is it about these things that make you think they will warm you?

What is thermal energy?

Hot cocoa, as shown in **Figure 12-9,** and a fire will warm you because they have high amounts of the form of energy known as thermal (THUR mul) energy. **Thermal energy** is the total amount of energy—both potential and kinetic—of the atoms that make up a material. The amount of thermal energy that a cup of cocoa contains is determined by the amount of cocoa in the cup and the amount of energy in the atoms that make it up.

Thermal energy moves from a warmer object to a cooler one. Thermal energy in the cup of hot cocoa will transfer to your body and you will become warmer.

FIGURE 12-9
Drinking warm cocoa transfers thermal energy from the cocoa to your body.

Program Resources

 Reproducible Masters
Activity Worksheets, pp. 5, 75-78 L2
Enrichment, p. 45 L3
Lab Manual, pp. 81-84 L2
Reinforcement, p. 45 L2
Science Integration Activities, pp. 61-62 L2
Study Guide, p. 45 L1

 **Transparencies**
Science Integration Transparency 12 L2
Section Focus Transparency 39 L2
Teaching Transparencies 23, 24 L2

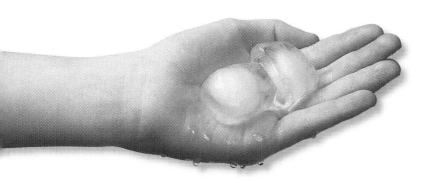

FIGURE 12-10

Thermal energy moves from a warmer object to a cooler one. *Is the air around the ice cubes colder than, warmer than, or the same temperature as the rest of the air in the room?*

How much thermal energy does it have?

An object's kinetic energy doesn't determine its thermal energy. For example, a ball at 25°C sitting in your backyard has the same amount of thermal energy as it does at 25°C heading through the air toward your friend. The energy the atoms have, not the energy of the object itself, determines the amount of thermal energy.

Different materials may have different thermal energies. If you get into a car that has been closed up on a sunny, summer day, where will you sit? If you can choose between sitting on a fabric-covered seat and sitting on a vinyl-covered seat, you'll find out that the vinyl has more thermal energy, even though the temperatures inside the cars are the same and the masses of the seats are similar. This difference is based on the way the atoms in each material are attracted to each other.

Heat and Thermal Energy

Thermal energy is what you may have been calling heat. Heat and thermal energy are related, but they are not the same. Look at **Figure 12-10.** Suppose you pick up an ice cube to place it in a drink. As you hold the ice cube, your hand becomes colder. Some of the ice melts because thermal energy from your hand transfers to the ice. The thermal energy that moves from a warmer object to a cooler one is **heat.**

Mini LAB

Observe a Strained Flame

See what happens when heat is transferred.

1. Light a candle.
2. Hold a metal strainer over the flame. Try to make the flame come through the wire mesh of the strainer.
3. Extinguish the candle.

Analysis

1. What happened to the flame when the strainer was in it?
2. In terms of heat transfer, suggest an explanation for your observations.

12-2 How do you use thermal energy?   329

Activity 12-2

PREPARE

Purpose

IS Kinesthetic Students will examine the relationship between thermal energy and kinetic energy. **L2 COOP LEARN**

Process Skills

observing, hypothesizing, designing an experiment, comparing, predicting

Time

45 minutes

Materials

Be sure balls are identical for each group of students. Racquetball balls and tennis balls are excellent for this activity because they bounce well.

Safety Precautions

Do not allow students to warm balls in a microwave or oven. Use alcohol, not mercury, thermometers in case of breakage.

Possible Hypotheses

Students may intuitively assume that removing thermal energy from a ball would negatively affect its kinetic energy. Most students' hypotheses will reflect that decreases in thermal energy will decrease kinetic energy; the cooler the balls, the less bounce they will have.

Activity Worksheets, pages 5, 75-76

Design Your Own Experiment

Does thermal energy affect kinetic energy?

You are in charge of dinner, and spaghetti sounds great. You put the dry spaghetti in a pot of boiling water and cover it. What soon happens? If you have ever seen steam lift the lid off a pot of boiling water, you've witnessed thermal energy transforming to energy of motion.

Possible Materials

- identical balls, at least 3
- meterstick
- thermometer
- paper
- pencil
- freezer

PREPARE

What You'll Investigate

Will the thermal energy in a ball affect how high it will bounce?

Form a Hypothesis

Think about what you know about how thermal energy affects matter, such as the air inside a ball. Remember that a measure of temperature indicates the amount of thermal energy the ball contains. In your group, **make a hypothesis** about how the temperature of a ball affects the height of its bounce.

Goals

Design an experiment that tests several balls to see whether the temperature of each ball when it is dropped makes it bounce to different heights.

Observe how energy is transformed from thermal energy to kinetic energy.

Safety Precautions

Wear goggles. Do not use any temperature beyond the range of what would normally be found in any safe area of your classroom.

Sample Data Table

Ball	Temp. (°C)	Height of Bounce (m)			
		Trial 1	Trial 2	Trial 3	Average
Hot	60	1.9	2.2	1.9	2
Room Temp.	20	1.5	1.2	1.2	1.3
Cold	10	0.39	0.41	0.40	0.40

PLAN

1. **Choose** a way to test your group's hypothesis. Remember to test only one thing.
2. **Decide** how many balls you are going to test and at which temperatures. Decide how long each ball will be kept at that temperature before you test it.
3. Pick a height from which to drop each ball. The height from which each ball is dropped must always be the same. **Decide** how to measure the bounce of the ball and how many times you are going to drop each ball.
4. **Prepare a data table** in your Science Journal that will show the temperature of each ball and the height of each bounce. **Decide** how to graph the results so the data are easy to see and compare.
5. **Read** over your experiment and see whether you can picture what is happening. Do any parts seem unclear? These parts will need more planning.

DO

1. Make sure your teacher has approved your plan and your data table before you proceed.
2. Carry out your plan.
3. While doing the experiment, **record** any observations and **complete** your data table.

CONCLUDE AND APPLY

1. **Graph** temperature versus height of bounce for each ball. **Compare** your graphs to those of other students in the class.
2. How did the temperature of the ball affect its bounce? **Summarize** your findings in a few sentences.
3. **Compare** your results and your hypothesis. From your results, what is the relationship between thermal energy in the ball and kinetic energy in the bounce?
4. **APPLY** Use your results to **predict** how the air temperature when you are playing tennis might affect how hard you have to swing the racket.

PLAN

Possible Procedures

Heat one ball, cool one ball, and leave one ball at room temperature. Hold the meterstick at a set position. Hold a thermometer against a ball and measure and record its temperature. In front of a certain point on the meterstick, drop the ball. Measure the height of the bounce. Repeat for several trials for each ball.

Teaching Strategies

Temperature differences can be achieved through use of a refrigerator, a sunny spot, and room temperature. Balls can also be held in warm water, room-temperature water, and ice water, so long as all three are placed in water and towel dried.

DO

Expected Outcome

The higher the temperature, the higher the bounce.

CONCLUDE AND APPLY

1. Sample graph:

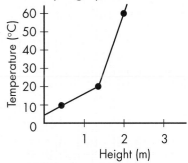

2. Height increases as temperature increases.
3. Student answers will vary. As thermal energy increases, so does kinetic energy.
4. On a hot day, the ball will have more bounce off the racket, and the players might not need to swing as hard.

Go Further

 **Intrapersonal** Have students find out why hockey pucks, which are made of solid disks of rubber, are frozen before a hockey game. Only frozen pucks are used during the game. A frozen hockey puck glides over the ice and rebounds off the boards, but it does not bounce as much. This means it can be angled off the boards. **L3**

✓ Assessment

Performance Design an experiment that would show whether or not heating a perfume will increase the kinetic energy of its molecules to the point where it moves more quickly through the air. Use the Performance Task Assessment List for Designing an Experiment in **PASC**, p. 23.

331

A scientist named Werner Heisenberg showed that certain physical properties, such as temperature, can never be accurately measured because the presence of the measuring instrument changes the state of the object slightly. Heisenberg's rule was named the Uncertainty Principle. Ask students how measuring the temperature of a cup of hot cocoa would change its temperature. The introduction of a thermometer, which is cooler than the liquid, will lower its temperature.

Demonstration

LS Kinesthetic To help students understand that *hot* and *cold* are relative terms, assemble three large containers of water: one with ice water, one with very warm—not hot—water, and one with water at room temperature. Have a student place one hand in the ice water and the other hand in the very warm water. Then, after about 30 s, have the student remove both hands and place them in the water that is at room temperature. That water will feel warm to the hand that was in the ice water and cold to the hand that was in the very warm water.

*inter*NET CONNECTION

The Glencoe Homepage at **www.glencoe.com/sec/ science** provides links connecting concepts from the student edition to relevant, up-to-date Internet sites.

Temperature and Thermal Energy

Temperature also is related to thermal energy and heat. You are aware of temperature every day. When you get up in the morning, the temperature of the air outside helps you decide what to wear. When you are not feeling well, you take your body temperature to find out whether you have a fever, as shown in **Figure 12-11.**

Temperature Is an Average

But what is temperature? If you have a sample of a material, that sample is made up of many atoms. Each of those atoms has a certain amount of kinetic energy and is moving. If you measured the kinetic energy of each atom in the sample and averaged them, this average kinetic energy would be the temperature of the sample. For example, normal human body temperature is 98.6°F. If the girl in **Figure 12-11** has a body temperature of 101.2, the atoms in her body have more kinetic energy than usual.

Hot and *cold* are terms that are used in everyday language to indicate temperature. They are not scientific words because they mean different things to different people. Water that seems hot to one person may seem just right to another. If you usually live in Florida but go swimming in the ocean off the coast of Maine while on vacation, you might find the water unbearably cold. Have you ever complained that a room was too cold when other people insisted that it was warm?

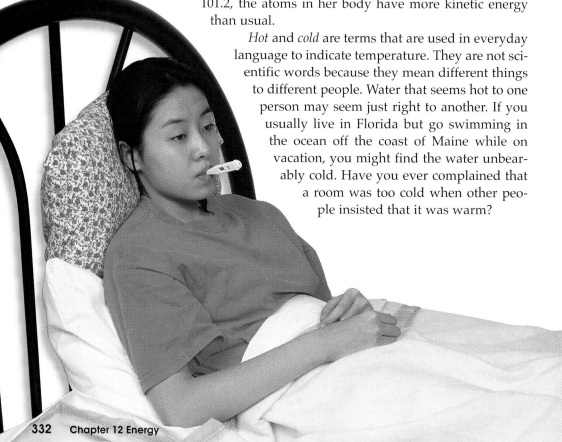

FIGURE 12-11

In the United States, thermometers that use the Fahrenheit scale are used to find body temperature.

Cultural Diversity

Alternate Fuels Before the widespread use of fossil fuels such as coal and oil, many people burned wood for energy. For some, however, wood is scarce, so alternate energy sources were used. For the Inuit, wood was not available in the Arctic environment, so most meals were eaten uncooked. Their primary fuels, which were fats and oils from hunted animals, were burned for heat and light, not for cooking. In India, firewood is also scarce. Dried cow dung provides most fuel for warmth and cooking fires for many families. Students can research alternate sources of fuel used by other peoples when wood and fossil fuels are scarce.

Temperature Scales

Because everyone experiences temperature differently, you cannot accurately measure temperature by how it "feels." Recall that temperature is the average kinetic energy of all the atoms in a material. But there is no way to measure the kinetic energy of each atom and then calculate an average. Instead, scientists use various temperature scales designed to measure the average kinetic energy of the particles in a material. These scales divide changes in kinetic energy in atoms into regular intervals. Look at **Figure 12-12.** The scales most commonly used in daily life are the Celsius (SEL see us) and Fahrenheit (FAYR un hite) scales.

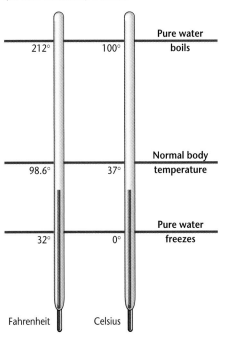

212°	100°	Pure water boils
98.6°	37°	Normal body temperature
32°	0°	Pure water freezes
Fahrenheit	Celsius	

FIGURE 12-12

These thermometers show at what temperatures water freezes and boils and normal human body temperature on the Fahrenheit and Celsius scales.

Mini LAB

Comparing Energy Content

Find out which has more kinetic energy, hot water or cold water.

1. Pour equal amounts of hot, cold, and room-temperature water into each of three transparent, labeled containers.
2. Measure and record the temperature of the water in each container.
3. Use a dropper to gently put a drop of food coloring in the center of each container.
4. Observe after 2 minutes to see how quickly the color has spread through each container.

Analysis

1. Based on the speed at which the food coloring spreads through the water, rank the containers from fastest to slowest. What can you infer about how water temperature affected the movement of the food coloring?
2. If air acts much like water, where on Earth would you expect to see the most energetic weather develop?

---

interNET CONNECTION

Go to the Glencoe Homepage at *www.glencoe.com/sec/science* for a link to a site that compares temperature scales and gives conversion factors.

Mini LAB

Purpose

LS **Visual-Spatial** Students will examine the kinetic energy of water molecules, observing how that energy changes with different temperatures. **L1** **ELL**

Materials

3 transparent containers, graduated cylinder, alcohol thermometer, dropper, food coloring, timer

Teaching Strategies

Safety Precautions Make sure all water temperatures are safe for student handling.

Troubleshooting Students will better be able to observe the movement of the food coloring if a piece of white paper is held behind the cups.

Activity Worksheets, pages 5, 78

Analysis

1. The slowest dispersal of color will be in the coolest water, and the fastest is with the highest temperature. Heat increases kinetic energy.
2. equatorial regions

Assessment

Portfolio Have students draw models of the water molecules in each of the three cups. Have them indicate the relative motion of each. Use the Performance Task Assessment List for Scientific Drawing in **PASC,** p. 55. **P**

Content Background

Kelvin is another temperature scale commonly used in laboratories and in industry. Absolute zero, the bottom of the Kelvin scale at 0 K, is a scientific concept that exists in theory, not reality. It is the theoretical temperature at which all kinetic molecular motion ceases.

Across the Curriculum

Language Arts *Absolute zero* is a term used frequently in science fiction writing. Have students write a one-page science fiction story that involves absolute zero, where there is no kinetic molecular motion. **LS** **L2** **P**

Using an Analogy

Conduction of heat occurs without transferring molecules from one material to another. The energy of one molecule increases the energy of another. This situation is similar to how a steel ball can hit a string of balls and make the one on the end move.

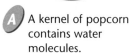

FIGURE 12-13

Popcorn can be cooked by either radiation or conduction.

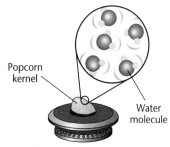

Popcorn kernel

Water molecule

A A kernel of popcorn contains water molecules.

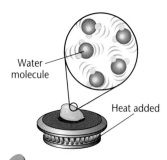

Water molecule

Heat added

B When heat is added to the kernel, the water molecules in the kernel move faster. They become water vapor and cause pressure inside the kernel.

334 **Chapter 12 Energy**

Thermal Energy on the Move

Thermal energy has many properties that affect our lives in important ways. What is the weather like today? Weather occurs because thermal energy causes air in the atmosphere to move. When you cook food, as in **Figure 12-13**, heat moves through the food, causing changes in it. Your home is heated and cooled by this same movement of thermal energy. Thermal energy moves in three distinct ways—by radiation, by conduction, and by convection.

Radiation

Radiation (ray dee AY shun) is energy that travels by waves in all directions from its source. The sun transfers energy to Earth through radiation. Have you ever been on a hike on a cool day and chosen to lean against a large boulder because it was warm from the sunlight? The boulder was heated by radiation. A microwave oven and a solar cooker cook food because of transfer of energy by radiation.

Conduction

Have you ever picked up a spoon that was in your soup and discovered that the spoon was hotter than it was when you placed it there? The spoon handle became hot because of conduction (kun DUK shun). **Conduction** is the transfer of kinetic energy from one molecule to another, much the same way kinetic energy is transferred from a bowling ball to the pins. When you put a pan on the stove to make a grilled-cheese sandwich, as shown in **Figure 12-14**, the heat from the stove's heating element transfers energy to the pan, making its molecules move

C When the pressure becomes great enough, the kernel explodes.

Popped kernel

Content Background

All molecules are in constant motion. To add thermal energy to something means to increase the motion of the molecules. Some materials, due to the density or character of their molecular structure, can acquire a great deal more thermal energy than other things. Three balls—the same size but made of metal, glass, and rubber—can illustrate this difference.

faster. Some of these molecules bump into the molecules in the bread and butter that are on the surface of the pan. Energy is transferred to the sandwich. This transferred energy causes the sandwich to cook. Even though conduction is a transfer of kinetic energy from particle to particle, the molecules involved don't travel from one place to another. They simply move in place, bumping into each other and transferring energy from faster-moving particles to slower-moving ones.

FIGURE 12-14

Energy transferred by conduction from the pan to the sandwich causes the cheese to melt and the bread to toast.

USING TECHNOLOGY

The Microwave Oven

When people's lives became busier, they needed faster ways to prepare food. The microwave oven became the answer for many households. It is easy to use, quick, and relatively inexpensive.

The microwave oven uses microwave radiation to transfer thermal energy to the inside of a food item, cooking it from the inside out. Microwaves increase the motion of the molecules of the food placed in the oven. The more mass something has, the more motion its molecules will have. This is why even large items can be cooked in just a little more time than smaller ones.

If you ever made microwave popcorn, you used radiation to speed up the water molecules inside the kernels, causing them to explode. If you don't use a microwave oven to make popcorn, you use conduction. Heat moves up from the bottom of the pan, through the metal, and into the kernels.

*inter*NET CONNECTION

Many meals require heat transfer in their preparation. What advantages and disadvantages are there to different methods? Visit the Glencoe Homepage at *www.glencoe.com/sec/science* for a link to the science of food preparation.

335

USING TECHNOLOGY

For more information and a diagram of the microwave oven, refer to page 152 of *The Way Things Work* by David Macaulay.

*inter*NET CONNECTION

Advantages can include that microwave ovens are a fast way to prepare food, and they fit a busy lifestyle. Conventional ovens usually produce better results on baked goods. Disadvantages of most methods include that they deplete fossil fuels.

? FLEX Your Brain

Use the Flex Your Brain activity to have students explore THERMAL ENERGY.

📁 **Activity Worksheets,** page 5

Discussion

Have students discuss what materials make good conductors of heat and what ones make good insulators. **L2**

GLENCOE TECHNOLOGY

💿 **Videodisc**

STVS: Physics
Disc 1, Side 1
Images of Heat (Ch. 8)

Solar Food Dryer (Ch. 12)

Content Background

Materials that have a high degree of thermal energy are *conductors* of heat. Materials that have low degrees of thermal energy are *insulators*.

If you look around your house, you might find that the heating vents are on the floor, frequently by outside walls or windows. With what you now know about heat convection, why does this arrangement make sense? *Warm air is forced up and throughout the room by cold air from the outside.*

Making a Model

Have students draw models that illustrate the three types of heat transfer: conduction, convection, and radiation. Captions should compare and contrast the methods.
P **L2**

3 Assess

Check for Understanding

Activity Place different fabrics and colors of material in the sun. Have students compare their ability to acquire thermal energy. Have them explain how this information could be useful in the clothing industry. **L2**

Reteach

Have students use the relationship between thermal energy and kinetic energy to explain why the temperature in a room increases when the room becomes full of people.

Extension

For students who have mastered this section, use the **Reinforcement** and **Enrichment** masters.

FIGURE 12-15

Even though Labrador is further south than Norway, Norway's ports are kept free of ice by the warm convection current that flows along its coast. Labrador is much colder because of the icy current that flows along its shore.

Convection

Some energy transfers involve molecules that do not stay in place but move from one place to another. **Convection** (kun VEK shun) transfers thermal energy by the movement of molecules from one place to another in a gas or a liquid. It is because of convection that thermal energy causes changes in the atmosphere that produce weather on Earth. Convection currents are also formed in oceans by cold water from the poles and warm, tropical water, as shown in **Figure 12-15**.

Have you ever seen an eagle or a hawk coasting high in the air? Look at **Figure 12-16**. The bird can fly even though it is not flapping its wings because it is held up by something called a "thermal." A thermal is a column

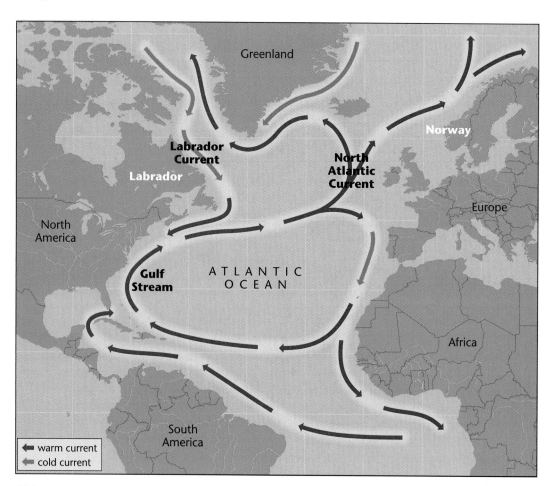

- ← warm current
- ← cold current

Content Background

The daily weather map published in the newspaper is a good illustration of the movement of air through the atmosphere by convection. Emphasize that Earth's weather is the result of this convection that results from uneven heating of Earth's atmosphere.

Theme Connection

Systems and Interactions

LS **Logical-Mathematical** Weather systems are closely tied to heat and energy transfer. The sun heats the surface of Earth unevenly. As cool air sinks and warm air is forced up, winds and weather fronts are created. Have students discuss how convection plays a part in the interaction of the forces of weather. **L1**

of warm air that is forced up as cold air around it sinks. A thermal is a convection current in the air.

You can see that thermal energy and its transfer affect you constantly. How many uses of thermal energy can you see around you right now?

FIGURE 12-16
A convection current holds this bird in the air. It is the same type of current used by people who are hang gliding. It keeps them in the air as long as possible.

Section Wrap-up

1. Popcorn can be cooked in a hot-air popper, in a microwave oven, or in a pan on the stove. Identify each method as using either convection, conduction, or radiation.

2. Use a glass of cold tea and a cup of hot tea to explain the differences among temperature, heat, and thermal energy.

3. **Think Critically:** On the Fahrenheit temperature scale, water boils at 212°F and freezes at 32°F. On the Celsius scale, water boils at 100°C and freezes at 0°C. Which scale do you think would be easier to use? Explain.

4. 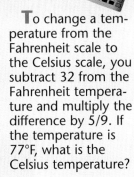 **Skill Builder**
   Making Models A thermos bottle includes a glass container with a shiny coating that is surrounded by a vacuum, which contains no air. This glass is then placed in a container that is the outer part of the thermos bottle. Make a model of what a cross-section of a thermos bottle looks like. If you need help, refer to Making Models on page 557 in the **Skill Handbook.**

USING MATH

To change a temperature from the Fahrenheit scale to the Celsius scale, you subtract 32 from the Fahrenheit temperature and multiply the difference by 5/9. If the temperature is 77°F, what is the Celsius temperature?

---

4 Close

•MINI•QUIZ•

Fill in each blank with one of these words: *conduction, convection, heat, radiation, temperature.*

1. **Thermal energy that moves from a warmer to a cooler area is _____.** *heat*

2. **The movement of energy by waves in all directions from a central point is _____.** *radiation*

3. **The process by which thermal energy moves through fluids such as air and water is _____.** *convection*

Section Wrap-up

1. Hot-air popping uses convection, a microwave oven uses radiation, and a pan on a stove uses conduction.

2. Temperature is the measure of average molecular energy in the tea. Hot tea has a greater temperature because its molecules have more kinetic energy. Heat is the movement of thermal energy from a warmer area to a cooler one. Hot tea will transfer heat to the air, and heat from the air will transfer to the cold tea. Thermal energy is the energy actually being transferred.

3. **Think Critically** Student answers will vary. Most students would believe that the Celsius scale would be easier to use because it is based on 100 units, not 180.

 **Skill Builder**

Making Models Students should be able to produce an accurate cross section of a thermos. It should include all of the parts mentioned, placed in the correct order. Parts should be labeled. Point out that the shiny coating in the glass container reflects heat back into the liquid, reducing the conduction of heat away from the contents.
P

✓Assessment

Oral Have students make oral presentations of their models of thermos bottles. Encourage students to propose improvements in the thermos bottles. Use the Performance Task Assessment List for Oral Presentation in **PASC,** p. 71. [LS] [L2]

USING MATH
$$\frac{5}{9} \times (77°F - 32°F) = 25°C$$

Science & Society

12●3 Sun Power

Bellringer

Before presenting the lesson, display **Section Focus Transparency 40** on the overhead projector. Assign the accompanying **Focus Activity** worksheet.

L2 ELL

Content Background

- In 1982, Australia celebrated its centennial with a solar-powered automobile race. The race, with its need for new technology, sparked global interest in solar energy. This interest led to a solar-powered automobile race called the Sunrayce in the United States. Participants travel from Indianapolis, Indiana, to Colorado Springs, Colorado. The purpose of the race is to encourage the study of solar energy possibilities.

- For more information on solar-powered cars and the Sunrayce, contact the University of Michigan's Solar Car Team, 3411 Beal Avenue, Ann Arbor, MI 48109.

What YOU'LL LEARN

- How using solar energy helps conserve Earth's limited resources
- How to make generalizations

Science Words:
solar energy

Why IT'S IMPORTANT

You will be able to generalize by putting together pieces of information to better understand an issue.

Solar Energy

Much of the energy we use comes from nonrenewable resources. As you learned in Chapter 7, nonrenewable resources cannot be replaced easily or quickly. Coal and oil are nonrenewable resources. Because these resources are limited, many people believe that we should use renewable resources to meet our energy needs. Natural processes can replace renewable resources fairly quickly. Energy from the sun, called **solar energy** (SOH lur • EN ur jee), is an example of a renewable resource.

The sun supplies Earth with a constant source of heat and light. Scientists are exploring ways to gather the sun's energy and transform it into electrical energy. This energy can provide heat and electricity—and even power cars!

One of the most ambitious schemes for providing solar energy on a large scale is to put giant solar collectors, or solar cells, in space. These solar cells, as shown in **Figure 12-17,** would collect energy and transfer it back to Earth using microwave radiation.

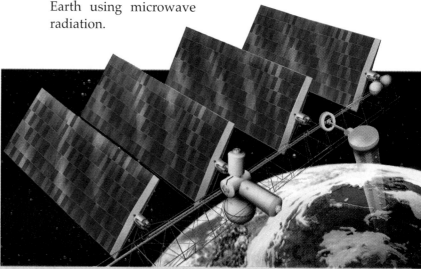

FIGURE 12-17
This is what solar collectors in space will probably look like.

Program Resources

 Reproducible Masters
Cross-Curricular Integration, p. 16 L2
Enrichment, p. 46 L3
Reinforcement, p. 46 L2
Study Guide, p. 46 L1

Transparencies
Section Focus Transparency 40 L2

Not all is sunny with this plan, however. To transform energy from the sun into electrical energy, solar cells require materials that are poisonous. In addition, microwaves can potentially harm humans. Solar energy itself can be unreliable. Light from the sun can be blocked by clouds, as in **Figure 12-18,** or air pollution. Even so, we can count on the sun's energy for a long, long time.

As you read material about topics such as solar energy, look for bits and pieces of information that hint at a main meaning—a generalization—about the topic. To make a generalization (jen rool luh ZAY shun), you put together bits of information to arrive at the big picture. For instance, if you knew that a football team was ranked first in the state last year and was unbeaten this season, you would make the generalization that the football team was great. This process will help you to form a better understanding of important issues.

FIGURE 12-18
Unless economical batteries that can store solar energy can be developed, using solar energy depends a lot on how much the sun shines.

Skill Builder: Making Generalizations

LEARNING the SKILL

1. Identify the general topic of what the reading is about.

2. Gather as many facts as you can about the subject. Be sure the information you gather is made up of facts, not opinions.

3. Identify similarities and patterns among the facts. Facts can fit more than one pattern.

4. Use the similarities and patterns to form some general ideas about the subject.

PRACTICING the SKILL

1. What is the main idea of this passage?

2. Which of the following generalizations is supported by the passage?
   a. The use of microwaves with solar cells must be planned and controlled.
   b. The use of microwaves with solar cells does not present a problem.

3. Identify one or two facts that support the correct generalization.

APPLYING the SKILL

Make a general statement that describes your class. Write three facts that support your generalization.

Teaching the Skill

Before starting the Skill Builder, provide students with supporting facts about a familiar event or thing. Then, have them make a generalization about it. For example, if a certain song is played every hour on the radio, most students know the words to the song, and the song is currently the top-selling single, a valid generalization could be that the song is currently popular. [L1]

Answers to Practicing the Skill

1. Solar energy has great possibilities as an energy source, but problems also are associated with it.

2. a

3. Student answers might include that microwaves can be harmful to humans.

Teaching the Content

- Show students pictures of several solar-powered race cars. Photos can be obtained on the Internet, by contacting organizers of the Sunrayce, or in magazines such as *Popular Science*. Have them compare and contrast the different designs. Point out any features that the cars have in common.

- Have students design and sketch a solar-powered automobile that uses the best design features of the solar-powered race cars. [L3]

- Challenge students to infer whether solar energy would be a good source of energy in their area. Students can research patterns of cloud cover and the level of air pollution in their area to support their conclusions. [L2]

- For more general information on solar energy, have students contact the Florida Solar Energy Center, a Research Institute of the University of Central Florida, 1679 Clearlake Road, Cocoa, FL 32922.

GLENCOE TECHNOLOGY

 Videodisc

STVS: Chemistry
Disc 2, Side 2
Solar House (Ch. 10)

Spiral Solar Concentrator (Ch. 12)

Solar Tower (Ch. 13)

339

Science & Language Arts

Source

London, Jack. *To Build a Fire and Other Stories.* Bantam Books: New York, 1982.

Biography

Jack London was born in San Francisco in January of 1876. His first book, *The Son of the Wolf,* was published in 1900. In the 40 years he lived, he wrote more than 40 books. His writing showed an obsession with adventure stories about men who were facing death. Although he made a fortune on his popular writings, he died penniless in November of 1916.

Teaching Strategies

- Demonstrate the need for oxygen in the burning process by placing a lit birthday or votive candle under a large, heat-proof beaker. Encourage students to hypothesize why the flame eventually flickers and goes out.

- During digestion, some of the energy stored in food is released as heat. Have students discuss why a high-calorie, high-fat diet would be the best choice for someone traveling on foot through the cold arctic environment.

To Build a Fire
by Jack London

He worked slowly and carefully, keenly aware of his danger. Gradually, as the flame grew stronger, he increased the size of the twigs with which he fed it. He squatted in the snow, pulling the twigs out from their entanglement in the brush and feeding directly to the flame. He knew there must be no failure. When it is seventy-five below zero, a man must not fail in his first attempt to build a fire . . .

Science Journal

Read Jack London's short story "To Build a Fire" to find out what happens to the man. What mistakes did he make that eventually put his life in danger? In your Science Journal, make a list of ways that heat, temperature, and thermal energy are important to the outcome of the story.

Jack London's story is set in the Yukon Territory in midwinter. A man hikes through the wilderness along a rarely traveled trail. He is alone except for a dog. It's so cold his spit crackles and freezes before it hits the ground.

The man has never spent a winter in the Arctic before. An old-timer cautioned him not to take this journey, saying that no one should travel alone when the temperature is colder than 50° below zero. But the man ignored the advice. Now, striding along the trail, he is still confident he can handle any problems that come along.

Then it happens. Thin ice covering a spring gives way. The man is soaked up to his knees. Almost instantly, his feet begin to freeze. He must build a fire to thaw his feet and dry his clothes.

But can he? The cold is frighteningly intense. The man's hands grow numb as he struggles with matches and wood. Perhaps he should have listened to the old man. Building a fire is now a matter of life or death.

Other Works

London, Jack. *The Call of the Wild* (originally published in 1903). Barry Moser, illustrator. Riverside, NJ: Simon & Schuster, 1995.

London, Jack. *To Build a Fire* audiocassette. (Story originally published in 1908). Old Greenwich, CT: Listening Library, 1992.

Science Journal Students might mention that the man did not dress warmly enough for the weather. A lot of heat was lost by the warmer man to the colder air. The temperature should have indicated to the man that it was not safe to be outside. The man's body did not produce thermal energy as quickly as heat was lost to the environment. LS P

Read the statements below that review the most important ideas in the chapter. Using what you have learned, answer each question in your Science Journal.

1. Energy is the ability to change your surroundings. Energy is found in many forms. The transformation of energy from one form to another allows humans to both store and use energy. *Compare and contrast potential energy and kinetic energy.*

2. Thermal energy can be transferred from place to place through convection, conduction, and radiation. Heat is thermal energy that is transferred from a warmer object to a cooler one. *What does temperature measure?*

3. Solar energy may be used to reduce use of nonrenewable energy resources, such as coal and petroleum. *Why isn't solar energy used more frequently?*

341

 **Assessment**

Portfolio Encourage students to place in their portfolios one or two items of what they consider to be their best work. Examples include:
- Flex Your Brain, p. 323
- Assessment, p. 326
- Making a Model, p. 336 **P**

Performance Additional performance assessments may be found in **Performance Assessment** and **Science Integration Activities.** Performance Task Assessment Lists and rubrics for evaluating these activities can be found in Glencoe's **Performance Assessment in the Science Classroom (PASC).**

Chapter 12

Review

Have students look at the illustrations on this page. Ask them to describe details that support the main ideas of the chapter found in the statement for each illustration.

Teaching Strategies

Divide students into three groups. Assign each group one of the illustrations. Have each group list all forms of energy and energy transformations they find in the photo. Ask them to share their lists and decide which illustration has the most examples. **L2**

Answers to Questions

1. Potential energy is stored energy and is acquired through the position or characteristics of matter. Kinetic energy is the energy of motion. It is energy that is being expended.
2. Temperature measures the average kinetic energy in the atoms of a particular sample of matter.
3. Student answers may include increased need for energy and the problems of pollution caused by fossil fuel use.

Science at Home

Intrapersonal Have students select several favorite cereals. Have them use the nutritional information on the label to make a poster that compares the amount of heat (calories) each provides per serving, before milk is added. This information can then be brought to school and displayed with other students' work for comparison. **L2**

Chapter 12 **Review**

Review

Using Key Science Words

1. potential energy
2. thermal energy
3. convection
4. conduction
5. energy

Checking Concepts

6. b   **7.** a   **8.** b
9. b   **10.** c

Thinking Critically

11. Answers will vary but can include: temperature, rubbing your hand on your arm; shape, mashing a ball of clay; speed, pedaling a bicycle faster; position, lifting a ball; direction, hitting one ball with another.

12. Over several hours, food is changed by the body to a usable form, which is then burned to become energy for muscles.

13. Setting up dominoes gives them potential energy. Tipping over a domino changes its potential energy to kinetic energy. If it hits another domino and knocks it over, it transfers kinetic energy.

14. Thermal energy moves from a place of high energy to a place of low energy. When you put cool cloths on a burn, heat moves from the skin to the cloth, minimizing damage to the skin.

15. As cool air in the room falls, warm air from the floor vents is forced up, creating even temperatures throughout the room. Cool air coming through the vents stays near the floor, and warm air accumulates near the ceiling.

Using Key Science Words

conduction   potential energy
convection   radiation
energy   solar energy
heat   thermal energy
kinetic energy

Match each phrase with the correct term from the list of Key Science Words.

1. stored energy
2. total kinetic and potential energy in the particles of a material
3. type of transfer of thermal energy that produces weather
4. transfer of thermal energy from a warmer object to a cooler one
5. ability to bring about change

Checking Concepts

Choose the word or phrase that completes the sentence.

6. Solar energy is _____.
   a. a nonrenewable resource
   b. a renewable resource
   c. an inexpensive source of electricity
   d. commonly used as a source of electricity

7. Potential energy is to kinetic energy as _____.
   a. water stored behind a dam is to electrical energy
   b. a flute is to a trumpet
   c. a tractor is to a sports car
   d. an ice cube is to a hot day

8. Thermometers measure _____.
   a. potential energy
   b. molecular motion
   c. mechanical energy
   d. coolness

9. Gravity changes potential energy to kinetic energy when _____.
   a. you push a child in a wagon.
   b. a skier starts down a slope
   c. you throw a ball to a friend
   d. a balloon floats upward

10. When you place a cooler object against a source of heat, you transfer heat by _____.
    a. convection   c. conduction
    b. radiation   d. microwave

Thinking Critically

Answer the following questions in your Science Journal using complete sentences.

11. Energy is the ability to change the temperature, shape, speed, position, or direction of something. Give an example of doing each of these things.

12. How can the chemical energy of your lunch today become the energy you need to deliver newspapers after school tomorrow?

13. How could you use a box of dominoes to represent potential energy, kinetic energy, and transfer of kinetic energy?

14. Use what you know about the movement of thermal energy to explain why you would place cool cloths on a burn.

Assessment Resources

Reproducible Masters
Chapter Review, pp. 27-28
Assessment, pp. 51-54
Performance Assessment, p. 50

Glencoe Technology
⦿ **Computer Test Bank**
▭ **MindJogger Videoquiz**

15. Explain why heat vents placed in the floor of a building would heat the building better than they would cool the building when the air conditioning is used.

Developing Skills

If you need help, refer to the description of each skill in the Skill Handbook.

16. **Observing and Inferring:** You observe a friend pick up a pot that has been on a lit burner, then quickly set it back down. In terms of thermal energy, infer what happened.

17. **Designing an Experiment:** Design an experiment that would test the ability of a raw egg to withstand being dropped. Point out the parts of the experiment that would represent each of the following:
    a. the controls in the experiment
    b. the dependent variable
    c. the data to be collected
    d. the egg when it has potential energy
    e. the egg when it has kinetic energy

18. **Comparing and Contrasting:** In the activities in this chapter, you made observations and collected data. Data and observations are not the same. Compare and contrast data and observations.

19. **Making Models:** Draw a diagram that shows the difference between the transfer of heat energy when you make popcorn in a microwave and when you make it on top of a stove.

20. **Concept Mapping:** Draw a concept map that shows the three ways that thermal energy is transferred. Include an example for each. Be prepared to explain your map to other students and your teacher.

Performance Assessment

1. **Designing an Experiment:** Use the following items to design an experiment to find out how quickly different materials absorb radiant energy: pockets made out of three different colors of construction paper, a thermometer for each pocket, and a sunny day or a heat lamp.

2. **Making Observations and Inferences:** Put an empty, glass soft-drink bottle in a freezer for about 15 minutes. Take the bottle out of the freezer. If possible, place it in the sun. Center a dime on top of the mouth of the bottle. Use your observations to infer how thermal energy affects the kinetic energy of molecules in the air.

Developing Skills

16. **Observing and Inferring** Conduction of heat from the burner to the bottom of the pan and from the bottom of the pan to the rest of the pan has made the handle of the pot hot.

17. **Designing an Experiment** Answers will vary but may include:
    a. the size of the eggs used and how the egg is protected
    b. the height from which they are dropped
    c. the number of eggs that withstand being dropped from specific heights
    d. when it is at the top of the place from which it is to be dropped
    e. when it is falling

18. **Comparing and Contrasting** Observations are made through using your senses, such as sight. Data are acquired through measurements and include numbers.

19. **Making Models** In the model, energy in the microwave moves through radiation in all directions toward the popcorn. When you make popcorn on the stove, energy from conduction moves from the heating element on the stove, to the pan, to the kernels of popcorn.

20. **Concept Mapping**

Sample map:

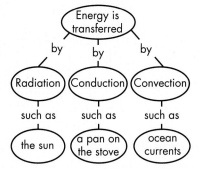

Performance Assessment

1. The experiment should show good use of variables and controls, data collection, a hypothesis, and a conclusion. Use the Performance Task Assessment List for Designing an Experiment in **PASC**, p. 23. **P**

2. As the dime moves, students should infer that increased thermal energy increases the kinetic energy of molecules in air. Use the Performance Task Assessment List for Making Observations and Inferences in **PASC**, p. 17. **P**

Chapter Organizer

Section	Objectives/Standards	Activities/Features
Chapter Opener		Explore Activity: Measure Magnet Motion, p. 345
13-1 **What is an electric charge?** (2 sessions, 1 block)*	1. **Describe** how the structure of the atom helps produce electricity. 2. **Classify** static electricity as a form of potential energy. National Science Content Standards: (5-8) UCP2, UCP3, A1, B2, E2, F3, F5	MiniLAB: Contrasting Static Charges, p. 348 Problem Solving: Static Image, p. 349 Skill Builder: Recognizing Cause and Effect, p. 350 Science Journal, p. 350
13-2 **How do circuits work?** (3 sessions, 1½ blocks)*	3. **Identify** what makes an electric circuit. 4. **Distinguish** between direct and alternating current. National Science Content Standards: (5-8) UCP1, UCP2, UCP3, A1, A2, E1, E2, F5	MiniLAB: Building a Model Circuit, p. 352 Activity 13-1: How long do batteries last?, pp. 356-35 Skill Builder: Designing an Experiment to Test a Hypothesis, p. 358 Science Journal, p. 358
13-3 **What is a magnet?** (2 sessions, 1 block)*	5. **Investigate** the relationship between magnetism and electricity. 6. **Describe** how a compass works. National Science Content Standards: (5-8) UCP1, UCP2, A1, A2, B1, E1, E2, F5	Activity 13-2: Lines of Magnetic Force, p. 361 Using Technology: Magnetic Signals at Work, p. 364 Skill Builder: Observing, p. 365 Using Math, p. 365
13-4 **Science and Society:** **Electricity Sources** (2 sessions, 1 block)*	7. **Compare and contrast** ways of generating electricity. 8. **Make generalizations.** National Science Content Standards: (5-8) UCP1, A2, B3, C4, E2, F2, F4, F5, G2	Skill Builder: Making Generalizations, p. 367 Science & Language Arts: Promise Me the Moon, p. 368 Science Journal, p. 368

* A complete Planning Guide that includes block scheduling is provided on pages 31T-33T.

Activity Materials

Explore	Activities	MiniLABs
page 345 bar magnets, graph paper	pages 356-357 D-cell battery with holder, bell, buzzer, wire, masking tape, lightbulb and holder page 361 D-cell battery with holder, small lightbulb with holder, bar magnets, 20-penny iron nail, paper clip, shaker, wire, sealable plastic bag, graph paper, iron filings	page 348 balloon, glass rod, silk, wool, string page 352 D-cell battery with holder, small lightbulb with holder, wire

Need Materials? Call Science Kit (1-800-828-7777).

Chapter 13 Electricity and Magnetism

Teacher Classroom Resources

Reproducible Masters	Transparencies	Teaching Resources
Activity Worksheets, pp. 5, 83 **Enrichment,** p. 47 **Reinforcement,** p. 47 **Study Guide,** p. 47	**Section Focus Transparency 41,** It's Alive…With Electricity	**Spanish Resources** **English/Spanish Audiocassettes** **Cooperative Learning Resource Guide** **Lab Partner** **Lab and Safety Skills** **Lesson Plans**
Activity Worksheets, pp. 5, 79-80, 84 **Enrichment,** p. 48 **Reinforcement,** p. 48 **Science and Society/Technology Integration,** p. 31 **Study Guide,** p. 48	**Science Integration Transparency 13,** Electricity and the Heart **Section Focus Transparency 42,** Circuit Splashdown! **Teaching Transparency 25,** Circuits	
Activity Worksheets, pp. 5, 81-82 **Cross-Curricular Integration,** p. 17 **Enrichment,** p. 49 **Lab Manual 25, 26** **Multicultural Connections,** pp. 29-30 **Reinforcement,** p. 49 **Science Integration Activities,** pp. 63-64 **Study Guide,** p. 49	**Section Focus Transparency 43,** Car-Sized Magnets **Teaching Transparency 26,** Magnetism	**Assessment Resources** **Chapter Review,** pp. 29-30 **Assessment,** pp. 55-58, 59-60 **Performance Assessment,** pp. 35-36, 51 **Performance Assessment in the Science Classroom (PASC)** **MindJogger Videoquiz** **Alternate Assessment in the Science Classroom** **Computer Test Bank**
Enrichment, p. 50 **Reinforcement,** p. 50 **Study Guide,** p. 50	**Section Focus Transparency 44,** Spin Doctor	

Key to Teaching Strategies

The following designations will help you decide which activities are appropriate for your students.

L1 Level 1 activities should be appropriate for students with learning difficulties.

L2 Level 2 activities should be within the ability range of all students.

L3 Level 3 activities are designed for above-average students.

ELL ELL activities should be within the ability range of English Language Learners.

LS These activities are designed to address different learning styles.

COOP LEARN Cooperative Learning activities are designed for small group work.

P These strategies represent student products that can be placed into a best-work portfolio.

GLENCOE TECHNOLOGY

The following multimedia resources are available from Glencoe.

Science and Technology Videodisc Series (STVS)
Physics
  Electric Heart
  Making Integrated Circuits
Chemistry
  Battery Science
  Solar House
  Wind Power
  Hydroelectric Power
  Geothermal Wells
Human Biology
  Detecting the Body's
    Magnetic Fields

The Infinite Voyage Series
Unseen Worlds

Glencoe Physical Science Interactive Videodisc
Motors

Glencoe Physical Science CD-ROM

This is a representation of key blackline masters available in the Teacher Classroom Resources.

Teaching Aids

Section Focus Transparencies

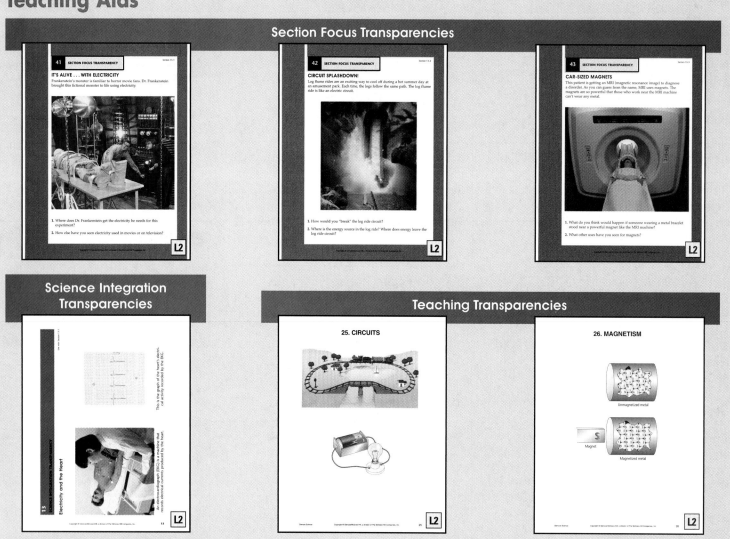

41 SECTION FOCUS TRANSPARENCY — Section 13-1

IT'S ALIVE . . . WITH ELECTRICITY

Frankenstein's monster is familiar to horror movie fans. Dr. Frankenstein brought this fictional monster to life using electricity.

1. Where does Dr. Frankenstein get the electricity he needs for this experiment?

2. How else have you seen electricity used in movies or on television?

L2

42 SECTION FOCUS TRANSPARENCY — Section 13-2

CIRCUIT SPLASHDOWN!

Log flume rides are an exciting way to cool off during a hot summer day at an amusement park. Each time, the logs follow the same path. The log flume ride is like an electric circuit.

1. How would you "break" the log ride circuit?

2. Where is the energy source in the log ride? Where does energy leave the log ride circuit?

L2

43 SECTION FOCUS TRANSPARENCY — Section 13-3

CAR-SIZED MAGNETS

This patient is getting an MRI (magnetic resonance image) to diagnose a disorder. As you can guess from the name, MRI uses magnets. The magnets are so powerful that those who work near the MRI machine can't wear any metal.

1. What do you think would happen if someone wearing a metal bracelet stood near a powerful magnet like the MRI machine?

2. What other uses have you seen for magnets?

L2

Science Integration Transparencies

This is the graph of the heart's electrical activity recorded by the EKG.

An electrocardiograph (EKG) is a machine that records electrical currents produced by the heart.

Electricity and the Heart

L2

Teaching Transparencies

25. CIRCUITS

L2

26. MAGNETISM

Unmagnetized metal

Magnet

Magnetized metal

L2

Meeting Different Ability Levels

Study Guide for Content Mastery

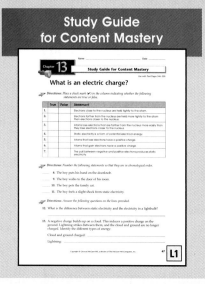

Chapter **13** — Study Guide for Content Mastery

What is an electric charge?

Directions: Place a check mark (✔) in the column indicating whether the following statements are true or false.

True	False	Statement
1.		Electrons close to the nucleus are held tightly to the atom.
2.		Electrons farther from the nucleus are held more tightly to the atom than electrons closer to the nucleus.
3.		Atoms lose electrons that are farther from the nucleus more easily than they lose electrons close to the nucleus.
4.		Static electricity is a form of potential electrical energy.
5.		Atoms that lose electrons have a positive charge.
6.		Atoms that gain electrons have a negative charge.
7.		The pull between negative and positive electrons produces static electricity.

Directions: Number the following statements so that they are in chronological order.

___ 8. The boy puts his hand on the doorknob.

___ 9. The boy walks to the door of his room.

___ 10. The boy pets the family cat.

___ 11. The boy feels a slight shock from static electricity.

Directions: Answer the following questions on the lines provided.

12. What is the difference between static electricity and the electricity in a lightbulb?

13. A negative charge builds up on a cloud. This induces a positive charge on the ground. Lightning strikes between them, and the cloud and ground are no longer charged. Identify the different types of energy.

Cloud and ground charged:

Lightning:

L1

Reinforcement

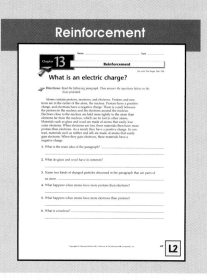

Chapter **13** — Reinforcement

What is an electric charge?

Directions: Read the following paragraph. Then answer the questions below on the lines provided.

Atoms contain protons, neutrons, and electrons. Protons and neutrons are in the center of the atom, the nucleus. Protons have a positive charge, and electrons have a negative charge. There is a pull between the protons in the nucleus and the electrons around the nucleus. Electrons close to the nucleus are held more tightly to the atom than electrons far from the nucleus, which can be lost to other atoms. Materials such as glass and wool are made of atoms that easily lose outer electrons. When electrons are lost, these materials then have more protons than electrons. As a result, they have a positive charge. In contrast, materials such as rubber and silk are made of atoms that easily gain electrons. When they gain electrons, these materials have a negative charge.

1. What is the main idea of the paragraph?

2. What do glass and wool have in common?

3. Name two kinds of charged particles discussed in the paragraph that are parts of an atom.

4. What happens when atoms have more protons than electrons?

5. What happens when atoms have more electrons than protons?

6. What is a nucleus?

L2

Enrichment Worksheets

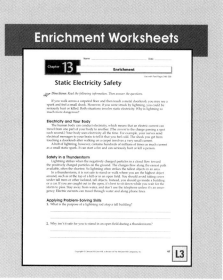

Chapter **13** — Enrichment

Static Electricity Safety

Directions: Read the following information. Then answer the questions.

If you walk across a carpeted floor and then touch a metal doorknob, you may see a spark and feel a small shock. However, if you were struck by lightning, you could be seriously hurt or killed. Both situations involve static electricity. Why is lightning so much more dangerous?

Electricity and Your Body

The human body can conduct electricity, which means that an electric current can travel from one part of your body to another. (The *current* is the charge passing a spot each second.) Your body uses electricity all the time. For example, your nerves send electrical messages to your brain to tell it that you feel cold. The shock you get from touching a doorknob after walking on a carpet involves a very small current.

A bolt of lightning, however, contains hundreds of millions of times as much current as a small static spark. It can start a fire and can seriously hurt or kill a person.

Safety in a Thunderstorm

Lightning strikes when the negatively charged particles in a cloud flow toward the positively charged particles on the ground. The charges flow along the easiest path available, often the shortest. So lightning often strikes the tallest objects in an area.

In a thunderstorm, it is not safe to stand or walk where you are the highest object around, such as at the top of a hill or in an open field. You should avoid taking cover under tall trees or other isolated, tall objects. Instead, you should go inside a building or a car. If you are caught out in the open, it's best to sit down while you wait for the storm to pass. Stay away from water, and don't use the telephone unless it's an emergency. Electric currents can travel through water and along phone lines.

Applying Problem-Solving Skills

1. What is the purpose of a lightning rod atop a tall building?

2. Why isn't it safe for you to stand in an open field during a thunderstorm?

L3

Chapter 13 Electricity and Magnetism

Hands-On Activities

Science Integration Activities
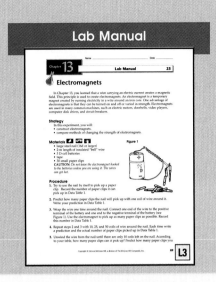

Lab Manual
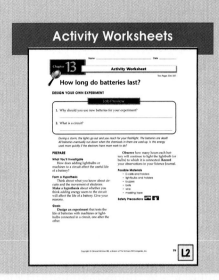

Activity Worksheets

Enrichment and Application

Cross-Curricular Integration
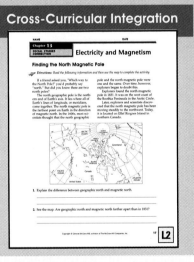

Multicultural Connections
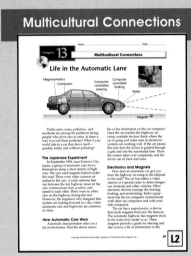

Science and Society/Technology Integration
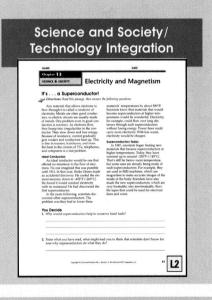

Assessment

Performance Assessment
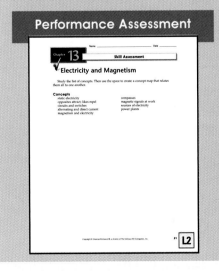

Chapter Review
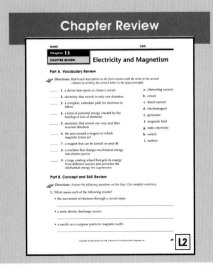

Assessment
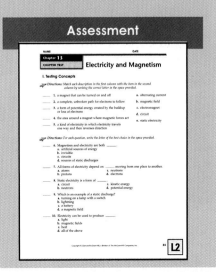

Electricity and Magnetism

CHAPTER OVERVIEW

Section 13-1 Students will review how the structure of the atom creates electrical force and how concepts of potential and kinetic energy apply to electrical energy. Charge interactions are introduced. Static electricity illustrates these concepts.

Section 13-2 This section explains circuits and the function of a switch in a circuit. Batteries are discussed, and direct and alternating current are defined.

Section 13-3 Magnetism is explained and the relationship between electric and magnetic fields is established. An electromagnet and a compass are introduced.

Section 13-4 Science and Society This section introduces the students to different ways of producing electrical energy. It also reviews how to make generalizations that enhance science understanding.

Chapter Vocabulary

electrical energy	switch
static electricity	magnetic field
circuit	electromagnet
	generator
	turbine

Theme Connection

Energy Electricity is presented as a source of energy, both potential and electrical. Students will see that electricity occurs naturally, and it can be produced by people to do work.

Chapter Preview

Section 13-1 What is an electric charge?

Section 13-2 How do circuits work?

Section 13-3 What is a magnet?

Section 13-4 Science and Society: Electricity Sources

Skills Preview

▶ **Skill Builders**
- recognize cause and effect
- design an experiment to test a hypothesis
- observe

▶ **MiniLABs**
- compare and contrast
- make a model

▶ **Activities**
- hypothesize
- observe and infer
- make and use graphs
- scientific drawing
- make a model

Learning Styles

Look for the following logo for strategies that emphasize different learning modalities. **LS**

Kinesthetic	Explore, p. 345; MiniLAB, pp. 348, 352; Enrichment, p. 354; Activity 13-1, p. 356; Activity, pp. 360, 362; Activity 13-2, p. 361; Reteach, p. 365
Visual-Spatial	Making a Model, pp. 347, 355; Visual Learning, pp. 347, 348, 352, 353, 355, 363; Demonstration, p. 353
Interpersonal	Community Connection, p. 354
Intrapersonal	Assessment, p. 365; Science at Home, p. 369
Logical-Mathematical	Brainstorming, p. 349; Inquiry Question, p. 350; Assessment, p. 350; Go Further, p. 357
Linguistic	Assessment, p. 348; Reteach, p. 350; Using Science Words, pp. 353, 362; Science Journal, pp. 361, 365

Chapter 13

Electricity and Magnetism

You hear music and feel a breeze on your skin, but you can't see the moving air that causes these things. Magnetism and electricity are like the air. They can't be seen. People learned about magnetic and electric forces by observing how these forces affected the world.

Invisible magnetic force can cause movement, which means magnets can do work. The following activity gives you a chance to test magnetic force.

EXPLORE ACTIVITY

Measure Magnet Motion

1. You'll need a sheet of graph paper and two magnets. Place one magnet at each end of the graph paper. Mark the positions of the magnets.
2. Predict how close the magnets must be to affect each other.
3. Slide one magnet forward one square.
4. Repeat Step 3 until one of the magnets moves without being touched. Measure the distance between the magnets.
5. Turn one magnet a half turn and repeat the activity. Then turn the other magnet and repeat again.
6. Was your hypothesis from Step 2 confirmed?

Science Journal

Imagine you are in a small, metal ship moving around one of the magnets. In your Science Journal, describe how your ship will be affected by the magnet's pull.

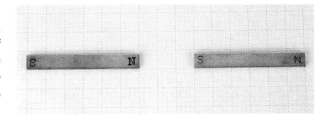

345

Prepare

Section Background
- Static electricity is found in intense forms, such as charged clouds, and less dangerous forms, such as charged socks coming out of a clothes drier. The intensity depends on the number of electrons available to be stored for later release.
- Friction allows the transfer of electrons from atoms that give up electrons easily to those that acquire electrons easily.

Preplanning
Refer to the Chapter Organizer on pages 344A-B.

1 Motivate

Bellringer

Before presenting the lesson, display **Section Focus Transparency 41** on the overhead projector. Assign the accompanying **Focus Activity** worksheet.

`L2` `ELL`

13•1 What is an electric charge?

What YOU'LL LEARN
- How the structure of the atom helps produce electricity
- Why static electricity is a form of potential energy

Science Words:
electrical energy
static electricity

Why IT'S IMPORTANT
Understanding static electricity will help you understand all the ways you use electricity.

FIGURE 13-1
Lightning is an example of electrical energy.

Static Electricity

Rap, blues, rock, classical—whatever your favorite style of music is, you can hear it any time by popping a compact disc or a cassette into a player. Hearing music from a CD or cassette player wouldn't be possible without electricity. Which type of energy are you using when you play a cassette?

Remember from Chapter 12 the difference between potential energy—stored energy—and kinetic energy? When a skier coasts down a hill, potential energy is transformed to kinetic energy—energy in use. A tape player changes potential electrical energy in the battery to **electrical energy,** the energy of charges in motion.

Where can you find examples of electrical potential energy? Have you ever seen a blinding bolt of lightning slice through a dark sky, as in **Figure 13-1?** What about seeing a spark and hearing a snap after you cross a carpeted floor? These examples of electrical energy are results of static electricity. It is a form of potential electrical energy.

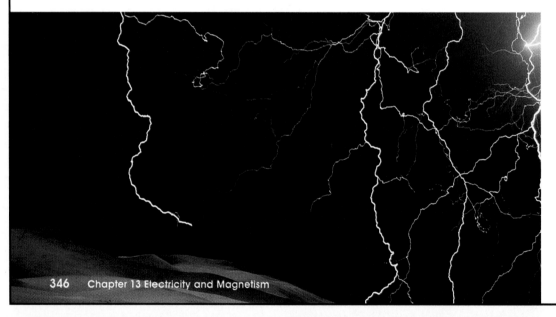

346   Chapter 13 Electricity and Magnetism

Program Resources

📁 **Reproducible Masters**
Activity Worksheets, pp. 5, 83 `L2`
Enrichment, p. 47 `L3`
Reinforcement, p. 47 `L2`
Study Guide, p. 47 `L1`

 Transparencies
Section Focus Transparency 41 `L2`

FIGURE 13-2

Objects with opposite charges attract. Objects with like charges repel.

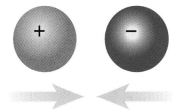

Opposite charges attract

Like charges repel   Like charges repel

To learn more about how you use electrical energy, you first need to understand the differences between electrical potential energy and electrical energy. In Chapter 12, potential and kinetic energy were explained using the example of a skier. The skier carried the energy. To understand electricity, you need to know what carries electrical energy.

Opposites Attract

Does the word *electricity* remind you of any other word? Think back to Chapter 8, where you learned about atoms and electrons, the negatively charged particles of an atom. Electrons and electricity are related. All forms of electricity depend on electrons moving from one place to another. An electron is like the skier that carries the energy.

Positively charged objects attract negatively charged objects, as shown in **Figure 13-2.** There is a pull between the negatively charged electrons around the nucleus and the positively charged protons in the nucleus. This means that electrons closer to the nucleus are held tightly to the atom. Electrons farther from the nucleus are held less tightly. **Figure 13-3** illustrates this.

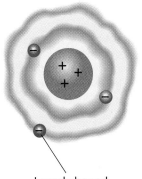

Loosely bound
outer electron

FIGURE 13-3

The electron farthest from the nucleus feels less of a pull than the electrons close to the nucleus.

13-1 What is an electric charge?   **347**

Mini LAB

Purpose

[LS] **Kinesthetic** Students will study static charges and their effects on charged and uncharged objects. [L2]

[ELL] [COOP LEARN]

Materials

two balloons, glass rod, silk, wool, string

Alternate Materials

Pith (foam) balls suspended from a string, a rubber wand, and felt cloth will work as well.

Teaching Strategies

Have students try both the silk and the wool on the balloons and compare the charges.

Safety Precautions Have students wear goggles.

Troubleshooting This experiment may work better on a dry day.

📁 **Activity Worksheets,** pages 5, 83

Analysis

1. neutral; negative

2. neutral; one now has a positive and the other a negative charge; the two objects attract one another

3. The glass rod gives up electrons and the silk cloth gains electrons, so the glass is positively charged and the cloth is negatively charged. They can attract and repel the balloons and charge them, which will cause the balloons to attract or repel each other.

4. Electrons are the only parts of the atom that can be exchanged because they are held loosely, outside the nucleus.

Mini LAB

Contrasting Static Charges

How can potential electrical energy be created and demonstrated?

1. Blow up two rubber balloons and hang them a few inches apart.

2. Rub a glass rod with a square of silk. Bring the glass rod close to one balloon. Observe what happens. Let the rod and balloon touch each another for a few seconds. What happens then?

3. Now, rub one balloon with a wool or silk cloth. Touch that balloon to the other balloon. How does it react? What happens when you touch the first balloon to the glass rod?

Analysis

1. What kinds of charges did the balloons have before rubbing? After?

2. What charge did the glass and cloths have before rubbing? After? How do you know?

3. What made the rod and balloons act as they did? Why do you think so?

4. Why do you think it would be correct to say that electrons are being exchanged, and not protons or neutrons?

FIGURE 13-4

Petting the family cat can be a lesson in static electricity.

Protons and electrons are much too small to see. But you can see evidence of the attraction between these positively and negatively charged particles. Because electrons farther from the nucleus aren't held as tightly, some atoms can lose those electrons. Materials such as glass and wool are made up of atoms that can lose outer electrons easily. When electrons are lost, these materials are left with more protons than electrons. As a result, they have a positive charge. In contrast, materials such as rubber and silk are made of atoms that easily gain electrons. The added electrons cause atoms to be negatively charged.

The buildup or loss of electrons creates a form of potential electrical energy called **static electricity.** Lightning and sparks result when many electric charges move all at once. This is sometimes called a static discharge, illustrated in **Figure 13-4.** What do you think would happen if you rubbed a balloon with a piece of wool?

Ⓐ As the boy pets the family cat, friction pulls electrons from the fur to his hand, giving his hand a negative charge. Then he walks to a door.

Assessment

Oral Have students describe what happens during this activity and why. Use the Performance Task Assessment List for Oral Presentation in **PASC,** p. 71. [LS]

Visual Learning

Figure 13-4 Point out that negative charges (electrons) are the only ones in motion. The front of the doorknob becomes positively charged because the atoms there lose electrons to the back of the doorknob. [LS] [ELL]

B The negative charge on his hand repels electrons near the surface of the metal knob. The surface of the knob now has a positive charge; his hand has a negative charge.

C Opposite charges attract. When his hand gets close enough, the extra electrons from his hand leap across the gap. The movement creates a spark and a snap, as electrons transfer energy to the air in the gap.

Problem Solving

Static Image

Photocopying machines work by giving a metal drum a negative charge. Then, a mirror and a bright light are used to put the image to be copied onto the drum. Where the light hits the drum, a positive charge forms, which balances the negative charge. Dark letters and dark parts of pictures don't reflect light. The drum keeps a negative charge where those parts of the image are.

Special dry ink powder called *toner* is brushed onto the drum. Toner sticks only to the charged areas where the letters or pictures were on the image. Then, the paper passes through a heated roller that melts the ink powder. That's why copies are warm when they first come out of the machine.

Solve the Problem:
You have to leave the copier cover up while you copy pages from a big spiral notebook. The edges of the copies are dark and fuzzy. Explain what you think is happening in terms of light and charges. How could you fix the problem?

Think Critically:
The drum has a negative charge. What kind of charge does the toner have? How do you know?

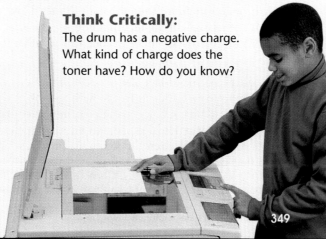

349

Problem Solving

Solve the Problem
When you can't close the lid of the photocopier completely, the light that should have hit the inside surface of the copier lid (they are always white) and reflected back to the drum travels into the room. Because the light doesn't hit the drum and remove the negative charge, toner sticks to the surface of the drum, making the edges of the copy dark.

Think Critically
The toner must have a positive charge to attract it to the negatively charged parts of the drum.

Reteach

[LS] **Linguistic** Have students explain why "Opposites Attract" is a good title for this part of the chapter. [L1]

Extension

📁 For students who have mastered this section, use the **Reinforcement** and **Enrichment** masters.

4 Close

Inquiry Question

[LS] **Logical-Mathematical** When an airplane lands and is being refueled, a grounding wire is attached to the plane to prevent sparks, which can ignite the fuel when it is pumped into the plane. **Why might an airplane have a static charge?** *The friction between the skin of the plane and the air and water molecules in the atmosphere gives the plane a static charge.*

Section Wrap-up

1. Sample answer: lightning and charged sock; friction builds up extra electrons on clouds or sock.

2. Sample answers might include blender, toaster, television, and CD player.

3. **Think Critically** In humid weather, electrons can more easily travel through the air, so they don't collect on surfaces.

Science Journal Encourage the students to think of applications outside of their schools and homes.

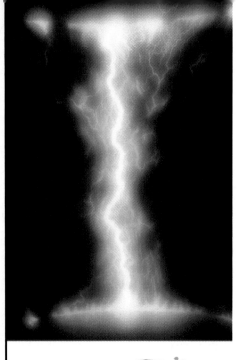

Static Electricity Problems

Lightning is one of the most impressive and dangerous examples of a static electric discharge. Each lightning strike transfers a huge amount of energy. Lightning strikes can start fires and damage trees and buildings. You also may have seen and heard the effect of these huge electrical discharges on your TV or radio.

Smaller static discharges, like that shown in **Figure 13-5**, also can be dangerous. They can harm computers and other electronic devices. Computer chips work on low electric power. The power of a static spark can be enough to overload and permanently burn out the circuits on a chip.

FIGURE 13-5
A spark discharges the static electrical charge built up between two metal objects.

Section Wrap-up

1. Static electrical charges happen in nature. They also can be made by humans. Give one example of natural static electricity and one of human-made static electricity. Explain how each occurs.

2. Electricity does work when it produces motion, heat, light, or sound. Give examples of appliances in your home that do each of those things.

3. **Think Critically:** It is easier for electrons to travel through humid air than through dry air. How does this explain the fact that you see the effects of static electricity more in cool, dry weather?

4. 📊 *Skill Builder*
   Recognizing Cause and Effect
   Demonstrate static electricity with a comb, a wool cloth, and bits of tissue paper. Explain what happens with a series of cause-and-effect statements. If you need help, refer to Recognizing Cause and Effect on page 551 in the **Skill Handbook.**

Science Journal

Write a story that tells how your life would change if you had to live without electricity for one day, one week, and one month. How does the length of time without electricity make a difference?

📊 *Skill Builder*
Recognizing Cause and Effect When you rub a comb with a wool cloth, the comb will pick up small pieces of tissue. Rubbing the comb with cloth is a cause; the wool losing electrons to the comb is an effect. Bringing a charged comb near bits of tissue is a cause; attracting the paper is an effect. Some may conclude that bringing a charged comb close to paper (cause) moves the electrons in the paper away from the comb (effect).

✔ **Assessment**

Portfolio Show the students how to read an electric meter and review the math for figuring its change. Have students measure the electricity used in their homes on a given day. Use the Performance Task Assessment List for Using Math in Science in **PASC,** p. 29. [L2] [P] [LS]

How do circuits work?

Creating a Path for Electrons

If you've watched lightning or touched a doorknob after walking across a carpeted room, you know that a static electric discharge is over in an instant. However, if you want to use electricity to do work, you probably need a nonstop, controlled flow of electrons. A device called a circuit controls the movement of electrons. A **circuit** (SUR kut) is a complete, unbroken path for electrons to follow. Usually, this path is through metal wires. When electrons move in this path, they can do work, as you will find out in the MiniLAB.

A circuit can have one path or many paths. In this book, we will study circuits with only one path. The racetrack in **Figure 13-6** is an example of a circuit with only one path.

FIGURE 13-6

This bicycle track is a model of an electrical circuit. The cyclists follow a complete, unbroken path around the track.

What YOU'LL LEARN

- What makes an electric circuit
- The difference between direct and alternating current

Science Words:
circuit
switch

Why IT'S IMPORTANT

Technology that uses electricity is a large part of your day-to-day life.

Section 13•2

Prepare

Section Background

- A circuit has a source of electrons, a pathway for the electrons, and a switch that controls their flow.
- Circuits that use batteries flow in only one direction and are called direct current (DC) circuits.
- Circuits that move energy back and forth in a continuous flow are called alternating current (AC) circuits.

Preplanning

Refer to the Chapter Organizer on pages 344A-B.

1 Motivate

Bellringer

Before presenting the lesson, display **Section Focus Transparency 42** on the overhead projector. Assign the accompanying **Focus Activity** worksheet.
L2 ELL

Tying to Previous Knowledge

Review with students cycles, such as the water and carbon cycles.

Program Resources

Reproducible Masters
Activity Worksheets, pp. 5, 79-80, 84 L2
Enrichment, p. 48 L3
Reinforcement, p. 48 L2
Science and Society/Technology
   Integration, p. 31 L2
Study Guide, p. 48 L1

Transparencies
Science Integration Transparency 13 L2
Section Focus Transparency 42 L2
Teaching Transparency 25 L2

Mini LAB

Purpose

LS **Kinesthetic** Students will visualize the actual movement of electrons through the path created. **L2** **ELL** **COOP LEARN**

Materials

D-cell battery, small light-bulb, piece of wire at least 20 cm long for each station

Teaching Strategies

After a few minutes of experimentation, let the students work in groups of three to discover all the ways that the circuit can work.

Safety Precautions Students should wear goggles.

📁 **Activity Worksheets,** pages 5, 84

Analysis

1. In a completed circuit, the lightbulb is touching either end of the battery. The wire must have one end touching the part of the bulb base not in contact with the battery, and the other end touching the opposite end of the battery.

2. If there is no closed loop to follow, electricity doesn't flow and the bulb won't light.

✔ Assessment

Process Ask the students to sketch at least two different ways to complete the circuit and explain how each one works. Use the Performance Task Assessment List for Scientific Drawing in **PASC,** p. 55. **P**

Mini LAB

Building a Model Circuit

Even simple circuits have many possible designs.

1. You'll need a battery, a small lightbulb, and one length of wire. Don't use holders.
2. Work with a partner to connect the battery, lightbulb, and wire in such a way that the bulb lights up.
3. Sketch how the circuit looked when you completed it.
4. Find a second way to make the circuit work. (Hint: The circuit can be completed in four possible ways.) Sketch your new circuit.

Analysis

1. How do the sketches of the successful circuits show that electrons move in a circuit?
2. Choose a design that did not work and infer what stopped a circuit from forming.

FIGURE 13-7

A simple circuit like this one allows electrons to move and in this case, produce light.

Controlling Circuits

As electrons move through a circuit, they don't stop at any particular spot and they don't get used up. They are simply moving from a place where extra electrons exist toward a place where there is a shortage of electrons. Electrons can cause movement and operate lights, machines, or appliances when these devices are connected to a circuit, as in **Figure 13-7.** The electrons then move in the devices, producing heat, light, motion, or magnetic signals.

Circuits control the nonstop movement of electrons in two basic ways: (1) they stop or start the movement of electrons, and (2) they have a source of energy. First, let's look at how a circuit stops or starts the movement of electrons. This is done with a **switch,** which is a device that opens or closes a circuit. A switch is like a movable bridge in a model railroad. If the bridge is closed, as in **Figure 13-8A,** the loop is complete. A train on the track could make a complete circle, ending up where it began. But, if the bridge is open, as in **Figure 13-8B,** the train can't go around the track.

Visual Learning

Figure 13-7 Draw on the chalkboard the following variations of the circuit shown in the figure. Ask students which circuit would light the bulb and why.

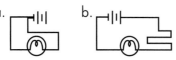

a. No, both ends of the battery must be included in the circuit.

b. Yes, the extra wire bends don't affect the continuous path. **LS**

A track is not a complete, closed loop if the bridge is open. Likewise, a circuit is not a complete, closed loop if a switch is open. Electrons cannot travel through the whole circuit if the switch is open.

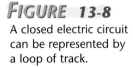

FIGURE 13-8
A closed electric circuit can be represented by a loop of track.

A When the circuit of track is connected, the model train can run. *What parts of the circuit do the train, track, and bridge represent?*

B An open switch can be represented by a break in the track. The model train cannot run through the open circuit, just as electrons cannot run through an open electric circuit.

---

2 Teach

Using Science Words

 Linguistic This section offers an opportunity to examine the Latin derivations of many science words. In the word *circuit*, point out the word part *circ*, meaning "a circle," and its function in such words as *circle, circumference, circumnavigate,* and *cycle.*

Visual Learning

Figure 13-8 What parts of a circuit do the train, track, and bridge represent? *electron, wire, switch*

Demonstration

 Visual-Spatial This is a good opportunity to show the components of a simple circuit. Use the same equipment students will need in Activity 13-1 on pages 356-357. Concentrate on the concept of a closed vs. an open circuit.

?FLEX Your Brain

Use the Flex Your Brain activity to have students explore CIRCUITS.

📁 **Activity Worksheets,** page 5

GLENCOE TECHNOLOGY

💿 **Videodisc**
STVS: Physics
Disc 1, Side 2
Electric Heart (Ch. 14)

‖‖‖‖‖‖‖‖‖‖‖

Making Integrated Circuits (Ch. 15)

‖‖‖‖‖‖‖‖‖‖‖

Content Background

- The movement of electrons can be contained in a circuit and made to do work. That work will be in the form of motion, heat, sound, or light.

- A circuit has three main types of measure: voltage, amperes, and ohms. Voltage measures potential energy in a battery or circuit. An ampere measures current, or the movement of electrons. An ohm measures resistance, the ease or difficulty with which electrons move through an object.

353

Batteries and Circuits

Now you've seen how a switch controls the movement of electrons in a circuit. You also know that circuits use a nonstop, controlled source of electrons. Where does the energy to move those electrons come from? A battery is one source of energy in circuits. A battery works by chemically separating positive and negative charges. This creates extra electrons at one end of the battery and extra positively charged atoms at the other. There is potential electrical energy in the position of the separated charges, just as there is potential energy in an object at the top of a hill, as in **Figure 13-9.** The only way for the negatively charged electrons to combine with the positively charged atoms is by moving outside the battery. If the ends of a battery are connected by a wire, electrons will move in the wire to recombine with the positive charges. This is similar to the uncontrolled way electric charges recombine during a static discharge, such as lightning.

Different types and sizes of batteries add different amounts of energy to electrons. The energy is measured in volts (V). Voltage measures how much potential energy

FIGURE 13-9

A battery can be represented by a hill. Your hand does work to move a model train up the hill. A battery does work to move electrons to one end of the battery. *How could you change the train model to represent a stronger battery? To allow more electrons per second?*

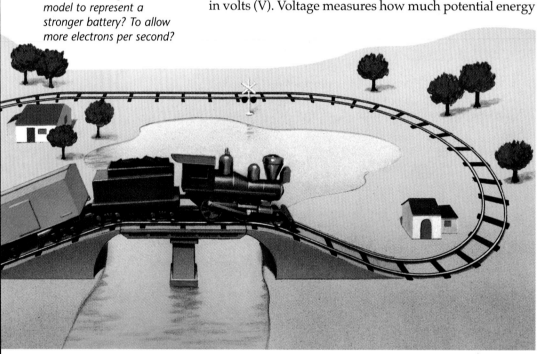

difference there is between the ends of a battery. A 9-V battery adds more energy to electrons than a 1.5-V battery. This is like the skier from Chapter 12 being lifted to different heights on a hill. More height is like more voltage.

Do you have a radio or tape player? How many batteries do they need to operate? Adding another battery to a circuit is like adding more height to the skier's hill. More potential electrical energy is available to be converted to electrical energy. With your parents' permission, compare the batteries needed to operate several different household items: a tape player, a flashlight, a smoke detector, and a television remote control. **Figure 13-10** shows the inside of one of these batteries. In the activity on the next pages, you can test how long these batteries last.

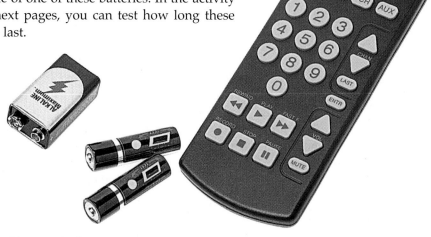

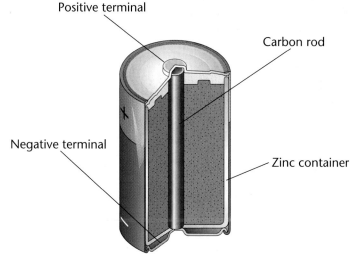

Positive terminal

Carbon rod

Negative terminal

Zinc container

FIGURE 13-10

A battery has two *terminals,* or connection points. When the two terminals of this battery are connected in a circuit, a chemical reaction causes the carbon rod to lose electrons, which accumulate in the zinc. The positively charged carbon rod is the positive terminal. The negatively charged zinc is the negative terminal.

13-2 How do circuits work?   **355**

Content Background

The D-cell battery shown in Figure 13-10 is a *dry-cell* battery. A car battery is an example of a *wet-cell* battery. It uses two connected plates of different metals (or metallic compounds) in an electrolyte solution. An *electrolyte solution* is a solution that easily separates into charged particles and conducts electricity well.

Making a Model

[LS] **Visual-Spatial** Have students draw a diagram of the arrangement of batteries in a multiple battery array. Note the connection of positive (+) to negative (−) ends. Ask students to illustrate what is happening in this arrangement. [L3] [P]

Discussion

In using the skier as an analogy, you may want to point out that the distance from the top of the hill to the bottom determines the potential energy. Skiers, representing electrons, can take a slow winding path or a straight, steep one. Ask the students what could represent resistance in this analogy.

3 Assess

Check for Understanding

Making a Model Have students draw a model diagram of a circuit with either an open or closed switch and tell whether or not electricity is moving through the circuit and why. [L2] **ELL**

Reteach

To reinforce the difference between the components of a circuit, ask students to visualize a reservoir high on a hill. The reservoir is filled with water, which represents voltage. Water flowing out represents current. Anything that slows down the water is resistance.

Extension

For students who have mastered this section, use the **Reinforcement** and **Enrichment** masters.

355

PREPARE

Purpose

LS **Kinesthetic** Students will design and carry out an experiment that would test whether or not giving a circuit more work to do will shorten the life of the battery. **L2** **ELL** **COOP LEARN**

Process Skills

controlling variables, predicting, collecting data, analyzing data

Time

Designing and setting up the experiment should take only one 45-minute class period. The execution of the experiment may take several days, depending on how the students choose to monitor the discharge of the batteries.

Safety Precautions

Students should wear safety goggles and aprons.

Possible Hypotheses

- Adding work will deplete the supply of electrons and drain the batteries faster.
- It is the size of the battery and not the work done that matters; adding work will not change the life of the battery.

Activity Worksheets, pages 5, 79–80

Activity 13-1

Design Your Own Experiment

How long do batteries last?

During a storm, the lights go out and you reach for your flashlight. The batteries are dead! All batteries eventually run down when the chemicals in them are used up. Is this energy used more quickly if the electrons have more work to do?

Possible Materials

- D-cells and holders
- lightbulbs and holders
- buzzers
- bells
- wire
- masking tape

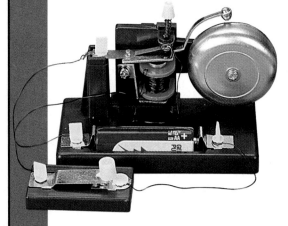

PREPARE

What You'll Investigate

How does adding lightbulbs or machines to a circuit affect the useful life of a battery?

Form a Hypothesis

Think about what you know about circuits and the movement of electrons. **Make a hypothesis** about whether you think adding energy users to the circuit will affect the life of a battery. Give your reasons.

Goals

Design an experiment that tests the life of batteries with machines or lightbulbs connected in a circuit, one after the other.

Observe how many hours each battery will continue to light the lightbulb (or bulbs) to which it is connected. Record your observations in your Science Journal.

Safety Precautions

356 Chapter 13 Electricity and Magnetism

Content Background

The circuits here and elsewhere in the chapter are connected in *series.* There is only one possible path. You can also connect a circuit in *parallel.* This provides two or more possible paths for electrons.

Because of the different ways lightbulbs add resistance, the series circuit will have less current than the parallel circuit and the bulbs will be dimmer.

PLAN

1. **Decide** how to set up the circuit.
2. **List** the things that you will not change during the experiment. List the things that you will change.
3. How will you measure the time each circuit works if you must be out of the classroom?
4. **Plan** how you will record data.

DO

1. Make sure your teacher has approved your plan, including how you will record data.
2. Carry out the experiment.
3. Keep careful records, and post the results where they can be shared with the other groups.

CONCLUDE AND APPLY

1. **Compare** the results of the experiment with your hypothesis. Was your hypothesis supported by the results? Why or why not?
2. **Graph** your experiment results on a class chart with the results of the other groups' experiments.
3. **APPLY** Discuss with the rest of the class the results of the experiment. Agree on how the work required of a battery affects it.

13-2 How do circuits work?   357

Go Further

LS **Logical-Mathematical** Ask the students what differences they would see if they added two more lightbulbs to the experiment. Possible answers: dimmer bulbs, battery has same lifetime

Assessment

Performance Have students demonstrate three different ways to empty the water from a paper cup using three different-sized holes in the bottom. In each case, the amount of water drained from the cup is the same, but the time it takes changes because of the "carrying" capacity of the hole. Use the Performance Task Assessment List for Carrying Out a Strategy and Collecting Data in **PASC**, p. 25. **L1**

Possible Procedures
  Connect a lightbulb in a holder to two connecting wires. Connect the other ends of the wires to the poles of a battery. Measure the length of time that the bulb stays lit.

Teaching Strategies
  The circuits should contain as little scaffolding as possible.

Tying to Previous Knowledge Students should review the components of an experiment and a circuit.

Troubleshooting All of the batteries must be unused and of the same size and brand (unless the students choose to compare brands).

DO

Expected Outcome
  The batteries in the circuits carrying the most lightbulbs will show a slightly longer life. (Ask students if the dimness of the bulbs might be important.)

CONCLUDE AND APPLY

1. Students who thought the circuit with the greatest workload would be drained first will find that their hypothesis is incorrect.
2. Student graphs should agree and confirm the expected outcome.
3. Because the results of this experiment tend to be counterintuitive, it is important for students to examine possible reasons for what happened.

357

4 Close

•MINI•QUIZ•

Use the Mini Quiz to check students' recall of chapter content.

1. **A circuit is a path for what form of energy?** *electrical*

2. **If electricity goes in only one direction, like electricity from a battery, it is called _____ current.** *direct*

3. **The part of a circuit that opens and closes the path for electrons is the _____.** *switch*

4. **Household current reverses its direction 60 times per second to keep electricity flowing at all times. This is called _____ current.** *alternating*

Section Wrap-up

1. Batteries are the power source in the circuit, storing potential energy.

2. A closed circuit is a complete path through which electrons can flow. It can do work. An open circuit is not a complete path.

3. **Think Critically** These appliances require more energy than a 120-V outlet can provide.

Science Journal Many rooms don't have two doors, so students will probably need to go outside for a circuit of several rooms.

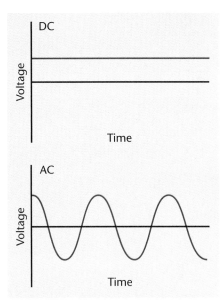

FIGURE 13-11

These graphs compare direct current (top) with alternating current (bottom). The voltage of alternating current changes in a regular pattern.

Alternating Current

You've learned about circuits that use electricity from batteries. But you also know that your TV, microwave oven, hair dryer, and many other appliances don't run on batteries. What kind of electricity are you using when you use these devices? Household devices use a kind of electricity called *alternating current.* With alternating current (AC), electricity travels one way for a while, then it reverses its direction. The direction alternates in a regular pattern. Battery-powered electricity is called *direct current* (DC) because the electricity travels in only one direction. See **Figure 13-11.**

The AC in your home (the electricity from wall sockets) can be dangerous. The voltage is almost 100 times greater than that of a D-cell battery. Most of the AC electricity used in homes in the United States is 120 V and changes direction 60 times each second. AC electricity can be generated at home, but usually comes from a power plant.

Section Wrap-up

1. What part do batteries play in a circuit?

2. What is the difference between a closed and an open circuit?

3. **Think Critically:** Why do electric dryers and ranges require 240 volts?

4. 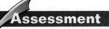 **Skill Builder**
   Designing an Experiment to Test a Hypothesis List the things you controlled in Activity 13-1. What was your dependent variable? Your independent variable? If you need help, refer to Designing an Experiment on page 553 in the **Skill Handbook.**

Science Journal
Draw a "circuit" through your home. You have to walk through as many rooms as possible without using any door more than once. Would you have to go outside your home to return to where you started?

Skill Builder
Designing an Experiment to Test a Hypothesis Controlled items were the freshness, size, and type of battery; the length of wires used; and the size of the lightbulbs. The dependent variable is the number of lightbulbs. The independent variables are the size and type of battery.

Assessment

Content Have students imagine what must happen when they flip a switch and an electric light goes on. They should then be able to explain how flipping a switch opens or closes a circuit. They may do this orally, in written form, or as a diagram. Use the Performance Task Assessment List for Making Observations and Inferences in **PASC,** p. 17.
L2

What is a magnet? 13•3

Magnets and Electricity

Did a school bell ring to tell you it was time for class? That ringing bell used a magnet and electricity. From school bells to cassette tapes, from your refrigerator to cabinet door latches, magnets are in use all around you. **Figure 13-12** shows a magnetic sculpture you might have experimented with. How does a magnet work?

When you pick up a paper clip with a magnet, one end of the magnet has to be close to the paper clip. The paper clip has to be close enough to the magnet to be affected by its magnetic field. A **magnetic field** is the area around a magnet where magnetic forces act. In a bar magnet, the magnetic field is strong at each end and weaker near the middle of the magnet.

If you have ever played with two bar magnets, you know that like ends repel and opposite ends attract. That means that two N (north) or two S (south) poles push away from each other. On the other hand, an N and an S pole will pull toward each other. In this way, poles (ends) of a magnet are similar to negative or positive charges.

What YOU'LL LEARN

- The relationship between magnetism and electricity
- How a compass works

Science Words:
magnetic field
electromagnet

Why IT'S IMPORTANT

Magnets are part of machines you use every day, such as computers, TVs, and radios.

FIGURE 13-12

The stronger magnet in the base of the sculpture magnetizes the metal diamond shapes. (See Figure 13-13 for an explanation.) When the shapes move out of the base's magnetic field, they lose their magnetism and fall.

13-3 What is a magnet? **359**

Prepare

Section Background

- Magnetism and electricity both have positive and negative aspects that interact similarly. Opposites attract and likes repel.
- Magnetism can produce an electrical current, and electricity traveling through a wire will create a magnetic field.
- Just as a lightbulb gives evidence of a complete circuit when it lights, a compass gives evidence of a magnetic field when the compass needle moves.

Preplanning

Refer to the Chapter Organizer on pages 349A-B.

1 Motivate

Bellringer

 Before presenting the lesson, display **Section Focus Transparency 43** on the overhead projector. Assign the accompanying **Focus Activity** worksheet.
L2 ELL

Program Resources

Reproducible Masters
Activity Worksheets, pp. 5, 81-82 L2
Cross-Curricular Integration, p. 17 L2
Enrichment, p. 49 L3
Lab Manual, pp. 89-94 L3 L2
Multicultural Connections, pp. 29-30 L2
Reinforcement, p. 49 L2
Science Integration Activities, pp. 63-64 L2
Study Guide, p. 49 L1

Transparencies
Section Focus Transparency 43 L2
Teaching Transparency 26 L2

Have students recall how like and unlike electric charges interact.

2 Teach

Activity

 Kinesthetic Have students use a good bar magnet to try to pick up as many paper clips as possible at one time. Have them see if they can transfer magnetic attraction through the paper clips by picking up one paper clip with another. L1 ELL

 GLENCOE TECHNOLOGY

 Videodisc

The Infinite Voyage: Unseen Worlds

Chapter 6

Magnetic Resonance Imaging: MRI a Medical Breakthrough

[barcode]

Chapter 7

Brain Tumor Surgery: Made Possible by MRI

[barcode]

 Videodisc

STVS: Human Biology

Disc 7, Side 1

Detecting the Body's Magnetic Fields

(Ch. 18)

[barcode]

 FIGURE 13-13

A strong magnet can magnetize a piece of metal.

 A Normally, the atom magnets in the piece of metal are arranged randomly.

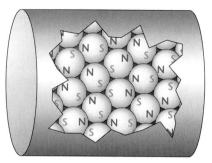

Unmagnetized metal

B A strong south pole attracts all the north poles of the atom magnets, making the metal a magnet.

 Magnet

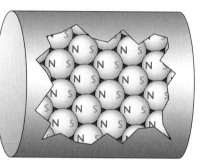

Magnetized metal

What makes a magnet?

Why are we talking about magnets in a chapter about electricity? Both electricity and magnetism involve electrons. The atoms that make magnets have certain numbers of electrons arranged in certain ways. Because of this, each atom within a magnetic material is like a small magnet, as shown in **Figure 13-13**. Normally, these atom magnets point in all directions so they cancel each other's magnetism. For a whole piece of material to act as a magnet, most of the atoms in the material must point in the same direction. When they point in the same direction, the individual magnetic forces of each atom add up to a much stronger magnetic force. In the activity on the next page, you can learn more about the relationship between electricity and magnetism.

360 Chapter 13 Electricity and Magnetism

Cultural Diversity

Electricity Top energy producers and consumers are the United States, Russia, and China. Japan is a major consumer even though it does not rank in the top ten producers. Mexico is the eighth-highest producer of energy, but it is not high on the consumer list.

Different societies also produce electricity by different methods. Iceland uses geothermal energy. Saudi Arabia relies on its extensive oil reserves. France uses much nuclear energy. Different parts of the United States use wind power, water, or fossil fuels such as coal.

Students can research energy production and consumption in different cultures using world almanacs and the Internet.

Lines of Magnetic Force

You usually can't see the lines of magnetic force. In this activity, you'll make the invisible become visible.

What You'll Investigate

How can you illustrate lines of force that show the relationship between magnetism and electricity?

Procedure 🥽 🧤 🧹

1. Put a sheet of graph paper over the bar magnet.
2. Sprinkle iron filings on the paper. **Observe** and **record** the pattern that forms.
3. **Build** a working circuit with a battery and lightbulb. Put a sheet of graph paper over one of the wires.
4. Repeat Step 2.
5. **Wrap** the nail with 20 turns of wire. Connect one end of the wire to a battery. **Touch** the other wire to the other terminal of the battery, and test to see whether the nail can pick up a paper clip. Put a sheet of graph paper over the nail. **CAUTION:** *Do not leave the battery connected unless you are using it. If you do, the wires may become hot enough to cause burns.*
6. Repeat Step 2.

Conclude and Apply

1. **Compare** and **contrast** your data for the bar magnet, circuit, and nail magnet.
2. **Infer** whether magnetic forces exist in all three of the objects that you tested. What is your evidence?
3. **Write** a step-by-step procedure to sketch the lines of attraction for a magnet.

Goals

- Build an electromagnet.
- Observe and compare magnetic fields around different types of magnets.

Materials

- bar magnet
- 20-penny iron nail
- lengths of wire
- D-cell and holder
- lightbulb and holder
- graph paper
- iron filings in a shaker container
- paper clips

Activity 13-2

Purpose

LS Kinesthetic This activity gives students a chance to see and compare the fields around magnets and fields created with electric current.
L2 ELL

Process Skills

observing, comparing, contrasting, inferring, drawing conclusions

Time

one 45-minute class period

📁 **Activity Worksheets,** pages 5, 81-82

Teaching Strategies

- Have the students first build a working circuit.
- Keep the number of iron filings to a minimum. The pattern will be easier to see.
- Using graph paper will allow students to measure the size of the field.

Troubleshooting

- The iron filings will stick to the magnet and be difficult to remove if you don't keep them separated by the paper or plastic.
- Magnets grow weaker over time. Check the magnets for strength before using them.

Answers to Questions

1. The nail magnet, circuit, and bar magnet will each have an observable magnetic field. Differences in their strengths will be obvious.
2. You can infer that a magnetic field exists because the iron filings arrange themselves according to lines of force.
3. Sprinkle iron filings on a covering sheet of graph paper and sketch the lines they form.

📓 *Science Journal* **Magnetism and Electricity** Have students compare and contrast magnetism and electricity in their Journals as follows: Make two columns marked *Magnetism* and *Electricity*. Down the side, make a list of characteristics. Put a check in each column to which that characteristic applies. **LS**

✔ **Assessment**

Performance Have students brainstorm situations in which it would be useful to be able to create a magnet electrically and then turn it off. Ask them to design what they think one of these electromagnets would look like. Use the Performance Task Assessment List for Model in **PASC**, p. 51. **L3**

FIGURE 13-14

The magnetic fields around bar magnets and electromagnets have the same shape. *Where would a paper clip be strongly attracted to the electromagnet's iron core? Weakly attracted?*

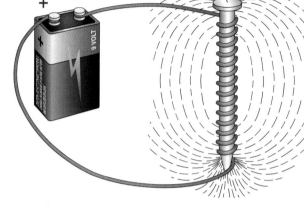

Battery

Iron core

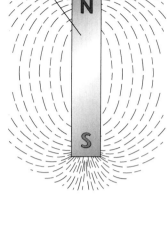

Magnet

N

S

Electricity and Magnetism

Magnets and current-carrying wires affect iron filings in much the same way. When a wire is wrapped around a nail, the pattern of iron filings looks more like the pattern made by a bar magnet. This shows an important relationship between electricity and magnetism: A magnetic field always surrounds an electric current.

A nail-and-wire magnet has a special name. An **electromagnet** is a magnet that can be turned on and off. As you can see in **Figure 13-14,** it is made by wrapping a current-carrying wire around an iron core. The strength of an electromagnet can be increased or decreased by changing the strength of the power source. Electromagnets are temporary because their magnetic field stops quickly when the circuit is opened. Electromagnets are used in many everyday devices from doorbells to computer disk drives.

Theme Connection

Stability and Change

The existence of magnetic fields represents disequilibrium of forces. Left alone, the atoms of matter in any object will tend to line up in random ways that create a neutral field. Therefore, creating a magnet involves the use of some force to rearrange the atoms in a way that creates two strong but opposing forces: positive and negative. *Entropy* tends to even out these forces over time and create another neutral field.

Compasses Are Magnets

You and your parents are in an unfamiliar part of town, and you suddenly realize you're not sure how to get home. You stop someone to ask directions. She tells you to go north and then hurries off before you can ask which way that is. If you only had a compass! How would that help? It's all a matter of magnetism.

In the activity, the iron filings and compass showed the lines of force around a magnet. The filings lined up along the force lines. The compass needle pointed along the lines. The compass works this way because its needle is a magnet.

Left undisturbed, the needle on a compass will point along the lines of force of the nearest or strongest magnet. Usually, this magnet is Earth, and the needle points toward the magnetic north pole of Earth.

FIGURE 13-15

Earth's magnetic field causes magnetic compass needles to line up along north-south lines.

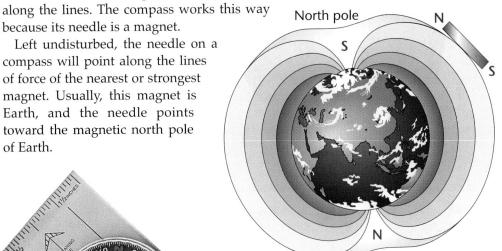

North pole

South pole

A The north magnetic pole is the pole that attracts the north end of magnets. The north magnetic pole corresponds to the south pole of a bar magnet.

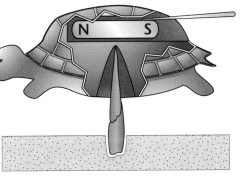

B Small factories in medieval China manufactured magnets for compasses like this one. The turtle's head points to magnetic north, allowing it to be used by travelers.

13-3 What is a magnet? **363**

Using an Analogy

Magnetic impulses on tape are like letters in a word. It is the order and rhythm of their appearance that conveys images. If you jumbled the order of letters on this page, there would be no way of telling what the message was. Likewise, if you jumble the order of the magnetic impulses, you lose the message those impulses were sending.

USING TECHNOLOGY

We know from using a tuning fork that the vibration of matter can create sound. Likewise, the electronic impulses from videotape can create sound and pictures. Television uses a screen that is sensitive to magnetic impulses to create pictures. A tape player uses magnetic impulses to create vibrations that produce sound. You can pick up the beat of a song from the feel of the speakers, without ever hearing the music because of this vibration. Using a magnet scrambles the imprinted signals on the videotape.

Troubleshooting

It will be fairly easy to erase the sound on the tape. Erasing the picture is more difficult. The stronger the magnet, the better. Electromagnets are best.

_inter_NET
CONNECTION

The Glencoe Homepage at **www.glencoe.com/sec/science** provides links connecting concepts from the student edition to relevant, up-to-date Internet sites.

Because of the connections between electricity and magnetism, they are usually studied together, as *electromagnetism.* Just as electric current can create a magnet, moving magnets can create an electric current.

FIGURE 13-16

A sound engineer uses this equipment to record music. It translates sound from air vibrations to electrical signals to magnetic impulses, and back again.

USING TECHNOLOGY
Magnetic Signals at Work

Audiotapes and videotapes are magnetic messages that are translated into sight and sound by machines. The messages are recorded as tape moves past an electromagnet. As the electric current in the magnet changes, the magnet becomes stronger or weaker. These changes are recorded as stronger and weaker magnetized areas on the tape. When the tape is played, a sensor detects and amplifies the changes. You hear and see the result.

To demonstrate the relationship between magnetism and electronic impulse, get an old videotape from home. *Be sure to get permission to use the tape, because it will be changed during your experiment.* Watch the tape, then rewind. Now, bring a magnet near the exposed part of the tape. The tape will be attracted to the magnet, proving that the tape is magnetic. Now play the tape. What do you hear and see? You have erased part of the tape by scrambling the magnetic code. Try to establish a pattern of blank spaces on the tape.

_inter_NET
CONNECTION

To learn more about magnetic recording, visit the Glencoe Homepage at **www.glencoe.com/sec/science** and follow the link for Chapter 13.

364

Content Background

- When sound hits a microphone, it is converted to varying voltage. For an analog audiotape, the strength of the magnetic field at each point on the tape corresponds to the strength of the voltage. For a digital audiotape, the signal is translated into areas of high and low magnetism on the tape corresponding to 1s and 0s, a binary code.

- A changing magnetic field induces an electric current in a wire. By rotating a loop of wire around a magnet, you induce an electric current that goes one way, then the other, as each pole passes by one side of the loop. This causes alternating current.

Harnessing Electromagnetic Energy

People first observed the effects of static electricity and magnetized metals thousands of years ago. It was not until the late 1800s that scientists learned to control the flow of electricity in a circuit and make electric batteries and motors. Now electromagnetic energy is a widely used tool. It powers lights and clocks, provides heat or cooling, and is used to record and transmit information, whether your favorite singer on the radio or the computer files of words and pictures used in publishing this book. As with all forms of energy, the more we learn about electromagnetism, the easier it is to use it.

Section ⟳ Wrap-up

1. A compass needle reacts to what type of field?

2. Magnets have poles that are like positive and negative charges. Describe how these poles act.

3. **Think Critically:** What are some possible uses for a magnet that can be turned on and off?

4. *Skill Builder*
   Observing Watch the behavior of a compass placed on a wire as the wire is connected and disconnected from a battery. Infer what would happen if you used a larger battery. If you need help, refer to Observing and Inferring on page 550 in the **Skill Handbook**.

USING MATH

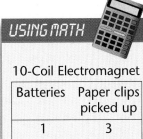

10-Coil Electromagnet

Batteries	Paper clips picked up
1	3
2	5
3	7

According to the table, how many paper clips would be picked up if four batteries and ten coils were used? How do you know?

Skill Builder
Observing The needle of a compass will move in response to the magnetic field created by current flowing through the wire. It will move quickly, and, with a larger battery, could be made to spin if the wire were connected and disconnected fast enough.

✓ Assessment

Portfolio Have students illustrate the structure and results of Activity 13-2 in a storyboard with captions that state the main ideas of the Activity. Use the Performance Task Assessment List for Poster in **PASC**, p. 73. L2 P LS

---

3 Assess

Check for Understanding

Science Journal Have students write a description of the relationship between magnetic and electrical forces. LS L2 P

Reteach

LS **Kinesthetic** Get several old doorbells or telephones, and have the students take them apart to try to identify the electromagnet. L1 ELL

Extension

📁 For students who have mastered this section, use the **Reinforcement** and **Enrichment** masters.

4 Close

•MINI•QUIZ•

Use the Mini Quiz to check students' recall of chapter content.

1. **Does electricity create or destroy a magnetic field?** *create*

2. **An iron core can become a magnet if it is surrounded by a(n) _____.** *electric current*

3. **A compass indicates direction because it is a(n) _____.** *magnet*

Section ⟳ Wrap-up

1. magnetic field
2. unlike attract and like repel
3. **Think Critically** Possible answers include lifting and dropping loads or making videotapes.

USING MATH

9; each battery adds two paper clips to the load.

365

13•4 Electricity Sources

What YOU'LL LEARN

- About different ways of generating electricity
- How to make generalizations

Science Words:
generator
turbine

Why IT'S IMPORTANT

You use electricity produced by generators every day.

FIGURE 13-17

Mechanical energy, such as that provided by these windmills, can be changed into electrical power by generators. The large generator above is powered by steam instead of wind.

366

A Typical Morning

Does your day start like this? You wake up to the bee-dee-BEEP of an electric alarm clock. You switch on the lights and stumble toward the bathroom, where you feel heat from an air duct. You take a hot shower and dress while listening to the radio. Then it's off to the kitchen for a glass of cold juice and some warm, buttered toast. Your day has just begun, and already you've used electricity at least seven times. Did you ever wonder where your electricity comes from?

Power Plants

Most electric power comes from power plants that use huge generators. A **generator** (JEN uh ray tur) is a machine that changes mechanical energy into electric power. One is shown in **Figure 13-17.** The mechanical energy comes from a **turbine** (TUR bun), a large rotating wheel that gets its energy from different sources of energy. As you read about these sources, find facts that hint at general statements about electricity. This will help you to better understand important issues.

Sources of Electricity

Coal, oil, and natural gas—fossil fuels—are the most commonly used sources of electricity in the United States. Fossil fuels are low in cost, but they can be replaced only through slow, natural processes that take millions of years. Plus, when fossil fuels are burned to produce electricity, they produce pollution.

The sun is another source of electricity. Solar energy is renewed every day. It does not pollute the air. What are the drawbacks? Places that receive little sunlight are not suited for solar-powered plants. These plants also use thousands of solar mirrors to reflect sunlight, so they require a lot of space.

Another source of electricity is nuclear energy, which harnesses the energy in the nuclei of certain atoms. It does not cause much air pollution, but the waste generated by the process is dangerous. This waste must be stored safely.

Wind and water are sources of electricity, too. Both are nonpolluting and renewable—but there are drawbacks. It's hard to build a windmill large enough to produce large amounts of electricity. Dams use water pressure to generate electricity. They create large lakes, destroying the habitats of the plants and animals that lived in the area before the dam was built.

 Skill Builder: *Making Generalizations*

LEARNING the SKILL

When you read, find facts that hint at general statements about the subject. This will help you to better understand and explain the reading.

1. To make generalizations (jen rul uh ZAY shuns), you first need to identify the subject matter.

2. Gather as many facts as you can about the subject.

3. Identify similarities and patterns among the facts.

4. Use the similarities and patterns to form some general ideas about the subject.

PRACTICING the SKILL

1. What is the subject matter of this passage?

2. What generalizations can you make about the sources of electricity we use?

3. Identify four facts that support your generalization.

APPLYING the SKILL

Make a general statement about the importance of electricity in your life. Then, identify three facts that support your statement.

Teaching the Content

- Reinforce the idea that electrical generation is the result of a magnet spinning in a coil of wire. Power plants are systems designed to spin the magnet in the most efficient way.

- Discuss how other systems (dams, windmills, solar-powered) are restricted by the immediate environment. You need an abundance of water, wind, or sun, respectively. No power systems are problem or pollution free.

Teaching the Skill

- Have the students copy what they believe is the most important sentence from each paragraph.

- They should use the copied sentences to rephrase and summarize the article, stating what they believe are the most important concepts.

- Using their summary, students should decide on an appropriate title for the selection.

GLENCOE TECHNOLOGY

 Videodisc

STVS: Chemistry
Disc 2, Side 2
Solar House (Ch. 10)

Wind Power (Ch. 14)

Hydroelectric Power (Ch. 15)

Geothermal Wells (Ch. 17)

Answers to Practicing the Skill

1. The subject matter is the many ways electricity may be generated, and the strengths and weaknesses of those ways.

2. We use fossil fuels and other natural resources to produce electricity, but all have problems connected with them.

3. Sample (a) Fossil fuels are the most commonly used sources of electricity in the United States. (b) When fossil fuels are burned, they also produce pollution. (c) Solar power is a source of electricity. We don't have to wait millions of years for it to be renewed when we use it up, but it requires thousands of solar mirrors and lots of sunlight. (d) Nuclear power is clean when it creates electricity but produces potentially dangerous wastes. (e) Wind and water are sources of electricity, but wind power may not have enough energy and dams may harm the surrounding environment. P

Source

Barnes, Joyce Annette. *Promise Me the Moon*. Dial Books: New York, 1997.

Biography

Joyce Annette Barnes is an assistant professor of English at Catonsville Community College in Maryland. She lives in Frederick, Maryland. The story of Annie Armstrong was begun in Barnes's first novel, *The Baby Grand, the Moon in July, and Me*. This book, published by Dial, was nominated for a Nebraska Golden Sower Award and was named a Child Study Children's Book of the Year 1994.

Teaching Strategies

- Ask students if they have ever had trouble with fuses or circuit breakers blowing.
- Ask students to comment on the quote "Discovery is the essence of science." Have they ever experienced this outside the classroom?

Science Journal Students will have a variety of answers. You may need to suggest where they can look for the fuse or circuit breaker box at home.

Promise Me the Moon

by Joyce Annette Barnes

I do a quick calculation. The total capacity of his system is only 1800 watts on each circuit, which is about enough to run an iron and a vacuum cleaner at the same time. Not nearly enough to supply all the new appliances. That's why the fuses keep blowing, shutting everything down. If lightning strikes his house, the surge of electricity could be too much for the circuit to handle. It's lucky the whole place hasn't gone up in flames . . .

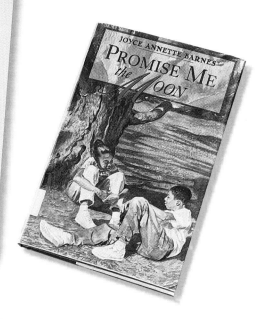

Science Journal

Explore your home like a curious scientist. In your *Science Journal*, list all the outlets, light fixtures, and electrical appliances you can find in two or three rooms during your investigation. With a parent's permission and help, visit the fuse or circuit breaker box and match your list of outlets and appliances to the circuits there. How many outlets are on a single circuit? Appliances? Lights?

Thirteen-year-old Annie Armstrong, who wants to be an astronaut, can't decide on a topic for her Enriched Science Class project. That is, not until a strange thing happens in her neighborhood. During a thunderstorm, the lights in Mr. Blackstone's stately old house go off. But the lights in the surrounding houses stay on. Annie wonders why. She reads up on electricity, forms a hypothesis, and then goes to investigate the fuse box in the dark, dusty basement of the Blackstone mansion.

Annie's curiosity and hard work pay off. She discovers that the mansion's electrical system is old and unsafe. And she understands what her science teacher means when he says, "Discovery is the essence of science."

Now Annie must discover something else—not about electricity but about herself. Does she have what it takes to get into McAllen High School and its special science and math program—one that can help her realize her dream of becoming an astronaut?

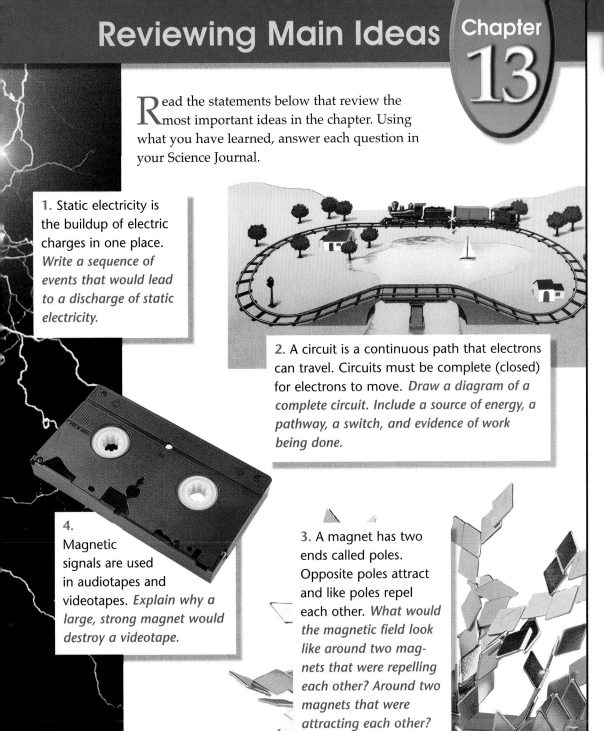

R̲ead the statements below that review the most important ideas in the chapter. Using what you have learned, answer each question in your Science Journal.

1. Static electricity is the buildup of electric charges in one place. *Write a sequence of events that would lead to a discharge of static electricity.*

2. A circuit is a continuous path that electrons can travel. Circuits must be complete (closed) for electrons to move. *Draw a diagram of a complete circuit. Include a source of energy, a pathway, a switch, and evidence of work being done.*

4. Magnetic signals are used in audiotapes and videotapes. *Explain why a large, strong magnet would destroy a videotape.*

3. A magnet has two ends called poles. Opposite poles attract and like poles repel each other. *What would the magnetic field look like around two magnets that were repelling each other? Around two magnets that were attracting each other?*

Chapter 13 Review **369**

Review

Ask students to describe details in the illustrations on this page that support the main ideas of the chapter.

Teaching Strategies

- Relate each review statement to the appropriate activity, experiment, illustration, or MiniLAB that demonstrates the concept.
- Have manipulative materials available that students may use to reconfirm or illustrate each statement.

Answers to Questions

1. Friction rubs electrons off a material whose electrons are loosely held. The result is an excess of electrons on the receiving material, giving it a negative charge. That charge will then jump to any material with an opposite, positive charge.
2. The diagram will most likely include a battery, wire, a closed switch, and a lighted lightbulb or ringing bell.
3. The lines of force that repel will fan out; those of attraction will be straight between the two poles.
4. A large magnet will alter the magnetic field and change the encoding on the tape.

✓Assessment

Portfolio Encourage students to place in their portfolios one or two items of what they consider to be their best work. Examples include:

- Model, p. 347
- MiniLAB, p. 352
- Skill Builder, p. 367 **P**

Performance Additional performance assessments may be found in **Performance Assessment** and **Science Integration Activities.** Performance Task Assessment Lists and rubrics for evaluating these activities can be found in Glencoe's **Performance Assessment in the Science Classroom (PASC).**

Science at Home

 **Intrapersonal** Guess the number of electric appliances in any room in your home. Then go in that room and make a list of everything there that uses electricity. How close was your guess? **L2**

Chapter 13 Review

Using Key Science Words

1. circuit
2. static electricity
3. magnetic field
4. electromagnet
5. switch

Checking Concepts

6. b   **9.** a
7. d   **10.** a
8. c

Thinking Critically

11. Their atomic structure will not allow them to line themselves up in positive and negative lines.

12. A static discharge of electricity in the form of lightning will be attracted to things that stick up above the ground. These are the objects that are most likely to have their electrons repelled by the excess electrons in the clouds, giving them a positive charge and making them attractive to the lightning discharge.

13. No, each lightbulb is part of the circuit. If a bulb is removed, the circuit is now open instead of closed and the electrons no longer have a path to follow.

14. The metal strip and the bottom of the maglev train must have the same magnetic pole (N or S) because they are repelling each other. (*Maglev* is short for "magnetic levitation.")

15. Answers may vary but should reflect regional resources and indicate that students understand the pros and cons of different types of power plants.

Chapter 13 Review

Using Key Science Words

circuit	magnetic field
electrical energy	static electricity
electromagnet	switch
generator	turbine

Match each phrase with the correct term from the list of Key Science Words.

1. closed path for electrons to follow
2. form of potential electrical energy
3. area where magnetic forces act
4. temporary magnet formed by electricity
5. gate in a circuit

Checking Concepts

Choose the word or phrase that completes the sentence.

6. Lightning represents a discharge of _____.
a. protons   c. neutrons
b. electrons   d. atomic nuclei

7. Static electricity moves from place to place because _____.
a. electrons are attracted to other electrons
b. potential energy is attracted to kinetic energy
c. electrons are attracted to Earth's natural magnetic poles
d. electrons are attracted to positive charges

8. Batteries represent _____.
a. a form of alternating current
b. the switch in a circuit
c. the source of electrical potential energy
d. a form of atomic energy

9. An electrical circuit requires _____.
a. a source of electrons, a pathway for the electrons, and a closed switch
b. a source of protons, a pathway for electrons, and a closed switch
c. a source of protons, a pathway for protons, and an open switch
d. a source of electrons, a pathway for electrons, and an open switch

10. All ways of generating electricity _____.
a. have benefits and drawbacks
b. are too expensive to use
c. depend on static
d. require positive charges

Thinking Critically

Answer the following questions in your Science Journal using complete sentences.

11. Aluminum and glass don't work as cores for electromagnets. Why do you think that is?

12. Playing golf or standing under a tall tree during a thunderstorm is dangerous. Why are these activities dangerous when lightning is present?

Assessment Resources

Reproducible Masters
Chapter Review, pp. 29-30
Assessment, pp. 55-58, 59-60
Performance Assessment, pp. 35-36, 51

Glencoe Technology
Computer Test Bank
MindJogger Videoquiz

13. A circuit has three lightbulbs connected as shown below. If one of the lightbulbs is removed, will the remaining lightbulbs stay lit? Explain.

14. A maglev train uses an electromagnet to raise a train above a metal strip and move it along a track. From what you know about magnetic poles, describe how you think a maglev train works.

15. Suppose you needed to establish a new electrical plant in your community. What power source would you use? Why?

Developing Skills

If you need help, refer to the description of each skill in the Skill Handbook.

16. Classifying: Make a list of 12 electrical appliances in your home. Classify them according to whether they are powered by DC or AC.

17. Making Models: Describe how you might make a model of a circuit using a garden hose and marbles.

18. Making and Using a Table: Make a table to compare the properties of five sources of electrical energy. Which source is best? Why? Would your answer change if you had different needs?

19. Measuring: Measure the distance at which a bar magnet will attract a compass needle. Now make the magnet an electromagnet by wrapping it with wire and connecting the wire to a battery. Measure the distance at which it will attract the compass needle now. **CAUTION:** *Disconnect the electromagnet as soon as possible or it will become hot and may cause burns.*

20. Sequencing: List the steps to magnetize a nail.

Performance Assessment

1. Write a Letter: In the Science and Society section, you compared and contrasted ways to produce electricity. Write a letter to your local power company and ask where and how the electricity you use is generated.

2. Calculating: Determine the number of hours in a typical Saturday that you and your family spend using electrical appliances. Pick out three different appliances. Keep a pad and pencil by each one so family members can write down when they are used. Then, total up the time used in just one day.

Developing Skills

16. Classifying Appliances that use AC will be any of those that plug into a wall; DC will be powered by batteries.

17. Making Models The marbles would represent electrons. They can go through the hose only when there is a closed path for the marbles to follow.

18. Making and Using a Table The table might include a grid with characteristics of energy sources on the vertical axis and five sources of energy on the upper horizontal axis. Students would put an x in squares that apply to each energy source. Answers will vary for best energy sources.

19. Measuring The students should use metric measurements and accurate lines of measurement to justify their data.

20. Sequencing Student answers should include either the steps that use a strong bar magnet as shown in Figure 13-13, or the steps that magnetize the nail by making it an electromagnet, as described in Step 5 on page 361.

Performance Assessment

1. Send the most representative letter to the power company for a response. Use the Performance Task Assessment List for Letter in **PASC**, p. 67. **P**

2. Encourage students to consult with their families about which appliances are used. When families consider this an invasion of privacy, students may select another electric item, such as a sign. Use the Performance Task Assessment List for Consumer Decision-Making Study in **PASC**, p. 43. **P**

Objectives

IS **Intrapersonal** Students will use Internet and other sources to research airplane design strategies. From their research, students will make and fly paper airplanes. They will then evaluate and adjust their designs according to their results. **L2**

Summary

Flight—whether it involves an actual airplane or a paper one—includes the same forces of gravity, lift, drag, and thrust. These forces act in pairs. When the downward force of gravity is less than the upward force of lift, the plane rises in the air. When thrust, pushing the plane forward, is greater than the drag holding the plane back, the plane moves forward.

Internet Students will evaluate paper airplane designs obtained from Internet sites that can be accessed through the Glencoe Homepage at **www.glencoe.com/sec/science** or by other similar Internet sites.

Non-Internet Sources If you do not have Internet access, use books on paper airplanes, such as those listed as references on page 373.

Time Required

One class period is required to research various designs and choose which to use. Two days should be allowed for experimenting with the designs and various materials. One-half class period is needed to decide on contest rules, and one class period should be used for the actual contest.

When the Wright brothers set out to make the first powered airplane, they spent time researching flight and studying designs that had failed, as well as gliders that had already flown. They considered variables such as distance, time, speed, force, and surface area. They recognized the forces involved in flight, such as gravity, lift, and drag (a form of friction). Consider all these variables and forces and help your team design a paper airplane that flies farther or longer than your classmates' airplanes.

A Paper Airplane Contest

Goals

• Research paper airplane design strategies.
• Design and build paper airplanes.
• Measure variables related to the airplane designs.

Researching Flight

Visit the Glencoe Homepage at **www.glencoe.com/sec/science** to find links to paper airplane and flight sites on the web. Learn about the physics of flight, and consider the benefits of different designs.

Procedure

Materials: paper, 50-m tape measure, metric ruler, stopwatch, balance, tape, stapler, paper clips, scissors

1. You may use a single sheet of any type of paper. You may also cut, fold, tape, glue, or staple the paper to form your airplane.

Preparation

• In addition to what the students bring to class, have supplies on hand. Different types of paper might be especially useful.

Internet If students will be using computers that might be shared with other classes, arrange access with the person in charge. Check out and become familiar with the links for paper airplane design on the Glencoe Homepage at **www.glencoe.com/sec/science.**

Non-Internet Sources Obtain reference books containing paper airplane designs for students to use.

2. Plan one or more designs. What type of paper will your airplane be made of? What shape of wings will you use? Make a sketch of your design and instructions on how to build it. Set up a data table like the one shown in your Science Journal.
3. Build your design. Use a balance to measure your airplane's mass. Record this mass in the data table.
4. Find an indoor testing area. It should be flat and open, such as a cafeteria or gymnasium.
5. Experiment with different ways of flying your airplane. Measure the distance and length of each flight. Record the data in the table in your Science Journal.
6. Make any modifications to your airplane that you think are necessary. Remember to change only one factor at a time. Record each modification in your Science Journal.
7. Tell your teacher when you have finished the airplane that you think will fly as long and as far as possible.
8. Hold a class contest to determine three categories: greatest time in the air, greatest distance flown from starting point, and the greatest overall flight (multiply flight distance and time). Your class will need to decide on the contest rules. Will teams get only one flight, or will they average the results of several? Who will judge flight time? Record the results in your Science Journal.

Conclude and Apply

1. Compare and contrast the designs your class came up with. What features did the winning planes have?
2. How did the planes that did well in the distance category differ from the planes that flew for a long time?

Go Further

Based on the results of your designs, what kind of features would a plane have that is designed to land on a target?

Sample Data

Data Table				
Flight	Mass (g)	Design Change	Flight Distance (m)	Flight Time (s)
1	5	N/A	3	2
2	10	added weight to nose	8	3
3	6	made wings wider	4	4

Unit 2 Internet Project   373

Teaching Strategies

• Encourage students who have Internet experience to help those students who do not.
• Encourage students to use paper clips as weights to adjust the flight of their airplanes. For example, a paper clip fastened on the nose of an airplane that loops up instead of flying straight will make the plane fly straighter.
• Remind students that only one variable should be changed at a time.
• Be sure students have proportions as well as general shape indicated in the sketches of their designs.

Answers to Conclude and Apply

1. Answers will vary, but the winning planes probably had narrow wings and were made of slick paper.
2. Planes that flew for a longer period of time but a shorter distance probably had wider wings that increased air resistance.

Go Further

Have students use the Internet to find out what airplane manufacturers do to reduce drag and increase lift.

Assessment

To assess the results of the project, use the Performance Task Assessment List for Model in **PASC**, p. 51.

References

Baker, Arthur. *Cut & Assemble Paper Airplanes That Fly.* Mineola, NY: Dover, 1982.

Mander, Jerry, George Dippel, and Howard Gossage. *The Great International Paper Airplane Book.* New York: Simon & Schuster, 1975.

Simon, Seymour. *The Paper Airplane Book.* New York: Viking Press, 1976.

Earth Science

In Unit 3, students are introduced to the fundamental concepts of Earth science, including the structure of Earth and the composition of the other planets and bodies found in the solar system. The physical and chemical properties of water and of the materials that comprise Earth, and their importance, are explored. The unit concludes with a discussion of Earth's atmosphere and its contribution to the ability of Earth to support life.

CONTENTS

Unit Contents

Unit 3 Project

The Internet Weather Project
pp. 522–523

374

interNET CONNECTION

The Glencoe Homepage at **www.glencoe.com/sec/ science** provides links to information related to topics in the chapters of Unit 3.

Science at Home

Making Soil Have students collect small pieces of sandstone. At home, have students open some newspaper on a table. Have students rub two pieces together hard enough to crumble the sandstone into sand. Time how long it takes to collect one full tablespoon of sand (soil) as a result of friction. Explain that this friction is similar to physical changes in rocks that result in the formation of soil. L2 ELL

Earth Science

What's happening here?

In the Toroweap section of the Grand Canyon, you can see the cones of extinct volcanoes. Lava from these volcanoes once spilled over the rim of the canyon, flowing almost a mile to the Colorado River below. The smaller photo is Io, one of Jupiter's moons. Volcanic activity is also changing the surface of Io, as shown in new images from NASA's *Galileo* spacecraft. Materials that make up Earth's structure also are found in meteorites, moons, and planets of our solar system. Could these same materials be found in solar systems of distant galaxies as well?

interNET CONNECTION

Visit the Glencoe Homepage at *www.glencoe.com/ sec/science* for links to information about our solar system. In your Science Journal, write about the materials that scientists hypothesize make up all of the planets and moons in our solar system.

375

Cultural Diversity

Myths about Earth's Origin Many cultures have myths about Earth and its features. Lakota myth states that first Earth was covered with water. The Creating Power sent a turtle under the waters. When it returned, it was covered with mud. The Creating Power spread this mud over the waters. He then cried, and his tears formed streams and lakes.

Unit Internet Project

The project for this unit guides students through using the Internet to investigate weather. In preparation for this project, found on pages 522 and 523, follow the suggestions indicated on page 522. Students can begin working on the project during Chapter 18.

INTRODUCING THE UNIT

What's Happening Here?

Have students look at the photos and read the text. Ask them to make a connection between the volcanic activity on Io and the evidence of extinct volcanoes in the Grand Canyon. Point out that all parts of our solar system were formed by the same forces and are made up of the same materials. Explain that in this unit, students will learn about the materials that make up Earth as well as other bodies in the solar system.

Background

- Areas of the Grand Canyon show evidence of lava flows dated to 100 000 years ago.
- Images from NASA's *Galileo* spacecraft indicate that Io is active volcanically. Scientists have identified some of the elements in the gases that Io's volcanoes emit. So far, they have discovered sodium, potassium, and sulfur—elements also found on Earth.

Previewing the Chapters

Before teaching Chapters 15 and 16, ask students to make a list of the types of rocks that can be found on Earth. Have them bring out their lists again when studying those chapters to see how their lists compare.

Tying to Previous Knowledge

Discuss with students what they know about Earth's moon. Students may have seen or touched pieces of moon rocks at museum exhibits. Ask students if they think they would be able to pick out a moon rock in a pile of Earth rocks.

375

Chapter Organizer

Section	Objectives/Standards	Activities/Features
Chapter Opener		Explore Activity: Estimate the Number of Stars, p. 3
14-1 **Earth's Place in Space** (3 sessions, 1½ blocks)*	1. **Explain** how the tilt of Earth's axis causes seasons. 2. **Analyze** what causes the phases of the moon. National Science Content Standards: (5-8) UCP1, UCP2, UCP3, D3	Activity 14-1: Moon Phases, p. 381 MiniLAB: Observing Distance and Size, p. 382 Skill Builder: Concept Mapping, p. 383 Using Math, p. 383
14-2 **The Solar System** (4 sessions, 2 blocks)*	3. **Recognize spatial relationships** between distances in space. 4. **Compare and contrast** objects in the solar system. National Science Content Standards: (5-8) UCP1, UCP3, D3, E1, E2, F2	Problem Solving: Distances in the Solar System, p. 386 Activity 14-2: Space Colony, pp. 390-391 Skill Builder: Developing Multimedia Presentations, p. 392 Using Math, p. 392 Science & Language Arts: Barbary, p. 393 Science Journal, p. 393
14-3 **Science and Society:** **Space Exploration:** **Boom or Bust?** (1 session, ½ block)*	5. **Analyze** the costs and benefits of space exploration. 6. **Recognize** bias in science articles. National Science Content Standards: (5-8) A2, E2, F5	Skill Builder: Recognizing Bias, p. 395
14-4 **Stars and Galaxies** (1 session, ½ block)*	7. **Discuss** how a star is born. 8. **Describe** the galaxies that make up the universe. National Science Content Standards: (5-8) UCP1, UCP2, A2, D3, E2, F5	MiniLAB: Modeling Constellations, p. 397 Using Technology: Space Probes, p. 398 Skill Builder: Formulating Models, p. 400 Science Journal, p. 400

* A complete Planning Guide that includes block scheduling is provided on pages 31T-33T.

Activity Materials

Explore	Activities	MiniLABs
page 377 black construction paper, white chalk or crayon, rice, spoon, ruler	page 381 drawing paper, softball, flashlight pages 390-391 drawing paper, markers, books about the planets	page 382 basketball, golf ball page 397 black construction paper, round cereal box (oatmeal), pencil, scissors, tape, flashlight

Need Materials? Call Science Kit (1-800-828-7777).

Teacher Classroom Resources

Reproducible Masters	Transparencies	Teaching Resources
Activity Worksheets, pp. 5, 85-86, 89 **Cross-Curricular Integration**, p. 18 **Enrichment**, p. 51 **Lab Manual 27** **Reinforcement**, p. 51 **Science Integration Activities**, pp. 65-66 **Study Guide**, p. 51	**Section Focus Transparency 45,** Autumn Flight **Teaching Transparency 27,** Moon Phases	**Spanish Resources** **English/Spanish Audiocassettes** **Cooperative Learning Resource Guide** **Lab Partner** **Lab and Safety Skills** **Lesson Plans**
Activity Worksheets, pp. 5, 87-88 **Enrichment**, p. 52 **Lab Manual 28** **Reinforcement**, p. 52 **Study Guide**, p. 52	**Section Focus Transparency 46,** Neptune **Teaching Transparency 28,** The Solar System	*Assessment Resources*
Enrichment, p. 53 **Reinforcement**, p. 53 **Science and Society/Technology Integration**, p. 32 **Study Guide**, p. 53	**Science Integration Transparency 14,** Smaller, Lighter, Cheaper **Section Focus Transparency 47,** Space Camp	**Chapter Review**, pp. 31-32 **Assessment**, pp. 61-64 **Performance Assessment**, p. 52 **Performance Assessment in the Science Classroom (PASC)** **MindJogger Videoquiz** **Alternate Assessment in the Science Classroom** **Computer Test Bank**
Activity Worksheets, pp. 5, 90 **Enrichment**, p. 54 **Multicultural Connections**, pp. 31-32 **Reinforcement**, p. 54 **Study Guide**, p. 54	**Section Focus Transparency 48,** Solar Flare-Up	

Key to Teaching Strategies

The following designations will help you decide which activities are appropriate for your students.

L1 Level 1 activities should be appropriate for students with learning difficulties.

L2 Level 2 activities should be within the ability range of all students.

L3 Level 3 activities are designed for above-average students.

ELL ELL activities should be within the ability range of English Language Learners.

LS These activities are designed to address different learning styles.

COOP LEARN Cooperative Learning activities are designed for small group work.

P These strategies represent student products that can be placed into a best-work portfolio.

GLENCOE TECHNOLOGY

The following multimedia resources are available from Glencoe.

Science and Technology Videodisc Series (STVS)
Earth & Space
Monitoring the Sun
Computerized Star Imaging
Modeling Black Holes
The Infinite Voyage Series
Sail On, Voyager

Glencoe Earth Science Interactive Videodisc
Introduction: Space Exploration
National Geographic Society Series
STV: The Solar System
Glencoe Earth Science CD-ROM

Teacher Classroom Resources

This is a representation of key blackline masters available in the Teacher Classroom Resources.

Teaching Aids

Section Focus Transparencies

AUTUMN FLIGHT
Each fall, geese fly south to warmer climates. They make their journey in large groups called flocks.

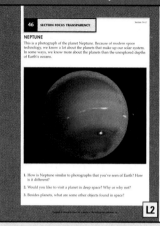

NEPTUNE
This is a photograph of the planet Neptune. Because of modern space technology, we know a lot about the planets that make up our solar system. In some ways, we know more about the planets than the unexplored depths of Earth's oceans.

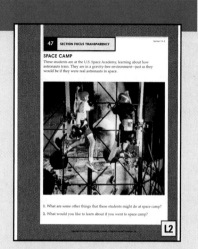

SPACE CAMP
These students are at the U.S. Space Academy, learning about how astronauts train. They are in a gravity-free environment—just as they would be if they were real astronauts in space.

Science Integration Transparencies

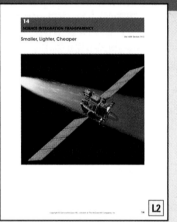

Smaller, Lighter, Cheaper

Teaching Transparencies

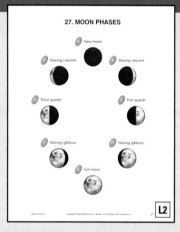

27. MOON PHASES

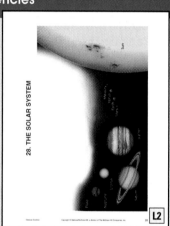

28. THE SOLAR SYSTEM

Meeting Different Ability Levels

Study Guide for Content Mastery

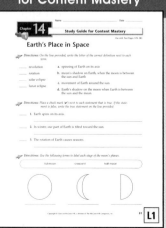

Earth's Place in Space

Reinforcement

Earth's Place in Space

Enrichment Worksheets

More Hours in a Day

Hands-On Activities

Science Integration Activities

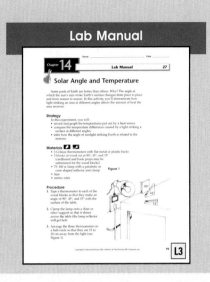

Angles of Light

Lab Manual

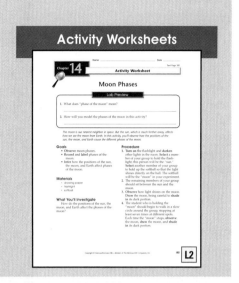

Solar Angle and Temperature

Activity Worksheets

Moon Phases

Enrichment and Application

Cross-Curricular Integration

The Solar System and Beyond

Fast Calendars and Slow Calendars

Multicultural Connections

The Other Side of the Sky

Science and Society/Technology Integration

The Solar System and Beyond

The Effects of Satellites on Society

Assessment

Performance Assessment

Make a Model of Earth, Sun, and Moon

Chapter Review

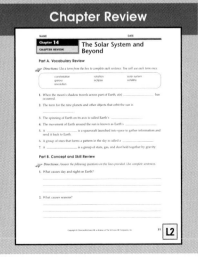

The Solar System and Beyond

Assessment

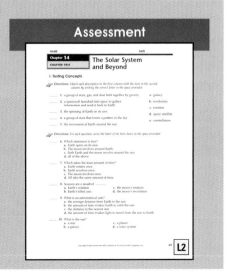

The Solar System and Beyond

The Solar System and Beyond

CHAPTER OVERVIEW

Section 14-1 This section describes the position and movement of Earth and the moon in relation to the sun and each other. Earth's axis as the determining factor in the seasons is discussed. Both lunar and solar eclipses are explained.

Section 14-2 The objects that make up our solar system are described in this section. Descriptions of comets, planets and their moons, and asteroids are given in sequence.

Section 14-3 Science and Society This section emphasizes the ability to recognize bias. Bias, motivation, and supporting statements are examined by presenting two points of view on the costs and benefits of space exploration.

Section 14-4 Students are introduced to the different sizes, temperatures, and lives of stars. Constellations, galaxies, and a look at the universe as a whole complete the chapter.

Chapter Vocabulary

rotation   satellite
revolution   constellation
eclipse   galaxy
solar system

Theme Connection

Scale and Structure The immensity, composition, and relatedness of objects in the solar system support this ongoing theme.

Chapter Preview

Skills Preview

▶ **Skill Builders**
- make a concept map
- develop multimedia presentations
- formulate models

▶ **MiniLABs**
- observe
- model

▶ **Activities**
- observe
- record
- infer
- classify
- draw

376

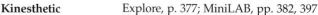

Learning Styles

Look for the following logo for strategies that emphasize different learning modalities. **LS**

Kinesthetic	Explore, p. 377; MiniLAB, pp. 382, 397
Visual-Spatial	Visual Learning, pp. 379, 382, 387, 388, 395, 397; Activity 14-1, p. 381; Demonstration, p. 397
Interpersonal	Check for Understanding, p. 399; Assessment, p. 400
Intrapersonal	Assessment, p. 383; Enrichment, p. 388; Reteach, p. 389; Go Further, p. 391; Science at Home, p. 401
Logical-Mathematical	Using Math, pp. 383, 392; Problem Solving, p. 386; Activity, 14-2, pp. 390-391
Linguistic	Using Science Words, p. 380; Assessment, pp. 381, 392; Science Journal, p. 386; Check for Understanding, p. 389

Chapter 14

The Solar System and Beyond

Here's a view of Earth taken from the moon by astronauts on the *Apollo 17* lunar mission. Do you ever gaze at the night sky? What do you see? On a clear night, it seems like the sky is full of sparkling points of lights. You can see dozens, no hundreds of these sparkles. Just how many stars are there?

EXPLORE ACTIVITY

Estimating the Number of Rice Grains

1. Divide a sheet of black construction paper into two-inch squares. Draw the lines with white crayon or chalk so that they show up clearly.
2. Spill a teaspoonful of rice onto the black paper.
3. Count the number of grains of rice in one square. Repeat this step with a different square. Add the number of grains of rice in the two squares, then divide this number by 2 to calculate the average number of grains of rice in the two squares.
4. Multiply this number by the total number of squares on the paper. This will give you an estimate of the total grains of rice on the paper.

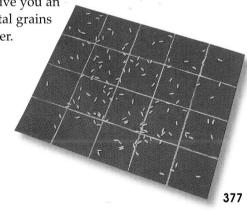

Science Journal

How might scientists use this same method to count the number of stars in the sky? In your Science Journal, describe the process scientists might use.

377

EXPLORE ACTIVITY

Purpose

LS **Kinesthetic** Use the Explore activity to introduce students to the concepts of random sampling and estimation. **L2**

Preparation

Review with students how to calculate averages.

Materials

black construction paper; ruler; white crayon or chalk; dry, uncooked rice; teaspoon

Teaching Strategies

• Have students test the accuracy of their estimate by counting the rice and comparing their estimate to the actual number.
• Discuss other areas where estimation is important (population, monthly costs for a family, the weight of people in an elevator).

Science Journal Scientists can divide the night sky into sections and use the method described to estimate the number of stars in the night sky.

✓ Assessment

Oral Have students use a parallel procedure to estimate the number of students in their school. Have them compare the accuracy of their estimates to the actual student population. Use the Performance Task Assessment List for Analyzing the Data in **PASC**, p. 27. **P**

Assessment Planner

Portfolio
Refer to page 401 for suggested items that students might select for their portfolios.

Performance Assessment
See page 401 for additional Performance Assessment options.
Skill Builders, pp. 383, 392, 395, 400
MiniLABs, pp. 382, 397
Activities 14-1, p. 381; 14-2, pp. 390-391

Content Assessment
Section Wrap-ups, pp. 383, 392, 400
Chapter Review, pp. 401-403
Mini Quizzes, pp. 383, 392, 400

Group Assessment
Opportunities for group assessment occur with Cooperative Learning Strategies and Flex Your Brain activities.

Prepare

Section Background

Earth completes one rotation on its axis every 24 hours, resulting in day and night. It completes one revolution around the sun every $365\frac{1}{4}$ days. One complete revolution is called a year. Different planets have years of different lengths.

Preplanning

Refer to the Chapter Organizer on pages 376A-B.

1 Motivate

Bellringer

 Before presenting the lesson, display **Section Focus Transparency 45** on the overhead projector. Assign the accompanying **Focus Activity** worksheet.

L2 ELL

Activity

Have students observe and record the shape of the moon for one month. This is a chance to visually verify moon phases. The drawings can be arranged on a sheet of paper or in a flip book. L2 ELL

378

14•1 Earth's Place in Space

What YOU'LL LEARN

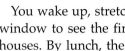

- How seasons are caused by the tilt of Earth's axis
- What causes the phases of the moon

Science Words:
rotation
revolution
eclipse

Why IT'S IMPORTANT

The movements of Earth cause night and day and the seasons.

FIGURE 14-1
The rotation of Earth on its axis causes night and day.

Axis

Rotation

Earth Moves

You wake up, stretch and yawn, then glance out your window to see the first rays of dawn peeking over the houses. By lunch, the sun is high in the sky. As you sit down to dinner that evening, the sun appears to sink below the horizon. It might seem like the sun moves across the sky. But it is Earth that is really moving.

Earth's Rotation

Earth spins in space like a dog chasing its tail—but not as fast! Our planet spins around an imaginary line called an axis. **Figure 14-1** shows this imaginary axis.

The spinning of Earth on its axis is called Earth's **rotation** (roh TAY shun). Earth rotates once every 24 hours. In the morning, as Earth rotates, the sun comes into view. In the afternoon, Earth continues to rotate, and the sun appears to move across the sky. In the evening, the sun seems to go down because the place where you are on Earth has rotated away from the sun.

You can see how this works by standing and facing the chalkboard. Pretend you are Earth and the chalkboard is the sun. Now turn around slowly in a counterclockwise direction. The chalkboard moves across your vision, then disappears. You rotate until finally you see the chalkboard again. The chalkboard didn't move—you did. When you rotated, you were like Earth, spinning in space so that different parts of the planet face the sun at different times. This movement of Earth, not the movement of the sun, causes night and day.

378   Chapter 14 The Solar System and Beyond

Program Resources

📁 **Reproducible Masters**
Activity Worksheets, pp. 5, 85-86, 89 L2
Cross-Curricular Integration, p. 18 L2
Enrichment, p. 51 L3
Lab Manual, pp. 95-97 L3
Reinforcement, p. 51 L2
Science Integration Activities, pp. 65-66 L2
Study Guide, p. 51 L1

🔦 **Transparencies**
Section Focus Transparency 45 L2
Teaching Transparency 27 L2

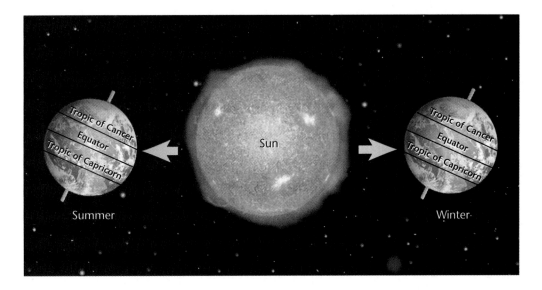

Summer

Sun

Winter

Earth's Revolution

You know that Earth rotates in space. It also moves in other ways. Like an athlete running around a track, Earth moves around the sun in a regular, curved path. This path is called an orbit. The movement of Earth around the sun is known as Earth's **revolution** (rev oh LEW shun). A year on Earth is the time it takes for Earth to revolve around the sun once. How many revolutions old are you?

Seasons

Who doesn't love summer? The days are long and warm. It's a great time to go swimming, ride a bike, or read a book just for fun. Why can't we have summer all year-round? Blame it on Earth's axis. The axis, that imaginary line that our planet spins around, is not straight up and down. It is tilted at an angle. It's because of this tilt that there are seasons in many areas on Earth. Why?

Look at **Figure 14-2.** The part of Earth that is tilted toward the sun receives more direct sunlight and more energy from the sun than the part of Earth that is tilted away from the sun. When the part of Earth that you live on is tilted away from the sun, you have winter. When the part of Earth that you live on is tilted toward the sun, you have summer.

FIGURE **14-2**

In the northern hemisphere on June 21 or 22, the sun's rays directly strike the Tropic of Cancer. On December 21 or 22, the sun's direct rays are over the Tropic of Capricorn. *When it's summer in the northern hemisphere, what season is it in the southern hemisphere?*

14-1 Earth's Place in Space   **379**

379

FIGURE 14-3

The moon is said to be waxing when it seems to be getting larger night by night. It is said to be waning when it seems to be getting smaller.

A New moon

B Waxing crescent

C First quarter

D Waxing gibbous

E Full moon

F Waning gibbous

G Third quarter

H Waning crescent

Movements of the Moon

Imagine a dog running in circles around an athlete who is jogging on a track. That's how you can picture the moon moving around Earth. As Earth revolves around the sun, the moon revolves around Earth. The moon revolves around Earth once every 27.3 days. But, as you have probably noticed, the moon does not always look the same from Earth. Sometimes it looks like a big, glowing disk. Other times, it's a thin sliver.

Moon Phases

How many different moon shapes have you seen? Round shapes? Half-circle shapes? The moon looks different at different times of the month, but it doesn't really change. What does change is the way the moon appears from Earth. We call these changes moon phases. **Figure 14-3** shows the different phases of the moon.

Light from the Sun

The moon phase you see on any given night depends on the positions of the moon, the sun, and Earth in space. Wait a minute. How can we see the different phases of the moon? Is someone shining a giant flashlight up there? No, the moon receives light from the sun, just as Earth does. And just as half of Earth experiences day while the other half experiences night, one half of the moon is lit by the sun while the other half is dark. As the moon revolves around Earth, we see different parts of the side of the moon that is facing the sun. This makes the moon appear to change shape.

Moon Phases

The moon is our nearest neighbor in space. But the sun, which is much farther away, affects how we see the moon from Earth. In this activity, you'll observe how the positions of the sun, the moon, and Earth cause the different phases of the moon.

What You'll Investigate
How do the positions of the sun, the moon, and Earth affect the phases of the moon?

Procedure

1. **Turn on** the flashlight and darken other lights in the room. **Select** a member of your group to hold the flashlight; this person will be the "sun." **Select** another member of your group to hold up the softball so that the light shines directly on the ball. The softball will be the "moon" in your experiment.
2. The remaining members of your group should sit between the sun and the moon.
3. **Observe** how light shines on the moon. **Draw** the moon, being careful to **shade in** its dark portion.
4. The student who is holding the "moon" should begin to **walk** in a slow circle around the group, stopping at least seven times at different spots. Each time the "moon" stops, **observe** the moon, **draw** the moon, and **shade in** its dark portion.

Conclude and Apply

1. **Compare** and **contrast** your drawings with those of other students. **Discuss** similarities and differences in the drawings.
2. In your own words, **explain** how the positions of the sun, the moon, and Earth affect the phase of the moon we see on Earth.
3. **Compare** your drawings with **Figure 14-3.** Which phase is the moon in for each drawing? **Label** each drawing with the correct moon phase.

381

Goals
- Observe moon phases.
- Record and label phases of the moon.
- Infer how the positions of the sun, the moon, and Earth affect phases of the moon.

Materials
- drawing paper
- softball
- flashlight

Purpose
Visual-Spatial Students will observe phases of the moon.

L2 ELL COOP LEARN

Process Skills
observing, comparing and contrasting, inferring, analyzing, modeling

Time
45 minutes

Alternate Materials
This activity can also be done using an unshaded lamp as a light source. Place the lamp on a desk and have students sit on the floor near the desk.

Activity Worksheets, pages 5, 85-86

Teaching Strategies
- The "terminator," the line that separates the moon's light and dark sides, should be drawn first. The darkened part of the sphere should be shaded.
- The student holding the softball should move in a counterclockwise motion to get the phases in correct order.

Troubleshooting This station activity works best in a fairly large, cleared area.

Answers to Questions
1. Answers will vary depending on drawings.
2. Answers will vary, but students should note that as the moon moves more nearly opposite the sun, we are able to see more of its lighted side from Earth.
3. Drawings will vary but should correspond to Figure 14-3.

✓ Assessment

Portfolio Many cultures have created myths and legends surrounding the phases of the moon. Have students research one myth, then write a short story or poem about the moon based on their research. They can use their moon sketches to illustrate their stories. Use the Performance Task Assessment List for Writer's Guide to Fiction in **PASC**, p. 83. P LS

3 Assess

Check for Understanding

Inquiry Question How can Earth experience the cold of winter, even though it is closer to the sun in the winter? *It is the tilt of Earth on its axis and not closeness to the sun that causes winter.*

Mini LAB

Purpose

LS **Kinesthetic** Students will observe the relationship between the distance and size of an object relative to another object. **L2** **COOP LEARN**

Materials

a basketball and a golf ball for each group

Teaching Strategies

- Have students work in groups.
- Before beginning the activity, ask students if they think that something small could hide something big. For example, "Could a peanut hide an elephant?"

Activity Worksheets, pages 5, 89

Analysis

1. Students' thumbs will cover the basketball when their arms are pulled approximately halfway toward their bodies. Their thumbs should block the view of the golf ball while their arms are outstretched.

2. Students should note that even though the moon is much smaller than the sun, it can block out light from the sun because it is much closer to Earth.

FIGURE 14-4
Only a small area of Earth ever experiences a total solar eclipse. *Why?*

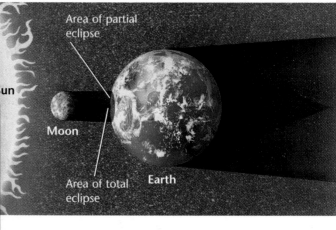

Area of partial eclipse

Sun

Moon

Area of total eclipse

Earth

Mini LAB

Observing Distance and Size

1. Place a basketball on a table at the front of the classroom. Then stand at the back of the room.
2. Extend your arm, close one eye, and try to block the ball from sight with your thumb.
3. Slowly move your thumb closer to you until it completely blocks the ball.
4. Repeat the experiment using a golf ball.

Analysis

1. In your Science Journal, describe what you observed. When did your thumb block your view of the basketball? When did your thumb block your view of the golf ball?
2. A small object can sometimes block a larger object from view. Explain how this relates to the moon, the sun, and Earth during a solar eclipse.

Eclipses

Have you ever tried to watch TV with someone standing between you and the screen? You can't see a thing! The light from the screen can't reach your eyes because someone is blocking it. Sometimes, the moon is like that person standing in front of the TV. It moves between the sun and Earth in a position that blocks sunlight from reaching Earth. The moon's shadow travels across parts of Earth. This event, shown in **Figure 14-4**, is called an **eclipse** (ee KLIHPS). Because it is an eclipse of the sun, it is known as a solar eclipse. The moon is much smaller than the sun, so not everywhere on Earth is in the moon's shadow. Sunlight is completely blocked only in the small area of Earth where the moon's shadow falls. In that area, the eclipse is said to be a total solar eclipse.

Lunar Eclipse

Sometimes, Earth can be like a person standing in front of the TV. It gets between the sun and the moon, blocking sunlight from reaching the moon. When

Assessment

Performance Have students draw a diagram of the relationship among their eye, their thumb, and the basketball. The lines should radiate straight from the eye past the thumb toward the basketball. Use the Performance Task Assessment List for Scientific Drawing in **PASC**, p. 55. **P**

Visual Learning

Figure 14-4 *During a total solar eclipse, sunlight is completely blocked only in the small area of Earth where the moon's shadow falls. The moon is much smaller than the sun. Its shadow falls on only a small portion of Earth.* **LS**

Earth's shadow falls on the moon, we have an eclipse of the moon, which is called a lunar eclipse. **Figure 14-5** shows a lunar eclipse.

Our Neighbors in Space

In this section, you've learned about what causes day and night and the seasons. You've also learned about the moon, Earth's nearest neighbor in space. Next, you'll look at our other neighbors in space—the planets that make up our solar system.

FIGURE 14-5

During a lunar eclipse, Earth moves between the sun and the moon.

Section Wrap-up

1. Explain the difference between Earth's revolution and its rotation.

2. Draw a picture showing the positions of the sun, the moon, and Earth during a solar eclipse.

3. **Think Critically:** Seasons are caused by the tilt of Earth's axis. What do you think seasons would be like if Earth's axis were not tilted?

4. *Skill Builder*
   Concept Mapping Make a cycle map showing the phases of the moon in sequence, beginning with the new moon. If you need help, refer to Concept Mapping on page 544 in the **Skill Handbook.**

USING MATH

Light travels 300 000 km per second. There are 60 seconds in a minute. If it takes eight minutes for the sun's light to reach us, how far is the sun from Earth?

Skill Builder

Concept Mapping The moon phases are as follows: new moon, waxing crescent, first quarter, waxing gibbous, full moon, waning gibbous, third quarter, waning crescent.

Assessment

Process Have students research and write a short paragraph explaining why the same side of the moon always faces Earth. The moon rotates once every 27.3 days, which is the same amount of time it takes to revolve around Earth. Thus, the same side of the moon always faces Earth. Use the Performance Task Assessment List for Writing in Science in **PASC**, p. 87. **L2** **P** **LS**

Reteach

Have students sketch and label the positions of the sun and Earth during summer and winter. **L1** **ELL** **P**

Extension

For students who have mastered this section, use the **Reinforcement** and **Enrichment** masters.

4 Close

•MINI•QUIZ•

Use the Mini Quiz to check students' recall of chapter content.

1. **The spinning of Earth on its axis is called _____.** *rotation*

2. **The movement of Earth around the sun is known as Earth's _____.** *revolution*

3. **A(n) _____ eclipse happens when the moon's shadow travels across part of Earth.** *solar*

Section Wrap-up

1. Rotation is the spinning of Earth on its axis; it produces night and day. Revolution is the movement of Earth around the sun. A complete revolution is a year.

2. Students should show the moon between the sun and Earth.

3. **Think Critically** There would be little difference between summer and winter, with fall and spring being absent.

USING MATH

144 000 000 km **LS**

Prepare

Section Background

- Comets are incredibly ancient objects in our solar system. They do not form their distinctive tails until they approach the sun. As the sun's warmth begins burning off the frozen gases, water, and dust that make up a comet, the "solar wind" (photons of light given off by the sun) blows the material away from the sun.

- Saturn's rings are the only ones large enough to be seen from Earth.

Preplanning

Refer to the Chapter Organizer on pages 376A-B.

1 Motivate

Bellringer

Before presenting the lesson, display **Section Focus Transparency 46** on the overhead projector. Assign the accompanying **Focus Activity** worksheet.

L2 ELL

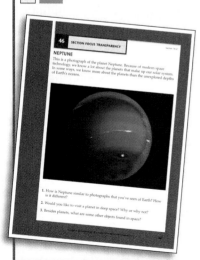

14•2 The Solar System

What YOU'LL LEARN

- About distances in space
- About the objects in our solar system

Science Words:
solar system

Why IT'S IMPORTANT

You'll learn more about how the planets, including Earth, were formed.

Distances in Space

Imagine that you are an astronaut living far in the future, doing research on a space station near the edge of our solar system. You've been working hard for a long time. You need a vacation. Where will you go? How about a tour of the solar system? The **solar system,** shown in **Figure 14-6,** is made up of the nine planets and numerous other objects that orbit the sun. How long do you think it would take you to cross the solar system?

Measuring Space

Distances in space are hard to imagine because space is so vast. Let's get back down to Earth for a minute. Suppose you had to measure your pencil, the hallway outside your classroom, and the distance from your home to school. Would you use the same units for each measurement? Probably not. You'd probably measure

Pluto

Neptune

Uranus

Saturn

Jupiter

384   Chapter 14 The Solar System and Beyond

Program Resources

Reproducible Masters
Activity Worksheets, pp. 5, 87-88 L2
Enrichment, p. 52 L3
Lab Manual, pp. 99-102 L2
Reinforcement, p. 52 L2
Study Guide, p. 52 L1

Transparencies
Section Focus Transparency 46 L2
Teaching Transparency 28 L2

your pencil in centimeters. You'd probably use something bigger to measure the length of the hallway, such as meters. You might measure the trip from your home to school in kilometers. We use larger units to measure longer distances. Imagine trying to measure the trip from your home to school in centimeters. If you didn't lose count, you'd end up with a very large number!

Astronomical Unit

Kilometers are fine for measuring long distances on Earth. But we need even bigger units to measure vast distances in space. One such measure is the astronomical (as truh NAHM uh kul) unit. An astronomical unit equals 150 million km, which is the average distance from Earth to the sun. It is abbreviated *AU*. If something is 3 AU away from Earth, it means that the object is three times as far away as Earth is from the sun.

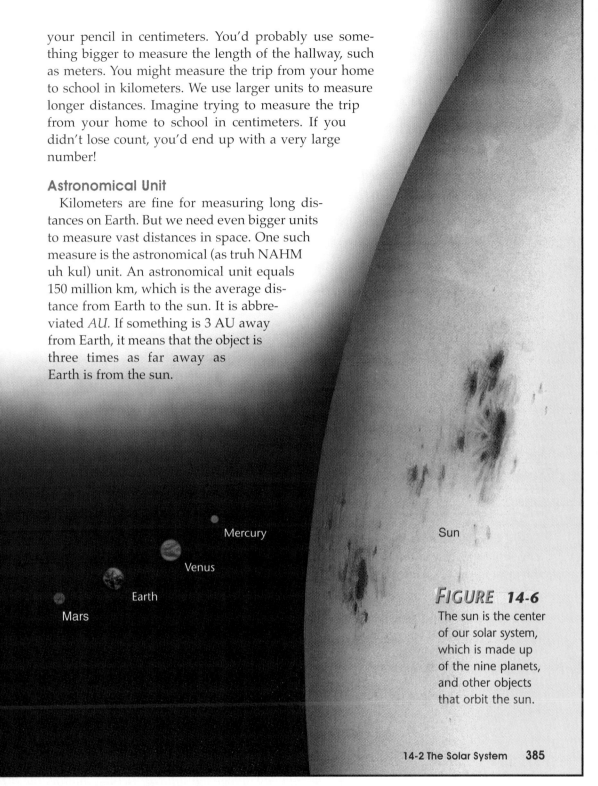

Mercury

Venus

Earth

Mars

Sun

FIGURE 14-6

The sun is the center of our solar system, which is made up of the nine planets, and other objects that orbit the sun.

14-2 The Solar System   **385**

Tying to Previous Knowledge

Remind students of the revolution of Earth around the sun. Point out that all objects influenced by the sun's gravity have similar motions. Gravity as a unifying force should be reviewed.

2 Teach

Discussion

Discuss why astronomers use the distance from Earth to the sun as a unit of measure. Point out that units of measure have to have meaning to the people who use them and be functional for the purpose they are created for. The Astronomical Unit fits this description.

? FLEX Your Brain

Use the Flex Your Brain activity to have students explore COMETS.

📁 **Activity Worksheets,** page 5

Activity

Remind students that Earth is the only planet where water exists as a gas, liquid, and solid. Encourage students to come up with additional ways that Earth differs from the other planets. As students learn about each planet, have them make an ongoing chart on the bulletin board or chalkboard that compares and contrasts each planet to Earth. **L2 COOP LEARN**

Problem Solving

Solve the Problem

Scientists know the most about Venus and Mars, the planets closest to Earth. This could be explained because of our ability to observe these planets from Earth, as well as the ability to get space probes to them in a relatively short amount of time.

Think Critically

The easiest way to make a model of the solar system would be to convert astronomical units to centimeters, using a meterstick. Given the vast distances from the outer planets to the sun, students may realize that some planets, such as Mercury, would be measured in centimeters, while others, such as Pluto, would be measured in meters. They may also note that the model would need to be built in an open area, such as a school gym. [IS]

Teacher F.Y.I.

The closer a comet gets to the sun, the longer its tail becomes. Issues of *Sky and Telescope* provide spectacular photographs of comets.

STV: The Solar System
Outer Planets
Unit 3
Jupiter

24512-27266

Problem Solving

Distances in the Solar System

The following table shows the distances in AU between the planets in our solar system and the sun. Study the distances carefully, then answer the questions below.

Planet	Distance from Sun
Mercury	0.38 AU
Venus	0.72 AU
Earth	1 AU
Mars	1.5 AU
Jupiter	5 AU
Saturn	9.5 AU
Uranus	19 AU
Neptune	30 AU
Pluto	39 AU

Solve the Problem:
Which planets do you think scientists know the most about? Explain your answer.

Think Critically:
Based on the distances shown in the table, how would you go about making a model of the solar system? What unit of measurement would you use to show the distances between the planets?

A Tour of the Solar System

Now you know how far you have to travel to tour the solar system, starting from your space station on the outer edge of the solar system. Strap yourself into your spacecraft. It's time to begin your journey. What will you see first as you enter the solar system?

Comets

What's this in **Figure 14-7**? A giant, dirty snowball? No, it's a comet—the first thing you see on your trip. Comets are made up of dust and frozen gases such as ice. From time to time, they swing close to the sun. When they do, the sun's radiation vaporizes some of the material. Gas and dust spurt from the comet, forming bright tails.

FIGURE 14-7

A comet's gas tail and dust tail are blown away from the sun by a combination of solar wind and the pressure of sunlight. Solar wind is a stream of charged particles from the sun.

Theme Connection

Scale and Structure

Compare and contrast the orbits of planets around the sun, multiple moons around the gaseous giants, and the movement of electrons around the nucleus of an atom. The scale changes but the structure remains basically the same.

Science Journal **Mythological Roots** Have students research and write about the mythological roots of the names Pluto and Charon in their Science Journals. In Greek mythology, Charon was the boatman who ferried dead souls across the river Styx to Hades. In Roman mythology, Pluto was the god of the underworld. [L3]

Outer Planets

Moving past the comets, you come to the outer planets. The outer planets are Pluto, Neptune, Uranus, Saturn, and Jupiter. Let's hope you aren't looking for places to stop and rest. Trying to stand on most of these planets would be like trying to stand on a cloud. That's because all of the outer planets except Pluto are huge balls of gas. Each may have a solid core, but none of them has a solid surface. The gas giants have lots of moons, which orbit the planets just like our own moon orbits Earth. They have outer rings made of dust and ice. In fact, the only outer planet that doesn't have rings is Pluto. Pluto isn't a gas giant. What does it look like? You'll soon find out.

Pluto

The first planet that you come to on your tour is Pluto, a small, rocky planet with a frozen crust. Pluto, the last planet discovered by scientists, is normally farthest from the sun. It is the smallest planet in the solar system, and the one we know the least about. Pluto, shown in **Figure 14-8A,** has no ring system. Its one moon, Charon, is nearly half the size of the planet itself.

Neptune

Neptune is the next stop in your space travel. Neptune, shown in **Figure 14-8B,** is the eighth planet from the sun most of the time. Sometimes, Pluto's orbit crosses inside Neptune's orbit during part of its voyage around the sun. When that happens, Neptune is the ninth planet from the sun. Neptune is the first of the big, gas planets with rings. Neptune's atmosphere is made of a gas called methane. Methane gives the planet a blue-green color.

Uranus

After Neptune, you come to the seventh planet from the sun, Uranus. Uranus needs a careful look because of the interesting way it spins on its axis.

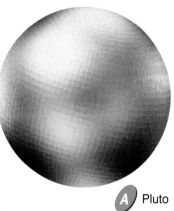

A) Pluto

FIGURE 14-8

The outer planets include Pluto and Neptune. Pluto is so small and far away that this is the best image current technology can produce.

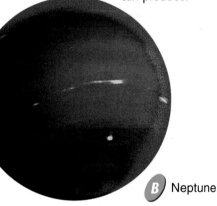

B) Neptune

14-2 The Solar System   **387**

Visual Learning

Figure 14-8 Why is Neptune normally the eighth planet from the sun, but sometimes the ninth? *Pluto's orbit sometimes crosses inside Neptune's orbit. When this happens, Neptune is the ninth planet from the sun.* LS

NATIONAL GEOGRAPHIC SOCIETY

Videodisc

STV: The Solar System
How the Solar System Began

00204-03596
The Nine Planets

03600-05294
Inner Planets
Unit 2
Mercury

11845-14069
Venus

14070-16477
Mars

18199-22050

GLENCOE TECHNOLOGY

 Videodisc

The Infinite Voyage: Sail On, *Voyager*
Introduction

Chapter 1
Preparation for the Grand Tour

Chapter 2
Technical Design and Capabilities

Chapter 3
Voyager 1: Jupiter

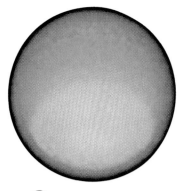

A Uranus

B Saturn

FIGURE 14-9

Uranus, Saturn, and Jupiter are gas giants. *In your Science Journal, list one unique characteristic about each of these planets.*

The axis of most planets, including Earth, is tilted just a little, somewhat like the hands of a clock when they are at 1 and 7. Uranus, shown in **Figure 14-9A,** has an axis that is tilted almost even with the plane of its orbit, as if the hands of the clock were at 3 and 9.

Saturn

You thought Uranus was unusual. Wait until you see Saturn, the sixth planet from the sun! You'll be dazzled by its rings, shown in **Figure 14-9B.** Saturn's several broad rings are made up of hundreds of smaller rings, which are made up of pieces of ice and rock. Some of these pieces are like specks of dust. Others are many meters across.

Jupiter

If you're looking for excitement, you'll find it on Jupiter, the largest planet in the solar system and the fifth from the sun. Watch out for a huge, red whirlwind rotating slowly around the middle of the planet. That's the Great Red Spot, a giant storm on Jupiter's surface. Jupiter, shown in **Figure 14-9C,** has 16 moons. Some are larger than Pluto! One of Jupiter's moons, Io, has more active volcanoes than anyplace else in the solar system.

Asteroid Belt

Look out for asteroids! On the next part of your trip, you must make your way through the asteroid belt that lies between Jupiter and the next planet, Mars. Asteroids are pieces of rock made of minerals similar to those which formed the planets. In fact, asteroids might have become planets if it weren't for that big giant, Jupiter. Jupiter's huge gravitational force probably kept any planets from forming in the area of the asteroid belt.

C Jupiter

Inner Planets

After traveling dozens of astronomical units, you finally reach the inner planets. These planets are solid and rocky. How do we know that? As with all the planets, much of what we know about planets comes from spacecraft that send data back to Earth to help us learn more about space. Look at **Figure 14-10A**. This photograph was taken by a spacecraft.

Mars

Hey! Has someone else been here? You see signs of earlier visits to Mars, the first of the inner planets. Tiny robot explorers have been left behind. But it wasn't a person who left them here. The roving robots were left by spacecraft sent from Earth to explore Mars's surface. If you stay long enough and look around, you may notice that Mars, shown in **Figure 14-10A,** has seasons and polar ice caps. There are signs that the planet once had liquid water. You'll also notice that the planet looks red. That's because the rocks on its surface contain iron oxide, which is what makes rust look red.

Earth

Home sweet home! You've finally reached Earth, the third planet from the sun. You didn't realize how unusual your home planet was until you saw the other planets. Earth's surface temperatures allow water to exist as a solid, a liquid, and a gas. Also, Earth's atmosphere works like a screen to keep ultraviolet (ul truh VI lut) rays from reaching the planet's surface. Ultraviolet rays are harmful rays from the sun. Because of Earth's atmosphere, life can thrive on the planet. You would like to linger on Earth, shown in **Figure 14-10B,** but you have two more planets to explore.

Venus

Maybe you should have stayed on Earth. You won't be able to see much at your next stop, shown in **Figure 14-10C.** Venus, the second-closest planet to the sun, is hard to see because its surface is surrounded by thick clouds.

FIGURE 14-10
Mars, Earth, and Venus are inner planets.

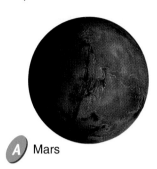

A Mars

B Earth

C Venus

14-2 The Solar System **389**

 Videodisc

STV: The Solar System
Outer Planets
Saturn

27379-29818

Uranus

29818-30810

Neptune

30833-34533

Pluto

34559-36600

3 Assess

Check for Understanding

Activity Have students make a Book of Planets listing various characteristics of the planets, such as size, number of moons, speed of rotation, and so on. **L2 IS**

Reteach

IS Intrapersonal Have students research why Venus is the most mapped planet in the solar system. Between 1990 and 1994, the *Magellan* space probe mapped almost the entire surface of Venus. **L2**

Extension

For students who have mastered this section, use the **Reinforcement** and **Enrichment** masters.

Content Background

The misconception that Mars has canals was actually the result of a linguistical error. An Italian astronomer, Schiaparelli, wrote of seeing "canali" on Mars, meaning "channels." This was written in English-speaking papers as "canals." There is an important difference. Channels can be made by the movement of any liquid. Canals, on the other hand, imply construction by intelligent beings, hence the beginning of the theories of life on Mars.

389

Activity 14-2

PREPARE

Purpose
 **Logical-Mathematical** Students will classify planetary surface conditions and relate these to the things humans need to survive in a space colony.

L2 **COOP LEARN** **P**

Process Skills
analyzing, comparing and contrasting, inferring, forming a hypothesis, designing an experiment

Time
three class periods: one for researching planets, one for designing and drawing space colonies, one for sharing designs and drawing conclusions

Possible Hypotheses
Space colonies will have to reflect the needs of humans for abundant amounts of oxygen, water, and energy. They will need to recycle each of these resources to have adequate supplies. Food will need to be produced and waste disposed of properly.

📁 **Activity Worksheets,** pages 5, 87-88

GLENCOE TECHNOLOGY

 Videodisc

Glencoe Earth Science Interactive Videodisc
Side 2, Lesson 8
Introduction: Space Exploration

31216-31963

Activity 14-2

Design Your Own Experiment
Space Colony

Have you ever seen a movie or read a book about astronauts from Earth living in space colonies on other planets? Some of these make-believe space colonies look awfully strange! So far, we haven't built a space colony on another planet. But if we did, what do you think it would look like?

Possible Materials
- drawing paper
- markers
- books about the planets

PREPARE

What You'll Investigate
How would conditions on a planet affect the type of space colony that might be built there?

Form a Hypothesis
Research a planet. Review conditions on the surface of the planet. **Make a hypothesis** about the things that would have to be included in a space colony to allow humans to survive on the planet.

Goals
Infer what a space colony might look like on another planet.

Classify planetary surface conditions.

Draw a space colony for a planet.

PLAN

1. **Select** a planet and **study** its surface conditions.
2. **Classify** the surface conditions in the following ways.
   a. solid or gas
   b. hot, cold, or changing temperatures
   c. heavy atmosphere or no atmosphere
   d. bright or dim sunlight
   e. special conditions unlike other planets

Planet: Venus   Sample list

Things Humans Need to Survive		
Air	Not available—must supply
Water	Not available—must supply
Food	Not available—must supply
Energy source	Can use sunlight

3. **List** the things that humans need to survive. For example, humans need air to breathe. Does your planet have air that humans can breathe, or would your space colony have to provide the air?
4. **Make a table** for the planet showing its surface conditions and the features the space colony would have to have so that humans could survive on the planet.
5. **Discuss** your decisions as a group to make sure they make sense.

DO

1. Make sure your teacher has approved your plan and your data table before you proceed.
2. **Draw** a picture of the space colony. **Draw** another picture showing the inside of the space colony. **Label** the parts of the space colony, and **explain** how they help humans to survive.
3. Present your drawing to the class. **Explain** the reasoning behind it.

CONCLUDE AND APPLY

1. **Compare and contrast** your space colony with those of other students who researched the same planet you did. How are they alike? How are they different?
2. Would you change your space colony after seeing other groups' drawings? If so, what changes would you make?
3. What was the most interesting thing you learned about the planet you studied?
4. **APPLY** Was your planet a good choice for a space colony? **Explain** your answer.

Assessment

Oral Have students present an oral explanation of the structure of their colonies. They should relate the structures to the unique conditions of the planet they studied. Use the Performance Task Assessment List for Oral Presentation in **PASC**, p. 71.

Go Further

LS **Intrapersonal** Have students research the *Mir* space station or *SkyLab* to see how the problems of human existence in an alien climate have been handled thus far.

Possible Procedures
Research the planets. Select one planet to include in the activity. Based on planetary conditions, determine what humans would need to survive in a space colony on the chosen planet. Draw a space colony for the planet.

Teaching Strategies
Brainstorm the needs for human habitation and possible solutions to the problems of meeting those needs.

Troubleshooting Keep the activity centered on analysis and evaluation. Otherwise, students may concentrate on the artistic aspects of the project rather than the scientific.

DO

Expected Outcome
Drawings will vary widely but should include strategies to deal with temperature control and availability of resources.

CONCLUDE AND APPLY

1. Answers will vary. Students may say that their drawings were either more detailed or less detailed than those of other groups.
2. Answers will vary. If no changes are proposed, students should explain why.
3. Answers will vary. Students should mention physical characteristics of the planet they studied, such as surface temperature or size.
4. Answers will vary. Students should support their explanation with facts, such as atmospheric conditions on the planet.

4 Close

Use the Mini Quiz to check students' recall of chapter content.

1. **Our solar system is made up of the planets, comets, and asteroids that orbit the _____.** *sun*

2. **Unlike the other outer planets, _____ is small and rocky.** *Pluto*

3. **The largest planet in the solar system is _____.** *Jupiter*

Section Wrap-up

1. Mercury, Venus, Earth, Mars, Jupiter, Saturn, Uranus, Neptune, and Pluto

2. The outer planets, with the exception of Pluto, are gaseous planets with rings and multiple moons. The inner planets are solid and rocky. All planets probably have a core of heavy metal. All rotate and revolve.

3. **Think Critically** Students will probably realize that scientists have developed small units of measurement to deal with the small distances between molecules.

USING MATH

Mercury, 3.8 cm; Venus, 7.2 cm; Earth, 10 cm; Mars, 15 cm; Jupiter, 50 cm; Saturn, 95 cm; Uranus, 190 cm; Neptune, 300 cm; Pluto, 390 cm **L3**

FIGURE 14-11
Mercury is the closest planet to the sun. Like our moon, its surface is scarred by meteorites.

Venus's clouds trap the energy that reaches the planet's surface from the sun. That energy causes surface temperatures on the planet to hover around 470°C. That's hot enough to bake a clay pot, and far too hot for you. You're on to your next stop.

Mercury

The last planet that you visit, and the one closest to the sun, is Mercury. Mercury, shown in **Figure 14-11,** is the second-smallest planet. Its surface is heavily scarred by craters made by meteorites. Meteorites are chunks of rock that fall from the sky when asteroids break up.

Section Wrap-up

1. List the nine planets in order from the sun, beginning with the planet closest to the sun.

2. In general, how are the outer planets different from the inner planets? How are they alike?

3. **Think Critically:** We use larger units of measure to deal with increasingly larger distances. How do you think scientists handle increasingly smaller distances, such as the distances between molecules?

4. **Skill Builder**
   Developing Multimedia Presentations
   Use your knowledge of the solar system to develop a multimedia presentation of the solar system. You may want to begin by drawing a poster that includes the sun, the planets, the asteroid belt, and comets. If you need help, refer to Developing Multimedia Presentations on page 564 in the **Technology Skill Handbook.**

USING MATH

Using the table in the Problem Solving on page 386, calculate the distances of the planets from the sun, using a scale of 10 cm = 1 AU.

Skill Builder
Developing Multimedia Presentations Student presentations should reflect unique characteristics of each object included in their displays. **P**

✓Assessment

Process The United States is hoping to launch a crewed mission to Mars in 2010. Have students research and write about the goals of the mission, as well as the obstacles NASA faces as it prepares for the mission. Use the Performance Task Assessment List for Writing in Science in **PASC,** p. 87. **L2** **P** **LS**

Barbary
by Vonda N. McIntyre

. . . But the stars were fantastic. Barbary thought she must be able to see a hundred times as many as on earth, even in the country where sky-glow and smog did not hide them. They spanned the universe, all colors, shining with a steady, cold, remote light. She wanted to write down what they looked like, but every phrase she could think of sounded silly and inadequate.

Science Journal

Have you ever dreamed of living on a space station? In your Journal, describe what you think life would be like at a remote outpost in space. What life-support systems would you need? How would you deal with weightlessness? What would you miss most about Earth?

Barbary is hoping to find a home for herself—and her mysterious stowaway—on the space station *Einstein.* On the way, Barbary learns from Jeanne Velory, the astronaut who is taking command of the station, that an alien ship is moving into the solar system. Some people think it may be abandoned; others aren't so sure.

The alien vessel is approaching on a path above the plane of the solar system. *Einstein,* circling Earth in a long orbit, is the ideal spot from which to observe, and hopefully contact, the mysterious ship.

Barbary explores the research station with her new sister, Heather. She learns how to "sly"—move gracefully in zero gravity—pilot a space raft, and do her homework with a very helpful computer. Along the way, she also learns to trust others and her own abilities.

Then without warning, Barbary finds herself using all her newfound skills to save her friends and discover the truth about the alien vessel . . . and its inhabitants.

14-2 The Solar System   **393**

Source
McIntyre, Vonda N. *Barbary.* Houghton Mifflin Company: Boston, 1986.

Background
The space station in *Barbary* is considerably more advanced than the Russian *Mir* space station that has been continually crewed and orbiting Earth since 1986. The sprawling *Mir* complex began with a single core module, but over the years a half dozen additional modules have been added. *Mir* has been the site of some of the world's most advanced space technologies and home to dozens of cosmonauts from many countries. The name *Mir* means "peace" or "world" in Russian.

Teaching Strategies
- Have students research the history of the *Mir* space station and write a report about it. L2 P
- If Internet access is available, encourage students to explore the web for information on space and spacecraft. A good place to start is NASA's Spacelink site. Spacelink will put students in touch with a wealth of information and images, and even allows them to take a virtual tour of the universe. L2

For More Information
Berliner, Don. *Living in Space.* Lerner Publications Co.: Minneapolis, 1993.

Kallen, Stuart A. *Space Stations (Giant Leaps).* Abdo & Daughters: Edina, 1996.

Kettelkamp, Larry. *Living in Space.* William F. Morrow & Company: New York, 1993.

14•3 Space Exploration: Boom or Bust?

What YOU'LL LEARN

- The costs and benefits of space exploration
- How to recognize bias in the things you read

Science Words:
satellite

Why IT'S IMPORTANT

You'll learn to judge the accuracy of what you read.

The Value of Space Exploration

As you sit at your desk reading about the solar system, space satellites and space probes are orbiting our planet or rocketing toward the edge of the solar system. Space **satellites** and space probes are spacecraft that are launched into space to gather information and send it back to Earth. **Figures 14-12** and **14-13** show examples of space technology. These and other kinds of space technology cost our country billions of dollars.

Has the money been well spent? That depends on who you ask. Some people think that the knowledge gained from space exploration has been well worth the cost. They point out that space technology is now used to study everything from the weather to the environment.

Others, though, think that the money would have been better spent elsewhere, and that the risk to human lives—12 astronauts have died—outweighs the benefits of space exploration.

Read the following passages about the space program. The people are fictional, which means they don't really exist. As you read, look for biases in the people's statements. Biases are viewpoints that influence the way people think. Recognizing bias will help you judge the accuracy of what you read.

Content Background

In addition to space satellites, other types of technology developed by the National Aeronautics and Space Administration (NASA) for space exploration are now being used on Earth. These technologies, called spin-offs, range from sunglasses that adjust to various light levels to the materials used in athletic shoes, bicycle seats, and blankets. Lightweight breathing systems and fire-resistant uniforms originally designed for astronauts are now used by firefighters.

FIGURE 14-12

The space shuttle is a reusable spacecraft that transports astronauts, satellites, and other materials to and from space. *Describe some advantages of using a spacecraft designed to make many trips into space.*

Program Resources

Reproducible Masters

Enrichment, p. 53 L3
Reinforcement, p. 53 L2
Science and Society/Technology
 Integration, p. 32 L2
Study Guide, p. 53 L1

Transparencies

Science Integration Transparency 14 L2
Section Focus Transparency 47 L2

Long Live the Space Program!

"The space program has been the best program this country has ever had. The technology developed for space is now used for many things on Earth. Space satellites are used to track weather patterns and to study the environment. Without the space program, computers would never have been developed."
—Maria Lopez, astronaut

Cut the Space Program!

"Since its beginning, the space program has been nothing but a drain on our country's resources. The billions of dollars spent on this program should have been used to build better roads, improve schools, and feed the hungry. The space program has also taken the lives of several astronauts. It has cost this country billions of dollars."
—Cholena Jones, highway worker

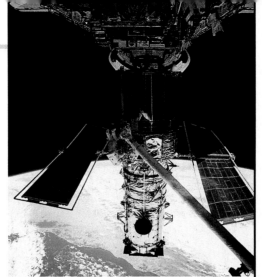

FIGURE 14-13

As part of space shuttle *Discovery's* mission, astronauts repaired the orbiting Hubble Space Telescope.

 Skill Builder: Recognizing Bias

LEARNING the SKILL

1. To recognize bias, you need to identify the speakers of the statements and examine their views and possible reasons for saying what they did. Determine how their biases influence the way they view an issue.

2. Sometimes biases are stated as facts, but in reality, they are opinions. Look for words that reflect an opinion, such as *never, best, nothing,* or *should.*

3. Examine the statements to see whether they are balanced. Do they mention other viewpoints?

4. Use your knowledge of the subject or research the subject to identify statements of facts.

PRACTICING the SKILL

1. Based on the material you read in the passages, list three facts about space exploration.

2. Identify the speakers of the statements. What might be some possible reasons for their opposing viewpoints?

3. Identify two facts and two biases in each of the speaker's statements.

APPLYING the SKILL

Look through the letters to the editor in your local newspaper. Write a report analyzing one of the letters for evidence of bias.

Visual Learning

Figure 14-12 Describe some advantages of using a spacecraft designed to make many trips into space. *A reusable spacecraft saves money and resources because it can be used again and again.* **LS**

Example of bias from Lopez: Without the space program, computers would not have been developed.

Example of fact from Jones: The space program has cost billions of dollars.

Examples of bias from Jones: The space program has been a drain on our resources.

Teaching the Content

Have students list costs and benefits of the space program. Encourage students to list items that are both directly and indirectly related to the program. Examples of direct costs and benefits include the cost of technology versus the knowledge gained by the technology. Examples of indirect costs and benefits include manufacturing jobs lost when human workers are replaced with robots developed using spin-off technology versus the benefits of spin-off technology. **L2**

Teaching the Skill

Give students examples to help them better understand the difference between an opinion and a fact. For instance, the statement "our soccer team is the best" is an opinion. The statement "our soccer team has gone undefeated this year" is a fact.

Sources NASA's Mission HOME project provides information about spin-off technology.

Answers to Practicing the Skill

1. Example facts: space satellites are used to track weather patterns and to study the environment; the technology developed for space is now used for many things on Earth.

2. Astronaut Maria Lopez wants the program to continue because her job depends on the space program; highway worker Cholena Jones may feel that the money spent on the space program should be spent on building better roads, which would increase her job security.

3. *Example of fact from Lopez:* Space technology is used for many things on Earth.

395

Prepare

Section Background

A black hole is extremely dense; its gravitational field is so strong that light cannot escape it. Evidence of black holes can be gathered by studying the emission rays of objects just before they are sucked into the outer edge of the black hole, an area called the "event horizon."

Preplanning

Refer to the Chapter Organizer on pages 376A-B.

1 Motivate

Bellringer

Before presenting the lesson, display **Section Focus Transparency 48** on the overhead projector. Assign the accompanying **Focus Activity** worksheet.

L2 ELL

Tying to Previous Knowledge

Have students compare and contrast electrons and solar systems. Both revolve.

14•4 Stars and Galaxies

What YOU'LL LEARN

• How a star is born
• About the galaxies that make up our universe

Science Words:
constellation
galaxy

Why IT'S IMPORTANT

Understanding the vastness of the universe helps us to understand Earth's place in space.

FIGURE 14-14

Find the Big Dipper in the constellation Ursa Major. *Why do you think people call it the Big Dipper?*

Stars

Every night, a whole new world opens to us. The stars come out. The fact is, stars are always in the sky. We just can't see them during the day because the sun's light is brighter than starlight. The sun is a star, too. It is the closest star to Earth. We can't see it at night because as Earth rotates, our part of Earth is facing away from it.

Constellations

Ursa Major, Orion, Taurus. Do these names sound familiar? They are **constellations** (kahn stuh LAY shunz), or groups of stars that form patterns in the sky. **Figure 14-14** shows some constellations.

Constellations are named after animals, objects, and people—real or imaginary. We still use many names that early Greek astronomers gave the constellations. But throughout history, different groups of people have seen different things in the constellations. In early England, people thought the Big Dipper, found in the constellation Ursa Major, looked like a plow. Native Americans saw it as a giant bear. To the Chinese, it looked like a governmental official and his helpers moving on a cloud. What does the Big Dipper look like to you?

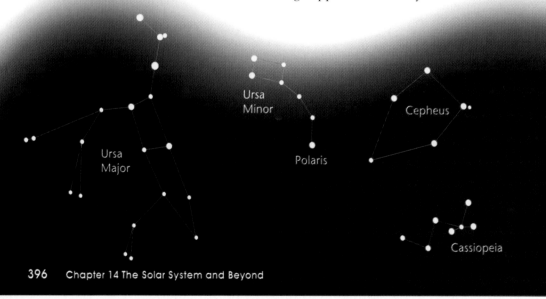

396   Chapter 14 The Solar System and Beyond

Program Resources

Reproducible Masters
Activity Worksheets, pp. 5, 90 L2
Enrichment, p. 54 L3
Multicultural Connections, pp. 31-32 L2
Reinforcement, p. 54 L2
Study Guide, p. 54 L1

Transparencies
Section Focus Transparency 48 L2

Starry Colors

When you glance at the sky on a clear night, the stars look like tiny pinpoints of light. It's hard to tell one from another, but stars come in different sizes and colors.

The larger a star is, the cooler it tends to be. How do you measure a star's temperature? You can't go there with a big thermometer. But you can use a star's color as a clue to its temperature. Red stars are the coolest. Yellow stars are of medium temperature. Bluish-white stars are the hottest. Our sun is a yellow, medium-sized star. The giant red star called Betelgeuse (BEE tul joos) is much bigger than the sun. If this huge star were in the same place as our sun, it would swallow Mercury, Venus, Earth, and Mars.

The Lives of Stars

You've grown up and changed a lot since you were born. You've gone through several stages in your life, and you'll go through many more. Stars go through stages in their lives, just as people do.

Scientists theorize that stars begin their lives as huge clouds of gas and dust. The force of gravity causes the dust and gases to move closer together. When this happens, temperatures within the cloud begin to rise. A star is formed when this cloud gets so dense and hot that it starts producing energy.

The stages a star goes through in its life depend on the star's size. When a medium-sized star like our sun uses up some of the gases in its center, it expands to become a giant. Our sun will become a giant in about 5 billion years. When the remaining gases are used up, our sun will contract to become a black dwarf.

Mini LAB

Modeling Constellations

1. Draw a dot pattern of a constellation on a piece of black construction paper. Choose a constellation from **Figure 14-14** or make your own pattern of stars.
2. With your teacher's help, cut off the end of a round, empty box. You now have a cylinder with both ends open.
3. Place the box over the constellation. Trace around the rim of the box. Cut the paper along the traced circle.
4. Tape the paper to the end of the box. Using a pencil, carefully poke holes through the dots on the paper.
5. Place a flashlight inside the open end of the box. Darken the room and observe your constellation on the ceiling.

Analysis

1. Turn on the overhead light and view your constellation again. Can you still see it? Why or why not?
2. The stars are always in the sky, even during the day. Knowing this, explain how the overhead light is similar to our sun.

Visual Learning

Figure 14-14 Why do you think people call it the Big Dipper? *Students will likely note that the stars are arranged in the shape of a dipper.* LS

2 Teach

Demonstration

LS **Visual-Spatial** Demonstrate the color and temperature of stars with a burning candle. The flame of a well-burning candle will burn red, yellow, and blue. The coolest temperatures are red, and the hottest are blue.

Mini LAB

Purpose

LS **Kinesthetic** Students will model constellations. L2

Materials
black construction paper; round, empty cornmeal or oats box; scissors; flashlight; pencil; tape

Teaching Strategies
Cut off the ends of the boxes for students.

Activity Worksheets, pages 5, 90

Analysis

1. No. The overhead light is brighter than the flashlight.
2. As the overhead light is brighter than the flashlight, our sun's light is brighter than starlight.

Assessment

Content Remind students that constellations are named after animals, objects, and people—real or imaginary. Have students make a cartoon about their constellation. The cartoon should illustrate what their constellation represents. Use the Performance Task Assessment List for Cartoon/Comic Book in PASC, p. 61. P

397

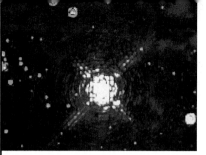

FIGURE 14-15
This photo, taken by the Hubble Space Telescope, shows a massive star in the Pistol Nebula.

Supergiants

When a huge star like the one shown in **Figure 14-15** begins to use up the gases in its core, it becomes a supergiant. Over time, the core of a supergiant collapses. Then the outer part of the star explodes and gets very bright. This is called a supernova. For a few brief days, the supernova might shine more brightly than a whole galaxy. The dust and gases released by this explosion may eventually form other stars.

Meanwhile, the core of the supergiant is still around. It is now called a neutron star. If the core is massive enough, it may rapidly become a black hole. Black holes are so dense that light shone into them would disappear.

USING TECHNOLOGY
Space Probes

Imagine what it would be like to explore a galaxy far away. It will be a long time before we can send astronauts millions of light-years into space. But we've already sent space probes to the outer limits of the solar system—some were launched 20 years ago and are still sending back data.

Cassini is one such space probe. It's about the size of a school bus. *Cassini* was launched in October 1997. Its mission is to study the planet Saturn, including Saturn's rings and moons. *Cassini* will reach Saturn in the year 2004. It will orbit the planet for four years, sending back data.

Afterward, if all systems are still working, *Cassini* might explore other planets. And in the future, who knows? Scientists may be able to design a space probe that can "boldly go where no space probe has gone before: to a galaxy far, far away."

*inter*NET CONNECTION

Go to the Glencoe Homepage at *www.glencoe.com/sec/science* to learn more about space probes.

398

Community Connection

Astronomer Invite a local professional or amateur astronomer to class. Have the person discuss his or her knowledge of astronomy and any astronomical events he or she may have witnessed.

Inclusion Strategies

Learning Disabled Show students the constellation Orion. Tell them that this constellation gets its name from a famous Greek warrior. Have them draw the figure of a warrior superhero over the constellation. **L1** **ELL**

Galaxies and the Universe

What do you see when you look at the night sky? If you live in a city, you may not see much. The glare from city lights makes it hard to see the stars. If you go to a dark place, far from the lights of towns and cities, you can see much more. In a dark area, with a powerful telescope, you might see dim clusters of stars grouped together. These clusters are galaxies (GAL uk seez). A **galaxy** is a group of stars, gas, and dust held together by gravity.

Light-Years

Do you remember what you learned earlier about astronomical units or AU? Distances between the planets are measured in AU. But to measure distances between galaxies, we need an even bigger unit of measure. Scientists use light-years to measure distances between galaxies. A light-year is the distance light travels in a year—about 9.5 trillion km. Light travels so fast it could go around Earth seven times in one second!

Have you ever wished that you could travel back in time? In a way, that's what you're doing when you look at a galaxy. The galaxy might be millions of light-years away. So the light that you see started on its journey long ago. You are seeing the galaxy as it was millions of years ago.

The Milky Way Galaxy

What's your address? You have a house number, a street, a city, a state, zip code, and country. Did you know that you have another address? Look at **Figure 14-16.** We all live in the Milky Way Galaxy. About 200 billion stars are in the Milky Way Galaxy, including our sun. Just as Earth revolves around the sun, stars revolve around the centers of galaxies. Our sun revolves around the Milky Way Galaxy about once every 240 million years.

FIGURE 14-16

The Milky Way Galaxy has spiral arms made up of dust and gas. Its inner region is an area of densely packed stars.

*inter*NET
CONNECTION

Explore the Glencoe Homepage at *www.glencoe.com/sec/science* to learn how telescopes have changed our view of space.

14-4 Stars and Galaxies   **399**

*inter*NET
CONNECTION

The Glencoe Homepage at **www.glencoe.com/sec/science** provides links to relevant, up-to-date Internet sites.

Discussion

Astronomers call the proliferation of light around populated areas light pollution. Ask students why they think astronomers use the word *pollution* to describe light. Can they think of a solution to this problem? For example, hooded lights are often used in parking lots and streetlights.

3 Assess

Check for Understanding

Activity Put students in groups of four. Have each group take a turn stating one fact about stars. Have the groups continue taking turns until they have covered what they have learned about stars. L2
COOP LEARN IS

Reteach

Use colored sticky dots to represent stars of different sizes and colors. Have students place them on a piece of black paper to represent a constellation. L1

Extension

For students who have mastered this section, use the **Reinforcement** and **Enrichment** masters.

Content Background

Galaxies, like stars, are grouped into clusters. The Milky Way belongs to a cluster called the Local Group. It is made up of approximately 30 galaxies of various types and sizes.

Content Background

There are three classes of galaxies: elliptical, spiral, and irregular. The most common type may be elliptical, which are shaped like footballs or spheres. Spiral galaxies look like pinwheels. Irregular galaxies don't fit into either classification. The Milky Way is a spiral galaxy. Our solar system is located on the inner edge of one arm of the spiral.

4 Close

•MINI•QUIZ•

Use the Mini Quiz to check students' recall of chapter content.

1. **The hottest stars are _____ in color.** *blue*
2. **A(n) _____ is a group of stars, gas, and dust held together by gravity.** *galaxy*
3. **We live in the _____ Galaxy.** *Milky Way*

Section Wrap-up

1. A constellation is a group of stars that form a pattern.
2. A medium-sized star such as our sun will use up the gases in its center, expanding to become a giant. When the remaining gases are used up, it contracts to become a black dwarf.
3. **Think Critically** Some stars are so far away that their light takes a long time to reach Earth. It is possible for us to be seeing the light from stars that have already burned out.

Science Journal Answers will vary, but students should explain why they selected a particular name for their constellation.

FIGURE 14-17
Stars are forming in the Orion Nebula.

A View from Within

We can see part of the Milky Way Galaxy as a band of light across the sky. But we can't see the whole Milky Way Galaxy. Why not? Think about it. When you're sitting in your classroom, can you see the whole school? No, you are inside the school and can see only parts of it. Our view of the Milky Way Galaxy from Earth is like the view of your school from a classroom. We can see only parts of our galaxy because we are inside it.

The Universe

Each galaxy probably has as many stars as the Milky Way Galaxy. Some may have more. And there may be as many as 100 billion galaxies. All these galaxies, with all their countless stars, make up the universe. Look at **Figure 14-17.** In this great vastness of revolving solar systems, exploding supernovae, and star-filled galaxies is one small planet called Earth. If you think about how huge the universe is, Earth seems like a speck of dust. Yet as far as we know, it's the only place where life exists.

Section Wrap-up

1. What is a constellation? Name three constellations.

2. Describe the life of a medium-sized star such as the sun.

3. **Think Critically:** Some stars may no longer be in existence, but we still see them in the night sky. Why?

4. **Skill Builder**
   Formulating Models The Milky Way Galaxy is 100 000 light-years in diameter. How would you build a model of the Milky Way Galaxy? If you need help, refer to Making Models on page 557 in the **Skill Handbook.**

Science Journal

Observe the stars in the night sky. In your Science Journal, draw the stars you observed. Now draw your own constellation based on those stars. Give your constellation a name. Why did you choose that name?

Skill Builder
Formulating Models Answers will vary. A scale of 1 cm = 1000 light-years is one possibility.

Assessment

Performance Have students perform a skit explaining the relationships among stars, solar systems, galaxies, and the universe. One student may portray the universe, while other students act as galaxies, and still others as solar systems and stars. Students should research galaxies and solar systems before performing their skits. Use the Performance Task Assessment List for Skit in **PASC**, p. 75.

L2 COOP LEARN IS

Read the statements below that review the most important ideas in the chapter. Using what you have learned, answer each question in your Science Journal.

1. The light-year and the astronomical unit or AU, are used to measure distances in space. *Why do we need special units of measurement for studying space?*

2. Seasons are created by the tilt of Earth's axis. *Explain why we have our warmest temperatures in summer.*

3. The inner planets and outer planets that make up our solar system revolve around the sun. *How is Pluto different from the other outer planets?*

4. The colors of stars tell us a lot about their temperature. *What color are the hottest stars? The coolest?*

Chapter 14 Review   **401**

Have students look at the illustrations on this page. Ask them to describe details that support the main ideas of the chapter found in the statement for each illustration.

Teaching Strategies

Divide students into groups of four. Have each group create a poster that illustrates a main idea of this chapter. **L2**
COOP LEARN

Answers to Questions

1. Distances in space are so vast.
2. Earth is tilted toward the sun and receives more direct (and therefore more intense) sunlight.
3. Pluto is a small, rocky planet, not a gaseous giant.
4. The hottest stars are blue; the coolest are red.

Science at Home

LS **Intrapersonal** Encourage students to keep a "moon diary." They should observe and sketch the moon for a month. Have them record when they made their observations and general weather conditions at the time observations were made. **L2**

✓ **Assessment**

Portfolio Encourage students to place in their portfolios one or two items of what they consider to be their best work. Examples include:
• Explore activity, p. 377
• Activity 14-1, p. 381
• MiniLab, p. 397

Performance Additional performance assessments may be found in **Performance Assessment** and **Science Integration Activities.** Performance Task Assessment Lists and rubrics for evaluating these activities can be found in Glencoe's **Performance Assessment in the Science Classroom** (PASC).

Using Key Science Words

1. eclipse
2. rotation
3. galaxy
4. constellation
5. satellite

Checking Concepts

6. d
7. b
8. b
9. c
10. d

Thinking Critically

11. advantages: technological innovation, employment, increased scientific knowledge; disadvantages: costs, danger to human life
12. Earth's surface temperatures allow water to exist as a solid, a liquid, and a gas. Also, Earth's atmosphere keeps ultraviolet rays from reaching the planet's surface.
13. Answers will vary. Students may say that Mars is most like Earth because of its temperature and atmosphere. Jupiter may be students' choice as most different. It is a gaseous giant, has no firm surface, and has multiple moons.
14. Eclipses are possible only when the moon is crossing Earth's orbital plane. By plotting the orbits of Earth and the moon, scientists can predict the time and day of a solar eclipse.
15. The constellations are only rough patterns of stars. They can be imagined to be almost anything, so different people would create animals and images that were familiar to them.

Chapter 14 Review

Using Key Science Words

constellation   rotation
eclipse   satellite
galaxy   solar system
revolution

Match each phrase with the correct term from the list of Key Science Words.

1. the shadow produced by the moon or Earth passing in front of the sun
2. the motion of Earth that produces day and night
3. a group of stars, gas, and dust held together by gravity
4. a group of stars that forms a pattern in the sky
5. an object that is launched into space to gather information and send it back to Earth

Checking Concepts

Choose the word or phrase that completes the sentence.

6. The tilt of Earth's _____ causes seasons.
   a. equator   c. moon
   b. oceans   d. axis
7. When the moon is waning, it appears to be _____.
   a. growing larger
   b. growing smaller
   c. a full moon
   d. a new moon

8. An astronomical unit is the distance from _____.
   a. Earth to the moon
   b. Earth to the sun
   c. Pluto to Mercury
   d. Pluto to the sun
9. Earth is the _____ planet from the sun.
   a. first   c. third
   b. second   d. fourth
10. There may be as many as _____ galaxies in the universe.
    a. 1 billion   c. 50 billion
    b. 10 billion   d. 100 billion

Thinking Critically

Answer the following questions in your Science Journal using complete sentences.

11. Describe some advantages and disadvantages of the U.S. space program.
12. What conditions on Earth allow life to thrive?
13. Which of the planets in our solar system seems most like Earth? Which seems most different? Explain your answer, using facts that you've learned about the planets.
14. How might a scientist predict the day and time of a solar eclipse?
15. Throughout history, different groups of people have viewed the constellations in different ways. Infer why this is true.

Assessment Resources

Reproducible Masters

Chapter Review, pp. 31-32
Assessment, pp. 61-64
Performance Assessment, p. 52

Glencoe Technology

■ **Computer Test Bank**

▭ **MindJogger Videoquiz**

Chapter 14 Skills Review

Developing Skills

If you need help, refer to the description of each skill in the Skill Handbook.

16. **Making and Using Tables:** Research the size, period of rotation, and period of revolution for each planet. Show this information in a table. How do tables help us to better understand information?

17. **Comparing and Contrasting:** Compare the inner planets with the outer planets. How are they alike? How are they different?

18. **Making a Model:** Based on what you have learned about the sun, the moon, and Earth, make a model of a lunar or a solar eclipse.

19. **Sequencing:** Sequence the following terms in order of smallest object to largest group: galaxy, inner planets, solar system, universe, Earth.

20. **Using a CD-ROM:** Choose one of the nine planets and, using a CD-ROM, research information about that planet. Write a report based on your research. In your report, describe the articles, photographs, pictures, or videos found in the CD-ROM.

Performance Assessment

1. **Model:** Based on the information provided in the Problem Solving table on page 386, work as a class to make a three-dimensional model of the solar system. Use the scale you developed from the Using Math activity on page 392 to accurately show the distances between the planets and the sun.

2. **Newspaper Article:** The photo below shows the remains of a supernova explosion. Write a newspaper article describing the life cycle of a star that becomes a supernova. Explain what will happen to the star afterwards.

Developing Skills

16. **Making and Using Tables** See sample table below. Periods of rotation and revolution are given in sidereal (Earth) equivalents.

17. **Comparing and Contrasting** The inner planets are rock; the outer ones (excepting Pluto) are gas. The inner planets are much smaller than the gaseous giants. The outer planets (again, excepting Pluto) have multiple moons.

18. **Making a Model** The lunar eclipse will place Earth between the sun and moon, casting a shadow over the moon. A solar eclipse will show the moon between Earth and the sun, casting a small shadow on one part of Earth.

19. **Sequencing** Earth, inner planets, solar system, galaxy, universe

20. **Using a CD-ROM** Answers will vary.

Performance Assessment

1. The scale of the model should follow the scale developed in the Using Math activity on page 392. Students can use string and paper to create their models. The planets can be drawn on paper, then hung along the appropriate positions on the string. Use the Performance Task Assessment List for Model in **PASC**, p. 51. **P**

2. Articles will vary. Use the Performance Task Assessment List for Newspaper Article in **PASC**, p. 69. **P**

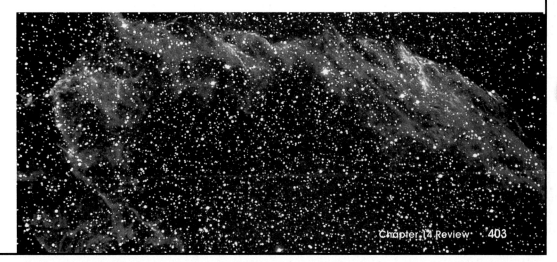

Chapter 14 Review 403

Solar System Information (Question 16)									
Planet	Mercury	Venus	Earth	Mars	Jupiter	Saturn	Uranus	Neptune	Pluto
Diameter (km)	4878	12 104	12 756	6794	142 796	120 660	51 118	49 528	2290
Period of rotation	59 d	243 d	24 h	24.5 h	10 h	10.4 h	11 h	16 h	7 d
Period of revolution	87.97 d	224.70 d	365.26 d	686.98 d	11.86 y	29.46 y	84.04 y	164.79 y	248.53 y
Number of known satellites	0	0	1	2	16	18	15	8	1
Known rings	0	0	0	0	1	thousands	11	4	0

Chapter Organizer

Section	Objectives/Standards	Activities/Features
Chapter Opener		Explore Activity: Observe a Rock, p. 405
15-1 **Minerals: Earth's Jewels** (4 sessions, 2 blocks)*	1. **Compare** and **contrast** minerals and rocks. 2. **Identify** minerals using physical properties. National Science Content Standards: (5-8) UCP1, A1, A2, B1, D1, C4, E2, F2, F5, G1	MiniLAB: Classifying Minerals, p. 411 Problem Solving: Identifying Mineral Origin, p. 412 Using Technology: Controlling the Waste from Mining Operations, p. 413 Skill Builder: Using a CD-ROM, p. 414 Science Journal, p. 414
15-2 **Igneous and Sedimentary Rocks** (2½ sessions, 1½ blocks)*	3. **Compare** and **contrast** extrusive and intrusive igneous rocks. 4. **Investigate** how different types of sedimentary rocks form. National Science Content Standards: (5-8) UCP3, A1, A2, B1, D1	Activity 15-1: Cool Crystals Versus Hot Crystals, pp. 418-419 MiniLAB: How do fossils form rocks?, p. 422 Skill Builder: Sequencing, p. 423 Science Journal, p. 423
15-3 **Science and Society: Monument or Energy?** (1 session, ½ block)*	5. **Investigate** Earth processes that produce landscapes and resources. 6. **Predict** the consequences of changes in land use on the economy of an area. National Science Content Standards: (5-8) UCP3, B1, C4, F2	Skill Builder: Predicting Consequences, p. 425
15-4 **Metamorphic Rocks and the Rock Cycle** (2½ sessions, 1 block)*	7. **Describe** the conditions needed to form metamorphic rocks. 8. **Analyze** how rocks change in the rock cycle. National Science Content Standards: (5-8) UCP1, UCP2, UCP3, A1, B1, G1	Activity 15-2: Gneiss Rice, p. 429 Skill Builder: Observing, p. 431 Using Computers, p. 431 People & Science, p. 432

* A complete Planning Guide that includes block scheduling is provided on pages 31T-33T.

Activity Materials

Explore	Activities	MiniLABs
page 405 basalt sample, hand lens, granite sample	pages 418-419 hot plate, jar, measuring cup, paper clip, igneous rocks, saucepan, wooden spoon, string, sugar page 429 rolling pin, uncooked rice, modeling clay, granite sample, gneiss sample	page 411 calcite sample, hornblende sample, jar, magnetite, quartz sample, plastic sealable bag, white vinegar, iron filings page 422 limestone with fossils, white glue, pie pan, macaroni

Need Materials? Call Science Kit (1-800-828-7777).

Teacher Classroom Resources

Reproducible Masters	Transparencies	Teaching Resources
Activity Worksheets, pp. 5, 95 **Cross-Curricular Integration**, p. 19 **Enrichment**, p. 55 **Lab Manual 29** **Multicultural Connections**, pp. 33-34 **Reinforcement**, p. 55 **Science and Society/Technology Integration**, p. 33 **Study Guide**, p. 55	**Science Integration Transparency 15**, A Closer Look at Minerals **Section Focus Transparency 49**, Earth's Treasures **Teaching Transparency 29**, Mohs Hardness Scale	**Spanish Resources** **English/Spanish Audiocassettes** **Cooperative Learning Resource Guide** **Lab Partner** **Lab and Safety Skills** **Lesson Plans**
Activity Worksheets, pp. 5, 91-92, 96 **Enrichment**, p. 56 **Lab Manual 30** **Reinforcement**, p. 56 **Study Guide**, p. 56	**Section Focus Transparency 50**, A Trip Through Time	
Enrichment, p. 57 **Reinforcement**, p. 57 **Science Integration Activities**, pp. 67-68 **Study Guide**, p. 57	**Section Focus Transparency 51**, Drilling with Diamonds	**Assessment Resources** **Chapter Review**, pp. 33-34 **Assessment**, pp. 65-68 **Performance Assessment**, p. 53 **Performance Assessment in the Science Classroom (PASC)** **MindJogger Videoquiz** **Alternate Assessment in the Science Classroom** **Computer Test Bank**
Activity Worksheets, pp. 5, 93-94 **Enrichment**, p. 58 **Reinforcement**, p. 58 **Study Guide**, p. 58	**Section Focus Transparency 52**, "Changed" Rock **Teaching Transparency 30**, Rock Cycle	

Key to Teaching Strategies

The following designations will help you decide which activities are appropriate for your students.

L1 Level 1 activities should be appropriate for students with learning difficulties.

L2 Level 2 activities should be within the ability range of all students.

L3 Level 3 activities are designed for above-average students.

ELL ELL activities should be within the ability range of English Language Learners.

LS These activities are designed to address different learning styles.

COOP LEARN Cooperative Learning activities are designed for small group work.

P These strategies represent student products that can be placed into a best-work portfolio.

GLENCOE TECHNOLOGY

The following multimedia resources are available from Glencoe.

The Infinite Voyage Series
The Future of the Past

Glencoe Earth Science Interactive Videodisc
Igneous Processes
Metamorphic Processes

Glencoe Earth Science CD-ROM

Teacher Classroom Resources

This is a representation of key blackline masters available in the Teacher Classroom Resources.

Teaching Aids

Section Focus Transparencies

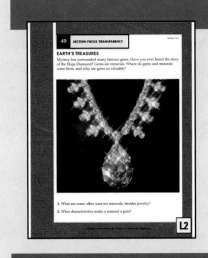

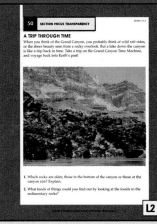

Science Integration Transparencies

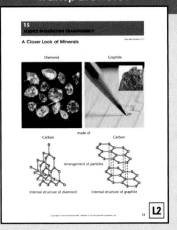

Teaching Transparencies

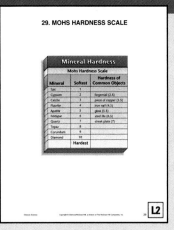

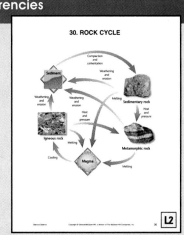

Meeting Different Ability Levels

Study Guide for Content Mastery

Reinforcement

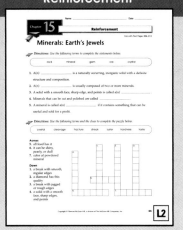

Enrichment Worksheets

Hands-On Activities

Science Integration Activities

Lab Manual

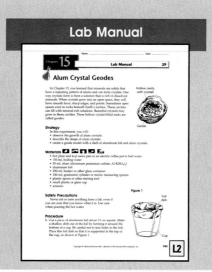

Activity Worksheets

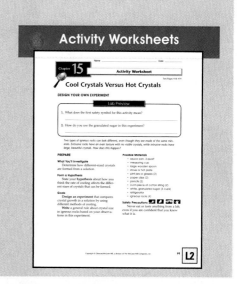

Enrichment and Application

Cross-Curricular Integration

Multicultural Connections

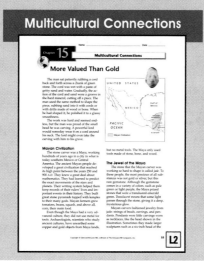

Science and Society/Technology Integration

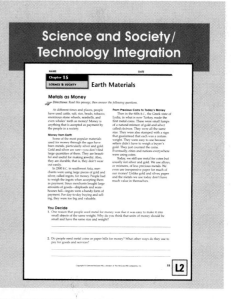

Assessment

Performance Assessment

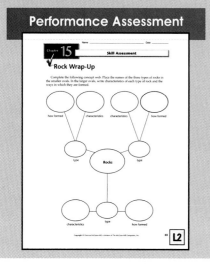

Chapter Review

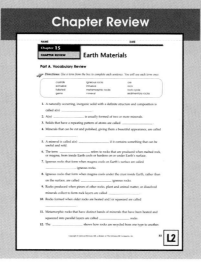

Assessment

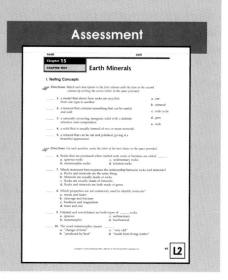

Chapter 15

Earth Materials

CHAPTER OVERVIEW

Section 15-1 This section describes minerals—how they form and how their properties are used to identify them.

Section 15-2 Igneous and sedimentary rocks are described. The classification of these two types of rocks is also presented.

Section 15-3 Science and Society This section compares the importance of the Grand Staircase-Escalante National Monument to the economy of Utah as a tourist attraction or as a reserve of fossil fuels. The skill of predicting consequences is also discussed.

Section 15-4 Metamorphic rocks and how they are classified are described. An overview of the rock cycle is presented.

Chapter Vocabulary

mineral	sedimentary
rock	rock
crystal	fossil fuel
gem	metamorphic
ore	rock
igneous rock	foliated
extrusive	non-foliated
intrusive	rock cycle

Theme Connection

Scale and Structure/Stability and Change The theme of Scale and Structure is presented, showing how the atomic arrangement inside atoms and other properties of minerals are related. The theme of Stability and Change is presented as students realize that rocks change from one kind to another through the rock cycle.

Chapter Preview

Skills Preview

▶ **Skill Builders**
- use a CD-ROM
- sequence

▶ **MiniLABs**
- classify
- infer

▶ **Activities**
- observe
- compare
- hypothesize
- collect data

404

Learning Styles

Look for the following logo for strategies that emphasize different learning modalities. **LS**

Kinesthetic	Explore, p. 405; Inclusion Strategies, p. 407; MiniLAB, p. 411; Activity 15-1, pp. 418-419; Check for Understanding, p. 421; Activity 15-2, p. 429
Visual-Spatial	Demonstration, pp. 408, 427; Reteach, p. 430; Visual Learning, pp. 420, 428
Interpersonal	Activity, p. 407; Assessment, p. 414; Brainstorming, p. 416
Intrapersonal	Across the Curriculum, p. 421; Science at Home, p. 433
Logical-Mathematical	Problem Solving, p. 412; Inquiry Question, p. 416; Using an Analogy, p. 417; Go Further, p. 419; Reteach, p. 421; MiniLAB, p. 422
Linguistic	Using Science Words, pp. 407, 428; Science Journal, p. 417
Naturalist	Community Connection, p. 427

Chapter 15

Earth Materials

You're at the top! You and a friend have been climbing Zane's Bluff. Now, after all the hard work of the climb, you are treated to a great view. You also take time to get a closer look at the rock you've been climbing. First, you notice that it sparkles in the sun because of the silvery specks that are stuck in the rock. Looking closer, you also see clear, glassy pieces and gray, irregular chunks. What is the rock made of? And how did it get here?

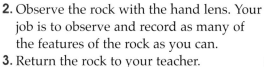

EXPLORE ACTIVITY

Observe a Rock

1. Obtain a sample of granite or basalt from your teacher. You will also need a hand lens.
2. Observe the rock with the hand lens. Your job is to observe and record as many of the features of the rock as you can.
3. Return the rock to your teacher.
4. Try to describe your rock so that other students could pick your rock out of the group of rocks.

Science Journal

In your Science Journal, describe and draw how the parts of the granite or basalt fit together to form the rock. Be sure to label your drawing.

405

Prepare

Section Background

- The most common group of minerals is the silicates. Silicates contain a silica tetrahedron structure that places four oxygen atoms around one silicon atom.

- Several "rocks" that we study don't actually fit the definition of a rock. Examples are coal and obsidian, neither of which contains minerals. For this reason, a more inclusive definition is used by geologists: A rock is a mixture of two or more minerals, mineraloids, glass, or organic materials.

Preplanning

Refer to the Chapter Organizer on pages 404A-B.

1 Motivate

Bellringer

Before presenting the lesson, display **Section Focus Transparency 49** on the overhead projector. Assign the accompanying **Focus Activity** worksheet.
[L2] [ELL]

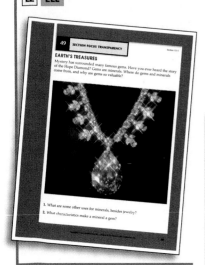

What YOU'LL LEARN

- The difference between a mineral and a rock
- Properties that are used to identify minerals

Science Words:
mineral
rock
crystal
gem
ore

Why IT'S IMPORTANT

You'll learn the value and uses of minerals and rocks.

FIGURE 15-1

You use minerals every day without realizing it. Minerals are used to make many common objects.

A Quartz is melted to form glass.

---

15 • 1 Minerals: Earth's Jewels

What is a mineral?

Get ready! You're going on an expedition to find minerals (MIH nuh rools). Where will you look? Do you think you'll have to crawl into a cave or brave the depths of a mine? Well, put away your flashlight and hard hat. You can find minerals right in your own home—in the salt shaker on your table; in the "lead" of your pencil; and in the metal pots and pans, glassware, and dishes in your kitchen. Minerals are everywhere around you, as shown in **Figure 15-1.**

What is a mineral? A **mineral** is an inorganic solid material found in nature. Inorganic substances are things that are not formed by plants or animals. A mineral must always have the same chemical makeup and an orderly pattern of atoms. For example, the mineral calcite always has the same chemical makeup. And the atoms in calcite are always arranged in the same way. Minerals are also the basic building blocks for all rocks. A **rock,** like the granite or basalt used in the Explore activity on page 405, is usually made of two or more minerals. More than 4000 different minerals have been identified. Each mineral has traits or characteristics you can use to identify it.

B The mineral gypsum is used to make drywall, a building material.

406 Chapter 15 Earth Materials

Program Resources

📁 Reproducible Masters
Activity Worksheets, pp. 5, 95 [L2]
Cross-Curricular Integration, p. 19 [L2]
Enrichment, p. 55 [L3]
Lab Manual, pp. 103-105 [L2]
Multicultural Connections, pp. 33-34 [L2]
Reinforcement, p. 55 [L2]

Science and Society/Technology
  Integration, p. 33 [L2]
Study Guide, p. 55 [L1]

📽 Transparencies
Science Integration Transparency 15 [L2]
Section Focus Transparency 49 [L2]
Teaching Transparency 29 [L2]

How do minerals form?

On your rock-climbing adventure, you started asking questions about how minerals and rocks form. Minerals can form in several ways from either melted rock or from solutions. The first way is from melted rock called magma. As magma cools, similar atoms combine to form minerals in the cooling rock. These minerals can form on Earth's surface or deep inside Earth. Minerals also form when solutions evaporate. As water leaves a solution, minerals are left behind. This is how the salt you put on your popcorn was formed. A third way minerals form is also from solution. If a solution is extremely rich in minerals, the minerals may settle out. A solution can hold only so much of the dissolved mineral. If more minerals are added, some of the minerals will settle out as solid materials.

You can tell how a mineral formed by how it looks. If the mineral grains fit together like a puzzle, it cooled from magma. If you see layers of different minerals, the minerals probably formed by evaporation. The beautiful crystal you see in **Figure 15-2** grew from a solution rich in dissolved minerals. To figure out how a mineral forms, you must consider the size of the mineral grain or crystal, the presence of layers, and the texture. Texture is the size of the crystal and how the crystals or mineral grains fit together. Next, you will learn ways to identify different minerals. Knowing how minerals form is interesting, but being able to identify them is more fun.

FIGURE 15-2
This fluorite crystal grew from a solution rich in dissolved minerals.

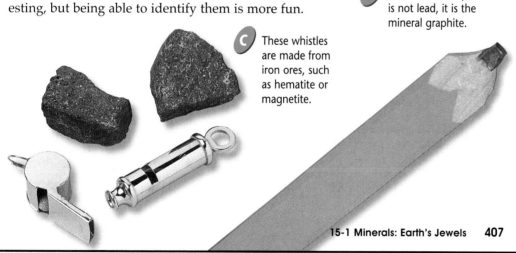

C These whistles are made from iron ores, such as hematite or magnetite.

D The "lead" in a pencil is not lead, it is the mineral graphite.

15-1 Minerals: Earth's Jewels   **407**

Revealing Preconceptions

Ask students to describe what comes to mind when they think of the word *crystal*. Many students will think that a mineral has a crystalline structure only if the sample shows the crystal shape, such as cubic salt crystals. Emphasize that a mineral's crystalline structure refers to the internal arrangement of atoms. Whether or not a mineral shows a crystal shape, it still has an internal crystalline structure.

Demonstration

Visual-Spatial Obtain samples of several crystals and share them with the class. Some examples to use are halite (cubic), corundum (hexagonal), and pyrite (cubic). Draw the internal atomic arrangement of a cubic crystal (NaCl) on the chalkboard and have students compare it with cubic crystals of salt.

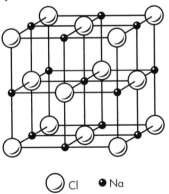

○ Cl   ● Na

FLEX Your Brain

Use the Flex Your Brain activity to have students explore MINERALS.

📁 **Activity Worksheets,** page 5

FIGURE 15-3

This beautiful amethyst has large crystals. Amethyst is a variety of the mineral quartz.

Properties of Minerals

Perched on your rocky outlook, you notice another person on a ledge some distance away. Is it someone you know? She's wearing a yellow shirt and has her long, dark hair in braids, just like a friend you saw this morning. You're only sure it's your friend when she turns, and you recognize her smile. You've identified your friend by physical properties that set her apart from other people—her clothing, hair color and style, and facial features. Each mineral, too, has a set of physical properties that you can use to identify it. No fancy equipment is needed. Most common minerals can be identified with items you have around the house and can carry in your pocket, like a penny and a nail. Now let's take a look at the properties that will help you identify minerals.

Crystals

All minerals have a definite atomic structure. The atoms that make up the mineral are arranged in a repeating pattern. Solid materials that have this pattern of atoms are called **crystals.** Some mineral samples, such as the amethyst pictured in **Figure 15-3,** have beautiful crystals. Crystals have smooth surfaces, sharp edges, and points. Some crystals look like tiny cubes. **Figure 15-4** gives a close look at table salt and reveals that each grain is a little cube. These cubes are crystals of the mineral halite.

FIGURE 15-4

Common table salt is the mineral halite, which is called rock salt. Halite crystals are tiny cubes.

Theme Connection

Scale and Structure

Minerals that have the same composition but have a different internal structure are called *polymorphs*. Comparing the structure of polymorphs at the atomic level reinforces the theme of Scale and Structure. An excellent example of this is diamond and graphite. Both minerals are composed only of carbon. In diamond, each carbon atom is bonded tightly with four other carbon atoms. Because of this, diamond is the hardest known mineral. In graphite, each carbon is bonded tightly to three other carbon atoms and bonded to one other carbon atom by weak atomic forces. This structure makes graphite one of the softest known minerals. Graphite is often used as a lubricant because of its softness.

Cleavage and Fracture

Another clue to a mineral's identity is the way it breaks. Minerals that split into pieces with smooth, regular edges and surfaces are said to have cleavage (KLEE vuj). The mineral mica in **Figure 15-5A** shows cleavage by splitting into thin sheets. Splitting one of these minerals along a cleavage surface is something like taking pieces of bread from a sliced loaf. Cleavage is caused by weaknesses in the arrangement of the atoms that make up the mineral.

Not all minerals have cleavage. Some break into pieces with jagged or rough edges. Instead of neat slices, these pieces are shaped more like hunks of bread torn from an unsliced loaf. Materials that break this way, such as quartz, have what is called fracture (FRAK chur). **Figure 15-5C** shows the hackly fracture of native copper.

FIGURE 15-5

Cleavage is one of the most useful properties of minerals. Some minerals have one or more directions of cleavage. If minerals do not break in a regular way, they have what is called fracture.

 Mica has one cleavage direction and can be peeled off in sheets.

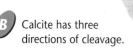 Calcite has three directions of cleavage.

 The smooth, curved fracture of obsidian is called conchoidal fracture.

 Native copper has a jagged, hackly fracture.

15-1 Minerals: Earth's Jewels **409**

Content Background

Cleavage planes occur along planes of weak atomic bonds in the internal structure of the crystal. The mineral mica, for example, has perfect cleavage in one direction. Sheets of silica tetrahedra are held together by weak bonds in between the sheets. Minerals with no planes of weak atomic bonds do not have cleavage planes. Instead, they fracture.

GLENCOE TECHNOLOGY

⦿ **Videodisc**

The Infinite Voyage: The Future of the Past

Chapter 2
The Cologne Cathedral: Preserving Stained Glass

Chapter 3
Preserving Medieval Glass with a Chemical Lacquer

Chapter 4
The Statue of Liberty Restoration Project

Chapter 7
Restoring the Parthenon

Content Background

Obsidian, or volcanic glass, illustrates conchoidal fracture better than most minerals. However, obsidian is an igneous rock not a mineral.

Across the Curriculum

History French chemist Marc Gaudin produced the first laboratory-grown ruby in 1837. He melted chemical compounds in the lab that decomposed to aluminum oxide. He added chromium to the aluminum oxide for color. Aluminum oxide is what rubies are made of. With the red color from the chromium, Marc Gaudin produced synthetic rubies.

Pyrite is often called fool's gold. Many prospectors were fooled by pyrite. Another form of fool's gold is chalcopyrite. Chalcopyrite is basically the same color but tends to tarnish more because of its copper content. Both minerals have dark-gray to black streaks and are shiny gold in color. One characteristic that can be used to distinguish the two minerals is hardness. Pyrite has a hardness of 6.5, whereas chalcopyrite is softer with a hardness of 3.5 to 4. Pyrite will scratch chalcopyrite.

Content Background

- Nonmetallic lusters are described in many ways. If a mineral looks like glass, it has a glassy luster. The brilliant luster of diamond is called adamantine. The sparkly luster of sphalerite is referred to as resinous. If the mineral looks earthy, its luster is earthy. Pearly minerals have pearly luster.

- Diamond chips are embedded on dentists' drills so they can drill into human teeth.

- Diamond wafers have been used as tiny windows for space probes. The *Pioneer* space probe to Venus used diamond wafers for this purpose because the crystal can withstand the extremes of heat and cold encountered in space.

- Diamond-tipped scalpels are excellent instruments for use in delicate eye surgery. Diamond is able to accept and maintain a sharp edge. A sharp edge is necessary for an instrument to slice tissue without tearing it.

FIGURE 15-6

Luster describes how a mineral looks. If it shines like a metal, it has metallic luster (galena). All other minerals have some type of nonmetallic luster (talc).

Galena

Talc

Streak and Luster

The reddish-gold color of a new penny shows that it's made of copper. The bright yellow color of sulfur is a valuable clue to its identity. Sometimes a mineral's color can help you figure out what it is. But color also can fool you. The common mineral pyrite (PI rite) has a shiny, gold color similar to real gold—close enough to disappoint many prospectors back in the Gold Rush days in the 1800s. Pyrite is called "fool's gold."

A test called the streak test will help identify a mineral, even if it looks a lot like a different mineral. Scratching a mineral sample across an unglazed, white tile (called a streak plate) produces a streak of color, as shown in **Figure 15-7**. The streak is not necessarily the same color as the mineral itself. This streak of powdered mineral is more useful for identification than the mineral's color. Gold prospectors could have saved themselves a lot of heartache if they had known about the streak test. Pyrite makes a greenish-black or brownish-black streak, while gold makes a yellow streak.

Shiny? Dull? Pearly? Words like these describe another property of minerals called luster. Luster is how light is reflected from a mineral's surface. If it shines like a metal, the mineral has metallic (muh TAL lihk) luster. Nonmetallic minerals can be described as having pearly, glassy, dull, or earthy luster, as shown in **Figure 15-6**. Using color, streak, and luster helps identify minerals. What other properties can be used?

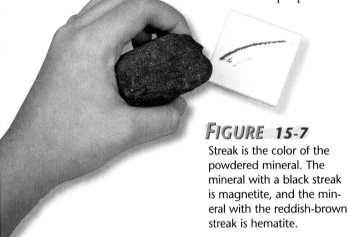

FIGURE 15-7

Streak is the color of the powdered mineral. The mineral with a black streak is magnetite, and the mineral with the reddish-brown streak is hematite.

Science Journal   Streak Tables Obtain samples of specular hematite, jasper, earthy hematite, and oolitic hematite. In their Journals, ask students to make a table to describe and contrast the appearance of each sample and to compare the streaks of each sample. Some hematites are red, others are steel gray, and yet others appear shiny metallic in color, but in all cases, the streak color is deep red-brown. **L2** **P**

Community Connection

Jeweler Invite a jeweler from your community to visit the class and share methods used to determine the quality of gemstones. The jeweler could explain how he or she uses tests such as hardness and specific gravity when determining the value of gems brought in by customers.

Table 15-1

Mineral Hardness		
Mohs scale		
	softest	**Hardness of common objects**
Talc	1	
Gypsum	2	fingernail (2.5)
Calcite	3	copper penny (3.5)
Fluorite	4	iron nail (4.5)
Apatite	5	glass (5.5)
Feldspar	6	steel file (6.5)
Quartz	7	streak plate (7)
Topaz	8	
Corundum	9	
Diamond	10	
	hardest	

Hardness

As you investigate different minerals, you'll find that some are harder than others. Some minerals, like talc, are so soft that they can be scratched with a fingernail. Others, like diamond, are so hard that they can be used to cut almost anything else.

In 1822, an Austrian geologist named Friedrich Mohs also noticed this property. He developed a way to classify minerals by their hardness. The Mohs scale, as shown in **Table 15-1**, classifies minerals from 1 (softest) to 10 (hardest). You can determine hardness by trying to scratch one mineral with another to see which is harder. For example, fluorite (4 on the Mohs scale) will scratch calcite (3 on the scale), but calcite will not scratch fluorite. You can also use a homemade mineral identification kit: a penny, a nail, and a glass plate with smooth edges. You just find out what scratches what. Is the mineral hard enough to scratch a penny? Will it scratch glass?

Mini LAB

Classifying Minerals

What other properties help to identify minerals?

1. Place your four samples in separate sealable, plastic bags, and dip each in a pile of iron filings. Record which mineral(s) attract the filings.
2. Place each sample in a small jar full of vinegar and record what happens.

Analysis

1. Identify each mineral—quartz, calcite, hornblende, and magnetite—by its magnetic or fizzing properties.
2. Describe, in a data table, the other physical properties of the four minerals.

Inclusion Strategies

Learning Disabled Provide students with specific instructions on how to dip samples in the iron filings and how to prepare the jars of vinegar. Help students prepare their data tables using an example on the chalkboard, or pair students with other students who better understand the type of data table to be used.

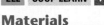

Mini LAB

Purpose

Kinesthetic Students will see how certain mineral properties can be used to identify the minerals. **L2**
ELL **COOP LEARN** **P**

Materials

4 sealable plastic bags, iron filings, quartz, calcite, hornblende, magnetite, small jar of vinegar

Teaching Strategies

Using groups of four students, have each student be responsible for one sample.

Safety Precautions Caution students to wear safety goggles and aprons.

📁 **Activity Worksheets,** pages 5, 95

Analysis

1. Magnetite is strongly magnetic. Calcite reacts with vinegar.
2. Data tables should include the following. Quartz is harder than glass; hornblende is black and has two directions of cleavage; calcite has three directions of cleavage and a hardness of 3; and magnetite is black with a black streak.

✓ Assessment

Oral Ask students what two tests can distinguish quartz from calcite. The two tests are hardness and reaction to vinegar. Quartz has a hardness of 7 and does not react to vinegar, whereas calcite has a hardness of 3 and does react to vinegar. Use the Performance Task Assessment List for Oral Presentation in **PASC**, p. 71.

Problem Solving

Solve the Problem

If mineral grains fit together like a puzzle, the minerals cooled from magma. If there are layers of different minerals, it probably formed from a solution by evaporation. Minerals with distinct, beautiful crystals often form from solution.

Think Critically

1. Students should separate the samples by how the mineral grains fit together and the general appearance of the grains. If mineral grains appear to fit together, the mineral formed from magma. Minerals formed from solution by evaporation will show layers of different minerals. Minerals with distinct crystals most likely formed from solution.

2. from magma: interlocking mineral grains; from solution by evaporation: layers of different minerals; from solution: distinct, beautiful crystals ▣

Content Background

A mineral's specific gravity (also known as heft) is a ratio of the mineral's weight to the weight of an equal volume of water. If a cubic centimeter of a mineral weighs four times more than a cubic centimeter of water, its specific gravity is 4.

Teacher F.Y.I.

If your hardness value derived by using a penny is different from what you expect, be sure you are using a copper penny, not a laminated one.

Problem Solving

Identifying Mineral Origin

At a neighbor's yard sale, you find a dusty old box filled with mineral samples. The neighbor, a former science teacher, says she'll sell you the whole thing for ten dollars. Then she gives you a challenge. If you can come back in a week with the samples organized by the way they formed, she'll give your money back. You rush home to start your detective work.

Solve the Problem:

Think about the traits of mineral crystals that are formed from magma and minerals that settle out of solution. Refer to page 407 for help.

Think Critically:

1. How will you identify and separate the samples by how they formed? What steps will you follow?
2. Determine the characteristics that you will look for in each group.

Gypsum

Calcite

Biotite

Content Background

A specific chemical reaction occurs when dilute HCl acid is placed on a sample of calcite or any rock that has calcite in it. The chemical reaction that occurs is as follows.

$$CaCO_3 + 2HCl \rightarrow H_2O + CO_2 + CaCl_2$$

Water is produced and carbon dioxide gas is released.

Other Properties

Some minerals have other unusual properties that can help identify them. Magnetism occurs in a few minerals, such as magnetite. The mineral calcite has two other unusual properties. It will fizz when it comes into contact with an acid like vinegar. And if you look through a clear calcite crystal, you will see a double image. Although, because of safety concerns, you should not try it, taste can be used by scientists to identify some minerals. Halite, also called rock salt, has a salty taste. Combinations of all of these properties are used to identify minerals. Learn to use them, and you can be a mineral detective.

Common Minerals

In the chapter opener, the rocks you scrambled up on your climb were made of minerals. But only a few of the more than 4000 minerals make up most rocks. The most common minerals found in rocks include quartz, feldspar, biotite mica, calcite, gypsum, hornblende, fluorite, augite, and hematite, among others. Most rocks on Earth are made up of these common minerals. Most other minerals, such as diamonds and emeralds, are rare.

Gems

Which would you rather win, a diamond ring or a quartz ring? A diamond ring would be more valuable. Why? The diamond in a ring is a kind of mineral called a gem. **Gems** are minerals that are rare and can be cut and polished, giving them a beautiful appearance, as shown in **Figure 15-8**. This makes them ideal for jewelry. Not all diamonds are gems. To be gem quality, a mineral must be clear with no blemishes or cracks. It also must have a beautiful color. Few minerals meet these standards. That's why the ones that do are so rare and valuable.

FIGURE 15-8

Gems such as these are rare and beautiful.

3 Assess

Check for Understanding

Discussion Lead students in a discussion of how to classify minerals using the properties they have learned.

Reteach

Display four different minerals that can be recognized easily by one or two characteristics. Help students note the characteristics and identify the minerals. Some minerals that may be used are quartz (hardness of 7), sulfur (yellow color), calcite (three directions of cleavage or reaction to acid), and hematite (red-brown streak). **L2**

Extension

For students who have mastered this section, use the **Reinforcement** and **Enrichment** masters.

Content Background

- About two dozen minerals make up most of the rocks that form on Earth. These are called the rock-forming minerals and include feldspar, quartz, olivine, mica, clay minerals, amphibole, pyroxene, calcite, and dolomite.

- The most common group of minerals is the silicates. The most common silicate mineral is feldspar. The name *feldspar* means "field stone," indicating how common this mineral is.

413

4 Close

FIGURE 15-9

To be profitable, ores must be found in large deposits or rich veins. Mining is expensive. This mine in Montana is producing copper.

Ores

A mineral is called an **ore** if it contains something that can be useful and sold for a profit. Many of the metals that we use come from ores. A copper mine is shown in **Figure 15-9.** For example, the iron used to make steel comes from the mineral hematite, lead for batteries is produced from galena, and the magnesium used in vitamins comes from dolomite.

Now you have a better understanding of minerals and their uses. Can you name five things in your classroom that come from minerals? You will find that you use lots of minerals every day. Next, you will look at rocks, which are Earth materials that are made up of combinations of minerals.

Section   Wrap-up

1. Explain the difference between a mineral and a rock. Name five common rock-forming minerals.

2. List five properties or traits that are used to identify minerals.

3. **Think Critically:** Would you want to live close to a working gold mine? Explain your answer.

4. **Skill Builder**
   Using a CD-ROM Use an electronic encyclopedia or Earth science CD-ROM to help create your own mineral identification chart. List the traits of at least ten different minerals. If you need help, refer to Using a CD-ROM on page 563 of the **Technology Skill Handbook.**

Science Journal

Select a mineral and write a story in your Journal about how it formed. Be creative, but everything must be scientifically correct. Include how the mineral forms and several of its characteristics. If you need help, refer to Appendix J on pages 538–540.

Igneous and Sedimentary Rocks

Earth's Fire

A rocky cliff, a jagged mountain peak, a huge boulder—they all look about as solid and permanent as anything can be. Rocks seem as if they've always been here and always will be. But things are constantly changing on Earth. New rocks form, and old rocks crack and wear away. These changes produce three main kinds of rocks—igneous, sedimentary, and metamorphic.

If you could travel down into Earth, you would find that the deeper you go, the higher the temperature and the greater the pressure. Deep inside Earth, it is hot enough to melt rock, as seen in **Figure 15-10. Igneous rocks** are produced when melted rock, or magma, from inside Earth cools. Igneous rocks can cool and harden on or under Earth's surface. When magma cools, it makes either an extrusive igneous (EKS trew sihv • IG nee us) rock or an intrusive (IN trew sihv) igneous rock.

FIGURE 15-10
This erupting volcano is Mount Etna in Italy.

What YOU'LL LEARN

- How extrusive and intrusive igneous rocks are different
- How different types of sedimentary rocks form

Science Words:
igneous rock
extrusive
intrusive
sedimentary rock

Why IT'S IMPORTANT

You'll learn about the rocks that form the land around you.

Prepare

Section Background

Sills and dikes can be made of very small mineral crystals. When they form close enough to Earth's surface, the magma cools quickly, thus producing small crystals.

Preplanning

Refer to the Chapter Organizer on pages 404A-B.

1 Motivate

Bellringer

Before presenting the lesson, display **Section Focus Transparency 50** on the overhead projector. Assign the accompanying **Focus Activity** worksheet.
L2 ELL

Tying to Previous Knowledge

Help students recall that rocks can be made of more than just minerals. Coal, made of organic material, and obsidian, made of volcanic glass, are considered rocks.

Program Resources

📁 **Reproducible Masters**
Activity Worksheets, pp. 5, 91-92, 96 L2
Enrichment, p. 56 L3
Lab Manual pp. 107-109 L2
Reinforcement, p. 56 L2
Study Guide, p. 56 L1

🔦 **Transparencies**
Section Focus Transparency 50 L2

415

2 Teach

Inquiry Question

LM **Logical-Mathematical** What might happen to lava that would cause it to cool so quickly that no crystals would form? *If the lava is shot through the air or flows into water, it can cool so rapidly that no crystals form.*

Brainstorming

LM **Interpersonal** Have students work in groups of three or four. Help them recall that minerals containing higher concentrations of silicon and oxygen are light in color, and minerals containing less silicon and oxygen and more iron and magnesium are dark in color. Show students samples of granite, rhyolite, gabbro, and basalt. Ask them to brainstorm within their groups and try to determine which minerals are contained in each of the rock samples. **L3**

COOP LEARN

 FLEX Your Brain

Use the Flex Your Brain activity to have students explore IGNEOUS ROCKS.

📁 **Activity Worksheets,** page 5

GLENCOE TECHNOLOGY

 **Videodisc**

Glencoe Earth Science Interactive Videodisc
Side 1, Lesson 1
Igneous Processes

‖‖‖‖‖‖‖‖‖‖‖‖

2252-2483

‖‖‖‖‖‖‖‖‖‖‖‖

2485-2793

416

Rocks from Lava

Extrusive igneous rocks form when magma cools on Earth's surface. Magma that reaches Earth's surface is called lava. Lava cools quickly, before large mineral crystals have time to form. That's why extrusive igneous rocks usually have a smooth, sometimes glassy appearance. They have a few or no visible crystals.

Extrusive igneous rocks can form in two ways. In one way, volcanoes erupt and shoot out lava and ash. Also, large cracks in Earth's crust, called fissures (FIHSH urs), can open up. When they do, the lava oozes out onto the ground. Oozing lava from a fissure or a volcano is called a lava flow. It's something like what happens when you put too much batter in the waffle iron. The batter oozes out the sides and flows onto the countertop. In Hawaii, lava flows are so common that you can see a volcanic eruption almost every day.

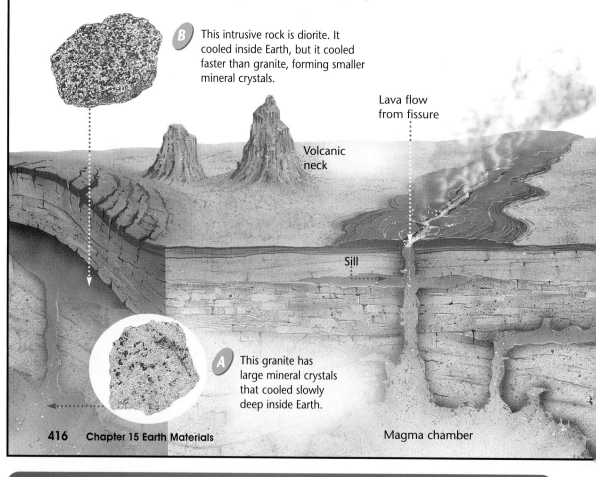

FIGURE 15-11

Intrusive igneous rocks form when molten rock or magma cools inside Earth. Extrusive igneous rocks form when lava cools at Earth's surface.

B This intrusive rock is diorite. It cooled inside Earth, but it cooled faster than granite, forming smaller mineral crystals.

Lava flow from fissure

Volcanic neck

Sill

Magma chamber

A This granite has large mineral crystals that cooled slowly deep inside Earth.

416 Chapter 15 Earth Materials

Content Background

- Intrusive rocks have a *phaneritic texture,* which refers to those rocks' large crystals. When cooling slowly, ions have lots of time to migrate and join with others, forming large crystals from the melt.

- In the case of extrusive rocks, ions don't have time to migrate. The cooling is so quick that many small crystals form, producing an *aphanitic texture.* Extrusive rocks form when lava cools on Earth's surface.

- When igneous rocks form quickly, no crystals form. This produces a rock with a *glassy texture,* such as obsidian, pumice, or scoria.

Rocks from Magma

What about magma that doesn't reach the surface? Can it form rocks, too? Yes, **intrusive** igneous rocks are produced when magma cools below the crust inside Earth, rather than on the surface, as shown in **Figure 15-11**.

When intrusive igneous rocks form, a huge glob of magma from inside Earth rises toward the surface. But it never quite breaks through to erupt onto the surface. It's similar to a helium balloon that is let loose in a gym and gets stuck on the ceiling. It doesn't make it outside—just to the roof. This hot mass of rock sits right under the surface and cools slowly over thousands of years until it is solid. The cooling takes so long that the minerals in the magma have time to form large crystals. The size of the mineral crystals is the main difference between intrusive and extrusive igneous rocks. *Intrusive* igneous rocks have large crystals that are easy to see. *Extrusive* igneous rocks do not have large crystals that you can see.

inter**NET** CONNECTION

What do different igneous rocks look like? Check out the link on the Glencoe Homepage. *www.glencoe.com/ sec/science*

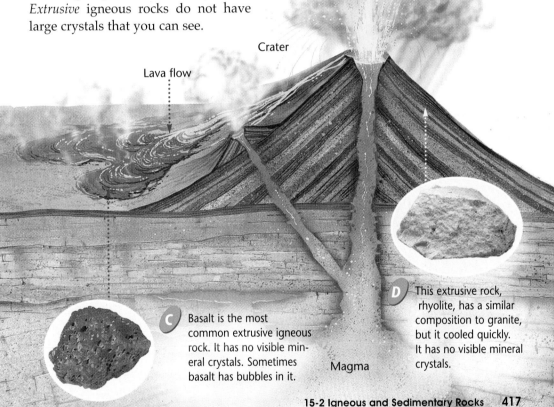

Crater

Lava flow

C Basalt is the most common extrusive igneous rock. It has no visible mineral crystals. Sometimes basalt has bubbles in it.

Magma

D This extrusive rock, rhyolite, has a similar composition to granite, but it cooled quickly. It has no visible mineral crystals.

Science Journal **Igneous Rock Formation** In your Science Journal, write a poem about what happens to lava and magma when igneous rocks form. Be sure to include where the cooling occurs, how long it takes, and what size the crystals are when they form. [LS] [L2] [P]

Using an Analogy

[LS] **Logical-Mathematical** To help students understand why crystal size in igneous rocks is affected by how much time passes during the cooling process, use the following analogy. Give each student one piece of paper made into an open-ended ring. Tell students that when you say "go," they are to take their loop of paper and attach it to another. They are to continue until you say "stop" or until all loops are connected. At first, give students very little time (this is an analogy of crystals forming quickly). They should note that a high number of small groups of loops have formed. Now have students return to their original position. Give students more time the second time (this is an analogy of crystals forming slowly). Students should note that fewer locations of larger groups of loops have formed. [L2] [ELL] [COOP LEARN]

Discussion

Discuss with students what they think would happen to magma that begins to cool slowly underground, then is erupted to Earth's surface as lava, and finishes cooling on Earth's surface. This causes a porphyritic texture. Large crystals begin to form when the magma is underground. When the lava is then erupted, it finishes cooling rapidly, forming little crystals. A rock forms that has large crystals contained within a mass of small crystals.

inter**NET** CONNECTION

The Glencoe Homepage at **www.glencoe.com/sec/ science** provides links connecting concepts from the student edition to relevant, up-to-date Internet sites.

Activity 15-1

Purpose

Kinesthetic Students will learn why some igneous rocks have no visible crystals whereas others have large, beautiful crystals. **L2** **ELL** **COOP LEARN**

Process Skills

observing and inferring, comparing and contrasting, recognizing cause and effect, making and using tables or graphs, predicting, interpreting data, separating and controlling variables, classifying, forming a hypothesis, designing an experiment to test a hypothesis

Time

20 minutes on one day for planning the experiment and 30 minutes on another day for carrying out the lab

Materials

Be sure hot plates are in good working order. Make up saturated sugar solutions before class time by adding sugar to water until no more will dissolve. An ice chest can be used instead of a refrigerator.

Safety Precautions

Caution students about hot surfaces (stove or hot plate and jars) and to be careful near the electric cord. Also, students should wear safety goggles and aprons.

Possible Hypotheses

If the supersaturated solution is cooled quickly, small crystals will form. If the solution is cooled slowly, large crystals will form.

📁 **Activity Worksheets,** pages 5, 91-92

Possible Materials

- saucepan, 3-quart
- measuring cup
- large wooden spoon
- stove or hot plate
- pint jars or glasses (2)
- paper clips (2)
- pencils (2)
- 3-cm piece of cotton string (2)
- white, granulated sugar (3 cups)
- refrigerator
- igneous rocks (4)

Design Your Own Experiment
Cool Crystals Versus Hot Crystals

Two types of igneous rocks can look different, even though they are made of the same minerals. Extrusive rocks have an even texture with no visible crystals, while intrusive rocks have large, beautiful crystals. How does this happen?

What You'll Investigate

Determine how different-sized crystals are formed from a solution.

Form a Hypothesis

State your **hypothesis** about how you think the rate of cooling affects the different sizes of crystals that can be formed.

Goals

Design an experiment that compares crystal growth in a solution by using different methods of cooling.

Write a general rule about crystal size in igneous rocks based on your observations in this experiment.

Safety Precautions

Never eat or taste anything from a lab, even if you are confident that you know what it is.

Inclusion Strategies

Learning Disabled Provide students with an overview of Activity 15-1 showing a higher degree of structure. You may wish to include a simple procedure to follow. Help students develop and write a workable hypothesis. Suggest ways that students can set up equipment for the experiment.

Solutions	Time for crystals to form
A (fast cooling)	(Answers will depend on temperature, humidity, amount of sunlight, etc.)
B (slow cooling)	

PLAN

1. As a group, agree upon and write out your **hypothesis**.
2. As a group, **list** the steps that you will take to test your hypothesis. Be specific, describing exactly how you are going to make different sizes of crystals. If you have questions about how to mix the solution, check with your teacher.
3. **List** all of the materials that you will need to complete your experiment.
4. Before you begin, **make a data table or graph** that will allow you to compare the size of the crystals that form in each solution.
5. **Read** over your entire experiment to make sure that all the steps are in logical order.
6. Be sure to double-check your list of materials and method for creating sizes of crystals.
7. Will you repeat any part of the experiment more than once to allow for human error?
8. Is your data table ready to handle the amount of data that you want to collect?
9. Make sure that your teacher approves your plan and that you have included any changes to the plan.

DO

1. Carry out the lab as planned and approved.
2. Be sure to **record** your observations in the data table as you **complete** each test.

CONCLUDE AND APPLY

1. **Compare** your findings with those of the other student groups.
2. Using your data table, **conclude** how this information could help you to classify igneous rocks.
3. **APPLY** Identify four igneous rock samples as either intrusive or extrusive.

✓ Assessment

Performance To further assess students' understanding of crystal formation, have them continue to work in threes. Encourage them to write a poem, song, or story that explains the process involved when atoms arrange themselves into crystals. Use the Performance Task Assessment List for Song with Lyrics in **PASC**, p. 79.

Go Further

LS **Logical-Mathematical** Hypothesize what would happen if a jar of the sugar solution were placed in the freezer rather than in the refrigerator. Because of the rapid cooling that would occur, tiny crystals or no crystals at all would form.

PLAN

Possible Procedures

Use two jars or glasses in which to cool identical solutions. Place one jar in a refrigerator or cooler to cool the solution quickly. Place the other on a shelf to allow for slow cooling. The only variable is the time allowed for the solution to cool.

Teaching Strategies

- Have students work in groups of three.
- The jars of sugar solution must be left undisturbed for at least a week. This allows large crystals to grow. When solutions are cooled suddenly, only small crystals will grow.

Troubleshooting Adding seed crystals on the thread may help the precipitation of sugar.

DO

Expected Outcome

Students will be able to describe the growth process of sugar crystals from solution. They will note that larger crystals grow when the solution cools slowly.

CONCLUDE AND APPLY

1. Answers will vary.
2. Data in student tables should show that rocks with large crystals probably cooled slowly and may be intrusive igneous rocks.
3. Answers will vary depending on samples used. Suggestions include granite, rhyolite, gabbro, and basalt.

Sedimentary Rocks

The second major group of rocks is called sedimentary (sed uh MEN tree) rocks. **Sedimentary rocks** are made when pieces of other rocks, plant and animal matter, or dissolved minerals collect to form rock layers, as shown in **Figure 15-12.** These pieces of rock and other materials are called sediments (SED uh muntz). Rivers, ocean waves, mudslides, and glaciers can carry sediments. Sediments can also be carried by the wind. When the sediments are dropped by or deposited by wind, ice, gravity, or water, they collect in layers. After the sediments are deposited, they begin the long process of becoming rock. Most sedimentary rocks form over thousands to millions of years. The changes that form sedimentary rocks are always happening in our world, as shown in **Figure 15-13.**

Broken Rocks

When most people talk about sedimentary rocks, they are usually talking about rocks such as sandstone. This type of rock is made up of broken pieces of other rocks. These rocks are deposited by water, ice, gravity, or wind into layers or piles. These piles are then cemented together by other minerals and squeezed or compacted into rock by the weight of sediments on top of them.

FIGURE 15-12

All of these sedimentary rocks are exposed at Cathedral Rocks in Arizona. The layers are different types of sedimentary rocks. *How are sediments carried to form sedimentary rocks?*

A Conglomerate is made up of large pebbles cemented together.

B Sandstone is made up of sand grains cemented together.

420 Chapter 15 Earth Materials

Cultural Diversity

Rocks and Minerals Rocks and minerals have been important resources for cultures for a very long time. The oldest stone tool from Ethiopia dates to more than 2.5 million years ago. Often, early people traveled miles to find hard rocks like flint and quartz for tools.

Obsidian, or volcanic glass, provided extremely sharp cutting edges, as well as beautiful artwork. Obsidian trade routes ran throughout the Americas by 200 B.C. and throughout Asia as early as 9000 B.C.

Delicate, decorative pieces were carved from the mineral mica. The earliest paintings, preserved in caves throughout the world, are masterpieces created with charcoal, chalk, and various shades of ochre.

These broken sedimentary rocks are identified by the size of the pieces or grains that make up the rock. Rocks made of the smallest grains, clay, are called shale. Silt grains, which are slightly larger than clay, make up siltstone. Sandstone is made of sand. Sand is larger than silt. The largest grains are called pebbles. Pebbles mixed and cemented together with other sediments make up rocks called conglomerates (kon GLAW muh rutz).

Chemical Rocks

Some sedimentary rocks are formed when seawater, loaded with dissolved minerals, evaporates. Chemical sedimentary rock can also form when mineral-rich water from geysers or hot springs evaporates. As the water evaporates, layers of minerals are left behind. If you've ever sat in the sun after swimming in the ocean, you may have noticed salt crystals on your skin. The seawater on your skin evaporated, leaving behind deposits of salt. The salt had been dissolved in the water. Something similar happens to form chemical rocks.

FIGURE 15-13

Waves deposit sand and move sand around on this beach in California.

C Siltstone is made up of grains that are smaller than sand grains.

D Shale is made of the smallest grains, clay.

15-2 Igneous and Sedimentary Rocks   421

3 Assess

Check for Understanding

Making a Model Check for student understanding of how broken rocks form by having them make a model of a sedimentary rock made of broken pieces of other rocks. Provide teams of students with small pieces of wood as a model of crushed rock. Give each team a bottle of wood glue or white glue. Have students mix the wood chips into the glue and then place the mixture between two sheets of waxed paper. Tell students to apply pressure to the mixture and then set it aside to dry. When dry, students will have a model of a broken rock. 🔢 L2 ELL

Reteach

🔢 **Logical-Mathematical** Ask students the following question. **How is the formation of deposits of salt forming on your skin after swimming in the ocean similar to the formation of chemical sedimentary rock?** *In both cases, deposits of salt form as water containing salt in solution evaporates.*

Extension

📁 For students who have mastered this section, use the **Reinforcement** and **Enrichment** masters.

Content Background

Remind students that the mineral calcite reacts with acid or vinegar. Chemical and biochemical rocks that contain calcite should also react. This reaction can be used as a test for limestone.

Across the Curriculum

Geography Have students research the location of the Bonneville Salt Flats. Ask them what processes occurred to form the salt flats. In addition, ask students why they think the Great Salt Lake is so salty when inland lakes usually contain freshwater. 🔢

421

FIGURE 15-14
There are a great variety of biochemical sedimentary rocks.

FIGURE 15-14
There are a great variety of biochemical sedimentary rocks.

A The White Cliffs of Dover, England, are made of chalk.

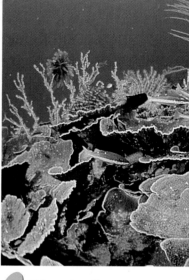

B This coral reef is located in New Guinea. The frame of the reef is limestone.

Mini LAB

How do fossils form rocks?

Create an artificial fossil rock sample of biochemical limestone.

1. Fill a small aluminum pie pan with an assortment of broken macaroni. These represent different kinds of fossils.
2. Mix three tablespoons of white glue into half a cup of water. Pour this solution over the macaroni and set it aside to dry.
3. When your fossil rock sample has set, pop it out of the pan and compare it with a real fossil limestone sample.

Analysis

1. Explain why you used the glue solution and what this represents in nature.
2. Using whole macaroni samples as a guide, match the macaroni pieces in your "rock" to the intact macaroni fossils. Then draw and label them in your Science Journal.

422   Chapter 15 Earth Materials

Biochemical Rocks

Would it surprise you to know that the chalk your teacher may be using on the chalkboard may also be a rock? So is the coal that is used as a fuel to produce electricity.

Chalk and coal are examples of the group of sedimentary rocks called biochemical rocks. Biochemical rocks form over millions of years. Living matter dies, piles up, and then is compressed into rock. If the rock is produced from layers of plants piled on top of one another, it is called coal. Biochemical sedimentary rocks can also form in the oceans and are most often classified as limestone. Many different kinds of limestone exist. Chalk is a kind of limestone made from the fossils of millions of tiny animals and algae, as shown in **Figure 15-14**. A fossil is the remains or trace of a once-living plant or animal. A dinosaur bone and footprint are both fossils.

Fossil limestone

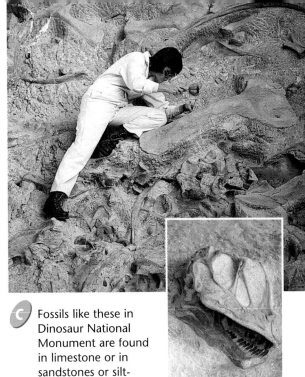

C Fossils like these in Dinosaur National Monument are found in limestone or in sandstones or silt-stones.

Camarasaurus skull

Section 3 Wrap-up

1. Describe the different ways that extrusive and intrusive igneous rocks form.

2. Diagram the way that each of the three kinds of sedimentary rocks is formed. List one sample of each kind of rock: broken, chemical, and bio-chemical.

3. **Think Critically:** If someone handed you a sample of an igneous rock and asked you if it was extru-sive or intrusive, what would you look for first? Explain.

4. **Skill Builder**

   Sequencing Describe one possible sequence of events that explains how coal is formed. If you need help, refer to Sequencing on page 543 of the **Skill Handbook.**

Science Journal

Select a national park or monument that has had volcanic activity. Read about the park and the features that you'd like to see. Then describe it in your Science Journal.

Skill Builder

Sequencing Plant material is buried in a swamp, plant material changes to peat in the absence of oxygen, peat changes to lignite (palm coal), and then lig-nite changes into bituminous coal.

✔ Assessment

Oral Ask students the following questions. **Why are rocks such as sandstone and shale called sedimentary rocks?** *These rocks form from sediment.* **Why is it easy to distinguish extrusive from intrusive rocks?** *The crystal sizes are different. Extrusive rocks contain crystals that can be seen with the unaided eye.* Use the Performance Task Assessment List for Oral Presentation in **PASC,** p. 71. P

4 Close

•MINI•QUIZ•

Use the Mini Quiz to check students' recall of chapter content.

1. **When magma cools slowly inside Earth, _____ igneous rocks form.** *intrusive*

2. **Igneous rocks that form on Earth's surface are _____ igneous rocks.** *extrusive*

3. **Remains or traces of once-living plants or ani-mals are called _____.** *fossils*

Section Wrap-up

1. Intrusive igneous rocks cool slowly from magma. Extrusive igneous rocks cool quickly from lava.

2. Broken rocks form from bits and pieces of other rocks, biochemical rocks form from the remains of once-living things, and chemical rocks form as water evaporates or set-tles out of liquids when conditions change. broken: sandstone; biochemical: coal; chemical: rock salt

3. **Think Critically** You would look at the crystals. If they were visible with the unaided eye, the rock would be intrusive.

Science Journal Answers will vary. For exam-ple, if Volcano Na-tional Park in Hawaii is chosen, descriptions should include the vol-cano surrounding the park and the fact that basalt is a rock common to the area. Other examples include Yel-lowstone National Park and Crater Lake in Oregon.

423

Science & Society

15●3 Monument or Energy?

Bellringer

 Before presenting the lesson, display **Section Focus Transparency 51** on the overhead projector. Assign the accompanying **Focus Activity** worksheet.

L2 **ELL**

Teaching the Content

• Have pairs of students use large chocolate-chip cookies and dissecting probes to "mine" the "coal" (chocolate chips) from the "land" (rest of the cookie). Have students describe how mining affects the land and suggest ways to reclaim the area once mining is complete. Students should not eat any cookies that have been used in the activity. **L1**

COOP LEARN **ELL**

• Review, if necessary, the processes of weathering and erosion. Weathering includes all the chemical and physical processes that change Earth's rocks into smaller pieces. Erosion is the transport of weathered materials. Agents of weathering and erosion are water, wind, ice, and gravity.

What YOU'LL LEARN

• That earth processes produce landscapes and resources
• How land use can change the economy of an area

Science Words:
fossil fuel

Why IT'S IMPORTANT

You will learn that there are consequences to the decisions you make.

FIGURE 15-15
Rock layers eroded to form a series of giant steps, which give the national monument its name.

In September of 1996, 1.7 million acres of Utah became a national monument. The Grand Staircase-Escalante (es kuh LAWN tay) National Monument is one of the most beautiful areas of the American west, as shown in **Figure 15-15.** For millions of years, running water has carved spectacular canyons into the rocks. Wind has blown bits of sand and silt against other rocks, wearing the rocks away. Water and wind have combined to form mesas (MAY sus), which are flat, tablelike rock structures that rise high above the surrounding land. Wind has also eroded rocks within the monument to form steep cliffs with what look like staircases carved into them.

On the Surface . . .

The monument also contains the remains of ancient cultures. One such group, the Anasazi (ah nuh SAW zee), lived in the caves and shelters that were carved into the rock by wind and water. The rocks of the Grand Staircase-Escalante National Monument also contain many fossils. Earlier, you learned that fossils are the remains of ancient plants and animals. Scientists use fossils to learn about ancient climates, how living things have changed over time, and about Earth's geography.

Beneath Your Feet . . .

Deep beneath the majestic cliffs and canyons of the largest and newest national monument in the United States are other treasures—buried treasures. Experts think that coal, natural gas, and oil valued at billions of dollars lie under the monument, as shown in **Figure 15-16.** These resources are fossil fuels. **Fossil fuels** are the remains of ancient plants and animals that can be burned to produce energy.

424   Chapter 15 Earth Materials

Program Resources

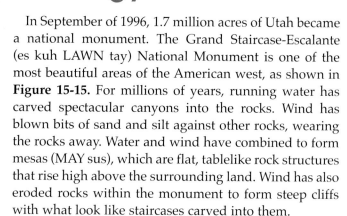 **Reproducible Masters**
Enrichment, p. 57 **L3**
Reinforcement, p. 57 **L2**
Science Integration Activities, pp. 67-68 **L2**
Study Guide, p. 57 **L1**

Transparencies
Section Focus Transparency 51 **L2**

Preserve or Reserves?

A lot of people are happy about the decision to create the monument. The desert ecosystem is fragile, they say, and should be protected. And, said one scientist, "It's the most beautiful spot in the U.S." But there is a price. Now that the area has been declared a national monument, none of the fossil fuels can be removed.

Because of that, some people worry about the state's economy. These natural resources could provide lots of money and hundreds of jobs for the state of Utah. This money could be used to improve roads and schools, provide career training, and pay for other valuable things. At the same time, the national monument will provide money and jobs because tourists will visit. It is also important to protect and preserve beautiful areas as parks and monuments.

FIGURE 15-16
Large deposits of coal and other fossil fuels are found in this region.

 Skill Builder: *Predicting Consequences*

LEARNING the SKILL

1. Identify the main idea.

2. Identify supporting details.

3. Study any photographs or illustrations that accompany the material.

4. Try to find out about similar decisions or situations. Then ask yourself what the consequences were of a similar action or decision.

PRACTICING the SKILL

1. Identify the main point of the article.

2. Why are some people in Utah worried about the new monument?

3. What points most strongly support the new monument?

APPLYING the SKILL

Former U.S. presidents Jimmy Carter, Theodore Roosevelt, and Franklin Roosevelt used the Antiquities Act to establish many monuments and national parks in the United States. Find out more about the act and the consequences that have resulted.

15-3 Monument or Energy?   **425**

425

Prepare

Section Background

- Tremendous heat, great pressure, and chemical reactions can cause rocks to change into different forms.
- Sedimentary rocks, igneous rocks, and even other metamorphic rocks can be changed into new metamorphic rocks.

Preplanning

Refer to the Chapter Organizer on pages 404A-B.

1 Motivate

Bellringer

Before presenting the lesson, display **Section Focus Transparency 52** on the overhead projector. Assign the accompanying **Focus Activity** worksheet.

L2 ELL

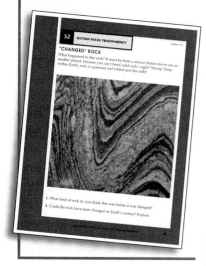

Tying to Previous Knowledge

Ask students to compare and contrast a sample of granite and a sample of gneiss, a metamorphic rock.

426

15•4 Metamorphic Rocks and the Rock Cycle

What YOU'LL LEARN

- The conditions needed for metamorphic rocks to form
- How all rocks are tied together in the rock cycle

Science Words:
metamorphic rock
foliated
non-foliated
rock cycle

Why IT'S IMPORTANT

Learning about metamorphic rocks and the rock cycle will help you see Earth as a constantly changing planet.

New Rocks from Old Rocks

When you wake up in the morning, does the world look different than it did the day before? Usually not. But even if you don't notice, Earth is constantly changing. Layers of sediment are piling up in the bottoms of lakes, landmasses are moving, and rocks are disappearing back under Earth's crust. Some of these changes cause sedimentary and igneous rocks to be heated and squeezed, as shown in **Figure 15-17**. In the process, new rocks form.

FIGURE 15-17

These mountains in the Swiss Alps were formed under great pressures and temperatures. The rocks were squeezed into these spectacular shapes.

Program Resources

Reproducible Masters
Activity Worksheets, pp. 5, 93-94 L2
Enrichment, p. 58 L3
Reinforcement, p. 58 L2
Study Guide, p. 58 L1

 Transparencies
Section Focus Transparency 52 L2
Teaching Transparency 30 L2

Metamorphic Rocks

Do you recycle your plastic milk jugs? After the jugs are collected, sorted, and cleaned, they are heated and squeezed into pellets. The pellets later can be made into useful products. Or material can be reused and made into art, as shown in **Figure 15-19.** Did you know that rocks get recycled, too? Rocks that form when older rocks are heated or squeezed are called **metamorphic rocks.** The word *metamorphic* (met uh MOR fihk) means "change of form." This is a good description of how some rocks take on a whole new look when under great temperatures and pressures, as shown in **Figure 15-18.**

FIGURE *15-19*
This dinosaur statue was made from recycled car parts.

Community Connection

Rock Uses Have students look around the school grounds and throughout the community for examples of metamorphic rocks uses. Some examples they might search for include slate on the rooftops of buildings, and marble as building stones or statues. LS L2 ELL

2 Teach

Enrichment

Remind students that the mineral calcite reacts with acid or vinegar. Metamorphic rocks formed from limestone would also contain calcite and should also react. This reaction can be used as a test for marble.

Demonstration

LS **Visual-Spatial** Present a demonstration that allows students to see samples of metamorphic rocks and the rocks from which they may form. Some examples to use are: slate forms from shale, quartzite forms from sandstone, marble forms from limestone, and gneiss forms from granite.

? FLEX Your Brain

Use the Flex Your Brain activity to have students explore METAMORPHIC ROCKS.

📁 **Activity Worksheets,** page 5

GLENCOE TECHNOLOGY

 Videodisc
Glencoe Earth Science Interactive Videodisc
Side 1, Lesson 1
Metamorphic Processes

5385-6351

6354

6356-6817

6819-7220

428

Using Science Words

Linguistic Non-foliated metamorphic rocks in which mineral grains only get bigger and generally do not rearrange into flat layers may be referred to as *recrystallized* metamorphic rocks.

Using Science Words

Linguistic The heat, pressure, and chemical reactions that cause rocks to change can be generated in several different ways. *Contact metamorphism* occurs when a body of hot magma comes into contact with surrounding rocks, baking them and possibly changing their composition. *Regional metamorphism* occurs when large areas of Earth are subjected to increased pressure and temperature, such as occurs during mountain building.

Content Background

Certain minerals produced by specific degrees of metamorphism can be used to indicate whether that degree of metamorphism has occurred. If these minerals are found in a rock, you could determine the degree of metamorphism that the rock underwent. Minerals of this type are referred to as *index minerals*. Some examples are chlorite, garnet, staurolite, and sillimanite.

Visual Learning

Figure 15-20 What force could cause the parallel layers in foliated rocks? *pressure directed at right angles to the layers* LS

Types of "Changed" Rocks

Metamorphic rocks are divided into two groups: foliated (FOH lee ay tud) and non-foliated, as shown in **Figure 15-20. Foliated** rocks have bands of minerals. These minerals have been heated and squeezed into parallel layers. Many foliated metamorphic rocks have bands of different-colored minerals. Slate, gneiss (NISE), phyllite (FIHL ite), and schist are all examples of foliated rocks. **Non-foliated** metamorphic rocks do not have distinct layers or bands. These rocks, such as quartzite, marble, and soapstone, usually have more even colors than foliated rocks.

Try the next activity to better understand how Earth recycles rocks from one kind to another.

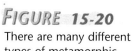

FIGURE 15-20

There are many different types of metamorphic rocks. *What force could cause the parallel layers in foliated rocks?*

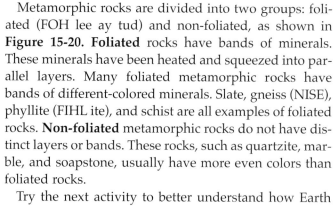

A This carved walrus is soapstone, a non-foliated metamorphic rock.

B This schist is a foliated metamorphic rock. The red grains are garnets.

C This statue from a fountain in Italy is made of marble, a non-foliated metamorphic rock.

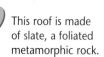

E Gneiss is a foliated metamorphic rock.

D This roof is made of slate, a foliated metamorphic rock.

Inclusion Strategies

Learning Disabled Help learning-disabled students better understand Activity 15-2 by making sure that members of each group of students evenly share the tasks. Provide independent guidance to each learning-disabled student throughout the activity.

Gneiss Rice

You will experiment with pressure to see how it can produce bands of minerals in metamorphic rocks. You will compare real rock samples with clay rocks that you will make.

What You'll Investigate

You will try to re-create conditions that cause an igneous rock to change into a banded rock.

Procedure

1. **CAUTION:** *Don't eat anything from the experiment.*
2. **Sketch** the granite specimen in your Science Journal.
3. **Pour** the rice onto the table. **Roll** the ball of clay in the rice. Some of the rice will stick to the outside of the ball. **Knead** the ball until the rice is spread out fairly evenly inside and out. Roll and knead the ball again, and repeat until your clay sample has lots of "minerals" distributed throughout it.
4. Using the rolling pin, **roll** the clay so that it is about 0.5 cm thick. Don't roll it too hard. The grains of rice should be pointing in different directions. **Draw** a picture of the clay in your Science Journal.
5. Take the edge of the clay closest to you and **fold** it toward the edge farthest from you. **Roll** the clay in the direction you folded it. Fold and roll the clay in the same direction several more times. Flatten the lump to 0.5 cm thickness again. **Draw** what you observe in your "rock" and in the gneiss sample in your Science Journal.

Conclude and Apply

1. What features did the granite and the first lump of clay have in common?
2. What force(s) caused the positions of rice grains in the lump of clay to change? How is this process similar to and also different from what happens in nature?

Goals

- Investigate ways rocks become changed.
- Construct a metamorphic rock.

Materials

- rolling pin
- lump of modeling clay
- uncooked rice (wild rice, if available), half cup
- sample of granite
- sample of gneiss

15-4 Metamorphic Rocks and the Rock Cycle **429**

Activity 15-2

Purpose

Kinesthetic Students will illustrate how pressure causes a realignment of mineral grains, forming a metamorphic rock. **L2** **COOP LEARN** **P** **ELL**

Process Skills

observing, inferring, recognizing cause and effect, interpreting scientific illustrations, making models

Time

30 minutes

Safety Precautions

Caution students to wear safety goggles and aprons during the entire activity. Tell them not to eat anything from the experiment.

Activity Worksheets, pages 5, 93-94

Teaching Strategies

- Have students work in groups of four.
- Have students compare their metamorphic rock with the rocks of other groups. If there are variations in how the model rocks appear, have students hypothesize as to the reason why.

Troubleshooting Roll small amounts of the clay in the rice. Put the smaller pieces together, and continue to knead the clay until the rice is thoroughly mixed. Use a clay ball the size of a tennis ball.

Answers to Questions

1. The granite has many mineral crystals arranged randomly throughout. The clay has many "minerals" distributed randomly throughout.
2. Compression on the clay forced the rice grains to change. In nature, compression could occur during mountain-building processes. In nature, there would also be heat and chemical change causing metamorphism.

Assessment

Process Provide students with three different metamorphic rocks and the rocks from which they formed. Have students determine which rock changed to form a foliated metamorphic rock, similar to the banded clay rock in the activity. Use the Performance Task Assessment List for Making Observations and Inferences in **PASC,** p. 17. **P**

3 Assess

Check for Understanding

Using Science Words To help students understand how metamorphic rocks are formed, have them research the origin of the word *metamorphic*. (*Meta* means "change"; *morph* means "form.") **L2**

Reteach

LS **Visual-Spatial** Have students sketch foliated and non-foliated metamorphic rocks. Ask individual students to describe what their sketches show. **L1**

Extension

For students who have mastered this section, use the **Reinforcement** and **Enrichment** masters.

4 Close

·MINI·QUIZ·

Use the Mini Quiz to check students' recall of chapter content.

1. Rocks that form when older rocks are heated or squeezed are called _____ rocks. *metamorphic*

2. Metamorphic rocks in which mineral grains are aligned in parallel layers are _____. *foliated*

3. Metamorphic rocks that do not have distinct layers are _____. *non-foliated*

4. An illustration of how different kinds of rocks are related to each other and how rocks change from one form to another is the _____. *rock cycle*

The Rock Cycle

Rocks can be recycled from one type to another. If you wanted to describe the process to someone, how would you do it? Would you use words or pictures? Scientists have created a diagram called the **rock cycle** to show the process. It shows how different kinds of rock are related to one another and how rocks change from one form to another. Each rock is on a continuing journey through the rock cycle, as shown in **Figure 15-21.**

FIGURE 15-21
This diagram of the rock cycle shows how rocks are constantly recycled from one kind of rock to another.

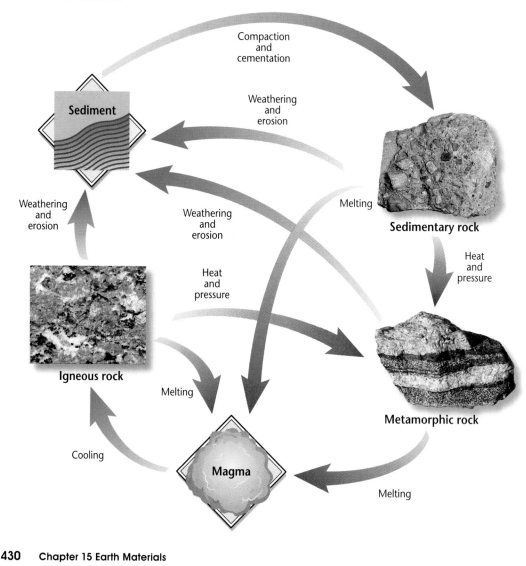

The Journey of a Rock

Pick any point on the diagram of the rock cycle in **Figure 15-21**, and you will see how the rock at that point could become any other kind of rock. Let's start with a blob of lava that oozes to the surface and cools, as shown in **Figure 15-22**. It forms an igneous rock. If that rock happens to fall into a stream, the water will gently wear off small pieces. These pieces of rock are now called sediment and will be washed downstream. In time, this sediment is piled up and cemented together. It becomes a sedimentary rock. But, pressure and heat inside Earth may change the sedimentary rock into a metamorphic rock. In this way, all rocks on Earth are reused and recycled over millions and millions of years. This process is happening right now.

FIGURE 15-22

This lava in Hawaii is flowing into the ocean and cooling rapidly, causing the steam.

Section Wrap-up

1. Identify two factors that combine to produce metamorphic rocks.

2. Name examples of foliated and non-foliated rock samples, and explain the difference between the two types of metamorphic rocks.

3. **Think Critically:** Trace the journey of a granite through the rock cycle. Explain how this rock could be changed from an igneous rock to a sedimentary and metamorphic rock.

4.  *Skill Builder*

   Observing Describe a part of the rock cycle you can observe occurring around you or see on television news. If you need help, refer to Observing and Inferring on page 550 in the **Skill Handbook**.

Using Computers

Using a spreadsheet program, create a data table to list and compare the properties of different rocks and minerals that you have studied in this chapter.

Section Wrap-up

1. heat, pressure, and chemical reactions

2. foliated: gneiss; nonfoliated: marble. Foliated rocks have minerals that have been rearranged into distinct parallel layers. Minerals in nonfoliated rocks are recrystallized and do not show parallel layers.

3. **Think Critically** Igneous rocks can be weathered into sediment. The sediment is carried and deposited where it is compacted and cemented together, forming a sedimentary rock. The sedimentary rock can be squeezed, heated, and changed chemically, forming a metamorphic rock.

Using Computers

Answers will vary but should include hardness, color, streak, luster, and cleavage and fracture for minerals. Answers should include intrusive and extrusive for igneous rocks; broken, chemical, and biochemical for sedimentary rocks; and foliated and nonfoliated for metamorphic rocks.

Skill Builder

Observing Answers will vary but could include erosion from floods or landslides and volcanic eruptions.

Assessment

Content Have students write a letter to you or another student in which they describe a rock that they "find" on their way home. The letter must describe the rock well enough without naming it so that you or the student can identify it. Use the Performance Task Assessment List for Writing in Science in **PASC**, p. 87. L2 P

People & Science

Background

- Composed primarily of silica and alumina, pipestone is also known as *catlinite*. The artist George Catlin (1796-1871) visited the pipestone quarries of southwestern Minnesota in 1836. He sent a sample of the stone to Boston to be analyzed, where it was identified as a new compound and named in his honor.

- Native Americans probably began quarrying pipestone in the 17th century, about the same time that they acquired metal tools from European traders. The Minnesota quarries came to be the preferred source of pipestone among Plains tribal peoples and were controlled by them until 1928. In 1937, the federal government created Pipestone National Monument. The monument is open to the public, but quarrying rights belong exclusively to Native Americans.

- The craft of pipestone carving nearly died out in the years following World War II. A handful of Dakota and Ojibwa families—including Alice Derby Erickson's—kept the craft alive.

Teaching Strategies

- Have students work in teams to research the history of the pipe in Native American culture. **L3** **COOP LEARN**

- Invite a local sculptor, potter, or wood carver to your class to demonstrate the techniques he or she uses to work with materials such as clay, stone, and wood.

- Have students write a fictional story about a pipestone carving that is traded from tribe to tribe in the early 1800s. **L2** **P**

Alice Derby Erickson, *Pipestone Carver*

Q **Ms. Erickson, tell us about yourself.**

A I work as a park ranger at the Pipestone National Monument in southwestern Minnesota. I'm a full-blooded Dakota Sioux and a pipestone carver.

Q **What is pipestone?**

A Pipestone is a soft, red stone made of silica and aluminum, also known as catlinite. It's about as hard as your fingernails.

Q **Why is pipestone important to Native Americans?**

A Many Native Americans believe that pipestone is the flesh and blood of our ancestors. For centuries, this sacred stone has been quarried here and carved into small figures and, of course, pipes. The pipe—called *channunpa* in Dakota—is used in most of our important ceremonies.

Q **How is the stone quarried?**

A Only Native Americans are allowed to quarry pipestone. All the work is done by hand. We use sledgehammers, picks, and shovels to remove the hard rock that lies over the pipestone. When we reach the pipestone layer, pieces of it are carefully pried out.

Q **Would you describe the carving process?**

A We first cut out a rough form, then shape it using various files, and smooth the surface with sandpaper. Finally, we heat the stone, apply beeswax, and put it in cold water. That sets the wax and brings out the rich, red color.

Q **What are your earliest memories of carving?**

A I started carving at about age ten. I helped my mother carve turtles. Turtles, to the Dakota people, are a symbol of long life. I'd cut the rough shapes of the turtles and then help my mother sand and finish them. The craft of pipestone carving passes from generation to generation. In my family, we are the fourth generation of carvers.

Career Connection

Think about careers that deal with natural materials. Interview another artist who uses rocks or minerals, and make a presentation to the class about him or her.

For More Information

Pipestone National Monument is administered by the National Park Service. Information about the quarrying process and the role of pipestone in Native American culture can be obtained by contacting: The Superintendent, Pipestone National Monument, P.O. Box 727, Pipestone, MN 56164.

Career Connection

Career Path Artists such as sculptors may have fine arts degrees. However, many artists have no formal art degrees.

R ead the statements below that review the most important ideas in the chapter. Using what you have learned, answer each question in your Science Journal.

1. All minerals occur naturally, are inorganic solids, and have a definite pattern of atoms and makeup. *List five properties that can be used to help identify a mineral.*

2. Intrusive igneous rocks form below Earth's surface and have large mineral crystals. *Why don't extrusive igneous rocks have large mineral crystals?*

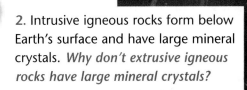

3. Sedimentary rocks can be formed from pieces of other rocks, organic material, and minerals that evaporate or settle out of solution. *How could the grains in a sedimentary rock like sandstone be moved and then deposited?*

4. Rocks form and change because of processes in the rock cycle. *Explain or draw the route through the rock cycle that granite could take, from an igneous rock to a metamorphic rock, to ending up as a sandstone.*

Chapter 15 Review   **433**

Review

Have students look at the illustrations on this page. Ask them to describe details that support the main ideas of the chapter found in the statement for each illustration.

Teaching Strategies

Divide the class into three groups. Have one group create a model showing why extrusive rocks don't have large crystals. Have another group use a stream table to show how sand is eroded and deposited by water. The final group can make a poster of the rock cycle. L2

Answers to Questions

1. hardness, streak, color, luster, and cleavage
2. They cool too rapidly for large crystals to grow.
3. As sandstone is weathered, sand particles are eroded by flowing water. As the water slows, sand grains are deposited.
4. Granite could be changed to a metamorphic rock by heat and pressure changes. The metamorphic rock could then be weathered, eroded, and deposited where it could be compacted and cemented into sandstone.

Science at Home

IS **Intrapersonal** Encourage students to find rocks of each kind used somewhere in or around their homes. Examples include sedimentary limestone used as driveway or landscaping stone, metamorphic slate used as roof tile or for flooring, and igneous scoria used as the rock in gas grills.

✔Assessment

Portfolio Encourage students to place in their portfolios one or two items of what they consider to be their best work. Examples include:
- Skill Builder, p. 414
- MiniLAB, p. 411
- Activity 15-2, p. 429 P

Performance Additional performance assessments may be found in **Performance Assessment** and **Science Integration Activities.** Performance Task Assessment Lists and rubrics for evaluating these activities can be found in Glencoe's **Performance Assessment in the Science Classroom (PASC).**

Chapter 15 Review

Using Key Science Words

1. intrusive
2. sedimentary rock
3. gem
4. foliated
5. rock cycle

Checking Concepts

6. b
7. d
8. b
9. a
10. c

Thinking Critically

11. No, sugar comes from plants. Minerals are non-living.
12. Answers could include that deposits are too small, the expense of mining in remote areas, and environmental concerns.
13. If gneiss, granite, and basalt were all exposed to erosion, pieces of all three could be moved and deposited to form a conglomerate.
14. No, the heat, pressure, and chemical changes that form metamorphic rocks would most likely destroy the dinosaur bone.
15. Quartz forms in igneous rocks as magma cools, and it is eroded, moved, and deposited in sedimentary rocks.

Using Key Science Words

crystal	metamorphic rock
extrusive	mineral
foliated	non-foliated
fossil fuel	ore
gem	rock
igneous rock	rock cycle
intrusive	sedimentary rock

Match each phrase with the correct term from the list of Key Science Words.

1. type of igneous rocks that are cooled below Earth's crust
2. kind of rock made of pieces of other rocks
3. a rare, precious mineral that can be cut and polished
4. type of metamorphic rocks with layers of different minerals
5. diagram that shows how the formation of igneous, sedimentary, and metamorphic rocks can be inter-related

Checking Concepts

Choose the word or phrase that completes the sentence.

6. Rocks are usually composed of two or more _____.
   a. pieces   c. fossil fuels
   b. minerals   d. foliations

7. Metamorphic rocks are formed when _____.
   a. volcanoes erupt
   b. fissures ooze lava
   c. globs of lava cool underground
   d. heat and pressure change rocks

8. Sedimentary rocks can be classified as _____.
   a. foliated or non-foliated
   b. biochemical, chemical, or broken
   c. extrusive or intrusive
   d. gems or ores

9. _____ rocks are composed of broken pieces of rock.
   a. Sedimentary   c. Old
   b. Banded   d. Extrusive

10. Natural gas, coal, and oil are all _____.
    a. minerals
    b. metamorphic rocks
    c. fossil fuels
    d. igneous rocks

Thinking Critically

Answer the following questions in your Science Journal using complete sentences.

11. Is a sugar crystal a mineral? Why or why not?
12. Metal deposits in Antarctica are not considered to be profitable. List some reasons for this.
13. How could pieces of gneiss, granite, and basalt all be found in one conglomerate?

Assessment Resources

Reproducible Masters
Chapter Review, pp. 33-34
Assessment, pp. 65-68
Performance Assessment, p. 53

Glencoe Technology
Computer Test Bank
MindJogger Videoquiz

Chapter 15 Skills Review

14. Would you expect to find a well-preserved dinosaur bone in a metamorphic rock like schist? Explain.

15. Explain how the mineral quartz could be in an igneous rock and in a sedimentary rock.

Developing Skills

If you need help, refer to the description of each skill in the Skill Handbook.

16. **Making and Using Tables:** Using the information that you collected in the MiniLAB on page 411, create a data table to compare and contrast the four minerals that you tested. Be sure to include color and luster in addition to the two tests that you performed.

17. **Comparing and Contrasting:** Compare and contrast the differences between intrusive and extrusive rock appearance.

18. **Interpreting Data:** Clem has several mineral samples and is doing his best to identify them. He has tested the hardness, streak, luster, and color of each of the samples but still can't figure out what the samples are. Suggest two other tests that may help. Explain why these tests would be useful.

19. **Observing and Inferring:** You are hiking in the mountains and as you cross a shallow stream, you spy an unusual rock. When you pick it up, you notice it is full of fossil shells. Your dad asks you what it is. What do you tell him and why?

Schist   Conglomerate   Granite

20. **Interpreting Scientific Pictures:** Review the pictures above and determine whether each is a sedimentary, igneous, or metamorphic rock.

Performance Assessment

1. **Display:** Create a display that shows the rock cycle using real rock specimens.

2. **Designing an Experiment:** You were able to make a model for metamorphism using clay and rice. Now experiment with baking soda and vinegar and design an experiment using these two materials to demonstrate how a volcano may erupt and produce igneous rocks. You may want to add red food coloring.

Developing Skills

16. **Making and Using Tables** Student data tables should include the following. Quartz is nonmagnetic, usually colorless or white, harder than glass, nonmetallic in luster, and does not react to acid. Hornblende is black, is nonmagnetic, has two directions of cleavage, is nonmetallic in luster, and does not react to acid. Calcite is nonmagnetic, white or colorless, has three directions of cleavage, a hardness of 3, is nonmetallic in luster, and reacts to acid. Magnetite is magnetic, is black, has a black streak, does not react to acid, and has a metallic luster.

17. **Comparing and Contrasting** Crystals usually form in both cases; however, the crystals in intrusive rocks are easily visible with the unaided eye. Crystals in the extrusive rocks are usually too small to be seen without a microscope or they do not form at all.

18. **Interpreting Data** He could use reaction to acid and cleavage. Reaction to acid can be used to test for calcite. Many minerals break with a specific number of cleavage planes, which can be used to identify them.

19. **Observing and Inferring** The rock is a biochemical sedimentary rock. It is a fossil-rich limestone.

20. **Interpreting Scientific Pictures** granite, igneous; schist, metamorphic; conglomerate, sedimentary

Performance Assessment

1. Student displays should show all parts of the rock cycle. In place of the names of the rocks, actual samples could be cemented to the display. Use the Performance Task Assessment List for Display in **PASC**, p. 63. [P]

2. Suggest that students add flour to the mixture to add substance. Caution students to use only one tablespoon of flour and one tablespoon of baking soda and about 250 mL of vinegar. The vinegar reacts with the baking soda, releasing carbon dioxide gas. The gas forces the red-colored liquid containing the flour out of the container. This material flows like lava. As it dries, it will become like igneous rock forming from lava erupted from a volcano. Use the Performance Task Assessment List for Designing an Experiment in **PASC**, p. 23. [P]

Chapter Organizer

Section	Objectives/Standards	Activities/Features
Chapter Opener		Explore Activity: Modeling Quakes, p. 437
16-1 **What's shaking?** (3½ sessions, 1½ blocks)*	1. **Determine** what happens during an earthquake. 2. **Investigate** how volcanoes erupt and where they occur. National Science Content Standards: (5-8) UCP2, UCP3, B3, D1, E2, F3, F5	Problem Solving: Identifying Epicenters, p. 441 MiniLAB: Constructing a Model Volcano, p. 442 Activity 16-1: Lava Landscapes, p. 444 Skill Builder: Building Models, p. 445 Using Math, p. 445
16-2 **Science and Society:** **Predicting Earthquakes** (1 session, ½ block)*	3. **Describe** where earthquakes occur in California. 4. **Explain** how to use library and Internet resources. National Science Content Standards: (5-8) UCP3, D1, E2, F5, G1	Skill Builder: Using Resources, p. 447
16-3 **A Journey to Earth's Center** (2½ sessions, 1½ blocks)*	5. **Describe** the interior of Earth. 6. **Investigate** the evidence for the theory of continental drift. National Science Content Standards: (5-8) UCP2, UCP3, A2, B3, D1, D2, G1, G3	Activity 16-2: Making a Seismograph, pp. 450-451 Using Technology: Satellite Tracking, p. 454 Skill Builder: Making a Model, p. 455 Science Journal, p. 455
16-4 **Crashing Continents** (4 sessions, 2 blocks)*	7. **Explain** how ocean crust forms. 8. **Determine** how Earth's surface is changed by plate tectonics. 9. **Discuss** how mountains are formed. National Science Content Standards: (5-8) UCP1, UCP2, UCP3, A1, A2, D1, D2	MiniLAB: Observing Convection, p. 458 Skill Builder: Sequencing, p. 461 Using Math, p. 461 Science & Language Arts: Pacific Crossing, p. 462 Science Journal, p. 462

* A complete Planning Guide that includes block scheduling is provided on pages 31T-33T.

Activity Materials

Explore	Activities	MiniLABs
page 437 spring toy	page 444 cookie sheet, spoon, timer, cornstarch, paper cups, white glue, dehydrated silica gel pages 450-451 ring stand with ring, wire coat hanger, metric ruler, wide rubber band, masking tape, fine-tip marker	page 442 1-L plastic bottle with cap, cookie sheet, long flat pan, baking soda, liquid soap, white vinegar page 458 10-gallon aquarium, bucket, stiff plastic, food coloring

Need Materials? Call Science Kit (1-800-828-7777).

Teacher Classroom Resources

Reproducible Masters	Transparencies	Teaching Resources
Activity Worksheets, pp. 5, 97-98, 101 **Cross-Curricular Integration**, p. 20 **Enrichment**, p. 59 **Lab Manual 31** **Multicultural Connections**, pp. 35-36 **Reinforcement**, p. 59 **Study Guide**, p. 59	**Science Integration Transparency 16**, Hot Bacteria **Section Focus Transparency 53**, Is California cracked? **Teaching Transparency 31**, Earthquake Model **Teaching Transparency 32**, Earthquakes and Volcanoes	**Spanish Resources** **English/Spanish Audiocassettes** **Cooperative Learning Resource Guide** **Lab Partner** **Lab and Safety Skills** **Lesson Plans**
Enrichment, p. 60 **Reinforcement**, p. 60 **Science and Society/Technology Integration**, p. 34 **Study Guide**, p. 60	**Section Focus Transparency 54**, San Francisco—1906	**Assessment Resources**
Activity Worksheets, pp. 5, 99-100 **Enrichment**, p. 61 **Reinforcement**, p. 61 **Study Guide**, p. 61	**Section Focus Transparency 55**, An Amazing Journey	**Chapter Review**, pp. 35-36 **Assessment**, pp. 69-72 **Performance Assessment**, p. 54 **Performance Assessment in the Science Classroom (PASC)** **MindJogger Videoquiz** **Alternate Assessment in the Science Classroom** **Computer Test Bank**
Activity Worksheets, pp. 5, 102 **Enrichment**, p. 62 **Lab Manual 32** **Reinforcement**, p. 62 **Science Integration Activities**, pp. 69-70 **Study Guide**, p. 62	**Section Focus Transparency 56**, A New Ocean	

Key to Teaching Strategies

The following designations will help you decide which activities are appropriate for your students.

L1 Level 1 activities should be appropriate for students with learning difficulties.

L2 Level 2 activities should be within the ability range of all students.

L3 Level 3 activities are designed for above-average students.

ELL ELL activities should be within the ability range of English Language Learners.

LS These activities are designed to address different learning styles.

COOP LEARN Cooperative Learning activities are designed for small group work.

P These strategies represent student products that can be placed into a best-work portfolio.

GLENCOE TECHNOLOGY

The following multimedia resources are available from Glencoe.

Science and Technology Videodisc Series (STVS)
Earth & Space
 Moving Continents
 Seismic Simulator
 Miniature Volcanoes
 Underwater Seismograph
 Indian Pompeii

The Infinite Voyage Series
To the Edge of the Earth
Living with Disaster
Restless Earth

Glencoe Earth Science CD-ROM

Teacher Classroom Resources

This is a representation of key blackline masters available in the Teacher Classroom Resources.

Teaching Aids

Section Focus Transparencies

53 SECTION FOCUS TRANSPARENCY

IS CALIFORNIA CRACKED?

You've probably heard that earthquakes will eventually make California fall into the ocean. That isn't true, but the San Andreas Fault in California is a lot like a big crack in Earth's crust. Earth's crust is constantly moving.

1. Do we feel all the earthquakes that occur? Explain.
2. Where was the largest recorded earthquake in the United States?

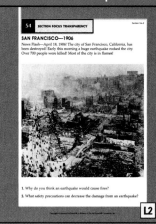

54 SECTION FOCUS TRANSPARENCY

SAN FRANCISCO—1906

News Flash—April 18, 1906! The city of San Francisco, California, has been destroyed! Early this morning a huge earthquake rocked the city. Over 700 people were killed! Most of the city is in flames!

1. Why do you think an earthquake would cause fires?
2. What safety precautions can decrease the damage from an earthquake?

55 SECTION FOCUS TRANSPARENCY

AN AMAZING JOURNEY

Imagine digging a hole through Earth. What do you think you might find? Jules Verne imagined such a journey in his book *Journey to the Center of the Earth.* The world that he imagined beneath Earth's surface is nothing like what scientists have discovered about the world beneath our feet!

1. Name another movie or book that shows an amazing journey like this.
2. What would you find at Earth's center?

Science Integration Transparencies

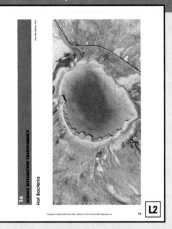

16 SCIENCE INTEGRATION TRANSPARENCY

Hot Bacteria

Teaching Transparencies

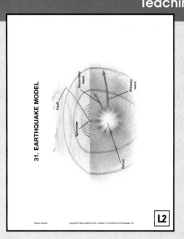

31. EARTHQUAKE MODEL

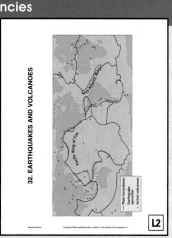

32. EARTHQUAKES AND VOLCANOES

Meeting Different Ability Levels

Study Guide for Content Mastery

Chapter 16 Study Guide for Content Mastery

What's shaking?

Reinforcement

Chapter 16 Reinforcement

What's shaking?

Enrichment Worksheets

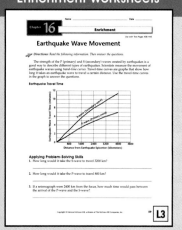

Chapter 16 Enrichment

Earthquake Wave Movement

436C

Hands-On Activities

Science Integration Activities

Lab Manual

Activity Worksheets

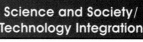

Enrichment and Application

Cross-Curricular Integration

Multicultural Connections

Science and Society/ Technology Integration

Assessment

Performance Assessment

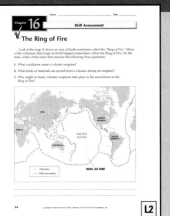

Chapter Review

Assessment

Chapter 16

Earth's Structure

CHAPTER OVERVIEW

Section 16-1 This section explains what causes earthquakes and volcanoes. The different types of volcanoes are also described.

Section 16-2 Science and Society Students are presented with the skill of doing research on how earthquakes might be predicted. Students develop skills in using the library and Internet resources.

Section 16-3 This section describes the layered structure inside Earth. Students are also presented with evidence in support of continental drift.

Section 16-4 This section describes how the movement of plates changes Earth's surface. The formation of new crust and mountain building are described.

Chapter Vocabulary

earthquake	continental
fault	drift
focus	plate
epicenter	seafloor
volcano	spreading
mantle	convection
inner core	current
outer core	plate tectonics

Theme Connection

Energy/Systems and Interactions
The theme of Energy is presented as forces inside Earth are shown to be responsible for earthquakes and volcanoes. Earth's moving plates generate forces inside Earth. These forces cause energy to build up inside Earth, which in turn causes earthquakes and volcanoes. A secondary theme is Systems and Interactions, dealing with the evidence in support of continental drift.

436

Chapter Preview

Section 16-1 What's shaking?

Section 16-2 Science and Society: Predicting Earthquakes

Section 16-3 A Journey to Earth's Center

Section 16-4 Crashing Continents

436

Skills Preview

▶ **Skill Builders**
  • make models
  • sequence

▶ **MiniLABs**
  • observe
  • make models

▶ **Activities**
  • make models
  • observe
  • hypothesize
  • collect data
  • design an experiment
  • compare results

Learning Styles

Look for the following logo for strategies that emphasize different learning modalities.

Kinesthetic	Explore, p. 437; Making a Model, pp. 441, 449; Inclusion Strategies, p. 441; MiniLAB, pp. 442, 458; Reteach, pp. 443, 454; Activity 16-1, p. 444; Activity 16-2, p. 450; Demonstration, p. 459
Visual-Spatial	Visual Learning, pp. 440, 443, 449; Inclusion Strategies, p. 442; Teaching the Content, p. 447; Demonstration, pp. 452, 459
Interpersonal	Enrichment, pp. 439, 457; Brainstorming, p. 440; Assessment, p. 458
Intrapersonal	Science at Home, p. 463
Logical-Mathematical	Problem Solving, p. 441; Assessment, p. 444; Using Math, pp. 445, 461; Inquiry Questions, p. 457
Linguistic	Science Journal, pp. 440, 454; Assessment, p. 445
Naturalist	Revealing Preconceptions, p. 439

Chapter 16

Earth's Structure

Next stop on your summer vacation: a Hollywood movie studio. You feel excited as you board a tram that will take you to see giant apes, spaceships, and shark attacks. As the tram starts, the track begins to shake, and buildings around you collapse. You're doomed, it seems. Just then, the tram starts to move and whisks you away to safety. You have just been through a fake earthquake. Thank goodness it was only make-believe. But for people who live in California and in many places around the world, the possibility of an earthquake is real, as in the case of the Loma Prieta earthquake in 1989 shown here.

EXPLORE ACTIVITY

Modeling Quakes 🥽

1. You and your lab partner are going to make two different kinds of waves, similar to those that happen during an earthquake. Stretch out a toy spring along the floor until the coils are about an inch apart. You and your lab partner should each hold one end.
2. Gather several coils together and then let them go. Observe the action of the wave as it travels down the spring.
3. Next, whip the end of the spring quickly to your right, keeping it on the floor. Note the difference in the shape of the wave.

Science Journal

In your Science Journal, draw pictures of the waves that you and your partner made with the spring. Describe the motion of the spring in each case.

437

EXPLORE ACTIVITY

Purpose

LS **Kinesthetic** Use the Explore activity to help students create models of waves produced during earthquakes. **COOP LEARN** **L2** **ELL**

Materials

spring toy or Slinky, pen

Alternate Materials

A string may be used instead of a Slinky to form waves.

Teaching Strategies

- To help students observe the movement of the wave through the spring, tie four or five short pieces of string along the length of the spring.
- When squishing the spring, instruct students to compress some coils and then let them go.

Science Journal Drawings will vary. By compressing the spring, a wave will be produced that travels parallel to the spring. When the end of the spring is moved quickly, the resulting wave moves perpendicular to the direction the wave travels.

✓ Assessment

Oral Ask students to explain how particles of Earth might move as both types of waves move through Earth. Use the Performance Task Assessment List for Oral Presentation in **PASC**, p. 71.

Assessment Planner

Portfolio
Refer to page 463 for suggested items that students might select for their portfolios.

Performance Assessment
See page 463 for additional Performance Assessment options.
Skill Builders, pp. 445, 447, 455, 461
MiniLABs, pp. 442, 458
Activities 16-1, p. 444; 16-2, pp. 450-451

Content Assessment
Section Wrap-ups, pp. 445, 455, 461
Chapter Review, pp. 463-465
Mini Quizzes, pp. 445, 455, 461

Group Assessment
Opportunities for group assessment occur with Cooperative Learning Strategies and Flex Your Brain activities.

Prepare

Section Background

- Although we normally think of Earth's crust as being broken into plates, the plates actually are composed of more than just the crust. The crust and the upper rigid portion of the mantle make a layer called the *lithosphere*. It is the lithosphere that is broken into plates.

- Compressional forces are applied to rocks where Earth's plates come together. Where plates pull apart, tensional forces are applied. When plates slide past each other, shearing forces are applied to rocks.

Preplanning

Refer to the Chapter Organizer on pages 436A-B.

1 Motivate

Bellringer

 Before presenting the lesson, display **Section Focus Transparency 53** on the overhead projector. Assign the accompanying **Focus Activity** worksheet.
`L2` `ELL`

16•1 What's shaking?

What YOU'LL LEARN

- What happens in an earthquake
- How volcanoes erupt and where they occur

Science Words:
earthquake
fault
focus
epicenter
volcano

Why IT'S IMPORTANT

When you see news reports of earthquakes and volcanic eruptions, you'll understand what's causing these events.

Earthquakes

The surface of Earth is always changing. The outer layer of Earth is called the crust. The crust is firm and seems solid, but it is broken up into large pieces. These pieces can move around. Because the pieces do move, energy builds up wherever two pieces meet. The energy builds up in Earth like it does when you bend a stick. As you bend the stick, energy builds until the stick breaks and the energy is released. When that energy is released rapidly, Earth shakes or vibrates.

An **earthquake** is the shaking of Earth produced by the fast release of energy. The energy travels as waves through Earth, as shown in **Figure 16-1.** The amount of damage caused by an earthquake depends upon the size of the waves. Most earthquakes are so small that you can't even feel them.

FIGURE 16-1
Earthquakes occur when sudden movement along a fault releases energy. The first earthquake wave is called the primary wave, and the second wave is called the secondary wave.

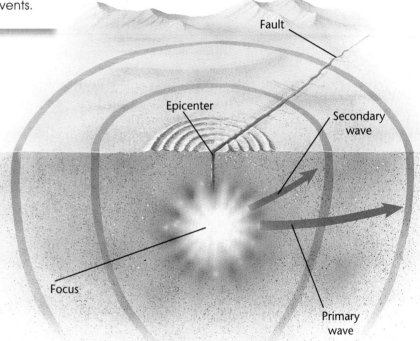

Fault

Epicenter

Secondary wave

Focus

Primary wave

Program Resources

 Reproducible Masters
Activity Worksheets, pp. 5, 97-98, 101 `L2`
Cross-Curricular Integration, p. 20 `L2`
Enrichment, p. 59 `L3`
Lab Manual, pp. 111-114 `L2`
Multicultural Connections, pp. 35-36 `L2`
Reinforcement, p. 59 `L2`
Study Guide, p. 59 `L1`

Transparencies
Section Integration Transparency 16 `L2`
Section Focus Transparency 53 `L2`
Teaching Transparencies 31, 32 `L2`

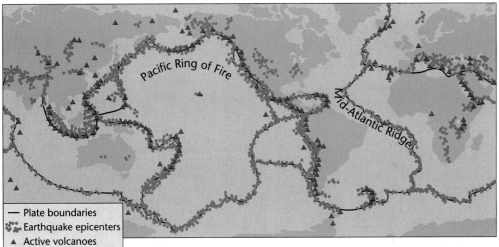

Plate boundaries
Earthquake epicenters
Active volcanoes

Earthquake Epicenters

Have you noticed that earthquakes seem more common in certain parts of the world than in others? Places such as Japan, China, and Mexico have many earthquakes. The pattern of where earthquakes happen has to do with how Earth's crust fits together. The outer layers of Earth, which include the crust, are broken into large pieces or sections called plates. As shown in **Figure 16-2,** earthquakes are more common where the edges of the giant plates meet. Scientists believe that these plates have been moving for millions and millions of years and continue to move today.

Faults

When the plates rub against one another, energy builds up along areas called faults. A **fault** is like a crack in the crust between the two pieces, where there has been movement. You can think of it like a crack in the sidewalk, but much bigger.

When there is an earthquake, energy moves away from the break in waves, like ripples on a pond. The exact spot inside Earth where the earthquake starts is called the **focus,** as shown in **Figure 16-1.** The waves travel in every direction including upward. The point where the waves hit the surface of Earth, directly over the focus, is called the **epicenter** (EP uh sen tur). A great deal of damage often occurs at the epicenter in a big earthquake.

FIGURE 16-2

Most of the earthquakes and active volcanoes occur along the edge of the Pacific Ocean. This area is known as the Pacific Ring of Fire.

Tying to Previous Knowledge

Help students recall recent earthquakes and discuss where they occurred and the damage done.

2 Teach

Teacher F.Y.I.

The plates in Earth's lithosphere (crust and upper mantle) move around on the asthenosphere (plasticlike layer in Earth's mantle).

Using Science Words

Faults occur when there is movement of rock layers along a fracture. A *normal fault* occurs when rocks above the fault move down in relation to rocks below the fault. A *reverse fault* occurs when rocks above the fault move up in relation to rocks below the fault. A *strike-slip* fault forms when rocks on one side of a fault slide past rocks on the other side of the fault.

Enrichment

IS **Interpersonal** Ask students to research the types of faults that can happen: normal fault, reverse fault, and strike-slip fault. Once students have researched faults, ask for volunteers to demonstrate to classmates how rocks move during each type of fault.
COOP LEARN **L2**

Revealing Preconceptions

IS **Naturalist** Students may think that earthquakes and volcanoes occur randomly around Earth. Their occurrence does follow a general pattern on Earth's surface. Their prediction, however, is not always possible.

Content Background

Compressional and shearing forces cause rock layers to bend or fold. If forces are great enough or sudden enough, breaks in the rock layers can occur. Tensional forces tend to cause rock layers to break.

FIGURE 16-3

Earthquakes are measured in two ways.

A This model shows how earthquake waves travel through Earth. The primary wave (P-wave) is recorded first because it travels faster. Secondary waves (S-waves) are slower and are recorded later.

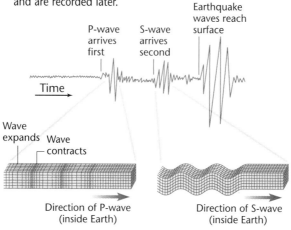

P-wave arrives first

S-wave arrives second

Earthquake waves reach surface

Time

Wave expands

Wave contracts

Direction of P-wave (inside Earth)

Direction of S-wave (inside Earth)

B A seismograph is an instrument that records the wave magnitudes from earthquakes.

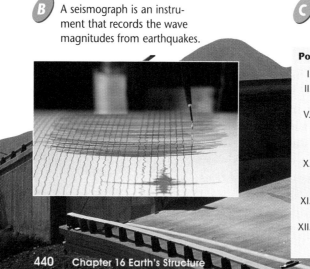

440 Chapter 16 Earth's Structure

C The Modified Mercalli scale is used to measure the damage caused by earthquakes.

Portion of the Modified Mercalli Scale

I. (1) Earth movement is not felt by people.

II. (2) A few people may notice movement if they are sitting still. Hanging objects may sway.

V. (5) Felt by almost everyone. Sleeping people are awakened. Some windows are broken and plaster cracked. Some unstable objects are overturned. Bells ring.

X. (10) Some well-built wooden structures are destroyed. Most masonry structures are destroyed. Ground is badly cracked.

XI. (11) Few, if any, structures remain standing. Broad open cracks appear in the ground.

XII. (12) Destruction is complete. Waves are seen on the ground surface.

Earthquake Magnitudes

Perhaps you've seen pictures of crumbled buildings and broken freeway overpasses in cities where strong earthquakes have happened. The strength or magnitude of an earthquake is usually measured in two ways: by the Modified Mercalli (mur KAH lee) scale as shown in **Figure 16-3,** and by the Richter (RICK tur) scale. The damage caused by an earthquake is measured using the Modified Mercalli scale, and the energy released is measured by the Richter scale.

Richter Scale

The first scale measures the size or magnitude of an earthquake. This scale was developed by Charles Richter. Richter's scale measures the size of the waves produced by an earthquake. He used an instrument called a seismograph (SIZE muh graf), as shown in **Figure 16-3B,** to measure the wave magnitudes. Stronger earthquakes make larger waves that can be measured by a seismograph. The Richter scale is used worldwide. News reports of earthquakes often mention where the earthquake registered on the Richter scale.

The most obvious way to measure the strength of an earthquake is to look at the damage caused by the quake. Scientists use the Modified Mercalli scale to measure damage. The scale ranges from a low of 1, where you barely notice any damage, to a 12, where buildings are totally destroyed. In an earthquake rated 12 on this scale, the ground moves in rolling waves, and objects on the ground can be sent flying through the air.

Problem Solving

Identifying Epicenters

When an earthquake happens, two types of waves, P (primary) and S (secondary) waves, are produced. P-waves travel faster than the S-waves. P-waves move faster because of how the wave moves through Earth, as seen in **Figure 16-3.** A seismograph records both kinds of waves, but the P-waves arrive first. The amount of time between when the two waves arrive at the station tells how far away the epicenter is. For example, if a P-wave reached the station at 9:00, and the S-wave arrived 4 minutes and 1 second later, from a graph you can calculate that the earthquake epicenter was 2000 km from the station. If you draw a 2000-km circle around that station, the epicenter would lie somewhere along that circle.

Solve the Problem:
Use the information given to find an earthquake epicenter. If the epicenter is 6000 km from station 3, is the epicenter at point A or at point B?

Think Critically:
Explain why you need recordings from three stations to pinpoint the epicenter. Why wouldn't just two stations work? Are four stations better?

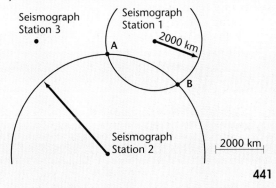

Seismograph Station 3

Seismograph Station 1

2000 km

A

B

Seismograph Station 2

2000 km

441

Kinesthetic Use two students to make a model of a seismograph. Have one student hold an opened file folder. Tell the second student to draw a straight black line from the top to the bottom of the file folder. Tell the first student to hold the file folder steady while the other student draws the line. During a second try, ask the student holding the file folder to shake it back and forth while the other student tries to draw the straight line. The pen is held steady while the folder is shaken, just as the seismograph would shake during an earthquake. [L1]
COOP LEARN

Problem Solving

Solve the Problem
The epicenter is at point B. By drawing a circle around station 3 with a radius of 6000 km, this circle intersects the other two at point B.

Think Critically
One station shows a circle on which the epicenter occurs. Two stations show two possible locations of the epicenter. A third station is needed to identify the exact location of the epicenter. Three stations are all that are needed. No, four stations don't make the location of the epicenter any more accurate.

Theme Connection
Energy
Describing the causes of the eruption of lava and gases from a volcano reinforces the theme of Energy. Energy inside Earth provides heat that melts rock material. The magma is less dense than surrounding rock material and is forced upward toward Earth's surface.

Inclusion Strategies
Visually Impaired To help your visually impaired students, provide a copy of the map used in the Problem Solving activity with the drawn circles and locations A and B marked with raised lines and letters. Match visually impaired students with other students who can help them understand what is being done. [L2]
COOP LEARN

Mini LAB

Mini LAB

Constructing a Model Volcano

How is molten rock forced up through Earth's crust?

1. Pour water into a 1-L plastic soda bottle until it is one-third full. Add a large squirt of liquid soap and 3 tablespoons of baking soda. Cap the bottle and shake it for 30 seconds.
2. Place the bottle in the middle of a cookie sheet. Remove the cap, and quickly add 125 mL of vinegar.
3. Observe the reaction.

Analysis

1. The mixture of vinegar and baking soda formed the gas carbon dioxide. Explain why the carbon dioxide gas erupted out of the mouth of the bottle instead of staying inside.
2. Compare and contrast the reaction that you observed with the formation of a real volcano.

FIGURE 16-4

There are three types of volcanoes: composite, shield, and cinder cone. *Which of these three types of volcanoes do you think would be the most dangerous to live near? Explain.*

A Anak Krakatau in Indonesia is a cinder cone volcano. Cinder cone volcanoes have steep sides and are made up of small blocks of solid lava.

442

Volcanoes

Can you think of another event on Earth that can cause such damage? What about an erupting volcano? Rivers of hot lava rush along, wiping out forests and destroying towns. Clouds of gas and ash spew into the air. Volcanoes, like earthquakes, are signs that Earth is active and moving. **Volcanoes** are places in the world where the hot, liquid magma, which is molten rock from below the crust, breaks through and flows onto the ground. Like earthquakes, volcanoes are also most common along the edges of Earth's plates.

Eruptions

Volcanic eruptions happen when magma is forced up through the surface by pressure inside Earth. The magma collects in a chamber just under the surface. When enough pressure builds up in the chamber, lava, which is what magma is called at the surface, erupts through an opening or vent. This eruption shoots lava, blocks of rock, cinders, ash, and hot gases into the air. These fall down around the vent, forming a small mound. As this happens over and over, a cone builds up around the opening, and a volcano is born.

Different Volcanoes

Scientists use several things to classify volcanoes. Some volcanoes are tall and pointy. Others have long, gentle slopes. Some are made of sticky lava, while some are mostly small chunks of solid lava called cinders. The kind of magma from which the volcano forms and the amount of water in the magma determine the type of volcano. But the main difference among them is how easily they flow. Some magmas are thick and sticky, like cooked oatmeal, and don't flow easily. They produce violent, explosive eruptions. Other magma is more fluid or liquid, like honey, and flows more easily. Magma that flows easily produces quieter lava flows.

Composite and Shield Volcanoes

Figure 16-4 shows three types of volcanoes: composite, shield, and cinder cone. Composite volcanoes—the tall, pointy ones—are made up of layers of thick, sticky lava and ash. Composite volcanoes, such as Mount St. Helens in Washington, are formed by alternating quiet and explosive eruptions. Shield volcanoes, such as Kilauea in Hawaii, are made up of long, sloping layers of rock that formed from quiet eruptions of fluid lava. The third type of volcano is known as a cinder cone. This type is formed by explosive eruptions. Krakatau in Indonesia is an example of a cinder cone volcano. When it erupted in 1883, the explosion was heard 2000 km away in Australia!

C Composite volcanoes like Mount Ruapehu in New Zealand form tall, pointed mountains. They are made of layers of thick, sticky lava and ash.

B Mauna Kea in Hawaii is a shield volcano, with long, sloping sides.

16-1 What's shaking? 443

GLENCOE TECHNOLOGY

 Videodisc

The Infinite Voyage: To the Edge of the Earth
Chapter 3
Exploring Volcanoes: To the Center of the Earth

The Infinite Voyage: Living with Disaster
Chapter 5
Earthquakes: A Turbulence Beneath the Earth

Chapter 6
Predicting Earthquakes: The Parkfield Experiment

3 Assess

Check for Understanding

Discussion Lead students in a discussion about volcanoes or volcanic eruptions they have read about or seen on the news.

Reteach

Kinesthetic Help students understand what earthquakes are by allowing them to experience a minor one. Take the class to the gymnasium where they can sit on the bleachers. Ask them to close their eyes and then start pounding their feet on the bleachers or floor. They should experience vibrations just as if they were in a small earthquake. **L2** **ELL**

Extension

For students who have mastered this section, use the **Reinforcement** and **Enrichment** masters.

Visual Learning

Figure 16-4 **Which of the three types of volcanoes do you think would be the most dangerous to live near? Explain.** *composite volcanoes or cinder cones because they have explosive eruptions*

Content Background

During explosive volcanic eruptions, lava is thrown high into the air. The lava cools and hardens into different-sized material. This material is called *tephra*. It ranges in size from volcanic ash, to cinders, to large rocks called bombs.

443

Activity 16-1

Purpose

LS **Kinesthetic** Students will model different types of lava that erupt from different types of volcanoes and compare how different types of lava flow. **L2** **COOP LEARN** **ELL**

Process Skills

observing, inferring, recognizing cause and effect, making models

Time

35 minutes

Safety Precautions

Caution students to wear an apron and safety goggles. Materials must be handled and disposed of properly. Throw the waste into the garbage; do not wash the material down a drain. Caution students not to eat any materials used in this activity.

📁 **Activity Worksheets,** pages 5, 97-98

Teaching Strategies

- Have students work in groups of four.
- Have students use the following directions in making their different lavas—rhyolite lava: 5 oz water and 1 oz dehydrated silica gel, no mixing or stirring; andesite lava: mix water and cornstarch until a thick, slow-running mixture is achieved (add white glue to the mixture to help it retain a thick texture); basalt lava: pile cornstarch, then slowly add water, mix, and squish it until you achieve a runny texture.

Troubleshooting Use small cups in which to mix each of the lavas. The basaltic lava model may need to be mixed in continually wet hands in order to achieve the correct texture.

Activity 16-1

Goals

- Model lavas.
- Observe how different lavas flow.

Materials

- cookie sheet
- 6-ounce paper cups (3)
- dehydrated silica gel
- cornstarch
- white glue
- spoon
- thick book
- stopwatch
- meterstick

Lava Landscapes

In this lab, you will experiment with common, everyday materials and use them to create models of different kinds of lava. You will compare these homemade lavas by racing them down a cookie sheet to help you understand why volcanic eruptions are different.

What You'll Investigate

How do the three kinds of lava flow?

Procedure 🧤 🥽 ☠️ 🚫

1. **Set** the book on your lab table, and **place** the cookie sheet on the book at an angle. You now have a long, gentle slope for your lava to flow down.
2. Each person in your group will make one kind of lava. You will then race the lava down the cookie sheet. Your teacher will provide the lava recipes for the rhyolite (RI uh lite), andesite (AN duh site), or basalt (buh SAWLT) lavas. Each homemade lava will model how real lava flows.
3. **Prepare a data table** where you **record** the distance the flow travels in centimeters. Also time the flow in seconds and **describe** the features of each kind of lava.
4. **Leave** the lavas in the cups until you are ready to race.
5. When all three lavas are prepared, you are ready to experiment.
6. **Place** all three kinds of lava at the top of the tilted cookie sheet. **Compare** the rates that they flow downhill. **Match** the lava with the different kinds of volcanic landforms that can be found around the world.

Conclude and Apply

1. Which lava model—rhyolite, andesite, or basalt—was the most resistant to flow?
2. Which model of lava would be likely to produce a gently sloping lava plain? **Explain** your answer.
3. The amount of silica (liquid glass) in lava helps control how it will flow. The more silica, the stickier the lava. **Sequence** the three kinds of lava from silica-rich to silica-poor based on your **observations**.

Answers to Questions

1. The model of rhyolite lava was most resistant to flow.
2. The basalt lava model would most likely produce a gently sloping lava plain. Lava of this type would flow long distances before cooling, thus creating large lava plains.
3. rhyolite, andesite, and basalt

✔ Assessment

Oral Ask students to determine which type of lava or lavas would be more likely to produce a steep-sided volcano. (rhyolite and andesite) Use the Performance Task Assessment List for Making Observations and Inferences in **PASC,** p. 17. **LS** **P**

Where do volcanoes occur?

Geologists have mapped the volcanoes and earthquakes of the world, as shown in **Figure 16-2** on page 439. Looking at the map, do you see any areas with a lot of volcanic activity? One especially active area circles the Pacific Ocean. It has been called the Ring of Fire. Most of the world's active volcanoes are found in this ring around the Pacific Ocean. This area of active volcanoes gives clues to the arrangement of Earth's plates. As with earthquakes, volcanoes occur more commonly where two of Earth's plates meet. What else can scientists learn from volcanoes? Volcanoes can also give information about the inside of Earth. The lava that erupts out of volcanoes, as shown in **Figure 16-5**, came from rock that melted deep inside Earth.

FIGURE 16-5

This eruption in the Galápagos Islands occurred along the Pacific Ring of Fire.

Section Wrap-up

1. Draw a diagram of an imaginary earthquake. Label the focus, epicenter, and fault on your illustration.

2. List the three kinds of volcanoes, and briefly describe why they look different.

3. **Think Critically:** A strong earthquake happens in an area without tall buildings or freeway structures. What other kinds of damage could you look for in order to rank the quake on the Modified Mercalli scale?

4. **Skill Builder**
   Building Models Make a model that shows a slice through a volcano. Explain how the volcano was formed and which type of magma was the likely building material. If you need help, refer to Making Models on page 557 in the **Skill Handbook**.

USING MATH

Mount Mazama exploded about 6900 years ago, creating Crater Lake in Oregon. It is estimated that the eruption produced 1 275 000 tons of ash and rock. Assume 1 ton of rock and ash takes up 25 m³ of space. How much total space did the material exploding out of Mount Mazama take up?

Skill Builder
Building Models Student models should show the outside and inside of a volcano. The vent down from the volcano's crater should be shown running down to the magma chamber. Layers of lava that have slowly built the volcano should be shown, as well. **L2** **P**

Assessment

Content Have students write a brief story that describes the journey taken by rock material that has been heated inside Earth so that it melts. Student stories should include getting pushed upward until the material erupts from a volcano as lava. Use the Performance Task Assessment List for Writer's Guide to Fiction in **PASC**, p. 83. **LS** **L2**

---

4 Close

•MINI•QUIZ•

Use the Mini Quiz to check students' recall of chapter content.

1. **The shaking of Earth produced by a fast release of energy is a(n) _____.** *earthquake*
2. **A fracture in rock along which there is movement is called a(n) _____.** *fault*
3. **The point on Earth's surface directly above an earthquake's focus is the _____.** *epicenter*

Section Wrap-up

1. Answers should show a break in rock along which there has been movement (fault). On this fault, students should indicate the focus (location inside Earth where the movement occurred) and the epicenter (the point on Earth's surface directly above the focus).

2. The difference in shape for these volcanoes is caused by the type of lava that erupts from each. Shield volcanoes are built from hot, easy-flowing basaltic lava; cinder cone volcanoes form from explosive, more silica-rich lavas; and composite volcanoes form when the silica-rich lava erupts explosively and then quietly.

3. **Think Critically** Look for geologic damage, such as large cracks in the rock layers. Major rivers and streams can have their flow altered, also.

USING MATH

1 275 000 tons × 25 m³/ton = 31 875 000 m³ **LS**

Science & Society

16•2 Predicting Earthquakes

Before presenting the lesson, display **Section Focus Transparency 54** on the overhead projector. Assign the accompanying **Focus Activity** worksheet.

L2 **ELL**

Teaching the Content

- Demonstrate how earthquake waves move using a large spring toy. Have a volunteer compress a dozen or so coils of the spring and let them go. Students will observe that energy is transferred horizontally along the spring toy. This movement mimics primary earthquake waves. Now have two volunteers hold either end of the spring toy. Instruct one volunteer to cause motion in the spring with a quick flick of the wrist. The up-and-down motion is similar to that caused by an earthquake's secondary waves. Ask students to hypothesize which type of wave causes the most damage. (secondary waves)

What YOU'LL LEARN

- About earthquakes in California
- How to use library and Internet resources

Why IT'S IMPORTANT

You'll be able to use and understand scientific resources.

FIGURE 16-6
This photo shows an aerial view of the San Andreas Fault in California.

If you wanted to learn more about earthquakes, where would you go? A library would be a great place to find out about how and where earthquakes happen. In an encyclopedia, you would learn that earthquakes are caused by movement of Earth's crust along faults. The fault you have probably heard the most about is the San Andreas Fault in California, as shown in **Figure 16-6.** This fault is part of a group of hundreds of faults that crisscross much of California. The thousands of earthquakes that happen in California each year are the result of movements along these faults. Scientists think that the next great earthquake in California will happen along this group of faults. When will it happen?

Shake, Rattle, and Roll

Of all the major earthquakes worldwide in the 20th century, about one fifth have taken place in California. In 1906, a great earthquake almost destroyed San Francisco. In 1971, an earthquake in San Fernando, California, left nearly 60 people dead. Another major earthquake struck California in 1994, at Northridge, causing freeway overpasses to collapse. If you wanted to know if any earthquakes had happened last month, where would you find out? Newspapers and magazines are good sources in which to find information about recent events.

Future Earthquakes

Many of the large California earthquakes have caused a lot of damage, as shown in **Figure 16-7.** But scientists and California residents are still waiting for an earthquake as big as the 1906 San Francisco quake. Scientists are trying to predict when and where the next big earthquake will happen. They study the land around the major faults, as well as California's earthquake history.

446 Chapter 16 Earth's Structure

Program Resources

 Reproducible Masters
Enrichment, p. 60 **L3**
Reinforcement, p. 60 **L2**
Science and Society/Technology Integration, p. 34 **L2**
Study Guide, p. 60 **L1**

Transparencies
Section Focus Transparency 54 **L2**

They think that a major quake—with a magnitude similar to the 1906 San Francisco quake—will happen in the next 20 to 30 years.

To predict major earthquakes, scientists are using various technologies. They use lasers and satellites to study small movements along faults. Scientists need to know where the faults are located and which ones are the most likely to cause major damage if they move. It is important to know whether hospitals, police departments, and fire stations will be able to operate after a big quake. This information will make people more prepared for a big earthquake.

Earthquake Information

Where could you get information on recent earthquakes directly—from city officials or from scientists? You can use electronic mail to talk with scientists, or use the Internet to learn about where the faults are located and about past quakes. You may be able to find out about a quake anywhere in the world on the same day it happens!

FIGURE 16-7
This damage was caused by the Loma Prieta earthquake in 1989.

Skill Builder: Using Resources

LEARNING the SKILL

1. Encyclopedias and other reference books in libraries have general information on most subjects.

2. Magazines, journals, and newspapers usually cover fewer topics in depth.

3. The Internet has search engines that locate information you identify.

PRACTICING the SKILL

1. Which of the resources listed would probably be a good source for information about faults?

2. Which resource—an encyclopedia or a newspaper—would have the most recent information about earthquakes? Explain.

3. Where would you look for details about the 1906 San Francisco quake?

APPLYING the SKILL

Find out more about California earthquakes using the Internet and at least two different types of library resources.

Prepare

Section Background

- Many earthquakes and volcanoes occur where Earth's plates are separating, coming together, or sliding past each other.
- Some volcanoes, such as those in Hawaii, occur over hot spots, which are areas of Earth's surface that are located over rising plumes of hot magma.

Preplanning

Refer to the Chapter Organizer on pages 436A-B.

1 Motivate

Bellringer

Before presenting the lesson, display **Section Focus Transparency 55** on the overhead projector. Assign the accompanying **Focus Activity** worksheet.

L2 ELL

Tying to Previous Knowledge

Have students recall that volcanoes form when magma is forced to the surface.

448

What YOU'LL LEARN

- What the inside of Earth looks like
- About the evidence that supports the theory of continental drift

Science Words:
mantle
inner core
outer core
continental drift

Why IT'S IMPORTANT

You'll increase your knowledge about what's inside Earth.

FIGURE 16-8
This scientist is studying the rocks from a volcano to learn more about Earth's interior. *Why can't scientists study Earth's interior directly?*

16•3 A Journey to Earth's Center

What's inside Earth?

Imagine that you could get into an elevator and ride down to Earth's center. What would you see? It's impossible to do such a thing—Earth's interior is much too hot, but you can learn a lot about the inside of Earth by studying earthquake waves. We also gather clues about the inside of Earth by looking at landforms such as volcanoes, deep-sea trenches, and mountains.

But why does anyone need to know what is inside Earth? Scientists study Earth's interior for many reasons. A major reason to study what's inside the planet is to learn more about earthquakes and volcanoes. Why do they happen and why do they happen where they do? Scientists hope to accurately predict earthquakes and volcanic eruptions to help save lives. Scientists also learn about how Earth was formed and why it looks the way it does today.

Using Volcanoes and Earthquakes

Scientists use information and materials from volcanoes and earthquakes to help them make a model of Earth's interior. Material shot out of volcanoes may have come from deep within Earth. It gives scientists clues to

448 Chapter 16 Earth's Structure

Program Resources

 Reproducible Masters
Activity Worksheets, pp. 5, 99-100 L2
Enrichment, p. 61 L3
Reinforcement, p. 61 L2
Study Guide, p. 61 L1

 Transparencies
Section Focus Transparency 55 L2

what the inside of the planet is made of, as shown in **Figure 16-8**. When an earthquake happens, scientists everywhere know about it because the waves are sent all over the world. Seismographs all over the world record the waves and give scientists information about the earthquake. The people studying earthquakes discovered that these waves do not travel through Earth in a regular way or even at the same speeds. At certain places inside Earth, the waves are bent. This is similar to how light waves are bent when they go from air into water. The bending of the earthquake waves gave scientists the first clue that Earth is made up of layers that have different thicknesses and physical properties.

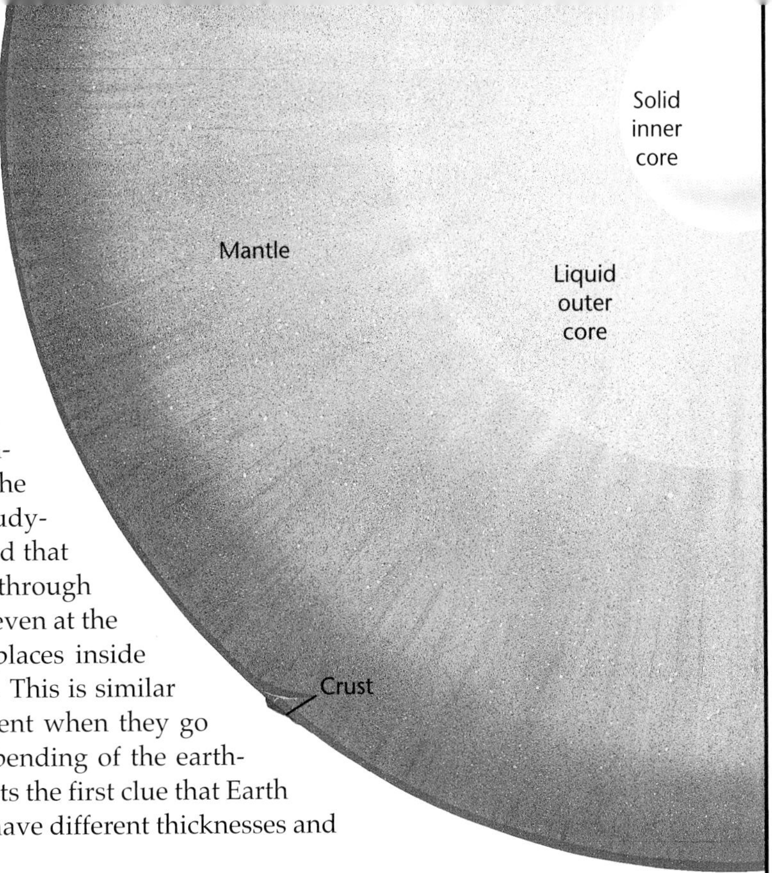

Earth's Layers

Primary (P) waves and secondary (S) waves, as shown in **Figure 16-3**, from earthquakes travel at different speeds through Earth. Both kinds of waves slow down in a plastic-like layer under the crust. This layer is weak and can move or flow like warm tar. This plasticlike layer is part of the thickest layer of Earth called the **mantle,** as shown in **Figure 16-9.** When earthquake waves are bent or slowed down, scientists know that the waves have hit a layer within Earth that has different properties. From these data, scientists are able to infer that Earth must have a solid **inner core** surrounded by a liquid **outer core.** Scientists can tell the outer core is liquid because S-waves are completely stopped at this layer. They already knew S-waves could not travel through fluids. It has taken many years to get a clear picture of the inside of Earth, and scientists learn more every day.

Solid inner core

Mantle

Liquid outer core

Crust

FIGURE 16-9
Earth's interior is made of four major layers: crust, mantle, outer core, and inner core.

*inter***NET**
CONNECTION

Go to the Glencoe Homepage, *www. glencoe.com/sec/ science,* for a link to Volcano World.

2 Teach

Teacher F.Y.I.

P-waves and S-waves are called seismic waves. There are other seismic waves, as well. The most destructive seismic waves are the surface waves.

Making a Model

LS **Kinesthetic** Have students research the thickness of each layer inside Earth. Using these data, have them make a model of Earth's interior. Students may wish to use modeling clay or other materials. Some students may want to use papier-mâché. Encourage students to be creative. Ask them to explain how scientists obtained the data they used in constructing their models. L2

*inter***NET**
CONNECTION

For more information about volcanoes, go to the Glencoe Homepage at **www.glencoe.com/sec/ science** for a link to Volcano World.

Revealing Preconceptions

Many students believe that all earthquakes occur at Earth's surface. Explain to them that in addition to Earth's surface, earthquakes also can occur deep inside Earth's crust or even in the upper parts of Earth's mantle. Earthquakes are labeled as shallow focus, intermediate focus, or deep focus.

Using an Analogy

Cut a peach in half lengthwise. Have students study the layers of the peach and draw an analogy to the layers inside Earth.

Visual Learning

Figure 16-8 Why can't scientists study Earth's interior directly? *The interior is too hot, and no technology exists to drill completely through Earth.* LS

Content Background

In a recent article, the boundary between Earth's mantle and outer core is described as "an upside-down landscape of anti-mountains" that hang downward toward Earth's outer core. Swirling currents of hot, molten iron are thought to flow rapidly through this landscape. See: Vogel, S. "Anti-matters." *Earth.* August 1995, pp. 42-49.

449

Activity 16-2

Purpose

[KS] **Kinesthetic** Students will understand how measurements of Earth vibrations are taken and recorded. [L2]

COOP LEARN [P]

Process Skills

observing and inferring, recognizing cause and effect

Time

40 minutes

Materials

Be sure to cut wire hooks from coat hangers and file ends so they are smooth.

Students may use the edge of a desk with paper being pulled over the floor.

Safety Precautions

Caution students to handle the wire hooks with care.

Possible Hypotheses

- As the paper is pulled under the pen, vibrations that occur will shake the ring stand and the pen. By pulling the paper, it is isolated from the vibration.
- The greater the magnitude of the vibrations, the larger the amplitude of the waves drawn on the paper.

📁 **Activity Worksheets,** pp. 5, 95-100

PLAN

Possible Procedures

Tape the ring stand down carefully. The pen must be attached to the stand just so its tip touches the sheet of paper. One person shakes the table, and another gradually pulls the paper under the pen. Measure the amplitude or height of the waves produced on the paper.

Activity 16-2

Design Your Own Experiment

Making a Seismograph

An earthquake is the vibrations caused by sudden movements of Earth. The sudden movements occur as energy that has built up in Earth's crust is released. The more energy that builds up, the greater the magnitude of the earthquake that occurs. A measure of the vibrations caused by the release of energy can be used to measure the size of the earthquake.

Possible Materials

- ring stand with ring
- wire hook from coat hanger
- masking tape
- sheet of paper
- piece of string
- 2 rubber bands
- fine-tip marker
- metric ruler

PREPARE

What You'll Investigate

How can you measure the size of vibrations?

Form a Hypothesis

If you pound on the top of your desk, you can feel vibrations in the desk. On a much larger scale, this is what happens when an earthquake occurs. If you tried to draw a straight line while a classmate pounded on the table, your line would end up wavy. Using this analogy, create a model that can measure vibrations, and **form a hypothesis** about how your model instrument could be used to measure these vibrations.

Goals

Design an instrument and use it in an experiment that models how vibrations produced by earthquakes are measured.

Safety Precautions

Sample Data Table

Data and Observations		
Trial	Height	Observations
1	3 mm	Frame moves a little
2	1 mm	Frame doesn't seem to move
3	6 mm	Frame moves noticeably

Content Background

In actual seismographs, paper is wrapped around a turning drum. The drum is connected to the frame of the instrument, which is sitting on Earth. The pen is suspended from a spring that absorbs all vibrations. As vibrations occur, the drum vibrates under the pen, and a wavy line is drawn that can be used as a measure of the earthquake's magnitude.

PLAN

1. Agree upon and design a **model** that will measure vibrations.
2. Agree upon and write out your **hypothesis** statement.
3. List the steps needed to build your model and to test your hypothesis. Be specific. Describe exactly what you will do.
4. **Prepare a data table** in your Science Journal.
5. **Read** over your plan to make sure your model can be used in your experiment to test your hypothesis.
6. What **variable** will you be testing? What **variable** will you need to control?
7. Make sure you **measure** and **record** carefully.
8. **Read** over the concluding questions to see if you have included all the necessary steps in your experiment.

DO

1. Make sure your teacher approves your plan and data table before you proceed.
2. **Build** your model and carry out the experiment as planned.
3. While doing the experiment, **record** any observations that you make in your Science Journal.

CONCLUDE AND APPLY

1. **Analyze** your data. Which trial resulted in the most wavy line on your model?
2. How did the movement of the suspended marker **compare and contrast** with the movement of the frame of your model instrument?
3. **Infer** whether the results of your experiment support your original **hypothesis.**
4. **Determine** the cause-and-effect relationship between the size of vibrations caused by pounding on the desk and how wavy the line was.

Teaching Strategies

Have students work in groups of three or four.

Troubleshooting Encourage students to pull the paper evenly with a smooth, slow motion to avoid distorting results, and be sure students hit the table from the sides so the line is wavy, not straight.

DO

Expected Outcome

The amplitude of the waves is greater when the table is hit with greater strength. The same relationship exists between the strength of waves generated by an earthquake and their amplitude waves.

CONCLUDE AND APPLY

1. the trial with the greatest force applied to the table
2. The pen is not totally isolated from the motion of the frame. The pen moves along with the frame. In a real seismograph, the pen is stationary while the frame moves with the vibrations.
3. Answers will vary depending on hypotheses.
4. As the magnitude of the vibrations increased, the amplitude or height of the wave peaks increased.

Assessment

Content Ask students to explain how the amplitude of the waves on a seismogram might be related to the amount of damage expected from an earthquake. The greater the amplitude, the more damage that can be expected. Use the Performance Task Assessment List for Analyzing the Data in **PASC**, p. 27.

Go Further

Hypothesize how the model would have to be changed in order for it to work like a true seismograph. (suspend the pen from a spring that absorbs vibrations)

Demonstration

IS **Visual-Spatial** Hold up a world map and ask students if they ever noticed how similar the coastlines of eastern South America and western Africa are. While holding the map, ask students to hypothesize about what other continents might have been connected at some time during the past. **L2**

? FLEX Your Brain

Use the Flex Your Brain activity to have students explore CONTINENTAL DRIFT.

📂 **Activity Worksheets,** page 5

Discussion

Discuss with students how difficult it can be to get your ideas across to someone else. This is especially true if your ideas are different from those of everyone else. Ask students to share experiences in which their ideas were not immediately accepted by others. Encourage them to think of the reasons why. Relate this discussion to the problems facing Alfred Wegener when he tried to convince people during the early 1900s about the idea of drifting continents.

FIGURE 16-10

The continents have moved over millions of years from their positions in Pangaea to their present-day positions.

A The fossil plant *Glossopteris* has been found on most of the continents that formed Pangaea.

B About 250 million years ago, the continents were joined to form the supercontinent Pangaea.

Continental Drift

Has anyone ever made fun of you for trying something unusual or for talking about a different idea? The same thing happens to scientists! Sometimes it takes a long time to find evidence to support new ideas. Many great discoveries and famous scientists were thought to be wrong at first. One example comes from the life of a German scientist named Alfred Wegener (VEG nur).

In 1912, he proposed the idea of a supercontinent that he named Pangaea (pan JEE uh). The word *Pangaea* means "all lands." He looked at a map of the continents as they were at that time and noticed that they could be arranged to fit together. Look at the eastern coastline of South America and the western coastline of Africa, as shown in **Figure 16-10C.** Don't they look like they could fit together as puzzle pieces? Wegener recognized this. He proposed that all the continents had once been connected to form Pangaea. Then, he said, Pangaea broke apart. The continents slowly moved to their present positions and are still moving. He called this the theory of continental drift. **Continental drift** is the movement of continents. Most other scientists at that time thought his theory was totally wrong.

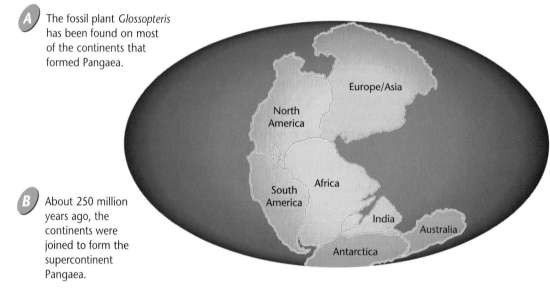

Theme Connection

Energy

The fact that Wegener's hypothesis of continental drift indicated that Earth's continents were once together and then drifted apart reinforces the theme of Energy. The energy needed to move continents great distances across the globe would be great. Even with all the evidence compiled by Wegener in support of his theory, the lack of a means to move the continents was its downfall. This, above all else, kept scientists of his time from accepting it as a possibility.

Fossil and Climate Clues

In addition to the puzzlelike fit of the continents, Wegener found other clues. He noticed that many similar fossils were found on the now-separated continents. For instance, fossils of the plant *Glossopteris* (glaws OP trus) were found in Africa, South America, Antarctica, and Australia. And fossils of the reptile *Mesosaurus* were found in Africa and South America. How did these organisms get to these widely separated places? *Mesosaurus* was a freshwater and land animal, so it is unlikely that it swam across oceans. And how did the plant get from one place to another? Using fossils from Antarctica, Wegener also found evidence that it had once had a tropical climate. He found fossils in cold, icy Antarctica of organisms that today live in warm, tropical climates. He also found that glaciers existed long ago in what are now Africa and South America. He thought the best way to explain these things was that the continents had once been together.

C This illustration shows the present-day positions of the continents.

D The fossils of the reptile *Mesosaurus* have been found in South America and Africa, landmasses that are now widely separated.

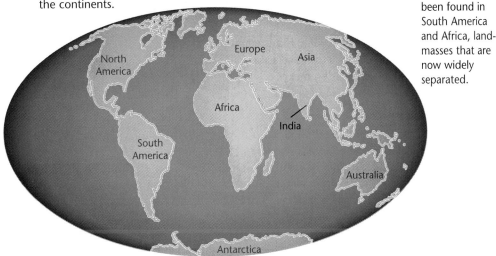

North America • Europe • Asia • Africa • India • South America • Australia • Antarctica

NATIONAL GEOGRAPHIC SOCIETY

 Videodisc

STV: Restless Earth

Continents today

52179

Faulting (art)

52204

Map of major plates

52174

Mid-Atlantic Ridge (art)

52185

Pangaea, 225 million years ago

52176

Pangaea, 125 million years ago

52177

Pangaea, 65 million years ago

52178

Rifts, ocean floor (art)

52183

Content Background

Mesosaurus was a freshwater reptile that lived in cool regions and probably hibernated during the winter. *Lystrosaurus* was a small reptile that lived about 200 million years ago. Fossils of *Lystrosaurus* have been found in South Africa, Antarctica, S.E. Asia, and India.

Theme Connection

Systems and Interactions

The secondary theme for this chapter is supported in the integration of Wegener's evidence for continental drift. Students can see how Wegener integrated his evidence—continental outlines, fossils, climate data, and rock structures—to support his hypothesis of continental drift.

FIGURE 16-11

Rocks and fossils in the Appalachian Mountains along the eastern coast of North America can be matched with similar rocks, fossils, and mountains across the Atlantic Ocean in Europe and Africa.

3 Assess

Check for Understanding

Discussion Lead students in a discussion that summarizes Wegener's evidence for continental drift.

Reteach

LS **Kinesthetic** Use small jigsaw puzzles with about 50 pieces. Without any indications of what the completed puzzles look like, hand out puzzles and ask groups of students to put them together. After 10 minutes, discuss the clues students used to fit the pieces together. Relate these results to how Wegener arrived at his hypothesis. **L2**

Extension

For students who have mastered this section, use the **Reinforcement** and **Enrichment** masters.

USING TECHNOLOGY
Satellite Tracking

Wouldn't it be great to find out exactly where you are anywhere on Earth—from Mount Everest to Miami Beach—at the touch of a button? A network of satellites called the Global Positioning System (GPS) lets you do just that.

No More Getting Lost

GPS has changed forever how people find their way. Hikers use GPS to keep from getting lost. Scientists use GPS to make accurate maps of fossil discoveries, rock outcroppings, and other land features.

Satellite Signals

Twenty-four GPS satellites orbit Earth twice a day. Each satellite sends a non-stop time and location signal. A handheld GPS unit uses signals from at least four satellites to find where a spot is located to within a few meters. It also tells how high above sea level the GPS unit is located.

454

Science Journal **Drifting Continents** Have students consider themselves to be the person presenting Alfred Wegener's ideas on continental drift to a meeting of geologists around 1912. Ask them to write their speech, including evidence they will use to support his ideas, in their Science Journals. **L2** **P** **LS**

Rock Clues

Wegener also found similar rocks on the different continents. You can walk along the hills in Scotland following a distinctive layer of rock. Then if you traveled to eastern Canada, you could find a similar rock layer in the hills of Newfoundland. About 250 million years ago, these two areas, which are now separated by the Atlantic Ocean, were joined together. Mountain ranges could be traced across oceans from one continent to another. In all, Wegener found a lot of evidence to support his idea—similar fossils, similar rocks, similar climates, and similar mountain ranges, as shown in **Figure 16-11.** The only problem was that he could not explain how these huge landmasses could have moved. Because he could not explain how it happened, few scientists took his idea seriously. It wasn't until years after his death that scientists using modern technology found new evidence of continental drift to support his ideas.

Today, his ideas form the basis for most of our knowledge of how the continents have changed through time. Next, we'll learn about the new evidence of continental drift.

Section Wrap-up

1. Explain how scientists used earthquakes and volcanoes to infer the structure of Earth's interior.

2. List and discuss the evidence Wegener used to support the theory of continental drift.

3. **Think Critically:** Where would you look in North America to find evidence for Wegener's theory of continental drift?

4. 🖉 *Skill Builder*
   Making a Model Build a model of the inside of Earth. If you need help, refer to Making Models on page 557 in the **Skill Handbook.**

Science Journal

In your Science Journal, write a short story about what you would see if you could take a trip to the center of Earth. Use your imagination, but be scientifically accurate.

🖉 *Skill Builder*
Making a Model Student models should show the four basic layers of Earth—the crust, the mantle, the outer core, and the inner core.

✔ Assessment

Performance Ask students to indicate on their models and in their stories about how far down into Earth each layer extends. Use the Performance Task Assessment List for Writing in Science in **PASC**, p. 87. [L2]

Science Journal
Student stories will differ, but all should include a description of the crust, the mantle, the outer core, and the inner core. [L2] [P]

---

4 Close

• MINI • QUIZ •

Use the Mini Quiz to check students' recall of chapter content.

1. **The thickest layer of Earth is the _____.** *mantle*
2. **The part of Earth's core that is solid is the _____ core.** *inner*
3. **The part of Earth's core that is liquid and lies just below the mantle is the _____ core.** *outer*

Section Wrap-up

1. Material from volcanoes may give clues as to the composition of the inside of Earth. The speed and angle of movement of waves produced by earthquakes indicate changes in density and state for material inside Earth.

2. Some continents look like they could fit together. Fossils of some organisms are widely spread over Earth. There are indications of a tropical climate in cold areas and indications of a cold climate in warm areas. Similar rocks have been found on different continents.

3. **Think Critically** The eastern coast seems to have been connected to Europe and Africa about 200 million years ago.

Prepare

Section Background

About 200 million years ago, Pangaea ("all Earth") contained all landmasses on Earth. Panthalassa ("all ocean") covered the rest of the planet. When Pangaea began to separate, it broke into two landmasses, Laurasia and Gondwanaland. These two landmasses were separated by the Tethys Sea. The Mediterranean Sea is a remnant of this prehistoric sea.

Preplanning

Refer to the Chapter Organizer on pages 436A-B.

1 Motivate

Bellringer

Before presenting the lesson, display **Section Focus Transparency 56** on the overhead projector. Assign the accompanying **Focus Activity** worksheet.

L2 **ELL**

16•4 Crashing Continents

What YOU'LL LEARN

- How new crust forms
- How plate tectonics changes the surface of Earth
- How mountains are formed

Science Words:
plate
seafloor spreading
convection current
plate tectonics

Why IT'S IMPORTANT

You'll understand how the continents move and mountains form.

FIGURE 16-12
Exploration of the oceans in the 1950s and 1960s revealed a system of ridges and valleys on the ocean floor.

Earth's Plates

Nearly 90 years after Wegener proposed the theory of continental drift, we have a much better understanding of his ideas. In that time, scientists have gathered more evidence to support his theory.

Our understanding of the upper layers of Earth changed a lot in the late 1900s. Scientists now think that Earth's crust and the upper part of the mantle are made up of 12 large sections called **plates.** The plates can move because they rest on a plasticlike layer within the mantle. The plates can move around on this layer. Imagine standing on pieces of plywood that have been tossed onto a muddy field. As you walk across the pieces of wood, they slip and bump into the pieces next to them. The plates on Earth are doing the same thing, but in slow motion.

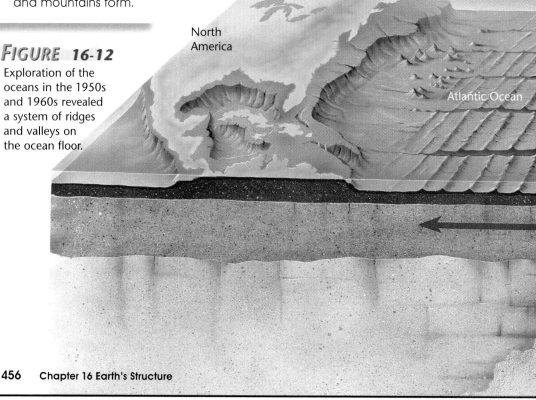

North America

Atlantic Ocean

456 Chapter 16 Earth's Structure

Program Resources

 Reproducible Masters
Activity Worksheets, pp. 5, 102 **L2**
Enrichment, p. 62 **L3**
Lab Manual, pp. 115-118 **L2**
Reinforcement, p. 62 **L2**
Science Integration Activities, pp. 69-70 **L2**
Study Guide, p. 62 **L1**

Transparencies
Section Focus Transparency 56 **L2**

Seafloor Spreading

Wegener had a problem. He couldn't explain how the continents moved. Scientists have now discovered how the plates move. Exploration of the ocean revealed areas where molten rock or magma oozes out onto the ocean floor. These areas, called mid-ocean ridges, have a gap or valley in the center. In this area, new ocean crust forms as the molten rock cools. As new magma moves up onto the surface, the two sides of the ridge are pushed apart. This process that forms new ocean crust is called **seafloor spreading.** If nothing filled the space in between the plates, a gap would be left. But as the plates get pushed farther apart, magma oozes up and hardens in the gap, adding to the edges of the plates. In **Figure 16-12,** you can see that the North American plate is moving slowly away from the European and African plates.

 The North American and South American plates are moving away from the European and African plates. The Atlantic Ocean is slowly getting wider at the rate of about 1.25 cm per year.

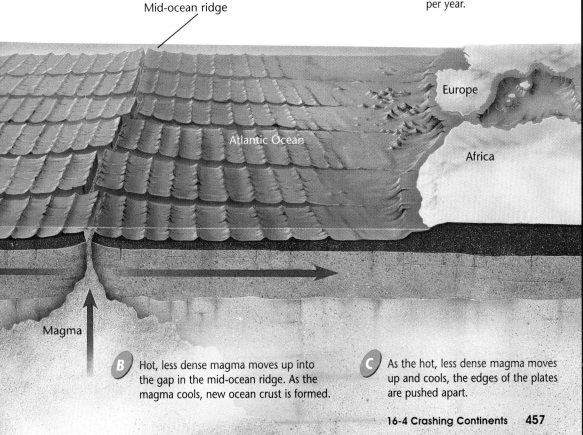

Mid-ocean ridge

Europe

Atlantic Ocean

Africa

Magma

B Hot, less dense magma moves up into the gap in the mid-ocean ridge. As the magma cools, new ocean crust is formed.

C As the hot, less dense magma moves up and cools, the edges of the plates are pushed apart.

16-4 Crashing Continents   457

Mini LAB

Purpose

IS Kinesthetic Students will observe the movement of convection currents. **L2**
ELL **COOP LEARN**

Materials

10-gallon aquarium with plastic divider, blue and red food coloring, two containers, ice water, hot water

Alternate Materials

A clear plastic container can be used in place of the aquarium. Use equal amounts of hot and cold water.

Teaching Strategies

Make the hot water as hot as is safely possible. Be careful of burns from water that is too hot. Place the water in the aquarium and allow it to become calm before removing the plastic divider.

Safety Precautions Caution students to handle the hot water carefully. They should wear aprons and handle glass with care.

Troubleshooting Be sure the plastic divider is placed securely. Attach it so that water cannot flow around it, but so that it can be removed. Remove the plastic divider carefully with a steady pulling motion.

📁 **Activity Worksheets,** pages 5, 102

Analysis

1. The denser, blue water flows into the red water near the bottom of the aquarium. The denser, blue water forces the less dense, red water to flow into the upper portion of the blue-water area.

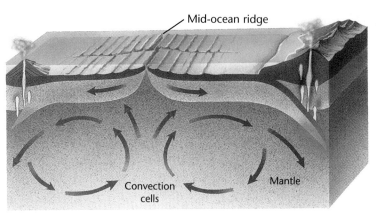

FIGURE 16-13
Convection cells form where hot, less dense magma is forced upward. As the magma cools, it becomes more dense and sinks back down into Earth.

Mid-ocean ridge

Convection cells

Mantle

Mini LAB

Observing Convection 🔧

1. Place a plastic divider in a large clear rectangular container or an aquarium.
2. Prepare a separate container with 10 liters of ice water. Add blue food coloring and stir.
3. Add red food coloring to 10 liters of hot water in a second container.
4. With the help of another student, pour both containers of water into the aquarium at the same time. Have another student hold the plastic divider in place.
5. Quickly remove the divider and observe what happens to the water.

Analysis

1. Draw and describe what happens when you remove the divider.
2. Explain why the warm, red water and the cooler, blue water moved as they did.
3. Using this lab as a model, explain how magma would rise and possibly produce volcanoes.

Convection Currents

The motion of magma in the plastic-like layer, just under the crust, causes the movement of the plates. Hot magma is forced upward toward the surface, becoming cooler and more dense as it rises. Then the cooler, dense magma starts to sink back down into the mantle. Deep in the mantle, the magma is heated again. This cycle repeats over and over and is called a **convection current,** as shown in **Figure 16-13.** A convection (kun VEK shun) current works just like hot air from a furnace. Hot air in a room gets pushed up by the cool air under it because the warm air is less dense. Then as it cools, it becomes more dense and it sinks to the floor. As convection currents move around in the plasticlike layer, they drag the stiff plates along with them.

Plate Tectonics

As scientists gather new data, they sometimes blend old and new ideas into new theories. That's what happened in the 1960s. Scientists combined the ideas about continental drift that Wegener

---

2. The warm, red water is less dense than the cool, blue water. The cool, blue water forces the warm, red water upward.

3. Hot magma is less dense than surrounding cooler rock. The cooler rock would force the less dense magma upward toward Earth's surface. Volcanoes would form as the hot magma flows onto Earth's surface.

Assessment

Content Have students work in groups of three and create poems that describe a convection current. Use the Performance Task Assessment List for Writer's Guide to Fiction in **PASC,** p. 83. **IS** **COOP LEARN** **P**

had discovered with the evidence of seafloor spreading and came up with the theory of **plate tectonics** (tek TAW nihks). Their new understanding of seafloor spreading and convection currents gave scientists an explanation of how the plates move. Volcanoes and earthquakes are the direct result of the movement of Earth's plates as the plates ride on the giant convection currents.

How do mountains form?

You have learned about how plate tectonics changes the ocean floor because of seafloor spreading. But how is the land changed by plate tectonics? Have you ever wondered how mountains are formed? Can we also use the theory of plate tectonics to explain how mountains form over many millions of years? From the rugged Rocky Mountains, as shown in **Figure 16-14,** in western North America to the gentle Appalachians along the eastern side of North America, mountains form where two of Earth's plates come together. The type of mountain that forms depends on what happens when plates meet.

FIGURE 16-14
Mountains form where two of Earth's plates come together. These mountains are part of the Rocky Mountains in Colorado.

459

3 Assess

Check for Understanding

Activity To help students understand how mountains form, have them use materials such as clay blocks, sheets of paper, or rubber. With these materials, students can model how the edges of the plates fold to form mountains. L2 COOP LEARN

Reteach

Demonstrate the three types of plate movement using your hands. Hold your hands level in front of the class. Bring them together slowly until they collide. Either your hands will bend up like a mountain, or one of your hands will slide under the other. Hold your hands level again, and slowly move them apart. Now slide one hand slowly past the other. This demonstrates basic ways that plates move.

Extension

For students who have mastered this section, use the **Reinforcement** and **Enrichment** masters.

460

FIGURE 16-15

There are four main kinds of mountains: folded, volcanic, fault-block, and upwarped. The kind of mountain depends on how the mountains were formed.

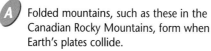

A Folded mountains, such as these in the Canadian Rocky Mountains, form when Earth's plates collide.

B Volcanic mountains, such as Mount Hood in Oregon, can form when one plate gets pushed under another. As the plate melts, the magma rises and volcanoes can form.

C Fault-block mountains form large sections of Earth's crust and are faulted and uplifted, such as the Sierra Nevada Mountains in California.

D Upwarped mountains, such as Mount Princeton in Colorado, form when Earth's crust is stretched and pushed up by forces inside Earth.

Cultural Diversity

Sacred Mountains Where geologists see mountains, many cultures see *sacred* mountains. For example, the Navajo in the American Southwest have four sacred mountains: Tsisnaajini (Mount Blanca), Sacred Mountain of the East; Tsoodzil (Mount Taylor), Sacred Mountain of the South; Doko'oosliid (San Francisco Peaks), Sacred Mountain of the West; and Dibentsaa (Mount Hesperus), Sacred Mountain of the North. Students could research traditional stories about various mountains. Some suggestions might be: Mount Kailas in western Tibet, the Black Hills in the Dakotas, and Mount Kilimanjaro in Kenya.

Kinds of Mountains

Scientists classify mountains by the way they form. Folded mountains are formed when plates collide. Volcanic mountains arise in areas where one plate gets pushed under another. As a plate gets pushed down into the mantle, the plate starts to melt. Magma is then pushed up through the crust and volcanoes form. This is happening now along the west coast of the state of Washington, forming volcanic mountains like Mount St. Helens. Fault-block mountains arise when whole sections of rock are faulted and broken. When this happens, some of the blocks get pushed up to form mountains. Upwarped mountains, as shown in **Figure 16-15D**, are formed when the crust is stretched by forces inside Earth, such as rising magma. Almost all mountains are found at or near the boundaries between two different plates. As you've learned, heat from within Earth drives the movements of the plates. This heat also forms mountains, opens oceans, and causes volcanic eruptions and earthquakes. Earth really is an active planet!

Section Wrap-up

1. What are Earth's plates, and how do they move?

2. List the four kinds of mountains, and describe how they were formed.

3. **Think Critically:** The highest mountain in the world is Mount Everest between Tibet and Nepal. Mount Everest is 8846 m high. Estimate the height of your school. How many "schools" high is Mount Everest?

4. *Skill Builder*

   Sequencing Volcanic mountains are produced near the boundary of two colliding plates, with one plate sliding under the other. Draw your own pictures showing the formation of a range of volcanic mountains. Put the pictures in their proper sequence to show the growth of the mountains. If you need help, refer to Sequencing on page 543 in the **Skill Handbook.**

USING MATH

Based on a seafloor-spreading rate of 5 cm/year, show how far apart two continents will move each 1000 years for the next 20 000 years.

Skill Builder

Sequencing Student pictures should be in the following order: one plate sliding under another, subducting material melts, melted material is forced upward toward the surface, upward moving magma flows onto Earth's surface, and lava flow after lava flow builds up into a range of volcanic mountains.

Assessment

Performance Ask students the following question. **Based on a seafloor spreading rate of 2.5 cm/year, how much further apart will North America and Africa be in 2 million years?** *50 km* Use the Performance Task Assessment List for Using Math in Science in **PASC,** p. 29.

---

4 Close

MINI·QUIZ

Use the Mini Quiz to check students' recall of chapter content.

1. **Earth's crust and rigid upper mantle are broken into sections called ____.** *plates*

2. **Heat transfer by ____ is thought to provide the energy required for plate tectonics.** *convection currents*

3. **The theory that explains how Earth's plates move is called ____.** *plate tectonics*

Section Wrap-up

1. Earth's plates are sections of the crust and rigid upper mantle that move by convection currents.

2. Folded mountains form when plates collide, volcanic mountains form from lava that builds up in layers, upwarped mountains form when Earth's crust is stretched by forces inside Earth, such as rising magma, and fault-block mountains form when whole sections of Earth's surface are faulted.

3. **Think Critically** Answers will vary depending on the height of your school. For example, if the building is 10 m high, then Mount Everest is 884.6 schools high.

USING MATH

Logical-Mathematical The separation will grow by 5 km every 1000 years; for example, 1000 years—5 km, 2000 years—10 km, etc., until at 20 000 years—100 km.

Science & Language Arts

Source

Soto, Gary. *Pacific Crossing.* New York: Harcourt, Brace and Jovanovich, 1992.

Biography

Gary Soto was born in Fresno, California, in 1952. His grandparents immigrated from Mexico during the depression. He has published many award-winning books of both prose and poetry.

Teaching Strategies

- Divide the class into two groups. Have one group research the San Francisco earthquake of 1906 and the other group the 1923 earthquake that occurred in Kwanto, Japan. Have each group present its findings. Then, as a class, have students compare and contrast the two events. **L2**

COOP LEARN

- Create a model of a subduction zone, where one crustal plate slides beneath another.

- Use the Internet as a tool to monitor earthquake or volcanic activity around the world.

Pacific Crossing
by Gary Soto

"Tokyo's like America," Lincoln said, smiling and trying to make conversation. "You know, we even have our own Cherry Blossom Festival. In San Francisco."

"Yes," Mr. Ono said, braking so hard that Lincoln had to hold on to the dashboard. "Yes, yes." A car was stalled in the left lane. Mr. Ono wiggled his steering wheel as he maneuvered dangerously into the next lane.

"I'm from San Francisco," Lincoln continued. "We're right on a bay like Tokyo."

"Yes, but America is very large," Mr. Ono said . . . "It is big as the sky."

Science Journal

Tokyo and San Francisco are known for their frequent earthquakes. Both cities lie along the Pacific Ring of Fire. In your Journal, draw a map of the Pacific, and show the locations of Tokyo and San Francisco. Label other cities in the Ring of Fire on your map.

Lincoln Mendoza is a 14-year-old Mexican-American boy from San Francisco who travels to Japan on an exchange program to study shorinji kempo, a martial art. For six weeks, Lincoln lives with his host family, Mr. and Mrs. Ono, and their son, Mitsuo, in the small farming village of Atami outside Tokyo. Both San Francisco and Tokyo are cities that have many earthquakes. As far as the number of earthquakes and their magnitude are concerned, Tokyo and San Francisco could be sister cities.

Before crossing the Pacific, Lincoln had imagined Japan would be all snow-capped mountains, women in kimonos, and temple-like dojos where he would practice kempo on neat straw mats. Instead, he finds that Japan is full of surprises. Tokyo is a lot like San Francisco. But beyond the city, Lincoln discovers small farms, public baths, hard-working people . . . and a dojo that looks like a driveway!

The Onos have misconceptions about the United States, too. Eager to share his culture, Lincoln teaches Mitsuo American slang, and succeeds—almost—in making a real Mexican dinner. But can he succeed in mastering kempo before he must cross the Pacific again?

Other Works

Baseball in April, Harcourt Brace: New York, 1990.

Chato's Kitchen, Putnam Publishing: New York, 1995.

Big Bushy Mustache, Random House: New York, 1998.

Science Journal Answers will vary but could include Manila, Philippines; Brisbane, Australia; Vancouver, Canada; and Santiago, Chile.

Read the statements below that review the most important ideas in the chapter. Using what you have learned, answer each question in your Science Journal.

1. Forces inside Earth cause earthquakes and volcanoes. *Where do most earthquakes occur?*

2. Information on earthquakes in California can be found in the library and on the Internet. *How can damage be minimized during an earthquake?*

3. Information from earthquakes and volcanoes helped scientists understand Earth's interior. *What are the four major layers of Earth?*

4. Mountains commonly form along Earth's plate boundaries. *What type of mountains are formed from the collision of two plates?*

Chapter 16 Review   463

Portfolio Encourage students to place in their portfolios one or two items of what they consider to be their best work. Examples include:
- Activity 16-1 assessment, p. 444
- Activity 16-2 experiment design and analysis, pp. 450-451
- Skill Builder, p. 445 **P**

Performance Additional performance assessments may be found in **Performance Assessment** and **Science Integration Activities.** Performance Task Assessment Lists and rubrics for evaluating these activities can be found in Glencoe's **Performance Assessment in the Science Classroom (PASC).**

Review

Ask students to describe details that support the main ideas of the chapter found in the statement for each illustration.

Teaching Strategies

Divide the class into four groups. Have the first group make a map of the world on which the locations of most earthquakes and volcanoes are shown. Have another group locate and share two sources of earthquake information from the library. Have another group make a poster showing the four major layers of Earth's structure. The final group can make a poster of folded mountains formed when plates collide. **L2**

COOP LEARN

Answers to Questions

1. Most earthquakes occur around the Pacific Ocean.
2. Answers could include to build stronger buildings, try to predict quakes, and don't build near a fault.
3. crust, mantle, outer core, and inner core
4. Folded mountains form when two plates collide.

Science at Home

IS Intrapersonal Using the illustration of Pangaea included here and the current positions of these continents, have students make a series of flash cards showing the major continents on either side of the present-day Atlantic Ocean as they break up and move. The last flash card should show the continents at their present locations. **L2**

Chapter 16 Review

Review

Using Key Science Words

1. epicenter
2. volcano
3. mantle
4. earthquake
5. plate

Checking Concepts

6. b
7. a
8. d
9. d
10. a

Thinking Critically

11. Movement of magma inside Earth can produce forces that cause earthquakes. Earthquakes can release pressure, thus causing a volcano to erupt. Volcanoes and earthquakes both occur where Earth's plates meet.

12. Alaska; one tectonic plate is sliding under another off the coast of Alaska.

13. The seabird could fly from one continent to another. Therefore, the presence of fossils on both continents can be explained even if the continents are far apart.

14. Scientists want to decrease loss of life, injuries, and damage from earthquakes. They use this information to make emergency plans in case of an earthquake.

15. It would be safer to live near a shield volcano. Lava flows from a shield volcano are nonexplosive. Although property could be damaged by it, easy-flowing lava is not as threatening to life. Humans could escape a typical shield volcano lava flow.

Using Key Science Words

continental drift	mantle
convection current	outer core
earthquake	plate
epicenter	plate tectonics
fault	seafloor spreading
focus	volcano
inner core	

Match each phrase with the correct term from the list of Key Science Words.

1. the surface of Earth directly above the focus of an earthquake
2. cone-shaped mountain formed by igneous activity
3. the thickest layer inside Earth
4. shaking of Earth produced by movement along a fault
5. large section of Earth's crust and upper mantle

Checking Concepts

Choose the word or phrase that completes the sentence.

6. The movement of Earth's plates over millions of years is called _____.
   a. a convection current
   b. plate tectonics
   c. an epicenter
   d. a fault

7. The movement of the plates is believed to be caused by _____.
   a. convection currents   c. earthquakes
   b. volcanoes   d. lavas

8. Earthquakes are caused by movement along a feature called a(n) _____.
   a. epicenter   c. landfill
   b. river   d. fault

9. Earthquake waves are produced at the _____.
   a. epicenter   c. extinct volcanoes
   b. plasticlike layer   d. focus

10. Volcanoes that are formed by quiet eruptions of fluid lava are _____ volcanoes.
    a. shield   c. cinder cone
    b. composite   d. earthquake

Thinking Critically

Answer the following questions in your Science Journal using complete sentences.

11. Why do volcanoes and earthquakes often occur in the same areas?

12. Would an earthquake be more likely in Nebraska or Alaska? Explain.

13. Why would the fossils of a seabird found on two different continents not be good evidence of continental drift?

14. Why would scientists want to know how many people live around faults, and how close water lines, gas lines, and hospitals are to faults?

Assessment Resources

Reproducible Masters
Chapter Review, pp. 35-36
Assessment, pp. 69-72
Performance Assessment, p. 54

Glencoe Technology
Computer Test Bank
MindJogger Videoquiz

15. Would it be safer to live near a shield volcano or a composite volcano? Explain.

Developing Skills

If you need help, refer to the description of each skill in the Skill Handbook.

16. Making and Using Tables: Research data about ten different earthquakes. Create a data table to list and record the magnitudes of the ten earthquakes. Plot the intensity of the earthquakes on a graph, and compare the strength of each earthquake.

17. Comparing and Contrasting: Compare and contrast the different layers of Earth.

18. Interpreting Data: Carlos is looking at a graph produced by a seismograph. He notices that the normally slightly wavy line had several peaks in it sometime early in the morning. What does this tell him?

19. Observing and Inferring: You are hiking in the desert of California, and you come to a wall of exposed rock. When you look at it, half of the wall appears to have shifted downward while the other half stayed in place. What are you probably looking at?

20. Interpreting Scientific Pictures: Review the pictures below and decide whether each is a composite, shield, or cinder cone volcano.

Performance Assessment

1. Display: Create a display about earthquakes. Include information on how and where earthquakes happen, and drawings or photos of earthquake damage.

2. Designing an Experiment: You created an experiment using baking soda and vinegar to make a model of a volcano erupting. Design an experiment where you use the same materials but a different setup to make a lava flow.

Developing Skills

16. Making and Using Tables Student data tables should include the location and magnitude of ten different earthquakes. Students may use a line graph, bar graph, or other type of graph on which they are to plot magnitude numbers (a measure of earthquake intensity or strength). Using the graph, students are to compare the strengths of each earthquake.

17. Comparing and Contrasting All layers are composed of Earth materials and make up the interior of Earth. Earth's crust is composed of solid rock with pockets of magma, the mantle has rigid and plasticlike layers, the outer core is liquid, and the inner core is solid.

18. Interpreting Data Several peaks on the graph indicate that an earthquake has occurred, and the seismograph registered the arrival of P- and S-waves. The difference in arrival times could be used to determine the earthquake's distance.

19. Observing and Inferring Most likely, you are looking at a fault in which rocks on one side have moved in relation to rocks on the other side.

20. Interpreting Scientific Pictures The upper photo is a composite volcano, Mt. Misti, Peru. The lower photo is of a cinder cone volcano.

Performance Assessment

1. Student displays should include illustrations explaining how and where most earthquakes occur. Displays should also include drawings or photos of earthquake damage. Use the Performance Task Assessment List for Display in **PASC,** p. 63. **P**

2. Students should obtain teacher approval before beginning the experiment. Answers will vary, but one idea is to build a volcano model with a small cup placed inside the crater of the model. Students might add red food coloring to the baking soda. When the reaction occurs, students will see red, frothy "lava" coming from the crater and flowing down the sides of the volcano. Use the Performance Task Assessment List for Designing an Experiment in **PASC,** p. 23. **P**

Chapter Organizer

Section	Objectives/Standards	Activities/Features
Chapter Opener		Explore Activity: Observe Water Filtering, p. 467
17-1 **Recycling Water** (2½ sessions, 1 block)*	1. **Investigate** how water moves through Earth and its atmosphere. 2. **Describe** what groundwater is and how it moves. National Science Content Standards: (5-8) UCP1, UCP3, A1, C4, D1, F2, F5	Activity 17-1: The Ground Is Soaked, pp. 472-473 Skill Builder: Comparing and Contrasting, p. 474 Using Computers, p. 474
17-2 **Earth Shaped by Water** (2 sessions, 1 block)*	3. **Explain** how water changes Earth's surface. 4. **Determine** how rivers and floods move sediment. National Science Content Standards: (5-8) UCP1, UCP2, UCP3, A1, A2, D1, F3	MiniLAB: Compare Stream Slopes, p. 477 Activity 17-2: Let's Get Settled, p. 478 Problem Solving: Interpreting Flooding Problems, p. 479 Skill Builder: Developing Multimedia Presentations, p. 480 Science Journal, p. 480
17-3 **Oceans** (2½ sessions, 1½ blocks)*	5. **Determine** why oceans are salty. 6. **Compare and contrast** currents, waves, and tides. National Science Content Standards: (5-8) UCP1, UCP2, UCP3, B2, C4, D1, D2, D3, E2, F2, F5	MiniLAB: Observing Waves, p. 484 Using Technology: Submersibles, p. 485 Skill Builder: Predicting, p. 487 Science Journal, p. 487
17-4 **Science and Society: Ocean Pollution** (1 session, ½ block)*	7. **Discuss** the importance of oceans as resources. 8. **Describe** various sources of ocean pollution. National Science Content Standards: (5-8) UCP1, C4, F2, F3, F5, G1	Skill Builder: Distinguishing Fact from Opinion, p. 489 People & Science, p. 490

* A complete Planning Guide that includes block scheduling is provided on pages 31T-33T.

Activity Materials

Explore	Activities	MiniLABs
page 467 ash, 600-mL beaker, graduated cylinder, sand, potting soil, coffee filter, salt	pages 472-473 600-mL beaker, graduated cylinder, gravel, sand, clay soil, timer, permanent marker, metric ruler page 478 2-L bottle, cup, graduated cylinder, gravel, sand, clay soil, spoon, timer, table salt, colored pencils, silt	page 477 250-mL beaker, long flat pan, potting soil page 484 Ping-Pong ball, cork, long flat pan, wooden rod, Styrofoam

Need Materials? Call Science Kit (1-800-828-7777).

Teacher Classroom Resources

Reproducible Masters	Transparencies	Teaching Resources
Activity Worksheets, pp. 5, 103-104 Cross-Curricular Integration, p. 21 Enrichment, p. 63 Lab Manual 33 Reinforcement, p. 63 Science Integration Activities, pp. 71-72 Study Guide, p. 63	Science Integration Transparency 17, Rainfall and Plants Section Focus Transparency 57, Old Faithful Teaching Transparency 33, Water Cycle	Spanish Resources English/Spanish Audiocassettes Cooperative Learning Resource Guide Lab Partner Lab and Safety Skills Lesson Plans
Activity Worksheets, pp. 5, 105-107 Enrichment, p. 64 Lab Manual 34 Multicultural Connections, pp. 37-38 Reinforcement, p. 64 Study Guide, p. 64	Section Focus Transparency 58, Rampaging River	**Assessment Resources**
Activity Worksheets, pp. 5, 108 Enrichment, p. 65 Reinforcement, p. 65 Science and Society/Technology Integration, p. 35 Study Guide, p. 65	Section Focus Transparency 59, Surf's Up Teaching Transparency 34, Ocean Motion	Chapter Review, pp. 37-38 Assessment, pp. 73-76 Performance Assessment, p. 55 Performance Assessment in the Science Classroom (PASC) MindJogger Videoquiz Alternate Assessment in the Science Classroom Computer Test Bank
Enrichment, p. 66 Reinforcement, p. 66 Study Guide, p. 66	Section Focus Transparency 60, Save the Beaches	

Key to Teaching Strategies

The following designations will help you decide which activities are appropriate for your students.

L1 Level 1 activities should be appropriate for students with learning difficulties.

L2 Level 2 activities should be within the ability range of all students.

L3 Level 3 activities are designed for above-average students.

ELL ELL activities should be within the ability range of English Language Learners.

LS These activities are designed to address different learning styles.

COOP LEARN Cooperative Learning activities are designed for small group work.

P These strategies represent student products that can be placed into a best-work portfolio.

GLENCOE TECHNOLOGY

The following multimedia resources are available from Glencoe.

Science and Technology Videodisc Series (STVS)
Earth & Space
  Artificial Waves
  Studying the Pacific Ocean

National Geographic Society Series
GTV: Planetary Manager
STV: Water

Glencoe Earth Science CD-ROM

This is a representation of key blackline masters available in the Teacher Classroom Resources.

Teaching Aids

Section Focus Transparencies

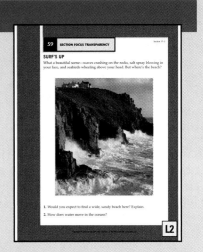

Science Integration Transparencies

Teaching Transparencies

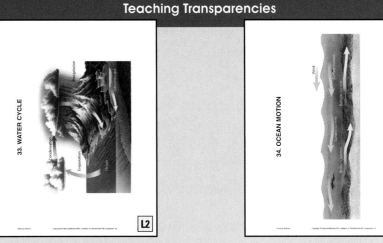

Meeting Different Ability Levels

Study Guide for Content Mastery

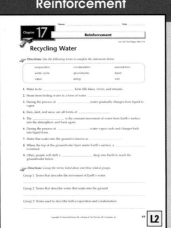

Reinforcement

Enrichment Worksheets

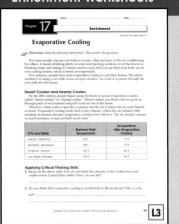

Hands-On Activities

Science Integration Activities

Soil Chemistry

Lab Manual

Comparing the Density of Salt Water and Freshwater

Activity Worksheets

The Ground Is Soaked

Enrichment and Application

Cross-Curricular Integration

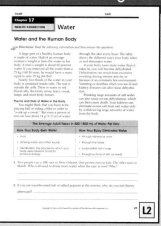

Water and the Human Body

Multicultural Connections

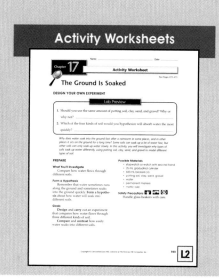

Water and Civilizations

Science and Society/ Technology Integration

The Ocean as a Source of Thermal Energy

Assessment

Performance Assessment

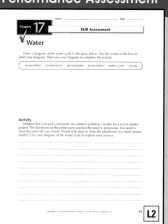

Water

Chapter Review

Water

Assessment

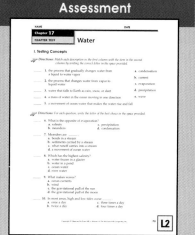

Water

Water

CHAPTER OVERVIEW

Section 17-1 This section describes how water is recycled through Earth and its atmosphere. How groundwater flows through rock and soil is described.

Section 17-2 This section describes how water moves sediment and changes Earth's surface. It describes how streams develop and explains what causes a river to flood.

Section 17-3 This section explains why ocean water is salty. It also describes the causes of currents, waves, and tides.

Section 17-4 Science and Society Various sources of ocean pollution are described. The importance of distinguishing fact from opinion is also discussed.

Chapter Vocabulary

evaporation	erosion
condensation	meander
precipitation	salinity
water cycle	current
groundwater	wave
deposition	tide
runoff	

Theme Connection

Systems and Interactions When precipitation falls to Earth, part of it eventually runs off to join Earth's river systems. Water that soaks into the ground flows through holes in rock and soil, forming the groundwater system. Interactions among ocean water density; winds in Earth's atmosphere; and the gravities of Earth, the moon, and the sun cause ocean currents, winds, and tides.

Chapter Preview

Skills Preview

▶ **Skill Builders**
- compare and contrast
- develop multimedia presentations
- predict

▶ **MiniLABs**
- compare
- observe

▶ **Activities**
- observe
- record
- hypothesize
- collect data
- compare and contrast
- model

466

Learning Styles

Look for the following logo for strategies that emphasize different learning modalities. **LS**

Kinesthetic	Explore, p. 467; Activity 17-1, p. 472; MiniLAB, pp. 477, 484; Inclusion Strategies, p. 484; Assessment, p. 487
Visual-Spatial	Visual Learning, pp. 469, 470; Activity, p. 476; Demonstration, p. 476; Activity 17-2, p. 478; Making a Model, p. 483
Interpersonal	Brainstorming, p. 470; Assessment, pp. 473, 474; Community Connection, p. 477; Theme Connection, p. 486; Teaching Strategies, p. 490
Intrapersonal	Activity, p. 486; Science at Home, p. 491
Logical-Mathematical	Inquiry Question, p. 469; Theme Connection, p. 472; Problem Solving, p. 479; Discussion, p. 482
Linguistic	Using an Analogy, p. 476; Science Journal, pp. 479, 482; Across the Curriculum, p. 483

Chapter 17 Water

Dolphins glide by in a graceful water ballet. Playful sea otters splash and twirl. It's fun to spend a day watching the animals in an aquarium. Don't you wonder what it's like to live in a world that's mostly water?

Well, guess what! You *do* live in a world that's mostly water. Some people call Earth the water planet. If you look at Earth from space, you'll see that water covers most of the planet's surface. But less than one percent of Earth's water is freshwater. This small amount of water we can use easily is important. It's the water we need to survive.

EXPLORE ACTIVITY

Observe Water Filtering

1. Mix some "pollutants"—sand, potting soil, ash, and salt— into a 600-mL beaker containing 300 mL of fresh tap water.
2. Carefully pour the polluted water into another beaker through a coffee filter.
3. Observe and record the kinds of and amount of pollutants that are filtered out.

Science Journal

In your Science Journal, discuss whether or not you think clear water is always clean. How do you think you could test to make sure water is clean enough to drink without tasting it?

467

Prepare

Section Background
- Plants also move water from the soil to the atmosphere through a process called transpiration.
- Groundwater passes through some rocks more easily than others. A rock layer in which water can collect is called an aquifer.

Preplanning
Refer to the Chapter Organizer on pages 466A-B.

1 Motivate

Bellringer
 Before presenting the lesson, display **Section Focus Transparency 57** on the overhead projector. Assign the accompanying **Focus Activity** worksheet.
L2 ELL

Tying to Previous Knowledge
Have students recall experiences they may have had during local flooding. What problems did the flooding cause?

17•1 Recycling Water

What YOU'LL LEARN
- How water moves through Earth and its atmosphere
- What groundwater is and how it moves

Science Words:
evaporation
condensation
precipitation
water cycle
groundwater

Why IT'S IMPORTANT
You will better understand the source of your water.

FIGURE 17-1
This family is enjoying a day at the park. For them and for us all, water is a vital resource.

The Water Cycle

The sun is out. The sky is clear and blue. It's a great day to meet your friends at the park for a picnic, as shown in **Figure 17-1.** You get there early and spread out a blanket for everyone to sit on. Soon, you notice that the blanket is damp because the grass is wet. You pick up the blanket and lay it on the sun-warmed picnic table to dry. Sure enough, when your friends show up with the food, both the blanket and the grass are dry. Where did the water go? It evaporated. **Evaporation** (ee vap uh RAY shun) is the process that gradually changes water from a liquid to water vapor. If you place a bowl of water in the hot sun all day, by the end of the day some of the water is gone. Where did the water go? It evaporated.

Later in the day, the sky gets cloudy and dark. You'd better pack up and head for home before it starts to rain! As you scramble to gather up everything, you wonder just where those clouds came from.

You see clouds almost every day. Let's look at how clouds are formed. Remember the water vapor that formed when liquid water on the grass and blanket evaporated? Water vapor rises through Earth's atmosphere, cooling as it rises.

468 Chapter 17 Water

Program Resources

Reproducible Masters
Activity Worksheets, pp. 5, 103-104 L2
Cross-Curricular Integration, p. 21 L2
Enrichment, p. 63 L3
Reinforcement, p. 63 L2
Science Integration Activities, pp. 71-72 L2
Study Guide, p. 63 L1

Transparencies
Science Integration Transparency 17 L2
Section Focus Transparency 57 L2
Teaching Transparency 33 L2

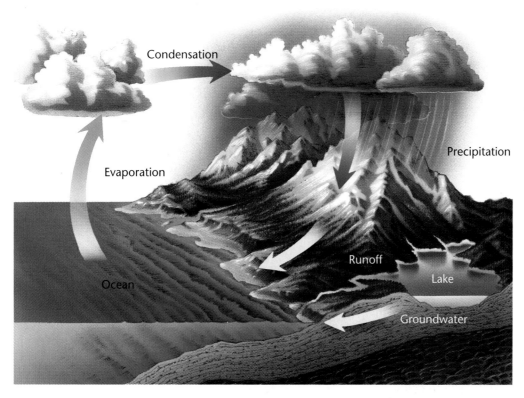

Condensation

Precipitation

Evaporation

Runoff

Ocean

Lake

Groundwater

During the process of **condensation** (kahn dun SAY shun), the water changes back into liquid form as the vapor cools. As water vapor condenses into tiny water droplets, clouds form.

The tiny droplets of water in the clouds bump into each other. When they do, they form larger drops. When the drops grow so large that they can no longer stay suspended in the clouds, drops of water fall to Earth as **precipitation** (pree sihp uh TAY shun). Rain, snow, hail, and sleet are all kinds of precipitation.

As your day in the park has shown, water constantly moves from Earth's surface into the atmosphere and then back to Earth's surface again. As shown in **Figure 17-2,** this constant movement of water is called the **water cycle.** Even though there is little water in the atmosphere at any one time, this water is important to life on Earth. This small amount of water is recycled through the water cycle to provide the precipitation needed for plants and crops.

FIGURE 17-2

The water cycle illustrates how water moves through the atmosphere, on the surface, and under the surface.

Visual Learning

Figure 17-2 The amount of water in and on Earth remains in balance with the amount of water in Earth's atmosphere through the water cycle. **What process changes water vapor into clouds?** *condensation* LS

ELL

Across the Curriculum

Geography Have students research and list any large bodies of water located near your school that would play a big part in the evaporation portion of Earth's water cycle. L2

2 Teach

Inquiry Question

LS **Logical-Mathematical** Explain why precipitation can occur in many forms: rain, snow, hail, or sleet. *Moisture high in the clouds is in the form of ice crystals. The form of precipitation depends on whether the ice crystals completely melt (rain), partially melt (sleet), or do not melt (snow). Hail forms when air currents carry water droplets up into the cloud, above the freezing level, where a coating of ice forms. The more times this occurs, the larger the hail grows.*

Revealing Preconceptions

Some students might think that streams carry water only from far-off locations. Remind them that streams often overflow their banks when the amount of rain is great locally. Tell students that streams get their water from all locations within the stream's drainage basin.

NATIONAL GEOGRAPHIC SOCIETY

 Videodisc

GTV: Planetary Manager
Water, Water Everywhere
Show 2.7, Side 2

38779-41076
Shall We Gather at the River?
Show 2.8, Side 2

41078-46201

How does water affect you?

Where do you get the water you use? You probably just turn on a faucet and water comes out. But where does that water come from? If you live in a city or town, your water may come from a nearby lake, reservoir, or river. In many places, people get water from wells drilled into the ground. If you don't know where the water in your area comes from, find out. Ask your teacher or parents, or check with your city water department.

How many ways have you used water today? The water you drink when you're thirsty is just a small part of all the water you use in a day, as shown in **Figure 17-3**. In the United States, the average family turns on the faucet 70 to 100 times a day. We use gallons and gallons of water to shower, brush our teeth, and flush the toilet. We use still more to cook, wash clothes and dishes, clean house, and water our plants. Can you think of other ways you and your family use water?

FIGURE **17-3**
You use water all the time for many things. Before you use water, it must be cleaned, tested, and treated to make sure it is safe. *List some other ways you use water.*

A We need lots of clean, safe water to drink.

B This pool requires a lot of water, and the water must be filtered and treated to be clean and safe.

470 Chapter 17 Water

Cultural Diversity

Valuable Water People who live in areas where water is scarce have had to be inventive to ensure a continuing water supply.

The Mayan culture of the Yucatan Peninsula in Mesoamerica flourished in a hot, humid area and had plenty of rainfall. But the soil layer was thin, and water quickly ran off. The Mayans used naturally occurring wells (cenotes) as water sources.

The San in Africa get liquid by pounding the moist insides of melons. They also eat soft, spiked cucumbers that retain water.

Today, grape growers in dry northern California wine country depend on the daily moisture from fog as it rolls in from the Pacific to help irrigate their vineyards.

Where is water found?

Have you ever stood at the edge of a big lake, looking across the water? Some lakes are so big you can't even see the other side. Many rivers wind along for miles and miles. Imagine how much water there must be in all the world's lakes and rivers! Yet the water in lakes and rivers makes up less than one percent of Earth's total freshwater supply. The rest is frozen in glaciers and the polar ice caps or underground.

Underground Water

If you have ever watched as rain falls on the ground, you have seen that sometimes the rain soaks in quickly. Some of the rainwater can run along the surface of the ground to flow into streams and lakes. Other times, the water sits on top of the ground and evaporates or soaks in slowly. The water that soaks into the ground becomes groundwater. **Groundwater** is water that soaks into the ground and collects in the small spaces between bits of soil and rock, as shown in **Figure 17-4.** If the small spaces are connected, then water can flow through layers of rock and soil. People drill down into these layers to make wells. They then pump the water to the surface for use as drinking water, for factories, or for watering crops and animals.

FIGURE 17-4

Groundwater is water that soaks into the ground and collects in the spaces between soil and rock particles.

 **D** We use water for our plants, lawns, and gardens.

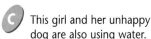 **C** This girl and her unhappy dog are also using water.

471

FLEX Your Brain

Use the Flex Your Brain activity to have students explore GROUNDWATER.

Activity Worksheets, page 5

NATIONAL GEOGRAPHIC SOCIETY

 **Videodisc**

STV: Water
Unit 1
Natural Filters

09425-10612
Hydrologic Cycle

10619-13542

3 Assess

Check for Understanding

Discussion Have students discuss where they think the drinking water in your area comes from and how it gets to their homes.

Reteach

Have students list all possible locations near your school that have concentrations of surface water. Ask students if they think any of the locations get some of their water from underground. **L2**

Extension

For students who have mastered this section, use the **Reinforcement** and **Enrichment** masters.

Content Background

- Flowing artesian wells occur where groundwater flows freely to a level above Earth's surface.
- The erosive action of groundwater is responsible for the formation of caves and caverns. As groundwater slowly seeps from the roof of a cavern and falls to the floor, formations known as stalactites and stalagmites form. Stalactites hang from the ceiling, and stalagmites form upward from the floor. When they meet, a column or pillar is formed.

Activity 17-1

PREPARE

Purpose

 Kinesthetic Students will predict, test, and compare how easily water flows through different soils.

L2 **ELL** **COOP LEARN** **P**

Process Skills

observing and inferring, comparing and contrasting, making and using tables, measuring in SI, predicting, interpreting data, separating and controlling variables, forming a hypothesis

Time

1 hour or more

Materials

Potting soil, clay, sand, and gravel can be obtained from a gardening store. Provide a set of materials for each student group.

Safety Precautions

Warn students to handle glass beakers with care.

Possible Hypotheses

Water will flow fastest through soil made mostly of gravel. Water will flow slowest through soil made mostly of clay. Water will flow fastest through layers of gravel, then potting soil, then sand, and finally clay.

📁 **Activity Worksheets,** pages 5, 103-104

PLAN

Possible Procedures

Place different types of soil in each of the four 600-mL beakers. Pour an equal amount of water into each of the beakers and measure the time it takes for the water to soak into the soil.

Activity 17-1

Design Your Own Experiment
The Ground Is Soaked

Why does water soak into the ground fast after a rainstorm in some places, and in other places it sits on the ground for a long time? Some soils can soak up a lot of water fast, but other soils can only soak up water slowly. In this activity, you will investigate why types of soils soak up water differently, using potting soil, clay, sand, and gravel to model different types of soil.

Possible Materials
- stopwatch or watch with second hand
- 25-mL graduated cylinder
- 600-mL beakers (4)
- potting soil, clay, sand, gravel
- water
- permanent markers
- metric ruler

PREPARE

What You'll Investigate
Compare how water flows through different soils.

Form a Hypothesis

Remember that water sometimes runs along the ground and sometimes soaks into the ground quickly. **Form a hypothesis** about how water will soak into different soils.

Goals

Design and **carry out** an experiment that compares how water flows through three different kinds of soil.

Compare and **contrast** how easily water soaks into different soils.

Safety Precautions

Handle glass beakers with care.

472 Chapter 17 Water

Inclusion Strategies

Learning Disabled Provide students with an overview of Activity 17-1 that shows a higher degree of structure. Help students develop and write a workable hypothesis. Suggest ways that students can set up equipment and data records for the experiment. **L1**

Theme Connection

Systems and Interactions

Ask students to consider how the different processes in the system of Earth's water cycle interact with each of them personally (especially when they are planning a social function or family trip). **LS**

472

PLAN

1. Agree upon and write out your **hypothesis** statement.
2. **List** the steps needed to test your hypothesis. Be specific. Describe exactly what you will do.
3. **Prepare a data table** in your Science Journal.
4. **Read** over your plan to make sure your experiment tests your hypothesis.
5. What **variable** will you be testing? What variables will you need to control?
6. Make sure you **measure** and **record** carefully.
7. **Read** over the concluding questions to see if you have included all the necessary steps in your experiment.

DO

1. Make sure your teacher approves your plan and data table before you proceed.
2. Carry out the experiment as planned.
3. While doing the experiment, **record** all observations that you make in your Science Journal.

CONCLUDE AND APPLY

1. **Infer** how your data and observations support or do not support your hypothesis. Explain your answer.
2. **Compare** and **contrast** how water soaked into each test soil.
3. **APPLY** If the soil is already soaked with water and you add more water, what will happen to the additional water?

17-1 Recycling Water   473

Content Background

The ability of soils to allow water to flow through them is called permeability. Soils or rock layers containing many openings or pores are considered to be porous. If the pores are connected, the layer is permeable and porous. Porous layers can also be impermeable if the pores are not connected.

Teaching Strategies
- Have students work in co-operative teams of four students.
- Data tables should have the soil types in the rows and columns for soil type, amount of water, and time for water to soak in.

Troubleshooting
- Students may need direction when creating a data table.
- Remind students to compare all the soils that they tested when answering the Conclude and Apply questions.

DO

Expected Outcome
Rockier soil (gravel and potting soil) will allow water to flow through more quickly.

CONCLUDE AND APPLY

1. If student hypotheses state that water will flow more quickly through soil containing gravel, then they should indicate that data support their hypothesis.
2. Answers will vary, but students should see that rockier soil tends to allow water to flow through more quickly.
3. If the soil is already soaked, additional water will not be able to soak into the wetted soil.

Go Further

Have students interpret how the ability of water to flow through different types of soil can be used to indicate how much runoff will occur with each soil type.

4 Close

Section Wrap-up

1. Thermal energy from the sun changes the liquid water into water vapor by the process of evaporation.

2. Yes; toxic liquids spilled on the ground may enter the groundwater supply and eventually flow into the town's reservoir.

3. **Think Critically** Yes; a rock layer may contain pores that are not connected, and water cannot flow through easily. This type of rock layer would stop or severely slow down the flow of groundwater.

FIGURE 17-5

This artesian well in Michigan flows without a pump. Most cities and towns have water wells that require pumps to bring the water to the surface.

The Importance of Groundwater

In some parts of the world, people rely on wells drilled down into groundwater for their everyday water needs. One type of well, an artesian (ar TEE zhun) well, is shown in **Figure 17-5.** In some places, this water is close to the ground surface. Springs form where the top of the groundwater layer meets Earth's surface. At springs, water flows from the ground.

In the United States, groundwater provides 40 percent of our public water supply. Industries and farms also use groundwater. Groundwater is the only source of water in many agricultural areas. You've learned how people depend on water and where they get the water they use. But water has other important roles on Earth. Next, you'll learn about those roles.

Section Wrap-up

1. Early in the morning, grass and other surfaces are often wet with dew. What happens to the water during the day?

2. Should you be concerned about the safety of your water supply if a large amount of a toxic liquid were spilled on the ground near the town's reservoir? Explain your answer.

3. **Think Critically:** Can a rock layer have small spaces but not allow water to flow through? How might this type of rock layer affect the flow of groundwater?

4. *Skill Builder*
   Comparing and Contrasting Compare and contrast the processes of evaporation and condensation. If you need help, refer to Comparing and Contrasting on page 550 in the **Skill Handbook.**

Using Computers

Model Using computer graphics or drawing software, make a diagram that shows how water circulates through Earth's water cycle.

Using Computers

Student models should include the processes of evaporation, condensation, and precipitation and the location of surface water and groundwater. Student models should also show where the processes occur.

Skillbuilder

Comparing and Contrasting Both processes deal with changing the state of water. In evaporation, liquid water is changed to water vapor, whereas in condensation, water vapor is changed to liquid water.

Assessment

Performance Assess students' abilities to compare and contrast concepts by having them work in pairs to critique what each compare-and-contrast answer should include. Use the Performance Task Assessment List for Group Work in **PASC,** p. 97. [IS] [L2] **COOP LEARN**

Earth Shaped by Water

Water Erosion and Deposition

You've gone camping at what seems like a perfect place. It is shady and just a few steps from a clear river. But now, you're getting worried. The level of water in the river is rising. And the water looks muddier than it did earlier. What's going on?

What you don't know is that large amounts of rain upstream from your campsite have made the water level rise. In some places, the river has risen higher than its banks. It has flooded the nearby land. The fast-moving water is stirring up sediment, making the water look muddy. You'd better pack up and come back another day!

The next weekend, you try again. The land around the river isn't flooded anymore. But what a mess! Everything is covered with mud and sand, as shown in **Figure 17-6.** Where did the mud and sand come from? The river left it there. Streams and rivers move a lot of mud, sand, and other sediments (SED uh munts). Moving water lays down sediments in a process called **deposition.** Deposition (dep uh ZIH shun) is one way that streams and rivers change Earth. But how do streams and rivers form in the first place? Let's take a look.

What YOU'LL LEARN

- How water changes Earth's surface
- How rivers and floods move sediment

Science Words:
deposition
runoff
erosion
meander

Why IT'S IMPORTANT

Water helps shape many features of our planet.

FIGURE 17-6

Floods carry mud, sand, and other debris over the area covered by water. This cornfield was flooded in 1993 by the Missouri River.

17-2 Earth Shaped by Water **475**

Prepare

Section Background

- The area of Earth's surface from which a river gets its water is the river's drainage basin. Drainage basins are separated by higher ground called divides.

- Stream valleys develop in three stages. Young streams have steep sides and flow swiftly through steep valleys. Mature streams flow less swiftly and begin to develop a floodplain. Old streams flow slowly through a broad, flat floodplain that it has carved.

Preplanning

Refer to the Chapter Organizer on pages 466A-B.

1 Motivate

Bellringer

 Before presenting the lesson, display **Section Focus Transparency 58** on the overhead projector. Assign the accompanying **Focus Activity** worksheet.
L2 ELL

Program Resources

Reproducible Masters
Activity Worksheets, pp. 5, 105-107 L2
Enrichment, p. 64 L3
Multicultural Connections, pp. 37-38 L2
Reinforcement, p. 64 L2
Study Guide, p. 64 L1

Transparencies
Section Focus Transparency 58 L2

Have students recall the discussion on sedimentary rocks.

2 Teach

Activity

LS **Visual-Spatial** Take students outside to look for evidence of water erosion or deposition on the school grounds. They might find small gullies near the corners of buildings or deposits of sediment left after a rainstorm. Encourage interested students to take photographs or make drawings. **L2**

Using an Analogy

LS **Linguistic** Explain erosion and deposition as the picking up and moving of sediment from one location to another. Ask a student to pick up books from your desk and move them to a table in the back of the room. The student picking up the books is similar to erosion; the student placing them on the table is similar to deposition.

Demonstration

LS **Visual-Spatial** Use a water hose and a bucket to demonstrate runoff. Take students outside and gently pour a bucket of water on a slightly sloping area of the school grounds. Ask students to comment on what happens to the water. Now use the water hose to thoroughly saturate the ground and repeat the demonstration. Ask students to comment on any differences they notice.

❓ FLEX Your Brain

Use the Flex Your Brain activity to have students explore STREAM DEVELOPMENT.

📁 **Activity Worksheets,** page 5

476

FIGURE 17-7

The amount of runoff from an area is affected by several factors. These include the amount of vegetation, how steep the land is, the amount of rain, the intensity of the rain, and whether the ground is already wet.

Stream Development

Streams and rivers can be big and mighty. But we're going to start small—with raindrops. Have you ever just sat by the window on a rainy day, watching where the rain goes? Some rainwater sinks right into the soil to become groundwater. Some rain collects on leaves and grass and in puddles. When the sun comes out, this rainwater will evaporate. But some rain doesn't sink in or stay in one place. Because it runs off the surface, it is called runoff. **Runoff** is water that flows over the ground surface and eventually flows into rivers, streams, or lakes.

Runoff is one of the causes of erosion (ee ROH zhun). **Erosion** occurs when soil or rock is loosened and then moved from one place to another. Some parts of the ground are more easily eroded than others. **Figure 17-7** shows some things that affect how much water runs off. The water begins to cut small grooves in some areas. Eventually, the flowing water makes the grooves deeper and wider, forming small valleys called gullies. Water in small gullies flows into streams. Water in streams flows into rivers and eventually into oceans.

Theme Connection

Systems and Interactions

Ask students to consider how the interactions of so many factors affect the amount of water that soaks into the ground to become groundwater and the amount of water that runs off to become part of Earth's surface-water system.

Meandering Stream

Deepest part of stream

Sandbar

Sandbar

Steep bank

FIGURE 17-8

When a stream begins to erode its banks more than the channel bottom, wide bends or meanders are formed. The stream cuts into the outside of the meanders and deposits sediment on the inside of the meanders.

Rivers as Earthmovers

In the MiniLAB, you will see that erosion can make a stream deeper or wider. The same thing happens in nature. As a stream is shaped by moving water, erosion cuts its channel deeper and deeper. In steep areas, streams erode the channel bottoms, making the channel deeper. But in flatter areas, streams flow slower and begin to erode the channel sides more than the bottom. When the stream can't get any deeper, it gets wider. Water in the stream moves sediments around, forming wide bends in the stream called **meanders** (mee AN durz). **Figure 17-8** shows a river with meanders.

Mini LAB

Compare Stream Slopes

How does stream slope affect erosion?

1. Fill three long, flat pans with moistened soil.
2. Level the soil in each pan, and gently pack it down.
3. Use a pencil to make a straight or slightly curved groove lengthwise in the soil in each pan.
4. Tilt each pan to a different angle.
5. Using a 250-mL beaker of water, begin pouring a steady stream of water into each pan.

Analysis

1. For each stream, describe where the water eroded the most sediment.
2. In each pan, did the water make the groove deeper or wider? Describe how the slope of the stream affects the type of erosion (deepening or widening). Also, describe how the slope of the pan affects where the most erosion occurs.

Community Connection

Flood Damage Using a map of the area around your school, have students locate the nearest stream or river. Help students determine the extent of the floodplain. Ask students how far from the stream or river they would expect flooding to occur. Then ask them if any structures might be in danger of flooding. **LS**

L2 **COOP LEARN**

Mini LAB

Purpose

LS **Kinesthetic** Students will see how the slope of the ground affects the amount of erosion. **COOP LEARN** **L2** **ELL**

Materials

3 long, flat pans; soil; pencil; three 250-mL beakers; water

Teaching Strategies

- Have students work in groups of at least three so they can pour the water into all the pans at the same time.
- Catch overflow in a bucket or pan. Dispose of properly.

Safety Precautions Caution students to handle glass beakers with care.

Troubleshooting Be sure that students pack the sediment in each pan the same amount. A slightly curved groove works best.

Activity Worksheets, pages 5, 107

Analysis

1. More sediment was eroded near where the water was poured in and on the outsides of the curves.
2. Lower slopes allow the water to carve wider channels. Steeper slopes cause the stream to cut deeper channels. The steeper the slope, the greater the erosion at the upper end of the pan. With decreases in slope, the erosion occurs farther and farther down the pan.

✓ Assessment

Oral Have students state a general relationship of stream slope to channel formation. Use the Performance Task Assessment List for Oral Presentation in **PASC**, p. 71. **P**

Activity 17-2

Purpose

LS **Visual-Spatial** Students will observe how different types of sediments settle out of water. **L2** **COOP LEARN**

Process Skills

measuring in SI, interpreting data, forming a hypothesis

Time

40 minutes

Alternate Materials

A soil sample containing a variety of particle sizes can replace the gravel, sand, clay, and silt. A quarter cup is approximately 50 mL.

Safety Precautions

Wear safety goggles when handling all materials.

📁 **Activity Worksheets,** pages 5, 105-106

Teaching Strategies

Discuss how rivers carry and deposit sediment and how changes in the speed of the water affect how sediments settle out.

Troubleshooting Provide trays to contain loose sediment.

Answers to Questions

1. Gravel and sand; salt, clay, and silt. Salt dissolved in the water, and the clay and silt remained in suspension.

2. The gravel was on the bottom overlain by the sand. Undissolved salt and some of the clay and silt formed the next layer. Some salt, clay, and silt remained in the water.

3. Gravel is deposited first, followed by sand. Salt remains dissolved in solution, and clay and silt are carried in suspension for long distances.

Goals

* Observe how sediments settle in water.
* Model how sediments move in streams.

Materials

* graduated cylinder
* clear plastic cups (5)
* 2-L bottle
* water
* spoon
* gravel (50 mL)
* salt (50 mL)
* sand (50 mL)
* clay (50 mL)
* silt (50 mL)
* colored pencils (3)
* stopwatch

Let's Get Settled

You have learned that streams move sediment around. There are different kinds of sediment—some heavy, some light. Streams carry light particles farther than heavy particles. This activity will help you understand how streams move sediments.

What You'll Investigate

How do sediments settle out of water?

Procedure 🥽 🧤 🚫

1. **Label** five cups *A, B, C, D,* and *E.*

2. Use the graduated cylinder to measure 25 mL of gravel. **Pour** the gravel into cup *A.* Do the same for each of the other materials, one type per cup.

3. **Pour** the remaining gravel, salt, sand, clay, and silt (fine sediment that feels like flour) into a 2-L bottle. **Add** water to the bottle, leaving 2-4 cm of space at the top of the bottle. Put the cap on the bottle.

4. **Add** water to each cup until it is about two-thirds full.

5. **Shake** the 2-L bottle and stir each cup thoroughly. **Observe** what happens to the sediment in the bottle. **Record** your observations in a data table.

6. **Observe** each cup for 5 seconds and then at 15-second intervals for 1 minute. Then observe each cup at 2-minute intervals for 16 minutes. **Record** your observations in your data table.

7. After shaking the 2-L bottle, let it settle for 16 minutes. **Observe** the bottle. Use the colored pencils to **draw** and **label** your observations of the 2-L bottle. Use a different color for each layer that forms in the bottle.

Conclude and Apply

1. Which two sediments settled out most quickly? Which three sediments did not settle out? **Explain.**

2. **Describe** the mixture in the 2-L bottle. Are the sediments layered?

3. Based on your observations in this experiment, **explain** what happens to gravel, salt, sand, clay, and silt carried by streams.

✓ Assessment

Oral Based on the results students noted in the 2-L bottle, ask them to describe how they would remove sediment still in suspension from their drinking water. Use the Performance Task Assessment List for Oral Presentation in **PASC,** p. 71.

Data and Observations

Cup A (gravel) settles out quickly; **Cup B (salt)** much dissolves with some settling out after 5 s; **Cup C (sand)** settles out quickly; **Cup D (clay)** some settles out; most remains suspended; **Cup E (silt)** some settles out after 5 s, more after 10 and 15 s; most remains suspended; **2-L bottle** Within 5 s, gravel and some sand settled. More sand settled after 15 s. Undissolved salt and some clay and silt formed the next layer. Some salt, clay, and silt remained in the water. **P**

Muddy Streams

What have you noticed about streams you have seen? Did they all look the same? Some were clear. Others may have been muddy. The information that you learned in Activity 17-2 should help you understand why the streams you have seen are not all alike. It should also help you understand that water moves different kinds of materials in different ways. Water can push or bounce grains of sediment along the bottom of a stream. Smaller grains of sediment can be carried suspended in the water. Other materials can be carried dissolved in the water. As it constantly moves sediment from one place to another, water truly is an earthmover, as shown in **Figure 17-9** on page 480.

Problem Solving

Interpreting Flooding Problems

The floodplain of a river is the area that is normally flooded during a time of high water. The floodplain is built from sediments deposited by the river. This area makes great farmland but is not the wisest place to build a city. The area of the floodplain is the mostly flat area that exists on both sides of a river. It stretches outward from the river's natural levees or banks. Look at the illustration of the Mississippi River shown here. Try to determine the extent of the river's floodplain.

Solve the Problem:

Interpret the illustration below as to whether any structures exist within the floodplain of this part of the Mississippi River. If you were planning a farm in an area like this, where would you plant crops?

Think Critically:

Based on what you interpreted from the illustration of the Mississippi River and its floodplain, where would you build the farmhouse and barn? Explain why you chose the sites you did for the planted fields, the farmhouse, and the barn.

Problem Solving

The extent of the river's floodplain is the relatively flat area that becomes flooded on both sides of the river.

Solve the Problem

Students should note any buildings, such as homes or barns. The soil in the floodplain should be plowed and planted.

Think Critically

The best location for any building would be at higher elevations beyond the floodplain. The planted fields were chosen because the floodplains of a river contain rich, fertile soil. **IS**

3 Assess

Check for Understanding

Discussion Have students discuss any nearby areas that have had flooding problems in the past. Students could discuss why the area floods and what could be done to decrease the flood risk. **L2**

Reteach

Students may have difficulty with the words *suspended* and *dissolved*. Explain that material carried in suspension is small enough that the water can prevent it from settling. Explain that material dissolved in the water is similar to sugar stirred into a drink. The sugar dissolves.

Extension

For students who have mastered this section, use the **Reinforcement** and **Enrichment** masters.

Content Background

Streams and rivers can carry larger volumes of sediment and other materials during floods because the water is moving at a high velocity. The higher the velocity, the larger size sediment and sediment volume the stream can carry.

Science Journal Flood Damage Prevention Write a paragraph in your Science Journal that describes how you could make your classroom or home more protected in case of a flood. **IS** **L2** **P**

4 Close

Section 3 Wrap-up

1. Larger sediment is pushed and bounced along the bottom of the stream channel. Smaller sediment is held up by the water in suspension. Some sediment dissolves and is carried in solution.
2. The steeper the slope, the greater the amount of runoff. If the soil is wet, there may not be any room for water to soak in, so runoff will be greater.
3. **Think Critically** Over time, erosion cuts down the land surface. As the land around the stream becomes less steep, the stream begins to erode its banks more than the channel bottom, and curves form.

Science Journal Answers will vary but should include information about how a stream erodes its banks and channel bottom and how a stream carries sediment.

FIGURE 17-9

This photo was taken during a flash flood of the Colorado River in the Grand Canyon. Notice how muddy the water looks. The river is carrying a lot of sediment. This is how, over millions of years, the relatively small Colorado River eroded the mighty Grand Canyon.

Section 3 Wrap-up

1. During stream erosion, how does the water carry material?

2. Explain how runoff might be affected by the slope of the land and how wet the soil is.

3. **Think Critically:** Why does a stream begin to curve as it develops over time?

4. **Skill Builder**
   Developing Multimedia Presentations
   Use posters, videos, photos, or computers to make a class presentation about how rivers form and develop. Be sure to include information on flooding. If you need help, refer to Developing Multimedia Presentations on page 564 in the **Technology Skill Handbook.**

Science Journal

Write a short story in your *Science Journal*, describing how a small stream can start in the mountains and flow down to the ocean, forming a wide river valley along the way.

Skill Builder
Developing Multimedia Presentations Student answers will vary but must include explanations of how rivers form, how rivers develop, and why rivers flood. **P** **L2**

Assessment

Oral Presentation Encourage students to share their Multimedia Presentations about rivers and flooding with other students. Use the Performance Task Assessment List for Oral Presentation in **PASC**, p. 71.

Oceans

Ocean Water

If you were getting ready to sail around the world, what would you take with you? You'd have to take food, clothes, and many supplies. You'd also have to take water. Doesn't that seem kind of strange? You'd be surrounded by water, so why would you need to take your own? The reason is that ocean water is salty. It is not good for people to drink because too much salt will make you sick.

Why is the ocean salty?

How does ocean water get so salty? In the last section, you learned that water flows from streams to rivers and from rivers to the oceans. The water in streams and rivers isn't salty. What happens along the way? Both streams and groundwater pick up elements such as calcium, magnesium, and sodium from rocks and minerals. These elements dissolve in the water and are carried to the oceans. In the oceans, they combine with other elements, such as sulfur and chlorine. These elements have been added to the oceans by volcanic activity. Volcanoes release these elements into the air and they later fall into the oceans. When the elements from the rivers and groundwater combine with the elements already in the ocean, they form salts, as shown in **Figure 17-10.** The most common combination of elements is sodium and chlorine—the same elements that make table salt. Over millions of years, these salts have built up to make the oceans salty.

What YOU'LL LEARN

- Why oceans are salty
- About currents, waves, and tides

Science Words:
salinity
current
wave
tide

Why IT'S IMPORTANT

You'll better understand the oceans, where most of Earth's water is found.

FIGURE 17-10

Rivers and streams carry many elements in solution. This material is carried to the ocean, where it combines with other elements to form salts.

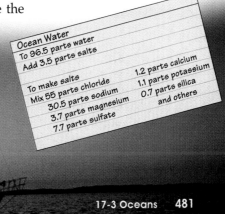

Ocean Water
To 96.5 parts water
Add 3.5 parts salts

To make salts
Mix 55 parts chloride
30.5 parts sodium
3.7 parts magnesium
7.7 parts sulfate
1.2 parts calcium
1.1 parts potassium
0.7 parts silica
and others

17-3 Oceans   **481**

Prepare

Section Background

- The study of the oceans, the interrelationships of ocean systems, and the varied ocean environments is oceanography.
- Even though salinity is defined as a measure of the dissolved salts in water, it is formally defined as the total amount of dissolved solids (parts per thousand) in seawater. These solids include other substances besides salts.

Preplanning

Refer to the Chapter Organizer on pages 466A-B.

1 Motivate

Bellringer

Before presenting the lesson, display **Section Focus Transparency 59** on the overhead projector. Assign the accompanying **Focus Activity** worksheet.

L2 ELL

Program Resources

📁 Reproducible Masters
Activity Worksheets, pp. 5, 108 L2
Enrichment, p. 65 L3
Lab Manual, pp. 119-126 L2
Reinforcement, p. 65 L2
Science and Society/Technology
  Integration, p. 35 L2
Study Guide, p. 65 L1

🔦 Transparencies
Section Focus Transparency 59 L2
Teaching Transparency 34 L2

Help students recall that running water erodes sediment. Some of the sediments are dissolved in the water. Dissolved sediments carried by rivers may eventually enter the oceans.

2 Teach

Discussion

 Logical-Mathematical Ask students to consider what Earth would be like if oceans had not formed. Suggest that they think about how Earth's surface would be different. Would there be any life as we know it? How might weather conditions be different on Earth if no oceans had formed? Could there be rain?

GLENCOE TECHNOLOGY

 Videodisc

STVS: Earth & Space
Disc 3, Side 2
Artificial Waves (Ch. 13)

Studying the Pacific Ocean (Ch. 14)

FIGURE 17-11
Many organisms in the ocean use elements in ocean water to build their shells and skeletons. These corals, in the Great Barrier Reef in Australia, have exoskeletons made of calcium carbonate.

A measure of the saltiness of water is called **salinity** (suh LIH nuh tee). The salinity of ocean water has stayed about the same for hundreds of millions of years. That must mean that the amount of salt added to the oceans over time has been balanced by the amount lost over the same time. How do oceans lose salt? Plants and animals that live in the oceans remove some from ocean water. They use elements, such as calcium and silicon, to build their shells and skeletons, as shown in **Figure 17-11.** Some of the elements in the ocean water just settle to the ocean floor. They become part of sedimentary rocks.

Now you know why the oceans are salty—and why you need to take water along when you sail around the world. Something else that could affect your sea trip is ocean motion.

Ocean Motion

Have you ever watched someone stir a big spoonful of sugar into a glass of iced tea? Before the sugar dissolves, you can see it swirling around in the glass. The sugar moves along with the tea because the stirring motion

Surface current

Surface current

Bottom current

Science Journal Poems from the Shore Ask students to write a poem in their Science Journals about what it might be like to live near an ocean. Suggest that they include topics such as walking along the beach, fishing on a large fishing boat, or visiting a working lighthouse. For students who live near the ocean, encourage them to try looking at the ocean as if it is the first time they have ever seen it. **LS** **P** **L3**

forms currents. A **current** is a mass of water (or tea) moving in one direction. Currents happen in the ocean. But it's not because a giant spoon is mixing it up. Let's see why currents form in the ocean.

Surface Currents

You can think of ocean currents as rivers of water flowing through the ocean. Some move along the ocean bottom. Bottom currents are caused mainly by the differences in temperature or density of different water masses. Others, called surface currents, move at or near the ocean surface, as shown in **Figure 17-12.** Wind powers surface currents. Friction between the water in the ocean and the air in the wind makes the water move. You can see how this works by blowing across the top of a bowl of water. Surface currents can carry seeds and plants from one continent to another. Sailors use currents to make sailing easier. People who fish in the ocean can catch more fish if they know about currents. That's because fish and other sea animals often follow currents. Fish follow currents because they eat tiny animals and plants that are carried by the currents.

Currents also can affect the climate. One well-known surface current called the Gulf Stream does just that. The warm water of the Gulf Stream makes the climate of Great Britain much milder than it would be otherwise.

interNET
CONNECTION

Do you want to know more about the oceans? Check out the Glencoe Homepage, *www.glencoe.com/sec/science*, for a link to questions and answers from an oceanographic institute.

FIGURE 17-12

Ocean water is always moving. Ocean currents move like rivers flowing through the ocean. Some currents flow near the surface and are caused by winds. Other currents flow along the bottom, or somewhere between the bottom and the surface.

Wind

Surface current

Bottom current

17-3 Oceans **483**

Teacher F.Y.I.

Water waves have many parts. The highest part of the wave is the wave crest. The length between successive crests of the wave is the wavelength. The lowest part of the wave is the wave trough.

? FLEX Your Brain

Use the Flex Your Brain activity to have students explore OCEAN CURRENTS.

Activity Worksheets, page 5

Making a Model

Visual-Spatial Have students make a model of a current. Ask them to fill a 250-mL beaker with cold water. Then direct them to fill another 250-mL beaker with warm water containing food coloring. Instruct the students to use a dropper to place a small amount of warm, colored water into the cold water. Ask students to record their observations in their Science Journals. Students should see that the warm, colored water moves upward toward the surface of the beaker, forming a density current. L2 ELL

Discussion

Ask students to discuss how winds blowing over the surface of the water might cause surface currents to form in the oceans.

Across the Curriculum

Literature Contact the reading teacher at your school to find out whether your students will be reading any stories about sea voyages or ocean life. If so, team plan with the reading instructor so the students will be reading about the oceans in their reading class while studying the oceans in science. LS

interNET
CONNECTION

The Glencoe Homepage at **www.glencoe.com/sec/science** provides links connecting concepts from the student edition to relevant, up-to-date Internet sites.

Mini LAB

Purpose

LS **Kinesthetic** Students will see how water waves move things. **L2** **ELL**

Materials

large, flat, plastic box or pan; a cork; a Ping-Pong ball; small bits of Styrofoam; wooden rod

Teaching Strategies

Place the pan over a large piece of paper so the waves are easier to see.

Troubleshooting When students tap the water with the wooden rod, be sure they tap only once so they can clearly see what the wave does to the floating objects.

📁 **Activity Worksheets,** pages 5, 108

Analysis

1. Objects floating in the water move in a circular path.
2. The movement of individual particles of water can be inferred from the movement of objects floating in the water. Individual particles of water move backward, then up, forward, and then down, staying in one location for the most part.

Assessment

Oral Ask students to explain whether water moves across the pan with the wave. As the wave goes by, what happens to the particles of water? Use the Performance Task Assessment List for Making Observations and Inferences in **PASC,** p. 17.

Mini LAB

Observing Waves 🖐️

Observe how waves move things in the water.

1. Fill a large, flat, plastic box or pan halfway with water.
2. Place some floating items in the water. These things might include a cork, a Ping-Pong ball, or small bits of Styrofoam.
3. Make waves in the pan by moving a wooden rod or large pencil up and down in the pan.

Analysis

1. Observe the objects floating in the pan, and describe how the objects are moved by the water waves.
2. How could you test how individual particles of water are moved by a wave?

FIGURE 17-13

In a wave, water particles move in a circular path. The deeper you go below the water's surface, the smaller the circles. Below a certain depth, little water movement occurs.

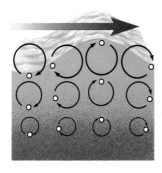

Ocean Waves

One after another, ocean waves crash onto the shore. Couldn't you just sit and watch them forever? No matter how long you sit and watch, the waves would keep coming. What is a wave? A **wave** is a movement of ocean water that makes the water rise and fall. Ocean water is constantly in motion. The movement of waves transmits energy. If you have ever tried to stand up to a big wave at the seashore or in a wave pool, you have felt that energy in the force of the moving water.

But waves in the ocean transmit energy without actually moving the water very far. To understand that, picture what happens when you're floating on a rubber raft in the ocean or at a wave pool. As each wave passes under your raft, the raft moves backward, then up and forward. But it ends up almost where it started. This same type of motion, shown in **Figure 17-13**, makes individual particles of water move in a circular path.

Most of the circular movement of water particles is near the surface of the ocean because that's where most waves are. The wave's energy spreads from one water particle to another through the particles' circular motion. Have you ever been at a sports stadium where fans did "the wave"? In the stadium, one group of people stands up and sits back down. Then the people right next to them do the same thing. Then the next group of people does it, and so on all around the stadium. If you're watching from far away, it looks like a huge wave of people is moving around and around the stadium. But it's really just lots of people standing up and sitting down, one by one. In the ocean, energy spreads from one water particle to another in a similar way.

What makes waves?

What makes waves in the ocean? Wind blows over the ocean's surface. Energy is transferred from the wind to the ocean water. That transfer of energy forms waves.

If you've watched waves roll in toward a beach, you've noticed that the waves "break" as they get close to the shore. That's because the water is shallower close to shore. As a wave comes into the shallow water, the wave doesn't have as much room to move the water around. Water in the wave piles up. As that happens, the top of the wave moves faster than the bottom, and the wave breaks against the shore. Surfers and boogie boarders like to ride those collapsing or breaking waves, as shown in **Figure 17-14**.

FIGURE 17-14

Ocean waves "break" as they move close to the shore.

USING TECHNOLOGY

Inside the craft, the diver is affected by the pull of gravity. Because of this, the Bio Shell part of this craft is designed to place the pilot in a position that is natural for creatures underwater. The pilot's body will be aligned with the axis of the craft—head down, face-down, and body lined up with the direction of movement through the water.

*inter*NET
CONNECTION

Visit the Glencoe Homepage, **www.glencoe.com/ sec/science**, for a link to information about the *Deep Flight* project.

Enrichment

Have a team of two or three students research the early dives of *Alvin* and report back to the class. Encourage them to produce posters and illustrations to help in their presentation. [L2] **COOP LEARN** P

USING TECHNOLOGY

Submersibles

Scientists use submersibles (sub MUR suh bulz) such as ALVIN to study the deep ocean. They can also study life in the oceans and human effects on the ocean environment.

In the past, submersibles have been made to sink or rise in the water by changing their weight. This was done by pumping water into or out of tanks in the sub and by using weights. A new type of submersible may change all of that.

Deep Flight I

The experimental craft now called *Deep Flight I* moves up and down in the water by its own power instead of using weights. This new kind of submersible moves in a way similar to the way an airplane flies in the air or a dolphin swims in the ocean. It can move through the water in any direction. To *Deep Flight I*, there is no up or down. It is as much at home in the water in level "flight" as it is preparing for a steep dive.

*inter*NET
CONNECTION

Go to the Glencoe Homepage, *www.glencoe.com/sec/science*, for a link to a site that tells more about ocean research vessels.

485

Content Background

When the depth of the water is less than one half of the wavelength of a wave, the wavelength of a wave becomes shorter and the wave height increases. Waves start to pile up and the ocean bottom applies friction to the water in the wave, slowing down the bottom of the wave. When this happens, the top of the wave outruns the bottom and eventually collapses. The crest of the wave falls, water tumbles over on itself, and the wave crashes onto shore, forming breakers.

3 Assess

Check for Understanding

Activity Have students research a major ocean current, such as the Gulf Stream, to help them understand how currents affect climate and ocean travel. ▣ ▣

Reteach

To help students understand how density currents form (currents that occur because of a difference in the density of ocean water), encourage them to do the following activity. Fill two 600-mL beakers halfway with water. Place food coloring in one of the beakers of water. Pour salt into the beaker containing the colored water, and stir it until all of the salt dissolves. Tell students to keep putting salt into the same beaker until no more can be dissolved. Carefully pour the contents of the two beakers into two sides of a flat pan that has a divider down the middle. Remove the divider and observe what happens. Students will see that the blue, denser water will flow into the less dense clear water. Because of this movement, the less dense, clear water will be forced into the other side of the pan. ▣ ▣

Extension

▣ For students who have mastered this section, use the **Reinforcement** and **Enrichment** masters.

Ocean Tides

The waves that constantly crash on the beach are caused by wind, as you just learned. But another kind of wave motion in the ocean is caused by something else. The effects of this water motion are called tides. **Tides** are changes in the level of ocean water during the course of a day. The level of ocean water usually reaches a high and low level twice a day. These high and low levels are called high and low tides, as shown in **Figure 17-15.** At high tide, water is pushed up onto the shore. As the low tide occurs, usually about six hours later, water is pulled away from shore.

What causes the tides? One cause is the pull of gravity between Earth, the moon, and the sun. The moon's gravity exerts a force on Earth. The water in the oceans responds to this force, forming a bulge of ocean water on the side of Earth facing the moon. Another bulge forms

FIGURE 17-15
The pull of gravity between Earth, the moon, and the sun is one cause of the tides in the ocean. In most places, there are two high tides and two low tides each day.

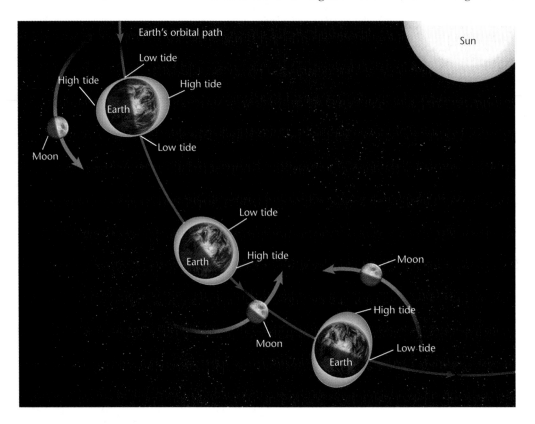

Theme Connection

Systems and Interactions

Ask a group of three students to research the most recent occurrence of El Niño. Encourage them to relate this occurrence to any changes in global weather systems. Suggest that they study the occurrence of hurricanes in the Pacific Ocean. ▣ ▣ ▣

COOP LEARN

on the opposite side of the planet. The bulge on the side away from the moon also occurs because the water is thrown outward as Earth and the moon revolve around a common point. The water in the oceans tries to travel away from Earth's orbit, but Earth's gravity keeps the water from flying off into space. This is similar to what happens if you spin a bucket of water around in a circle, as shown in **Figure 17-16.** The water in the bucket tries to travel in a straight line and is thrown outward toward the bottom of the bucket. The bottom of the bucket keeps the water contained, and the water can't flow out unless you stop spinning.

The bulges that form in the oceans are the highest parts of the tides. The sun's gravity also has an effect on the tides, but because the sun is much farther away, its effect is less than that of the moon.

You have learned how important water is. Next, you will learn about some of the problems of ocean pollution.

FIGURE 17-16
The bulge of ocean water that forms the tides is thrown outward as Earth and the moon revolve around the sun, like the water in the bucket.

Section 🌊 Wrap-up

1. Describe how ocean waves get energy from Earth's atmosphere.

2. Describe what causes tides.

3. **Think Critically:** You are swimming in a wave pool at a water park. The wave passes you and moves to the other side of the pool. Why doesn't the wave carry you and the water around you from one side of the pool to the other?

4. **Skill Builder**
   Predicting A surface current moves colder, denser water into an area of warm, less dense water. Predict what will happen to the ocean water in this area. Explain your prediction. If you need help, refer to Predicting on page 560 in the **Skill Handbook.**

Science Journal

In winter, the water of the ocean around Antarctica is saltier and denser than any other ocean water. Find out the reason for this, and write about it in your Science Journal.

Skill Builder
Predicting The colder, denser water that is moved into the area of warm, less dense water by a surface current will sink toward the bottom of the ocean.

✓ **Assessment**

Performance Give students two different liquids, with one denser than the other (e.g., water and syrup). Ask them to hypothesize what will happen when the two liquids are combined. Have the students combine the liquids and observe what happens. Use the Performance Task Assessment List for Formulating a Hypothesis in **PASC**, p. 21. [LS]
[L2] [ELL]

4 Close

• MINI • QUIZ •

Use the Mini Quiz to check students' recall of chapter content.

1. **A measure of the amount of dissolved salts in ocean water is called _____.** *salinity*

2. **A mass of water moving in one direction is a(n) _____.** *current*

3. **A movement of ocean water that makes the water level rise and fall is a(n) _____.** *wave*

Section 🌊 Wrap-up

1. As winds flow over the surface of the oceans, energy due to friction is transferred from the air to the water in the ocean.

2. The tides are caused by the interaction of gravity within the Earth-moon-sun system. Also, Earth is spinning, causing the water to be thrown outward forming a bulge.

3. **Think Critically** While the energy of a wave in water is transmitted across the body of water, the individual particles of water are not. The circular motion of the water particles transmits the energy of the wave from one particle to another.

Science Journal As water in the Antarctic Ocean freezes during winter, the salt is left in the unfrozen water. This makes the water denser, and it flows toward the ocean floor where it spreads along the ocean bottom toward the equator—an area of less dense ocean water. [L2] [P]

Science
& Society

17•4 Ocean Pollution

Bellringer

Before presenting the lesson, display **Section Focus Transparency 60** on the overhead projector. Assign the accompanying **Focus Activity** worksheet.

[L2] [ELL]

What YOU'LL LEARN

- The importance of oceans as resources
- Various sources of ocean pollution

Why IT'S IMPORTANT

Distinguishing fact from opinion helps you make better decisions.

FIGURE 17-17

Beaches, like this one in Costa Rica, can become polluted from ocean dumping, shipping accidents, and careless people.

The Blue Planet

The oceans are the homes for millions of different kinds of life, from tiny floating plants and animals to the largest known creature—the blue whale. And, while humans don't live *in* the oceans, about half of the world's population lives within 100 km of the oceans. Oceans and their beaches give humans some of their food, mineral resources, and recreation.

Sources of Ocean Pollution

Because humans use Earth's oceans for so many things, the oceans can become polluted. The oceans are huge, but they are also fragile resources. Tankers that carry oil have had accidents and spilled oil into the oceans. Beaches and animal and bird populations that live near the shore can suffer greatly from these spills. Oil from other kinds of seagoing ships can also pollute our oceans.

Ships can pollute the oceans by dumping their wastes overboard. Some trash is allowed to be dumped at sea. But things such as glass and metal must first be crushed before they are thrown overboard. The dumping of plastics of any kind is illegal. Along one shore, the Gulf of Mexico, wastes dumped by ships totaled 65 percent of the trash collected during one year.

Ocean beaches also can be polluted by people on the beach, such as the beach shown in **Figure 17-17.** On Padre Island in Texas, during just one of the national beach cleanups that occur several times a year in the United States, volunteers, such as those shown in **Figure 17-18,** collected more than 7 million pieces of trash including 25 000 plastic six-pack holders, 40 000 rubber balloons, 300 000 plastic and glass beverage bottles, and 200 000 metal cans!

488   Chapter 17 Water

Teaching the Content

- [LS] **Kinesthetic** Have students attempt to clean up an "oil spill" using a variety of materials including paper towels, spoons, liquid dishwashing soap, and sponges. Make the spill using a shallow, plastic box; a quarter of a cup of cooking oil; and water. [L2] [COOP LEARN] [ELL]

- Some students may not understand how actions far from the ocean can pollute ocean waters. Use a physiographic map of the United States to show students how something they pour into the soil in the backyard eventually reaches the ocean. Help students sequence the events, if necessary. The substance seeps into the ground and enters the groundwater system. From there, the substance is carried to local streams and rivers. All rivers ultimately empty into the ocean.

Program Resources

Reproducible Masters
Enrichment, p. 66 [L3]
Reinforcement, p. 66 [L2]
Study Guide, p. 66 [L1]

Transparencies
Section Focus Transparency 60 [L2]

Solutions to Ocean Pollution

For a long time, many people, even some scientists, thought that Earth's oceans were big enough to resist most of the effects of pollution. Now, however, scientists think that the oceans and ocean organisms are suffering from the effects of pollution.

Most people don't realize that many of the things that are poured down drains, thrown away, or dumped out in the backyard can end up in the ocean. Properly disposing of toxic items such as paint, motor oil, and household chemicals can prevent pollution. Reducing and recycling are two other ways to prevent and minimize all forms of pollution. Can you think of other solutions to ocean pollution or any other kind of pollution?

FIGURE 17-18
These students are helping in a beach cleanup.

 ## Skill Builder: Distinguishing Fact from Opinion

LEARNING the SKILL

1. Facts are statements that can be checked for accuracy and proven. Facts often answer the reporter's five questions: *who, what, where, when,* and *why?*

2. Opinions, on the other hand, express personal beliefs and feelings. Opinions can't be proved or disproved. Opinions often include the following words: *I believe, I think, probably, it seems to me, best, worst,* and *greatest,* among others.

PRACTICING the SKILL

1. Reread the first paragraph. Are there any opinions in it? If so, identify them.

2. List at least five facts from magazines or Internet sources on ocean pollution.

APPLYING the SKILL

The statements below were taken from magazine articles on ocean pollution. Identify each as either a fact or an opinion.

1. Fishing should be banned in contaminated areas.

2. About 1 million tons of military equipment were dumped into an ocean trench off the Scottish coast between 1945 and 1976.

3. During 1992 and 1993, pollution caused almost 5000 beach closings.

4. Human activities are killing our beaches.

Content Background

- The tanker *Exxon Valdez* accidentally spilled more than 40 million liters of crude oil into Prince William Sound in 1989. The hardest-hit marine mammals were probably the sea otters. Most were poisoned by the oil itself, while others died from hypothermia as their oil-laden fur lost its insulating ability.

- Cigarette butts are the most common item of trash found along beaches.
- Some countries currently use the oceans to dispose of radioactive wastes.

People & Science

Background

- Invertebrates are animals without backbones.

- The *Johnson-Sea-Link* submersibles are operated by Harbor Branch Oceanographic Institution in Florida. They can carry four people down to a maximum depth of about 910 m. They are equipped with hydraulic arms and suction tubes for collecting samples to study.

- One of the best-known American submersibles is the chubby yellow *Alvin*, built in 1964 for Woods Hole Oceanographic Institution in Massachusetts. *Alvin* is currently able to descend to 4500 m.

- The world's deepest-diving, manned submersible is Japan's *Shinkai 6500*. The *Shinkai* can carry three people down to 6500 m. Japan also operates the world's deepest-diving, unmanned, remotely operated vehicle, *Kaiko*. In 1995, *Kaiko* succeeded in coming close to the bottom in the Marianas Trench north of New Guinea, the deepest point on Earth at 11 022 m below sea level.

Teaching Strategies

LS Interpersonal Have students work in pairs to design hypothetical deep-sea dwellers. Have team members present their animal and describe its adaptations to the class. **L2**

ELL **COOP LEARN** **P**

Isidro Bosch, Aquatic Biologist

Q When and how did you first get interested in the ocean?

A As a child living in Cuba, some of my earliest memories are of trips with my family to the beach. There was one danger where we went swimming—*los erizos*, the long-spined sea urchins of the Caribbean. As a child, I feared them. Isn't it funny that as a grown-up, I study those same animals?

Q Dr. Bosch, tell us about your research.

A I mostly study ocean invertebrates—sea urchins, sea slugs, sponges, and so forth. I'm interested in how these animals live in tough environmental conditions, such as cold polar oceans and the deep sea with its darkness and high pressure.

Q How have you explored the oceans?

A In everything from huge research ships to small, inflatable boats. I've dived beneath the sea ice in Antarctica, and explored tropical coral reefs and giant kelp forests. One of my greatest experiences has been exploring the deep sea in tiny submarines that go down into deep parts of the ocean. I've been down to 900 meters in subs.

Q What's it like to be at those great depths?

A As the submersible drifts down through the darkness, some organisms have natural luminescence and light up in the water around you. They flash like fireworks on the Fourth of July. On the bottom, you see a world few other people have seen. This world is inhabited by strange animals that have adapted to the crushing pressures, total darkness, and extreme cold.

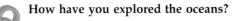

Q Recently, you've also been doing research on freshwater lakes. Why?

A Lakes are threatened habitats, and the biologists who study them have a very important and urgent job. Research in lakes has shown clearly how all life-forms depend on each other—how adding or taking away just one organism can change the entire community of organisms that live there.

Career Connection

Interview someone in your area who works with the oceans, or contact someone at an oceanographic institute on the Internet. Present your findings to the class.

Career Connection

Career Path Aquatic biologists usually have a master's degree or Ph.D. in marine or freshwater biology or ecology.

- High School: science and math
- College: biology, chemistry, and physics
- Post-College: Master's degree or Ph.D. in marine/aquatic biology or larval ecology.

For More Information

The National Oceanic and Atmospheric Administration's (NOAA) mission is to conserve and manage the coastal and marine resources of the United States. They maintain a website that contains useful information on marine organisms.

Read the statements below that review the most important ideas in the chapter. Using what you have learned, answer each question in your Science Journal.

1. Water moves through Earth's water cycle by evaporation, condensation, and precipitation. *How does condensation of water vapor in Earth's atmosphere eventually lead to precipitation?*

2. Water that does not run off will soak into the ground. *What is this water that fills and moves through holes in soil and rock called?*

3. Runoff is affected by the amount of rain and its intensity, the slope of the land, the amount of vegetation, and whether the ground is already wet. *Describe how two of these affect runoff.*

4. Streams and groundwater have dissolved elements such as calcium, magnesium, and sodium from rocks and minerals. *How do these elements end up as salts in the oceans?*

Chapter 17 Review **491**

Chapter 17

Review

Have students look at the illustrations on this page. Ask them to describe details that support the main ideas of the chapter found in the statement for each illustration.

Teaching Strategies

Divide the class into three groups. Have the first group make a poster that shows Earth's water cycle. Have another group locate an area on the school grounds at which they can demonstrate water soaking into the ground. Have another group make a bulletin board that shows how at least two of the listed items affect runoff. L2 **COOP LEARN**

Answers to Questions

1. Water condenses into clouds. When water droplets are heavy enough, they fall as precipitation.

2. groundwater

3. If the amount of rain is great and intense, water can't soak in and more of it will run off. If the slope is steep, water will run off before it has time to soak in. Vegetation will hold water on Earth's surface, enabling it to soak in. If the ground is already wet, no more can soak in and runoff increases.

4. The dissolved materials are carried to the ocean in streams and rivers.

✓ Assessment

Portfolio Encourage students to place in their portfolios one or two items of what they consider to be their best work. Examples include:

• Activity 17-1 model and experiment design, p. 472
• MiniLAB analysis, p. 477
• Activity 17-2 data and observations, p. 478
P

Performance Additional performance assessments may be found in **Performance Assessment** and **Science Integration Activities.** Performance Task Assessment Lists and rubrics for evaluating these activities can be found in Glencoe's **Performance Assessment in the Science Classroom (PASC).**

Science at Home

IS **Intrapersonal** Ask students to identify areas in their homes where evaporation and condensation occur. Have them make posters that show where these processes occur. L2

Chapter 17 Review

Using Key Science Words

1. runoff
2. meander
3. salinity
4. deposition
5. condensation

Checking Concepts

6. a   7. c   8. b
9. d   10. a

Thinking Critically

11. Cities have a lot of pavement. The pavement does not let water soak into the ground. The water level gets higher than in the country, where water can soak into the ground.

12. It would be better to drill for water in a layer of sand. Groundwater flows into and through openings in the sand. The openings between particles in clay are so small that little or no water can enter.

13. When the cities were built, waterways were important for commerce. Rivers and lakes were also important sources of freshwater.

14. 3.5 kg, 1.75 kg

15. The forest fire destroys all or most of the vegetation. Without vegetation, there is nothing to hold the soil. With heavy rains, the soil will wash off as mud, perhaps causing a mudslide. Flooding is more likely to occur because little water will soak into the ground. Without vegetation, the water runs off before it has a chance to soak in. Increased runoff could lead to flooding.

Using Key Science Words

condensation	precipitation
current	runoff
deposition	salinity
erosion	tide
evaporation	water cycle
groundwater	wave
meander	

Match each phrase with the correct term from the list of Key Science Words.

1. water flowing over the land
2. a wide curve in a stream channel
3. saltiness of ocean water
4. when moving water drops sediment
5. water vapor changing back into a liquid

Checking Concepts

Choose the word or phrase that completes the sentence.

6. The process in which water changes from a liquid to water vapor is
   _____.
   a. evaporation
   b. condensation
   c. runoff
   d. precipitation

7. Earth's _____ is the continual cycling of water between Earth's surface and the atmosphere.
   a. rock cycle   c. water cycle
   b. evaporation d. runoff

8. Rivers change Earth by dropping sediments in a process called
   _____.
   a. runoff
   b. deposition
   c. condensation
   d. evaporation

9. Currents that move large masses of water across the ocean, at or near the surface, are called _____.
   a. waves   c. hurricanes
   b. tides   d. surface currents

10. The chlorine in ocean water comes from _____.
    a. volcanoes   c. rain
    b. storms   d. rivers

Thinking Critically

Answer the following questions in your Science Journal using complete sentences.

11. Why do you think that floods are often more severe in cities than in the countryside?

12. Would it be better to drill for water in a layer of clay or a layer of sand? Explain.

13. If you look at a map of the United States or the world, you will see that most major cities are located near bodies of water. Explain why you think cities are located near rivers, lakes, and the ocean.

14. If the salinity of ocean water is 3.5 percent, how many grams of solids (salts) are dissolved in 100 kg of water? In 50 kg of water?

Assessment Resources

Reproducible Masters
Chapter Review, pp. 37-38
Assessment, pp. 73-76
Performance Assessment, p. 55

Glencoe Technology
Computer Test Bank
MindJogger Videoquiz

15. Why do people worry about mudslides and flooding near hillsides after there has been a forest fire?

Developing Skills

If you need help, refer to the description of each skill in the Skill Handbook.

16. **Comparing and Contrasting:** Compare and contrast the processes of evaporation and condensation.

17. **Recognizing Cause and Effect:** What is runoff? How does the amount of vegetation on the ground affect runoff?

18. **Hypothesizing:** Hypothesize why fine particles are carried farther by a stream than heavier particles.

19. **Using Variables, Constants, and Controls:** Explain how you could measure undissolved water pollution by comparing how clear the water looks.

20. **Comparing and Contrasting:** Compare and contrast waves and currents. How are they similar? How are they different?

Performance Assessment

1. **Making Observations and Inferences:** Experiment to find out the amount of space between particles. Use 100 mL of sand. Measure the amount of water needed to just cover the sand. Repeat your experiment to measure the spaces when you mix 50 mL of gravel and 50 mL of sand. Repeat your experiment to measure the spaces when you mix 50 mL of clay and 50 mL of sand.

2. **Observing and Inferring:** Study the photo below and infer how water has affected the landscape and organisms shown. Discuss how erosion and deposition have affected the mountains and the land around the river.

Chapter 17 Review **493**

Developing Skills

16. **Comparing and Contrasting** Both processes deal with changing physical state. During evaporation, liquid is changed to vapor, whereas during condensation, vapor is changed to liquid.

17. **Recognizing Cause and Effect** Runoff is water that flows over Earth's surface. More vegetation holds the water in place, allowing more of it to soak into the ground.

18. **Hypothesizing** Fine-grain particles can be carried by water in suspension, whereas heavier particles are often pushed or bounced along the bottom of the stream channel. The finer the particles, the farther the water can carry the particles.

19. **Using Variables, Constants, and Controls** Make a chart that has five blocks that are colored in with a colored pencil. Each block should be progressively darker from clear to completely colored. Water samples could be held up next to the color chart. The amount of pollution could be given in numbers from 0 to 4, depending on how clouded the water is. Zero would indicate no pollution.

20. **Comparing and Contrasting** Waves and some currents (surface currents) are generated by energy from the wind blowing over the ocean's surface. Waves transmit energy without moving particles of the ocean across the body of water. Currents transmit energy and move water from one location to another.

Performance Assessment

1. Student inferences should indicate the volume of water that is absorbed into the pore space in 100 mL of sand, 50 mL of gravel, and 50 mL of sand. Students should also infer which materials had the greatest pore space. Use the Performance Task Assessment List for Making Observations and Inferences in **PASC**, p. 17. P

2. Answers will vary but should include that water allows plants to survive, water has deposited sediment along the river, and it has eroded mountains. Use the Performance Task Assessment List for Making Observations and Inferences in **PASC**, p. 17. P

Chapter Organizer

Section	Objectives/Standards	Activities/Features
Chapter Opener		Explore Activity: Observe the Mass of Air, p. 495
18-1 **What's in the air?** (3 sessions, 1½ blocks)*	1. **Analyze** the makeup of the atmosphere. 2. **Identify** the parts of the atmosphere. 3. **Determine** the effects of air pressure. National Science Content Standards: (5-8) UCP2, UCP3, A1, A2, B1, C4, D1, F2, F3, G1	MiniLAB: Observing the Effects of Holes in the Ozone Layer, p. 498 Problem Solving: Air Pressure Egg-zample!, p. 500 Activity 18-1: Air Pressure, p. 501 Skill Builder: Making and Using Graphs, p. 502 Science Journal, p. 502
18-2 **Weather** (3½ sessions, 2 blocks)*	4. **Discuss** the causes of weather. 5. **Compare and contrast** different types of weather. National Science Content Standards: (5-8) UCP1, UCP2, UCP3, A1, A2, B2, D3, F2, F3	MiniLAB: Inferring, p. 504 Activity 18-2: Blowing in the Wind, pp. 508-509 Skill Builder: Comparing and Contrasting, p. 511 Using Math, p. 511
18-3 **Science and Society:** **Forecasting the** **Weather** (1 session, ½ block)*	6. **Explain** how scientists forecast the weather. 7. **Develop** note-taking skills. National Science Content Standards: (5-8) UCP2, UCP3, E2, F5, G1	Skill Builder: Note-Taking Skills, p. 513
18-4 **Climate** (1½ sessions, ½ block)*	8. **Describe** climate. 9. **Analyze** how people affect Earth's climate. National Science Content Standards: (5-8) UCP1, UCP2, UCP3, A2, C4, D2, F2, G1, G2	Using Technology: GLOBE Project, p. 516 Skill Builder: Recognizing Cause and Effect, p. 517 Science Journal, p. 517 Science & History: How Typhoons Saved Japan, p. 518

* A complete Planning Guide that includes block scheduling is provided on pages 31T-33T.

Activity Materials

Explore	Activities	MiniLABs
page 495 balance, balloon	page 501 glass soft-drink bottle, balloon pages 508-509 directional compass, straight pin, stiff plastic, scissors, drinking straws, clear tape, pencil	page 498 flashlight, globe, waxed paper page 504 balloon

Need Materials? Call Science Kit (1-800-828-7777).

Teacher Classroom Resources

Reproducible Masters	Transparencies	Teaching Resources
Activity Worksheets, pp. 5, 109-110, 113 **Enrichment,** p. 67 **Lab Manual 35** **Reinforcement,** p. 67 **Study Guide,** p. 67	**Science Integration Transparency 18,** Altitude and Air Pressure **Section Focus Transparency 61,** Don't Look Down!	**Spanish Resources** **English/Spanish Audiocassettes** **Cooperative Learning Resource Guide** **Lab Partner** **Lab and Safety Skills** **Lesson Plans**
Activity Worksheets, pp. 5, 111-112, 114 **Cross-Curricular Integration,** p. 22 **Enrichment,** p. 68 **Lab Manual 36** **Reinforcement,** p. 68 **Science Integration Activities,** pp. 73-74 **Study Guide,** p. 68	**Section Focus Transparency 62,** Cloudy Weather **Teaching Transparency 35,** Heat from the Sun **Teaching Transparency 36,** Fronts	

Assessment Resources

Reproducible Masters	Transparencies	Assessment Resources
Enrichment, p. 69 **Multicultural Connections,** pp. 39-40 **Reinforcement,** p. 69 **Study Guide,** p. 69	**Section Focus Transparency 63,** Twister	**Chapter Review,** pp. 39-40 **Assessment,** pp. 77-80, 81-82 **Performance Assessment,** pp. 37-38, 56 **Performance Assessment in the Science Classroom (PASC)** **MindJogger Videoquiz** **Alternate Assessment in the Science Classroom** **Computer Test Bank**
Activity Worksheets, p. 5 **Enrichment,** p. 70 **Reinforcement,** p. 70 **Science and Society/Technology Integration,** p. 36 **Study Guide,** p. 70	**Section Focus Transparency 64,** Ice Sheet	

Key to Teaching Strategies

The following designations will help you decide which activities are appropriate for your students.

L1 Level 1 activities should be appropriate for students with learning difficulties.

L2 Level 2 activities should be within the ability range of all students.

L3 Level 3 activities are designed for above-average students.

ELL ELL activities should be within the ability range of English Language Learners.

LS These activities are designed to address different learning styles.

COOP LEARN Cooperative Learning activities are designed for small group work.

P These strategies represent student products that can be placed into a best-work portfolio.

GLENCOE TECHNOLOGY

The following multimedia resources are available from Glencoe.

Science and Technology Videodisc Series (STVS)
Earth & Space
The Infinite Voyage Series
Crisis in the Atmosphere
Glencoe Earth Science Interactive Videodisc
Introduction: Carl Chases a Thunderstorm
Spring: Time for Thunderstorms

Thunderstorm Formation
Thunder and Lightning
Lightning Rods
Conclusion: What You Should Do in a Thunderstorm
National Geographic Society Series
STV: Atmosphere
Glencoe Earth Science CD-ROM

Teacher Classroom Resources

This is a representation of key blackline masters available in the Teacher Classroom Resources.

Teaching Aids

Section Focus Transparencies

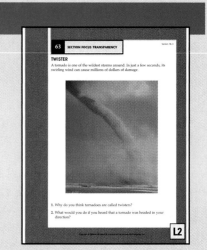

Science Integration Transparencies

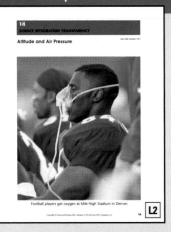

Teaching Transparencies

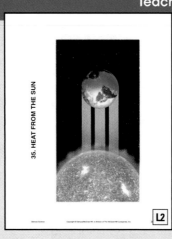

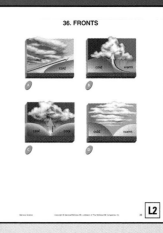

Meeting Different Ability Levels

Study Guide for Content Mastery

Reinforcement

Enrichment Worksheets

Hands-On Activities

Science Integration Activities

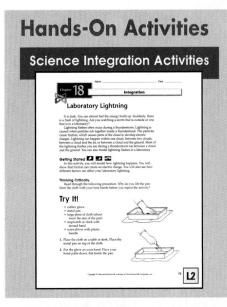

Lab Manual

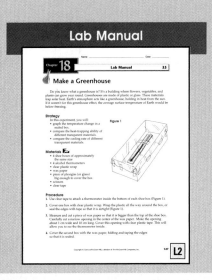

Activity Worksheets

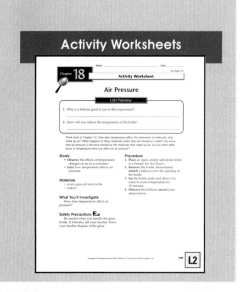

Enrichment and Application

Cross-Curricular Integration

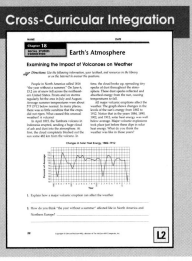

Multicultural Connections

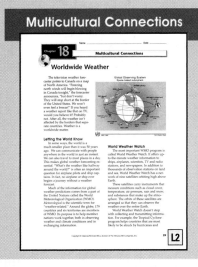

Science and Society/Technology Integration

Assessment

Performance Assessment

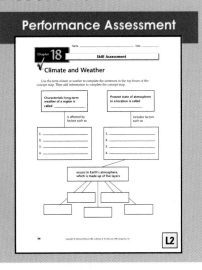

Chapter Review

Assessment

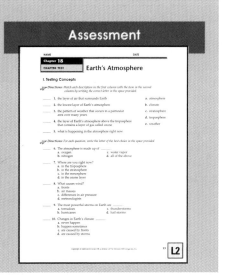

Earth's Atmosphere

CHAPTER OVERVIEW

Section 18-1 The composition and structure of Earth's atmosphere are discussed, as well as the effects of air pressure.

Section 18-2 This section presents factors that cause Earth's weather. Different types of weather are also compared and contrasted.

Section 18-3 Science and Society Methods used by scientists to forecast the weather are described, as are ways to improve student note-taking skills.

Section 18-4 This section discusses climate and describes different types of climate. Climatic change is discussed, and the effect human activity has on global warming is also presented.

Chapter Vocabulary

atmosphere	front
troposphere	meteorologist
stratosphere	climate
weather	global
wind	warming
air mass	

Theme Connection

Scale and Structure/Stability and Change The theme of Scale and Structure is emphasized by a discussion of the composition and structure of Earth's atmosphere. The theme of Stability and Change is emphasized by comparing and contrasting different types of weather. The Stability and Change theme is also supported by the fact that many factors affect the climate of an area.

Chapter Preview

Skills Preview

► **Skill Builders**
- make and use graphs
- compare and contrast
- recognize cause and effect

► **MiniLABs**
- observe
- infer

► **Activities**
- observe
- model
- infer
- record

494

Learning Styles Look for the following logo for strategies that emphasize different learning modalities. **LS**

Kinesthetic	Explore, p. 495; MiniLAB, pp. 498, 504; Inclusion Strategies, p. 498; Activity, p. 505; Activity 18-2, p. 508
Visual-Spatial	Visual Learning, pp. 497, 504, 505, 506, 512, 515; Demonstration, p. 499; Activity 18-1, p. 501; Reteach, p. 510; Inclusion Strategies, p. 516
Interpersonal	Community Connection, p. 497; Assessment, p. 511
Intrapersonal	Inclusion Strategies, p. 500; Assessment, p. 504; Go Further, p. 509; Check for Understanding, p. 510; Science at Home, p. 519
Logical-Mathematical	Student Text Question, p. 497; Problem Solving, p. 500; Community Connection, p. 506; Using Math, p. 511
Linguistic	Using Science Words, p. 499; Science Journal, p. 510; Activity, p. 510

Chapter 18

Earth's Atmosphere

Whoosh! What was that? A big airplane just took off right over your head. Don't you wonder how that giant jet stays in the sky, held up by nothing but air? Air is more than it seems to be. Air is matter. Even though we can't see it, air has mass and takes up space. Air can exert pressure on objects. That's why it's able to provide lift for airplanes. You've felt this pressure when the wind blows. What other properties of air can you observe?

EXPLORE ACTIVITY

Observe the Mass of Air

1. Place an empty balloon on a balance and measure its mass.
2. Blow up the balloon with air. Tie the end so that the air will not get out.
3. Place the blown-up balloon on a balance and determine its mass.

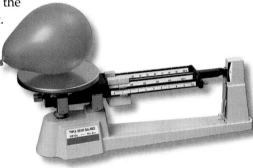

Science Journal

What was the mass of the empty balloon? What was the mass of the blown-up balloon? In your Science Journal, infer why the two measurements were different. What property of air did you observe?

495

Prepare

Section Background

Earth's atmosphere has not always been composed of the same gases as today's atmosphere. Over time, volcanoes have added water vapor, carbon dioxide, and nitrogen. Oxygen has been added by plants through the process of photosynthesis.

Preplanning

Refer to the Chapter Organizer on pages 494A-B.

1 Motivate

Bellringer

 Before presenting the lesson, display **Section Focus Transparency 61** on the overhead projector. Assign the accompanying **Focus Activity** worksheet.
L2 ELL

Tying to Previous Knowledge

Help students recall that matter is anything that has mass and takes up space, as they learned in Chapter 8.

496

18●1 What's in the air?

What YOU'LL LEARN

- What the atmosphere is made of
- The parts of the atmosphere
- The effects of air pressure

Science Words:
atmosphere
troposphere
stratosphere

Why IT'S IMPORTANT

Your life and the lives of all living things on Earth depend upon the atmosphere.

FIGURE 18-1

This graph shows the composition of Earth's atmosphere.

- ● Nitrogen 78%
- ● Oxygen 21%
- ● Other gases 1%

The Air Around You

On a hazy day, you might say that you can see the air. When it's breezy, you feel the air. But most of the time, you can't see the air that is all around you every second of your life. How can you describe something that you can't see? One way is to study what it's made of.

What makes up air?

Take a deep breath. What did you take into your body with that breath? Most people say they breathe oxygen when they take a breath of air. That's partly true. But as you can see from **Figure 18-1**, air is more than oxygen.

Air is a mixture of gases with a tiny amount of solids, such as dust particles, and liquids, like water, in its makeup. A layer of air surrounds Earth like a blanket. This blanket of air is called Earth's **atmosphere** (AT muh sfeer). The most common gas in Earth's atmosphere is nitrogen. Nitrogen makes up about 78 percent of the air we breathe. Oxygen makes up about 21 percent. That leaves about one percent for everything else. The amount of water vapor in the air changes from place to place and from time to time. It can be as low as nearly zero to as high as four percent.

Other Substances in the Atmosphere

We said that the atmosphere is like a blanket. On some days, it looks like a blanket that needs a spin in the washing machine! What could cause the brown layer of air that covers some cities like a dirty blanket?

Look at **Figure 18-2**. Hazy or smoky air is caused by pollutants (puh LEW tuntz) in the atmosphere. Pollutants are materials that harm living things by interfering with life processes. Some pollutants are released by natural

496 Chapter 18 Earth's Atmosphere

Program Resources

 Reproducible Masters
Activity Worksheets, pp. 5, 109-110, 113
Enrichment, p. 67 L3
Lab Manual, pp. 127-130
Reinforcement, p. 67 L2
Study Guide, p. 67 L1

Transparencies
Science Integration Transparency 18
Section Focus Transparency 61 L2

 A The burning of fossil fuels, such as coal, releases pollutants into the air.

B Coal scrubbers dissolve the gases in the smoke in water. This is one way that power plants can reduce pollution.

C The exhaust from cars, buses, and trucks adds polluting chemicals to the air.

D All new cars are now equipped with catalytic converters, which change harmful chemicals into harmless compounds. But vehicles are still a major source of air pollution. *Why?*

FIGURE 18-2

Most air pollution is caused by human activities.

events, such as when ashes from volcanic eruptions and forest fires enter the atmosphere. But most pollutants are produced by human activities. Do you do anything that causes air pollution? In one way or another, almost everyone does.

Air Pollution

When you hop in a car or catch a bus, you're usually just thinking about where you're going. But stop and think about how cars and buses affect the air. When cars and buses burn gasoline, they release pollutants into the atmosphere. Pollutants are also given off when coal and other fossil fuels are burned to make electricity. **Figure 18-2** shows how these and other human activities cause air pollution. It also shows some of the things that we use to keep air clean.

Visual Learning

Figure 18-2 Vehicles are still a major source of air pollution. Why? *Many older-model vehicles are still in use today. These vehicles are not equipped with pollution-reducing technology.* LS

Community Connection

Air Pollution Invite to class a person responsible for monitoring air quality such as a local EPA official. Encourage students to ask questions about dangerous levels of pollution and possible causes of air pollution near your school or in your town.

COOP LEARN LS

---

2 Teach

Student Text Question

What did you take into your body with that breath? *Most of the air is nitrogen and oxygen, with traces of other gases.* LS

FLEX Your Brain

Use the Flex Your Brain activity to have students explore COMPOSITION OF AIR.

Activity Worksheets, page 5

Teacher F.Y.I.

At times in Hawaii, *vog alerts* are given when volcanically produced smog is at dangerous levels.

Inquiry Questions

How does Earth's atmosphere act like a blanket? *The atmosphere keeps Earth warm and protects life on Earth from harmful radiation.* **What gas makes up most of the air you breathe?** *nitrogen*

GLENCOE TECHNOLOGY

Videodisc

The Infinite Voyage: Crisis in the Atmosphere
Chapter 5
Chlorofluorocarbons and Their Effect on the Ozone Layer

Chapter 6
NASA: Studying the Ozone

Chapter 7
The Ozone and Its Effect on the Ecosystem

Mini LAB

Purpose

LS **Kinesthetic** Students will make a model of Earth's ozone layer and observe effects of holes in that layer. **L2** **ELL** **COOP LEARN**

Materials

waxed paper, a globe, flashlight

Teaching Strategies

- Have one student in each group hold the waxed paper, another hold the flashlight, and the third student observe the globe.
- Inform students that in this analogy, the waxed paper is affecting visible light similar to the way ozone would affect ultraviolet light.

📁 **Activity Worksheets,** pages 5, 113

Analysis

1. The amount of light increased. It was more concentrated.
2. The waxed paper absorbs light in much the same way that ozone absorbs ultraviolet radiation.

✓ Assessment

Oral Ask students to explain how a thinning of Earth's ozone layer might affect the health of humans on Earth's surface. If additional ultraviolet radiation reaches Earth's surface, more cases of skin cancer may occur. Use the Performance Task Assessment List for Asking Questions in **PASC**, p. 19.

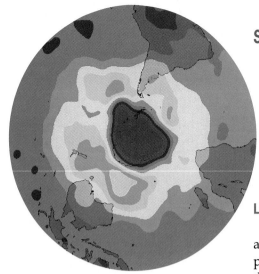

FIGURE 18-3

Every year, the ozone layer thins out over Antarctica by almost 50 percent. The dark pink color shows areas of very low ozone concentration.

Mini LAB

Observing the Effects of Holes in the Ozone Layer

1. Hold a piece of waxed paper a few inches above a globe.
2. Shine a flashlight on the waxed paper from above. Observe how the light shines on the globe.
3. Tear a small hole in the waxed paper. Shine the flashlight on the waxed paper again. Observe how the light shines on the globe.

Analysis

1. How did the amount of light on the globe change after a hole was made in the waxed paper?
2. How is the waxed paper similar to Earth's ozone layer?

Structure of the Atmosphere

In Chapter 14, you went on a tour of the solar system. You learned that Neptune has an atmosphere made of methane and that Saturn is surrounded by rings of rock and ice. But we don't have to go into outer space to find a planet with interesting surroundings. Let's take a trip through Earth's atmosphere to see what's around us.

Layers of the Atmosphere

We'll start close to home. You know the city and state you live in. But do you know what part of the atmosphere you live in? You live in the **troposphere,** the lowest layer of Earth's atmosphere. The troposphere (TROH puh sfeer) is where you'll find weather, clouds, and air pollution. It contains about 75 percent of atmospheric gases.

Going up!

As you rise above the troposphere, you come to the stratosphere (STRAT uh sfeer). The **stratosphere** contains a layer of gas called ozone. This important layer of gas protects life on Earth from ultraviolet rays, which are harmful rays from the sun. Photographs from space satellites, such as the one in **Figure 18-3,** show that the ozone layer is thinning and developing holes. Why? Many scientists think that pollutants in the atmosphere cause the ozone layer to thin. Remember, this is the layer that protects us from the sun's ultraviolet rays. If the ozone layer thins, we could see an increase in skin cancer.

Above the stratosphere are the mesosphere, thermosphere, and exosphere. Beyond the exosphere is space. There is no clear boundary between the atmo-

---

Inclusion Strategies

Visually Impaired Pair up your visually impaired students with other students who can relate what the light looks like shining through the waxed paper and what happens to the light from the flashlight once the hole is cut into the waxed paper. **LS**

Content Background

Every October, the ozone layer thins over Antarctica by almost 50 percent. This thinning occurs during certain seasons and disappears during others. One idea about the thinning of the ozone layer is that it is caused by pollutants in our atmosphere. These pollutants break down the ozone molecules forming regular oxygen, which cannot absorb ultraviolet radiation.

sphere and space. If you traveled upward through the exosphere, you would encounter fewer and fewer molecules, until eventually, for all practical purposes, you would be out of Earth's atmosphere and in space. **Figure 18-4** gives more information about Earth's atmospheric layers.

Air Pressure

Why doesn't Earth's atmosphere just drift away? What keeps it wrapped around the planet like a cozy blanket? Consider what you already know about air. You know that air is matter. In Chapter 8, you learned that matter has mass. In Chapter 10, you learned that matter has a gravitational attraction to other forms of matter. The gravitational attraction between Earth and the gases in the atmosphere causes the gases to be pulled toward Earth. The weight of the gases at the top of the atmosphere presses down on the gases below so that the molecules that make up air near Earth's surface are squeezed together. It's sort of like stacking a pile of heavy blankets on top of a fluffy quilt. The quilt stuffing gets pressed down and becomes more dense. Air at the bottom of the

FIGURE **18-4**

In the mesosphere, air temperature decreases with increasing height. In the thermosphere, air temperature increases with height. Beyond the exosphere is space.

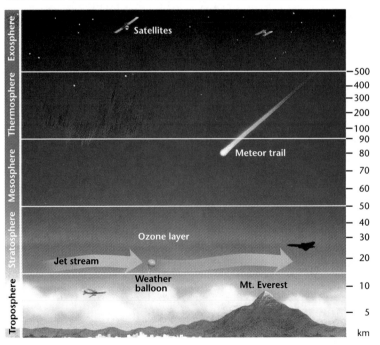

Using Science Words

IS **Linguistic** Air pressure is measured using a *mercury barometer* or an *aneroid* (which means "without liquid") *barometer*. Air pressure measured using a barometer is referred to as *barometric pressure*.

Demonstration

IS **Visual-Spatial** Display a mercury or an aneroid barometer and allow students to collect barometric pressure readings each day for a week. Demonstrate how changes in the barometric pressure are associated with changes in the weather.

NATIONAL GEOGRAPHIC SOCIETY

Videodisc

STV: Atmosphere
Unit 1
What Is the Atmosphere?
Introduction

00400-03109
Atmospheric Gases

03109-07851
Layers of the Atmosphere

07854-13145

Content Background

An aneroid barometer can provide continuous readings of air pressure. This instrument consists of metal chambers that have partial vacuums inside. A spring is employed within the chambers to keep them from collapsing. The metal chambers are susceptible to changes in air pressure. As air pressure increases, the metal chambers are compressed and change shape slightly. In like manner, the chambers expand as air pressure decreases. These changes in shape of the metal containers can be recorded on a graph in an instrument called a barograph.

Problem Solving

Solve the Problem
The egg fell into the jar because of a difference in air pressure between the inside of the jar and the outside of the jar. As air inside the jar was heated, it expanded and the air pressure dropped. Because the egg was somewhat pliable, the higher pressure outside of the jar pushed on the egg, forcing it to fall into the jar.

Think Critically
After the paper was set on fire, the air inside the jar was heated. The air expanded and the air pressure inside the jar dropped. Your ears pop because of a similar difference between the higher air pressure inside your ear and the lower air pressure outside your ear. **LS**

3 Assess

Check for Understanding
Brainstorming Ask students to brainstorm about ways they have been affected by changes in air pressure, such as flying in a plane or feeling air rush out of a deflating balloon.

Reteach
Have students draw a cross section of Earth's atmosphere showing each layer. Ask them to hypothesize on why the thermosphere is so named. Temperature steadily increases throughout this layer. **L1** **P**

Extension
For students who have mastered this section, use the **Reinforcement** and **Enrichment** masters.

Problem Solving

Air Pressure Egg-zample!
One day, you walk into science class and your teacher announces that she's going to conduct an experiment. Your job is to observe and infer what caused the results of the experiment. Your teacher puts a peeled, hard-boiled egg on the top of an open glass jar. The opening of the jar is smaller than the egg, so the egg won't fit in the jar. Your teacher puts the egg aside, then lights a piece of paper and drops it into the jar. She puts the egg back on top of the jar. The egg wiggles, then, presto! The egg falls into the jar.

Solve the Problem:
Why did the egg fall into the jar? Explain your answer in terms of air pressure.

Think Critically:
What happened to the air in the jar after the paper was set on fire? How is this egg-zample related to your ears popping as you drive up a mountain?

500 Chapter 18 Earth's Atmosphere

atmosphere is more dense than air at the top. Dense air pushes on objects more than less dense air does. Would the air pressure be greater in the troposphere or exosphere?

Effects of Air Pressure
If you guessed the troposphere, you're right. Overall, air pressure is greater in the troposphere than in upper layers of the atmosphere. Why? In the upper layers, there are fewer molecules exerting less pressure. Would you expect air pressure to be the same everywhere in the troposphere? It isn't. Air pressure changes from place to place. It generally gets lower as you go higher. That's why your eardrums pop when you go up a steep mountain in a car. As you go higher, there is less air pressure around you. The popping is your eardrum trying to balance the higher air pressure inside your ear with the lower air pressure outside your ear.

Changes in air pressure affect humans in other ways, too. Consider a mountain climber. As she climbs higher, she takes in less oxygen with each breath. Why? Because air in high places is less dense and exerts less pressure.

In the following activity, you'll see how temperature affects air pressure.

Inclusion Strategies

Gifted Have students make a barometer. Instruct them to stretch a circle cut from a balloon over the open end of a coffee can, attach the balloon to the can with a thick rubber band, then tape a straw to the top of the stretched balloon so that it lays flat and protrudes over the edge of the can about 5 cm. The straw shows changes in air pressure by pointing slightly up or down. Have students attach a card to the side of the can to calibrate rises or drops in air pressure. **L3** **ELL** **LS**

Air Pressure

Think back to Chapter 12. How does temperature affect the movement of molecules that make up air? What happens to these molecules when they are heated or cooled? You know that air pressure is the force exerted by the molecules that make up air. Do you think differences in temperature have any effect on air pressure?

What You'll Investigate
How does temperature affect air pressure?

Procedure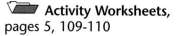
1. **Place** an open, empty soft-drink bottle in a freezer for two hours. Be careful when you handle the glass bottle. If it breaks, tell your teacher and have your teacher dispose of the glass.
2. **Remove** the bottle. Immediately **stretch** a balloon over the opening of the bottle.
3. **Set** the bottle aside and allow it to come to room temperature for 15 minutes.
4. **Observe** the balloon; **record** your observations.

Conclude and Apply
1. What happened to the balloon after it was stretched over the glass bottle?
2. What property of air was demonstrated by this activity?
3. **Describe** the movement of air in the bottle after the bottle was taken out of the freezer.
4. Does temperature affect air pressure? **Explain** your answer.

4. Yes, in an open environment, increases in temperature decrease air pressure by making the molecules move farther apart.

Goals
- Observe the effects of temperature changes on air in a container.
- Infer how temperature affects air pressure.

Materials
- empty glass soft-drink bottle
- balloon

501

Purpose
Visual-Spatial Students will observe how temperatures affect air pressure. **L2**
COOP LEARN **P**

Process Skills
observing and inferring, recognizing cause and effect, separating and controlling variables, interpreting data

Time
30 minutes

Alternate Materials
Any glass container with a small enough neck for a balloon to be stretched over it can be used in place of the 1-L bottle.

Safety Precautions
Caution students to handle glass carefully.

Activity Worksheets, pages 5, 109-110

Teaching Strategies
- Have students work in groups of three or four.
- If a refrigerator is not available, the bottle can be placed in ice for cooling.

Troubleshooting Stretch the balloons prior to class to minimize breakage, and have extra balloons available.

Answers to Questions
1. As the glass bottle and the air inside became warmer, molecules of air moved farther apart and moved out of the bottle into the balloon, inflating it.
2. Air expands and takes up a larger space when warmed.
3. As the air inside the bottle became warmer, the molecules moved around faster and moved farther apart from each other.

✓ Assessment

Oral Ask students to explain what they think might happen if the empty bottle is closed with a cap and removed from the refrigerator. Air molecules in the bottle would move around faster and because the volume is controlled, air pressure would increase inside the bottle. Use the Performance Task Assessment List for Formulating a Hypothesis in **PASC**, p. 21.

501

4 Close

1. **The layer of air that surrounds Earth is the _____.** *atmosphere*
2. **Haze and smoke are _____ in the air.** *pollutants*
3. **The lowest layer of Earth's atmosphere is the _____.** *troposphere*

Section 3 Wrap-up

1. All layers of Earth's atmosphere contain air. The air pressure in each layer becomes less as you go up from Earth's surface. The troposphere contains weather, clouds, and pollution. The stratosphere contains the ozone layer. Temperatures drop with altitude in the troposphere and mesosphere, but rise with altitude in the stratosphere and thermosphere.
2. exhaust from cars and buses, burning fossil fuels to generate electricity
3. **Think Critically** People's eardrums pop when they move up a steep mountain. Lower air pressure at higher elevations makes it difficult for people to breathe. It can also tire people.

Science Journal Encourage students to be creative when writing their poems. Student poems should discuss what air is made of or ways air affects life on Earth.

The Importance of the Atmosphere

On a cool but sunny day, you head to the beach with a blanket under your arm. You can wrap it around yourself to keep warm and to keep from getting too much sun. Look at **Figure 18-5.** Earth's blanket, the atmosphere, does the same things. It keeps temperatures on Earth just right to support life. It protects living things from the sun's harmful rays. In the next section, you'll learn how the atmosphere affects something that you deal with every day: the weather.

FIGURE 18-5
This photograph, taken from a satellite in space, shows how Earth's atmosphere covers the planet.

Section 3 Wrap-up

1. Compare and contrast the different layers of Earth's atmosphere.
2. Describe some of the human activities that cause air pollution.
3. **Think Critically:** Describe some ways air pressure affects humans.
4. **Skill Builder**
   Making and Using Graphs Make a circle graph that shows the composition of gases in Earth's atmosphere. If you need help, refer to Making and Using Graphs on page 547 in the **Skill Handbook.**

Science Journal
In your Science Journal, write a poem about air. Write about the things that make up air or about the ways that air affects life on Earth.

Skill Builder
Making and Using Graphs Student graphs should include the following compositions: 78% nitrogen; 21% oxygen; 1% all other gases. (Water vapor can be as low as nearly zero or as high as 4%.)

✓ Assessment

Oral Ask a student to explain how the ozone layer is important to life on Earth. Use the Performance Task Assessment List for Oral Presentation in **PASC,** p. 71.

Weather

What is weather?

Do you like to walk in the rain? Or would you rather watch the clouds in a bright blue sky? We use words like *rain, snow, blue skies,* and *clouds* to describe the weather. But what are we really talking about? **Weather** is what is happening in the atmosphere right now. Take a look at **Figure 18-6**. What's the weather like in these photographs?

Causes of Weather

Look outside. What do you see? Rain or sunshine? Trees blowing in the breeze? What do you think causes the type of weather you see right now? Recall what you learned about Earth's water cycle from Chapter 17. Where does the energy come from that causes the water to evaporate, condense into clouds, and eventually fall to Earth once again? It comes from the sun. Energy from the sun powers Earth's water cycle. The water cycle, in turn, forms the basis of our weather. But there's more to weather than just water.

What YOU'LL LEARN

- The causes of weather
- How to compare and contrast different types of weather
  Science Words:
  weather
  wind
  air mass
  front

Why IT'S IMPORTANT

Changes in the weather affect you and your family every day.

FIGURE 18-6
Weather is the current state of the atmosphere. *Describe the weather in these three photographs.*

Program Resources

📁 Reproducible Masters
Activity Worksheets, pp. 5, 111-112, 114
Cross-Curricular Integration, p. 22
Enrichment, p. 68 L3
Lab Manual, pp. 131-133
Reinforcement, p. 68 L2
Science Integration Activities, pp. 73-74
Study Guide, p. 68 L1

📽 Transparencies
Section Focus Transparency 62 L2
Teaching Transparencies 35, 36

Prepare

Section Background
- Warm air is forced upward by cooler, denser air. As it cools, water vapor begins to condense, forming clouds.
- Condensing water vapor forms droplets around small dirt particles in Earth's atmosphere. When large enough, they may fall as rain or other precipitation.

Preplanning
Refer to the Chapter Organizer on pages 494A-B.

1 Motivate

Bellringer

 Before presenting the lesson, display **Section Focus Transparency 62** on the overhead projector. Assign the accompanying **Focus Activity** worksheet.
L2 ELL

Tying to Previous Knowledge
Help students to recall what they learned about the water cycle in Section 17-1.

Encourage students to research the early use of kites. Have them make drawings of kites from around the world.

L3

Mini LAB

Purpose

LS **Kinesthetic** Students will observe the cause of wind.

L2 **COOP LEARN**

Materials

balloons

Teaching Strategies

Have students work in pairs. One student can deflate the balloon while the other student observes its effects.

Safety Precautions Caution students not to blow up the balloon too much.

📁 **Activity Worksheets,** pages 5, 114

Analysis

1. Air pressure was greater inside the inflated balloon.
2. Air was flowing out of the balloon as wind. Air inside the balloon flowed from a high-pressure area to a lower-pressure area. This is the same process that occurs with air in Earth's atmosphere as air from cooler, denser areas flows into low-pressure areas.

✓Assessment

Performance Have students research and write about why a hot-air balloon rises and descends. Use the Performance Task Assessment List for Writing in Science in **PASC**, p. 87. P LS

504

FIGURE 18-7

The sun's rays strike different parts of Earth at different angles. *Which area receives the most direct sunlight?*

Equator

Other Factors That Affect Weather

If you wanted to vacation in a sunny, warm place, would you head for Alaska or Mexico? Your chances of finding the weather you want would be better in Mexico. You know that some parts of Earth are warmer than other parts. But do you know why? Look at **Figure 18-7.** The round shape of Earth causes the sun's rays to strike different parts of Earth at different angles. This means that some parts of Earth, such as the equator, receive more sun energy—and more heat—than other parts of Earth.

Air Pressure and Wind

Temperature isn't the whole story. A day can be warm or cold. But it can also be cloudy or sunny, rainy or dry. What causes these differences? Look at **Figure 18-7** again. In warmer areas, air pressure is lower because the molecules that make up air are farther apart. In cooler areas, air pressure is higher. The air from high-pressure areas moves into low-pressure areas. This movement of air due to differences in pressure is called **wind.** Wind, air pressure, temperature, and water all work together to cause weather.

Mini LAB

Inferring

What causes wind?

1. Blow up a balloon and hold the end closed with one hand.
2. Place your other hand, palm down, a few inches above the end of the balloon.
3. Slowly let the air out of the balloon.

Analysis

1. Air flows from areas of high pressure to areas of low pressure. Where was the air pressure greater—inside or outside of the inflated balloon?
2. What did you feel on your hand when you opened the end of the balloon? In your Science Journal, explain how this movement of air is similar to the way air moves on a blustery day.

504 Chapter 18 Earth's Atmosphere

Content Background

Global wind patterns are caused by unequal global heating. Large convection cells form in the atmosphere as air moves from cool, high-pressure areas to warmer, low-pressure areas. As the air moves, its path is deflected due to the rotation of Earth. This sets up major wind systems that control global weather patterns.

Visual Learning

Figure 18-6 Describe the weather in these three photographs. *Sunny, snowy, and cloudy are possible answers.*

Figure 18-7 Which area receives the most direct sunlight? *equator* LS

Weather Patterns

The sun was out when you decided to go to an afternoon movie. But now the movie is over, and something has changed. It's still daytime, but the sky is as dark as night. Thunder is crashing. Lightning is flashing. The wind is blowing hard. Heavy raindrops start to splatter on the sidewalk. And of course you didn't bring an umbrella! How did you get into this mess? Let's take a look at what may have caused the change in the weather.

Air Masses

The sudden weather change that left you trapped in the movie theater lobby was caused by movements of air masses. An **air mass** is a large body of air that has the same properties as the area of Earth's surface over which it develops and moves. In other words, if that part of Earth's surface is hot, the air mass will be hot. If the air mass develops over a cold, wet area on Earth, the air mass will be cold and wet. **Figure 18-8** shows the different kinds of air masses that affect weather in the United States.

FIGURE 18-8

Weather in the United States is affected by six major air masses. *Why do air masses have different temperatures and different levels of moisture?*

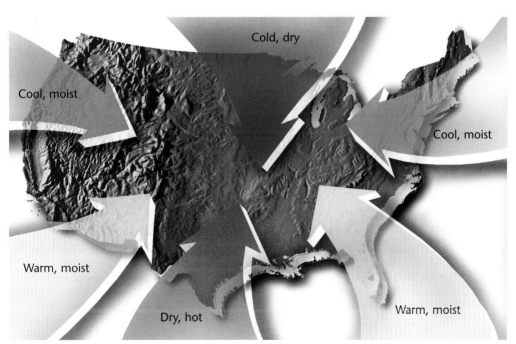

Cold, dry

Cool, moist

Cool, moist

Warm, moist

Dry, hot

Warm, moist

18-2 Weather **505**

Community Connection

Winds, Kites, and Flying Invite a local pilot, model airplane enthusiast, or kite maker into class to discuss how winds affect flying airplanes, model airplanes, or kites.

Theme Connection

Stability and Change

The theme of Stability and Change is supported in the study of changing weather patterns and the stability of the characteristics of air masses. Cold, dry air masses are more stable than warm, moist air masses because cold air is more dense than warm air and will tend to remain at Earth's surface.

? FLEX Your Brain

Use the Flex Your Brain activity to have students explore FRONTS.

Activity Worksheets, page 5

Visual Learning

Figure 18-8 Why do air masses have different temperatures and different levels of moisture? *They form over different areas of Earth—they have the same properties as the areas over which they form.*

Activity

 Kinesthetic Encourage students to build kites. Schedule time in class for students to take their kites outside and attempt to fly them in the wind. **L2** **ELL**

GLENCOE TECHNOLOGY

Videodisc

STVS: Earth & Space
Disc 3, Side 1
World's Worst Weather
(Ch. 15)

Studying Thunderstorms
(Ch. 16)

Reducing Hail Damage
(Ch. 17)

505

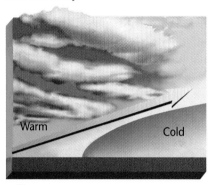
Fronts

Look at **Figure 18-8** again. The air masses come from different directions. Sometimes they bump into each other. They'll slide over one another or push each other out of the way. The place where air masses come together is called a **front**. The weather conditions that occur at a front are caused by interactions between the air masses. **Figure 18-9** shows weather conditions that occur at different kinds of fronts. Look at the picture and tell what kind of weather occurs along a stationary front.

FIGURE 18-9
There are four different types of fronts: warm, cold, occluded, and stationary.

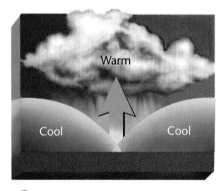

A In a warm front, less dense warm air slides over departing cold air. High, feathery clouds often form along warm fronts.

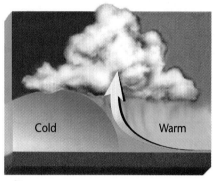

B In a cold front, cold air pushes up warm air, causing a narrow band of violent storms. *What kind of clouds would you expect to see with cold fronts?*

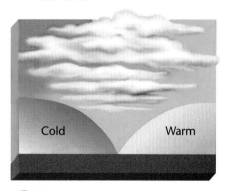

C In an occluded front, two cool air masses meet and force warmer air between them to be pushed up. Often, this causes strong winds and heavy precipitation.

D In a stationary front, a warm air mass or a cold air mass stops moving because of pressure differences. The weather might be rainy and slightly windy for several days.

506   Chapter 18 Earth's Atmosphere

Severe Weather

We're back in the theater lobby, with thunder and lightning crashing and flashing outside. Should you start walking home in the storm? Not a good idea. Lightning bolts can reach temperatures of around 30 000°C—that's five times hotter than the surface of the sun! Obviously, lightning can be dangerous if it strikes an object or person. What causes the lightning, thunder, and heavy rains of a thunderstorm?

Thunderstorms take place inside warm, moist air masses and at fronts. The warm air is pushed up rapidly, creating an updraft of moist air that forms huge clouds like the one shown in **Figure 18-10.** As the air rises, it cools, and the moisture in the air changes into water droplets. When the droplets become heavier than the air, they begin to fall. The falling droplets collide with other droplets, forming heavy raindrops. The heavy raindrops create downdrafts of air as they fall. This, in turn, creates the strong winds associated with thunderstorms.

Thunderstorms can be dangerous. Lightning can damage property and people. Heavy rains can cause flooding. Strong winds can knock down power lines or tree branches. Sometimes lumps or balls of ice called hail fall from the sky. The hail can damage crops and make dents in cars. Thunderstorms are the most common type of severe weather. But they are not the only type.

FIGURE 18-10.
Huge clouds called cumulonimbus clouds can unleash violent thunderstorms. Some cumulonimbus clouds are 18 000 m high.

Revealing Preconceptions

Prior to the class discussion on thunderstorms and lightning, ask students what they know about each of these phenomena. Ask them whether lightning comes from the clouds downward to Earth or from Earth's surface up to the clouds. Some students will believe that all lightning goes from clouds to the ground. Inform them that lightning can go from cloud to cloud, cloud to Earth, and Earth to cloud. Explain that in less than 1/10 of a second, a single lightning discharge goes back and forth many times between the cloud and Earth's surface.

GLENCOE TECHNOLOGY

 Videodisc

Glencoe Earth Science Interactive Videodisc
Side 1, Lesson 3
Thunder and Lightning

34719-37099
Lightning Rods

37101-39629
Conclusion: What You Should Do in a Thunderstorm

39631-41631

Cultural Diversity

Weather Patterns Weather patterns influence the way people live. In Southeast Asia, people produce and store extra food during the dry season to carry them through the monsoon season.

Some weather patterns appear unpredictably. El Niño causes the eastern coast of the Pacific Ocean to warm, affecting parts of Australia, Africa, and South and North America. El Niño is Spanish for "the child."

It is so named because it usually appears around Christmastime every three to seven years.

A sirocco is a hot, dusty wind that blows from north African deserts through the Mediterranean. Its name is Arabic for "east," the direction from which it comes.

Have students research names and explanations for other weather systems.

PREPARE

Purpose
 Kinesthetic Students will design an instrument to measure wind direction and an experiment to check on the instrument's operation.

L2 **COOP LEARN** **P**

Process Skills
making models, observing and inferring, making and using tables, recognizing cause and effect, forming a hypothesis, designing an experiment, using numbers, separating and controlling variables, interpreting data

Time
30 minutes to make and check plan; 30 minutes to do the experiment; 15 minutes to analyze and apply

Safety Precautions
Caution students to handle the straight pin and scissors with care.

Possible Hypotheses
Most students will hypothesize that there is a relationship between wind direction and weather.

📁 **Activity Worksheets,** pages 5, 111-112

PLAN

Possible Procedures
Cut an arrow out of the stiff plastic. Using glue or tape, mount the stiff plastic arrow onto the drinking straw. Using the straight pin, mount the wind vane onto the eraser of a pencil.

Activity 18-2

Design Your Own Experiment
Blowing in the Wind

Hold onto your hat! In this activity, you'll learn more about the movement of air called wind. You'll build a simple instrument to measure wind direction. Then you'll compare your measurements to weather conditions in your area to see if there's a relationship between wind direction and weather.

Possible Materials
- compass
- drinking straw
- a piece of stiff plastic or poster board
- straight pin
- clear tape or glue
- scissors
- pencil

PREPARE

What You'll Investigate
Is there a relationship between wind direction and weather?

Form a Hypothesis
Remember what you've learned about weather patterns and wind. **Make a hypothesis** about how wind direction might affect the weather in your area.

Goals
- **Design** and **construct** an instrument to measure wind direction.
- **Observe** and **record** wind direction and weather conditions.
- **Infer** how wind direction and weather are related.

Safety Precautions 🧤 🥽
Take care when handling the straight pin and scissors.

508   Chapter 18 Earth's Atmosphere

	M	T	W	TH	F
Wind direction	no wind	west	west	west	no wind
Weather conditions	sunny	slightly cloudy	thunderstorms	rain	slightly cloudy

1. **Study** the materials provided by your teacher. Also, study the photograph of the wind vane shown on this page. Wind vanes measure wind direction. As a group, **agree** upon a way to use the materials to **build** a wind vane to measure wind direction.
2. **List** the steps you will take to **build** your wind vane. How will you **cut** and **mount** the plastic so it can **show** wind direction? How will you use the compass to **determine** wind direction?
3. **Make a data table** in your Science Journal to **record** your measurements of wind direction and weather conditions in your area. **Decide** how long your experiment will last. When and where will you make your measurements? Who will **record** the information?

DO

1. Make sure your teacher has approved your plan and your data tables before you proceed.
2. Build the instrument and carry out the experiment as planned.
3. Carefully **measure** and **record** your observations.

CONCLUDE AND APPLY

1. **Explain** how your wind vane works. Why does the arrow of the wind vane point into the wind? Why is the arrow long and narrow?
2. **Compare and contrast** your measurements with other groups' measurements that were taken on the same day. How might you **explain** any differences in the measurements?
3. Does the wind change often in your area? From which direction did the wind blow most often?
4. **APPLY** Is there a relationship between the wind direction and the weather in your area? **Explain** your answer.

LS **Linguistic** Have students keep a weather diary over the course of ten days. Have them compare the actual weather recorded in their diaries with weather forecasts from TV, radio, or newspapers. Students should note that the weather forecasts were generally reliable several days in advance, but that the reliability decreased with long-term forecasts. **L2**

3 Assess

Check for Understanding

Enrichment Have students research the anemometer. Encourage them to draw a poster that explains each part of the instrument. Have students relate the anemometer to the wind direction instrument they made in Activity 18-2. **L3** **P** **LS**

Reteach

LS **Visual-Spatial** Have students review weather conditions associated with fronts. Ask them to draw flash cards of each of the four types of fronts: warm, cold, occluded, and stationary. On the back of each flash card, students should list all weather conditions associated with each front. **L1**

Extension

For students who have mastered this section, use the **Reinforcement** and **Enrichment** masters.

FIGURE 18-11

Differences in wind height and speed cause a tornado's distinctive funnel cloud.

Tornadoes

Cows go flying through the air. Cars and trucks are picked up and dropped like toys. Even houses can be whisked off their foundations. Tornadoes are powerful and frightening. What causes them? In very severe thunderstorms, winds blow at different heights and at different speeds. Look at **Figure 18-11.** This difference in wind height and speed can cause a funnel cloud to form. The result? A tornado! Tornadoes are violent, whirling winds that move in a narrow path over land. These twisting winds can travel as fast as 500 km per hour. Tornadoes can destroy lives and property in a matter of seconds.

Hurricanes

Tornadoes are awesome to behold, but they aren't the most powerful storms that develop on Earth. That title goes to hurricanes. Hurricanes are large, swirling, low-pressure systems that form over tropical oceans. They usually begin when several small thunderstorms come together. Strong, high winds that blow in tropical ocean areas cause these thunderstorms to start spinning as one. The heat from the tropical waters gives the storm energy to spin even faster. When the winds of the storm reach at least 120 km per hour, the storm is called a hurricane.

510   Chapter 18 Earth's Atmosphere

Science Journal **Tornadoes vs. Hurricanes** Have students compare and contrast hurricanes and monsoons in their Science Journals. They should describe similarities and differences between the two, including how the storms form and the types of damage they cause. **L2** **P** **LS**

Content Background

The rapid upward rush of air in cumulonimbus clouds generates static electricity, called lightning. Atoms rushing past one another create friction that strips electrons from the atoms. Atoms in one place lose electrons, while atoms in another place gain electrons. When enough of a charge differential forms, an arc of electricity is generated.

If you watch worldwide weather reports, you can follow the movement of hurricanes. Look at **Figure 18-12.** When hurricanes strike land, they cause high waves and severe thunderstorms. If the wind is strong enough, the thunderstorms caused by hurricanes can cause tornadoes! Most hurricanes in the United States hit the Atlantic Coast or the Gulf of Mexico. The National Weather Service tracks these storms so there is enough time to warn people to move out of their way.

FIGURE 18-12

Hurricanes, the most powerful of storms, cause torrential rains and strong winds.

The Weather and You

Severe or not, the weather affects you and your family every day. It affects what you wear and what you do. On a larger scale, it affects crops and outdoor events, such as parades or sporting events. When you're trying to make plans, it helps to know what the weather is going to be like a few days in the future. In the next section, you'll learn how scientists predict the weather.

Section Wrap-up

1. What causes wind?
2. How do thunderstorms form?
3. **Think Critically:** Describe what happens when a cold air mass meets a warm air mass. What type of weather is associated with this event?
4. *Skill Builder*
   Compare and Contrast Compare and contrast tornadoes and hurricanes. If you need help, refer to Comparing and Contrasting on page 550 in the **Skill Handbook.**

USING MATH

On the radio, you hear that a tornado is moving toward a nearby town at a speed of 150 km per hour. The storm front is 100 km from the town. How long will it take for the storm front to reach the town?

Skill Builder

Compare and Contrast Both are severe storms that include high-speed, whirling winds. Tornadoes are violent, whirling winds that move in a narrow path over land. Hurricanes are large, swirling, low-pressure systems that form over tropical oceans.

Assessment

Oral Have students work in teams of two to research tornadoes and hurricanes. Ask each member of the team to draw an illustration of either a tornado or a hurricane. Have students indicate on their drawings where the fastest and slowest winds occur. Use the Performance Task Assessment List for Scientific Drawing in **PASC**, p. 55. [L2]

COOP LEARN P LS

---

4 Close

•MINI•QUIZ•

Use the Mini Quiz to check students' recall of chapter content.

1. **Words like** *rain, snow, cloudy,* **and** *sunny* **describe _____.** *weather*
2. **Movement of air caused by differences in pressure is called _____.** *wind*
3. **A large body of air called a(n) _____ has the same properties as the area over which it develops.** *air mass*

Section Wrap-up

1. Unequal heating of Earth produces areas of different air pressures in Earth's atmosphere. Wind is air that flows from a high-pressure to a low-pressure area.
2. Warm air that rises rapidly creates an updraft of moist air that forms huge clouds. As the air rises, it cools, and the moisture in the air changes into water droplets. Once the water droplets are heavy enough, they fall. Heavy raindrops create downdrafts of air as they fall. This causes the strong winds associated with thunderstorms.
3. **Think Critically** The cold air mass pushes under the warm air mass, forcing the warm air up. As the warm air rises and cools, clouds form. A narrow band of violent storms occurs along the front.

USING MATH

150 km/1 h = 100 km/? h
? h = 100 km/150 km/h = 0.67 h or 40 min LS

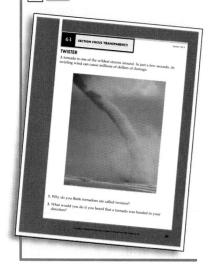

Teaching the Content

To help students understand how weather conditions can fluctuate unpredictably over short periods of time, have them look out the window to observe how the wind changes (leaves flutter, then are still) or how clouds can temporarily block out the sunlight. Help them to relate these small changes to larger changes in weather conditions. [L2]

Visual Learning

Figure 18-13 Why do meteorologists use a combination of technologies to forecast the weather? *They need a complete picture of the weather to make accurate forecasts.* [LS]

Science & Society

18●3 Forecasting the Weather

What YOU'LL LEARN

- How scientists forecast the weather
- How to improve your note-taking skills
- **Science Words:** meteorologist

Why IT'S IMPORTANT

Good note-taking skills will help you improve your study habits.

FIGURE 18-13

This satellite photograph shows weather patterns in the United States. *Why do meteorologists use a combination of technologies to forecast the weather?*

The Weather and You

Quick—list all the things you do outside: basketball, bike riding, football, kite flying, swimming, running, walking, Rollerblading. Can you do these things if the weather is bad? Probably not. That's one reason people read or listen to weather forecasts in the newspaper or on TV. They want to plan for picnics in the park or trips to the beach. Think back to what you learned about hurricanes. People also need to know about severe weather ahead of time so they can get to safe places.

Who forecasts the weather?

Weather forecasts are made by meteorologists. A **meteorologist** is a scientist who studies weather patterns to predict daily weather. In the past, people used their observations to make the predictions. If they saw a cow lying in a pasture, they might say that rain was on the way. Why? Because people thought that cows could feel moisture in the air and that the cows wanted to lie down in the dry grass before it got wet.

While these kinds of predictions are interesting, they're not very accurate. Today, meteorologists still use observation to predict the weather. But they combine their observations with modern technology to come up with accurate forecasts.

Modern Weather Forecasting

Modern meteorologists use airplanes, weather balloons, weather ships, satellites, computers, and data from weather stations located around the world to make

---

Program Resources

 Reproducible Masters
Enrichment, p. 69 [L3]
Multicultural Connections, pp. 39-40
Reinforcement, p. 69 [L2]
Study Guide, p. 69 [L1]

Transparencies
Section Focus Transparency 63 [L2]

their forecasts. **Figures 18-13** and **18-14** show some examples of weather technology. Why do meteorologists use so many things? They need a complete picture of the weather to make accurate forecasts. They need weather balloons and planes to measure temperature, wind speed, and air pressure from the atmosphere. They need ships to measure ocean currents and water temperatures. They need satellites to send back pictures of shifting clouds. And they need people at weather stations to take local measurements of temperature and rainfall.

All these measurements are constantly updated and recorded by a global network of weather stations. Every three hours, these stations send their data to the world's 13 main weather centers. Meteorologists at these centers enter the data into powerful computers. Your local meteorologist gets his or her information from these computers. Then, based on local weather conditions, your meteorologist makes a forecast.

FIGURE 18-14
This researcher is releasing a weather balloon into the atmosphere.

 Skill Builder: *Note-Taking Skills*

LEARNING the SKILL

1. To take good notes, first read the material to identify the main ideas. The heads and subheads in the material are clues to main ideas.

2. Look for *italicized* or **boldfaced** words in the material. These are also clues to important ideas.

3. Identify details or sentences in the material that support the main ideas.

4. Using the main ideas and the details or sentences that support them, take notes about the material.

PRACTICING the SKILL

1. List three main ideas in this passage.

2. What is a meteorologist? What are some things that meteorologists use to forecast the weather?

3. How did people forecast weather in the past? How is this different from the way we forecast weather now? How is it the same?

APPLYING the SKILL

Choose a subject about the weather that interests you. Research the topic in an encyclopedia, CD-ROM, or other resource. Take notes about the topic, and share what you've learned with the class.

18-3 Forecasting the Weather   **513**

GLENCOE TECHNOLOGY

 Videodisc

STVS: Earth & Space
Disc 3, Side 2
Computerized Weather Forecasting (Ch. 3)

Global Weather Forecasting (Ch. 4)

Balloons in Science (Ch. 5)

Teaching the Skill

To help students improve their note-taking skills, go over Section 18-1 or 18-2 in class. With student input, write the main ideas of the chosen section on the chalkboard.

Answers to Practicing the Skill

1. Some examples involve forecasting the weather, who forecasts the weather, and modern weather forecasting.

2. A meteorologist is a scientist who studies weather patterns to predict the weather. Meteorologists use airplanes, weather balloons, weather ships, satellites, computers, and data from weather stations around the world to make their forecasts.

3. In the past, people relied on observation to predict the weather. Modern meteorologists still use observation, but they combine observation with modern technology.

Content Background

The World Meteorological Organization is made up of 10 000 permanent weather stations around the world and 13 main weather centers. The computers at the centers plot the water and air temperature, air pressure, wind speed and direction, humidity, and other relevant data onto thousands of points located on a grid of Earth. Each half hour, the computers calculate how these weather factors will change for each point on the grid. The calculations are repeated continuously to produce forecasts several days in advance.

513

Prepare

Section Background

The Köppen classification of world climates is divided into six major groups based on differences in temperature and precipitation. Each group is identified by name: tropical, dry, mild, continental, polar, and high elevation.

Preplanning

Refer to the Chapter Organizer on pages 494A-B.

1 Motivate

Bellringer

Before presenting the lesson, display **Section Focus Transparency 64** on the overhead projector. Assign the accompanying **Focus Activity** worksheet.

[L2] [ELL]

Tying to Previous Knowledge

Help students recall that because Earth is round, the sun's rays strike Earth at different angles. In this section, they will learn how this fact affects the occurrence of various climates.

514

18•4 Climate

What YOU'LL LEARN

- How to describe climate
- How people affect Earth's climate

Science Words:
climate
global warming

Why IT'S IMPORTANT

You'll learn how human activities affect climate.

FIGURE 18-15

There are five main types of climate: polar, subarctic, temperate, mild, and tropical. *What type of climate does your area have?*

What is climate?

You've probably noticed that the weather in other parts of the country is different from where you live. If you could see a map of the world's weather, you'd see even greater differences. Some parts of the world always seem rainy. Others are cold every winter. **Figure 18-15** shows different patterns of climate (KLIME ut). **Climate** is the pattern of weather that occurs in a particular area over many years. To determine the climate of a region, scientists figure out the average weather conditions over a period of 30 years or more. They look at the average temperature, precipitation, air pressure, humidity, and the number of days of sunshine in each area.

What causes different climates?

Different parts of Earth receive different amounts of sunlight. The areas nearest the equator receive the most sunlight. They have a tropical climate with warm to hot temperatures year-round. Polar climates are found near the poles, where sunlight strikes Earth at a low angle.

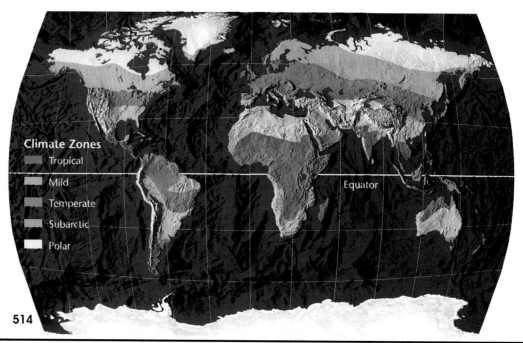

Climate Zones
- Tropical
- Mild
- Temperate
- Subarctic
- Polar

Equator

514

Program Resources

Reproducible Masters
Activity Worksheets, p. 5 [L2]
Enrichment, p. 70 [L3]
Reinforcement, p. 70 [L2]
Science and Society/Technology Integration, p. 36
Study Guide, p. 70 [L1]

Transparencies
Section Focus Transparency 64 [L2]

These places can get very cold. Some are always covered in ice. Subarctic, temperate (TEM prut), and mild climates lie between the tropics and the poles. Temperatures in these areas vary, but on average, they are not as hot as tropical climates and not as cold as polar climates. Is location the only factor that affects climate? No. Oceans, mountains, wind patterns, and even large cities can affect local climate.

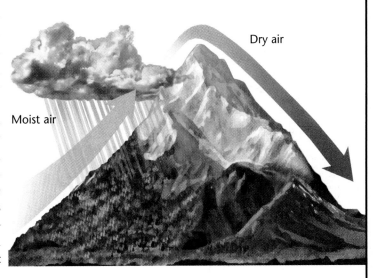
Dry air

Moist air

Local Effects on Climate

Imagine you're at the beach on a summer day. You take a moment to enjoy the cool, refreshing breeze that blows in from the ocean. Ocean winds and ocean currents affect the climate along the coasts. These areas are often cooler in the summer and warmer in the winter than areas located just a short distance inland.

As **Figure 18-16** shows, mountains also affect local climate. On the side of the mountain that faces the wind, the climate is generally cool and wet. That's because air cools as it moves up the mountain. As it cools, it drops its moisture as rain or snow. By the time the air reaches the top of the mountain, it is dry. The dry air continues over the mountain, heating up as it goes down the side of the mountain that faces away from the wind.

Cities can affect climate, too, especially in the summer. If you've ever walked barefoot on a hot street, you know that the street absorbs heat. So do parking lots and buildings. Some of the heat they absorb is sent back into the surrounding air. That makes the air in a city hotter.

Location, oceans, mountains, cities—now you know the things that cause different climates. Do the climates of regions ever change? They have in the past. Some people think they might change again in the future.

FIGURE 18-16
The climate on the side of the mountain that faces away from the wind is generally dry and hot. *Why?*

18-4 Climate **515**

Content Background

Local conditions can affect the local climate, setting up microclimates. For example, a house or an apartment can create microclimates around it. Buildings can cause differences in temperature, wind speed, relative humidity, and precipitation.

? FLEX Your Brain

Use the Flex Your Brain activity to have students explore GLOBAL WARMING.

📁 **Activity Worksheets,** page 5

Visual Learning

Figure 18-15 What type of climate does your area have? *Answers will vary depending on student's location.*
Figure 18-16 Why? *As air moves up the side of the mountain that faces the wind, it cools and drops its moisture. The dry air continues over the mountain, heating up as it descends, making the climate on the side of the mountain that faces away from the wind hot and dry.* **LS**

Teacher F.Y.I.

Cities increase precipitation as they affect microclimate formation. Studies indicate that precipitation can be 5% to 15% higher in cities and in areas downwind of cities than in surrounding locations.

Theme Connection

Stability and Change

The theme of Stability and Change is supported by the method used to determine the climate of an area; the overall stability of general weather conditions is studied for more than 30 years. The theme is also supported by the daily changes of weather conditions that produce so much variety within a specific climate.

Cultural Diversity

Worldwide Weather El Niño, which originates in the tropical waters of the Pacific Ocean, causes climate change. It can cause floods in some areas and droughts in others. Have students research El Niño during a recent three-year period. Each group should report on a different section of the world that El Niño affected. **L2** **P**

*inter*NET
CONNECTION

The Glencoe Homepage at **www.glencoe.com/sec/science** provides links to up-to-date Internet sites.

3 Assess

Check for Understanding

Using an Analogy Climates can be wet or dry. Temperatures can also vary a great deal from one climate to another. This effect can be experienced at an indoor pool and sauna. The air surrounding the indoor pool will be wet. In the sauna, temperatures will be much higher and the air much drier.

Reteach

Explain that the climate of an area is an average of weather conditions over a long period of time (usually 30 years or more). Ask students to record weather conditions for their town for one week. Encourage them to describe the "climate" their town might experience based on their limited records. ☐L2☐

Extension

📁 For students who have mastered this section, use the **Reinforcement** and **Enrichment** masters.

FIGURE 18-17
This glacier in Argentina is similar to the ones that once covered large parts of North America.

Climate Changes

Look at **Figure 18-17**. About 20 000 years ago huge sheets of ice called glaciers covered much of Canada and parts of the United States. What caused this type of climate change? Scientists are not sure. Some think that the tilt of Earth's axis or the path of Earth's orbit may change over a long period of time. Others think that a huge volcanic eruption or a meteorite collision with Earth may have caused the climate change.

Some people think we're headed for another big climate change. But this time, the change will be caused by people, not by volcanoes or meteorites.

USING TECHNOLOGY

GLOBE Project

Global warming is a big environmental problem—so big that scientists need all the help they can get! That's why more than 4000 schools in 55 countries are part of the GLOBE project. GLOBE stands for Global Learning and Observation to Benefit the Environment. It's a World Wide Web program that helps scientists keep track of changes in the environment. Who's doing the tracking? Students just like you.

In GLOBE, students collect environmental data near their schools. They might measure air temperature or test water samples for pollutants. They carefully record their data, then report their findings on the Internet. Scientists use the data to help them study environmental problems such as global warming.

*inter*NET
CONNECTION

Check out the Glencoe Homepage at *www.glencoe.com/sec/science* for a tour of the GLOBE project.

516   Chapter 18 Earth's Atmosphere

Will Earth's climate change again?

Look at **Figure 18-18.** Some human activities, such as burning fossil fuels to produce electricity and cutting down rain forests for farmland, increase the amount of carbon dioxide in the atmosphere. Carbon dioxide traps heat in the atmosphere. When there's more carbon dioxide, more heat is trapped, so temperature increases around the world. An increase in temperatures all over the world is called **global warming.**

Effects of Global Warming

Scientists are not sure what the effects of global warming will be. Some think rising temperatures might cause ice caps to melt. This, in turn, could cause a sudden rise in sea level and flooding along coastal areas. Other scientists aren't convinced that global warming is a problem. All scientists, though, warn against tampering with Earth's climate. We can help protect Earth from harmful climate change by using less electricity and recycling products to reduce our use of fossil fuels.

FIGURE 18-18
Cutting down rain forests may increase the amount of carbon dioxide in the atmosphere. This, in turn, may lead to global warming.

Section Wrap-up

1. What factors cause climate?

2. List three types of climate. Describe each.

3. **Think Critically:** Can humans affect climate? How?

4. **Skill Builder**
   Recognizing Cause and Effect
   Temperature is one of the weather conditions used to determine a region's climate. Describe how your life might change if the average monthly temperatures in your area decreased by 10°C. If you need help, refer to Recognizing Cause and Effect on page 551 in the **Skill Handbook.**

Science Journal

Using a newspaper, magazine, or the Internet, research global warming. Write about changes that are being made around the world to reduce levels of carbon dioxide and other gases that contribute to global warming.

4 Close

Use the Mini Quiz to check students' recall of chapter content.

1. **The pattern of weather that occurs in a particular area over many years is the area's _____.** *climate*

2. **Climates near Earth's equator are called _____ climates.** *tropical*

3. **One of the gases that traps heat in Earth's atmosphere is _____.** *carbon dioxide*

Section Wrap-up

1. temperature, precipitation, air pressure, humidity, and the number of sunny days averaged over 30 years

2. tropical—warm to hot temperatures year-round; polar—cold; temperate—temperatures vary, not as hot as tropical nor as cold as polar

3. **Think Critically** Yes, human activities such as burning fossil fuels and cutting down rain forests increase the amount of carbon dioxide in the atmosphere, which can lead to global warming.

Science Journal Journal entries should list several things being done to cut down on carbon dioxide in the atmosphere. For example, more fuel-efficient cars are being produced, countries are working together to preserve rain forests, and many people are using renewable energy sources other than fossil fuels.

Skill Builder
Recognizing Cause and Effect Student answers will vary but should be based on scientific principles. One possible answer follows. Winters would be much colder and last longer. Summers would be cooler with normal highs 10° to 15° lower than usual.

Assessment

Performance Ask students to develop a global plan that would encourage nations to cut back on cutting down rain forests. Encourage them to find out reasons for saving the rain forests other than for reducing carbon dioxide in the atmosphere. Use the Performance Task Assessment List for Carrying Out a Strategy and Collecting Data in **PASC,** p. 25. L3 P

517

Source

Watson, Lyall. *Heaven's Breath: A Natural History of the Wind.* William Morrow and Company: New York, 1984.

Background

- Kublai Khan launched his first invasion of Japan in 1274, with 1000 ships and 40 000 soldiers. The Japanese fought fiercely, holding the Mongolians to the beaches. When darkness fell, the invaders retreated to the safety of their ships. During the night, a typhoon struck, driving most of the ships out to the open sea, where they sank.

- Kublai Khan launched a second assault in 1281—this time with 4500 ships and 150 000 warriors. The second typhoon destroyed the majority of the fleet. The Japanese came to call the winds that had saved them *kamikaze,* or "divine winds."

Teaching Strategies

- Arrange a field trip to a branch of the National Weather Service or the weather center at a local television station so that students can see how weather forecasting is done and how severe storms are tracked.

- Have students research important historical weather events in their region or state, and write short plays about these events and their impact on people's lives, crops, and homes. **L2** **P**

interNET
CONNECTION

Other fleets have been destroyed by hurricanes through the years. Visit the Glencoe Homepage at *www.glencoe.com/ sec/science* for a link to this history.

How Typhoons Saved Japan

Weather affects our lives every day. It can also change the course of history. One of the most dramatic events caused by weather happened in Japan during the 13th century.

When Kublai Khan, Genghis Khan's successor, came to power in China, he set his sights on conquering Japan. In 1274 and again in 1281, the Khan sent a huge fleet of ships to Japan carrying thousands of soldiers. But on both occasions, the Mongolian forces were utterly destroyed by typhoons. Without warning, these great windstorms swept in from the sea, crushed the ships that had anchored off the Japanese coast, and saved the country from invasion.

Typhoon, hurricane, and *cyclone* are all different names for the same thing—a powerful tropical storm that begins as a low-pressure area over warm seas. Gradually, it grows into a huge, spiraling mass of moisture-rich air, spinning at speeds of more than 120 km per hour. People living near the western Pacific Ocean and the China Sea call such storms typhoons, from *ty fung,* meaning "great wind." People living near the Atlantic Ocean call the storms hurricanes, after *Huracan,* a West Indian storm god. And for people living near the Indian Ocean, the storms are called cyclones, from the Greek word *kuklos,* meaning "circular."

As a typhoon moves across a body of water, it creates waves that can be 20 to 30 meters high. Violent winds and enormous waves make a typhoon one of the most destructive forces on Earth—as Kublai Khan found out.

518   Chapter 18 Earth's Atmosphere

Bibliography

Longshore, David. *Encyclopedia of Hurricanes, Typhoons & Cyclones.* Facts on File: New York, 1998.

Souza, Dorothy M. *Hurricanes.* Lerner Pub: Minneapolis, 1996.

interNET
CONNECTION

The Glencoe Homepage at **www.glencoe.com/sec/ science** provides links connecting concepts from the student edition to relevant, up-to-date Internet sites.

Read the statements below that review the most important ideas in the chapter. Using what you have learned, answer each question in your Science Journal.

1. Air is mainly a mixture of gases, plus small amounts of solids and liquids. *What percentage of Earth's atmosphere is made of nitrogen? What percentage is made of oxygen?*

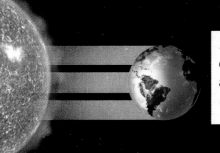

2. The sun's rays strike different parts of Earth at different angles. *How does this affect weather?*

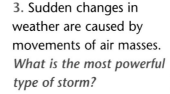

3. Sudden changes in weather are caused by movements of air masses. *What is the most powerful type of storm?*

4. Climate is the pattern of weather that occurs in a particular area over many years. *List three things that affect local climate.*

Chapter 18 Review   519

Review

Have students look at the illustrations on this page. Ask them to describe details that support the main ideas of the chapter found in the statement for each illustration.

Teaching Strategies

Divide the class into four groups. Assign each group a different project from the following list: make a three-dimensional pie graph showing the composition of Earth's atmosphere; model the angles at which the sun's rays strike different areas of Earth's surface; use colored, inflated balloons to model air masses; make a poster illustrating factors that affect local weather conditions. **L2 COOP LEARN**

Answers to Questions

1. 78% N, 21% O
2. When the sun's rays strike at direct angles, Earth's surface receives more energy and heats up. When the sun's rays strike at glancing angles, the surface is cooler. Differences in temperature affect several aspects of the weather.
3. hurricanes
4. location on Earth, closeness to a large body of water, mountains, cities

Science at Home

LS **Intrapersonal** Have students repeat the Explore activity for Chapter 8. Have them relate the activity to air pressure. Air exerts pressure and takes up space. The air in the cup kept the water out.

✔ Assessment

Portfolio Encourage students to place in their portfolios one or two items of what they consider to be their best work. Examples include:
- Activity 18-1, p. 501
- MiniLAB, p. 504
- Activity 18-2, pp. 508-509 **P**

Performance Additional performance assessments may be found in **Performance Assessment** and **Science Integration Activities.** Performance Task Assessment Lists and rubrics for evaluating these activities can be found in Glencoe's **Performance Assessment in the Science Classroom (PASC).**

Chapter 18 Review

Chapter 18 Review

Using Key Science Words

1. climate
2. front
3. global warming
4. atmosphere
5. troposphere

Checking Concepts

6. c
7. a
8. c
9. b
10. d

Thinking Critically

11. No, solar power is a clean source of fuel. No harmful exhaust would be released.

12. The amount of oxygen in the air is less as you climb because the air is thinner. They carry oxygen to make up for the lower amount.

13. As air heats up, it expands. The molecules move farther apart, thus producing an area of low pressure.

14. By tracking hurricanes, the National Weather Bureau has enough time to warn the population to move out of the approaching storm's way.

15. When rain forests are cut down, the number of trees available to take in carbon dioxide is less. The burning of the trees releases carbon dioxide into the atmosphere. Carbon dioxide is a gas that traps heat, leading to global warming.

Using Key Science Words

air mass	meteorologist
atmosphere	stratosphere
climate	troposphere
front	weather
global warming	wind

Match each phrase with the correct term from the list of Key Science Words.

1. the pattern of weather that occurs in an area over many years
2. the place where two air masses meet
3. an increase in temperatures all over the world
4. the layer of air that surrounds Earth like a blanket
5. the lowest layer of Earth's atmosphere

Checking Concepts

Choose the word or phrase that completes the sentence.

6. The _____ is the outermost layer of Earth's atmosphere.
   a. troposphere
   b. stratosphere
   c. exosphere
   d. thermosphere

7. A(n) _____ front forms when a warm air mass slides up and over a cold air mass.
   a. warm   c. stationary
   b. cold   d. occluded

8. One of the gases that contributes to global warming is _____.
   a. helium   c. carbon dioxide
   b. hydrogen   d. oxygen

9. The amount of water vapor in the air ranges from _____.
   a. ten percent to 25 percent
   b. zero to four percent
   c. zero to ten percent
   d. 25 percent to 50 percent

10. _____ is a gas found in the stratosphere that protects life on Earth from the sun's ultraviolet rays.
    a. Carbon dioxide   c. Hydrogen
    b. Nitrogen   d. Ozone

Thinking Critically

Answer the following questions in your Science Journal using complete sentences.

11. Would a solar-powered car cause pollution? Explain your answer.

12. Some mountain climbers carry bottles of oxygen when they climb high mountains. Why?

13. Air in warmer places has a lower pressure than air from cooler places. Explain why this is true.

14. Meteorologists use airplanes, weather balloons, satellites, and computers to track weather. Why does the National Weather Bureau track hurricanes?

15. You know that changes in climate can be triggered by human activities. How does cutting down rain forests contribute to global warming?

Assessment Resources

Reproducible Masters

Chapter Review, pp. 39-40
Assessment, pp. 77-80, 81-82
Performance Assessment, pp. 37-38, 56

Glencoe Technology

Computer Test Bank

MindJogger Videoquiz

Developing Skills

If you need help, refer to the description of each skill in the Skill Handbook.

16. **Making and Using Graphs:** Review the composition of Earth's atmosphere shown in **Figure 18-1** on page 496. Make a bar graph showing the amounts of the major gases found in the atmosphere.

17. **Comparing and Contrasting:** Compare and contrast a stationary front and an occluded front.

18. **Recognizing Cause and Effect:** Based on what you know about movement of air masses, can a meteorologist be correct 100 percent of the time? Why or why not?

19. **Interpreting Scientific Illustrations:** What kind of weather is associated with the front shown below? Describe how the air masses move in this type of front.

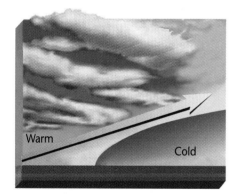

Warm

Cold

20. **Using a CD-ROM:** Using a CD-ROM, research a weather topic, such as weather forecasting, clouds, or severe storms. Write a report based on your research. In your report, describe the articles, photographs, pictures, or videos found in the CD-ROM.

Performance Assessment

1. **Model:** Design and build a rain gauge, an instrument to measure the amount of rain that falls in a given area. Place the rain gauge near your home or at school. Measure the amount of rain collected over a period of five days. Compare your measurements with those of other students. As a class, discuss some of the things that may have caused different measurements.

2. **Poster:** Review the causes of air pollution on page 497. Then make a poster showing some of the things that you and your family can do to help reduce air pollution.

Developing Skills

16. **Making and Using Graphs** Student graphs should show nitrogen at 78%, oxygen at 21%, water vapor from 0% to 4%, and other gases at less than 1%.

17. **Comparing and Contrasting** An occluded front occurs when two cool air masses merge and force warm air between them to rise. In a stationary front, air masses remain in the same place for several days, causing rain and slight winds.

18. **Recognizing Cause and Effect** No, meteorologists base their predictions on the movement of air masses. If an air mass changes its direction of movement, it may carry the predicted weather to another location. Another air mass may generate weather conditions not originally forecast.

19. **Interpreting Scientific Illustrations** The illustration shows a warm front. Warm fronts develop when warm air slides over cold air. High feathery clouds and sunny weather are sometimes associated with warm fronts.

20. **Using a CD-ROM** Student answers will vary depending on the CD-ROM used and the topic chosen by each student. Student answers should include information on a weather topic.

Performance Assessment

1. Student models should collect and measure the amount of rain that falls in a given amount of time. Students should include illustrations that explain how their rain gauges work. Students should also prepare a data table that can be compared to data from other students. Use the Performance Task Assessment List for Model in **PASC**, p. 51. P

2. Student posters should include illustrations and explanations of what family members can do to help reduce air pollution. Examples might be: make fewer trips by car or bus or combine many trips together, cut down on using fossil fuels by setting the home thermostat lower, or turn off lights in rooms not in use to save electricity and thus cut down on fossil fuel use. Use the Performance Task Assessment List for Poster in **PASC**, p. 73. P

Objectives

IS **Intrapersonal** Students will search for, gather, and analyze information from the Internet, newspaper, radio, and television on weather. They will make weather forecasts based on their data. They will present their results to the class. **L1** **COOP LEARN**

Summary

Internet Students will gather data from Internet sites that can be accessed through the Glencoe Homepage at **www. glencoe.com/sec/science.** Students can post their data on the site and get data from other schools around the country. Just click on your region of the country on the United States map. This will open a database of weather data for that region. The site will provide links to weather sites where data are available. Weather and prediction data tables are also provided on the site. Students can print out the tables or use them as a model and make weather maps using the map on the site.

Non-Internet Sources If you do not have access to the Internet, use weather data available from newspapers, radio, and television. Use a map of the United States with an overlay to post data. Record data in a table such as the one on page 523.

Time Required

One month (two weeks of data collection and two weeks of weather forecasts) are required.

UNIT 3 Internet Project

Earth Science

It's raining cats and dogs!
Red sky at night, sailor's delight.

These sayings are about the weather. You may check out the weather forecast to decide if a concert will be rained out or if you need to wear a coat to the park. Knowing what the weather will be like is important. But how do scientists predict or forecast the weather? They collect weather data every day all over the country and try to find a pattern in the data. Why do they get it wrong sometimes? Let's find out.

Goals

• Collect weather data available on the Internet, in newspapers, and on television.
• Produce a weather forecast based on the data.

Researching Weather

1. Make a data table like the one on the next page.
2. Collect data in your area every day for at least two weeks.
3. Your weather data should include temperature, barometric pressure, wind speed, wind direction, amount and type of precipitation, and cloud cover. Be sure to include the date and your location.

Data Sources

Go to the Glencoe Homepage at *www.glencoe. com/sec/science* to find links to weather data on the Internet. You can post your weather data on the Homepage and collect data from other schools around the country. You can use these data to make your own weather maps. Print the map from the Glencoe

Preparation

Internet Access the Glencoe Homepage at **www.glencoe.com/sec/science** to run through the steps the students will follow.

Non-Internet Sources Obtain a large United States map and a thin paper or clear plastic overlay. Have examples of weather data from the local newspaper.

site or post your data on a large map of the United States, using an overlay of tissue paper or plastic. *If you do not have access to the Internet,* you can find weather data on television news shows, in newspapers, or on the radio. You can also make your own weather station. You'll need a thermometer, a barometer, a rain gauge, and a wind vane.

Predicting the Weather

1. After you have collected at least two weeks of weather data for your area, it's time to make your forecast.
2. Using your data, predict what the weather will be like each day for the next two weeks. For each day, make a prediction of the temperature, amount of cloud cover, wind, and whether there will be precipitation. Record your predictions in another data table.
3. Each day, record what the weather was really like and compare it to your prediction for that day.

Conclude and Apply

1. How close did your predictions come to the actual weather? Were your forecasts for the first few days more accurate than the later days' forecasts? Explain.
2. How could you make your predictions more accurate? Would data from other areas help? Explain your answer.

Go Further

Look up weather data from your area for the last ten years. Find out the record high and low temperatures and amounts of precipitation. How could historical weather data help you make better forecasts?

Weather Data Collection Table					
Date					
Location					
Temperature					
Barometric Pressure					
Wind Speed					
Wind Direction					
Type of Precipitation					
Amount of Precipitation					
Cloud Cover					

Teaching Strategies

- To make their forecasts, students will use the data they collected. Record the actual weather each day to determine the accuracy of the students' forecasts.
- Show students examples of weather maps from the Internet, newspapers, and other sources to help them prepare their own maps.
- If the school does not have Internet access, many libraries provide it. There may be a fee for logging onto the Internet.
- Some students may have Internet access at home. These students could work in pairs with students who do not have access.
- Post data on a large United States map with a thin paper or plastic overlay.
- **Weather Station** Access the weather station directions on the website or have students research how to build their own weather stations.

Going Further

Historical weather data is available from the National Weather Service or from local meteorologists at television stations or universities. Newspapers also publish weather records. Historical weather data can help to establish trends that will increase the accuracy of weather forecasts.

Assessment

To assess the results of the project, use the Performance Task Assessment List for Poster in **PASC**, p. 73.

References

- *Glencoe Science Interactions Course 4,* Design a School Weather Station, page 108.
- Williams, Jack. *The USA Today Weather Book.* New York: Vantage Books, 1992.

Appendices

Appendix A

SI Units of Measurement

Table A-1

SI Base Units					
Measurement	**Unit**	**Symbol**	**Measurement**	**Unit**	**Symbol**
length	meter	m	temperature	kelvin	K
mass	kilogram	kg	amount of substance	mole	mol
time	second	s			

Table A-2

Units Derived from SI Base Units		
Measurement	**Unit**	**Symbol**
energy	joule	J
force	newton	N
frequency	hertz	Hz
potential difference	volt	V
power	watt	W
pressure	pascal	Pa

Table A-3

Common SI Prefixes					
Prefix	**Symbol**	**Multiplier**	**Prefix**	**Symbol**	**Multiplier**
Greater than 1			Less than 1		
mega-	M	1 000 000	*deci-*	d	0.1
kilo-	k	1 000	*centi-*	c	0.01
hecto-	h	100	*milli-*	m	0.001
deka-	da	10	*micro-*	μ	0.000 001

Appendix B

SI/Metric to English Conversions

	When you want to convert:	To:	Multiply by:
Length	inches	centimeters	2.54
	centimeters	inches	0.39
	feet	meters	0.30
	meters	feet	3.28
	yards	meters	0.91
	meters	yards	1.09
	miles	kilometers	1.61
	kilometers	miles	0.62
Mass and Weight*	ounces	grams	28.35
	grams	ounces	0.04
	pounds	kilograms	0.45
	kilograms	pounds	2.2
	tons (short)	tonnes (metric tons)	0.91
	tonnes (metric tons)	tons (short)	1.10
	pounds	newtons	4.45
	newtons	pounds	0.23
Volume	cubic inches	cubic centimeters	16.39
	cubic centimeters	cubic inches	0.06
	cubic feet	cubic meters	0.03
	cubic meters	cubic feet	35.30
	liters	quarts	1.06
	liters	gallons	0.26
	gallons	liters	3.78
Area	square inches	square centimeters	6.45
	square centimeters	square inches	0.16
	square feet	square meters	0.09
	square meters	square feet	10.76
	square miles	square kilometers	2.59
	square kilometers	square miles	0.39
	hectares	acres	2.47
	acres	hectares	0.40
Temperature	Fahrenheit	5/9 (°F − 32)	Celsius
	Celsius	9/5 (°C + 32)	Fahrenheit

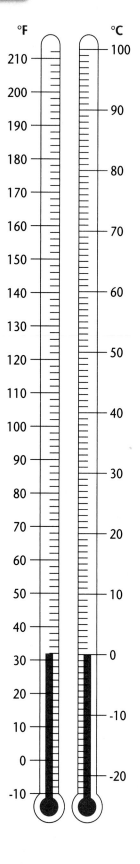

*Weight as measured in standard Earth gravity

Appendix C

Safety in the Classroom

1. Always obtain your teacher's permission to begin an investigation.
2. Study the procedure. If you have questions, ask your teacher. Be sure you understand any safety symbols shown on the page.
3. Use the safety equipment provided for you. Goggles and a safety apron should be worn during an investigation.
4. Always slant test tubes away from yourself and others when heating them.
5. Never eat or drink in the lab, and never use lab glassware as food or drink containers. Never inhale chemicals. Do not taste any substances or draw any material into a tube with your mouth.
6. If you spill any chemical, wash it off immediately with water. Report the spill immediately to your teacher.
7. Know the location and proper use of the fire extinguisher, safety shower, fire blanket, first aid kit, and fire alarm.
8. Keep all materials away from open flames. Tie back long hair and loose clothing.
9. If a fire should break out in the classroom, or if your clothing should catch fire, smother it with the fire blanket or a coat, or get under a safety shower. NEVER RUN.
10. Report any accident or injury, no matter how small, to your teacher.

Follow these procedures as you clean up your work area.
1. Turn off the water and gas. Disconnect electrical devices.
2. Return all materials to their proper places.
3. Dispose of chemicals and other materials as directed by your teacher. Place broken glass and solid substances in the proper containers. Never discard materials in the sink.
4. Clean your work area.
5. Wash your hands thoroughly after working in the laboratory.

Table C-1

First Aid	
Injury	**Safe Response**
Burns	Apply cold water. Call your teacher immediately.
Cuts and bruises	Stop any bleeding by applying direct pressure. Cover cuts with a clean dressing. Apply cold compresses to bruises. Call your teacher immediately.
Fainting	Leave the person lying down. Loosen any tight clothing and keep crowds away. Call your teacher immediately.
Foreign matter in eye	Flush with plenty of water. Use eyewash bottle or fountain.
Poisoning	Note the suspected poisoning agent and call your teacher immediately.
Any spills on skin	Flush with large amounts of water or use safety shower. Call your teacher immediately.

Appendix D

SAFETY SYMBOLS

SAFETY SYMBOLS	HAZARD	PRECAUTION	REMEDY
Disposal	Special disposal required	Dispose of wastes as directed by your teacher.	Ask your teacher how to dispose of laboratory materials.
Biological	Organisms that can harm humans	Avoid breathing in or skin contact with organisms. Wear dust mask or gloves. Wash hands thoroughly.	Notify your teacher if you suspect contact.
Extreme Temperature	Objects that can burn skin by being too cold or too hot	Use proper protection when handling.	Go to your teacher for first aid.
Sharp Object	Use of tools or glassware that can easily puncture or slice skin	Practice common sense behavior and follow guidelines for use of the tool.	Go to your teacher for first aid.
Fumes	Potential danger from smelling fumes	Must have good ventilation and never smell fumes directly.	Leave foul area and notify your teacher immediately.
Electrical	Possible danger from electrical shock or burn	Double-check setup with instructor. Check condition of wires and apparatus.	Do not attempt to fix electrical problems. Notify your teacher immediately.
Irritant	Substances that can irritate your skin or mucous membranes	Wear dust mask or gloves. Practice extra care when handling these materials.	Go to your teacher for first aid.
Chemical	Substances (acids and bases) that can react with and destroy tissue and other materials	Wear goggles and an apron.	Immediately flush with water and notify your teacher.
Toxic	Poisonous substance	Follow your teacher's instructions. Always wash hands thoroughly after use.	Go to your teacher for first aid.
Fire	Flammable and combustible materials may burn if exposed to an open flame or spark	Avoid flames and heat sources. Be aware of locations of fire safety equipment.	Notify your teacher immediately. Use fire safety equipment if necessary.

Eye Safety
This symbol appears when a danger to eyes exists.

Clothing Protection
This symbol appears when substances could stain or burn clothing.

Animal Safety
This symbol appears whenever live animals are studied and the safety of the animals and students must be ensured.

Mitosis

Mitosis is the process by which a nucleus divides into two nuclei, each containing the same number of chromosomes that the original cell had. Usually the cytoplasm then also divides.

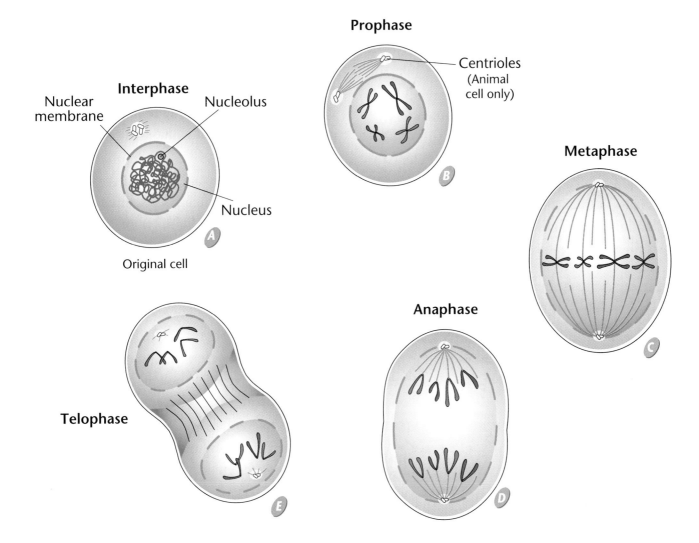

Prophase
Centrioles
(Animal cell only)

Interphase
Nuclear membrane
Nucleolus
Nucleus

A

Original cell

Metaphase

C

Anaphase

D

Telophase

E

Ⓐ	Interphase	The chromosomes duplicate.
Ⓑ	Prophase	Duplicated chromosomes become visible.
Ⓒ	Metaphase	Duplicated chromosomes line up at the equator of the cell.
Ⓓ	Anaphase	Duplicated chromosomes separate.
Ⓔ	Telophase	The cytoplasm separates. Two new cells contain same number of chromosomes as the original cell.

Diversity of Life: Classification of Living Organisms

Scientists use a six-kingdom system for the classification of organisms. These kingdoms are Kingdom Eubacteria, Kingdom Archaebacteria, Kingdom Fungi, the Plant Kingdom, and the Animal Kingdom.

Kingdom Eubacteria (True Bacteria)

Phylum Cyanobacteria one-celled organisms without a true nucleus; make their own food; contain chlorophyll; some species form colonies; most are blue-green

Bacteria one-celled organisms without a true nucleus; most absorb food from their surroundings, some are photosynthetic; many are parasites; round, spiral, or rod shaped

Kingdom Archaebacteria (Ancient Bacteria)

Archaebacteria one-celled prokaryotes found in extreme environments; phyla found in salt ponds, hot sulfur springs, and deep ocean thermal vents

Kingdom Protista

Phylum Euglenophyta one celled; can photo-synthesize or take in food; most have one flagellum; euglenoids

Phylum Chrysophyta most are one-celled; make their own food through photosynthesis; golden-brown pigments mask chlorophyll; diatoms

Phylum Pyrrophyta one celled; make their own food through photosynthesis; contain red pigments and have two flagella; dinoflagellates

Phylum Chlorophyta one celled, many celled, or colonies; contain chlorophyll and make their own food; live on land, in freshwater, or in salt water; green algae

Phylum Rhodophyta most are many-celled and photosynthetic; contain red pigments; most live in deep saltwater environments; red algae

Phylum Phaeophyta most are many-celled and photosynthetic; contain brown pigments; most live in saltwater environments; brown algae

Phylum Phaeophyta
Kelp *Macrocystis pyrifera*

Phylum Chlorophyta
Volvox

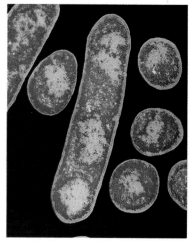

Bacteria
Clostridium botulinum
× 57 811

Appendix F

Phylum Sarcodina one-celled; take in food; move by means of pseudopods; free-living or parasitic; amoebas

Phylum Mastigophora one-celled; take in food; have two or more flagella; free-living or parasitic; flagellates

Phylum Ciliophora one-celled; take in food; have large numbers of cilia; ciliates

Phylum Sporozoa one-celled; take in food; no means of movement; parasites in animals; sporozoans

Phyla Myxomycota and Acrasiomycota one- or many-celled; absorb food; change form during life cycle; cellular and plasmodial slime molds

Kingdom Fungi

Phylum Zygomycota many-celled; absorb food; spores are produced in sporangia; zygote fungi; bread mold

Phylum Ascomycota one- and many-celled; absorb food; spores produced in asci; sac fungi; yeast

Phylum Basidiomycota many-celled; absorb food; spores produced in basidia; club fungi; mushrooms

Phylum Deuteromycota members with unknown reproductive structures; imperfect fungi; penicillin

Lichens organisms formed by symbiotic relationship between an ascomycote or a basidiomycote and a green alga or a cyanobacterium

Plant Kingdom
Spore Plants

Division Bryophyta nonvascular plants that reproduce by spores produced in capsules; many-celled; green; grow in moist land environments; mosses and liverworts

Division Lycophyta many-celled vascular plants; spores produced in cones; live on land; are photosynthetic; club mosses

Division Sphenophyta vascular plants with ribbed and jointed stems; scalelike leaves; spores produced in cones; horsetails

Division Pterophyta vascular plants with feathery leaves called fronds; spores produced in clusters of sporangia called sori; live on land or in water; ferns

Division Pterophyta
Bracken fern *Pteridium aquilinum*

Phylum Basidiomycota
Russula rubescens

Lichens
British soldier
lichen × 3

Appendix F **531**

Appendix F

Seed Plants

Division Ginkgophyta deciduous gymnosperms; only one living species called the maidenhair tree; fan-shaped leaves with branching veins; reproduces with seeds; ginkgoes

Division Cycadophyta palmlike gymnosperms; large compound leaves; produce seeds in cones; cycads

Division Coniferophyta deciduous or evergreen gymnosperms; trees or shrubs; needlelike or scalelike leaves; seeds produced in cones; conifers

Division Gnetophyta shrubs or woody vines; seeds produced in cones; division contains only three genera; gnetum

Division Anthophyta dominant group of plants; ovules protected at fertilization by an ovary; sperm carried to ovules by pollen tube; produce flowers and seeds in fruits; flowering plants

Animal Kingdom

Phylum Porifera aquatic organisms that lack true tissues and organs; they are asymmetrical and sessile; sponges

Phylum Cnidaria radially symmetrical organisms with a digestive cavity with one opening; most have tentacles armed with stinging cells; live in aquatic environments singly or in colonies; includes jellyfish, corals, hydra, and sea anemones

Phylum Platyhelminthes bilaterally symmetrical worms with flattened bodies; digestive system has one opening; parasitic and free-living species; flatworms

Phylum Annelida
Sabillid worms
Feather duster

Division Anthophyta
Oak tree
Quercus robur

Division Anthophyta
Viola

Phylum Cnidaria
Rose Anemone
Tealia lofotensis

Division Coniferophyta
Red pine cone
Pinus resinosa

Phylum Nematoda round, bilaterally symmetrical body; digestive system with two openings; some free-living forms but mostly parasitic; roundworms

Phylum Mollusca soft-bodied animals, many with a hard shell; a mantle covers the soft body; aquatic and terrestrial species; includes clams, snails, squid, and octopuses

Phylum Annelida bilaterally symmetrical worms with round segmented bodies; terrestrial and aquatic species; includes earthworms, leeches, and marine polychaetes

Phylum Arthropoda large phylum of organisms that have segmented bodies with pairs of jointed appendages and a hard exoskeleton; terrestrial and aquatic species; includes insects, crustaceans, spiders, and horseshoe crabs

Phylum Echinodermata saltwater organisms with spiny or leathery skin; water-vascular system with tube feet; radial symmetry; includes sea stars, sand dollars, and sea urchins

Phylum Chordata organisms with internal skeletons, specialized body systems, and paired appendages; all at some time have a notochord, dorsal nerve cord, gill slits, and a tail; includes fish, amphibians, reptiles, birds, and mammals

Phylum Chordata
Red-headed woodpecker

Phylum Chordata
Gray wolf
Canis lupus

Phylum Arthropoda
Euphaedra uganda

Phylum Echinodermata
Asterias forbesi

Phylum Chordata
Enneacanthus obesus

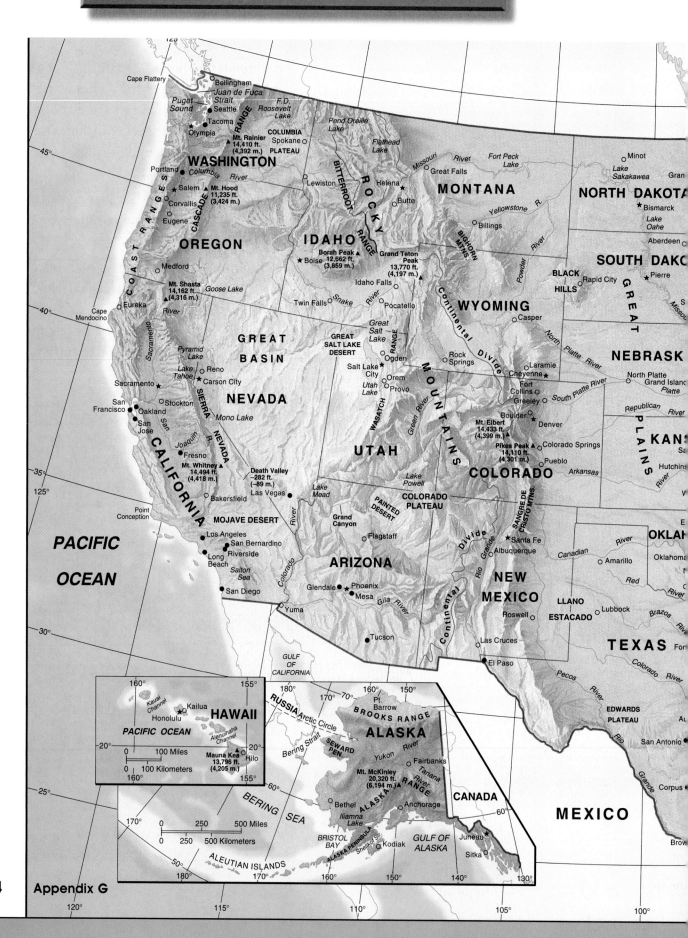

Appendix H

Care and Use of a Microscope

Eyepiece   Contains a magnifying lens you look through

Arm   Supports the body tube

Low-power objective   Contains the lens with low-power magnification

Stage clips   Hold the microscope slide in place

Coarse Adjustment   Focuses the image under low power

Fine Adjustment   Sharpens the image under high and low magnification

Base   Provides support for the microscope

Body tube   Connects the eyepiece to the revolving nosepiece

Revolving nosepiece   Holds and turns the objectives into viewing position

High-power objective   Contains the lens with the highest magnification

Stage   Supports the microscope slide

Light source   Allows light to reflect upward through the diaphragm, the specimen, and the lenses

Care of a Microscope

1. Always carry the microscope holding the arm with one hand and supporting the base with the other hand.
2. Don't touch the lenses with your fingers.
3. Never lower the coarse adjustment knob when looking through the eyepiece lens.
4. Always focus first with the low-power objective.
5. Don't use the coarse adjustment knob when the high-power objective is in place.
6. Store the microscope covered.

Using a Microscope

1. Place the microscope on a flat surface that is clear of objects. The arm should be toward you.
2. Look through the eyepiece. Adjust the diaphragm so that light comes through the opening in the stage.
3. Place a slide on the stage so that the specimen is in the field of view. Hold it firmly in place by using the stage clips.

4. Always focus first with the coarse adjustment and the low-power objective lens. Once the object is in focus on low power, turn the nosepiece until the high-power objective is in place. Use ONLY the fine adjustment to focus with the high-power objective lens.

Making a Wet-Mount Slide

1. Carefully place the item you want to look at in the center of a clean glass slide. Make sure the sample is thin enough for light to pass through.
2. Use a dropper to place one or two drops of water on the sample.
3. Hold a clean coverslip by the edges and place it at one edge of the drop of water. Slowly lower the coverslip onto the drop of water until it lies flat.
4. If you have too much water or a lot of air bubbles, touch the edge of a paper towel to the edge of the coverslip to draw off extra water and force out air.

Appendix I

Rocks

Rock Type	Rock Name	Characteristics
Igneous (intrusive) Photos on pages 416, 430.	Granite	Large mineral grains of quartz, feldspar, hornblende, and mica. Usually light in color.
	Diorite	Large mineral grains of feldspar, hornblende, mica. Less quartz than granite. Intermediate in color.
	Gabbro	Large mineral grains of feldspar, hornblende, and mica. No quartz. Dark in color.
Igneous (extrusive) Photos on page 417.	Rhyolite	Small mineral grains of quartz, feldspar, hornblende, and mica or no visible grains. Light in color.
	Andesite	Small mineral grains of feldspar, hornblende, mica or no visible grains. Less quartz than rhyolite. Intermediate in color.
	Basalt	Small mineral grains of feldspar, hornblende, mica or no visible grains. No quartz. Dark in color.
	Obsidian	Glassy texture. No visible grains. Volcanic glass. Fracture looks like broken glass.
	Pumice	Frothy texture. Floats.
Sedimentary (broken, detrital) Photos on pages 420-421, 430.	Conglomerate	Coarse-grained. Gravel or pebble-sized grains.
	Sandstone	Sand-sized grains 1/16 to 2 mm in size.
	Siltstone	Grains are smaller than sand but larger than clay.
	Shale	Smallest grains. Usually dark in color.
Sedimentary (biochemical) Photos on pages 422-423.	Limestone	Major mineral is calcite. Usually forms in oceans, lakes, and rivers. Often contains fossils.
	Coal	Occurs in swampy, low-lying areas. Compacted layers of organic material, mainly plant remains.
Sedimentary (chemical)	Rock Salt	Commonly formed by the evaporation of seawater.
Metamorphic (foliated) Photos on pages 427-429.	Gneiss	Well-developed banding because of alternating layers of different minerals, usually of different colors. Common parent rock is granite.
	Schist	Well-defined parallel arrangement of flat, sheet-like minerals, mainly micas. Common parent rocks are shale, phyllite.
	Phyllite	Shiny or silky appearance. Looks wrinkled. Common parent rocks are shale, slate.
	Slate	Harder, denser, and shinier than shale. Common parent rock is shale.
Metamorphic (non-foliated) Photos on pages 428, 430.	Marble	Interlocking calcite or dolomite crystals. Common parent rock is limestone.
	Soapstone	Composed mainly of the mineral talc. Soft with a greasy feel.
	Quartzite	Hard and well cemented with interlocking quartz crystals. Common parent rock is sandstone.

Appendix J

Minerals

Mineral (formula)	Color	Streak	Hardness	Breakage pattern	Uses and other properties
graphite (C)	black to gray	black to gray	1-2	basal cleavage (scales)	pencil lead, lubricants for locks, rods to control some small nuclear reactions, battery poles
silver (Ag)	silvery white, tarnishes to black	light gray to silver	2.5	hackly	coins, fillings for teeth, jewelry, silverplate, wires; malleable and ductile
galena (PbS)	gray	gray to black	2.5	cubic cleavage perfect	source of lead, used in pipes, shields for X rays, fishing equipment sinkers
gold (Au)	pale to golden yellow	yellow	2.5-3	hackly	jewelry, money, gold leaf, fillings for teeth, medicines; does not tarnish
bornite (Cu_5FeS_4)	bronze, tarnishes to dark blue, purple	gray-black	3	uneven fracture	source of copper; called "peacock ore" because of the purple shine when it tarnishes
copper (Cu)	copper red	copper red	3	hackly	coins, pipes, gutters, wire, cooking utensils, jewelry, decorative plaques; malleable and ductile
chalcopyrite ($CuFeS_2$)	brassy to golden yellow	greenish black	3.5-4	uneven fracture	main ore of copper
chromite ($FeCr_2O_4$)	black or brown	brown to black	5.5	irregular fracture	ore of chromium, stainless steel, metallurgical bricks
pyrrhotite (FeS)	bronze	gray-black	4	uneven fracture	often found with pentlandite, an ore of nickel; may be magnetic
hematite (specular) (Fe_2O_3)	black or reddish brown	red or reddish brown	6	irregular fracture	source of iron; roasted in a blast furnace, converted to "pig" iron, made into steel
magnetite (Fe_3O_4)	black	black	6	conchoidal fracture	source of iron, naturally magnetic, called lodestone
pyrite (FeS_2)	light, brassy, yellow	greenish black	6.5	uneven fracture	source of iron, "fool's gold," alters to limonite

Minerals

Mineral (formula)	Color	Streak	Hardness	Breakage pattern	Uses and other properties
talc ($Mg_3(OH)_2$ Si_4O_{10})	white, greenish	white	1	cleavage in one direction	easily cut with fingernail; used for talcum powder; soapstone; is used in paper and for tabletops
kaolinite ($Al_2Si_2O_5$ $(OH)_4$)	white, red, reddish, brown, black	white	2	basal cleavage	clays; used in ceramics and in china dishes; common in most soils; often microscopic-sized particles
gypsum ($CaSO_4$ $\cdot 2H_2O$)	colorless, gray, white, brown	white	2	basal cleavage	used extensively in the preparation of plaster of paris, alabaster, and dry wall for building construction
sphalerite (ZnS)	brown	pale yellow	3.5-4	cleavage in six directions	main ore of zinc; used in paints, dyes, and medicine
sulfur (S)	yellow	yellow to white	2	conchoidal fracture	used in medicine, fungicides for plants, vulcanization of rubber, production of sulfuric acid
muscovite ($KAl_3Si_3O_{10}$ $(OH)_2$)	white, light gray, yellow, rose, green	colorless	2.5	basal cleavage	occurs in large flexible plates; used as an insulator in electrical equipment, lubricant
biotite ($K(Mg, Fe)_3$ $AlSi_3O_{10}$ $(OH)_2$)	black to dark brown	colorless	2.5	basal cleavage	occurs in large flexible plates
halite (NaCl)	colorless, red, white, blue	colorless	2.5	cubic cleavage	salt; very soluble in water; a preservative
calcite ($CaCO_3$)	colorless, white, pale blue	colorless, white	3	cleavage in three directions	fizzes when HCl is added; used in cements and other building materials
dolomite ($CaMg$ $(CO_3)_2$)	colorless, white, pink, green, gray, black	white	3.5-4	cleavage in three directions	concrete and cement; used as an ornamental building stone

Appendix J

Minerals

Mineral (formula)	Color	Streak	Hardness	Breakage pattern	Uses and other properties
fluorite (CaF_2)	colorless, white, blue, green, red, yellow, purple	colorless	4	cleavage	used in the manufacture of optical equipment; glows under ultraviolet light
limonite (hydrous iron oxides)	yellow, brown, black	yellow, brown	5.5	conchoidal fracture	source of iron; weathers easily, coloring matter of soils
hornblende ($CaNa(Mg, Al,Fe)_5(Al,Si)_2 Si_6O_{22}(OH)_2$)	green to black	gray to white	5-6	cleavage in two directions	will transmit light on thin edges; 6-sided cross section
feldspar (orthoclase) ($KAlSi_3O_8$)	colorless, white to gray, green and yellow	colorless	6	two cleavage planes meet at 90° angle	insoluble in acids; used in the manufacture of porcelain
feldspar (plagioclase) ($NaAlSi_3O_8$) ($CaAl_2Si_2O_8$)	gray, green, white	colorless	6	two cleavage planes meet at 86° angle	used in ceramics; striations present on some faces
augite ($(Ca, Na)(Mg, Fe, Al)(Al, Si)_2O_6$)	black	colorless	6	2-directional cleavage	square or 8-sided cross section
olivine ($(Mg, Fe)_2 SiO_4$)	olive green	colorless	6.5	conchoidal fracture	gemstones, refractory sand
quartz (SiO_2)	colorless, various colors	colorless	7	conchoidal fracture	used in glass manufacture, electronic equipment, radios, computers, watches, gemstones
garnet ($(Mg, Fe,Ca)_3 (Al_2Si_3O_{12})$)	deep yellow-red, green, black	colorless	7.5	conchoidal fracture	used in jewelry; also used as an abrasive
topaz ($Al_2SiO_4 (F, OH)_2$)	white, pink, yellow, pale blue, colorless	colorless	8	basal cleavage	valuable gemstone
corundum (Al_2O_3)	colorless, blue, brown, green, white, pink, red	colorless	9	fracture	gemstones: ruby is red, sapphire is blue; industrial abrasive

Skill Handbook

Table of Contents

Organizing Information

Communicating

Being able to explain ideas to other people is an important part of our everyday lives. Whether reading a book, writing a letter, or watching a television program, people everywhere are giving their opinions and sharing information with each other.

Science Journal One way to record information, and express how you think about a topic in science is by writing in your Science Journal. It also lets you show how much you know about a subject.

There are many different kinds of Science Journal assignments. You may be asked to pretend you are a scientist, a TV reporter, or a committee member of a local environmental group and write from that point of view. Maybe you will be communicating your opinions to a member of Congress, a doctor, or to the editor of your local newspaper. Sometimes, you will summarize information, make an outline or a diagram, or write a letter or a paragraph in your Science Journal.

FIGURE 1

A Science Journal entry

FIGURE 2

Classifying dishes

Classifying

You may not realize it, but you make things orderly in the world around you. If you hang all your shirts together in the closet or if your favorite CDs are stacked together by performer, you have used the skill of classifying. Classifying is the process of sorting objects or events into groups based on things they have in common. When classifying, first look closely at the objects or events to be classified. Then, select one characteristic or feature that is shared by some members in the group but not by all. Place those members that share that feature into their own group. You can classify members into smaller and smaller groups based on similar characteristics.

What do you do with the dishes after they are washed? You classify them as you put them away, as shown in **Figure 2.** You separate the silverware from the plates and glasses. Forks, spoons, and knives each have their own place in the drawer. You would keep separating the dishes until all are classified and put away. Remember that all the smaller groups still share the common feature of being an eating utensil.

Sequencing

A sequence is an arrangement of things or events in a certain order. When you are asked to sequence objects or events, decide what comes first, then think about what should come next. Continue to choose objects or events until all are in order. Then, go back over the sequence to make sure each thing or event in your sequence logically leads to the next.

A sequence you are familiar with is alphabetical order. Another example of sequence would be the steps in a recipe, as shown in **Figure 3.** Think about following a recipe. Steps in a recipe for chocolate-chip cookies have to be followed in order for the cookies to turn out right.

FIGURE 3

A recipe for chocolate-chip cookies contains a sequence of steps.

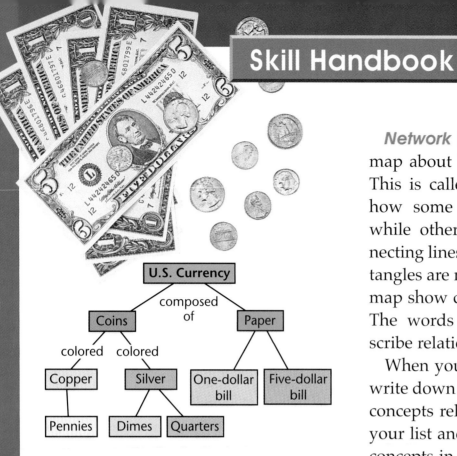

FIGURE 4

Network tree describing U.S. currency

Concept Mapping

If you were taking a trip in a car, you would probably take along a road map. The road map shows you where you are, where you are going, and other places along the way.

A concept map is similar to a road map. But a concept map shows relationships among ideas (or concepts) rather than places. A concept map is a diagram that shows how concepts are related visually. Because the concept map shows relationships among ideas, it can make the meanings of ideas and terms clear and help you understand better what you are studying.

Three types of concept maps are described here: a network tree, an events chain, and a spider map.

Network Tree Look at the concept map about U.S. currency in **Figure 4.** This is called a network tree. Notice how some words are in rectangles while others are written across connecting lines. The words inside the rectangles are main ideas. The lines in the map show connections between ideas. The words written on the lines describe relationships between concepts.

When you construct a network tree, write down the topic and list the major concepts related to that topic. Look at your list and begin to put the ideas or concepts in order from general to specific. Branch the related concepts from the major concept and describe the relationships on the lines.

Events Chain An events chain map is used to describe concepts in order.

FIGURE 5

Events chain of a typical morning routine.

In science, an events chain can be used to describe a sequence of events, the steps in a procedure, or the stages of a process.

When making an events chain, first find the one event that starts the chain. This event is called the initiating event. Then, find the next event in the chain and continue until you reach an outcome. Suppose you are asked to describe what happens when your alarm clock rings. An events chain map describing the steps might look like **Figure 5.**

Cycle Map A cycle concept map is a special type of events chain map. In a cycle concept map, the series of events does not produce a final outcome. The last event in the chain relates back to

the initiating event. Because there is no outcome and the last event relates back to the initiating event, the cycle repeats itself. Look at the cycle map describing the relationship between day and night in **Figure 6.**

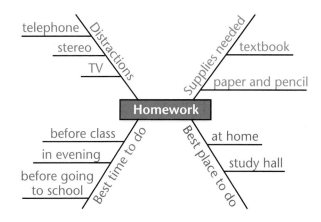

FIGURE 7

Spider map about homework.

Spider Map A fourth type of concept map is the spider map. This is a map that you can use for brainstorming. Once you brainstorm ideas from a central idea, you may find you have a jumble of ideas. Many of these ideas are related to the central idea but are not necessarily clearly related to each other. As illustrated by the spider map in **Figure 7,** you may begin to separate and group unrelated terms so that they become more useful by writing them outside the main concept.

FIGURE 6

Cycle map of day and night.

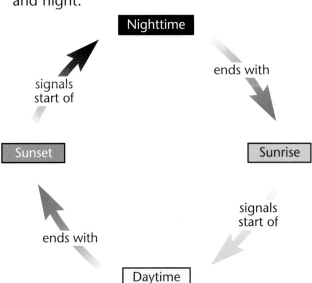

Making and Using Tables

Browse through your textbook and you will notice tables in the text and in the activities. In a table, data or information is arranged in a way that makes it easier for you to understand. Activity tables help organize and interpret the data you collect during an activity.

Most tables have a title. The title tells you what the table is about. A table is divided into columns and rows. The first column lists items to be compared. In **Figure 8,** a collection of recy-

Recycled Materials			
Day of Week	Paper (kg)	Aluminum (kg)	Plastic (kg)
Mon.	4.0	2.0	0.5
Wed.	3.5	1.5	0.5
Fri.	3.0	1.0	1.5

FIGURE 8

Table of recycled materials.

clable materials is being compared in a table. The row across the top lists the specific characteristics being compared. Collected data are recorded within the grid of the table. To make a table, list the items to be compared down in columns and the characteristics to be compared across in rows.

The title of the table in **Figure 8** is "Recycled Materials." What is being compared? This table shows the different materials being recycled and on which days they are recycled. To find out how much plastic, in kilograms, is being recycled on Wednesday, locate the column labeled "Plastic (kg)" and the row "Wed." The datum in the box where the column and row intersect gives the answer. Did you answer "0.5"? How much aluminum, in kilograms, is being recycled on Friday? If you answered "1.0," you understand how to use the parts of the table.

Making and Using Graphs

After scientists organize data in tables, they often show the data in a graph. A graph is a diagram that shows the relationship of one item or variable to another. A graph makes interpretation and analysis of data easier. There are three basic types of graphs used in science: the line graph, the bar graph, and the circle graph.

Line Graphs A line graph is used to show the relationship between two variables. The variables being compared go on two axes of the graph. The independent variable always goes on the horizontal axis, called the *x*-axis. The independent variable is the condition that is being changed. The dependent variable always goes on the vertical axis, called the *y*-axis. The dependent variable is any change that results from the changes in the independent variable.

Suppose your class started to record the amount of materials they collected in one week for their school to recycle. The collected information is shown in **Figure 9.**

Materials Collected During Week		
Day of Week	Paper (kg)	During Week (kg)
Mon.	5.0	4.0
Wed.	4.0	1.0
Fri.	4.0	2.0

FIGURE 9

Amount of recyclable materials collected during one week.

You could make a graph of the materials collected over the three days of the school week. The three weekdays are the independent variables and are placed on the *x*-axis of your graph. The amount of materials collected is the dependent variable and would go on the *y*-axis. After drawing your axes, label each with a scale. The *x*-axis lists the three weekdays. To make a scale of the amount of materials collected on the *y*-axis, look at the data values. Because the lowest amount collected was 1.0 and the highest was 5.0, you will have to start numbering at least at 1.0 and go through 5.0.

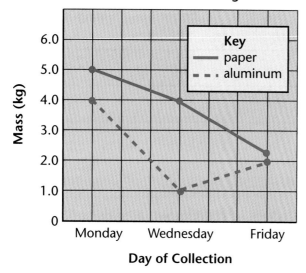

Material Collected During Week

Key
— paper
- - - aluminum

Mass (kg) — vertical axis: 6.0, 5.0, 4.0, 3.0, 2.0, 1.0, 0

Day of Collection — Monday, Wednesday, Friday

FIGURE 10

Line graph of materials collected during week.

Next, plot the data points for collected paper. The first pair of data you want to plot is Monday and 5.0 kg of paper. Locate "Monday" on the *x*-axis and locate "5.0" on the *y*-axis. Where an imaginary vertical line from the *x*-axis and an imaginary horizontal line from the *y*-axis would meet, place the first data point. Place the other data points the same way. After all the points are plotted, connect them with the best smooth curve. Repeat this procedure for the data points for aluminum. Use continuous and dashed lines to distinguish the two line graphs. The resulting graph should look like **Figure 10.**

Bar Graphs Bar graphs are similar to line graphs. They compare data that do not continuously change. In a bar graph, vertical bars show the relationships among data.

To make a bar graph, set up the *x*-axis and *y*-axis as you did for the line graph. The data are plotted by drawing vertical bars from the *x*-axis up to a point where the *y*-axis would meet the bar if it were extended.

Look at the bar graph in **Figure 11** comparing the mass of aluminum collected over three weekdays. The *x*-axis is the days on which the aluminum was collected. The *y*-axis is the mass of aluminum collected, in kilograms.

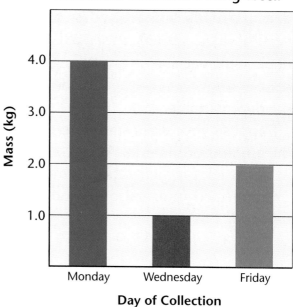

Aluminum Collected During Week

Mass (kg) — vertical axis: 4.0, 3.0, 2.0, 1.0

Day of Collection — Monday, Wednesday, Friday

FIGURE 11

Bar graph of aluminum collected during week.

Circle Graphs A circle graph or pie graph uses a circle divided into sections to show data. Each section represents part of the whole. All the sections together equal 100 percent.

Suppose you wanted to make a circle graph to show the number of seeds that sprouted or grew from a package of seeds. You count the total number of seeds. You find that there are 143 seeds in the package. This represents 100 percent, the whole circle.

You plant the seeds, and 129 seeds sprout. The seeds that sprouted will make up one section of the circle graph, and the seeds that did not sprout will make up the remaining section.

To find out how much of the circle each section should take, divide the number of seeds in each section by the total number of seeds. Then multiply your answer by 360, the number of degrees in a circle. Round to the nearest whole number. The section of the circle graph in degrees that represents the seeds sprouted is figured below.

$$\frac{129}{143} \times 360 = 324.75 \text{ or } 325 \text{ degrees}$$

To plot these data on a circle graph, you need a compass and a protractor. Use the compass to draw a circle. It will be easier to measure the part of the circle representing the seeds that did not sprout, so subtract 325° from 360° to get 35°. Draw a straight line from the center of the circle to the edge of the circle. Place your protractor on this line and use it to mark a point at 35°.

Use this point to draw a straight line from the center of the circle to the edge. This is the section for the group of seeds that did not grow. The other section represents the group of 129 seeds that did grow. Label the sections and title the graph, as shown in **Figure 12.**

FIGURE 12

Circle graph of seed germination

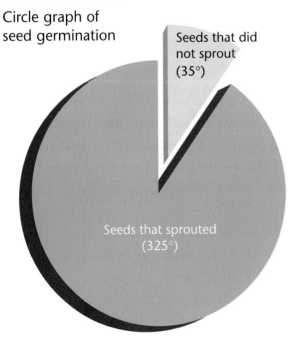

Seeds that did not sprout (35°)

Seeds that sprouted (325°)

Thinking Critically

Observing and Inferring

Observing Scientists try to make careful and accurate observations. When possible, they use instruments such as microscopes, thermometers, and balances to make observations. Measurements with a balance or thermometer provide numerical data that can be checked and repeated. Other observations are made using your senses. The basis of all scientific inquiry is observation.

Inferring Scientists often make inferences based on their observations. An inference is a conclusion about what was observed.

When making an inference, be certain to use correct data and observations. Analyze all of the data that you've collected. Then, based on everything you know, draw a conclusion about what you've observed. If possible, investigate further to find out if your inference was correct.

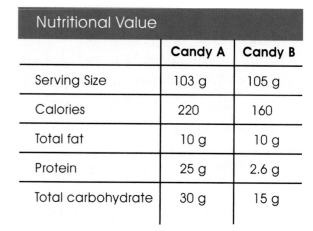

Nutritional Value		
	Candy A	**Candy B**
Serving Size	103 g	105 g
Calories	220	160
Total fat	10 g	10 g
Protein	25 g	2.6 g
Total carbohydrate	30 g	15 g

FIGURE 13

Table comparing the nutritional value of *Candy A* and *Candy B*.

When you drank a glass of orange juice after the volleyball game, you observed that the orange juice was cold. You might infer or conclude that the juice was cold because it had been made earlier in the day and had been kept in the refrigerator. The only way to be sure which inference is correct is to investigate further.

Comparing and Contrasting

Observations can be analyzed by looking at the similarities and the differences between two or more objects or events that you observe. When you look at objects or events to see how they are similar, you are comparing them. Contrasting is looking for differences in objects or events. Compare and contrast the nutritional value of two candy bars in **Figure 13**.

Recognizing Cause and Effect

Have you ever watched something happen and tried to figure out how or why it came about? If so, you have observed an effect and inferred a reason for the event. The event is an effect, and the reason for the event is the cause.

Suppose that every time your teacher fed the fish in a classroom aquarium, he or she tapped the food container on the edge of the aquarium. Then, one day your teacher just happened to tap the edge of the aquarium with a pencil. You observed the fish swim to the surface of the aquarium to feed, as shown in **Figure 14**. What is the effect, and what would you infer to be the cause? The effect is the fish swimming to the surface of the aquarium. You might infer the cause to be the teacher tapping on the edge of the aquarium. In determining cause and effect, you have made a logical inference based on your observations.

Perhaps the fish swam to the surface because they reacted to the teacher's waving hand or for some other reason. When scientists are unsure of the cause of a certain event, they plan experiments to determine what causes the event. Although you have made a logical conclusion about the behavior of the fish, you would have to perform an experiment to be certain that it was the tapping that caused the effect you observed.

FIGURE 14

What cause-and-effect situations are occurring in this aquarium?

Practicing Scientific Processes

Scientists use an orderly approach to learn new information and to solve problems. The methods scientists may use include observing to form a hypothesis, designing an experiment to test a hypothesis, separating and controlling variables, and interpreting data.

Forming Operational Definitions

Operational definitions define an object by showing how it functions, works, or behaves. Such definitions are written in terms of how an object works or how it can be used; that is, what its job or purpose is.

Some operational definitions explain how an object, such as the car in **Figure 15,** can be used.

- A car is a vehicle that can move things from one place to another.

Or such a definition may explain how an object works.

- A car is a vehicle that can move from place to place.

FIGURE 16

What hypothesis could be made about these plants?

Forming a Hypothesis

Hypotheses A hypothesis is a prediction, based on observation, that can be tested. Hypotheses are often stated as if-and-then statements. A hypothesis needs to be testable. For example, a scientist has observed in **Figure 16** that plants that are fertilized grow taller than plants that are not. A scientist may form a hypothesis that says: If plants are fertilized, then they will grow taller. This hypothesis can be tested by an experiment.

FIGURE 15

What observations can be made about this car?

Designing an Experiment to Test a Hypothesis

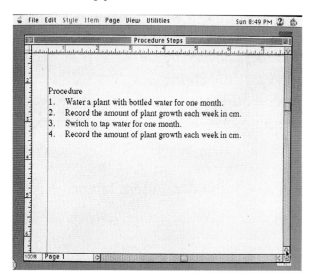

FIGURE 17

Possible procedural steps.

Once you have stated a hypothesis, you probably want to find out whether or not it explains an event or observation. In order to test a hypothesis, you must perform an experiment. When conducting an experiment, it is best to begin by writing out a procedure. A procedure is the plan that you follow in your experiment. A procedure tells you what materials to use and how to use them. After following the procedure, data are obtained. From this data, you can then draw a conclusion and make a statement about your results.

If the conclusion you draw from the data supports your hypothesis, then you can say that your hypothesis is reliable. Reliable means that you can trust your conclusion. If it did not support your hypothesis, then you would have to make new observations and state a new hypothesis—just make sure that it is one that you can test.

Planning a Procedure Suppose you hypothesize that a houseplant will grow better if watered with bottled water than with tap water. Let's figure out how to conduct an experiment to test the hypothesis: If a plant is watered with bottled water, then it will grow better and look healthier. An example procedure is seen in **Figure 17.** The data generated from this procedure are shown in **Figure 18.** The data show that the plant grew the same amount each month regardless of the type of water used. This conclusion does not support the original hypothesis made.

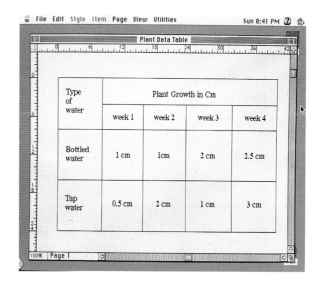

FIGURE 18

Data generated from procedural steps.

Separating and Controlling Variables

In any experiment, it is important to keep everything the same except for the item you are testing. The one factor that you change is called the independent variable. The independent variable in the experiment on the previous page was the type of water. The factor that changes as a result of the independent variable is called the dependent variable. The dependent variable in the experiment was the plant height. Always make sure that there is only one independent variable. If you have more than one, you will not know what caused the changes you observe in the independent variable. Many experiments have a control. A control is a treatment or an experiment that you can compare with the results of your test groups.

In the experiment with the plants, you made everything the same except the type of water being used. The soil, the amount of water given, the amount of light, and the temperature of the room should remain the same throughout the entire experiment. By doing so, you made sure that at the end of the experiment, any differences were the result of the type of water being used—bottled or tap. The type of water was the independent factor, and the height of the plant was the dependent factor.

Interpreting Data

The word *interpret* means "to explain the meaning of something." Look at the problem being explored in the plant experiment and find out what the data show. When you looked at the data you collected, you were checking to see if the variable had an effect. You were looking for an explanation. If there are differences in the data, the variable being tested may have had an effect.

If there is no difference between the control and the test groups, the variable being tested apparently has had no effect.

Look back at **Figure 18** on page 553, which shows the results of this experiment. In this example, the use of tap water to water the plant was the control, while watering the plant with bottled water was the test. Data showed no difference in the amount of plant growth over a one-month period.

What are data? In the experiment described on these pages, measurements were taken so that at the end of the experiment, you had concrete numbers to interpret. Not every experiment that you do will give you data in the form of numbers. Sometimes data will be in the form of a description. At the end of a chemistry experiment, you might have noted that one solution turned yellow when treated with a particular chemical, and another remained clear when treated with the same chemical. Data, therefore, are stated in different forms for different types of scientific experiments.

Are all experiments alike? Keep in mind as you perform experiments in science that not every experiment makes use of all of the parts that have been described on these pages. For some situations, it may be difficult to design an experiment that will always have a control. Other experiments are complex enough that it may be hard to have only one dependent variable. Scientists use many variations in their methods of performing experiments. The skills in this handbook are here for you to use and practice. In real situations, their uses will vary.

Representing and Applying Data

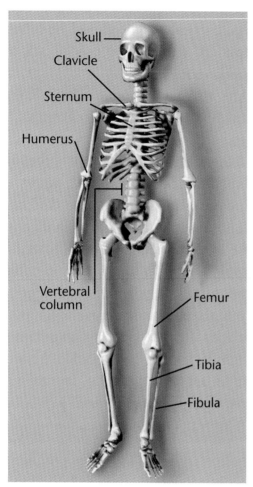

FIGURE 19

Interpreting Scientific Illustrations

As you read a science textbook, you will see many drawings, diagrams, and photographs. Illustrations help you to understand what you read. Some illustrations are included to help you understand an idea that you can't see easily by yourself. For instance, we can't see atoms, but we can look at a diagram of an atom that helps us to understand

some things about atoms. Seeing something often helps you remember more easily. Illustrations also provide examples that clarify difficult concepts or give additional information about the topic you are studying. Maps, for example, help you to locate places that may be described in the text.

Most illustrations have captions. A caption identifies or explains the illustration. Some captions are short; others are longer and more descriptive. Diagrams often have labels that identify parts of the organism or the order of steps in a process, such as the labels in **Figure 19.**

Learning with Illustrations An illustration of an organism shows that organism from a particular side. In order to understand the illustration, you may need to identify the front (anterior) end, the tail (posterior) end, the underside (ventral), and the back (dorsal) side, as shown in **Figure 20.**

FIGURE 20

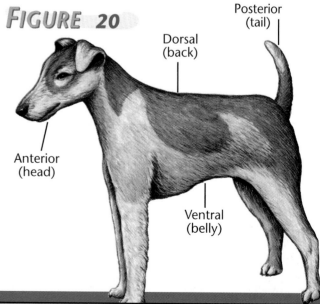

Making Models

Have you ever worked on a model car, plane, or rocket? Models look, and sometimes work, much like the real thing, but they are often smaller or larger. In science, models are used to help simplify processes or structures that otherwise would be difficult to see and understand.

To make a model, you first have to get a basic idea about the structure or process involved. For example, make a model to show the differences in size of arteries, veins, and capillaries. First, read about these structures. All three are hollow tubes. Arteries are round and thick. Veins are flat and have thinner walls than arteries. Capillaries are small.

Now, decide what you can use for your model. Common materials are often best and cheapest to work with when making models. The different kinds and sizes of pasta shown in **Figure 21** might work for these models. Different sizes of rubber tubing might do just as well. Cut and glue the different noodles or tubing onto thick paper so the openings can be seen. Then label each. Now you have a simple, easy-to-understand model showing the differences in size of arteries, veins, and capillaries.

What other scientific ideas might a model help you to understand? A model of a molecule can be made from gumdrops (using different colors for the different elements present) and toothpicks (to show different chemical bonds). A working model of a volcano can be made from clay, a small amount of baking soda, vinegar, and a bottle cap. Other models can be devised on a computer.

FIGURE 21

Different types of pasta may be used to model blood vessels.

Measuring in SI

The International System (SI) of Measurement is accepted as the standard for measurement throughout most of the world. Four of the base units in SI are the meter, liter, kilogram, and second.

The size of the unit can be determined from the prefix used with the base unit name. Look at **Figure 22** for some common metric prefixes and their meanings. The prefix *kilo-* attached to the unit *gram* is kilogram, or 1000 grams. The prefix *deci-* attached to the unit *meter* is decimeter, or one-tenth (0.1) of a meter.

The metric system is convenient because its unit sizes vary by multiples of 10. When changing from smaller units to larger units, divide by 10. When changing from larger units to smaller units, multiply by 10. For example, to convert millimeters to centimeters, divide the millimeters by 10. To convert 30 millimeters to centimeters, divide 30 by 10 (30 millimeters equal 3 centimeters).

Metric Prefixes

Prefix	Symbol	Meaning	
kilo-	k	1000	thousand
hecto-	h	200	hundred
deka-	da	10	ten
deci-	d	0.1	tenth
centi-	c	0.01	hundredth
milli-	m	0.001	thousandth

FIGURE 22

Common metric prefixes

The meter is the SI unit used to measure length. A baseball bat is about one meter long. When measuring smaller lengths, the meter is divided into smaller units called centimeters and millimeters. A centimeter is one-hundredth (0.01) of a meter. A millimeter is one-thousandth of a meter (0.001).

Most metric rulers have lines indicating centimeters and millimeters, as shown in **Figure 23.** The centimeter lines are the longer, numbered lines; the shorter lines are millimeter lines. When using a metric ruler, line up the 0-centimeter mark with the end of the object being measured and read the number of the unit where the object ends.

FIGURE 23

Metric ruler showing centimeter and millimeter divisions.

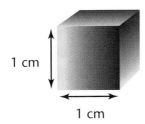

FIGURE 24

A square centimeter

Surface Area Units of length are also used to measure surface area. The standard unit of area is the square meter (m^2). A square that's one meter long on each side has a surface area of one square meter. A square centimeter, (cm^2), shown in **Figure 24,** is one centimeter long on each side. The surface area of an object is determined by multiplying the length times the width.

Volume The volume of a rectangular solid is also calculated using units of length. The cubic meter (m^3) is the standard SI unit of volume. A cubic meter is a cube one meter on each side. You can determine the volume of rectangular solids by multiplying length times width times height.

Liquid Volume During science activities, you will measure liquids using beakers and graduated cylinders marked in milliliters, as illustrated in **Figure 25.** A graduated cylinder is a cylindrical container marked with lines from bottom to top. Liquid volume is measured using a unit called a liter. A liter has the volume of 1000 cubic centimeters. Because the prefix *milli-* means thousandth (0.001), a milli-

liter equals one cubic centimeter. One milliliter of liquid would completely fill a cube measuring one centimeter on each side.

Mass Scientists use balances to find the mass of objects in grams. You will use a triple beam balance similar to the one shown in **Figure 26** on the next page. Notice that on one side of the balance is a pan and on the other side is a set of beams. Each beam has an object of a known mass, called a rider, that slides along the beam.

Before you find the mass of an object, set the balance to zero by sliding all the riders back to the zero point. Check the pointer on the right to make sure it swings an equal distance above and below the zero point on the scale. If the swing is unequal, find and turn the adjusting screw until you have an equal swing.

FIGURE 25

A volume of 79 mL is measured by reading at the lowest point of the curve.

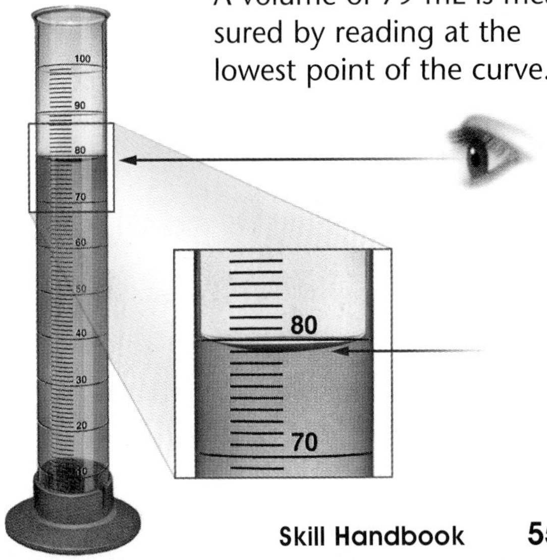

Place an object on the pan. Slide the rider with the largest mass along its beam until the pointer drops below zero. Then move it back one notch. Repeat the process on each beam until the pointer swings an equal distance above and below the zero point. Add the masses on each beam to find the mass of the object.

You should never place a hot object or pour chemicals directly onto the pan. Instead, find the mass of a clean beaker or a glass jar. Place the dry or liquid chemicals in the container. Then find the combined mass of the container and the chemicals. Calculate the mass of the chemicals by subtracting the mass of the empty container from the combined mass.

Predicting

When you apply a hypothesis, or general explanation, to a specific situation, you predict something about that situation. First, you must identify which hypothesis fits the situation you

FIGURE 27

The daily high temperature is predicted every day.

are considering. People use prediction to make decisions every day. Based on previous observations and experiences, you may form a hypothesis that if it is wintertime, then temperatures will be low. From weather data in your area, temperatures are lowest in February. You may then use this hypothesis to predict specific temperatures and weather for the month of February. Someone could use these predictions to plan to set aside more money for heating bills during that month.

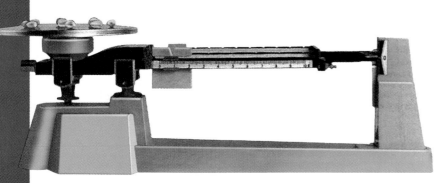

FIGURE 26

A beam balance is used to measure mass.

Using Numbers

When working with large populations of organisms, scientists usually cannot observe or study every organism in the population. Instead, they use a sample or a portion of the population. To sample is to take a small number of organisms of a population for research. Information discovered with the small sample may then be applied to the whole population. For example, scientists may take a small number of mice from a field to study the effects of day length on reproductive rate. This information could be applied to the population as a whole.

Estimating   Scientific work also involves estimating. To estimate is to make a judgment about the size of something or the number of something without actually measuring or counting every member of a population. Here is a familiar example. Have you ever tried to guess how many kernels of popcorn were in a sealed jar?

If you did, you were estimating. What if you knew the jar of popcorn held one liter (1000 mL)? If you knew that 60 popcorn kernels would fit in a 100-milliliter jar, how many kernels would you estimate to be in the one-liter jar? If you said about 600 kernels, your estimate would be close to the actual number of popcorn kernels.

Scientists use a similar process to estimate populations of organisms from bacteria to buffalo. Scientists count the actual number of organisms in a small sample and then estimate the number of organisms in a larger area. For example, if a scientist wanted to count the number of black-eyed Susans, the field could be marked off in a large grid of 1-meter squares. To determine the total population of the field, the number of organisms in one square-meter sample can be multiplied by the total number of square centimeters in the field.

Using a Computerized Card Catalog

When you have a report or paper to research, you go to the library. To find the information, skill is needed in using a computerized card catalog. You use the computerized card catalog by typing in a subject, the title of a book, or an author's name. The computer will list on the screen all the holdings the library has on the subject, title, or author requested.

A library's holdings include books, magazines, databases, videos, and audio materials. When you have chosen something from this list, the computer will show whether an item is available and where in the library to find it.

Example You have a report due on dinosaurs, and you need to find three books on the subject. In the library, follow the instructions on the computer screen to select the "Subject" heading. You could start by typing in the word *dinosaurs.* This will give you a list of books on that subject. Now you need to narrow your search to the kind of dinosaur you are interested in, for example, *Tyrannosaurus rex.* You can type in *Tyrannosaurus rex* or just look through the list to find titles that you think would have information you need. Once you have selected a short list of books, click on each selection to find out if the library has the books. Then, check on where they are located in the library.

Using a CD-ROM

What's your favorite music? You probably listen to your favorite music on compact discs (CDs). But, there is another use for compact discs, called CD-ROM. CD-ROM means Compact-Disc–Read Only Memory. CD-ROMs hold information. Whole encyclopedias and dictionaries can be stored on CD-ROM discs. This kind of CD-ROM and others are used to research information for reports and papers. The information is accessed by putting the disc in your computer's CD-ROM drive and following the computer's installation instructions. The CD-ROM will have words, pictures, photographs, and maybe even sound and videos on a wide range of topics.

Example Load the CD-ROM into the computer. Find the topic you are interested in by clicking on the Search button. If there is no Search button, try the Help button. Most CD-ROMs are easy to use, but refer to the Help instructions if you have problems. Use the arrow keys to move down through the list of titles on your topic. When you double-click on a title, the article will appear on the screen. You can print the article by clicking on the Print button. Each CD-ROM is different. Click the Help menu to see how to find what you want.

Developing Multimedia Presentations

It's your turn—you have to present your science report to the entire class. How do you do it? You can use many different sources of information to get the class excited about your presentation. Posters, videos, photographs, sound, computers, and the Internet can help show our ideas. First, decide the most important points you want your presentation to make. Then sketch out what materials and types of media would be best to illustrate those points. Maybe you could start with an outline on an overhead projector, then show a video, followed by something from the Internet or a slide show accompanied by music or recorded voices. Make sure you don't make the presentation too complicated, or you will confuse yourself and the class. Practice your presentation a few times for your parents or brothers and sisters before you present it to the class.

Example Your assignment is to give a presentation on bird-watching. You could have a poster that shows what features you use to identify birds, with a sketch of your favorite bird. A tape of the calls of your favorite bird or a video of birds in your area would work well with the poster. If possible, include an Internet site with illustrations of birds that the class can look at.

Using E-Mail

It's science fair time and you want to ask a scientist a question about your project, but he or she lives far away. You could write a letter or make a phone call. But you can also use the computer to communicate. You can do this using electronic mail (E-mail). You will need a computer that is connected to an E-mail network. The computer is usually hooked up to the network by a device called a *modem*. A modem works through the telephone lines. Finally, you need an address for the person you want to talk with. The E-mail address works just like a street address to send mail to that person.

Example There are just a few steps needed to send a message to a friend on an E-mail network. First, select Message from the E-mail software menu. Then, enter the E-mail address of your friend. Next, type your message. Make sure you check it for spelling and other errors. Finally, click the Send button to mail your message and off it goes! You will get a reply back in your electronic mailbox. To read your reply, just click on the message and the reply will appear on the screen.

Using an Electronic Spreadsheet

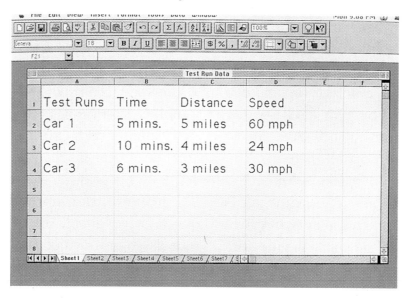

Your science fair experiment has produced lots of numbers. How do you keep track of all the data, and how can you easily work out all the calculations needed? You can use a computer program called a *spreadsheet* to keep track of data that involve numbers. A spreadsheet is an electronic worksheet. Type in your data in rows and columns, just as in a data table on a sheet of paper. A spreadsheet uses some simple math to do calculations on the data. For example, you could add, subtract, divide, or multiply any of the values in the spreadsheet by another number. Or you can set up a series of math steps you want to apply to the data. If you want to add 12 to all the numbers and then multiply all the numbers by 10, the computer does all the calculations for you in the spreadsheet.

Example Let's say that to complete your project, you need to calculate the speed of the model cars in your experiment. Enter the distance traveled by each car in the rows of the spreadsheet. Then enter the time you recorded for each car to travel the measured distance in the column across from each car. To make the formula, just type in the equation you want the computer to calculate, in this case, *speed 5 distance 3 time.* You must make sure the computer knows what data are in the rows and what data are in the columns so the calculation will be correct. Once all the distance and time data and the formula have been entered into the spreadsheet program, the computer will calculate the speed for all the trials you ran. You can even make graphs of the results.

English Glossary

This glossary defines each key term that appears in **bold type** in the text. It also shows the page number where you can find the word used.

Pronunciation Key

a...b**a**ck (bak)	oh...g**o** (goh)	sh...**sh**elf (shelf)
ay...d**ay** (day)	aw...s**o**ft (sawft)	ch...na**t**ure (nay chur)
ah...f**a**ther (fahth ur)	or...**or**bit (or but)	g...**g**ift (gihft)
ow...fl**ow**er (flow ur)	oy...c**oi**n (coyn)	j...**g**em (jem)
ar...**car** (car)	oo...f**oo**t (foot)	ing...si**ng** (sing)
e...l**e**ss (les)	ew...f**oo**d (fewd)	zh...vi**s**ion (vihzh un)
ee...l**ea**f (leef)	yoo...p**u**re (pyoor)	k...**c**ake (kayk)
ih...tr**i**p (trihp)	yew...f**ew** (fyew)	s...**s**eed, **c**ent (seed, sent)
i (i + con + e)...**i**dea	uh...comm**a** (cahm uh)	z...**z**one, rai**s**e (zohn, rayz)
(i dee uh), l**i**fe (life)	u (+ con)...flow**er** (flow ur)	

abiotic (AY bi AH tihk) **factors:** all the non-living parts of an ecosystem, including air, soil, water, and temperature. (Chap. 6, p. 153)

acceleration (ak sel uh RAY shun): a change in speed or direction; depends on the mass of an object and the force pushing or pulling the object. (Chap. 10, p. 272)

acid rain: damaging acidic rain or snow formed when gases released by burning oil and coal mix with water in the air; can kill fish when it falls into rivers and lakes and can kill plants and trees when it falls to the ground. (Chap. 7, p. 193)

adaptation: any characteristic of an organism—a body shape, body process, or behavior—that helps the organism to survive in its environment and carry out life processes. (Chap. 5, p. 120)

air mass: a large body of air with the same properties as the area of Earth's surface over which it develops and moves. (Chap. 18, p. 505)

allele (uh LEEL): a different form that a gene may have for a specific trait. (Chap. 4, p. 106)

alternating current (AC): a kind of electricity that regularly reverses its direction, usually is generated by a power plant, and is used to run household appliances. (Chap. 13, p. 358)

anatomy: the physical makeup of a living organism. (Chap. 5, p. 132)

ancestry: a line of descent of a living organism. (Chap. 5, p. 132)

Archimedes (ar kih MEE deez): a Greek mathematician, inventor, and philosopher whose ideas included a compound machine used to pump water for irrigation more than 2000 years ago. (Chap. 11, p. 310) artesian well: a well drilled down into the groundwater layer of Earth's surface. (Chap. 17, p. 474)

asexual (ay SEK shuhl) **reproduction:** a type of reproduction in which an organism produces a new organism from one parent, and all the DNA of the new organism comes from that one parent. (Chap. 4, p. 92)

asteroids: pieces of rock found mostly in the asteroid belt that lies between Mars and Jupiter and that are made up of minerals similar to those that formed the planets. (Chap. 14, p. 388)

astronomical (as truh NAHM uh kul) unit (AU): a unit of measure equal to 150 million km, which is the average distance from Earth to the sun, and that is used to measure the distances between the planets. (Chap. 14, p. 385)

atmosphere (AT muh sfeer): the layer of air that surrounds Earth like a blanket, is made up of a mixture of gases plus small amounts of tiny solids and liquids, and keeps temperatures on Earth just right to support life. (Chap. 18, p. 496)

atom: an extremely small particle that is the basic unit of matter and is composed mainly of protons, neutrons, and electrons. (Chap. 8, p. 212)

average speed: describes the movement for an entire trip by dividing measured distance by measured time. (Chap. 10, p. 270)

axis: the imaginary line that Earth spins around and whose tilt is responsible for Earth's seasons. (Chap. 14, p. 379)

balance: a small, precise tool that is used to measure the mass of an object. (Chap. 10, p. 267)

battery: a device that converts internally stored chemical energy into electric current. (Chap. 13, p. 354)

biochemical rock: a type of sedimentary rock, such as coal or limestone, formed from dead organic matter that piled up and was compressed into rock over millions of years. (Chap. 15, p. 422)

biodiversity (bi oh duh VUR suh tee): the measure of the number of different species in a specific area. (Chap. 5, p. 119)

biosphere (BI oh sfeer): the part of Earth in which organisms can live; includes the topmost layer of Earth's crust, the surrounding atmosphere, and all the rivers, lakes, and oceans. (Chap. 6, p. 151)

biotic (bi AH tihk) **factors:** all the organisms that make up the living parts of an ecosystem, including trees, birds, bacteria, and frogs. (Chap. 6, p. 152)

captive breeding: the breeding of endangered species in captivity in order to build up the number of individuals in a population. (Chap. 5, p. 134)

cell: the smallest unit of life in a living thing; cells take in materials and release energy and waste products. (Chap. 2, p. 36)

cell membrane: the flexible structure that holds a plant or animal cell together, gives it shape, provides a boundary between the cell and its environment, and helps control what enters and exits the cell. (Chap. 3, p. 68)

cellular respiration: the process in which food and oxygen combine in the mitochondria to make carbon dioxide and water and release energy to do all of the cell's work. (Chap. 3, p. 70)

chemical change: the change of materials into other, new materials with different properties. (Chap. 9, p. 255)

chemical formula: a formula composed of symbols and numbers that tell what elements make up a compound and what their ratios are. (Chap. 8, p. 227)

chemical property: a characteristic of a substance that permits its change to a new substance—for example, the ability to burn and the ability to react with oxygen. (Chap. 9, p. 243)

chloroplasts (KLOR uh plasts): plant cell organelles that give plants their green color and that trap energy from sunlight and turn it into food during the process of photosynthesis. (Chap. 3, p. 72)

cichlid (SI klud): a type of African fish whose numerous species show a variety of differences yet still share many similarities. (Chap. 5, p. 132)

cinder cone volcano: a cone-shaped volcano whose steep sides are made up of blocks of solid lava that formed by explosive eruptions. (Chap. 16, p. 443)

cinders: small chunks of solid lava produced by a volcano. (Chap. 16, p. 443)

circuit (SUR kut): a complete, unbroken path that electrons can travel to do work. (Chap. 13, p. 351)

classification (klas if uh KAY shun): the grouping of objects based on common traits; for living organisms, can be based on common body parts, similarities in the materials that make up the bodies, or chemicals found inside the organisms' bodies. (Chap. 2, p. 45)

clay: a type of soil formed from the smallest grains of broken sedimentary rocks; holds moisture and is often found at the mouths of rivers. (Chap. 17, p. 472)

cleavage (KLEE vuj): property of some minerals to split into pieces with smooth, regular edges and surfaces. (Chap. 15, p. 409)

climate (KLIME ut): the pattern of weather that occurs in a particular area over many years based on such things as average temperature, precipitation, and humidity. (Chap. 18, p. 514)

cloning: the process of creating an organism that has exactly the same DNA as another organism; a clone receives all of its DNA from one parent and is genetically identical to its parent. (Chap. 4, p. 100)

comet: an object in the solar system made up of ice, dust, and frozen gases that develops a tail of light when it swings close to the sun and is pushed on by the solar wind. (Chap. 14, p. 386)

common ancestor (AN ses tur): the shared ancestor of new, different species that arose from one population. (Chap. 5, p. 129)

community (kuh MYEW nuh tee): all the populations living in an area who depend on each other for food, shelter, and other assorted needs. (Chap. 6, p. 158)

composite volcano: a volcano made of layers of thick, sticky lava and ash that is formed from alternating quiet and explosive eruptions; forms tall, pointed mountains. (Chap. 16, p. 443)

compound (KAHM pownd): a form of matter made by combining two or more different kinds of elements. (Chap. 8, p. 225)

compound machine: a machine made from the combination of two or more simple machines—for example, hand can openers and bicycles. (Chap. 11, p. 307)

condensation: the process in which water vapor changes back into its liquid form as the water vapor cools. (Chap. 17, p. 469)

conduction (kun DUK shun): the transfer of kinetic energy from faster-moving molecules to slower-moving molecules that do not travel from one place to another, but move in place. (Chap. 12, p. 334)

conglomerates (kon GLAW muh rutz): rocks composed of large pebbles mixed and cemented together with other sediments. (Chap. 15, p. 421)

constellation (kahn stuh LAY shun): a group of stars, such as the Big Dipper, that forms a pattern in the sky and may be named after a real or imaginary animal, person, or object. (Chap. 14, p. 396)

consumer: an organism that eats other organisms. (Chap. 6, p. 168)

continental drift: a theory proposed by Wegener to explain the movement of the continents over millions of years from Pangaea to their present-day positions. (Chap. 16, p. 452)

convection (kun VEK shun): the transfer of thermal energy by the movement of molecules from one place to another in a liquid or gas; the type of transfer of thermal energy that produces weather. (Chap. 12, p. 336)

convection current: the cycle in which hot magma is forced upward toward Earth's surface, becomes cooler and denser as it rises, then starts to sink back down into the mantle, where it is heated again; as convection currents move around in the plastic-like layer just under the crust, they cause movement of the plates. (Chap. 16, p. 458)

crust: Earth's outer layer, which is firm and seems solid but is broken up into giant plates that are able to move around. (Chap. 16, p. 438)

crystals: solid materials, such as quartz and halite, whose atoms are arranged in repeating patterns and that have smooth surfaces, sharp edges, and points. (Chap. 15, p. 408)

curator: an individual who is responsible for the care of something—for example, a person in charge of a zoo. (Chap. 6, p. 165)

current: a mass of water moving in one direction in the ocean, either at the surface or flowing along the bottom or between the bottom and the surface; can carry seeds and plants from one continent to another and can affect the climate. (Chap. 17, p. 483)

decomposers: organisms, such as bacteria or fungi, that feed on dead organisms and the waste material of other organisms. (Chap. 6, p. 168)

density: a physical property of matter that can be found by dividing the mass of a material by its volume. (Chap. 9, p. 239)

deposition (dep uh ZIH shun): process in which rivers and streams change Earth by dropping sediments. (Chap. 17, p. 475)

desert: a hot, dry environment with low biodiversity. (Chap. 5, p. 119)

development: all the changes that take place during the life of an organism. (Chap. 2, p. 38)

direct current (DC): battery-powered electricity that travels in only one direction. (Chap. 13, p. 358)

dissolved: describes materials that have passed into solution. (Chap. 17, p. 479)

diversity: the variety of species shown by Earth's living organisms. (Chap. 5, p. 118)

DNA: a chemical that chromosomes are made of and that is found in the nucleus of every cell of all living things—the "genetic blueprint" that provides instructions for how organisms look and function. (Chap. 4, p. 103)

earthquake: the shaking of Earth produced when sudden movement occurs along a fault and quickly releases energy. (Chap. 16, p. 438)

eclipse (ee KLIHPS): an event that happens when the moon passes between the sun and Earth (a solar eclipse) or Earth passes between the sun and the moon (a lunar eclipse), producing a shadow. (Chap. 14, p. 382)

ecologist: a scientist who studies the interactions of organisms and their environment. (Chap. 6, p. 151)

ecology (ee KAH luh jee): the study of all the interactions between living and nonliving parts of an ecosystem. (Chap. 6, p. 151)

ecosystem (EE koh sihs tum): a working unit made up of organisms interacting with each other and with nonliving factors. (Chap. 6, p. 148)

electrical energy: the energy of charges in motion. (Chap. 13, p. 346)

electricity: electric current or power; all forms depend on electrons moving from one place to another. (Chap. 13, p. 346)

electromagnet: a temporary magnet made from a current-carrying wire wrapped around an iron core, whose strength can be increased or decreased by changing the strength of the power source. (Chap. 13, p. 362)

electrons (ee LEK trahn): negatively charged particles that are the lightest of the three main particles found in an atom and that form a cloud around the atom's nucleus. (Chap. 8, p. 219)

element: natural or synthetic matter that is made up of only one type of atom; there are 112 known elements—for example, gold, aluminum, and oxygen. (Chap. 8, p. 223)

embryo: a fertilized egg that has begun dividing into more cells. (Chap. 4, p. 100)

endangered species: a species with so few living members that it is in danger of becoming extinct. (Chap. 6, p. 166)

energy (EN ur jee): the ability to bring about change in shape, temperature, speed, position, or direction of something; occurs in several different forms, such as heat energy and chemical energy, and often changes from one form to another. (Chap. 12, p. 320)

environment (en VI ur munt): everything in an organism's surroundings including other organisms, weather, water, sound, light, temperature, soil, and rocks. (Chap. 2, p. 35)

epicenter: the point at which the waves of an earthquake hit the surface of Earth, directly over the focus. (Chap. 16, p. 439)

erosion (ee ROH zhun): the wearing away of Earth's surface when soil or rock is loosened—for example, by wind and water—and then moved from one place to another. (Chap. 17, p. 476)

evaporation (ee vap uh RAY shun): the process that gradually changes water from a liquid state to water vapor. (Chap. 17, p. 468)

extinction: the dying out of a whole species. (Chap. 5, p. 141)

extrusive (EKS trew sihv): a type of igneous rock with an even, smooth texture and few or no visible crystals that forms when magma or melted rock from inside Earth cools at Earth's surface. (Chap. 15, p. 416)

fault: a feature of Earth's crust where plates rub against each other and build up energy that is released in waves during an earthquake. (Chap. 16, p. 439)

fault-block mountain: a type of mountain that is formed when whole sections of Earth's crust are faulted and broken and then some sections are uplifted. (Chap. 16, p. 461)

fertilization: the process in which sperm and egg unite, resulting in a new individual with a full set of chromosomes. (Chap. 4, p. 97)

fissures (FIHSH urs): large cracks in Earth's crust that can open up and allow lava to ooze out onto the ground and form extrusive igneous rocks. (Chap. 15, p. 416)

floodplain: a fertile, mostly flat area formed from sediments deposited by a river and that normally becomes flooded during a time of high water. (Chap. 17, p. 479)

focus: the exact spot inside Earth where an earthquake starts. (Chap. 16, p. 439)

folded mountain: a type of mountain formed when Earth's plates collided. (Chap. 16, p. 461)

foliated (FOH lee ay tud): a type of metamorphic rock, such as gneiss and slate, with bands of minerals that have been heated and squeezed into parallel layers. (Chap. 15, p. 428)

force: a push or a pull—for example, gravity and friction; can be balanced or unbalanced and can be found mathematically by multiplying mass times acceleration. (Chap. 10, p. 275)

forecast: a prediction of weather conditions based on observation and technology such as weather balloons, satellites, computers, and data from weather stations located around the world. (Chap. 18, p. 512)

fossil fuel: a fuel (coal, oil, natural gas) formed from the remains of ancient plants and animals that can be burned to produce energy. (Chap. 15, p. 424)

fossil record: all of the fossils that scientists have recovered from the ground; provides scientists with strong evidence that life on Earth has changed over time. (Chap. 5, p. 136)

fossils: the traces or remains of ancient plants and animals, which provide a history for life on Earth and help scientists form a picture of the past. (Chap. 5, p. 130)

fracture (FRAK chur): property of some minerals to break into pieces with irregular, jagged, or rough edges. (Chap. 15, p. 409)

freshwater: water in lakes, rivers, frozen in glaciers and polar ice caps, or underground and that is necessary for our survival. (Chap. 17, p. 467)

friction (FRIHK shun): the push or pull that opposes motion when two touching surfaces are sliding on each other. (Chap. 10, p. 273)

front: the place at which air masses meet and produce different kinds of weather conditions. (Chap. 18, p. 506)

fulcrum: the pivot point around which a lever turns. (Chap. 11, p. 299)

galaxy (GAL uk see): a group of stars, gas, and dust held together by gravity—for example, the Milky Way Galaxy, which contains 200 billion stars. (Chap. 14, p. 399)

gem: a rare, precious mineral that is clear, with no blemishes or cracks, and that can be cut and polished, making it ideal for jewelry. (Chap. 15, p. 413)

gene (JEEN): a small section of a chromosome that determines a trait; genes control all the traits of all organisms and provide all the information for the growth and life of a species. (Chap. 4, p. 103)

generator (JEN uh ray tur): a machine that changes mechanical energy into electric power. (Chap. 13, p. 366)

genetics (juh NET ihks): the study of how traits—the physical characteristics of an organism—are passed from parent to offspring. (Chap. 4, p. 102)

geologic time scale: a "diary" for life on Earth that is divided into time periods and that helps scientists keep track of when important events happened in Earth history, such as the appearance or disappearance of a species. (Chap. 5, p. 138)

geometric shapes: the shapes formed from straight lines, curves, and angles. (Chap. 11, p. 297)

global warming: an increase in temperatures all over the world. (Chap. 18, p. 517)

gravel: loose, rounded rock fragments. (Chap. 17, p. 472)

gravity (GRA vuh tee): a pull that every object exerts on every other object and that pulls all things toward Earth in the same way. (Chap. 10, p. 264)

groundwater: water that soaks into the ground and collects in the spaces between soil and rock particles. (Chap. 17, p. 471)

habitat (HAB uh tat): the place in which an organism lives out its life and where resources such as food, space, and shelter are shared among the different species. (Chap. 6, p. 164)

heat: thermal energy that is transferred from a warmer object to a cooler object. (Chap. 12, p. 329)

hemoglobin (HEE muh gloh bun): the blood protein that carries oxygen to the body cells. (Chap. 5, p. 137)

hypothesis (hi POTH uh sus): a statement about a problem that can be tested. (Chap. 1, p. 18)

igneous (IG nee us) **rock:** a type of rock produced when magma from inside Earth cools and hardens on Earth's surface (extrusive igneous rock) or under Earth's surface (intrusive igneous rock). (Chap. 15, p. 415)

inclined plane: a simple machine with a sloped surface that makes it easier to lift objects—for example, a moving van ramp. (Chap. 11, p. 296)

inertia (ih NUR shuh): resistance to a change in motion. (Chap. 10, p. 276)

inference (IHN fuh runtz): an explanation of why something happened. (Chap. 1, p. 10)

inner core: the solid, innermost layer of Earth. (Chap. 16, p. 449)

inorganic: describes a substance that is not formed by plants or animals. (Chap. 15, p. 406)

intrusive (IN trew sihv): a type of igneous rock with large, beautiful crystals that is produced when magma cools under the crust inside Earth. (Chap. 15, p. 417)

invertebrate: an animal without a backbone, such as a sea urchin or sponge. (Chap. 17, p. 490)

joule (J): unit of measure used for work; a joule is equal to one newton-meter. (Chap. 11, p. 293)

kinetic (kih NET ihk) **energy:** the energy that comes from the motion of an object and that depends on the object's mass and speed. (Chap. 12, p. 323)

kingdom: a large group of organisms that share certain features; Earth's species are classified into six kingdoms—Archaebacteria, Eubacteria, Protista, Fungi, Plant, and Animal. (Chap. 2, p. 47)

landfill: an area in which garbage or trash is deposited, covered with a thin layer of dirt, and then watered down to keep the deposited materials from blowing around; 80 percent of U.S. trash and garbage goes to landfills. (Chap. 7, p. 187)

lava: the name given to magma when it erupts through an opening or vent at Earth's surface. (Chap. 16, p. 442)

law of conservation of matter: states that mass is never created or destroyed when things react with one another. (Chap. 8, p. 214)

levees: the natural banks of a river. (Chap. 17, p. 479)

lever: a simple machine made from a bar or rod that is free to rotate around a fulcrum; examples include seesaws and fishing rods. (Chap. 11, p. 299)

life-cycle analysis: a tool to help figure out the environmental impact of a product through its entire life, beginning with getting the natural resources to make the product and ending with disposal of the product in a landfill or by burning. (Chap. 7, p. 200)

light-year: a unit equal to the distance that light travels in a year, which is used to measure the distances between galaxies. (Chap. 14, p. 399)

limiting factor: a factor that controls the size of a population—for example, the availability of food or amount of rainfall. (Chap. 6, p. 159)

luster: a property of minerals that describes how light is reflected from the surface of a mineral—can be either shiny like a metal (metallic) or pearly, dull, or earthy (nonmetallic). (Chap. 15, p. 410)

magma: melted rock in which similar atoms combine as the rock cools, forming minerals whose grains fit together like a puzzle. (Chap. 15, p. 407)

magnet: a device that produces a magnetic field; has two poles that repel or attract and are similar to positive and negative charges. (Chap. 13, p. 359)

magnetic field: the area around a magnet in which magnetic forces act. (Chap. 13, p. 359)

magnitude: the size of something—for example, the size of a wave produced by an earthquake. (Chap. 16, p. 441)

mantle: the thickest layer inside Earth, located beneath the crust and above the outer core. (Chap. 16, p. 449)

mass: the amount of matter in an object. (Chap. 10, p. 266)

mass extinction: a large-scale dying out of many species over a short period of time. (Chap. 5, p. 141)

matter: anything that has mass and takes up space—for example, air; its basic unit is the atom. (Chap. 8, p. 212)

meanders (mee AN durz): wide bends in a stream formed when a stream begins to erode its banks more than the channel bottom. (Chap. 17, p. 477)

meiosis (mi OH sus): the process of sex cell formation during which the chromosome number in each sex cell becomes half the number found in other body cells. (Chap. 4, p. 96)

mesas (MAY sus): flat-topped, tablelike, natural rock structures that rise high above the surrounding land and were formed by the erosion of rock layers. (Chap. 15, p. 424)

metamorphic (met uh MOR fihk) **rock:** a type of rock, either foliated or non-foliated, that is formed when older rocks are heated and squeezed. (Chap. 15, p. 427)

meteorologist: a scientist who studies weather patterns to produce a forecast, or prediction of daily weather. (Chap. 18, p. 512)

mid-ocean ridge: an area where molten rock or magma oozes out onto the ocean floor. (Chap. 16, p. 457)

mineral (MIH nuh rool): an inorganic solid found in nature, having a definite, orderly pattern of atoms that are always arranged in the same way and that have the same chemical makeup throughout. (Chap. 15, p. 406)

mitochondria (mi tuh KAHN dree uh): the sausage-shaped cell organelles in which energy is released from food, providing power for all of the cell's activities. (Chap. 3, p. 70)

mitosis (mi TOH sus): the process of nuclear division that results in two nuclei containing identical information. (Chap. 4, p. 91)

mixture: two or more substances put together that do not combine to form a compound; can be the same throughout or made of several different items that can be seen. (Chap. 8, p. 228)

model: a representation or working version created to understand something that is too big, too small, or too complicated to understand otherwise. (Chap. 8, p. 214)

Modified Mercalli (mur KAH lee) scale: a scale used to measure the damage caused by earthquakes; ranges from a low of 1, where movement is not felt by people, to a high of 12, where buildings are completely destroyed. (Chap. 16, p. 440)

multiple alleles: two or more alleles that control a specific trait in an organism. (Chap. 4, p. 110)

mutation (myew TAY shun): a change in a gene or chromosome resulting from an error in mitosis or meiosis or an environmental factor; adds variation to the genes of a species and can be beneficial, harmful, or neutral. (Chap. 4, p. 111)

natural resources: things found in nature that organisms use to meet their needs; examples include vegetables, trees, oil, and minerals. (Chap. 7, p. 177)

natural selection: process in which organisms that are best suited to their environment survive, reproduce, and pass on their traits to their offspring; a mechanism for species' change over time. (Chap. 5, p. 125)

negatively charged: describes materials whose atoms easily gain electrons. (Chap. 13, p. 348)

neutron (NEW trahn): a particle without a charge that is found in the nucleus of an atom. (Chap. 8, p. 218)

newton (N): unit of force. (Chap. 11, p. 293)

niche (NIHCH): the role, or job, of an organism in an ecosystem. (Chap. 6, p. 164)

non-foliated: a type of metamorphic rock, such as marble and soapstone, that lacks distinct bands of minerals. (Chap. 15, p. 428)

nonrenewable (NAHN ree new uh bul) **resources:** resources, such as coal and oil, that form slowly over long periods of time and cannot be replaced by natural processes within 100 years. (Chap. 7, p. 182)

nucleus (NEW klee us): a cell organelle that contains chromosomes and that controls most of the cell's activities. (Chap. 3, p. 69)

observation: careful watching that can include measurements and descriptions and sketches or drawings of exactly what is seen. (Chap. 1, p. 10)

ore: a mineral that contains something that can be useful and sold for a profit. (Chap. 15, p. 414)

organ (OR gun): in many-celled organisms, a structure made up of two or more different types of tissue that work together to keep an organism alive—for example, the stomach is an organ that has muscle tissue, nerve tissue, and blood tissue. (Chap. 3, p. 80)

organ system: a group of different organs that work together to do a specific job—for example, a digestive system can be made up of a mouth, stomach, intestines, and liver. (Chap. 3, p. 81)

organelle: structure within the cytoplasm in cells that breaks down food, moves waste, and stores materials. (Chap. 3, p. 68)

organism: a living thing that has all five traits of life—it responds, moves, shows organization, reproduces, and grows and develops. (Chap. 2, p. 34)

outer core: the liquid layer inside Earth, located beneath the mantle and above the inner core. (Chap. 16, p. 449)

Pangaea (pan JEE uh): the supercontinent made up of all the continents joined together about 250 million years ago, before they broke apart and slowly moved to their present positions. (Chap. 16, p. 452)

particle: a tiny fragment; the three main particles in an atom are protons, neutrons, and electrons. (Chap. 8, p. 212)

periodic (peer ee AW dihk) **table:** a chart that organizes all known elements by the number of protons in the nucleus of each element and gives information such as the mass of an atom, the pattern of electrons, and whether an element is a solid, liquid, or gas. (Chap. 8, p. 223)

phase: the change in appearance of the moon when seen from Earth, depending on the position of the moon, the sun, and Earth in space. (Chap. 14, p. 380)

photosynthesis (foh toh SIHN thuh sus): the food-making process of plants, green algae, and many types of bacteria; in green plants, this process occurs inside the chloroplasts, where the sun's energy and water and carbon dioxide are converted into food and oxygen. (Chap. 3, p. 72)

physical change: a change in the shape, size, form, or state of matter that can be observed without changing the identity of the matter. (Chap. 9, p. 252)

physical property: a characteristic of matter that can be observed by the senses—for example, shape, size, taste, texture, color, and form. (Chap. 9, p. 238)

plate: a large section of Earth's crust and upper mantle that moves around on the plasticlike layer within the mantle. (Chap. 16, p. 456)

plate tectonics (tek TAW nihks): the theory developed to explain how Earth's plates move, based on evidence of seafloor spreading, convection currents, and Wegener's ideas about continental drift. (Chap. 16, p. 459)

pollutants (puh LEW tuntz): any materials that harm living things by interfering with life processes. (Chap. 7, p. 187); some are released by natural events such as erupting volcanoes, and others are produced by human activities such as burning fossil fuels. (Chap. 18, p. 497)

population: a group of the same type of organisms living in the same place at the same time. (Chap. 6, p. 157)

population density (DEN suh tee): the size of a particular population compared to the size of the area it occupies. (Chap. 6, p. 158)

positively charged: describes materials whose atoms easily lose electrons and as a result are left with more protons than electrons. (Chap. 13, p. 348)

potential (poh TEN shul) **energy:** the energy that is stored and comes from position or condition; can be transformed into kinetic energy when something acts to release the stored energy. (Chap. 12, p. 323)

precipitation (pree sihp uh TAY shun): water droplets that fall from the clouds to Earth—for example, rain and snow. (Chap. 17, p. 469)

predation (pruh DAY shun): the act of one organism feeding on another organism. (Chap. 6, p. 162)

predator (PRED uh tur): an organism that captures and eats other organisms. (Chap. 6, p. 162)

producer: an organism, such as a plant, that is able to use energy from the sun to make its own food. (Chap. 6, p. 168)

properties: the physical and chemical characteristics of matter. (Chap. 9, p. 238)

protein: a type of chemical that is used in the building of bones, muscles, and skin, and helps living things grow, digest food, and fight disease. (Chap. 5, p. 131)

proton (PROH tahn): a positively charged particle located in the nucleus of an atom. (Chap. 8, p. 218)

pulley: a simple machine composed of a surface with a chain or rope going around it and that can be either fixed or movable. (Chap. 11, p. 305)

radiation (ray dee AY shun): the energy that travels by waves in all directions from its source. (Chap. 12, p. 334)

ratio (RAY shee oh): a simple fraction that compares the amounts of two items. (Chap. 8, p. 214)

recycling (ree SIKE ling): reusing materials after they have been changed into another form, which often saves energy, water, and other resources. (Chap. 7, p. 198)

regeneration (ree jen uh RAY shun): the process by which some organisms can reproduce asexually or regrow missing body parts that have been lost because of an injury. (Chap. 4, p. 93)

renewable (ree NEW uh bul) **resources:** resources, such as water and trees, that can be replaced by natural processes in 100 years or less. (Chap. 7, p. 181)

reproduction: process by which organisms make more of their own kind—for example, by giving birth to live offspring, laying eggs, or producing flowers. (Chap. 2, p. 37)

revolution (rev oh LEW shun): the movement of Earth in a regular, curved path around the sun, which takes one year to complete. (Chap. 14, p. 379)

Richter (RICK tur) **scale:** a scale used to measure the amount of energy released by an earthquake. (Chap. 16, p. 440)

rock: a material, such as granite, that is usually composed of two or more minerals. (Chap. 15, p. 406)

rock cycle: a diagram that shows how the formation of igneous, sedimentary, and metamorphic rocks can be interrelated and how rocks are constantly recycled from one kind of rock to another. (Chap. 15, p. 430)

rotation (roh TAY shun): the spinning of Earth on its imaginary axis, which occurs once every 24 hours and causes day and night as Earth rotates toward or away from the sun. (Chap. 14, p. 378)

runoff: water that flows over the land and eventually flows into lakes, rivers, or streams; affected by the intensity and amount of rain, amount of vegetation, slope of the land, and whether the ground is already wet. (Chap. 17, p. 476)

salinity (suh LIH nuh tee): the saltiness of ocean water. (Chap. 17, p. 482)

salts: useful compounds formed along with water when acids and bases react with each other. (Chap. 9, p. 248)

sand: a kind of soil made up of fine particles of rock that have been weathered by glaciers, wind, or water. (Chap. 17, p. 472)

satellite: a spacecraft that is launched into space to gather information and send it back to Earth. (Chap. 14, p. 394)

science: the process of trying to understand the world around you; a set of steps that can be followed to help find out about something or to solve a problem. (Chap. 1, p. 8)

scientific law: a theory about the natural world that has been tested many times and produces the same results. (Chap. 10, p. 275)

screw: an inclined plane wrapped around a rod; this simple machine makes it easier to hold objects together. (Chap. 11, p. 297)

seafloor spreading: the process that forms new ocean crust by magma oozing out onto the ocean floor through a mid-ocean ridge. (Chap. 16, p. 457)

sedimentary rock: a type of rock formed in layers from pieces of other rocks, plant and animal matter, and minerals that evaporate or settle out of solution, usually over a period of thousands to millions of years. (Chap. 15, p. 420)

sediments (SED uh muntz): materials composed of pieces of rock, dissolved minerals, and organic matter that is carried and deposited or dropped by wind, water, ice, or gravity and collects in layers. (Chap. 15, p. 420)

seismograph (SIZE muh graf): an instrument that measures the wave magnitudes of earthquakes and records both faster P-waves (primary waves) and slower S-waves (secondary waves). (Chap. 16, p. 440)

sex cells: specialized cells involved in reproduction; eggs are female sex cells and sperm are male sex cells. (Chap. 4, p. 96)

sexual (SEK shul) **reproduction:** a type of reproduction in which a new organism is produced from two parents whose DNA combines to produce the new individual with its own new DNA. (Chap. 4, p. 96)

shield volcano: a volcano with long, sloping rock layers formed from quiet eruptions of fluid lava. (Chap. 16, p. 443)

silica: a material (liquid glass) found in lava that helps determine how the lava flows—the more silica, the stickier and slower the lava. (Chap. 16, p. 444) silt: a type of loose sedimentary rock made up of tiny rock grains that are slightly larger than clay. (Chap. 17, p. 478)

simple machine: a device that does work with only one movement—changes the size or direction of a force; a pulley, an inclined plane, a wheel and axle, and the three classes of levers. (Chap. 11, p. 296)

soil: the topmost surface layer of Earth in which plants grow and that is made up of a combination of minerals, water, air, and the decaying parts of plants and animals. (Chap. 17, p. 472)

solar (SOH lur) energy (EN ur jee): the energy from the sun, which is a renewable resource that can be used to reduce the need for nonrenewable energy resources such as coal and petroleum. (Chap. 12, p. 338)

solar system: the nine planets (Earth, Venus, Mars, Mercury, Jupiter, Saturn, Uranus, Neptune, and Pluto) and numerous other objects, such as asteroids, that circle the sun. (Chap. 14, p. 384)

solid waste: things that people throw away that are in solid or near-solid forms; examples include old newspapers, old plastic toys, and scrap metal. (Chap. 7, p. 195)

species: a group of organisms that can breed with one another and produce healthy offspring. (Chap. 2, p. 45)

speed: a measure of how far an object moves in a given amount of time. (Chap. 10, p. 269)

spring: a water source that forms where the top of the groundwater layer meets Earth's surface. (Chap. 17, p. 474)

star: dense, massive celestial body composed of gas; produces energy and has a life cycle. (Chap. 14, p. 398)

state of matter: a physical property that describes matter as a solid, a liquid, or a gas. (Chap. 9, p. 241)

static discharge: uncontrolled recombination of electric charges, producing, for example, lightning. (Chap. 13, p. 348)

static electricity: a form of potential electrical energy caused by the buildup or loss of electrons. (Chap. 13, p. 348)

stationary: describes a warm front or a cold front that has stopped moving. (Chap. 18, p. 506)

stratosphere (STRAT uh sfeer): the layer of Earth's atmosphere located above the troposphere; contains the ozone layer, which protects life on Earth from the sun's harmful ultraviolet rays. (Chap. 18, p. 498)

streak test: a test to help identify a mineral that involves scratching a mineral sample across a streak plate and producing a streak of color. (Chap. 15, p. 410)

submersibles (sub MUR suh bulz): underwater craft used to study the oceans. (Chap. 17, p. 485)

subscript: a little number written below and immediately after a letter in a chemical formula that tells how many atoms of an element are present. (Chap. 8, p. 227)

suspended: describes materials that are kept from falling or sinking. (Chap. 17, p. 479)

switch: a device that opens or closes a circuit to stop or start the movement of electrons. (Chap. 13, p. 352)

synthetic (sihn THET ihk): describes elements that are made in particle accelerators and are not found in nature—for example, technetium and neptunium. (Chap. 8, p. 223)

technology: the use of knowledge learned by science; examples include computers, video games, and robots. (Chap. 1, p. 9) temperature: a measure of the kinetic energy of all the atoms of a material; commonly is measured by Celsius and Fahrenheit scales. (Chap. 12, p. 332)

terminals: the positive and negative connection points of a battery. (Chap. 13, p. 355)

texture: describes the size of the mineral crystal and how the crystals fit together. (Chap. 15, p. 407)

thermal (THUR mul) **energy:** the total amount of kinetic energy and potential energy in the atoms that make up a material; can be transferred from one place to another through conduction, convection, and radiation. (Chap. 12, p. 328)

threatened species: a species that is in need of protection but is not in immediate danger of extinction. (Chap. 6, p. 166)

tides: the changes in the level of ocean water during the course of a day—most places have two high tides and two low tides daily; one cause of tides is the pull of gravity between Earth, the moon, and the sun. (Chap. 17, p. 486)

tissue (TIH shew): in many-celled organisms, a group of similar cells that work together to do the same sort of work—for example, bone tissue is made up of just bone cells, and nerve tissue is made up of just nerve cells. (Chap. 3, p. 80)

toxic: describes something poisonous. (Chap. 17, p. 474)

trait: a specific feature of something, such as eye color, hair color, or height. (Chap. 2, p. 34)

transformation: a change of energy from one form to another with the total amount of energy staying the same—no energy is lost or gained. (Chap. 12, p. 321)

trilobites (TRI luh bites): the extinct relatives of today's lobsters, crabs, and insects. (Chap. 5, p. 130)

tropical rain forest: an environment with high biodiversity due to warm and steady temperatures and high rainfall, and which provides many resources for living organisms. (Chap. 5, p. 119)

troposphere (TROH puh sfeer): the lowest layer of Earth's atmosphere, which contains about 75 percent of the air we breathe and also is where weather, clouds, and air pollution are found. (Chap. 18, p. 498)

turbine (TUR bun): a large wheel that rotates and provides mechanical energy to a generator. (Chap. 13, p. 366)

upwarped mountains: mountains formed when Earth's crust was stretched and pushed up by forces inside Earth. (Chap. 16, p. 461)

vaccine: a preparation made from dead or weak viruses that is given by mouth or by a needle and which causes the body to make substances that resist particular viruses. (Chap. 2, p. 55)

variable: a factor that can change in an experiment. (Chap. 1, p. 19)

variations (vayr ee AY shuns): the different ways a certain inherited trait appears—for example, differences in height. (Chap. 4, p. 110)

virus (VI rus): a disease-causing particle that has some things in common with both living and nonliving things—it reproduces (in living cells) but it doesn't eat, grow, or respond to its environment. (Chap. 2, p. 54)

volcanic mountain: a mountain that forms when one plate is pushed beneath another plate, and as the plate melts, the magma rises and a volcano is created. (Chap. 16, p. 461)

volcano: a place where hot, liquid magma is forced up through the surface by pressure inside Earth and flows onto the ground; occurs most commonly along the boundaries of Earth's plates. (Chap. 16, p. 442)

voltage: a measure of the potential energy difference between the ends of a battery. (Chap. 13, p. 354)

volume: the amount of space taken up by matter. (Chap. 9, p. 239)

waning: the change in the moon's appearance when it seems to be getting smaller from night to night. (Chap. 14, p. 380)

water cycle: the constant cycling of water between Earth's surface and the atmosphere through evaporation, condensation, and precipitation, which is powered by energy from the sun and forms the basis for Earth's weather. (Chap. 17, p. 469)

wave: a movement of ocean water in which the water particles move in a circular path that makes the water rise and fall. (Chap. 17, p. 484)

waxing: the change in the appearance of the moon when it appears to be getting larger from night to night. (Chap. 14, p. 380)

weather: what is happening in the atmosphere right now—for example, rain, snow, clouds, and blue skies. (Chap. 18, p. 503)

weathering: the chemical and physical changes that alter the surface of Earth over time. (Chap. 9, p. 254)

wedge: an inclined plane with one or two sloping sides; examples of this simple machine include knives and chisels. (Chap. 11, p. 298)

weight: a measure of how much Earth's gravity pulls down on an object. (Chap. 10, p. 266)

wetland: a natural habitat whose soil contains much moisture and that may be completely or partially covered by water at different times of the year. (Chap. 7, p. 186)

wheel and axle: a simple machine that is made with two wheels of different sizes that are connected and turn together—for example, doorknobs and fishing reels. (Chap. 11, p. 304)

wind: the movement of air from warm, high-pressure areas into cooler, low-pressure areas due to differences in pressure. (Chap. 18, p. 504)

work: exerting a force on an object over a distance in the same direction as the object's motion. (Chap. 11, p. 292)

Glossary/Glosario

Este glosario define cada término clave que aparece en **negrillas** en el texto. También muestra el número de página en donde puedes encontrar la palabra usada.

abiotic factor / factor abiótico Cualquier cosa inanimada en un ecosistema. Los factores abióticos afectan el tipo y número de organismos que pueden sobrevivir en un ecosistema. (Cap. 6, pág. 153)

acceleration / aceleración Cambio en velocidad o dirección. (Cap. 10, pág. 272)

acid rain / lluvia ácida Ocurre cuando los gases que se liberan de la combustión del petróleo y del carbón se mezclan con agua en el aire, produciendo nieve o lluvia ácida. Este tipo de precipitación causa mucha contaminación y puede matar plantas y peces. (Cap. 7, pág. 193)

adaptation / adaptación Cualquier forma, proceso corporal o comportamiento que le permite a un organismo sobrevivir en su ambiente y llevar a cabo los procesos vitales. (Cap. 5, pág. 120)

air mass / masa de aire Gran cantidad de aire que tiene las mismas propiedades del área sobre la cual se forma y se mueve. (Cap. 18, pág. 505)

allele / alelo Diferente forma que presentan los genes que controlan un rasgo. (Cap. 4, pág. 106)

alternating current / corriente alterna Tipo de electricidad que usan los artefactos electrodomésticos, en el cual la electricidad viaja en una dirección por un rato y luego cambia de dirección. La dirección se alterna en un patrón regular. (Cap. 13, pág. 358)

Archimedes / Arquímedes Matemático, filósofo e inventor griego, que hace más de 2000 años propuso un sistema de bombeo de agua que usaba máquinas compuestas. (Cap. 11, pág. 310)

artesian well / pozo artesiano Tipo de pozo en que el agua fluye naturalmente hacia la superficie sin la ayuda de una bomba. (Cap. 17, pág. 474)

asexual reproduction / reproducción asexual Método de reproducción en que se crea un nuevo organismo a partir de un solo progenitor. En la reproducción asexual todo el DNA del organismo proviene de un solo progenitor. (Cap. 4, pág. 92)

asteroid / asteroide Pedazo de roca. Está compuesto de minerales similares a los que formaron los planetas. (Cap. 14, pág. 388)

astronomical unit (AU) / unidad astronómica (UA) Unidad que se usa para medir distancias en el espacio. Equivale a 150 millones de kilómetros, lo cual es la distancia promedio de la Tierra al sol. (Cap. 14, pág. 385)

atmosphere / atmósfera Capa de aire que rodea a la Tierra como si fuera una manta. El gas más común en la atmósfera es el nitrógeno. (Cap. 18, pág. 496)

atom / átomo Pequeña partícula que compone casi todo tipo de materia. (Cap. 8, pág. 212)

average speed / rapidez promedio Describe el movimiento de un viaje completo. (Cap. 10, pág. 270)

axis / eje Línea imaginaria, ligeramente inclinada, sobre la cual gira nuestro planeta. (Cap. 14, pág. 379)

balance / báscula Aparato que sirve para medir la masa. (Cap. 10, pág. 267)

bank / ribera Zona a ambos lados de un río. (Cap. 17, pág. 479)

battery / batería Fuente de energía en circuitos, la cual funciona mediante la separación química de las cargas positivas y negativas. (Cap. 13, pág. 354)

biochemical rock / roca bioquímica Tipo de roca que se forma cuando los materiales vivos se mueren, se apilan y se comprimen formando rocas. La tiza y el carbón son dos ejemplos. (Cap. 15, pág. 422)

biodiversity / biodiversidad Medida del número de diferentes especies en un área. Un bosque pluvial tropical tiene una biodiversidad alta, pero un desierto tiene una biodiversidad baja. (Cap. 5, pág. 119)

biosphere / biosfera La parte de la Tierra donde los organismos pueden vivir. Incluye la parte superior de la corteza terrestre, todos los océanos, ríos y lagos y la atmósfera que los rodea. La biosfera está compuesta de la combinación de todos los ecosistemas de la Tierra. (Cap. 6, pág. 151)

biotic factor / factor biótico Organismo que conforma la parte viviente de un ecosistema. (Cap. 6, pág. 152)

captive breeding / reproducción en cautiverio Apareamiento y reproducción en cautiverio de especies en peligro de extinción. (Cap. 5, pág. 134)

cell / célula Unidad básica de la vida en un ser viviente. (Cap. 2, pág. 36)

cell membrane / membrana celular Estructura flexible que mantiene unida a la célula y le da forma. (Cap. 3, pág. 68)

cellular respiration / respiración celular Proceso en el cual el alimento y el oxígeno se mezclan para formar dióxido de carbono y agua y así liberar energía. Este proceso ocurre dentro de la mitocondria. (Cap. 3, pág. 70)

chemical change / cambio químico Cambio que se produce cuando un material se convierte en otro con propiedades diferentes. Por ejemplo, el cambio que sufre una moneda de centavo brillante al volverse opaca y deslustrada. (Cap. 9, pág. 255)

chemical formula / fórmula química Se construye con símbolos o números que indican qué elementos hay en un compuesto y en qué razón se encuentran dichos elementos. (Cap. 8, pág. 227)

chemical property / propiedad química Característica de una sustancia que le permite convertirse en una nueva sustancia. La capacidad que tiene la madera de quemarse es un ejemplo de propiedad química. (Cap. 9, pág. 243)

chloroplast / cloroplasto Organelo verde que atrapa energía de los rayos solares y la convierte en alimento mediante el proceso de fotosíntesis. (Cap. 3, pág. 72)

cinder cone volcano / volcán de cono de carbonilla Tipo de volcán formado de erupciones explosivas de lava, las cuales pueden oírse a muchos kilómetros de distancia. (Cap. 16, pág. 443)

cinders / carbonilla Pequeños trozos de lava sólida. (Cap. 16, pág. 443)

circuit / circuito Trayecto completo y cerrado que siguen los electrones. Por lo general, este trayecto es a través de alambres metálicos. (Cap. 13, pág. 351)

classification / clasificación Agrupamiento de objetos o información basándose en rasgos comunes. (Cap. 2, pág. 45)

cleavage / crucero Característica de un mineral que hace que al romperse, los pedazos del mineral presenten superficies uniformes, suaves y con bordes regulares. (Cap. 15, pág. 409)

climate / clima Patrón del tiempo que ocurre en un área en particular, durante muchos años. (Cap. 18, pág. 514)

cloning / clonación Significa la creación de un organismo, el cual contiene exactamente el mismo DNA que otro organismo. (Cap. 4, pág. 100)

comet / cometa Masa enorme compuesta de hielo, polvo y gases congelados. (Cap. 14, pág. 386)

common ancestor / antepasado común Individuo del cual provienen otros individuos de varias especies diferentes. Los individuos de cada especie poseen sus propios rasgos específicos y no pueden producir progenie al aparearse con los miembros de las otras especies, pero todos los miembros de las especies comparten algunos rasgos que heredaron del antepasado común. (Cap. 5, pág. 129)

community / comunidad Todas las poblaciones que viven en un área. Los miembros de una comunidad dependen mutuamente de los alimentos, refugio y otras necesidades. (Cap. 6, pág. 158)

composite volcano / volcán compuesto Tipo de volcán alto y escarpado compuesto de capas de lava viscosa y cenizas.

Los volcanes compuestos se forman de erupciones sin ruidos y explosivas, alternativamente. (Cap. 16, pág. 443)

compound / compuesto Forma de materia que resulta de la combinación de dos o más elementos. (Cap. 8, pág. 225)

compound machine / máquina compuesta Dos o más máquinas simples que funcionan juntas. (Cap. 11, pág. 307)

condensation / condensación Proceso en el cual el agua pasa nuevamente al estado líquido a medida que se enfría el vapor de agua. (Cap. 17, pág. 469)

conduction / conducción Transmisión de energía cinética de una molécula a otra. (Cap. 12, pág. 334)

conglomerate / conglomerado Roca sedimentaria formada por la consolidación de guijarros y otros sedimentos. (Cap. 15, pág. 421)

constelation / constelación Grupo de estrellas que forma un patrón en el firmamento. (Cap. 14, pág. 396)

consumer / consumidor Organismo que se alimenta de otros organismos vivos. (Cap. 6, pág. 168)

continental drift / deriva continental Movimiento de los continentes. (Cap. 16, pág. 452)

convection / convección Proceso de transmisión de energía térmica dentro de un líquido o un gas mediante el movimiento de moléculas de un lugar a otro. (Cap. 12, pág. 336)

convection current / corriente de convección Ciclo en que el magma caliente se enfría y se vuelve más denso a medida que asciende hacia la superficie. Luego, este magma más frío y denso comienza a hundirse de nuevo en el manto, en donde es calentado nuevamente. Este ciclo se repite continuamente. (Cap. 16, pág. 458)

crust / corteza Capa externa de la Tierra. La corteza es firme y parece sólida, pero está dividida en trozos más grandes. (Cap. 16, pág. 438)

crystal / cristal Cuerpo sólido que tiene una estructura atómica con un patrón repetitivo. (Cap. 15, pág. 408)

curator / conservador Persona encargada de cuidar algo. (Cap. 6, pág. 165)

current / corriente Masa de agua que se mueve en una dirección. (Cap. 17, pág. 483)

decomposer / descomponedor Organismo que se alimenta de organismos muertos y de los materiales de desecho de otros organismos. (Cap. 6, pág. 168)

density / densidad Relación de la masa de un objeto con la cantidad de espacio que ocupa. (Cap. 9, pág. 239)

deposition / depositación Proceso en que el agua en movimiento deposita sedimentos. Es una de las maneras en que los ríos y arroyos cambian el relieve de la Tierra. (Cap. 17, pág. 475)

development / desarrollo Todos los cambios durante la vida de un organismo. (Cap. 2, pág. 38)

direct current / corriente directa Tipo de electricidad que viaja en una sola dirección y que permite el funcionamiento de una batería. (Cap. 13, pág. 358)

diversity / diversidad La variedad de la vida sobre la Tierra. (Cap. 5, pág. 118)

DNA / DNA Sustancia química que se encuentra en el núcleo de casi todas las células. Es el material del cual están hechos los cromosomas. Toda la información en el DNA de los cromosomas se llama información genética. (Cap. 4, pág. 103)

earthquake / terremoto Estremecimiento de la Tierra causado por un desatamiento rápido de energía. La energía viaja en forma de ondas a través de la Tierra y la cantidad de daño depende del tamaño de las ondas. (Cap. 16, pág. 438)

eclipse / eclipse Acontecimiento en que un cuerpo luminoso proyecta una sombra sobre otro cuerpo luminoso. Un eclipse puede ser lunar o solar. (Cap. 14, pág. 382)

ecologist / ecólogo Persona que se dedica al estudio de la ecología. (Cap. 6, pág. 151)

ecology / ecología Estudio de la relación entre los organismos vivientes y las partes inanimadas de un ecosistema. (Cap. 6, pág. 151)

ecosystem / ecosistema Está compuesto de organismos que interaccionan entre sí y con los factores inanimados del ambiente para formar una unidad funcional. (Cap. 6, pág. 148)

electrical energy / energía eléctrica Energía de las cargas eléctricas en movimiento. (Cap. 13, pág. 346)

electricity / electricidad Tipo de energía que depende del movimiento de electrones de un lugar a otro y que se usa para hacer funcionar artefactos, como por ejemplo, un reproductor de discos compactos. (Cap. 13, pág. 346)

electromagnet / electroimán Tipo de imán que se puede activar y desactivar. Se hace al enrollar un alambre que conduce corriente eléctrica alrededor de un núcleo de hierro. (Cap. 13, pág. 362)

electron / electrón Partícula del átomo la cual posee carga eléctrica negativa. (Cap. 8, pág. 219)

element / elemento Materia formada por un solo tipo de átomo. (Cap. 8, pág. 223)

embryo / embrión Huevo fecundado que ha comenzado a dividirse en más células. (Cap. 4, pág. 100)

endangered species / especie en peligro de extinción Se considera que una especie está en peligro de extinción, cuando la cantidad de miembros vivos es tan mínima que la especie completa puede desaparecer. (Cap. 6, pág. 166)

energy / energía Capacidad para causar cambio. La energía puede cambiar la temperatura, la forma, la velocidad, la posición o la dirección de un objeto. (Cap. 12, pág. 320)

environment / ambiente Todo lo que rodea a un organismo. Incluye a otros organismos, el agua, el tiempo, la temperatura, el tipo de suelo, las rocas, los sonidos y la luz; es decir, cualquier cosa con la que el organismo entre en contacto. (Cap. 2, pág. 35)

epicenter / epicentro Punto donde las ondas chocan contra la superficie terrestre, directamente sobre el foco de un terremoto. A menudo el epicentro es el lugar donde ocurre el peor daño, en un terremoto. (Cap. 16, pág. 439)

erosion / erosión Ocurre cuando se afloja el suelo o las rocas y son transportados de un lugar a otro. (Cap. 17, pág. 476)

evaporation / evaporación Proceso que convierte gradualmente el agua del estado líquido al estado gaseoso. (Cap. 17, pág. 468)

extintion / extinción La desaparición de una especie completa. (Cap. 5, pág. 141)

extrusive igneus rock / roca ígnea extrusiva Roca ígnea que se forma cuando la lava se enfría sobre la superficie terrestre. Las rocas ígneas intrusivas poseen cristales pequeños. (Cap. 15, pág. 416)

fault / falla Especie de grieta en la corteza, entre dos placas, donde ha habido movimiento. (Cap. 16, pág. 439)

fault-block mountain / montaña de bloque de falla Se forma cuando secciones completas de roca se convierten en fallas y se rompen; y algunas de estas secciones se elevan para formar una montaña. (Cap. 16, pág. 461)

fertilization / fecundación Proceso en el cual un óvulo y un espermatozoide se unen para formar un nuevo individuo, con un juego completo de 46 cromosomas. (Cap. 4, pág. 97)

fissure / grieta Hendidura de gran tamaño en la corteza terrestre. (Cap. 15, pág. 416)

floodplain / llanura aluvial Área que por lo general se inunda durante períodos de mucha precipitación. La llanura aluvial está formada por los sedimentos depositados por el río. (Cap. 17, pág. 479)

focus / foco Punto exacto donde comienza un terremoto. (Cap. 16, pág. 439)

folded mountain / montaña plegada Tipo de montaña que se forma del choque de las placas. (Cap. 16, pág. 461)

foliated rock / roca foliada Tipo de roca que presenta bandas de minerales que han sido calentados y compactados formando capas paralelas. (Cap. 15, pág. 428)

force / fuerza Cualquier empuje o atracción. Por ejemplo, la gravedad y la fricción son fuerzas. (Cap. 10, pág. 275)

fossil / fósil Resto o huella de una vida antigua. Provee a los científicos pruebas directas de que las especies cambian con el paso del tiempo. (Cap. 5, pág. 130)

fossil fuel / combustible fósil Restos de plantas y animales antiguos que podemos quemar para producir energía. (Cap. 15, pág. 424)

fossil record / récord fósil Todos los fósiles que los científicos han recobrado del suelo. Los fósiles de casi cada grupo principal de plantas o animales forman parte del récord fósil. (Cap. 5, pág. 136)

fracture / fractura Característica de un mineral que hace que al romperse, los pedazos del mineral presenten superficies desiguales, con bordes toscos e irregulares. (Cap. 15, pág. 409)

freshwater / agua fresca Agua que utilizamos para beber, cocinar y otras actividades. Es el agua que necesitamos para sobrevivir. (Cap. 17, pág. 467)

friction / fricción Es el resultado de movimientos opuestos entre dos superficies en contacto. (Cap. 10, pág. 273)

front / frente Lugar donde las masas de aire se juntan. Las interacciones entre las masas de aire ocasionan las condiciones del tiempo en el lugar del frente. (Cap. 18, pág. 506)

fulcrum / fulcro Punto de apoyo de una palanca. (Cap. 11, pág. 299)

genetics / genética Ciencia que estudia la manera en que los padres transfieren sus rasgos a la progenie. (Cap. 4, pág. 102)

geologic time scale / escala del tiempo geológico Especie de diario de la vida sobre la Tierra. Ayuda a los científicos a mantener un registro de cuándo una especie apareció y desapareció de la faz de la Tierra. Esta escala está dividida en cuatro largos lapsos de tiempo llamados eras y cada era está subdividida en períodos. (Cap. 5, pág. 138)

global warming / calentamiento global Aumento en las temperaturas en todo el globo terráqueo. (Cap. 18, pág. 517)

gravity / gravedad Fuerza de atracción mutua que existe entre todos los objetos. (Cap. 10, pág. 264)

ground water / agua subterránea Agua que absorbe la superficie terrestre y que se va acumulando en pequeños espacios entre las piedras y el suelo. (Cap. 17, pág. 471)

galaxy / galaxia Conjunto de estrellas, gases y polvo que se mantienen unidos debido a la gravedad. (Cap. 14, pág. 399)

gem / gema Mineral raro que se puede cortar y pulir dándole una apariencia bella. Son ideales para la joyería. (Cap. 15, pág. 413)

gene / gene Pequeña sección de un cromosoma, que determina un rasgo. (Cap. 4, pág. 103)

generator / generador Máquina que convierte la energía mecánica en energía eléctrica. (Cap. 13, pág. 366)

habitat / hábitat Lugar donde vive un organismo. Diferentes especies comparten un hábitat. Los organismos comparten los recursos del hábitat, tales como alimentos, espacio y refugio. (Cap. 6, pág. 164)

heat / calor Energía térmica que se mueve desde un objeto caliente a uno más frío. (Cap. 12, pág. 329)

hypothesis / hipótesis Enunciado que se puede comprobar acerca de un problema. (Cap. 1, pág. 18)

igneus rock / roca ígnea Tipo de roca que se forma cuando la roca derretida o magma se enfría dentro de la Tierra. (Cap. 15, pág. 415)

inclined plane / plano inclinado Superficie inclinada que se usa para facilitar el traslado de la carga cuesta arriba. (Cap. 11, pág. 296)

inertia / inercia Resistencia al cambio de movimiento. Es decir, la tendencia de los objetos en reposo de permanecer en reposo y de los objetos en movimiento de continuar moviéndose en línea recta hasta que una fuerza actúe sobre ellos. (Cap. 10, pág. 276)

inference / inferencia Explicación, con base en la observación, de por qué sucedió algo. (Cap. 1, pág. 10)

inner core / núcleo interno La capa sólida más interna de la Tierra. (Cap. 16, pág. 449)

inorganic substance / sustancia inorgánica Sustancia que no está compuesta ni de material vegetal ni animal. (Cap. 15, pág. 406)

intrusive igneus rock / roca ígnea intrusiva Roca ígnea que se forma cuando el magma se enfría dentro de la corteza, en lugar de enfriarse sobre la superficie terrestre. Las rocas ígneas intrusivas poseen cristales grandes. (Cap. 15, pág. 417)

joule (J) / julio (J) Equivale a un newton-metro (N·m). (Cap. 11, pág. 293)

kinetic energy / energía cinética Energía que posee un objeto en movimiento. Dicha energía depende de la masa y de la velocidad del objeto. (Cap. 12, pág. 323)

kingdom / reino Grupo de gran tamaño formado por organismos que comparten ciertos rasgos comunes. (Cap. 2, pág. 47)

landfill / vertedero controlado Área donde se deposita la basura. (Cap. 7, pág. 187)

lava / lava Magma que ha llegado a la superficie terrestre. (Cap. 16, pág. 442)

law of conservation of matter / ley de conservación de la materia Ley que dice que la materia no puede ser ni creada ni destruida cuando una sustancia reacciona con otra. (Cap. 8, pág. 214)

lever / palanca Barra que gira libremente alrededor de un punto de apoyo llamado fulcro. (Cap. 11, pág. 299)

life-cycle analysis / análisis del ciclo de vida Una forma de averiguar el impacto de un producto, durante su existencia, en el ambiente. (Cap. 7, pág. 200)

light-year / año-luz Distancia que la luz viaja en un año. (Cap. 14, pág. 399)

limiting factor / factor limitante Factor que limita el crecimiento de una población; como por ejemplo, la cantidad de lluvia o alimento. (Cap. 6, pág. 159)

luster / lustre Manera en que la luz se refleja desde la superficie de un mineral. (Cap. 15, pág. 410)

magma / magma Roca caliente derretida. (Cap. 15, pág. 407)

magnet / imán Tipo de material que posee cierto número de electrones arreglados en ciertas maneras. La mayoría de los electrones apuntan en la misma dirección. (Cap. 13, pág. 359)

magnetic field / campo magnético Área alrededor de un imán donde actúan las fuerzas magnéticas. (Cap. 13, pág. 359)

magnitude / magnitud La fuerza destructiva de un terremoto. (Cap. 16, pág. 441)

mantle / manto La capa más gruesa de la Tierra. Posee características parecidas al plástico y puede moverse o fluir como si fuera brea caliente. (Cap. 16, pág. 449)

mass / masa Cantidad de materia en un objeto. La masa no cambia de un lugar a otro. (Cap. 10, pág. 266)

mass extinction / extinción masiva La desaparición en gran escala de muchas especies en un corto período de tiempo. (Cap. 5, pág. 141)

matter / materia Término que se usa para describir cualquier cosa que tenga masa y que ocupe espacio. Todo aquello que puedas tocar, saborear u oler es materia. (Cap. 8, pág. 212)

meander / meandro Serpenteo ancho que forman las corrientes de agua cuando no pueden adquirir más profundidad. La corriente de agua erosiona la parte de afuera del meandro y deposita los sedimentos en la parte de adentro. (Cap. 17, pág. 477)

meiosis / meiosis Proceso de la formación de células sexuales. Durante la meiosis, el número de cromosomas en cada célula es disminuido a la mitad. (Cap. 4, pág. 96)

mesa / meseta Estructura plana y rocosa con forma de mesa, más elevada que el resto del terreno que la rodea. (Cap. 15, pág. 424)

metamorphic non-foliated rock / roca metamórfica no foliada Tipo de roca que no presenta capas o bandas distintivas. (Cap. 15, pág. 428)

metamorphic rock / roca metamórfica Roca que se forma cuando las rocas más antiguas se calientan o se compactan. La palabra metamórfica significa "cambiar de forma". (Cap. 15, pág. 427)

meteorologist / meteorólogo Científico que estudia los patrones del tiempo para pronosticar el estado del tiempo diariamente. (Cap. 18, pág. 512)

mid-ocean ridges / dorsal medioceánica Áreas del océano por donde salen rocas derretidas o magma. Estas áreas tienen unas brechas o valles en el centro, en donde se forma el nuevo suelo oceánico a medida que se enfría la roca derretida. (Cap. 16, pág. 457)

mineral / mineral Sustancia química como el sodio y el cloro, que se encuentran en el aire, en el suelo y en el agua. (Cap. 2, pág. 44)

mineral / mineral Material sólido inorgánico que se encuentra en la naturaleza. Cada mineral posee rasgos o características que puedes usar para identificarlo. (Cap. 15, pág. 406)

mitochondrion / mitocondria Organelo que provee la energía necesaria para las funciones celulares. (Cap. 3, pág. 70)

mitosis / mitosis Proceso de división del núcleo que resulta en la formación de dos núcleos, los cuales poseen exactamente la misma información genética. (Cap. 4, pág. 91)

mixture / mezcla Resulta de juntar dos o más sustancias que no se combinan para formar un compuesto. (Cap. 8, pág. 228)

model / modelo Versión pequeña de algo más grande. (Cap. 8, pág. 214)

Modified Mercali scale / escala modificada Mercali Escala con que se mide el daño causado por un terremoto. Un 1 en esta escala indica que el terremoto no ha causado casi ningún daño. Un 12 indica que el terremoto ha causado la destrucción total de los edificios. (Cap. 16, pág. 440)

moon phases / fases lunares Cambios en la apariencia de la luna. Dependen de la posición de la luna en relación con el sol y con la Tierra. (Cap. 14, pág. 380)

mutation / mutación Cambio en un gene o en un cromosoma debido a un error en la meiosis o en la mitosis, o debido a un factor ambiental. Muchas mutaciones ocurren aleatoriamente, pero algunas otras son causadas por influencias externas, tales como los rayos X o las sustancias químicas peligrosas en el ambiente. (Cap. 4, pág. 111)

natural resource / recurso natural Todo lo que se encuentre en la naturaleza y que usen los seres vivos. (Cap. 7, pág. 177)

natural selection / selección natural Proceso en que los organismos que poseen rasgos que los hacen más aptos para un ambiente sobreviven, se reproducen y pasan esos rasgos a su progenie. (Cap. 5, pág. 125)

negative charge / carga eléctrica negativa Ocurre cuando un material gana electrones. (Cap. 13, pág. 348)

neutron / neutrón Partícula del átomo, la cual no posee carga eléctrica. (Cap. 8, pág. 218)

newton (N) / newton (N) Unidad de medida para la fuerza. (Cap. 11, pág. 293)

niche / nicho Papel que juega un organismo en su ecosistema. (Cap. 6, pág. 164)

nonrenewable resource / recurso no renovable Recurso que no se puede reemplazar mediante procesos naturales en un período de 100 años o menos. (Cap. 7, pág. 182)

nucleus / núcleo Organelo que controla la mayor parte de las funciones celulares. (Cap. 3, pág. 69)

nucleus / núcleo Centro del átomo. Constituye casi toda la masa del átomo. (Cap. 8, pág. 218)

observation / observación Acción de mirar algo cuidadosamente para poder anotar exactamente lo que ocurre. (Cap. 1, pág. 10)

ore / mena Cualquier mineral que contiene algo que pueda ser útil y que se pueda vender para obtener una ganancia. Por ejemplo, el hierro que se usa para hacer acero proviene de la mena del mineral hematita. (Cap. 15, pág. 414)

organ / órgano Estructura formada por dos o más tipos de tejidos que trabajan conjuntamente. (Cap. 3, pág. 80)

organ system / sistema de órganos Grupo de órganos que trabajan conjuntamente para realizar cierta función. (Cap. 3, pág. 81)

organelle / organelo Parte de la célula que realiza funciones especializadas. Todos los organelos de la célula se mueven dentro del citoplasma. (Cap. 3, pág. 68)

organism / organismo Ser viviente que presenta todos los rasgos de la vida, tales como ser capaz de responder a estímulos, de moverse, de producir progenie, de crecer y de desarrollarse, además de mostrar organización,. (Cap. 2, pág. 34)

outer core / núcleo externo Capa líquida que rodea el núcleo interno de la Tierra. (Cap. 16, pág. 449)

pangaea / pangaea Idea propuesta por el científico alemán Alfred Wegener, quien propuso en 1912 que todos los continentes estuvieron una vez unidos formando un supercontinente, que al separarse formó los continentes actuales. (Cap. 16, pág. 452)

particle / partícula Parte diminuta que forma la materia. (Cap. 8, pág. 212)

periodic table / tabla periódica Tabla que organiza los elementos de acuerdo con el número de protones en el núcleo de cada elemento. (Cap. 8, pág. 223)

photosyntesis / fotosíntesis Proceso mediante el cual las plantas, las algas verdes y otros tipos de bacterias atrapan la energía solar y la convierten en alimento. (Cap. 3, pág. 72)

physical change / cambio físico Cualquier cambio de un material en tamaño, forma o estado físico, en que la identidad de la materia no cambia. (Cap. 9, pág. 252)

physical property / propiedad física Término que usan los científicos para describir todas las características de la materia que se puedan detectar con los sentidos. (Cap. 9, pág. 238)

plate / placa Cada una de las 12 secciones enormes que forman la parte superior del manto y la corteza terrestre. Las placas pueden moverse porque descansan sobre una capa del manto que parece plástico. (Cap. 16, pág. 456)

plate tectonics / tectónica de las placas Teoría que resultó de la combinación de ideas sobre la deriva continental y el desplazamiento del fondo oceánico y la cual explica el movimiento de las placas. (Cap. 16, pág. 459)

pollutant / contaminante Todo material que pueda causar daño a los seres vivos, al interferir con los procesos vitales. (Cap. 7, pág. 187) (Cap. 18, pág. 497)

population / población Grupo del mismo tipo de organismos que vive en el mismo lugar, al mismo tiempo. (Cap. 6, pág. 157)

population density / densidad demográfica Comparación del tamaño de una población con relación al tamaño del área en que vive. (Cap. 6, pág. 158)

positively charged / cargado positivamente Ocurre cuando un material pierde electrones y se queda con más protones que electrones. (Cap. 13, pág. 348)

potential energy / energía potencial Energía almacenada que posee un objeto en reposo. Esta energía depende de la posición o condición del objeto. (Cap. 12, pág. 323)

precipitation / precipitación Caída a la Tierra de las gotas grandes de agua suspendidas en las nubes. La lluvia, la nieve, el granizo y el aguanieve son todos ejemplos de precipitación. (Cap. 17, pág. 469)

predation / predación Acto mediante el cual un organismo se alimenta de otro organismo. (Cap. 6, pág. 162)

predator / predador Animal que captura y come otros animales. (Cap. 6, pág. 162)

producer / productor Organismo que produce su propio alimento, como por ejemplo una planta. (Cap. 6, pág. 168)

property / propiedad Característica que describe alguna cosa. (Cap. 9, pág. 238)

protein / proteína Sustancia química que lleva a cabo una variedad de funciones en los seres vivos. Algunas proteínas se usan en la construcción de material viviente, como los músculos, los huesos y la piel. Otras proteínas ayudan a los seres vivos a crecer, a digerir los alimentos y a combatir enfermedades. (Cap. 5, pág. 131)

proton / protón Partícula del átomo, la cual posee carga eléctrica positiva. (Cap. 8, pág. 218)

pulley / polea Superficie con una cuerda o cadena a su alrededor. (Cap. 11, pág. 305)

radiation / radiación Energía que viaja, desde su fuente, en todas direcciones por medio de ondas. (Cap. 12, pág. 334)

ratio / razón Fracción simple que compara las cantidades de dos artículos. (Cap. 8, pág. 214)

recycling / reciclaje Significa volver a usar un material después de que dicho material ha sido convertido en un material diferente. (Cap. 7, pág. 198)

regeneration / regeneración Proceso mediante el cual algunos organismos son capaces de reemplazar las partes corporales perdidas debido a lesiones. (Cap. 4, pág. 93)

renewable resource / recurso natural renovable Recurso que se puede reemplazar mediante procesos naturales en un período de 100 años o menos. (Cap. 7, pág. 181)

reproduction / reproducción La capacidad de un organismo de producir progenie. (Cap. 2, pág. 37)

revolution / traslación Movimiento de la Tierra alrededor del sol. La Tierra demora un año en trasladarse, una vez, alrededor del sol. (Cap. 14, pág. 379)

Richter scale / escala Richter Escala con que se mide la energía liberada durante un terremoto. Esta escala mide el tamaño de las ondas producidas por un terremoto. (Cap. 16, pág. 440)

rock / roca Sustancia hecha de dos o más minerales. (Cap. 15, pág. 406)

rock cycle / ciclo de las rocas Proceso de cambio en el cual un tipo de roca se convierte en otro tipo de roca. (Cap. 15, pág. 430)

rotation / rotación Movimiento de la Tierra sobre su eje. Cada rotación dura 24 horas. (Cap. 14, pág. 378)

runoff / agua de escorrentía Agua que fluye sobre la superficie terrestre y, que a la larga, desemboca en los ríos, arroyos o lagos. El agua de escorrentía es una de las causas de la erosión. (Cap. 17, pág. 476)

salinity / salinidad Medida de la cantidad de sal en el agua. (Cap. 17, pág. 482)

salt / sal Compuesto que resulta de la reacción entre un ácido y una base. (Cap. 9, pág. 248)

sand / arena tipo de suelo compuesto de partículas finas de rocas que han sido meteorizadas por los glaciares, el viento o el agua. (Cap. 17, pág. 472)

satellite / satélite Nave espacial lanzada al espacio que recoge información y la transmite a la Tierra. (Cap. 14, pág. 394)

science / ciencia Es el proceso mediante el cual tratamos de comprender el mundo que nos rodea. Esto quiere decir que la ciencia es un conjunto de pasos que podemos seguir para averiguar más acerca de algo o para resolver un problema. (Cap. 1, pág. 8)

scientific law / ley científica Ley que resulta después de que una teoría ha sido probada muchas veces y ha dado los mismos resultados. Es una descripción exacta de algo importante en la naturaleza. (Cap. 10, pág. 275)

screw / tornillo Plano inclinado enrollado alrededor de una barra. (Cap. 11, pág. 297)

seafloor spreading / desplazamiento del fondo oceánico Proceso mediante el cual se forma el nuevo suelo oceánico. A medida que el magma llega a la superficie, este empuja y separa los dos lados de la dorsal. (Cap. 16, pág. 457)

sediment / sedimento Pedazos de rocas y otros materiales arrastrados por los ríos, los océanos, las olas, los glaciares y el viento. (Cap. 15, pág. 420)

sedimentary rock / roca sedimentaria Tipo de roca que se forma cuando los pedazos de rocas, plantas y animales o minerales disueltos se acumulan formando capas rocosas. (Cap. 15, pág. 420)

seismograph / sismógrafo Aparato que registra la magnitud de las ondas de un terremoto. (Cap. 16, pág. 440)

sex cell / célula sexual Célula especializada que está involucrada en la reproducción. Las células sexuales femeninas se llaman óvulos y las masculinas se llaman espermatozoides. (Cap. 4, pág. 96)

sexual reproduction / reproducción sexual Método de reproducción en que dos progenitores producen un nuevo organismo. En este proceso, el DNA de ambos padres se combina para formar un individuo con su propio DNA. (Cap. 4, pág. 96)

shale / esquisto arcilloso Roca sedimentaria formada por partículas pequeñísimas de arcilla. (Cap. 15, pág. 421)

shield volcano / volcán de escudo Volcán formado de largas capas inclinadas de rocas que se acumulan debido a las erupciones sin ruidos de lava fluida. (Cap. 16, pág. 443)

silt / roca de cieno y de arcilla Roca sedimentaria formada por partículas más grandes que la arcilla, pero más pequeñas que la arena. (Cap. 17, pág. 478)

simple machine / máquina simple Dispositivo que realiza trabajo mediante un solo movimiento. Es la forma más básica de una herramienta útil. (Cap. 11, pág. 296)

solar energy / energía solar Energía que proviene del sol. Es un ejemplo de un recurso renovable. (Cap. 12, pág. 338)

solar system / sistema solar Está formado por nueve planetas y numerosos objetos que giran en órbitas alrededor del sol. (Cap. 14, pág. 384)

solid waste / desecho sólido Cualquier objeto en forma sólida o casi sólida que una persona bote. (Cap. 7, pág. 195)

species / especie Grupo de organismos que pueden aparearse entre sí y producir progenie saludable. (Cap. 2, pág. 45)

speed / rapidez Una medida del grado de velocidad con que se mueve un objeto, en un tiempo dado. (Cap. 10, pág. 269)

spring / manantial Flujo de agua fresca que brota de la tierra cuando la capa superior de las aguas subterráneas llega a la superficie. (Cap. 17, pág. 474)

star / estrella Inmensas nubes compuestas de polvo y gases que se calientan y producen energía. Son cuerpos celestes con luz propia. (Cap. 14, pág. 398)

state of matter / estado de la materia Indica si una muestra de materia es sólida, líquida o gaseosa. Por ejemplo, un cubo de hielo es agua en estado sólido. (Cap. 9, pág. 241)

static discharge / descarga estática Resulta cuando muchas cargas eléctricas se mueven al mismo tiempo. Los rayos son un ejemplo. (Cap. 13, pág. 348)

static electricity / electricidad estática Forma de energía eléctrica potencial que se presenta cuando se acumulan o se pierden electrones. (Cap. 13, pág. 348)

stationary / estacionario Tipo de frente que se ha detenido sobre una región. (Cap. 18, pág. 506)

stratosphere / estratosfera Capa de la atmósfera que contiene el gas ozono. Esta capa importante de gas protege la vida en la Tierra de los dañinos rayos ultravioletas del sol. (Cap. 18, pág. 498)

streak test / prueba de la veta Prueba que ayuda a identificar un mineral aunque el mineral parezca otro mineral diferente. La veta no necesariamente es del mismo color del mineral. (Cap. 15, pág. 410)

submersibles / sumergibles Naves submarinas experimentales, tales como ALVIN que se utilizan para el estudio del fondo oceánico. También pueden estudiar la vida en el océano y los efectos humanos sobre el ambiente marino. (Cap. 17, pág. 485)

subscript / subíndice Número pequeño que se escribe un poco más abajo en una fórmula y el cual indica el número de átomos del elemento. (Cap. 8, pág. 227)

switch / interruptor Dispositivo que abre o cierra un circuito. (Cap. 13, pág. 352)

synthetic / sintético Tipo de elemento que no se encuentra en la naturaleza, pero que se fabrica en máquinas llamadas aceleradores de partículas. (Cap. 8, pág. 223)

technology / tecnología Es el uso de los conocimientos aprendidos a través de la ciencia. (Ch. 1, pág. 9)

temperature / temperatura Promedio de la energía cinética de todos los átomos de un material. (Cap. 12, pág. 332)

terminal / terminal Cada uno de dos puntos de conexión de una batería. (Cap. 13, pág. 355)

texture / textura Se refiere al tamaño del cristal y a la manera cómo encajan los granos del cristal o mineral. (Cap. 15, pág. 407)

thermal energy / energía térmica Cantidad total de energía, tanto cinética como potencial, de los átomos que forman un material. (Cap. 12, pág. 328)

threatened species / especie amenazada de extinción Especie bajo protección pero que no está en peligro inmediato de extinción. (Cap. 6, pág. 166)

tide / marea Cambio en el nivel del agua del océano durante el curso de un día. Normalmente, el nivel del agua del océano alcanza un nivel alto y un nivel bajo dos veces al día. (Cap. 17, pág. 486)

tissue / tejido Grupo de células semejantes que realizan el mismo tipo de trabajo. (Cap. 3, pág. 80)

trait / rasgo Característica específica de algo. Por ejemplo, algunos rasgos humanos incluyen ojos marrones, cabello rojizo y la capacidad de caminar. (Cap. 2, pág. 34)

transformation / transformación Cambio de la energía de una forma a otra. (Cap. 12, pág. 321)

trilobite / trilobita Parientes extintos de animales actuales, tales como los cangrejos, las langostas y los insectos. (Cap. 5, pág. 130)

troposphere / troposfera La capa más baja de la atmósfera terrestre. Es donde ocurren las condiciones del tiempo, las nubes y la contaminación del aire. Esta capa contiene aproximadamente un 75% del aire que respiramos. (Cap. 18, pág. 498)

turbine / turbina Rueda giratoria gigante que recibe energía de diferentes fuentes y que provee energía mecánica a un generador. (Cap. 13, pág. 366) upwarped mountain / montaña plegada anticlinal Tipo de montaña que se forma cuando el magma ascendente hace que la corteza terrestre se estire. (Cap. 16, pág. 461)

vaccine / vacuna Medicamento hecho de virus muertos o debilitados. Se administra por vía oral o por inyección. Las vacunas hacen que el cuerpo produzca sustancias que resisten ciertos virus. (Cap. 2, pág. 55)

variable / variable Factor que se puede cambiar en un experimento. (Cap. 1, pág. 19)

variation / variación Las diferentes formas en que puede presentarse cierto rasgo. (Cap. 4, pág. 110)

variation / variación Diferencias en los rasgos de los miembros de una especie y que los distingue a unos de los otros. (Cap. 5, pág. 124)

virus / virus Partícula que comparte características tanto con los seres vivos como con los seres inanimados. Se parece a los seres vivos porque es capaz de reproducirse. Se parece a los seres inanimados porque no crece, no come y no responde a su ambiente. (Cap. 2, pág. 54)

volcanic mountain / montaña volcánica Se forma en áreas donde una placa es empujada debajo de otra placa. A medida que una placa es empujada dentro del manto, la placa comienza a derretirse, empujando magma a través de la corteza y formando volcanes. (Cap. 16, pág. 461)

volcano / volcán Lugar en la Tierra donde el magma caliente y líquido, el cual es roca que se ha derretido debajo de la corteza, sube y sale a la superficie terrestre. (Cap. 16, pág. 442)

voltage / voltaje Mide la diferencia de energía potencial que hay entre los extremos de una batería. (Cap. 13, pág. 354)

volume / volumen Espacio que ocupa un objeto. (Cap. 9, pág. 239)

waning moon / luna menguante Cuando la luna parece que disminuye de tamaño noche tras noche. (Cap. 14, pág. 380)

water cycle / ciclo del agua Movimiento constante del agua desde la superficie terrestre hasta la atmósfera y nuevamente a la superficie terrestre. (Cap. 17, pág. 469)

wave / ola Movimiento del agua oceánica que hace que el agua suba y caiga. Dicho movimiento transmite energía. (Cap. 17, pág. 484)

waxing moon / luna creciente Cuando la luna parece que aumenta de tamaño noche tras noche. (Cap. 14, pág. 380)

weather / tiempo Lo que sucede en la atmósfera en todo momento. Comprende los cambios en la atmósfera causados por la presión atmosférica, los vientos, la temperatura y el agua. (Cap. 18, pág. 503)

weather forecast / pronóstico del tiempo Estudio que hace un meteorólogo sobre las condiciones del tiempo. (Cap. 18, pág. 512)

weathering / meteorización Cambio físico responsable de la mayor parte de la forma de la superficie terrestre. (Cap. 9, pág. 254)

wedge / cuña Máquina simple compuesta de uno o dos lados inclinados. (Cap. 11, pág. 298)

weight / peso Una medida de la fuerza que ejerce la gravedad sobre un objeto. El peso cambia de un lugar a otro, de acuerdo con la fuerza de gravedad. (Cap. 10, pág. 266)

weight / peso Una medida de la fuerza de gravedad. (Cap. 11, pág. 294)

wetland / ciénaga Área parcialmente cubierta con agua. (Cap. 7, pág. 186)

wheel and axle / rueda y eje Máquina simple que consiste en dos ruedas de diferentes tamaños, conectadas por un eje, y las cuales giran al mismo tiempo. (Cap. 11, pág. 304)

wind / viento Movimiento del aire que resulta de las diferencias en presión atmosférica. (Cap. 18, pág. 504)

work / trabajo Es el producto de una fuerza ejercida sobre un objeto, a través de una distancia, cuando la fuerza y el movimiento son en la misma dirección. (Cap. 11, pág. 292)

Index

The index for *Glencoe Science* will help you locate major topics in the book quickly and easily. Each entry in the Index is followed by the numbers of the pages on which the entry is discussed. A page number given in **boldface type** indicates the page on which that entry is defined. A page number given in *italic type* indicates a page on which the entry is used in an illustration or photograph. The abbreviation *act.* indicates a page on which the entry is used in an activity.

B

reaction with oxygen, 216
in sugar, *225*
in water, 214, *214*

Hypothesis, **18**, 29, 145, 493, 552

Ice, and salt, 250

Igneous rocks, **415**-419
extrusive, 415, **416**, *417*, *act.* 418-419
intrusive, 415, *416*, **417**, *act.* 418-419
from lava, 416, *417*, 431
from magma, *416*, 417
in rock cycle, *430*, 431

Illustrations and pictures, interpreting, 145, 235, 261, 435, 521, 556

Inclined plane, **296**, *296*, 297, *act.* 302-303, 305, *305*, 312

Inertia, **276**-277, *276*, 282, 283, *283*

Inferring, 10, *act.* 11, 36, 59, 87, 122, 145, 158, 173, 235, 261, 268, *act.* 291, 325, 343, 435, 465, 493, 504, 549

Inner core, of Earth, **449**, *449*

Insects
adaptation of, *120*
diversity of, 117, *118*, 122
in ecosystem, 148, *148*, *162*, 163
monarch butterflies, 159, *159*
See also individual types of insects

Intrusive igneous rocks, 415, *416*, **417**, *act.* 418-419

Io (moon of Jupiter), 388

Iodine, 230

Iron, 414
chemical properties of, 244, *244*
chemical symbol for, 224
comparing gravel and, 216
density of, 240
as element, 223
as natural resource, *178*

Iron ore, *407*

Joule (unit of measure), **293**

Joule, James, 293

Journal. *See* Science journal

Judo, *292*

Jupiter, *384*, 386, 388, *388*
gravity of, 265, 266

Kasparov, Garry, 42

Kelp, 50, *50*

Kidneys, 80

Kilauea volcano, 443, *443*

Kilogram, 268

Kilometer, 269

Kinetic energy, 322, **323**, *323*, 327
comparing content of, 333

thermal energy and, *act.* 330-331
transfer of, 324, 325, *325*, *act.* 326
transformation of potential energy to, 324

King, Martin Luther, Jr., 232

Kingdoms, **47**, 118
animals, 45-46, *45*, *46*, 52, *52*, *53*
bacteria, 49, *49*
fungi, 50-51, *50*
plants, 51, *51*
protists, 50, *50*

Krakatau volcano, 443

Land, use of, 186-187, *186*, *187*, *act.* 188-189

Landfill, **187**, *187*, 195

Language arts, 56, 123, 340, 368, 393, 462

Lava, *431*, 442, 443, *act.* 444
rocks from, 416, *417*, 431

Lavoisier, Antoine, 213

Law. *See* Scientific laws

Law of conservation of matter, 214

Lead, *407*, 414

Leaves, *act.* 71, 72, *73*, *77*, *258*

Levers, 298-301, **299**, *299*, *act.* 302-303, 304
in compound machine, *307*

Measurement
  of acidity, *act.* 246-247
  of earthquakes, 440-441,
    440, act. 450-451
  of gravity, 266
  of magnetic motion, *act.*
    345
  of magnetism, 371
  of mass, 267-268, *267*
  in SI, 115, 261, 557
  of space, 384-385, 386,
    399
  of speed, 269-270, *269*
  of temperature, *332, 333,*
    333
  of thermal energy, 329
  units of. *See* Units of
    measure
  of weight, 268
  of work, 293, 294
Meiosis, **96**-97, *97*
Mendeleev, Dmitri, 223
Mercalli scale, modified,
  440-441
Mercury, *384,* 386, 392, *392*
Mesosaurus, 453, *453*
Mesosphere, 498, *499*
Metals, reaction with acid,
  245, *245. See also indi-*
  vidual metals
Metamorphic rocks, **427-**
  429, *427, 428, act.* 429,
  430, 431
Meteorologist, **512**
Methane, 387
Mica, *409,* 412
Microscopes, 61, 64, *64*
Microwave oven, 334, 335,
  335
Microwave radiation, 338-
  339
Migration, *121*

Milky Way Galaxy, 399-400
Millenium Seed Bank, 98,
  98
Minerals, **406**-414, *406*
  as basic need of living
    things, 44
  classification of, 411
  common, 412
  formation of, 407, *407*
  gems, 413, *413*
  hardness of, 411
  identifying origin of, 412,
    412
  mining, *177,* 179, 413,
    413, 414
  ores, 414
  properties of, 408-412,
    408, 409, 410
Mining, *177,* 179, 413, *413,*
  414
Mississippi River, *479*
Mitochondrion, *66, 67,* **70,**
  70, 72
Mitosis, **91,** *91*
Mixture, **228**-229, *228, 229*
Model
  of atom, 214-219, *217, 218*
  of cell, 69
  of circuit, 352
  computer, 170, 474
  Dalton's model of atom,
    216
  of earthquake, *act.* 437
  of energy flow in ecosys-
    tem, 169
  formulating, 400
  making, 196, 222, 337,
    343, 371, 445, 521, 556
  purpose of, 214, 215
  using, 215, *215*
  of volcano, 442
Modified Mercalli scale,
  440-441

Mohs, Friedrich, 411
Mohs scale, 411
Mold, 50, *50*
Monarch butterflies, 159, *159*
Monera, 49
Monument design, 232, *232*
Moon, 380-383
  gravity of, 265, 266, 486-
    487, *486*
  in lunar eclipse, 383, *383*
  phases of, 380, *380, act.*
    381
  in solar eclipse, 382, *382*
Mosses, *149*
Motion
  acceleration and, 272-273,
    274, 277-278, *278*
  friction and, 273-274, *274,*
    282, 283
  magnetic, *act.* 345
  Newton's laws of, 276-
    279, *277, 278, 279, act.*
    280-281
  of oceans, 482-483, *482-*
    483
  speed of, 269-274, *269,*
    270, act. 271
  stopping, 273
Mountains, *454*
  climate and, 515, *515*
  formation of, *426,* 459,
    460
  kinds of, *460,* 461
Movement
  of organisms, *35, 36, 36,*
    37
  of thermal energy, 334-
    337, *334, 335, 336, 337*
Multimedia presentations,
  53, 164, 194, 258, 392,
  480, 563
Muscle cells, *74, 80*

vi Dr. Gopal Murti/SPL/Photo Researchers; vii (tl)Sinclair Stammers/SPL/Photo Researchers, (tr)Betty Barford/Photo Researchers, (bl)Phil Degginger/Color-Pic; (br)Philip & Karen Smith/Tony Stone Images; viii (t, br)Dr. Mitsu Ohtsuki/ SPL/Photo Researchers, (bl)Henry Groskinsky/Peter Arnold; ix (t)Tim Davis/ Tony Stone Images, (b)Phil Jude/SPL/Photo Researchers; x E.R. Degginger/ Bruce Coleman, (tr)NASA/JPL/TSADO/Tom Stack & Associates, (c)C.Dani-I. Jeske/Earth Scenes, (bl)Mark Jones/Minden Pictures, (br)Tui De Roy/Minden Pictures; xi (t)Tracy Aiguier/The Picture Cube, (c)I.M. House/Tony Stone Images, (b)Charles D. Winters/Photo Researchers; xii (t)Glencoe photo; (b) Cabisco/Visuals Unlimited; xiv (t)Matt Meadows, (b)Morrison Photography; xv (t)Matt Meadows, (b)Morrison Photography; xvi (t)Morrison Photography, (b)Francois Gohier/Photo Researchers; xvii Morrison Photography; xviii (t c)Matt Meadows, (b)Breck P. Kent/Earth Scenes; xix (l)Phil Degginger/Color-Pic, (c)Morrison Photography, (r)E.R. Degginger/Color-Pic; xx (t)Leonard Lessin/FBPA, (b)David Parker/SPL/Photo Researchers; xxii (t)Mike Rustad Photography, (c)Courtesy Dr. Isidro Bosch, (b)Steve Skjold; xxiii (t)Historical Picture Archive/Corbis, (c)Morrison Photography, (b) Courtesy Addison Bain; 2 Ralph Lee Hopkins/The Wildlife Collection; 4 (l)Karl Weatherly/Corbis, (r)David Schultz/Tony Stone Images; 5 Pete Saloutos/Tony Stone Images; 6 Lawrence Migdale/Stock Boston; 7 Stephen Frisch/Photo 20-20; 8 Morrison Photography; 9 (t) David Parker/PL/Photo Researchers, (b)Jeff Greenberg/PhotoEdit; 10 Morrison Photography; 12 Will & Deni McIntyre/Photo Researchers; 13 Courtesy Amanda Shaw & FocusOne; 15 Jack Demuth; 16 (t)Morrison Photography, (b)Tony Freeman/PhotoEdit; 18 Donald Johnston/Tony Stone Images; 19 20 22 23 24 25 Morrison Photography; 26 Jack Demuth; 27 (l)Morrison Photography, (r)Ralph Lee Hopkins/The Wildlife Collection; 30-31 Karen Tweedy-Holmes/ Corbis; 31 Morrison Photography; 32 Cabisco/Visuals Unlimited; 33 Morrison Photography; 34 E.R. Degginger/Color-Pic; 35 Shin Yoshino/Minden Pictures; 39 (t)Picture Perfect, (c,b)Zefa/The Stock Market; 41 Morrison Photography; 42 Jonathan Elderfield/Gamma Liaison; 43 (t)Morrison Photography, (bl, br) E.R Degginger/Color-Pic; 45 (t) Pat & Tom Leeson/Photo Researchers, (b)E.R. Degginger/Color-Pic; 46 (t)Dr.G.J. Chafaris/Color-Pic, (b) H.Reinhard/ Okapia/Photo Researchers; 47 (t)Flip Nicklin/Minden Pictures, (b)Dave B. Fleetham/Tom Stack & Associates; 48 Morrison Photography; 49 (t, c) David M. Phillips/Visuals Unlimited, (b)Dr.Gopal Murti/ SPL/ Photo Researchers; 50 (tl, bl)Andrew Syred/SPL/Photo Researchers, (tr, br)Paul Skelcher/Rainbow; 51 (t, c) Picture Perfect, (b)Aaron Haupt; 52 (t)Zefa/The Stock Market, (c) Glencoe photo, (b)K.G. Preston Mafham/Animals Animals; 53 Zefa-Ziesler/The Stock Market; 54 Chris Bjornberg/Photo Researchers; 55 Matt Meadows; 56 Doug Martin; 57 (tl) E.R. Degginger/Color-Pic, (tr) Cabisco/Visuals Unlimited, (bl)Picture Perfect, (br)Dr.Gopal Murti/SPL/Photo Researchers; 59 Dietrich Gehring/The Wildlife Collection; 60 Manfred Kage/Peter Arnold; 61 Matt Meadows; 62 (l)Brian Parker/Tom Stack & Associates, (tr)Susan Van Etten/PhotoEdit, (br)John Cancalosi/Peter Arnold; 63 (l)David M.Dennis/Tom Stack & Associates, (c)Eric Grave/Photo Researchers, (r)David Young-Wolff/ PhotoEdit; 64 (t)Morrison Photography, (c)Dr.Tony Brain/ SPL/Photo Researchers, (b)Leonard Lessin/FBPA; 68 Biophoto Associates/Photo Researchers; 69 Michael Abbey/Photo Researchers; 71 (t)Robert & Linda Mitchell, (b)John Walsh/SPL/Photo Researchers; 72 Doug Sokell/Tom Stack & Associates; 74 (l)M.Abbey/Photo Researchers, (r) Biodisc; 74-75 Wendy Shattil & Bob Rozinski/Tom Stack & Associates; 75 (t)Dr.Dennis Kunkel/Phototake, (b)Andrew Syred/Tony Stone Images; 76 (l)Larry Mulvehill/Photo Researchers; 76-77 Morrison Photography; 77 (t)Carolina Biological Supply Co./Oxford Scientific Films/Earth Scenes, (c)R.Kessel-G.Shih/Visuals Unlimited, (b)Bruce Iverson; 78 Morrison Photography; 81 H.H. Sharp/Photo Researchers; 82 Courtesy Genzyme Corporation; 84 Courtesy Eyes On You Magazine; 85 (t)John Walsh/SPL/Photo Researchers, (bl)Dr.Dennis Kunkel/Phototake, (br)Morrison Photography; 86 Mike Eichelberger/Visuals Unlimited; 88 J-C Carton/Bruce Coleman; 89 Morrison Photography; 90 N. Smythe/Photo Researchers; 92 Holt Studios International/Photo Researchers; 93 (t)Betty Barford/Photo Researchers, (bl)Biophoto Associates/Photo Researchers, (br)Jerome Wyckoff/Earth Scenes; 94 95 Matt Meadows; 96 (t)Prof. P. Motta/SPL/Photo Researchers, (b)Andrew Syred/SPL/Photo Researchers; 98 Heather Angel Photography; 100 Dwight R. Kuhn; 101 AP/Wide World Photos; 102 Walter Hodges/Tony Stone Images; 103 Biophoto Associates/Photo Researchers; 104 Matt Meadows; 106 (t)Donald Specker/Earth Scenes, (b)Jane Grushow/Grant Heilman Photography; 108 Mark E. Gibson; 110 (t)Matt Meadows, (b)Morrison Photography; 111 (l)Michael P. Gadomski/Earth Scenes, (r)Gregory K. Scott/Photo Researchers; 112 (l)Bob Daemmrich/Stock Boston, (r)Derrick Ditchburn/Visuals Unlimited; 113 (tl)Betty Barford/Photo Researchers, (tr)AP/ Wide World Photos, (bl)Walter Hodges/Tony Stone Images, (br)Gregory K. Scott/Photo Researchers; 114 E.R.Degginger/ Color-Pic; 115 Michelle Garrett/Corbis; 116 Gerry Ellis/ENP Images; 117 Morrison Photography; 119 (t)Lynn M.Stone/Earth Scenes, (c)Mike Bacon/Tom Stack & Associates, (b)Kennan Ward; 120 (t)E.R. Degginger/Color-Pic, (c) Gregory G. Dimijian/Photo Researchers, (bl)Rod Planck /Photo Researchers, (br)Leonard Lee Rue III/Photo Researchers; 122 E.R. Degginger/Color-Pic; 123 Maslowski/Photo Researchers; 126 127 Morrison Photography; 130 (l)Sinclair Stammers/SPL/Photo Researchers, (c)Francois Gohier/Photo Researchers, (r) Gary Retherford/Photo Researchers; 131 (t)George E. Jones III/Photo Researchers, (b)Michael Durham/ENP Images; 134 Tom Brakefield/Bruce Coleman; 135 (t)Wendy Shatti/Bob Rozinski, (b) Wendy Shattil/Bob Rozinski/Tom Stack & Associates; 136 Patrick Aventurier/Gamma Liaison; 137 Richard T. Nowitz/Photo Researchers; 139 Morrison Photography; 142 Tom McHugh/Photo Researchers; 143 (t)Lynn M. Stone/Earth Scenes, (b)Gerry Ellis/ENP Images; 146 David M. Dennis; 147 Aaron Haupt; 148 (t) David M. Dennis/Tom Stack & Associates, (b)Todd Gipstein/Photo Researchers; 148-149 Carr Clifton/Minden Pictures; 149 (l)Colin Milkins/Oxford Scientific Films/ Animals Animals, (tr)Harold R.Hungerford/Photo Researchers, (br)Ed Reschke/Peter Arnold; 150 Matt Meadows; 151 (t)Morrison Photography, (b)SSEC/University of Wisconsin, Madison; 154 (l) F. Stuart Westmorland/ Photo Researchers, (r)Jim Zipp/Photo Researchers; 155 Francois Gohier/Photo Researchers; 156 (t)David Woodfall/ENP Images, (b)Gerry Ellis/ENP Images; 157 Flip Nicklin/Minden Pictures; 158 Fred Bavendam/Minden Pictures; 159 Glencoe photo; 160 Matt Meadows; 161 Doug Martin; 163 Lynn M. Stone; 164 Matt Meadows; 165 Courtesy Sea World of Ohio; 166 Raymond Gehman/Corbis; 167 Rosemary Calvert/Tony Stone Images; 170 Glencoe photo; 171 (t)David M. Dennis, (b)Kit Latham/FPG International; 173 Morrison Photography, (t)Jim Zipp/Photo Researchers, (b)Mark Newman/Photo Researchers; 174 Nancy Linden; 175 Morrison Photography; 176 Doug Armand/Tony Stone Images; 177 (t)Morrison Photography, (bl)David R. Frazier Photolibrary, (bc)John Mead/ SPL/Photo Researchers, (br)Philip & Karen Smith/Tony Stone Images; 178 (t)Steven Weinberg/Tony Stone Images, (bl)World View/Tony Stone Images, (bc)E.R. Degginger/Color-Pic, (br)Camcar; 179 (l)Mark E. Gibson, (r)Jack Demuth; 180 Michael Dwyer/Stock Boston; 181 (t)E.R. Degginger/Color-Pic, (b)Nancy Ross-Flanigan; 182 (clockwise from top)Kristin Finnegan/Tony Stone Images, Simon Fraser/SPL/Photo Researchers, Phil Degginger/Color-Pic;

185 David R. Frazier Photolibrary; 186 Jeff Greenberg/David R. Frazier Photolibrary; 187 Lonnie Duka/Tony Stone Images; 188 189 Morrison Photography; 190 (l)Elaine Comer-Shay, (r)Morrison Photography; 193 Jack Demuth; 194 Michael Giannechini/Photo Researchers; 196 Jack Demuth; 197 (l)Morrison Photography, (r)Fred Bavendam/Peter Arnold; 198 (l)Morrison Photography; (c)David R. Frazier Photolibrary, (r)Will McIntyre/Photo Researchers; 199 (l)E.R. Degginger/Color-Pic, (cl)F.Pedrick/The Image Works, (tr)Hank Morgan/Photo Researchers, (cr)Morrison Photography; 200 Morrison Photography; 201 John Elk III/Stock Boston; 202 Mark E. Gibson/Visuals Unlimited; 203 (tl) David R. Frazier Photolibrary, (tr)John Elk III/Stock Boston, (bl)Jeff Greenberg/David R. Frazier/Photolibrary, (br)Jack Demuth; 206 Johnny Johnson; 207 Susan Marquart; 208-209 George Steinmetz; 209 Courtesy Prof. H. Kazerooni, University of California at Berkeley; 210 Kaz Mori/The Image Bank; 211 Morrison Photography; 212 Ancient Art & Architecture Collection; 213 David R. Frazier Photolibrary; 215 Morrison Photography; 216 David R. Frazier Photolibrary; 217 (t)Dr. Mitsuo Ohtsuki/SPL/Photo Researchers, (c) ESA/TSADO/Tom Stack & Associates, (b)Morrison Photography; 220 221 222 Morrison Photography; 223 Fermi National Accelerator Laboratory/Photo Researchers; 225 Sandy King/The Image Bank; 227 228 Morrison Photography; 229 Ronnie Kaufman/The Stock Market; 230 Morrison Photography; 231 Simon Fraser/SPL/Photo Researchers; 232 Todd Gipstein/Corbis; 233 (t)Ancient Art & Architecture Collection, (b)Kaz Mori/The Image Bank; 236 Norbert Rosing/Earth Scenes; 237 Breck P. Kent/Earth Scenes; 238 239 Morrison Photography; 240 (t)E.R. Degginger/Color-Pic, (b)Morrison Photography; 241 (t)E.R. Degginger/ Color-Pic, (b)Morrison Photography; 242 Morrison Photography; 243 (l)E.R. Degginger/Color-Pic, (r)Alan L. Detrick/Color-Pic; 244 (t)Morrison Photography, (b)Bob Daemmrich/Stock Boston; 245 (t)Morrison Photography, (b)Simon Fraser/SPL/Photo Researchers; 247 248 Morrison Photography; 249 Courtesy Addison Bain; 250 Mark Joseph/Tony Stone Images; 251 Phil Degginger/Color-Pic; 252 Morrison Photography; 253 (tl) Art Montes de Oca/ FPG International, (tr) Morrison Photography,(bl)E.R. Degginger/Color-Pic, (br) Amy C. Etra/PhotoEdit; 254 (t)C.C. Lockwood/Earth Scenes, (b)Brenda Tharp/Photo Researchers; 255 Gerry Ellis/ENP Images; 256 257 Morrison Photography; 258 Charles Benes/FPG International; 259 (t, c)Morrison Photography, (b)Gerry Ellis/ENP Images; 261 (t, cl, cr)Morrison Photography, (b)Michael Nelson/FPG International; 262 William Sallaz/Duomo; 263 Morrison Photography; 264 Henry Groskinsky/Peter Arnold; 265 Werner H. Muller/Peter Arnold; 266 Michael Newman/PhotoEdit; 267 (t)Morrison Photography, (b)Matt Meadows; 268 (t)NASA, (c)Morrison Photography, (b)BLT Productions; 269 (t)Matt Meadows, (b)Chuck Kuhn/The Image Bank; 270 David Young-Wolff/ PhotoEdit; 271 272 273 274 Morrison Photography; 275 Lawrence Migdale/ Stock Boston; 276 (t)NASA, (b)Scott Markewitz/FPG International; 279 Matt Meadows; 280 281 Morrison Photography; 282 (l)NASA, (r)David Young-Wolff/PhotoEdit; 283 David Young-Wolff/PhotoEdit; 285 Insurance Institute for Highway Safety; 286 Steve Skjold; 287 (t)Werner H. Muller/Peter Arnold, (c) Insurance Institute for Highway Safety, (b)Scott Markewitz/FPG International; 290 Erick Bakke/Allsport; 291 Morrison Photography; 292 (t)Peter Steiner/The Stock Market, (b)Amwell/Tony Stone Images; 293 Peter Arnold/Peter Arnold; 296 Bob Daemmrich/Stock Boston; 300 Wiley & Wales/Adventure Photo & Film; 301 Mark C. Burnett/Photo Researchers; 302 303 304 307 Morrison Photography; 308 (t)Morrison Photography, (b)Sarah Putnam/The Picture Cube; 309 Matt Meadows; 310 Stephen R. Brown/The Picture Cube; 311 Bill Gallery/ Stock Boston; 312 Scott Robinson/Tony Stone Images; 313 Clive Brunskill/ Allsport; 314 Rube Goldberg,™ and © of Rube Goldberg Inc. Distributed by United Media; 315 (t)Amwell/Tony Stone Images, (b)Mark C. Burnett/Photo Researchers; 316 Morrison Photography; 318 Mark Junak/Tony Stone Images; 319 Morrison Photography; 320 Jim Cummins/FPG International; 322 (t) Morrison Photography, (b)Tim Thompson/Corbis; 323 (t)Tim Davis/Tony Stone Images, (b)Chris Harvey/Tony Stone Images; 324 SW.Productions; 325 (l)SW Productions, (r)John Eastcott/Photo Researchers; 326 Morrison Photography; 327 L & M Photos/FPG International; 328 Steve Skjold/PhotoEdit; 329 330 331 332 335 Morrison Photography; 337 Tom & Pat Leeson/Photo Researchers; 339 Miro Vintoniv/Stock Boston; 340 (t)David R. Frazier Photolibrary, (b)Michael Giannechini/Photo Researchers; 341 (l)Steve Skjold/ PhotoEdit, (r)John Eastcott/Photo Researchers; 342 343 Morrison Photography; 344 David W. Hamilton/The Image Bank; 345 Matt Meadows; 346 E.R. Degginger/Color-Pic; 349 Matt Meadows; 350 Phil Jude/SPL/Photo Researchers; 351 Ben Van Hook/Duomo; 355 Morrison Photography; 356 357 Matt Meadows; 359 363 Morrison Photography; 364 (t)Bob Kramer/The Picture Cube, (b)Morrison Photography; 366 (t)E.R. Degginger/Color-Pic, (b)Russell D.Curtis/Photo Researchers; 367 (t)Mark C.Burnett/Photo Researchers, (b)John Mead/SPL/PhotoResearchers; 368 369 Morrison Photography; 372 Jack Demuth; 373 Morrison photography; 374-375 Adriel Heisey/Photographers/Aspen; 375 NASA/JPL; 376 NASA; 377 381 Morrison Photography; 382 NASA; 383 Jerry Lodriguss/Photo Researchers; 386 TSADO/NOAO/Tom Stack & Associates; 387 (t)Alan Stern(Southwest Research Institute), Marc Buie(Lowell Observatory), ESA and NASA; (b)NASA/JPL/Tom Stack & Associates; 388 (t) E.K.Karkoschka (LPL) and NASA, (c)NASA/JPL/TSADO/Tom Stack & Associates, (b) NASA/ Photo Researchers; 389 (t)USGS/NASA/TSADO/Tom Stack & Associates, (c,b)NASA; 390 Morrison Photography; 391 Pat Rawlings; 392 Mark S. Robinson, Northwestern University; 393 NASA; 394 NASA/Photo Researchers; 395 NASA; 398 (t)D.Figer (UCLA) and NASA, (b)NASA/SPL/Photo Researchers; 400 Mike O'Brine/Tom Stack & Associates; 401 (l) D. Figer (UCLA) and NASA, (r) Alan Stern (Southwest Research Institute), Marc Buie (Lowell Observatory), ESA and NASA; 403 Bill & Sally Fletcher/Tom Stack & Associates; 404 Galen Rowell/Mountain Light; 405 Matt Meadows; 406 through 410 Morrison Photography; 412 (t)Morrison Photography, (bl) Runk/ Schoenberger/Grant Heilman Photography, (br)E.R. Degginger/Color-Pic; 413 (t)E.R. Degginger/Bruce Coleman, (b)David R.Frazier Photolibrary; 414 Harvey Lloyd/Peter Arnold; 415 Otto Hahn/Peter Arnold; 416 Morrison Photography; 417 (l)Morrison Photography, (r)Breck P. Kent/Earth Scenes; 418 Matt Meadows; 420 Morrison Photography; 420-421 Jim Corwin/Photo Researchers; 421 (t)Bob Evans/Peter Arnold, (b)Morrison Photography; 422 (l)Alan D. Carey/Photo Researchers, (c)Morrison Photography; 422-423 Fred Bavendam/Minden Pictures; 423 (tr)James L. Amos/Photo Researchers, (bl)Mark C. Burnett/Photo Researchers, (br)E.R. Degginger/Color-Pic; 424 Milton Rand/Tom Stack & Associates; 425 Steve Hamblin/Liaison International; 426 Derek Karp/Earth Scenes; 427 (t)Phil Degginger/Color-Pic, (b)Alice Q. Hargrave/Gamma Liaison; 428 (t)E.R. Degginger/Color-Pic, (cl)Breck P. Kent/ Earth Scenes, (cr)Jerome Wyckoff/Earth Scenes, (bl)Ancient Art & Architecture Collection, (br)E.R. Degginger/Color-Pic; 429 Matt Meadows; 430 (l)G.I. Bernard/Earth Scenes,(tr)E.R. Degginger/Color-Pic, (br)Jerome Wyckoff/Earth Scenes; 431 Magrath Photography/SPL/Photo Researchers; 432 Mike Rustad Photography; 433 (tl)Morrison Photography, (tr)Otto Hahn/Peter Arnold, (bl) E. R. Degginger/Color-Pic, (br)Matt Meadows; 435 (c,r)Morrison Photography, (l)Doug Martin; 436 Lysaght/Gamma Liaison; 437 Matt Meadows; 440 Jean-Marc Giboux/Gamma Liaison; 440-441 John T. Barr/Gamma Liaison; 442 C.Dani-I. Jeske/Earth Scenes; 442-443 David Muench Photography; 443 Mark Jones/Minden Pictures; 445 Tui de Roy/Minden Pictures; 446 Francois Gohier/ Photo Researchers; 447 David Weintraub/Photo Researchers; 448 Krafft/Photo Researchers; 450 Matt Meadows; 452 Martin Land/SPL/Photo Researchers; 454 (t)Carr Clifton/Minden Pictures, (b)David Parker/SPL/Photo Researchers; 459 Ann Duncan/Tom Stack & Associates; 460 (tl)Jeff Lepore/Photo Researchers, (tr)Michael Durham/ENP Images, (bl)E.R. Degginger/Color-Pic, (br)John Lemker/Earth Scenes; 463 (t)Lysaght/Gamma Liaison, (c)Jean-Marc Giboux/ Gamma Liaison,

Art Credits

Permissions

PERIODIC TABLE OF THE ELEMENTS

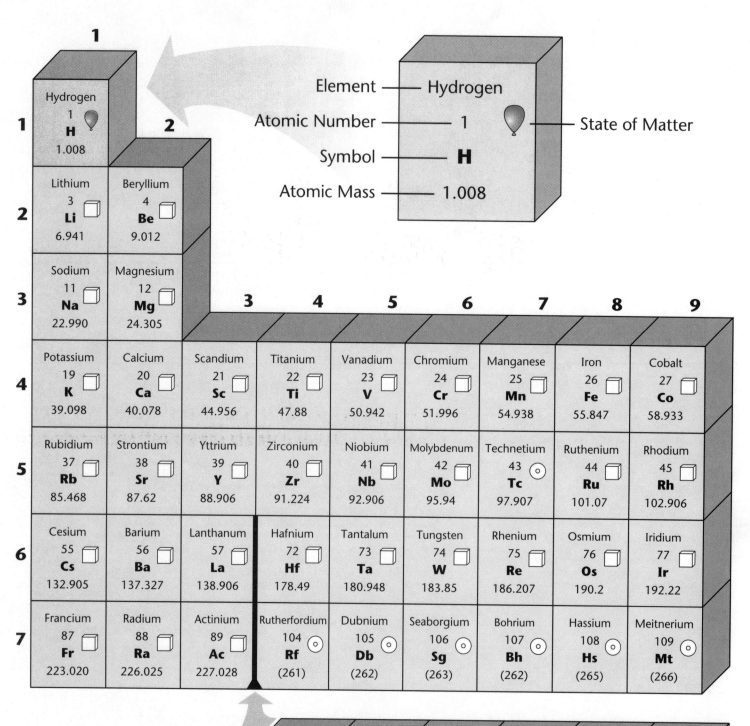